张直明

张直明文选

Selected Works of
Zhang Zhiming

上海大学出版社
·上海·

图书在版编目(CIP)数据

张直明文选 / 张直明著. —上海：上海大学出版社，2021.3
ISBN 978-7-5671-3811-7

Ⅰ.①张… Ⅱ.①张… Ⅲ.①转子—轴承系统—文集 Ⅳ.①TH13-53

中国版本图书馆 CIP 数据核字(2021)第 017759 号

责任编辑　王悦生
封面设计　柯国富
技术编辑　金　鑫　钱宇坤

张直明文选

张直明　著

上海大学出版社出版发行
(上海市上大路 99 号　邮政编码 200444)
(http://www.shupress.cn　发行热线 021-66135112)
出版人　戴骏豪

*

南京展望文化发展有限公司排版
上海华业装潢印刷厂有限公司印刷　各地新华书店经销
开本 787mm×1092mm　1/16　印张 45.25　字数 1136 千字
2021 年 3 月第 1 版　2021 年 3 月第 1 次印刷
ISBN 978-7-5671-3811-7/TH·11　定价 480.00 元

版权所有　侵权必究
如发现本书有印装质量问题请与印刷厂质量科联系
联系电话：021-56475919

序

我原籍浙江海宁,父亲张正学,母亲周大一。我1931年出生于上海,求学于上海:1937年入太华小学,1942年入民立中学,1946年转入大同附中二院,1948年入交通大学造船系,1949年转入机械系,1952年毕业于交通大学机械系。毕业后留校任助教;1954年升任讲师;1957年随校迁至交通大学西安部分;1959年学校更名为西安交通大学;1963年我开始招收"教师研究生";1979年开始招收硕士研究生;1979年升为副教授;1980年10月初至1981年底由西安交大派至联邦德国卡尔斯鲁厄大学做访问学者;1984年调至上海工业大学任职;1986年升为教授;1990年被国务院批准成立机械学博士点,我任博士生导师,并开始招收博士生;1994年学校改名为上海大学;1998年底我退休于上海大学。

我于1951年在上海虬江机器厂实习时,学习了雷天觉的两项关键技术:用可倾瓦轴承支撑磨床主轴,刮制镗床精密导轨。亲见这两项摩擦学技术将该厂提升为一个精密机床制造厂,成为后来的上海机床厂。由此,我深感摩擦学这门学科的重要性。

1958年,西安交大派我承担三峡水轮机大型推力轴承的设计计算工作。此工作如今已是易事,但当时却算难事。我从西安交大图书馆技术资料获知,当时只有Kingsbury用电解槽来模拟润滑油膜,成功地求解了雷诺方程,这可算是最先进的方法。我们学懂了此法,并正打算用这个方法来解决大推力轴承的设计计算时,三峡工程项目下马了。这是我参与摩擦学任务的初尝。

几年后,大学里开始提倡科学研究。我与几个年轻同事组成了"轴承兴趣小组",共同学习和交流,以增进这方面知识和能力。此事得到主管科研的副校长陈大燮教授的赞同,因他深知许多机器都有轴承问题,很需要这方面的研究工作。

在20世纪60年代初期,我国成立了许多"学科组",其中有一个"摩擦、磨损、润滑学科组"(那时尚无"摩擦学"之名)。首次学科组大会在兰州召开,学校派我参会。会议最后宣布,本学科组共六个组员(相当于后来的学会理事),其中包括我。"文革"后,学科组恢复活动,并更名为机械工程学会摩擦学分会,我仍任学会各届理事直至退休。

此后,计算工具和工程计算语言迅速发展,解算雷诺方程的方法主要采用差分方程的超松弛迭代法。我们不断地学习和紧跟世界潮流,并在一些细节技术上略有创新。那时我国开始生产汽轮发电机组,我们研究的主要应用对象,也改为汽轮机和发电机的径向轴承。

接着,一些重点大学接受了培养"教师研究生"的任务。我被指派每年培养一二名以轴承研究为方向的教师研究生。但培养了两届后,这种工作就停止了,让位于多年的政治运动。我在其中受到冲击,尤其在"文革"中的冲击更为强烈,毫无资格再继续从事研究工作。眼看着研究工作多年停滞,我深感损失极大,研究水平远远地落后于世界先进水平,极为可惜。

20世纪70年代末期,我恢复了摩擦学研究工作,并开始注重轴承的动力特性问题。有一家发电厂引进了瑞士BBC的一个5万千瓦的汽轮机,它装有三油楔轴承,运转非常平稳,我国主管这方面技术的机械研究所希望弄清楚其机理。这个问题必须从轴承的动力特性才能回答。我就着手进行轴承动力特性的理论工作,用小振动条件下表达振动的复数方式,将雷诺方程对复振幅求偏导的方法得出适用于动态增量的复数雷诺方程,由此求出油膜的四个刚度和四个阻尼,合称为八个动力特性。此法与 Lund 用有限小扰动以导出扰动雷诺方程有相同效果,但我的方式更加紧凑易懂。我就用此方式为三油楔轴承编制了电子计算机程序,以计算其静特性和动力特性。这是当时中国的第一个滑动轴承静、动特性的大型程序。

后来在我国开始广泛采用椭圆轴承。我又编制了椭圆轴承的计算机程序。并且无偿地送给哈尔滨汽轮机厂使用。1978年,我和一位助手由本校出路费到哈尔滨,将程序装入他们的计算机,教会他们的技术人员使用程序,讲了好几堂课的理论知识,哈尔滨电机厂和大电机研究所的许多人都来听。

由此开始了1978—1979年我被邀请的几次讲学:除了哈尔滨外,还有广州的机床研究所和上海交通大学。那些都是几十个小时的系统讲课。在上海交大讲学时,我全部录了音,交给西安交大轴承研究组任其参考使用。

那段时间,中国有两个大发电厂的汽轮发电机组发生极严重的事故,整台机组主轴和轴承系统在发生巨响后即刻断裂爆飞,所幸未伤人,但事故已毁了整套机组,直接损失何止千万。别的国家也发生过此种事故。因此转子-轴承系统的动力稳定性问题就成为世界上的一个热门课题。其原因是这个系统由于动力失稳而发生自激振动,振动频率与轴系的某个固有频率发生同步,构成了严重共振,从而引发灾难,亦称"油膜振荡"。所以这个问题也成为我的重要研究问题,并推动了我实现更完整的理论体系。

同时我也从单纯的轴承研究进入到转子-轴承系统的动力学研究。作为第一步,仅从单跨对称转子-轴承系统来分析理解自激振动的机理和推导出避免失去稳定性的条件。至于多轴承的大型多跨转子系统的计算软件,则是后来获 Glienicke 教授赠与,回国后在解算方法上加以改进而成的。

此外,汽轮机和发电机不断向大功率发展,轴颈在轴承里的线速度越来越高,润滑油膜的流态已失去层流,进入 Taylor 涡旋,再进入紊流状态。原来的润滑理论已不适用,需改用紊流润滑理论。紊流状态下的流体运动是一种随机运动,不可能精确描述,只能描述其时均值。紊流润滑理论即用以计算这些时均值的。世界上已经有好几种紊流润滑理论,均为半理论-半经验的实用方法。美国哥伦比亚大学的 H. T. Pan(潘弘道)教授的 Ng-Pan 紊流润滑理论是被最为普遍应用的一种。20世纪70年代末,荷兰的 Hirs 又提出了一种 Hirs 紊流润滑理论,似有被人们认可的迹象。我仔细阅读了他的文章,发现其基础站不住脚。我就写了一篇短评"论 Hirs 方法的理论基础",发表于《西安交通大学学报》。恰逢潘弘道来华,我与他在车里相遇,我给他看了我的这页意见。他表示赞同,并问我有无自己的紊流润滑理论。这完全出乎我的意外,我当时是觉得根本没有条件研制自己的紊流润滑理论。不过这句问话竟在我心里埋下一颗种子,多年后于条件成熟时真的研发了自己的紊流润滑理论。

那时我们虽已用计算机来分析计算轴承性能,但计算速度很慢。我在教研室里提出了"特征矢量法",其计算速度比超松弛迭代法平均快7倍。我将它编入静特性计算程序里,并准备将它推广用于动力特性计算程序里。在招收了"文革"后第一批研究生后,安排一名研究生的硕士论文来完成这个任务。此法后来实现于一套软件里,用于计算汽轮机轴承的静、动特性。我调回上海后,西安交大的此软件被一机部采纳为汽轮机轴承的标准计算法。

1980年,西安交大派我到联邦德国卡尔斯鲁厄大学 Glienicke 教授处做访问学者,我在那待了12个月后,又被 Glienicke 教授出资邀请延长了3个月,他第二次邀请我延长时我不再接受而回国了。Glienicke是著名的高速滑动轴承专家。我从接触到的德国人,和看到的德国情况,长了不少见识,得到了不少启示,也完成了一个课题。回国时,我向Glienicke要转子动力学的程序,他慷慨地给了我。回国后,我将它们在解算方法上做了一些改进,有效地提高了它们的计算效果。以此为基础,我后来完成了上海汽轮机厂的课题,提供他们一套验算大型转子-轴承系统动力稳定性的软件,以及一套验算其运转平稳性的软件。

我调到上海工业大学后,成立了轴承研究室,与上海的汽轮机厂、电机厂、汽轮发电机组之成套所、711研究所、杨浦发电厂、南京高速齿轮箱厂等等企业联系密切,多次研讨或完成他们提出的轴承和转子动力学课题。

接着,国务院批准了在上海工业大学设立以我为博士生导师并以摩擦学为主的机械学博士点。

此后,我曾邀请日本东京大学的教授崛幸夫(Hori)来校讲学,其助手加藤孝久(Kato)与他同来。加藤认为,滑动轴承参数太多,难以建立数据库以减轻设计计算。同时,他又进行了转子-轴承系统的非线性动力学仿真计算。其中,为了减轻轴承瞬时动力特性的计算量,他用"准雷诺边界条件"代替原来的雷诺边界条件,以减轻轴承力计算的工作量,得到了一些初步的结果。即使如此,其轴承瞬时动力的计算量仍很大,因此他只得到了很少量的仿真结果。

在20世纪80年代末期或90年代初期,我根据轴承结构参数的特殊性,用相似理论分析,提出了单块瓦的非定常油膜瞬时力的数据库建立方法,只需一个三维的表格式数据库,即可从数据库插值求出每一块瓦的瞬时力,接着组装成整个轴承的瞬时力。我指导一位研究生实现此法,并构成易于使用的软件。比加藤的方法快了400多倍,并且不用简化边界条件。我指导的一位博士生借助此法,实现了多跨转子-轴承系统的非线性动力学仿真技术和软件,得出了很丰富的仿真结果。后来我应崛幸夫的邀请,在东京大学作专题讲座,向他们介绍了我们的技术。不久后,在西安交大主办的"轴承-转子动力学"国际学术会议上,崛幸夫教授做主题报告列举各国进展情况时,单提了我们这套技术作为中国的最新发展。接着,西安交大和哈工大分别从我们这里引进了这套技术,用于转子-轴承系统的非线性动力学仿真计算。

那时,国内有一个大轴承在工作了一段时间后,它相对于轴承座发生了逆于轴转动方向的反转,反转的力量很大,把用来固定它位置的销子也挤断了,从而使轴承发生了损坏。此事件引起了关注,并在一次国内会议上有人对其机理做出了一些不着边际的猜测。我知道此事后,认为其机理在于轴的残余不平衡对轴承施加了一个旋转的压力,在轴承与轴承座之间的过盈不够大的情况下,使轴承与轴承座之间在这个动态压力的反方向处发生局部的微小滑移,甚至发生局部的微小松开。轴承与轴承座配合处,受压逐渐增大时引起径向和周向的压缩变形,受压逐渐减小时则发生回弹的伸长变形;而轴承座与轴承配合处也发生类似的径向变形,但其周向变形却方向相反。因此在配合面上有了局部周向滑移的趋势,而当它胜过摩擦力时就形成实际滑移。随着这个局部滑移的部位沿着轴转动方向变位,轴承相对于轴承座的局部滑移必定逆于轴转动方向。轴旋转一周后,轴承与轴承座的配合处就发生了整整一周的微小反向滑移,也就是整个轴承有了反向旋转。在每分钟3 000转的转动速度下,发生了每分钟3 000次的微小反向变位,其结果就可想而知了。我指导一位硕士生将这种机理分析作为论文,并且制作一套演示的模型,能清楚地观察到旋转压力引起轴承反转的现象。为了更慎重起见,我还写信给崛幸夫,问他认为轴承反转是何原因,他直接回答,是蠕变。可见这早就被人家洞察了,而我国因为长期关门闭塞,落后于人。

1989年末,加拿大多伦多大学的Leutheusser教授应邀来我室做3个月的学术休假。他在摩擦学方面曾有一项非常独特的贡献。他是流体力学方面的实验研究大家,能做大尺度的实验研究。他把轴承里的油膜放大了1 000倍,做成了空气膜的模型,在参数上使压缩性很小,可以代表不可压缩的油膜。实验时,将它在很长的轨道上高速移动,使气膜里发生紊流,用热膜法测出气膜里的流速分布。这是其他人无法得到的重要数据。我了解了他的这项成果后,敏锐地感觉到这是可以用来评价当前各种紊流润滑理论正确性的权威数据,也可以借此创制更好的紊流润滑理论。我先后指导过2位博士生做这项工作,很好地实现了我的要求,并且成功地提出了我们自己的紊流润滑理论(Zhang-Zhang理论)。

我退休后,1999年我到女儿家探亲时,在女婿的帮助下,用Visual Basic软件开发了一套各种常用径向滑动轴承的设计计算软件,将雷诺方程、能量方程、温粘关系联合求解,得到很切合实际工况的结果,使用上又像傻瓜照相机那样十分方便,使用者无须高深的理论知识即可应用。提供给一些滑动轴承厂使用,很受上海汽轮机厂等企业的欢迎。后来又应用户们的特殊需要,与合作者共同扩编成适应单个用户的便用软件,供应上海汽轮机厂、上海电机厂、四川的东方汽轮机厂等。

美国纽约州的Binghamton大学教授孙大成(D. C. Sun)早就与我相识,他退休后常来我室交流和合作。在他的密切参加并理论指导下,我们全室一起完成了一项挤压油膜的实验研究。这是一项基础性的研究,目的在于弄清动态情况下油膜破裂和复原的规律。这个规律与静态条件下的有很大不同,所以对动态仿真计算有重要的基础性意义。由本研究室同人建立的实验技术,以及在实验中观察到的一些现象,由孙先生与本室同人撰文发表后,被美国通用汽车公司等赏识,并提出该方向上的一项课题,委托本室一位教授进行研究。

我还曾受德国Atlas Copco压缩机公司聘为高级顾问,解决他们与上海电机厂之间的一些难协调的技术问题,并分析计算了电机的整套转子-轴承系统的稳定性和运转平稳性,发现当它安装在刚度较低的基础上时会发生振幅超标,所以必须安装在刚度好的基础上。在电动机试车时证实了我的结论。

2010年,我还曾受郑州大学邀请,作关于浮环轴承动力特性计算法的专题讲座。

20世纪80年代后期,一次与美国西北大学国家摩擦学中心郑绪云教授谈天时,他告诉我,弹流理论中计算固体变形均采用半空间理论,但近物体边缘处其实应当用四分之一空间理论。不过这种理论很难。他们校内有一位Keer教授在做,他的方法连郑教授也看不懂。此话也在我心里又埋下一颗种子。退休后,我常在早晨天未亮时醒来,躺在床上思考问题,此时脑筋特别清楚,多次解决了一些疑难或关键问题。在一个清晨,我忽然想出了一个解决四分之一弹性空间力学计算的矩阵方法,起身后即将它成文写下。我曾在室内向多名年轻同人讲解过我的方法,希望某个年轻同人能在一台高级电脑上实现这个方法。但无人有动于衷,可能因为没有正式立题,也没有课题经费支持之故吧,不能怪他们。

大约8年后,一位年轻老师知道了我的想法,帮我购置一台较好的个人电脑,并装设了较新的Fortran软件,使我的电脑也有能力可以解算四分之一空间的弹性问题。于是我自己实现了我的计算方法,并用算例验证了此法的正确性。那位老师又查阅了国际上已有的两种方法:Hetenyi方法和Keer方法。我的方法与它们对比,显示了它比原有方法具有更好的实用性和效率,以及更便于在工程应用上推广。我与合作者共同成文后,投稿学术杂志,获得发表。现在已经至少有国内外两位作者用此法解决问题。2013年,我受西安交大邀请作此专题的学术讲座。

此时我已经认识到我的矩阵方法可以推广到解算楔形空间的弹性力学问题上。接着我推导完成了它的理论系统。但楔形空间弹性计算当然比四分之一空间的技术难得多，而且需要存储量更大并且处理速度更快的计算机，所以迟迟未能成文。多年后，从 Keer 手下一位博士后交流的情况知道，已经有了切向载荷下的 Ahmadi 解，与原来已有的法向载荷下的 Love 解搭配起来，提供了解决楔形空间问题的完备基础。我于是完成了此问题的数学框架，由软件技术很强的年轻合作者将它实现为软件，并用一系列算例证实了此法的正确性、实用性和效率。投稿学术杂志后，两位评审人和学报主编都做出了高度评价，认为解决了多年来许多人未能解决的重要问题，接受发表。

以上是我从 1958 年以来参加摩擦学研究 60 年的概况。本书内容即选自我与合作者历年发表的文章。

回顾这 60 年，我要感谢前辈陈大燮、郑林庆、雷天觉的支持和提携。感谢上海大学前校长钱伟长推荐我获得王宽诚奖学金去香港大学和香港理工大学讲学，由此使我与香港城市大学黄柏林博士建立了多年的合作关系。感谢上海大学机械系前主任牟致中和总支书记李祖齐对我申请博士点的大力支持。感谢我系和我室许多同事们的合作和支持，以及我历年许多研究生和博士生的努力工作，我的业绩里包含着他们的贡献，他们的名字见于本书各篇文章。感谢陶德华，他从化学方向与我从力学方向共同支撑本校的摩擦学研究和教育。感谢老友毛谦德在我访德期间以及回国后给予我的合作和帮助。感谢孙大成、Glienicke、Leutheusser、崛幸夫等著名学者对我真诚的友谊和使我颇为得益的学术交流。

此外，我要特别感谢我的妻子吴西柳。我们结婚 61 年来，她一直支持我的工作。我在交大和上大均承担繁重的教学和科研任务，夹杂着受运动冲击而不能工作也无力顾家的时段。西柳除了完成自己的教学工作外，一直挑起繁杂的家务担子，悉心照护三个女儿的成长和教育，尽心照料两对年迈父母亲的生活和残病、直至四老尽享近九十之天年。她尽己所能创造条件让我能专心工作。如无贤妻的无怨无悔、不离不弃地与我同欢乐共患难，我的生涯必将大幅修改。

最后，我要感谢本校老教授协会、陶德华、王小静的大力支持，使本书得以出版。

张直明
2020 年 6 月 1 日

目 录

期 刊 论 文

轴颈偏斜时 360°径向滑动轴承中的油压分布和承载能力的理论探讨 ············ 3
论紊流润滑理论中的 Hirs 法的理论基础 ············ 15
动压式滑动轴承的稳定性问题 ············ 18
一种新型滑动轴承—隧道孔轴承性能的初步分析 ············ 26
滑动轴承—转子系统的系统阻尼值与稳定裕度的相互关系 ············ 36
计入弹性动变形的圆柱形径向滑动轴承动力特性的研究 ············ 46
圆柱滚子轴承固有频率计算方法的改进 ············ 60
有限宽隧道孔轴承的动力特性和稳定性 ············ 67
EHL Analysis of Rib-Roller End Contact in Tapered Roller Bearings ············ 75
圆锥滚子轴承挡边接触副弹性流体动力润滑分析 ············ 87
粘弹性轴瓦对滑动轴承动力特性的影响 ············ 96
小间隙浮动理论研究 ············ 102
关于高速内圆磨头主轴-滑动轴承系统稳定性的讨论 ············ 110
浮环轴承支承的转子系统的静态特性和动态稳定性研究 ············ 116
复套式转子-轴承系统的固有复频率计算 ············ 123
负压型磁头的设计分析 ············ 132
Analysis of Cylindrical Journal Bearing With Viscoelastic Bush ············ 138
负压型磁头的静动态特性 ············ 147
径向滑动轴承轴瓦反转的现象和机理分析 ············ 153
计入弹性动变形的单块径向轴承可倾瓦动特性 ············ 160
油叶型轴瓦性能数据库研究 ············ 168
计入弹性变形的可倾瓦轴承和转子系统的动力性态 ············ 175
Non-newtonian Elastohydrodynamic Lubrication Analysis of Rib-roller End Contact
　in Tapered Roller Bearings ············ 183
油叶型轴承非线性油膜力数据库 ············ 197
应用高级紊流模式的紊流润滑理论分析 ············ 203
应用高级紊流模式对雷诺润滑方程的解算方法 ············ 209
高速重载轴承紊流润滑特性分析 ············ 215
Thermal Non-Newtonian EHL Analysis of Rib-Roller End Contact in Tapered
　Roller Bearings ············ 220
一种紊流润滑理论分析新方法——复合型紊流模式理论 ············ 238
核电水泵在热冲击下瞬态热效应的研究和分析（Ⅰ）

——热弹性问题微分控制方程的建立和分析 ·· 243
核电水泵在热冲击下瞬态热效应的研究和分析(Ⅱ)
——三维时变有限元分析和计算 ·· 250
大自由度的转子-滑动轴承系统非线性动力学分析 ·· 257
大自由度的转子-滑动轴承系统非线性动力学分析(Ⅱ) ·· 268
Partial EHL Analysis of Rib-roller End Contact in Tapered Roller Bearings ············ 278
不平衡质量的大小和分布对柔性转子-轴承系统稳定性的影响 ································ 291
Approximate Tangent Plane Method for Calculating Surface Deformation in Elastic
　　Contact Problems ·· 297
弹性金属塑料瓦推力轴承的滑移问题研究 ·· 307
大扰动情况下滑动轴承内瞬态油膜分布的研究 ··· 314
360°动、静载荷滑动轴承油膜分布实验台的设计及实验研究 ··································· 321
有限长线接触弹流润滑研究的现状与展望 ·· 329
弹簧支承式推力轴承的热弹流研究 ··· 337
多跨转子—滑动轴承系统非线性动力学仿真 ·· 345
用梁单元表达转子刚度时阶梯处过渡段的最佳等效参数 ··· 352
Experimental Study of Active Magnetic Bearing on a 150 M^3 Turbo Oxygen Gas
　　Expander ·· 356
Effect of Geometry Change of Rough Point Contact due to Lubricated Sliding Wear
　　on Lubrication ·· 360
Jeffcot 转子-滑动轴承系统不平衡响应的非线性仿真 ·· 377
Improving the Performance of Spring-supported Thrust Bearing by Controlling Its
　　Deformations ·· 384
计入非牛顿效应的曲轴轴承的混合润滑分析 ·· 394
计入入口冲击压力的弹簧支承式推力轴承热弹流研究 ··· 401
复合型紊流润滑理论模式的研究 ·· 408
电磁轴承在透平膨胀机中的应用研究进展 ·· 413
A Comparison of Flow Fields Predicted by Various Turbulent Lubrication Models
　　With Existing Measurements ··· 418
THD Analysis of High Speed Heavily Loaded Journal Bearings Including Thermal
　　Deformation, Mass Conserving Cavitation, and Turbulent Effects ··················· 423
Application of the Non-Stationary Oil Film Force Database ·································· 434
Analysis on Dynamic Performance for Active Magnetic Bearing-Rotor System ······ 440
主动磁轴承转子系统动力学特性的研究 ··· 447
电磁轴承及其系统设计方法 ·· 457
周隙密封紊流润滑研究 ··· 465
Turbulence Models of Hydrodynamic Lubrication ··· 471
Theory of Cavitation in an Oscillatory Oil Squeeze Film ·· 485
Experimental Study of Cavitation in an Oscillatory Oil Squeeze Film ···················· 503
径向浮环动静压轴承稳定性研究 ·· 518

An Explicit Solution for the Elastic Quarter-space Problem in Matrix Formulation ········ 527
An Explicit Matrix Algorithm for Solving Three-dimensional Elastic Wedge Under Surface Loads ········ 535
Surface Normal Deformation in Elastic Quarter-space ········ 556
Study on the Free Edge Effect on Finite Line Contact Elastohydrodynamic Lubrication ········ 569

国际会议、全国会议论文集论文

Effects of Pad Elastic Deformation on Tilting Pad Bearing Properties and Rotor System Behavior ········ 589
System Damping Factor of an Elastic Rotor Supported on Tilting-Pad Bearings with Elastic and Damped Pivots ········ 599
The Effect of Dynamic Deformation on Dynamic Properties and Stability of Cylindrical Journal Bearings ········ 609
复套式转子-滑动轴承系统动力学计算 ········ 616
Effect of Bush Viscoelasticity on Journal Bearing Dynamic Properties and Stability ········ 625
Calculation of Hydrodynamic Lobe Type Journal Bearings Aided by Database of Properties of Single Bush Segment ········ 634
Calculation of Journal Dynamic Locus Aided by Database of Non-stationary Oil Film Force of Single Bush Segment ········ 644
Calculation of Tilting Pad Bearings with Elastically Deformable Pads and Pivots Aided by Database of Single Deformable Tilting Pad ········ 653
Nonlinear Simulation of Lateral Vibration of an Experimental Rotor-Journal Bearing System ········ 663
Analysis of Crankshaft Bearings in Mixed Lubrication Including Mass Conserving Cavitation ········ 671
Nonlinear Dynamic Analysis of Multi-span Rotor-journal Bearing-foundation System: Part I ········ 681
Nonlinear Dynamic Analysis of Multi-span Rotor-journal Bearing-foundation System: Part II ········ 694
A General Matrix Solution for the Elastic Quarter Space ········ 710

期刊论文

轴颈偏斜时 360°径向滑动轴承中的油压分布和承载能力的理论探讨[*]

摘 要：在径向滑动轴承中，轴颈偏斜对油压分布和承载能力的影响，尚少全面的理论分析．目前轴承计算法对轴颈偏斜的考虑，也没有充分的理论根据．

在本文中，将油压表为偏心率和偏斜率的幂级数，由雷诺方程解出了油压分布，并从而求出轴颈中心的轨迹和承载能力．由此推荐了一套计算方法，亦推荐了缺乏专门数据时较易行的简化计算法．

本文还没有能全部解决这一情况下的理论问题，因为当偏心率或偏斜率愈近于 1 时，级数的收敛愈慢，而且边界条件方面也做了简化的假设．但是作为研究这个问题的初步尝试，本文仍试图表达自己的观点．

一、引言

在液体摩擦的径向滑动轴承中，当轴颈由于安装误差和受力变形而发生偏斜时，对轴承的工作起着很不利的影响：在同样的径向力下，使轴承和轴颈间的最小间隙量减小，或在同样的最小间隙下使承载能力降低．有时因之而使油膜在轴承一端遭到破坏，使轴颈和轴承在此处产生"边缘摩擦"而引起损伤和失效．

一般的液体摩擦轴承计算法中，是在最小油膜厚度的临界值中加以轴颈的变形值，来考虑其影响的[1]．但是此种做法，并非建立在流体动力学的基础上的，而且是过分安全的．

从流体动学观点来全面分析这个问题的文章目前为数尚不很多，而且多半是针对个别情况进行的．G. B. Dubois 等曾做了大量试验[2]，并得出曲线，但是缺乏理论分析．几年前，苏联 Е. М. Гутьяр 曾在全苏第三次摩擦和磨损会议时提出有必要解决轴颈偏斜时的轴承计算问题[3]．

本文试图用一种级数解法来处理这个问题．

二、轴颈在轴承中的位置

轴颈在轴承中之所以会有偏斜（或其他变形），其原因很多，如安装误差、受力变形等，而所发生的偏斜情况亦各不相同．一般情况下，主要由两部分组成，一部分是轴颈轴线在轴承长度中央的切线方向不平行于轴承轴线[图 1(a)]，另一部分是轴颈轴线的挠曲[图 1(b)]．当采用自位轴承时，或对于某些中轴颈（也包括一些端轴颈，视具体情况而定），前一部分变形的影响由于轴承自动调位或其他原因而大为降低，主要是后一部分发生影响．但在采用非自位轴承时，特别是长度较大的端轴颈中，常常是前一部分占主要成分．

---

[*] 原发表于《西安交通大学学报》，1963(1)：50—61．

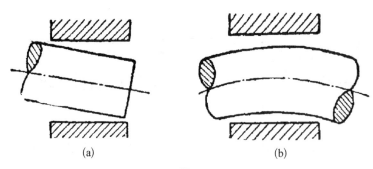

图 1

本文主要讨论在轴端采用非自位的 360° 轴承时，轴颈由于受力变形而发生偏斜的影响．但就解题的方法而言，则亦适用于许多其他情况．

当端轴颈由于受力变形偏斜时，轴颈在轴承中的位置如下（图 2）．在轴承长度的中央截面内，轴颈中心 O_1' 相对于轴承中心 O_2 有一偏心距 $e = \overline{O_2 O_1'}$，其偏位角为 θ．而在其他截面中，由于轴颈受力偏斜，在这个偏心距之外附加上一个偏心距 $e' = \overline{O_1' O_1}$，其方向大致平行于轴承径向载荷 P 的方向，其大小则正比于该截面与中截面间之距离．设 E 为端截面内的偏斜距，则 $E \approx \dfrac{L}{2}\lambda$，其中 λ——轴颈的偏斜角，L——轴承长度．

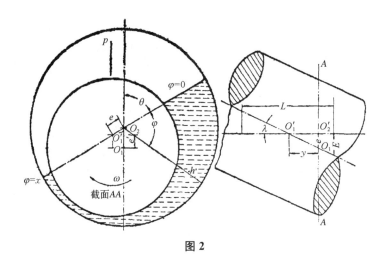

图 2

采用相对值，取中截面的偏心率 $\chi \approx \dfrac{2e}{\Delta}$，端截面的倾斜附加偏心率 $\varepsilon = \dfrac{2E}{\Delta}$，其中 Δ——直径间隙，即轴承孔直径与轴颈直径之差．取任一截面与中截面间之无量纲距离 $w = \dfrac{2y}{L}$，则该截面内的偏心率为 χ 与倾斜偏心率 εw 的几何和．

因此，如以中截面内最大间隙处作为起始角位，则在距中截面的任一截面内在 φ 角处的间隙为 $h \approx \dfrac{\Delta}{2} + e\cos\varphi + E\dfrac{2y}{L}\cos(\theta+\varphi)$．如以 $\bar{h} = \dfrac{2h}{\Delta}$ 表示局部的相对间隙，则

$$\bar{h} \approx 1 + \chi\cos\varphi + \varepsilon w \cos(\theta+\varphi) \tag{1}$$

三、方程和边界条件

当液体摩擦轴承中的油膜处在层流定常运动情况下时,描述油压 p 分布的雷诺方程为

$$\frac{\partial}{\partial x}\left(\frac{h^3}{\mu}\frac{\partial p}{\partial x}\right)+\frac{\partial}{\partial y}\left(\frac{h^3}{\mu}\frac{\partial p}{\partial y}\right)=6U\frac{\partial h}{\partial x} \tag{2}$$

式中 x——在轴颈相对于轴承运动方向上的坐标,即 $x=\frac{d}{2}\varphi$,其中 d——轴颈直径;y——轴向坐标;U——轴颈的圆周速度,即 $U=\frac{d}{2}\omega$,其中 ω——轴颈的角速度;μ——油的绝对黏度.

如不考虑黏度随压力和温度变化,并取无量纲油压 $\bar{p}=\frac{\psi^2 p}{2\mu\omega}$,其中 $\psi=\frac{\Delta}{d}$——相对间隙,则可将式(2)化为如下的无量纲形式:

$$\frac{\partial}{\partial\varphi}\left(\bar{h}^3\frac{\partial\bar{p}}{\partial\varphi}\right)+\left(\frac{d}{L}\right)^2\frac{\partial}{\partial w}\left(\bar{h}^3\frac{\partial\bar{p}}{\partial w}\right)=3\frac{\partial\bar{h}}{\partial\varphi} \tag{3}$$

对于油膜的始末端位置和形状,在轴颈不偏斜的360°径向轴承中,常可取最大间隙处 $(\varphi_1=0)$ 的轴向线作为油膜始端,略后于最小间隙 $(\varphi_2>\pi)$ 的某个位置作为油膜终端,但亦有取最小间隙处 $(\varphi_2=\pi)$ 作为油膜终端的[4]. 在轴颈有偏斜时,油膜始、末端可能有所改变,但尚乏充分实验资料. 作为初步近似,仍取 $\varphi_1=0$ 和 $\varphi_2=\pi$. 因此,无压力供油时,油压 $\bar{p}$ 的边界条件为

$$\varphi=0 \text{ 和 } \pi \text{ 时}, \quad \bar{p}=0;$$
$$w=\pm 1 \text{ 时}, \quad \bar{p}=0 \tag{4}$$

(以大气压作为参考压力).

四、$\bar{p}$ 的级数解

当轴颈有偏斜时,解方程(3)的主要困难在于 $\bar{h}$ 既是 φ 的函数又是 w 的函数,且无法采用分离变量法. 因此采用一种近似解法或数值解法较为方便. 数值解法和某些近似解法,每一次只能解一种具体 χ 和 ε 值下的 $\bar{p}$ 分布. 为能将 $\bar{p}$ 对于任何 ω 和 ε 一次解出,采用如下的级数解(所谓微扰法).

将 $\bar{p}$ 表为小参数 χ 和 ε 的幂级数:

$$\bar{p}=p_0+\chi p_1+\varepsilon p_2+\chi^2 p_3+\chi\varepsilon p_4+\varepsilon^2 p_5+\cdots \tag{5}$$

式中 $p_0,p_1,\cdots$ 都是坐标 φ 和 ω 的函数,且其边界条件同(4).

将式(5)代入式(3),得

$$\left[\frac{\partial^2 p_0}{\partial\varphi^2}+\left(\frac{d}{L}\right)^2\frac{\partial^2 p_0}{\partial w^2}\right]+\chi\left[\frac{\partial^2 p_1}{\partial\varphi^2}+\left(\frac{d}{L}\right)^2\frac{\partial^2 p_1}{\partial w^2}\right]+\varepsilon\left[\frac{\partial^2 p_2}{\partial\varphi^2}+\left(\frac{d}{L}\right)^2\frac{\partial^2 p_2}{\partial w^2}\right]+$$

$$\chi^2\left[3\frac{\partial}{\partial\varphi}\left(\cos\varphi\frac{\partial p_1}{\partial\varphi}\right)+\frac{\partial^2 p_3}{\partial\varphi^2}+3\left(\frac{d}{L}\right)^2\frac{\partial}{\partial w}\left(\cos\varphi\frac{\partial p_1}{\partial w}\right)+\left(\frac{d}{L}\right)^2\frac{\partial^2 p_3}{\partial w^2}\right]+$$

$$\chi\varepsilon\left\{3\frac{\partial}{\partial\varphi}\left[w\cos(\varphi+\theta)\frac{\partial p_1}{\partial\varphi}\right]+3\frac{\partial}{\partial\varphi}\left(\cos\varphi\frac{\partial p_2}{\partial\varphi}\right)+\frac{\partial^2 p_4}{\partial\varphi^2}+3\left(\frac{d}{L}\right)^2\cdot\right.$$

$$\left.\frac{\partial}{\partial w}\left[w\cos(\varphi+\theta)\frac{\partial p_1}{\partial w}\right]+3\left(\frac{d}{L}\right)^2\frac{\partial}{\partial w}\left(\cos\varphi\frac{\partial p_2}{\partial w}\right)+\left(\frac{d}{L}\right)^2\frac{\partial^2 p_4}{\partial w^2}\right\}+$$

$$\varepsilon^2\left\{3\frac{\partial}{\partial\varphi}\left[w\cos(\varphi+\theta)\frac{\partial p_2}{\partial\varphi}\right]+\frac{\partial^2 p_5}{\partial\varphi^2}+3\left(\frac{d}{L}\right)^2\frac{\partial}{\partial w}\left[w\cos(\varphi+\theta)\frac{\partial p_2}{\partial w}\right]+\right.$$

$$\left.\left(\frac{d}{L}\right)^2\frac{\partial^2 p_5}{\partial w^2}\right\}+\cdots=-3\chi\sin\varphi-3\varepsilon w\sin(\varphi+\theta). \tag{6}$$

由于此方程应对于任何 χ 和 ε 都为正确,故式左、右 χ 和 ε 的同幂项之系数必须各各相等,因此得一组方程:

$$\frac{\partial^2 p_0}{\partial\varphi^2}+\left(\frac{d}{L}\right)^2\frac{\partial^2 p_0}{\partial w^2}=0; \tag{7}$$

$$\frac{\partial^2 p_1}{\partial\varphi^2}+\left(\frac{d}{L}\right)^2\frac{\partial^2 p_1}{\partial w^2}=-3\sin\varphi; \tag{8}$$

$$\frac{\partial^2 p_2}{\partial\varphi^2}+\left(\frac{d}{L}\right)^2\frac{\partial^2 p_2}{\partial w^2}=-3w\sin(\varphi+\theta); \tag{9}$$

$$\frac{\partial^2 p_3}{\partial\varphi^2}+\left(\frac{d}{L}\right)^2\frac{\partial^2 p_3}{\partial w^2}=-3\frac{\partial}{\partial\varphi}\left(\cos\varphi\frac{\partial p_1}{\partial\varphi}\right)-3\left(\frac{d}{L}\right)^2\frac{\partial}{\partial w}\left(\cos\varphi\frac{\partial p_1}{\partial w}\right); \tag{10}$$

$$\frac{\partial^2 p_4}{\partial\varphi^2}+\left(\frac{d}{L}\right)^2\frac{\partial^2 p_4}{\partial w^2}=-3\left\{\frac{\partial}{\partial\varphi}\left[w\cos(\varphi+\theta)\frac{\partial p_1}{\partial\varphi}\right]+\left(\frac{d}{L}\right)^2\frac{\partial}{\partial w}\left[w\cos(\varphi+\theta)\frac{\partial p_1}{\partial w}\right]+\right.$$

$$\left.\frac{\partial}{\partial\varphi}\left(\cos\varphi\frac{\partial p_2}{\partial\varphi}\right)+\left(\frac{d}{L}\right)^2\frac{\partial}{\partial w}\left(\cos\varphi\frac{\partial p_2}{\partial w}\right)\right\}; \tag{11}$$

$$\frac{\partial^2 p_5}{\partial\varphi^2}+\left(\frac{d}{L}\right)^2\frac{\partial^2 p_5}{\partial w^2}=-3\left\{\frac{\partial}{\partial\varphi}\left[w\cos(\varphi+\theta)\frac{\partial p_2}{2\varphi}\right]+\right.$$

$$\left.\left(\frac{d}{L}\right)^2\frac{\partial}{\partial w}\left[w\cos(\varphi+\theta)\frac{\partial p_2}{\partial w}\right]\right\}; \tag{12}$$

······

逐个解出这些方程,代入式(5),即得 $\bar{p}$. 当 χ 和 ε 不大时,取头六项已经足够.

今解得

$$p_0=0; \tag{13}$$

$$p_1=3\sin\varphi\left(1-\frac{\operatorname{ch}\frac{L}{d}w}{\operatorname{ch}\frac{L}{d}}\right); \tag{14}$$

$$p_2=-3\cos\theta\sin\varphi\frac{\operatorname{sh}\frac{L}{d}w}{\operatorname{sh}\frac{L}{d}}+3w\left[\sin(\varphi+\theta)+\left(\frac{2\varphi}{\pi}-1\right)\sin\theta\right]+$$

$$\sum_{k=1}^{\infty} A_{2k} \sin 2k\varphi \frac{\operatorname{sh} 2k \frac{L}{d} w}{\operatorname{sh} 2k \frac{L}{d}}; \tag{15}$$

式中

$$A_{2k} = \frac{-6\sin\theta}{\pi k (4k^2 - 1)}; \tag{16}$$

$$p_3 = -\frac{3}{4} \sin 2\varphi \left(3 - 2 \frac{\operatorname{ch} \frac{L}{d} w}{\operatorname{ch} \frac{L}{d}} - \frac{\operatorname{ch} 2 \frac{L}{d} w}{\operatorname{ch} 2 \frac{L}{d}} \right); \tag{17}$$

$$p_4 = p'_4 + \sum_{k=1}^{\infty} B_k \sin\left[\frac{k\pi}{2}(w+1)\right] \frac{\operatorname{ch}\left[\frac{d}{L}\frac{k\pi}{4}(2\varphi - \pi)\right]}{\operatorname{ch}\left(\frac{d}{L}\frac{k\pi^2}{4}\right)}, \tag{18}$$

式中

$$p'_4 = \frac{9}{2}\sin\theta \left[\frac{d}{L}\operatorname{th}\frac{L}{d}\left(\frac{\operatorname{sh}\frac{L}{d}w}{\operatorname{sh}\frac{L}{d}} - w\right) + w\left(1 - \frac{\operatorname{ch}\frac{L}{d}w}{\operatorname{ch}\frac{L}{d}}\right)\right] - \frac{18}{\pi}\sin\theta$$

$$\sin\psi\left(w - \frac{\operatorname{sh}\frac{L}{d}w}{\operatorname{sh}\frac{L}{d}}\right) - \frac{\cos\theta}{2}\sin 2\psi \left[9w - \left(3 - \frac{d}{L}\operatorname{th}\frac{L}{d}\right)\frac{\operatorname{sh}\frac{L}{d}w}{\operatorname{sh}\frac{L}{d}} - \right.$$

$$\left. 3w\frac{\operatorname{ch}\frac{L}{d}w}{\operatorname{ch}\frac{L}{d}} - \left(3 + \frac{d}{L}\operatorname{th}\frac{L}{d}\right)\frac{\operatorname{sh}2\frac{L}{d}w}{\operatorname{sh}2\frac{L}{d}}\right] - \frac{\sin\theta}{2}\cos 2\varphi \left[9w + \right.$$

$$\left. \frac{d}{L}\operatorname{th}\frac{L}{d}\frac{\operatorname{sh}\frac{L}{d}w}{\operatorname{sh}\frac{L}{d}} - 3w\frac{\operatorname{ch}\frac{L}{d}w}{\operatorname{ch}\frac{L}{d}} - \left(6 + \frac{d}{L}\operatorname{th}\frac{L}{d}\right)\frac{\operatorname{sh}2\frac{L}{d}w}{\operatorname{sh}2\frac{L}{d}}\right] -$$

$$3\sum_{k=1}^{\infty}\left\{\frac{kA_{2k}}{4k+1}\sin(2k+1)\varphi\left[\frac{\operatorname{sh}2k\frac{L}{d}w}{\operatorname{sh}2k\frac{L}{d}} - \frac{\operatorname{sh}(2k+1)\frac{L}{d}w}{\operatorname{sh}(2k+1)\frac{L}{d}}\right] + \right.$$

$$\left. \frac{kA_{2k}}{4k-1}\sin(2k-1)\varphi\left[\frac{\operatorname{sh}2k\frac{L}{d}w}{\operatorname{sh}2k\frac{L}{d}} - \frac{\operatorname{sh}(2k-1)\frac{L}{d}w}{\operatorname{sh}(2k-1)\frac{L}{d}}\right]\right\} \tag{19}$$

$$B_k = 3\sin\theta \frac{1 + (-1)^k}{k\pi}\left\{\frac{-6}{1 + \left(\frac{d}{L}\frac{k\pi}{2}\right)^2} + \frac{6 + \frac{d}{L}\operatorname{th}\frac{L}{d}}{\left[1 + \left(\frac{d}{L}\frac{k\pi}{2}\right)^2\right]\left[1 + \left(\frac{d}{2L}\frac{k\pi}{2}\right)^2\right]} - \right.$$

$$\left.\frac{4\dfrac{d}{L}\operatorname{th}\dfrac{L}{d}}{\left[1+\left(\dfrac{d}{L}\dfrac{k\pi}{2}\right)^2\right]^2}\right\}; \tag{20}$$

$$p_5 = p_5' + \sum_{i=1}^{\infty} \sin\frac{i\pi}{2}(w+1)\left\{C_i\frac{\operatorname{ch}\left[\dfrac{d}{L}\dfrac{i\pi}{4}(2\varphi-\pi)\right]}{\operatorname{ch}\left(\dfrac{d}{L}\dfrac{i\pi^2}{4}\right)} + D_i\frac{\operatorname{sh}\left[\dfrac{d}{L}\dfrac{i\pi}{4}(2\varphi-\pi)\right]}{\operatorname{sh}\left(\dfrac{d}{L}\dfrac{i\pi^2}{4}\right)}\right\}, \tag{21}$$

式中 $p_5' = W_1 + \sin2\varphi \cdot W_2 + \cos2\varphi \cdot W_3 + \cos\varphi \cdot W_4 + \varphi\sin\varphi \cdot W_5 + \sin\varphi \cdot W_6 +$

$\varphi\cos\varphi \cdot W_7 + \sum_{k=1}^{\infty} kA_{2k}[\cos\theta\sin(2k+1)\varphi \cdot W_{k,8} + \cos\theta\sin(2k-1)\varphi \cdot W_{k,9} +$

$\sin\theta\cos(2k+1)\varphi \cdot W_{k,10} + \sin\theta\cos(2k-1)\varphi \cdot W_{k,11}]$, \qquad (22)

而
$$W_1 = \frac{9}{4}\sin2\theta\left(\frac{d}{L}\frac{\operatorname{ch}\dfrac{L}{d}w - \operatorname{ch}\dfrac{L}{d}}{\operatorname{sh}\dfrac{L}{d}} + 1 - w\frac{\operatorname{sh}\dfrac{L}{d}w}{\operatorname{sh}\dfrac{L}{d}}\right), \tag{23}$$

$$W_2 = -\frac{9}{4}\sin2\theta \cdot w^2 - \frac{1}{2}\frac{d}{L}\cos^2\theta\frac{\operatorname{ch}\dfrac{L}{d}w}{\operatorname{sh}\dfrac{L}{d}} + \frac{3}{2}\cos^2\theta \cdot w\frac{\operatorname{sh}\dfrac{L}{d}w}{\operatorname{sh}\dfrac{L}{d}} +$$

$$\left[\cos^2\theta\left(3 + \frac{1}{2}\frac{d}{L}\operatorname{cth}\frac{L}{d}\right) - \frac{9}{4}\right]\frac{\operatorname{ch2}\dfrac{L}{d}w}{\operatorname{ch2}\dfrac{L}{d}}; \tag{24}$$

$$W_3 = \sin2\theta\left[-\frac{9}{4}w^2 - \frac{1}{4}\frac{d}{L}\frac{\operatorname{ch}\dfrac{L}{d}w}{\operatorname{sh}\dfrac{L}{d}} + \frac{3}{4}w\frac{\operatorname{sh}\dfrac{L}{d}w}{\operatorname{sh}\dfrac{L}{d}} + \left(\frac{3}{2} + \frac{1}{4}\frac{d}{L}\operatorname{cth}\frac{L}{d}\right)\frac{\operatorname{ch2}\dfrac{L}{d}w}{\operatorname{ch2}\dfrac{L}{d}}\right]; \tag{25}$$

$$W_4 = \frac{18}{\pi}\sin^2\theta\left\{-4\left(\frac{d}{L}\right)^2 - w^2 - \frac{d}{L}\frac{w\operatorname{sh}\dfrac{L}{d}w}{\operatorname{ch}\dfrac{L}{d}} + \left[1 + \frac{d}{L}\operatorname{th}\frac{L}{d} + 4\left(\frac{d}{L}\right)^2\right]\frac{\operatorname{ch}\dfrac{L}{d}w}{\operatorname{ch}\dfrac{L}{d}}\right\} + 9\left(\frac{d}{L}\right)^2\sin\theta\cos\theta\left(-1 + \frac{\operatorname{ch}\dfrac{L}{d}w}{\operatorname{ch}\dfrac{L}{d}}\right); \tag{26}$$

$$W_5 = -\frac{18}{\pi}\left(\frac{d}{L}\right)^2\sin^2\theta\left(1 - \frac{\operatorname{ch}\dfrac{L}{d}w}{\operatorname{ch}\dfrac{L}{d}}\right); \tag{27}$$

$$W_6 = \frac{18}{\pi}\sin\theta\cos\theta\left\{-4\left(\frac{d}{L}\right)^2 - w^2 - \frac{d}{L}\frac{w\,\text{sh}\frac{L}{d}w}{\text{ch}\frac{L}{d}} + \left[1 + \frac{d}{L}\text{th}\frac{L}{d} + \right.\right.$$

$$\left.\left. 4\left(\frac{d}{L}\right)^2\right]\frac{\text{ch}\frac{L}{d}w}{\text{ch}\frac{L}{d}}\right\} + 9\left(\frac{d}{L}\right)^2\sin^2\theta\left(1 - \frac{\text{ch}\frac{L}{d}w}{\text{ch}\frac{L}{d}}\right); \qquad (28)$$

$$W_7 = \frac{18}{\pi}\left(\frac{d}{L}\right)^2\sin\theta\cos\theta\left(1 - \frac{\text{ch}\frac{L}{d}w}{\text{ch}\frac{L}{d}}\right); \qquad (29)$$

$$W_{k,8} = W_{k,10} = \frac{3}{4k+1}\left[-w\frac{\text{sh}2k\frac{L}{d}w}{\text{sh}2k\frac{L}{d}} + \frac{1}{4k+1}\frac{d}{L}\frac{\text{ch}2k\frac{L}{d}w}{\text{sh}2k\frac{L}{d}} + \right.$$

$$\left. \left(1 - \frac{1}{4k+1}\frac{d}{L}\text{cth}2k\right)\frac{\text{ch}(2k+1)\frac{L}{d}w}{\text{ch}(2k+1)\frac{L}{d}}\right]; \qquad (30)$$

$$W_{k,9} = -W_{k,11} = \frac{3}{4k-1}\left[-w\frac{\text{sh}2k\frac{L}{d}w}{\text{sh}2k\frac{L}{d}} + \frac{1}{4k-1}\frac{d}{L}\frac{\text{ch}2k\frac{L}{d}w}{\text{sh}2k\frac{L}{d}} + \right.$$

$$\left. \left(1 - \frac{1}{4k-1}\frac{d}{L}\text{cth}2k\frac{L}{d}\right)\frac{\text{ch}(2k-1)\frac{L}{d}w}{\text{ch}(2k-1)\frac{L}{d}}\right]; \qquad (31)$$

又 $$C_i = -\int_{-1}^{+1}\left(W_1 + W_3 \cdot \frac{\pi}{2}W_7\right)\sin\frac{i\pi}{2}(w+1)\mathrm{d}w = \frac{\sin 2\theta}{2i\pi}[1-(-1)^i]\cdot$$

$$\left\{\frac{-72}{(i\pi)^2} + \frac{-6+18\left(\frac{d}{L}\right)^2 - 4\frac{d}{L}\text{cth}\frac{L}{d}}{1+\left(\frac{i\pi d}{2L}\right)^2} + \frac{12\frac{d}{L}\text{cth}\frac{L}{d}}{\left[1+\left(\frac{i\pi d}{2L}\right)^2\right]^2} + \frac{6+\frac{d}{L}\text{cth}\frac{L}{d}}{1+\left(\frac{i\pi d}{4L}\right)^2}\right\}; \qquad (32)$$

$$D_i = \int_{-1}^{+1}\left[W_4 + \frac{\pi}{2}W_7 + \sum_{k=\lambda}^{\infty}kA_{2k}\sin\theta(W_{k,10} + W_{k,11})\right]\sin\frac{i\pi}{2}(w+1)\mathrm{d}w$$

$$= \frac{36\sin^2\theta}{\pi^2}\frac{1-(-1)^i}{i}\left\{\frac{8}{(i\pi)^2} - \frac{1+2\left(\frac{d}{L}\right)^2}{1+\left(\frac{i\pi d}{2L}\right)^2} - \frac{2\left(\frac{d}{L}\right)^2}{\left[1+\left(\frac{i\pi d}{2L}\right)^2\right]^2} - \right.$$

$$\sum_{k=1}^{\infty}\frac{1}{4k^2-1}\left[\frac{1}{1+\left(\frac{i\pi d}{4kL}\right)^2}\left(\frac{-2}{16k^2-1} + \frac{16k\frac{d}{L}\text{cth}2k\frac{L}{d}}{(16k^2-1)^2}\right) - \right.$$

$$\frac{1}{4k+1} \times \frac{1 - \dfrac{d}{L}\dfrac{\operatorname{cth}2k\dfrac{L}{d}}{4k+1}}{1 + \left[\dfrac{i\pi d}{2(2k+1)L}\right]^2} + \frac{1}{4k-1} \times \frac{1 - \dfrac{d}{L}\dfrac{\operatorname{cth}2k\dfrac{L}{d}}{4k-1}}{1 + \left[\dfrac{i\pi d}{2(2k-1)L}\right]^2} -$$

$$\left.\left.\frac{2d}{kL}\operatorname{cth}\left(2k\frac{L}{d}\right)\frac{1}{16k^2-1}\left(\frac{1}{1+\left(\dfrac{i\pi d}{4kL}\right)^2} - \frac{1}{\left[1+\left(\dfrac{i\pi d}{4kL}\right)^2\right]^2}\right)\right]\right\}. \tag{33}$$

五、对级数解的一种校核

对于轴颈不偏斜的无限长轴承,在定常运动下,设 μ 为常数,且油膜始于 $\varphi_1 = 0$ 而终于 $\varphi_2 = \pi$,则油压 $\bar{p}_\infty$ 很易由雷诺方程直接积分得到,为[4]:

$$\bar{p}_\infty = -\frac{1}{(1-\chi^2)\sqrt{1-\chi^2}}(B\sin\eta + C\sin 2\eta), \tag{34}$$

其中 η——"双极坐标"中的一个,即:

$$\eta = 2\tan^{-1}\left(\sqrt{\frac{1-\chi}{1+\chi}}\tan\frac{\varphi}{2}\right) - \pi;$$

而

$$B = 3\chi\frac{2-\chi^2}{2+\chi^2}; \qquad C = \frac{3}{2}\frac{\chi^2}{2+\chi^2}.$$

如将式(34)的 $\bar{p}_\infty$ 对 χ 展开为幂级数,则得:

$$\bar{p}_\infty = 3\chi\sin\varphi - \frac{9}{4}\chi^2\sin 2\varphi + \frac{3}{2}\chi^3\sin 3\varphi + \cdots \tag{35}$$

而按本文的级数解,在轴承的中截面($w=0$)内,当轴承无限长$\left(\dfrac{L}{d} \to \infty\right)$而轴颈不偏斜($\varepsilon = 0$)时,由式(13)~(33)及(5)亦可得:

$$\bar{p}_\infty = 3x\sin\varphi - \frac{9}{4}x^2\sin 2\varphi + \cdots,$$

恰好与式(35)相符.

六、中截面内的轴颈中心轨迹及承载能力

因油压的合力必须与轴颈上的外力相平衡,故大致处在轴颈偏斜平面中(当略去油层对轴颈转动的阻力时),即:

$$\tan\theta = -\frac{\int_0^\pi \int_{-1}^{+1} \bar{p}\sin\varphi\, dw\, d\varphi}{\int_0^\pi \int_{-1}^{+1} \bar{p}\cos\varphi\, dw\, d\varphi}. \tag{36}$$

由此即可定出给定 χ 和 ε 下的 θ 值.

$$\int_0^\pi \int_{-1}^{+1} \bar{p}\sin\varphi \,\mathrm{d}w\mathrm{d}\varphi = 3\pi\chi\left(1-\frac{d}{L}\text{th}\frac{L}{d}\right) + \varepsilon^2\sin2\theta\left[7-\frac{367}{12}\left(\frac{d}{L}\right)^2 + \right.$$

$$8\frac{d}{L}\text{th}\frac{L}{d} + \frac{99}{2}\left(\frac{d}{L}\right)^3\text{th}\frac{L}{d} + 9\left(\frac{d}{L}\right)^2\text{th}^2\frac{L}{d} - 19\frac{d}{L}\text{cth}\frac{L}{d} - \frac{d}{L}\text{th}2\frac{L}{d} + $$

$$\frac{1}{2}\frac{d}{L}\text{cth}2\frac{L}{d} + \frac{1}{3}\left(\frac{d}{L}\right)^2\text{cth}2\frac{L}{d}\text{th}\frac{L}{d} - \frac{1}{6}\left(\frac{d}{L}\right)^2\text{cth}\frac{L}{d}\text{th}2\frac{L}{d}\bigg] + $$

$$\varepsilon^2\frac{4}{\pi}\sum_{i=1}^{\infty}C_i\frac{1-(-1)^i}{i\left[1+\left(\frac{i\pi d}{2L}\right)^2\right]}; \tag{37}$$

$$\int_0^\pi\int_{-1}^{+1}\bar{p}\cos\varphi\,\mathrm{d}w\mathrm{d}\varphi = -\chi^2\left(6-4\frac{d}{L}\text{th}\frac{L}{d}-\frac{d}{L}\text{th}2\frac{L}{d}\right)+\varepsilon^2\bigg[-2-\frac{16}{3}\left(\frac{d}{L}\right)^2+$$

$$4\frac{d}{L}\text{cth}\frac{L}{d}+\left(1+\frac{2}{3}\frac{d}{L}\text{cth}\frac{L}{d}\right)\frac{d}{L}\text{th}2\frac{L}{d}\bigg]+\varepsilon^2\sin^2\theta\bigg[-2-\frac{449}{6}\left(\frac{d}{L}\right)^2-$$

$$20\frac{d}{L}\text{th}\frac{L}{d}+81\left(\frac{d}{L}\right)^3\text{th}\frac{L}{d}+18\left(\frac{d}{L}\right)^2\text{th}^2\frac{L}{d}-4\frac{d}{L}\text{cth}\frac{L}{d}-$$

$$\left(4+\frac{2}{3}\frac{d}{L}\text{cth}\frac{L}{d}\right)\frac{d}{L}\text{th}2\frac{L}{d}-\frac{d}{L}\text{cth}2\frac{L}{d}-\frac{2}{3}\left(\frac{d}{L}\right)^2\text{cth}2\frac{L}{d}\text{th}\frac{L}{d}\bigg]-$$

$$2\varepsilon^2\frac{d}{L}\sum_{i=1}^{\infty}D_i\frac{1-(-1)^i}{1+\left(\frac{d}{L}\frac{i\pi}{2}\right)^2}\text{cth}\left(\frac{d}{L}\frac{i\pi^2}{4}\right). \tag{38}$$

当 $\frac{L}{d}=1$ 时,得

$$\tan\theta = \frac{1.124\chi^2 + 0.1016\varepsilon^2\sin2\theta}{0.995\chi^2 + 0.1366\varepsilon^2 - 0.259\varepsilon^2\sin^2\theta}, \tag{39}$$

由此可计算不同 χ 及 ε 下的 θ,如表 1. 图 3 表示 $\varepsilon=0$ 及 $\varepsilon=0.5$ 时由 $\chi=0$ 至 $\chi=0.5$ 的中截面轴颈中心轨迹线.

将 χ 和 ε 作几何和,即可求得轴承一端(左端或右端)的最大偏心率 χ_{max}:

$$\chi_{max} = \sqrt{\chi^2+\varepsilon^2+2\chi\varepsilon\cdot|\cos\theta|} \tag{40}$$

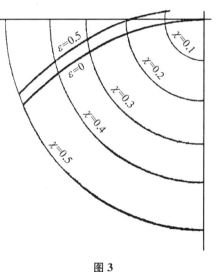

图 3

表 1 $\frac{L}{d}=1$ 时,各种 χ 和 ε 下的 θ 值

ε \ χ	0.1	0.2	0.3	0.4	0.5
0	84°56′	79°57′	75°7′	70°30′	66°7′
0.1	85°33′	80°15′	75°19′	70°38′	66°13′
0.2	87°28′	81°81′	75°54′	71°2′	66°36′
0.3	90°33′	82°45′	76°51′	71°42′	67°
0.4	94°50′	84°49′	78°12′	72°34′	67°41′
0.5	100°35′	87°40′	80°1′	73°54′	68°37′

当 $\frac{L}{d}=1$ 时，χ_{max} 值如表2.

表2 $\frac{L}{d}=1$ 时，各种 χ 和 ε 下的 χ_{max} 值

ε \ χ	0.1	0.2	0.3	0.4	0.5
0	0.1	0.2	0.3	0.4	0.5
0.1	0.147 5	0.238	0.34	0.443	0.549
0.2	0.227 5	0.304	0.399	0.502	0.608
0.3	0.317 5	0.381	0.47	0.57	0.676
0.4	0.420 5	0.463	0.546	0.645	0.751
0.5	0.527 5	0.545 5	0.723	0.723	0.827

以无量纲的承载量系数表示承载能力：

$$\zeta = \frac{P}{Ld} \cdot \frac{\psi^2}{\mu_0 \omega} = \frac{1}{2\sin\theta}\int_0^\pi \int_{-1}^{+1} \bar{p}\sin\varphi \,\mathrm{d}w\,\mathrm{d}\varphi$$

或

$$-\frac{1}{2\cos\theta}\int_0^\pi \int_{-1}^{+1} \bar{p}\cos\varphi \,\mathrm{d}w\,\mathrm{d}\varphi, \tag{41}$$

则当 $\frac{L}{d}=1$ 时，得

$$\zeta = \frac{1.124\chi + 0.101\,6\varepsilon^2 \sin2\theta}{\sin\theta}. \tag{42}$$

按此式算出的 ζ 值如表3. 图4表示 $\zeta \sim \chi_{max}$ 的关系，可作为轴承计算的依据.

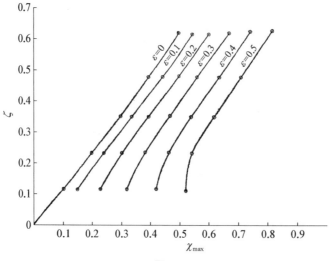

图4

表 3 $\frac{L}{d}=1$ 时,各种 χ 和 ε 下的 ζ 值

ε \ χ	0.1	0.2	0.3	0.4	0.5
0	0.112 8	0.228 5	0.349	0.476	0.615
0.1	0.112 9	0.228 5	0.349	0.478	0.615
0.2	0.113 5	0.228 5	0.35	0.478	0.615
0.3	0.112 4	0.229	0.351	0.379	0.617
0.4	0.11	0.228 5	0.351	0.481	0.619
0.5	0.105	0.228 5	0.351	0.482	0.621

七、几点讨论

由式(39)及图3可见,轴颈偏斜率 ε 对于轴颈中截面中心偏位角 θ 的影响较 χ 为小,当 χ 数值大时,ε 对 θ 的影响几可不计,这与 G. B. Dubois 等的试验结果[2]很符合.

由式(42)及表3又可见,在同样的 χ 下,ε 对 ζ 的影响很小,这亦符合于 G. B. Dubois 等的试验结果. 当然,在同样的 χ_{max} 下,则 ε 愈大将使 ζ 愈小,如图4.

可按图4的曲线对 $\frac{L}{d}=1$ 的轴承进行计算(当 χ 和 ε 不超出 0.5 时). 此时,先取定轴承的尺寸及间隙 Δ,再按轴的受力确定轴颈的偏斜角 λ(不计油压对轴颈的力偶作用),从而得出 ε,按轴承径向力计算 ζ,按图4中相应的曲线查出轴承一端的 χ_{max},并计算相应的最小油膜厚度:

$$\bar{h}_{min}=1-\chi_{max},$$

而
$$h_{min}=\bar{h}_{min}\frac{\Delta}{2}, \tag{43}$$

然后将算出的 h_{min} 与临界值 $h_{\pi p}$(在 $h_{\pi p}$ 中已无须计入轴颈的受力变形了)比较,并对轴承设计作必要的修改和重算.

鉴于 ε 对 θ 和 ζ 的影响较小,而且轴颈偏斜的其余情况下数据缺乏(例如 $\frac{L}{d}\neq 1$ 或 χ 或 $\varepsilon<0.5$),亦可作如下的简化计算,而无须采用曲线图4. 可按轴颈不偏斜时的一般结果,根据 ζ 查出 χ 和 θ,近似地认为此 χ 及 θ 即等于轴颈有偏斜时的数值,则代入式(40)求出 χ_{max},从而求出 h_{min}. 此与 Dubois 等根据试验结果所推荐的方法在性质上是一致的.

必须指出本文的一些缺点:级数解法只宜用于 χ 和 ε 不大时(例如不大于 0.5),否则收敛慢,需计入的项数增多,计算更繁,对于油膜边界形状作了很简化的假设,尚需专门的试验观察.

参 考 文 献

[1] [苏联]H. C. Aчеркан 主编. 机械制造者手册. 卷 4. 胡汉章,等译. 中国工业出版社,1962.
[2] Dubois G B. Properties of misaligned journal bearings. Transactions of the A. S. M. E. 1957: 79.
[3] Гутьяр Е М. Современные направления в развитии гидродинамической теории смазка, Труды третьей всесоюзной

Конференции по трению и износу в мащинах. Том Ⅲ, АН СССР, 1960.

[4] Коровчинский М. В. Теорзтичэские осиовы работы подшидников скольжения. Машгиз, 1959.

A Theoretical Study of the Oil Pressure Distribution and Load-Carrying Capacity of 360° Sliding Bearings with Inclined Journals

Abstract: The subject concerning the influences of the inclination of shaft journal on the pressure distribution and load-carrying capacity in a journal bearing is as yet not sufficiently treated with theoretically. As to what measure should be taken to account for the journal inclination, the methods of bearing calculations now commonly used also lack a hydrodynamic basis.

In this paper, the Reynolds' equation is solved for the oil pressure distribution, with the latter expressed as a power series of eccentricity ratio and inclination ratio, and hence the trajectory of the journal center and load-carrying capacity are determined. A method of bearing calculation considering journal inclination is recommended, and in addition, a simplified method to be used in the absence of special data is also introduced.

The problem discussed in the paper must not be considered as ultimately solved theoretically; it is because the nearer the eccentricity ratio or inclination ratio approaches unity, the more slowly will the power series converge; moreover, simplified suppositions for the boundary conditions are made in the paper.

论紊流润滑理论中的 Hirs 法的理论基础[*]

摘　要：本文对于 G. G. Hirs 提出的关于紊流润滑的"整体流动理论(bulk flow theory)"提出一些不同的看法. Hirs 认为, 在剪切流动和压力流动的任何一种组合下, 都可以用同一个简单的关系来表达作用在壁面上的剪应力 τ 与相对于该壁面的平均流速 u_m 之间的关系, 并从而得出他的计算方法. 但是本文指出了一些明显不能适用这种关系的剪切流和压力流的组合情况, 而这些组合情况往往是发生在油膜的最重要的部位上, 于是就造成了一定的误差.

70 年代初期, G. G. Hirs 在其博士论文中提出了一种关于流体动力润滑油膜在紊流工况下的计算方法, 并名之为"整体流动理论(bulk flow theory)", 后来又发表在美国机械工程学会学报上[1]. 这个方法的基础如下：

当两个壁面间发生紊流的剪切流动(即 Couette 流动)时, 或发生紊流的压力流动(即 Poiseuille 流动)时, 每一壁面上的剪应力与相对于该壁面的平均流速间的关系都可用同一种简单关系来表示, 即如下的经验公式：

$$\frac{\tau}{\frac{1}{2}\rho u_m^2} = n\left(\frac{\rho u_m h}{\mu}\right)^m \tag{1}$$

式中 u_m——相对于该壁面的平均流速；τ——该壁面上的剪应力. Hirs 进一步认为：在剪切流和压力流的任何一种组合下, 也都可用式(1)来表达 τ 与 u_m 的关系. 从这个基点出发, 他导出了用来计算润滑油膜工作性能的公式.

但实际上不难指出, 式(1)并不能适用于剪切流和压力流的一切组合, 至少在下述组合下是无法用式(1)来描述某一壁面上的情况的.

设在两个平行壁面间有着剪切流动, 同时又加上反向的压力流动(此时沿剪切流方向的压力梯度为正). 只要适当调节压力梯度的大小, 总可使流速时均值 $\bar{u}$ 沿膜厚方向 y 的分布成为如图 1 所示, 即在固定壁面处的 $\frac{\partial \bar{u}}{\partial y}=0$. 此时, 紧贴固定壁面的液流层间没有 U 方向的动量交换, 因此固定壁面上在此种复合流动情况下的剪应力 τ 显然为 0(这也符合 Bousinesq 概念). 但相对于固定壁面的平均流速 u_m 显然不为 0. 如果将式(1)套用到固定壁面上, 将得 $n=0$ 或 $m=-\infty$, 当然无法应用. 而如果用 Hirs 推荐的任何一种 n 和 m 值来计算, 都会使固定壁面白白地多受一个沿 U 方向的剪应力, 从而使力平衡关系变得不准确.

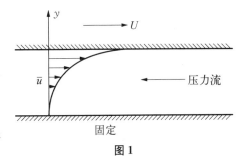

图 1

---

[*] 原发表于《西安交通大学学报》, 1978(4)：25—28.

如果进一步增大逆向压力流,将可达到图 2 所示的流动情况,即相对于固定壁面的平均流速为 0(这就是 Hirs 在[2]的 1113 目中引用的实验情况). 显然,这时固定壁面所受的剪应力(方向反于 U)不可能为 0. Hirs 所引的实验[3]并未测定固定壁面单独所受的剪力(不包括试验轴承的挡边的受力),而 Hirs 在论证中却随意地取之为 0,这当然是不妥的. 显然,式(1)对此时的固定壁面也是无法应用的. 如果仍用 Hirs 方法来计算,将使固定壁面所受的剪应力白白地卸去,也使力平衡关系不准确.

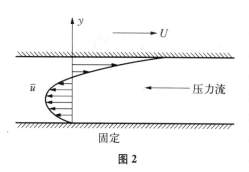

图 2

在这两种情况之间的各种复合流动下,将会发生相对于固定壁面的 u_m 与该壁面所受 τ 反向的情况,当然也是式(1)所无法表达的(按 Hirs 原意,n 是正值,u_m 与 τ 总是同号——即同向的).

在这些情况下,如果 Hirs 的式(1)准确地描述了运动壁面的情况,那么固定壁面上的上述误差,都将使流体动力油膜中算出的该处压力梯度偏低. Hirs 在[2]的 1123 和 1143 目中曾将他对于一些复合流动情况计算的压力值与实验结果[4,5]比较,发现计算值偏低,就是在 1113 目中也有偏低的结果. Hirs 将此种偏低归因于别人的实验误差和惯性影响等. 但上述问题所造成的误差,难道不是更明白的吗?

以上所讨论的两种情况,在实际轴承(无论是动压的或静压的)中都有可能发生于油膜的某些部位(往往是最重要的部位)上,或发生近乎这样的工况. 因此,在这类工况下来讨论问题,是有实际意义的.

如果把这两种情况中的运动壁面认为固定的,而把固定壁面改为以反向的 U 运动,并保持压力梯度不变,就得到了压力流与剪切流同向的情况. 由于此时相对运动没有改变,所以上述讨论对于此时的运动壁面亦显然有效.

总之,Hirs 方法作为一种在一定工况范围内实用的经验方法,并非不可行. 但如果认为它可适用于任何工况,或认为它是一种基础坚实的理论,恐怕是不行的.

参 考 文 献

[1] Hirs, G. G. A Bulk-Flow Theory for Turbulence in Lubricant Films. Trans. ASME, Series F, Apr. 1973: 137.
[2] Hirs, G. G. A Systematic Study of Turbulent Film Flow. Trans. ASME, Series F, Jan. 1974: 118.
[3] Shinkle, J. W., Hornung, K. G. Frictional Characteristics of Liquid Hydrostatic Journal Bearings, Trans. ASME, Series D, Mar. 1965: 163.
[4] Burton, R. A. Fundamental Investigation of Liquid-Metal Lubricated Lournal Bearings, second topical report SwRI-1238 P8-32, final report SwRI-1228P 8-30, Southwest Research Instute, San Antonio, Texas.
[5] Pan, C. H. T, Vohr. J. Super-Laminar Flow in Bearings and Seals, Bearing and Seal Design in Nuclear Power Machinery. ed. R. A. Burton, ASME, New York, N. Y., 1967.

On the Theoretical Basis of the Hirs Meteod in the Theory of Lubrication

Abstract: Some viewpoints different from the so-called Bulk Flow Theory for the turbulence in the lubricant film by G. G. Hirs are put forward in this paper. Hirs said that under any kinds of combinations of the shear flow and pressure flow it is possible to express the interconnections between the shear stress exerted on the wall and the mean velocity u_m relative to the wall by a simple relationship, thus deriving the calculating method. But as pointed out in this article, there are several evident examples of the combinations of the shear flow and pressure flow which do not justify the above relationship. Since such kinds of combinations might often happen in the most important regions of the oil film, a given error can appear in the Hirs calculation.

动压式滑动轴承的稳定性问题[*]

现代的机器日益向高速发展,它的振动问题和减振问题也相应地日益重要.在用滑动轴承支撑的转子上,除了由于不平衡所造成的同步振动(振动频率与转子转速相同)外,还会发生一种由于轴承中的油膜丧失稳定性而造成的自激振动,它与同步振动(同步涡动)在本质上和现象上都不相同.

由不平衡力引起的振动,是一种强迫振动,它除了在频率上有与转子转速同步的特点外,在振幅上还有这样的特点:当转子转速上升(或下降)而通过某些临界值时,振幅迅速增大,随后又迅速减小.

而动压式滑动轴承油膜失稳时发生的自激振动则有下列特点:它的振动频率(涡动频率)等于或低于转子转速的一半,而且以转子的一阶临界转速为上限;一旦发生这种涡动,是不能依靠提高转速来冲过去的,通常,转速越高,涡动越严重,所以,这个速度的上限,称为稳定界限转速.

当转速达到和超出这个界限值,从而油膜失稳,并发生自激振动时,如果涡动频率还达不到转子的一阶临界转速并有相当距离时,称为"油膜涡动"(轻载圆柱轴承中此种涡动频率常等于或近于转速的一半,故称"半速涡动"),这时的振幅还不一定大到立即造成损坏的程度.而如果失稳时涡动频率已接近转子一阶临界转速,或发生"油膜涡动"后转速再升高,以致涡动频率接近一阶临界转速时,则振幅会急剧增大,这种突发性的振动特别危险,称为"油膜振荡"(或"油膜共振").

现举两个典型的实验现象[1].

图 1 为转子在不同转速下的振动频率和振幅,它的轴承是带有两条轴向槽(位于与载荷方向成 90°处)的圆柱轴承,直径 $D=2$ 英寸,间隙 $C=0.0025$ 英寸,转子重量引起轴承载荷约 8 磅/英寸^2,轴的一阶临界转速为 6 100 r/min. 由图可见,在约 11 650 r/min 时,开始出现半速涡动,因这时涡动频率已很接近一阶临界转速,所以振幅增加得很快,随着转速的进一步提高而立即转入强烈振荡.与此同时,涡动频率却变化甚少,但愈来愈接近一阶临界转速的数值.图中还示出各种转速下测得的轴心涡动轨迹:5 970 r/min 时,只有同步涡动;12 720 r/min 时,则有较小的半速涡动与较大的同步涡动;14 300 r/min 时,由于发生台架共振而出现了颇大的同步涡动,从而暂时抑制了自激涡动;16 500 r/min 时,则有很大的亚半速自激振荡与较小的同步涡动.

图 2 为另一个转子在不同转速下的振动频率.由图 2 可见,当转速很低时,就已发生半速涡动;在转速达到约 2 倍一阶临界转速以前,涡动频率正比于转速而增加;当转速达到约 2 倍一阶临界转速时,涡动频率颇近于一阶临界转速,从而转入剧烈的油膜振荡;此后,涡动频率变化甚少.

油膜失稳的力学原因、油膜稳定性的判断及失稳界限转速的计算,都要从分析油膜动力

---

[*] 原发表于《润滑与密封》,1980(1):41—48.

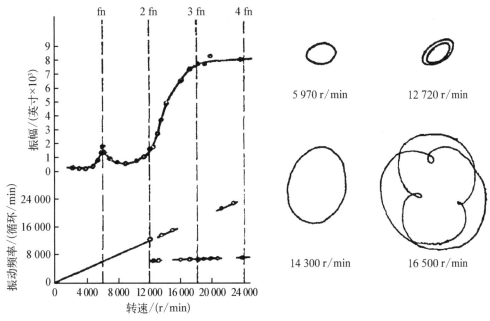

图1 转子在不同转速下的振动频率和振幅

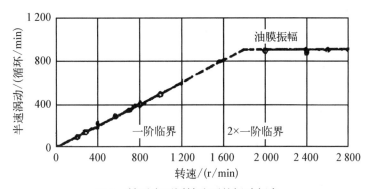

图2 转子在不同转速下的振动频率

特性着手. 关于这个问题的比较形象化和简化的分析,可阅[2,3]等文献. 这里用比较一般的动力学方法来分析一下.

一、油膜的刚度和阻尼

在一定的转速和载荷等工况参数下,轴颈中心 O_j 在径向滑动轴承内有一定的平衡位置,这个位置可以用轴颈中心相对于轴承中心 O_b 的偏心距 e 和偏位角 θ 来表示,如图 3(a)上的 O_e. 当轴颈中心静止地处在这个位置上运转时,轴颈所受油膜压力的合力 F_O 与外载荷 W 相平衡.

大家知道,平衡位置可以是稳定平衡的、不稳定平衡的、随遇平衡的,这要看物体在平衡位置附近有扰动时所受力的变化特性而定.

轴颈中心 O_j 可以有偏离平衡位置的位移扰动 x 和 y,如图 3(b),也可以有在平衡位置上的速度扰动 $\dot{x}$ 和 $\dot{y}$,如图 3(c);也可以有位移扰动和速度扰动组合的一般情况,如图

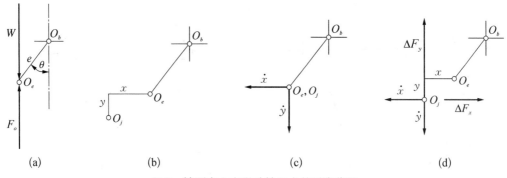

图 3 轴颈中心在滑动轴承内的平衡位置

3(d). 在这些扰动下,油膜力将发生变化而不再与载荷平衡,也就是说,油膜力与载荷加起来不等于零,它可以有水平方向和垂直方向的分力 ΔF_x 和 ΔF_y[图 3(d)],它们的数值通常是位移和速度扰动量的函数

$$\Delta F_x = \Delta F_x(x, y, \dot{x}, \dot{y}) \quad \Delta F_y = \Delta F_y(x, y, \dot{x}, \dot{y}) \tag{1}$$

这是一种非线性的函数关系.用这种非线性关系来计算是比较繁复的.由于我们只要判断平衡位置是否稳定,所以,可以在平衡位置附近取非常小的扰动来研究油膜力变化,因此,可把这两个分力 ΔF_x 和 ΔF_y 看成是扰动量的线性函数,也就是等于各扰动量 $x, y, \dot{x}, \dot{y}$ 分别乘上不同的常系数后再相加:

$$\begin{aligned} \Delta F_x &= k_{xx}x + k_{xy}y + d_{xx}\dot{x} + d_{xy}\dot{y} \\ \Delta F_y &= k_{yx}x + k_{yy}y + d_{yx}\dot{x} + d_{yy}\dot{y} \end{aligned} \tag{2}$$

式中四个常系数 k 称为刚度系数,表示单位位移扰动引起的油膜力;四个常系数 d 称为阻尼系数,表示单位速度扰动引起的油膜力;k 和 d 所带的两个标号中,第一个表示所产生的油膜力方向,第二个表示引起这个力的扰动方向.例如 k_{xy} 表示 y 方向的单位位移扰动引起的 x 方向油膜力;d_{xx} 表示 x 方向的单位速度扰动引起的 x 方向油膜力.

常用轴承的阻尼和刚度系数数值可查阅[1,4~7].计算时,用无量纲的刚度系数和阻尼系数更为方便

$$\begin{aligned} K_{xx} &= k_{xx}\frac{\psi^3}{\mu l \Omega}, K_{xy} = k_{xy}\frac{\psi^3}{\mu l \Omega}, K_{yx} = k_{yx}\frac{\psi^3}{\mu l \Omega}, K_{yy} = k_{yy}\frac{\psi^3}{\mu l \Omega} \\ D_{xx} &= d_{xx}\frac{\psi^3}{\mu l}, D_{xy} = d_{xy}\frac{\psi^3}{\mu l}, D_{yx} = d_{yx}\frac{\psi^3}{\mu l}, D_{yy} = d_{yy}\frac{\psi^3}{\mu l} \end{aligned} \tag{3}$$

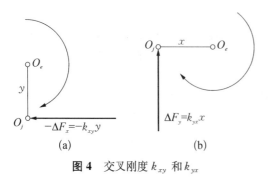

图 4 交叉刚度 k_{xy} 和 k_{yx}

式中 ψ——相对间隙;μ——动力黏度;l——轴承长度;Ω——轴角速度.

k_{xy} 和 k_{yx} 表示位移扰动所引起的同它相垂直的方向上的油膜力,这两个系数称为交叉刚度,它们代表着引起轴颈涡动的力学因素.例如由对于轻载的圆柱轴承计算结果可知,按图 3(d)的位移、速度和力的坐标系统,k_{xy} 是负的,即向下的位移扰动会引起向左的油膜力[图 4(a)];而 k_{yx} 是正的,即向左的位移扰动会引

起向上的油膜力[图 4(b)]. 可见,这时交叉刚度代表的油膜力将促使轴心绕着平衡位置 O_e 而涡动,涡动的方向与轴的转动方向一致. 这就是造成油膜失稳的力因素.

二、轴颈涡动时油膜力所作的机械功

除了交叉刚度外,其他的刚度、阻尼系数都起着一定的稳定化或不稳化的作用,在它们的共同作用下,轴心在平衡位置附近的涡动轨迹或者不断扩大——失稳,或者不断缩小——稳定,或者一次一次重复——界限状态(图 5).

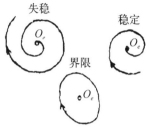

图 5 涡动轨迹

涡动轨迹的变化,意味着振动能量的相应变化. 而振动能量的增减,是由油膜力对涡动着的轴颈所作的功引起的. 在一个振动循环中,当油膜力对轴颈作正的功时,会使振动能量增大,从而促使涡动轨迹扩大. 当油膜力作负的功时,就会吸收掉振动能量,使轨迹缩小. 而在稳定的界限状态下,油膜力作的功等于零.

从这个观点来看,当轴颈绕平衡位置作微小的椭圆形涡动时,在每一循环中,交叉刚度 k_{xy} 和 k_{yx} 一般是作正功的,其功正比于 $(k_{yx}-k_{xy})$ 及轴的角速度 Ω;阻尼系数 d_{xx} 和 d_{yy} 则作负功,所以是稳定因素,它们的功分别正比于 D_{xx} 或 D_{yy} 以及涡动角频率 ω;刚度系数 k_{xx} 和 k_{yy} 作的功等于零,但对涡动频率 ω 及轨迹型式有很大影响,k_{xx} 和 k_{yy} 越大则 ω 越高,因而有助于 d_{xx} 和 d_{yy} 发挥有益作用;交叉阻尼 d_{xy} 和 d_{yx} 的数值常常较小,一般不起主要作用.

三、油膜在界限状态下的综合特性

在界限状态下,正负功相抵消,而正功大致正比于 Ω,负功正比于 ω,所以不难想象,油膜开始失稳(即界限状态下)时有一定的 ω/Ω 比值,称为失稳涡动比 γ_{st},它取决于各无量纲刚度、阻尼特性的相对关系. 此时的油膜力既不作正功,亦不作负功,所以油膜的动力特性综合起来表现为单纯的弹性,它的综合刚度系数可用 K_{eq} 表示,相应的无量纲综合刚度 $K_{eq}=\dfrac{k_{eq}\psi^3}{\mu l \Omega}$. 由分析结果可知,油膜在开始失稳时的 K_{eq} 和 γ_{st} 值可按下列公式计算:

$$K_{eq}=\frac{K_{xx}D_{yy}+K_{yy}D_{xx}-K_{xy}D_{yx}-K_{yx}D_{xy}}{D_{xx}+D_{yy}} \tag{4}$$

$$\gamma_{st}^2=\frac{(K_{eq}-K_{xx})(K_{eq}-K_{yy})-K_{xy}K_{yx}}{D_{xx}D_{yy}-D_{xy}D_{yx}} \tag{5}$$

从稳定性观点来看,显然 γ_{st} 越小越好,而 K_{eq} 越大越好. 由于是以无量纲形式表达,通常是取定偏心比 ε 而进行这种计算的.

根据 Routh-Hurwitz 稳定性准则可知,如果算出的 K_{eq} 为负值,则这个 ε 下油膜必为不稳;而如果 K_{eq} 为正值,同时 γ_{st}^2 却为负值时,则通常在这个 ε 下油膜总是稳定的;如果 K_{eq} 和 γ_{st}^2 均为正值,则在这个 ε 下,转速低为稳定,转速高则失稳,亦即存在某个失稳转速.

例如剖分式圆柱轴承（瓦张角150°，载荷垂直向下），按[7]的数据可得表1所示之 K_{eq} 及 γ_{st} 值．表中亦列出了相应的承载量系数 $\dfrac{P\psi^2}{\mu Ul}$ 值，以便查用．

表 1 剖分式圆柱轴承的 $\dfrac{P\psi^2}{\mu Ul}$ 值、K_{eq} 值及 γ_{st} 值

l/d	ε	0.1	0.2	0.3	0.4	0.5	0.6	0.7	0.8	0.9
0.4	$\dfrac{P\psi^2}{\mu Ul}$	0.0458	0.0983	0.1657	0.2611	0.4103	0.6695	1.1936	2.5150	8.1585
	K_{eq}	0.0642	0.140	0.240	0.363	0.552	0.878	1.600	恒稳	恒稳
	γ_{st}	0.502	0.509	0.525	0.514	0.492	0.459	0.406		
0.6	$\dfrac{P\psi^2}{\mu Ul}$	0.0927	0.1980	0.3314	0.5151	0.7931	1.2581	2.1467	4.2397	11.9894
	K_{eq}	0.115	0.249	0.427	0.665	1.005	1.574	5.655	5.299	恒稳
	γ_{st}	0.502	0.506	0.515	0.527	0.519	0.496	0.404	0.253	
1.0	$\dfrac{P\psi^2}{\mu Ul}$	0.1959	0.4122	0.6757	1.0254	1.5206	2.2960	3.6638	0.6275	16.3598
	K_{eq}	0.197	0.425	0.714	1.111	1.641	2.537	4.042	7.563	恒稳
	γ_{st}	0.501	0.504	0.511	0.519	0.505	0.498	0.427	0.079	

四、失稳转速的计算

当 K_{eq} 为正，γ_{st}^2 亦为正时（即 γ_{st} 为实数时），存在一个失稳角速 Ω_{st}．当工作角速 Ω 低于它时，油膜为稳定．当工作角速高于它时，则为不稳，此时或发生油膜涡动，或发生油膜振荡，视涡动角频率 ω（界限状态下的涡动角频率 ω_{st} 等于 $\gamma_{st}\Omega_{st}$）是否接近转子一阶临界角速而定．

以对称单质量弹性转子支撑在一对相同的滑动轴承上的情况为例[图6(a)]，设转子的质惯矩相对地较小，因而主要考虑柱形涡动．在开始失稳时，油膜呈现无阻尼的弹性，因而系统可简示如图6(b)．轴与油膜的串联刚度为

$$\frac{k_{eq}k}{k+k_{eq}}$$

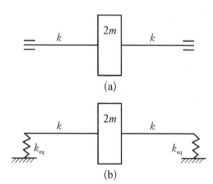

图 6 失稳时系统状态示意图

因此，自然振动角频率（即失稳时的涡动角频率）为

$$\omega_{st}=\sqrt{\dfrac{\dfrac{k_{eq}k}{k+k_{eq}}}{m}}$$

以上　$2m$——转子质量；$2k$——轴的横向刚度；

k_{eq}——每个轴承的油膜综合刚度．

将 $\omega_{st}=\Omega_{st}\gamma_{st}$ 及 $k_{eq}=K_{eq}\dfrac{\mu l \Omega_{st}}{\psi^3}$ 代入,可得失稳角速度为

$$\Omega_{st}=\omega_0\left[-\frac{\omega_0 m\psi^3}{2K_{eq}Ul}+\sqrt{\left(\frac{\omega_0 m\psi^3}{2K_{eq}\mu l}\right)^2+\frac{1}{\gamma_{st}^2}}\right] \tag{7}$$

式中　$\omega_0=\sqrt{k/m}$——转子的临界角频率.

例如有一转子,重 46 吨,一阶临界转速为 978 r/min,工作转速为 3 000 r/min,装在两个圆柱滑动轴承上,其 $l/d=1$,$\psi=0.0015$,$\mu=22.528\times10^{-4}$ kg·s/m²,直径 $d=420$ mm. 问在工作情况下是否会失稳?

先计算承载系数

$$\frac{P\psi^2}{\mu Ul}=\frac{23\,000\times(0.001\,5)^2}{22.528\times10^{-4}\times65.97\times0.420}=0.829$$

式中　U——轴颈圆周速度;$U=\pi dn/60$[其中 n——转速(r/min)].

根据此承载系数,按表 1 用插入法查得平衡位置的偏心比为 $\varepsilon=0.344$,相应的 K_{eq} 为 0.889,γ_{st} 为 0.515.

用式(7)计算此 ε 下的失稳角速为

$$\begin{aligned}\Omega_{st}=102.4\Bigg[&-\frac{102.4\times4\,689\times(0.001\,5)^3}{2\times0.889\times22.528\times10^{-4}\times0.42}\\&+\sqrt{\left(\frac{102.4\times4\,689\times(0.001\,5)^3}{2\times0.889\times22.528\times10^{-4}\times0.42}\right)^2+\frac{1}{(0.515)^2}}\,\Bigg]\\=&123.3(\text{弧度}/\text{秒})\end{aligned}$$

此 ε 下的失稳转速为

$$n_{st}=\frac{60\Omega_{st}}{2\pi}=\frac{60\times123.3}{2\pi}=1\,178(\text{r/min})$$

以上,ω_0 以弧度/s 计,m 以 kg·s²/m 计. 即:$\omega_0=\dfrac{2\pi n_1}{60}$,$m=w/g$,式中 n_1 是一阶临界转速(r/min),w 是重量(kg),$g=9.81$ m/s².

将工作转速(3 000 r/min)与同一 ε 下的失稳转速相比,可见此转子在工作转速下将失稳,应改用稳定性更好的轴承.

五、轴承参数对稳定性的影响

某些参数的改变究竟对稳定性的影响如何,不能一概而论. 例如,加大间隙而不改变其他条件,则 ε 将增大,从而使 K_{eq} 增大和 γ_{st} 减小;流量增多使温度降低,从而使 μ 增高,另一方面也使 ψ 增大,从式(7)看来,Ω_{st} 究竟提高了还是降低了,不是立刻可看出的. 实际上也是如此. 其他一些参数对稳定性的影响也常常有这种两面性.

不过,有些参数的影响的倾向性比较明显,或者影响比较单纯,可以对其有个大体上的估计:降低润滑油黏度、减少油量、减短轴承长度、加大轴承载荷(不是加大转子重量),都有

利于提高失稳转速.

从分析可知,在其他条件不变下,加大 ε 是提高稳定性的十分有效和基本的途径. 一些参数对稳定性的影响,一些轴承结构型式为何能提高稳定性,都同 ε 有关.

六、提高稳定性的轴承结构型式

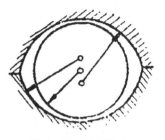

图 7 椭圆轴承示意

圆柱轴承在轻载时,ε 接近于 0,稳定性极差. 如果将上、下瓦的圆弧中心错开一个距离,成为图 7 所示的那种在载荷方向上略扁的轴承(通常称为椭圆轴承),则即使在空载因而轴颈中心处在轴承中心(即上、下瓦弧中心间的中点)上时,轴颈中心相对于瓦弧中心仍有相当大的相对偏心率,因而稳定性比圆柱轴承好得多.

三油叶轴承、四油叶轴承、多油楔轴承、横向偏置轴承、上瓦带油坝轴承等都是利用类似的原理(图 8).

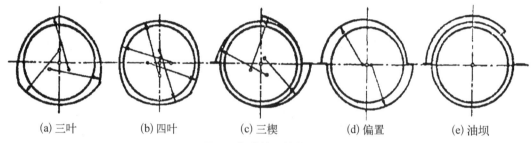

(a) 三叶   (b) 四叶   (c) 三楔   (d) 偏置   (e) 油坝

图 8 各种轴承结构型式

采用一些非固定瓦型的轴承,是一种极有效的提高稳定性的方法. 例如可倾瓦径向轴承的稳定性,早已得到公认. 当略去瓦惯性时,每块瓦上油膜压力的合力的动力作用都近似相当于装设在轴心与瓦支点间的一个弹簧和一个阻尼器(非线性的),如图 9. 弹簧代表一个保

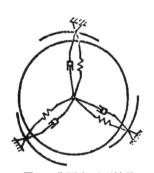

图 9 非固定瓦型轴承

守力场,它不会促进涡动,斜置弹簧有 $k_{xy}=k_{yx}$ 之特点. 因此,整个可倾瓦轴承不论其轴瓦和支点的结构和布置怎样,一般近似地有 $k_{xy}=k_{yx}$ 之特点(略去瓦惯性时),因此这两个交叉刚度在轴颈作椭圆轨迹涡动时作的功近于零,就没有了这方面的失稳因素. 所以这种轴承极为稳定. 当然,实际上轴瓦有一定惯性,而且支点亦非理想铰链,再加上其他一些因素,所以总还有一些不稳定因素. 但这种轴承的主要动力问题,一般不是稳定性问题,而是瓦块共振问题[9,10]. 属于同一类型的结构很多,例如弹性支片式轴承[11]及液支可倾瓦轴承[12]. 螺旋槽轴承[13]、浮环轴承[14]、薄膜式轴承[15] 等也有很好的稳定性.

将轴承装在弹性、阻尼支座上以提高稳定性,是近年来越来越广泛应用的方法[16,17]. 如果支座只有弹性而无阻尼,则常常会降低失稳转速. 但如果支座有较低的刚度,同时又有适当大小的阻尼,则可大大提高稳定性,如果支座的弹性和阻尼匹配得好,理论上可做到在任何转速下不失稳,而且通过临界速度的性能也极好.

参 考 文 献

[1] O. Pinkus, B. Sternlicht. Theory of Hydrodynamic Lubrication, 1961: 450-452.
[2] H. Poritsky. Contribution of the Theory of Oil Whip. Trans. ASME, 1958, 75: 1153-1161.
[3] D. Robertson. Whirling of a Journal in a Sleeve Bearing. Phil. Mag. Ser. 7, Vol. 15: 113-130, Jan., 1933.
[4] 西安交大轴承研究小组. 三油楔轴承性能计算. 1975.
[5] 西安交大轴承研究小组. 椭圆轴承性能计算. 1974.
[6] 西安交大轴承研究小组. 紊流工况下椭圆轴承性能计算. 1977.
[7] 西安交大轴承研究小组. 圆柱和椭圆轴承性能计算资料. 1978.
[8] 西安交大78—139号科技报告. 用滑动轴承支撑的转子失稳角速度计算法的补充条件.
[9] J. W. Lund. Spring and Damping Coefficients of the Tilting Pad Journal Bearing. Trans. ASLE, 1964, 7(4): 342.
[10] 西安交大78—067号科技报告. 关于可倾瓦径向滑动轴承油膜动力特性的讨论(一).
[11] A. Тондть. Экспериментальное исследование самовозбуждающихся колебаний роторов возникающих в результате воздействия смазочного слоя в полшипниках скольжения. сб. Развитие Гидродинамической теории смазки подшипников Быстроходных машин, 129-173.
[12] D. V. Nelson, L. W. Hollingsworth. The Fluid Pivot Journal Bearing. Trans. ASME, Ser. F, 1977: 122-127.
[13] F. T. Schuller. Experiments on the Stability of Various Water-Lubricated Fixed Geometry Hydrodynamic Journal Bearings at Zero Load. ASME, F, 1973, 95(4): 434-446.
[14] M. Tanaka, Y. Hori. Stability Characteristic of Floating Bearings. ASME, F, 1972, 94,(3): 248-259.
[15] T. Barnum. The Geometrically Irregular Foil Bearing. ASME, F, 1974, 96(2): 224-227.
[16] Lund. The Stability of an Elastic Rotor in Journal Bearings with Flexible, Damped Supports. ASME, E, 1965, 32(4): 911-920.
[17] Glienicke. Bauelemente zur äußeren Lagerdampfung. MTZ, 1974, 35(7): 205-210.

一种新型滑动轴承——隧道孔轴承性能的初步分析*

摘　要：本文用无限长轴承的近似方法计算了在不同位置上开隧道孔的轴承的静特性及动特性．由计算结果可知：在适当位置上开隧道孔，可大大改善轴承稳定性而不降低承载能力，因此有希望成为一种结构简单、性能良好、工作原理新颖的轴承型式，值得进一步研究．

符号与术语

C——轴承的半径间隙

r——轴承半径

e——轴承中心相对于轴颈中心的偏心距

ψ——相对间隙，$\psi=c/r$

ε——相对偏心，$\varepsilon=e/c$

μ——油的动力黏度

f——摩擦系数

ω——轴的角速度

θ——偏位角

ϕ——由垂直向上线算起的角度

φ——由最大间隙算起的角度

ϕ_a，ϕ_b——隧道孔二端的角位置

ϕ_1，ϕ_2——轴瓦起始和终止边的角位置

h——油膜厚度

H——无量纲油膜厚度，$H=h/c$

I_1，I_2，I_3，I_4——第一、第二、第三、四种 Sommerfeld 积分

k_{ik}——油膜刚度系数（每单位长度的）($i,k=\varepsilon,\theta$)

K_{ik}——无量纲油膜刚度，$K_{ik}=\dfrac{k_{ik}\psi^3}{\mu\omega}(i,k=\varepsilon,\theta)$

K_{eq}——无量纲相当油膜刚度

M_{kr}——无量纲临界质量

γ_{st}——界限涡动比

d_{ik}——油膜阻尼系数（每单位长度的）($i,k=\varepsilon,\theta$)

D_{ik}——无量纲阻尼系数，$D_{ik}=\dfrac{d_{ik}\psi^3}{\mu}(i,k=\varepsilon,\theta)$

p——油膜压力

P——无量纲油膜压力

P_S——无量纲油膜压力对无量纲扰动 S 的偏导数，$P_S=\dfrac{\partial P}{\partial S}$，$S=\varepsilon,\varepsilon_0\theta,\varepsilon',\varepsilon_0\theta'$．其中 $\varepsilon_0=$ 静平衡位置的偏心率 $\varepsilon'=\dfrac{1}{\omega}\dfrac{d\varepsilon}{dt}$，$\theta'=\dfrac{1}{\omega}\dfrac{d\theta}{dt}$

Q_1，Q_2，Q_3——无量纲流量，依次为入口流量、隧道孔流量及中部流量

引言

高速动压滑动轴承的严重问题之一，是有可能失去动力稳定性而发生油膜涡动或油膜振荡．迄今已找到许多提高油膜稳定性的方法，其中第一类方法主要是通过各种措施增大轴颈相对于主要承载瓦的相对偏心，从而改善油膜稳定性．例如截短轴承的长度，开中央环形卸载槽，采用多油叶轴承、多油楔轴承、横偏置轴承以及上瓦加油坦等都属于此类原理．这类轴承在改善动特性的同时，通常都不可避免地带来使承载能力等静特性变差的缺点，而且有些轴承的结构和制造工艺也较复杂．另一类方法虽在提高稳定性方面十分有效，但其结构更

---

* 本文合作者：吴西柳．原发表于《上海工业大学学报》，1985(2)：89-98．

为复杂,例如可倾瓦轴承、浮环轴承、轴承外加阻尼支承等.

油膜失稳是由油膜压力分布的特性所造成,如能从原理上改变压力分布规律,一定可引起油膜动力特性的相应变化,如措施得当,可收到改善轴承的性能的效果. 作者根据这样的设想,假定在油膜压力区内对称于载荷作用方向,开一隧道孔,将高压区与低压区连通,使高压区内的压力稍有下降,低压区内的压力稍有提高(见图1),此时轴承的承载能力并无多大变化,而动力特性(特别是交叉刚度)却将发生明显变化. 这是一种结构并不复杂,而能使动特性发生本质变化的方法. 为了在理论上探

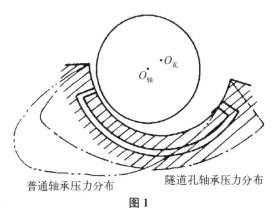

普通轴承压力分布　　隧道孔轴承压力分布

图 1

讨这种设想的效果,作者在180°无限长径向轴承上加设隧道孔,来计算其动特性和稳定性. 由于这种情况用分析法而无须用数值法解算 Reynolds 方程,所以较方便地得出了结果. 根据这种近似理论得出的结果,即可初步评价这种设想的有效性,从而确认作进一步研究的必要性.

一、基本关系

研究对象如图 2 所示的部分轴承,开有隧道孔以连通 ϕ_a 及 ϕ_b. 设轴承长度为无穷大,且全长都有隧道孔.

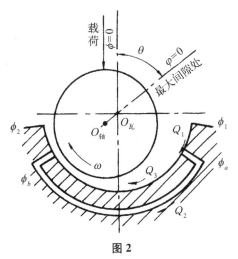

图 2

描述无限长油膜压力分布的无量纲形式的 Reynolds 方程为:

$$\frac{d}{d\varphi}\left(H^3 \frac{dP}{d\varphi}\right) = -3\varepsilon \sin\varphi \qquad (1)$$

起始边上的边界条件为

$$\phi = \phi_1 \text{ 处}, P = 0 \qquad (2)$$

终止边上的边界条件为

$$\phi = \phi_2 \text{ 处}, P = 0$$

如油膜在轴承下游边以前即自然破裂,则终止边的边界条件为

$$\phi = \phi_c \text{ 处}, P = 0 \text{ 且 } \frac{dP}{d\varphi} = 0 \qquad (3)$$

由隧道孔带来的条件为:隧道孔两端压力相等(设隧道孔无节流作用)

$$P_{\phi_a} = P_{\phi_b} \qquad (4)$$

隧道内流量守恒

$$H_{\phi_a}^3\left(\frac{\mathrm{d}P}{\mathrm{d}\varphi}\bigg|_{\phi_a^+} - \frac{\mathrm{d}P}{\mathrm{d}\varphi}\bigg|_{\phi_a^-}\right) + H_{\phi_b}^3\left(\frac{\mathrm{d}P}{\mathrm{d}\varphi}\bigg|_{\phi_b^+} - \frac{\mathrm{d}P}{\mathrm{d}\varphi}\bigg|_{\phi_b^-}\right) = 0 \tag{5}$$

根据式(1)~(5),解得无量纲压力分布为

对于 $\phi_1 \leqslant \phi \leqslant \phi_a$, $P = 3I_2(\varphi) + c_1 I_3(\varphi) + c_2$ (6)

对于 $\phi_a \leqslant \phi \leqslant \phi_b$, $P = 3I_2(\varphi) + c_3 I_3(\varphi) + c_4$ (7)

对于 $\phi_b \leqslant \phi \leqslant \phi_2$, $P = 3I_2(\varphi) + c_5 I_3(\varphi) + c_6$ (8)

且

$$c_3 = -3\frac{I_2(\varphi_a) - I_2(\varphi_b)}{I_3(\varphi_a) - I_3(\varphi_b)} \tag{9}$$

$$c_1 = c_5 = -3\frac{I_2(\varphi_1) - I_2(\varphi_a) + I_2(\varphi_b) - I_2(\varphi_c)}{I_3(\varphi_1) - I_3(\varphi_a) + I_3(\varphi_b) - I_3(\varphi_c)} \tag{10}$$

$$c_2 = -3I_2(\varphi_1) - c_1 I_3(\varphi_1) \tag{11}$$

$$c_6 = -3I_2(\varphi_c) - c_5 I_3(\varphi_c) \tag{12}$$

$$c_4 = P_{\phi_a} - 3I_2(\varphi_a) - c_3 I_3(\varphi_a) \tag{13}$$

$$P_{\varphi_a} = P_{\varphi_b} = -3[I_2(\varphi_1) - I_2(\varphi_a)] - c_1[I_3(\varphi_1) - I_3(\varphi_a)] \tag{14}$$

又,各 Sommerfeld 积分为

$$I_1(\varphi) = \frac{2}{\sqrt{1-\varepsilon^2}} \tan^{-1}\left(\sqrt{\frac{1-\varepsilon}{1+\varepsilon}} \tan\frac{\varphi}{2}\right) \tag{15}$$

$$I_2(\varphi) = \frac{1}{(1-\varepsilon^2)^{3/2}}\left[2\tan^{-1}\left(\sqrt{\frac{1-\varepsilon}{1+\varepsilon}} \tan\frac{\varphi}{2}\right) - \varepsilon\sqrt{1-\varepsilon^2}\,\frac{\sin\varphi}{1+\varepsilon\cos\varphi}\right] \tag{16}$$

$$I_3(\varphi) = \frac{1}{(1-\varepsilon^2)^{5/2}}\bigg[(2+\varepsilon^2)\tan^{-1}\left(\sqrt{\frac{1-\varepsilon}{1+\varepsilon}} \tan\frac{\varphi}{2}\right)$$
$$-2\varepsilon\sqrt{1-\varepsilon^2}\,\frac{\sin\varphi}{1+\varepsilon\cos\varphi} + \frac{\varepsilon^2\sqrt{1-\varepsilon^2}}{2}\,\frac{\sin\varphi(\varepsilon+\cos\varphi)}{(1+\varepsilon\cos\varphi)^2}\bigg] \tag{17}$$

$$I_4(\varphi) = \frac{1}{(1-\varepsilon^2)^{7/2}}\left[\left(1+\frac{3\varepsilon^2}{2}\right)\Psi - 3\varepsilon\left(1+\frac{\varepsilon^2}{4}\right)\sin\Psi + \frac{3\varepsilon^2}{4}\sin2\Psi - \frac{\varepsilon^3}{12}\sin3\Psi\right] \tag{18}$$

式中

$$\Psi = 2\tan^{-1}\left(\sqrt{\frac{1-\varepsilon}{1+\varepsilon}} \tan\frac{\varphi}{2}\right) \tag{19}$$

φ_c 的值则由条件(3)决定,即 $3H_{\varphi_c} + C_5 = 0$ (20)

根据 P 分布,即可求出无量纲油膜合力:

$$F_x = -\int_{\phi_1}^{\phi_2} P\sin\phi\,\mathrm{d}\varphi = \frac{1}{\varepsilon}(Q\sin\theta + R\cos\theta) \tag{21}$$

$$F_y = -\int_{\phi_1}^{\phi_2} P\cos\mathrm{d}\phi = \frac{1}{\varepsilon}(Q\cos\theta - R\sin\theta) \tag{22}$$

式中 $Q = 3(H_{\phi_c}^{-1} - H_{\phi_1}^{-1}) + \dfrac{C_1}{2}(-H_{\phi_1}^{-2} + H_{\phi_a}^{-2} - H_{\phi_a}^{-2} + H_{\phi_c}^{-2}) + \dfrac{C_3}{2}(-H_{\phi_a}^{-2} + H_{\phi_b}^{-2})$ (23)

$$\begin{aligned}R = &3[I_1(\varphi_1) - I_1(\varphi_2) - I_2(\varphi_1) + I_2(\varphi_2)] + c_1[I_2(\varphi_1) - I_2(\varphi_a) \\ &+ I_2(\varphi_b) - I_2(\varphi_c) - I_3(\varphi_1) + I_3(\varphi_a) - I_3(\varphi_b) + I_3(\varphi_c)] \\ &+ c_3[I_2(\varphi_a) - I_2(\varphi_b) - I_3(\varphi_a) + I_3(\varphi_b)]\end{aligned}$$ (24)

在各种 ε 值下,θ 的大小由

$$F_x = 0 \tag{25}$$

的条件决定.

如此,即可求得各种 ε 下的平衡偏位角 θ 及无量纲力 F_y. 此外可求出阻力系数 f/ψ 及流量 Q_1、Q_2、Q_3. 其结果见附表.

将 Reynolds 方程对 ε 取偏导[1]得

$$\frac{d}{d\varphi}\left(H^3 \frac{dP_\varepsilon}{d\varphi}\right) = -3\sin\varphi + \frac{9\varepsilon\sin\varphi\cos\varphi}{H} + 3H\sin\varphi \frac{dP}{d\varphi} \tag{26}$$

P_ε 的边界条件为

$$\varphi = \phi_1 处, P_\varepsilon = 0 \tag{27}$$

$$\phi = \phi_c 处, P_\varepsilon = 0 \tag{28}$$

$$P_\varepsilon\Big|_{\phi_a} = P_\varepsilon\Big|_{\phi_b} \tag{29}$$

$$H_{\phi_a}^3\left(\frac{dP_\varepsilon}{d\varphi}\Big|_{\phi_a^+} - \frac{dP_\varepsilon}{d\varphi}\Big|_{\phi_a^-}\right) + H_{\phi_b}^3\left(\frac{dP_\varepsilon}{d\varphi}\Big|_{\phi_b^+} - \frac{dP_\varepsilon}{d\varphi}\Big|_{\phi_b^-}\right)$$
$$+ 3H_{\phi_a}^2 \frac{\partial H_{\phi_a}}{\partial \varepsilon}\left(\frac{dP_\varepsilon}{d\varphi}\Big|_{\phi_a^+} - \frac{dP_\varepsilon}{d\varphi}\Big|_{\phi_a^-}\right) + 3H_{\phi_b}^2 \frac{\partial H_{\phi_b}}{\partial \varepsilon}\left(\frac{dP_\varepsilon}{d\varphi}\Big|_{\phi_b^+} - \frac{dP_\varepsilon}{d\varphi}\Big|_{\phi_b^-}\right) = 0 \tag{30}$$

与前相似地分三段积分求出 P_ε,求出两个无量纲油膜刚度

$$K_{\varepsilon\varepsilon} = -\int_{\varphi_1}^{\varphi_c} P_\varepsilon \cos\varphi d\varphi \tag{31}$$

$$K_{\theta\varepsilon} = -\int_{\varphi_1}^{\varphi_c} P_\varepsilon \sin\varphi d\varphi \tag{32}$$

将 Reynolds 方程对 θ 取偏导[1]得

$$\frac{d}{d\varphi}\left(H^3 \frac{dP_\theta}{d\varphi}\right) = 3\frac{\partial^2 H}{\partial\theta\partial\varphi} - \frac{9}{H}\frac{\partial H}{\partial\theta}\frac{\partial H}{\partial\varphi} - 3H^3 \frac{dP}{d\varphi}\frac{\partial}{\partial\varepsilon}\left(\frac{1}{H}\frac{\partial H}{\partial\theta}\right) \tag{33}$$

P_θ 的边界条件与 P_ε 相同.

解出 P_θ 后,可求出另两个无量纲油膜刚度

$$K_{\varepsilon\theta} = -\int_{\varphi_1}^{\varphi_c} \frac{P_\theta}{\varepsilon_a}\cos\varphi d\varphi \tag{34}$$

附表 $\phi_1 = 90°$, $\phi_2 = 270°$ 轴承的计算结果

I $\phi_a = 175°$, $\phi_b = 185°$ 的隧道孔

ε	θ	F_y	f/ψ	Q_1	Q_2	Q_3	K_{xx}	K_{xy}	K_{yx}	K_{yy}	D_{xx}	D_{xy}	D_{yx}	D_{yy}	K_{eq}	γ_{st}^2	M_{kr}	注
0.1	72.88°	0.459	7.059	2.902	2.912	−0.009 2	1.567	−4.251	3.018	6.048	1.771	0.545	0.545	9.769	2.313	0.591	3.915	
0.2	61.71°	0.883	3.941	2.680	2.715	−0.003 52	3.247	−5.153	2.943	4.411	2.126	1.144	1.144	10.645	3.638	0.697	5.221	
0.3	54.69°	1.301	2.944	2.405	2.479	−0.074 4	4.073	−4.875	3.335	3.749	2.702	1.913	1.895	11.969	4.208	0.568	7.404	
0.4	50.87°	1.759	2.437	2.104	2.241	−0.137	4.410	−4.509	3.626	4.027	3.641	2.962	2.957	13.746	4.479	0.397	11.291	
0.5	48.08°	2.290	2.139	1.782	1.994	−0.212	4.543	−3.875	4.181	4.900	4.154	3.730	3.730	15.662	4.556	0.315	14.479	
0.6	44.80°	2.968	1.933	1.440	1.717	−0.277	4.758	−2.889	5.428	6.611	4.546	4.578	4.578	18.650	4.620	0.250	18.478	
0.7	40.82°	3.964	1.749	1.085	1.403	−0.318	5.024	−1.279	8.167	10.386	5.237	6.064	6.066	24.351	4.456	0.145	31.500	稳
0.8	35.68°	5.766	1.535	0.722	1.039	−0.317	4.720	2.635	15.356	21.123	6.808	9.481	9.484	37.921	3.403	−0.102		
0.9	28.26°	10.643	1.216	0.358	0.605	−0.247	−8.070	23.830	49.430	76.842	11.179	20.797	20.821	88.689	−13.827	−1.176		不稳

II $\phi_a = 155°$, $\phi_b = 205°$ 的隧道孔

ε	θ	F_y	f/ψ	Q_1	Q_2	Q_3	K_{xx}	K_{xy}	K_{yx}	K_{yy}	D_{xx}	D_{xy}	D_{yx}	D_{yy}	K_{eq}	γ_{st}^2	M_{kr}	注
0.1	74.46°	0.460	7.024	2.904	2.917	−0.013 3	0.919	−2.854	2.838	7.013	1.059	0.295	0.284	9.636	1.520	0.477	3.209	
0.2	64.67°	0.889	3.896	2.678	2.732	−0.053 5	1.932	−3.637	2.122	5.938	1.266	0.599	0.585	10.114	2.453	0.474	5.179	
0.3	59.63°	1.306	2.903	2.392	2.517	−0.125	2.009	−2.996	1.772	5.533	1.633	0.957	0.952	10.653	2.572	0.221	11.640	
0.4	56.94°	1.722	2.455	2.070	2.290	−0.220	1.520	−1.993	1.354	5.609	2.247	1.463	1.462	11.222	2.272	0.008 3	274.44	
0.5	55.82°	2.134	2.246	1.728	2.061	−0.333	0.719	−0.759	1.289	5.445	3.058	2.077	2.078	11.728	1.622	−0.078 4		稳
0.6	54.28°	2.575	2.162	1.375	1.798	−0.423	−0.210	1.129	2.454	4.483	3.455	2.484	2.485	12.362	0.252	−0.129		不稳
0.7	51.50°	3.156	2.122	1.020	1.476	−0.456	−2.191	4.675	5.432	3.308	4.142	3.295	3.300	14.147	−2.767	−0.459		不稳
0.8	46.85°	4.187	2.041	0.669	1.069	−0.400	−9.608	14.709	14.649	2.046	5.422	5.083	5.096	19.124	−13.120	−2.085		不稳
0.9	37.98°	7.479	1.679	0.325	0.536	−0.211	−72.277	69.941	65.207	8.845	7.166	9.179	9.215	38.812	−86.699	−16.461		不稳

Ⅲ $\phi_a = 135°$, $\phi_b = 225°$ 的隧道孔

ε	θ	F_y	f/ψ	Q_1	Q_2	Q_3	K_{xx}	K_{xy}	K_{yx}	K_{yy}	D_{xx}	D_{xy}	D_{yx}	D_{yy}	K_{eq}	γ_{st}^2	M_{kr}	注
0.1	75.70°	0.462	6.992	2.904	2.918	−0.0136	0.505	−1.760	2.786	7.678	0.510	0.130	0.126	9.566	0.854	0.519	1.644	稳
0.2	66.26°	0.896	3.863	2.674	2.725	−0.0511	1.218	−2.552	1.747	6.973	0.611	0.269	0.263	9.888	1.572	0.427	3.685	稳
0.3	61.09°	1.315	2.876	2.377	2.486	−0.109	1.118	−1.804	1.118	6.840	0.794	0.438	0.437	10.225	1.554	−0.0374		稳
0.4	58.03°	1.742	2.422	2.047	2.218	−0.171	0.443	−0.511	0.507	7.365	1.107	0.691	0.691	10.697	1.093	−0.336		稳
0.5	56.05°	2.201	2.182	1.762	1.923	−0.221	−0.487	1.098	0.611	7.688	1.375	0.926	0.927	11.240	0.278	−0.435		不稳
0.6	50.29°	2.864	1.979	1.369	1.542	−0.173	−1.092	2.448	1.258	9.520	1.304	1.083	1.089	13.314	−0.421	−0.626		不稳
0.7	38.02°	4.196	1.663	1.114	1.093	0.0207	2.107	0.729	−3.836	22.066	1.660	2.123	2.120	22.177	3.774	−0.857		稳
0.8	29.01°	6.582	1.373	0.891	0.717	0.174	16.489	−4.836	−23.851	52.395	3.884	7.005	7.004	46.567	23.236	−2.367		稳
0.9	24.48°	12.234	1.127	0.615	0.369	0.246	64.481	−24.042	−107.766	162.551	16.953	37.239	37.226	147.380	104.464	−4.417		稳

Ⅳ $\phi_a = 115°$, $\phi_b = 245°$ 的隧道孔

ε	θ	F_y	f/ψ	Q_1	Q_2	Q_3	K_{xx}	K_{xy}	K_{yx}	K_{yy}	D_{xx}	D_{xy}	D_{yx}	D_{yy}	K_{eq}	γ_{st}^2	M_{kr}	注
0.1	68.76°	0.452	7.199	2.904	2.887	0.0172	1.548	−3.741	1.855	7.915	0.542	0.211	0.202	9.770	1.918	0.898	2.135	稳
0.2	56.54°	0.860	4.079	2.703	2.650	0.0532	3.235	−4.586	0.960	6.879	0.672	0.447	0.446	10.602	3.597	0.461	7.796	稳
0.3	47.74°	1.254	3.086	2.481	2.361	0.121	4.294	−4.383	−0.134	7.432	0.886	0.805	0.789	12.012	4.786	−0.189		稳
0.4	42.94°	1.709	2.537	2.245	2.059	0.186	4.977	−4.170	−2.170	9.748	1.252	1.346	1.340	13.983	5.928	−0.807		稳
0.5	39.21°	2.264	2.190	2.010	1.742	0.268	6.265	−4.480	−5.131	13.343	1.838	2.253	2.246	17.079	8.096	−1.238		稳
0.6	36.28°	3.008	1.932	1.767	1.417	0.351	8.389	−5.271	−8.393	18.234	2.910	3.965	3.961	22.343	11.668	−1.334		稳
0.7	34.22°	4.118	1.721	1.505	1.082	0.422	12.187	−7.214	−15.140	28.200	5.348	7.862	7.861	32.475	19.097	−1.539		稳
0.8	33.11°	6.0176	1.541	1.213	0.740	0.474	19.313	−12.294	−31.234	52.087	12.343	18.930	18.909	56.542	37.143	−1.913		稳
0.9	34.05°	10.452	1.380	0.844	0.381	0.462	30.218	−30.524	−94.400	149.273	48.800	72.202	72.192	148.298	105.455	−3.052		稳

V $\phi_a = 105°$, $\phi_b = 225°$ 的隧道孔

ε	θ	F_y	f/ψ	Q_1	Q_2	Q_3	K_{xx}	K_{xy}	K_{yx}	K_{yy}	D_{xx}	D_{xy}	D_{yx}	D_{yy}	K_{eq}	γ_{st}^2	M_{kr}	注
0.1	68.75°	0.451	7.204	2.911	2.888	0.0228	1.775	−4.177	2.099	7.326	0.895	0.348	0.332	9.814	2.300	0.707	3.253	
0.2	56.94°	0.861	4.072	2.724	2.655	0.0689	3.531	−4.932	1.610	6.027	1.111	0.723	0.722	10.741	3.967	0.617	6.428	
0.3	48.73°	1.258	3.072	2.517	2.372	0.145	4.567	−4.671	1.141	6.182	1.446	1.269	1.248	12.252	5.058	0.296	17.068	
0.4	44.35°	1.714	2.525	2.295	2.074	0.222	5.233	−4.455	0.0439	7.855	2.017	2.064	2.056	14.362	6.110	−0.0540		稳
0.5	41.21°	2.641	2.185	2.072	1.760	0.312	6.316	−4.761	−2.184	11.110	2.971	3.392	3.388	17.615	8.151	−0.388		稳
0.6	39.16°	2.992	1.942	1.843	1.435	0.407	8.125	−5.782	−5.890	16.652	4.732	5.810	5.810	22.926	12.035	−0.697		稳
0.7	37.99°	4.052	1.755	1.594	1.099	0.494	10.722	−7.750	−10.419	25.393	8.490	10.869	10.867	32.722	18.536	−0.841		稳
0.8	37.99°	5.850	1.600	1.302	0.750	0.552	14.615	−12.440	−21.172	46.973	19.003	24.329	24.345	54.929	33.995	−1.140		稳
0.9	39.71°	9.921	1.474	0.939	0.386	0.553	14.357	−28.423	−62.595	134.742	70.453	84.832	84.806	136.662	92.510	−2.088		稳

VI $\phi_a = 180°$, $\phi_b = 260°$ 的隧道孔

ε	θ	F_y	f/ψ	Q_1	Q_2	Q_3	K_{xx}	K_{xy}	K_{yx}	K_{yy}	D_{xx}	D_{xy}	D_{yx}	D_{yy}	K_{eq}	γ_{st}^2	M_{kr}	注
0.1	59.78°	0.183	17.873	3.010	2.738	0.272	2.622	−4.156	0.456	3.293	1.917	1.116	1.110	4.877	3.415	0.245	13.916	
0.2	53.08°	0.365	9.551	2.926	2.457	0.469	3.369	−3.955	1.489	1.823	2.306	1.733	1.724	5.859	3.451	0.572	6.032	
0.3	48.96°	0.560	6.758	2.786	2.173	0.613	3.994	−3.878	2.722	0.846	2.936	2.555	2.561	7.177	3.374	0.619	5.453	
0.4	45.91°	0.771	5.425	2.604	1.888	0.716	4.702	−4.019	4.175	0.0119	3.899	3.777	3.777	9.025	3.242	0.577	5.622	
0.5	43.91°	1.021	4.616	2.393	1.601	0.792	5.733	−4.573	6.063	−0.828	5.483	5.696	5.697	11.809	3.162	0.541	5.848	
0.6	42.63°	1.344	4.063	2.159	1.311	0.847	7.388	−5.834	8.783	−1.801	8.301	9.018	9.019	16.413	3.226	0.552	5.843	
0.7	41.96°	1.824	3.613	1.903	1.015	0.888	10.309	−8.635	12.996	−2.620	14.021	15.592	15.593	25.133	3.943	0.645	6.117	
0.8	41.96°	2.754	3.114	1.627	0.706	0.921	16.109	−15.825	18.209	1.101	28.684	31.899	31.902	45.744	9.306	0.787	11.830	
0.9	43.14°	5.719	2.358	1.314	0.375	0.940	24.553	−40.747	−34.590	96.170	92.682	98.911	98.922	124.158	89.530	−1.069		稳

$$K_{\theta\theta} = -\int_{\varphi_1}^{\varphi_c} \frac{P_\theta}{\varepsilon_a} \sin\varphi \, d\varphi \tag{35}$$

经坐标转换可得 x, y 坐标系中的刚度

$$\begin{bmatrix} K_{xx} & K_{xy} \\ K_{yx} & K_{yy} \end{bmatrix} = \begin{bmatrix} \sin\theta & \cos\theta \\ \cos\theta & -\sin\theta \end{bmatrix} \begin{bmatrix} K_{\varepsilon\varepsilon} & K_{\varepsilon\theta} \\ K_{\theta\varepsilon} & K_{\theta\theta} \end{bmatrix} \begin{bmatrix} \sin\theta & \cos\theta \\ \cos\theta & -\sin\theta \end{bmatrix} \tag{36}$$

将 Reynolds 方程对 ε' 取偏导得

$$\frac{d}{d\varphi}\left(H^3 \frac{dP'_\varepsilon}{d\varphi}\right) = 6\cos\varphi \tag{37}$$

P'_ε 的边界条件与 P_ε 同.

解出 P'_ε,求出两个无量纲油膜阻尼

$$D_{\varepsilon\varepsilon} = -\int_{\varphi_1}^{\varphi_c} P'_\varepsilon \cos\varphi \, d\varphi \tag{38}$$

$$D_{\theta\varepsilon} = -\int_{\varphi_1}^{\varphi_c} P'_\varepsilon \sin\varphi \, d\varphi \tag{39}$$

另两个无量纲阻尼为

$$D_{\varepsilon\theta} = -2F_y \cos\theta/\varepsilon_0 \tag{40}$$

$$D_{\theta\theta} = 2F_y \sin\theta/\varepsilon_0 \tag{41}$$

换算成 x, y 坐标系中的阻尼

$$\begin{bmatrix} D_{xx} & D_{xy} \\ D_{yx} & D_{yy} \end{bmatrix} = \begin{bmatrix} \sin\theta & \cos\theta \\ \cos\theta & -\sin\theta \end{bmatrix} \begin{bmatrix} D_{\varepsilon\varepsilon} & D_{\varepsilon\theta} \\ D_{\theta\varepsilon} & D_{\theta\theta} \end{bmatrix} \begin{bmatrix} \sin\theta & \cos\theta \\ \cos\theta & -\sin\theta \end{bmatrix} \tag{42}$$

油膜在界限状态下的无量纲相当刚度为

$$K_{eq} = \frac{K_{xx}D_{yy} + K_{yy}D_{xx} - K_{xy}D_{yx} - K_{yx}D_{xy}}{D_{xx} + D_{yy}} \tag{43}$$

界限状态下的涡动比平方值为

$$\gamma_{st}^2 = \frac{(K_{eq} - K_{xx})(K_{eq} - K_{yy}) - K_{xy}K_{yx}}{D_{xx}D_{yy} - D_{xy}D_{yx}} \tag{44}$$

刚性转子临界质量为

$$M_{kr} = K_{eq}/\gamma_{st}^2$$

二、计算结果及讨论

以半圆轴瓦 ($\phi_1 = 90°$, $\phi_2 = 270°$) 为对象,取不同的 ϕ_a 和 ϕ_b 值,计算了 $\varepsilon = 0.1 \sim 0.9$ 时的偏位角 θ、承载量系数 F_y、八个无量纲动特性 K_{ik} 及 D_{ik} ($i, k = x, y$),界限情况下的无量纲相当刚度 K_{eq} 及涡动比平方 γ_{st}^2 以及刚性转子临界无量纲质量 M_{kr}. 并注明了"绝对稳定"

或"绝对不稳定"的情况(这里的"绝对稳定"或"绝对不稳"是一种通俗的提法,即不论无量纲质量多大总是稳定或不稳的).

在计算结果中,选出一些值得注意的结果列于附表.

由这些结果可知,隧道孔极大地改变了轴承的动特性及稳定性. Ⅰ、Ⅱ、Ⅲ 三种情况均为对称于载荷方向开设隧道孔. Ⅰ 中隧道孔中两出口离载荷方向只有 5°,在 ε=0.1~0.8 的范围内,其稳定性变化趋势与不开隧道孔时的普通轴承情况相似,即临界无量纲量随 ε 而提高,到 ε=0.8 时变为"绝对稳定". 但有趣的是,当 ε 再增大到 0.9 时,反而变成了"绝对不稳". 它的直接原因是,无量纲相当刚度 K_{eq} 不像普通轴承那样随 ε 而单调上升,而是到了 ε=0.7 以后反而下降,在 ε=0.8 和 0.9 之间变成了 0,然后变成了负值,所以进入了"绝对不稳"的情况. Ⅱ 中隧道孔的两出口离载荷方向为 25°,由此可见,"绝对稳定"在 ε=0.5 时已出现了,但 ε 进一步增大时,出现了很大的"绝对不稳区"(ε=0.6~0.9 均是),其不稳的直接原因,ε=0.6 时是由于 $K_{xx}K_{yy}-K_{xy}K_{yx}<0$,ε=0.7~0.9 时则除了这以外同时也由于 $K_{eq}<0$. Ⅲ 中隧道孔的两出口离载荷方向为 45°,"绝对稳定"在 ε=0.3 时就已实现,但当 ε 仅增大到 0.5 时,又变成"绝对不稳";更有趣的是,当 ε 增大到 0.7 时,又出现了第二个"绝对稳定"区,换言之,"绝对不稳"区的 ε 范围缩小了. Ⅳ 中隧道孔的两出口离载荷方向的角度大到了 65°,亦即离轴瓦上游边及下游边分别为 25°,此时十分值得注意的现象出现了,即在 ε=0.3 时即进入"绝对稳定",而 ε 进一步增大时并不出现不稳区,直到 ε=0.9 时都是稳定的. 在如此大的 ε 范围内(亦即承载量范围内)能够实现"绝对稳定",这是各类常用固定瓦轴承(圆柱、椭圆、三叶、三油楔轴承)所不易做到的优异性质. Ⅴ 是进一步把隧道孔的两口扩大开在离载荷 75°处,结果情况略差,"绝对稳定"要到 ε=0.4 才发生. 可以想象,如再使两出口离开载荷方向,亦即更近于轴瓦上、下游边,则情况越接近普通轴承,即"稳定"要到 ε=0.8 以上才发生. 由上可见,在对称于载荷而开设隧道孔时,情况Ⅳ为最佳,即出口离载荷约 60°时. 且此时静承载能力并不降低. Ⅵ 是隧道孔开在载荷方向下游,稳定性和承载能力都明显下降.

三、结论

轴瓦上开隧道孔,可使轴承性能发生根本变化. 如果开设的位置不适当,会使承载能力和稳定性均变劣. 而开设得恰当时,则可在承载能力不下降的情况下大大提高稳定性. 这就意味着一种以新颖原理为依据的结构简单而性能极佳的高稳定轴承,值得从理论上和实验上进一步深入研究.

虞烈同志对本文曾提出宝贵意见,在此表示感谢.

参考文献

[1] 张直明. 流体动压轴承润滑理论. 1979.

A Preliminary Analysis of Properties of a New Type of Sliding Bearing-Tunnel Bearing

Abstract: This article calculates the static and dynamic properties of a bearing with a "tunnel" cut in each position using the approximate method based on the theory for infinitely long bearings. The results of calculation show that the stability of the bearing can be significantly improved without impairing its load capacity when a tunnel is cut in a suitable position. Hence a new type of sliding bearing with simple configurations and promising properties can be expected to develop on such a rather novel idea, and can deserve further investigations.

滑动轴承—转子系统的系统
阻尼值与稳定裕度的相互关系*

摘 要：本文以滑动轴承支撑的对称单质量转子为例，计算了系统阻尼值及系统抗交叉刚度和负阻尼的能力。由对比可知，按目前例行办法用系统阻尼值来表征系统的动力稳定裕度，常是不恰当的。建议可能时用相应的减稳因素的界限值来作为稳定裕度的尺度。

引言

众所周知，用滑动轴承支撑的转子，在工作转速很高时，有可能丧失动力稳定性。所以它的工作转速必须低于失稳转速。不仅如此，整个系统还应具有足够的稳定裕度，才能对付实际情况中的各种减稳因素（例如透平机械中的汽隙激振、间隙式密封的流体动压激振、轴材料内摩擦和压合座配合面摩擦激振等）和参数偏差（制造和安装误差、油特性的偏差以及工作参数的变化等）（例如[1]、[2]、[3]）。

近年来，不少研究者以系统阻尼值[4]的大小来衡量系统的稳定裕度。[5]即认为系统阻尼值"对一名义上稳定的转子提供了衡量其稳定裕度的尺度"，并且还建议了系统应具有的相对阻尼值。

诚然，根据系统阻尼值来衡量稳定裕度比单纯根据失稳转速是要进了一步。但究其实质，系统阻尼值（正比于对数衰减率）是系统运动方程组特征值的负实部，它只表达了系统受扰动偏离其平衡位置后不再受扰动外因的情况下回复原位的迅速程度。它究竟能否代表系统对付上述各种恒定起作用的减稳因素的能力，则是一个颇值得探究的问题。从这类转子系统的动力稳定优化设计来说，也只有弄清了这个问题，才能确立优化的目标。

为此，本文将系统对抗汽隙激振（实际上是一个交叉刚度）的能力与相应的系统阻尼值进行对比，来判断二者是否一致。此外，又假设了转子受一个负阻尼的情况，来判断系统阻尼值与系统对抗负阻尼激振因素的能力是否一致。文中的计算对比是针对支撑在两个圆柱轴承上的对称单质量转子进行的。所用的轴承参数为：直径 $D=420$ mm，长度 $L=210$ mm，相对间隙 $\psi=0.0015$，润滑油动力黏度 $\mu=22.528\times10^{-4}$ kg·s/m^2。该轴承在各种偏心率 ε_0 下的无量纲承载力 $\overline{W}$、无量纲刚度 K_{ij} 和无量纲阻尼 D_{ij} 如表1所示。考虑了三种转子参数：a) 重转子，重量 46 000 kg，轴刚度为无穷大或 $5\,791\times10^4$ kg/m；b) 中等转子，重量 4 600 kg，轴刚度为无穷大或 $1\,158\times10^4$ kg/m；c) 轻转子，重量 920 kg，轴刚度为无穷大或 $1\,158\times10^4$ kg/m。此外，也计算了这些转子受不同程度减稳因素作用时之系统阻尼值。

---

* 本文合作者：虞烈。原发表于《上海工业大学学报》，1985(4)：11-20。

一、无外加减稳因素时之系统阻尼值

刚性轴的简图如图 1(a),弹性轴如图 1(b). 图 2(a)和(b)相应地表示端视图中圆盘中心和轴颈中心的瞬时位置. 图中$(\varepsilon_0, \theta_0)$表示轴颈中心的静平衡位置;$(x, y)$为轴颈中心偏离静平衡位置的距离;$(\xi, \eta)$为轴的弯曲挠度;$(\Delta F_x, \Delta F_y)$为油膜力相对于静平衡油膜力的增量.

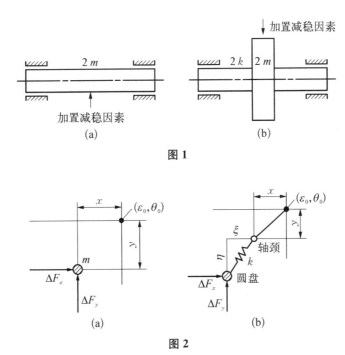

图 1

图 2

刚性转子的运动方程为[7]:

表 1 圆柱轴承的无量纲性能值[6]

ε_0	0.05	0.1	0.2	0.3	0.4	0.5	0.6	0.7	0.8	0.9	0.95
$\overline{W}$	.0335	.0681	.1459	.2451	.3833	.5965	.9597	1.6744	3.4093	10.2712	27.8700
K_{xx}	.0576	.1174	.2576	.4487	.6539	1.0021	1.5834	2.6683	5.5162	15.7608	39.6925
K_{xy}	−.1451	−.2715	−.3138	−.3894	−.3906	−.4054	−.3589	−.0495	1.0485	12.6741	70.8114
K_{yx}	.6768	.7114	.8532	1.1121	1.5036	2.2209	3.5433	6.5698	15.2536	56.3159	193.7993
K_{yy}	.0311	.1155	.2607	.4742	.9075	1.7163	3.4569	8.2396	24.6699	146.8495	806.7754
D_{xx}	.5245	.5526	.6694	.8852	1.0384	1.2612	1.6702	2.2657	3.8729	6.3956	10.2718
D_{xy}	.0485	.1200	.2745	.5116	.7444	1.1295	1.8188	3.0658	6.6133	16.1682	35.1833
D_{yx}	.0155	.1166	.2701	.5030	.7464	1.1172	1.8499	3.0657	6.5777	15.1829	32.6793
D_{yy}	1.3524	1.4179	1.6867	2.1753	2.9023	4.2009	6.7544	12.2137	28.2808	100.1107	328.3816

$$m\ddot{x} + k_{xx}x + k_{xy}y + d_{xx}\dot{x} + d_{xy}\dot{y} = 0 \left.\begin{matrix}\\\\\end{matrix}\right\} \quad (1)$$
$$m\ddot{y} + k_{yx}x + k_{yy}y + d_{yx}\dot{x} + d_{yy}\dot{y} = 0$$

以 $x = x_0 e^{\nu t}$ 及 $y = y_0 e^{\nu t}$ 代入,得特征方程:

$$a_0\nu^4 + a_1\nu^3 + a_2\nu^2 + a_3\nu + a_4 = 0 \tag{2}$$

如以 $\lambda = \nu/\omega$ 代入，则得关于无量纲特征值 λ 的方程：

$$A_0\lambda^4 + A_1\lambda^3 + A_2\lambda^2 + A_3\lambda + A_4 = 0 \tag{3}$$

式中

$$\left.\begin{aligned}
A_0 &= M^2 \\
A_1 &= M(D_{xx} + D_{yy}) \\
A_2 &= M(K_{xx} + K_{yy}) + D_{xx}D_{yy} - D_{xy}D_{yx} \\
A_3 &= K_{xx}D_{yy} + K_{yy}D_{xx} - K_{xy}D_{yx} - K_{yx}D_{xy} \\
A_4 &= K_{xx}K_{yy} - K_{xy}K_{yx} \\
M &= m\omega\psi^3/(\mu L)
\end{aligned}\right\} \tag{4}$$

式(2)或(3)有四个根。四个 ν_i 根的负实数中的最小的一个称为系统阻尼 μ，即 $\mu = \min(-\mathrm{Re}(\nu_1), \cdots, -\mathrm{Re}(\nu_4))$。对于刚性转子，定义无量纲系统阻尼为

$$U = \mu/\omega_0 \tag{5}$$

此处 ω_0——参考角频率，$\omega_0 = \sqrt{g/c}$。 \hfill (6)

类似地，对称单质量弹性转子的运动方程组为

$$\left.\begin{aligned}
m(\ddot{x} + \ddot{\xi}) + k\xi &= 0 \\
m(\ddot{y} + \ddot{\eta}) + k\eta &= 0 \\
k\xi &= k_{xx}x + k_{xy}y + d_{xx}\dot{x} + d_{xy}\dot{y} \\
k\eta &= k_{yx}x + k_{yy}y + d_{yx}\dot{x} + d_{yy}\dot{y}
\end{aligned}\right\} \tag{7}$$

其特征方程及无量纲特征方程依次为

$$a_0\nu^6 + a_1\nu^5 + a_2\nu^4 + a_3\nu^3 + a_4\nu^2 + a_5\nu + a_6 = 0 \tag{8}$$

$$A_0\lambda^6 + A_1\lambda^5 + A_2\lambda^4 + A_3\lambda^3 + A_4\lambda^2 + A_5\lambda + A_6 = 0 \tag{9}$$

式中

$$\left.\begin{aligned}
A_0 &= M^2 A_d \\
A_1 &= M^2[K(D_{xx} + D_{yy}) + A_g] \\
A_2 &= M[MK^2 + MK(K_{xx} + K_{yy}) + MA_k + 2KA_d] \\
A_3 &= MK[K(D_{xx} + D_{yy}) + 2A_g] \\
A_4 &= K[MK(K_{xx} + K_{yy}) + 2MA_k + KA_d] \\
A_5 &= K^2 A_g \\
A_6 &= K^2 A_k \\
A_d &= D_{xx}D_{yy} - D_{xy}D_{yx} \\
A_k &= K_{xx}K_{yy} - K_{xy}K_{yx} \\
A_g &= K_{xx}D_{yy} + K_{yy}D_{xx} - K_{xy}D_{yx} - K_{yx}D_{xy} \\
K &= k\psi^3/(\mu\omega L)
\end{aligned}\right\} \tag{10}$$

系统阻尼 u 为六个 ν_i 根中最小的一个负实部. 弹性转子的无量纲系统阻尼定义为

$$U = \mu/\omega_k \tag{11}$$

而
$$\omega_k = \sqrt{k/m} \tag{12}$$

刚性重转子 a、中转子 b、轻转子 c 的计算结果如图 3 所示. 在 500～2 500 r/min 转速范围内, 刚性转子的系统阻尼均取决于第三、四个 ν_i 根(一对共轭复根).

弹性重转子 a 在约 1 600 r/min 前, 系统阻尼取决于第 3 对 ν_i 根;更高速时则取决于第 2 对 ν_i 根(图4). 弹性中转子 b 在约 2 200 r/min 前取决于第 3 对根, 其后取决于第 1 对根(图5). 弹性轻转子 c 则在全部计算范围内取决于第 1 对根(图6).

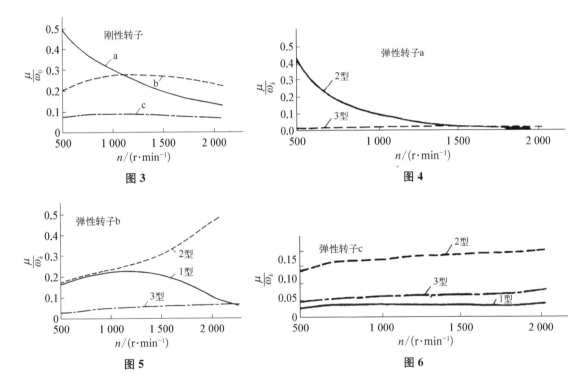

图 3

图 4

图 5

图 6

二、抗各向同性交叉刚度的能力

汽隙激振, 实际上就是在透平转子与定子间外加了一个交叉刚度. 当转子与定子近于处在同心位置时, 这是一个各向同性的交叉刚度. 这个交叉刚度有使转子顺着旋转方向涡动的趋势. 当这个交叉刚度足够大时, 亦即达到某个界限值时, 系统就失稳而发生自激涡动. 这个交叉刚度界限值, 就是该系统抗此类减稳因素的能力. 换言之, 在名义上没有这种减稳因素时, 系统抵抗实际上存在的此减稳因素的稳定储备(裕度)就应以这个界限值来标志.

当刚性转子与定子间的长度中央上加有各向同性交叉刚度 $2k_{st}$ 时, 由运动方程组可得

$$\left.\begin{array}{l}(m\nu^2 + d_{xx}\nu + k_{xx})x_0 + (k_{xy} - k_{st} + \nu d_{xy})y_0 = 0 \\ (k_{yx} + k_{st} + d_{yx}\nu)x_0 + (m\nu^2 + d_{yy}\nu + k_{yy})y_0 = 0\end{array}\right\} \tag{13}$$

存在非平凡解的条件为

$$\begin{vmatrix} m\nu^2+d_{xx}\nu+k_{xx} & d_{xy}\nu+k_{xy}-k_{st} \\ d_{yx}\nu+k_{yx}+k_{st} & m\nu^2+d_{yy}\nu+k_{yy} \end{vmatrix}=0 \quad (14)$$

当 k_{st} 大到其界限值时,系统处在由稳定到不稳的转变状态,此时 ν 为一纯虚数,可令 $\nu=i\Omega$,其中 Ω 为涡动角频率. 代入式(14),令其虚部和实部分别等于零,得

$$\Omega^2=\frac{k_{xx}d_{yy}+k_{yy}d_{xx}-(k_{xy}-k_{st})d_{yx}-(k_{yx}+k_{st})d_{xy}}{m(d_{xx}+d_{yy})} \quad (15)$$

$$(-m\Omega^2+k_{xx})(-m\Omega^2+k_{yy})-\Omega^2(d_{xx}d_{yy}-d_{xy}d_{yx})-(k_{xy}-k_{st})(k_{yx}+k_{st})=0 \quad (16)$$

将式(15)代入式(16)得

$$B_2 M K_{st}^2+(B_{i1}M-A)K_{st}+(\nu_{st}^2 M-K_{eq})=0 \quad (17)$$

式中

$$\left. \begin{aligned} & K_{st}=k_{st}\psi^3/(\mu\omega L) \\ & K_{eq}=\frac{A_g}{D_{xx}+D_{yy}},\ A_g \text{ 见式}(10) \\ & \gamma_{st}^2=\frac{(K_{eq}-K_{xx})(K_{eq}-K_{yy})-K_{xy}K_{yx}}{A_d},\ A_d \text{ 见式}(10) \\ & B_1=\frac{A(2K_{eq}-K_{xx}-K_{yy})}{A_d} \\ & B_2=\frac{1+A^2}{A_d} \\ & A=\frac{D_{yx}-D_{xy}}{D_{xx}+D_{yy}} \end{aligned} \right\} \quad (18)$$

由式(17)即可求出系统能抗的交叉刚度界限值 k_{st},对于刚性转子可定义其相对值为 $k_{st}\psi^3/(\mu\omega_0 L)$. 对于刚性重转子 a、中转子 b、轻转子 c 的计算结果如图 7 所示. 比较图 3 和图 7 可见,对于重转子 a,系统阻尼和交叉刚度界限值随转速的变化是相似的,都是随着转速的提高而单调下降;但对于中转子 b 和轻转子 c,则二者的变化规律却不同,随着转速的提高,系统阻尼先是增大然后降低,交叉刚度界限值却单调降低. 可见,就刚性转子而言,系统阻尼值随转速的变化情况并不与系统抗交叉刚度的能力互相一致. 换言之,系统阻尼值的

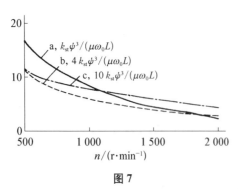

图 7

大小不能如实反映系统对付汽隙激振这类减稳因素的能力.

对于对称单质量弹性转子,当转子与定子间在长度中央上有交叉刚度 k_{st} 时,由运动方程组可得

$$\left.\begin{array}{l} m\nu^2(x_0+\xi_0)-k_{st}(y_0+\eta_0)+k\xi_0=0 \\ k_{st}(x_0+\xi_0)+m\nu^2(y_0+\eta_0)+k\eta_0=0 \\ (d_{xx}\nu+k_{xx})x_0+(d_{xy}y+k_{xy})y_0-k\xi_0=0 \\ (d_{yx}\nu+k_{yx})x_0+(d_{yy}y+k_{yy})y_0-k\eta_0=0 \end{array}\right\} \tag{19}$$

式中 x_0, y_0——轴颈复振幅; ξ_0, η_0——转子挠度复振幅.

由存在非平凡解的条件可得关于 ν 或其无量纲值 λ 的六次方程:

$$A'_0\lambda^6+A'_1\lambda^5+A'_2\lambda^4+A'_3\lambda^3+A'_4\lambda^2+A'_5\lambda+A_6=0$$

式中 $A'_i=A_i+B_i$,而 A_i 由式(10)确定, B_i 值如下:

$$B_0=B_1=B_2=B_3=0$$

$$B_4=A_d K_{st}^2$$

$$B_5=K^2 K_{st}(D_{yx}-D_{xy})+A_g K_{st}^2+KK_{st}^2(D_{xx}+D_{yy})$$

$$B_6=K^2 K_{st}(K_{yx}-K_{xy})+A_k K_{st}^2+KK_{st}^2(K_{xx}+K_{yy})+K^2 K_{st}^2$$

当 k_{st} 大到其界限值对, ν 及 λ 为纯虚数, 由此可算出 k_{st} 的界限值. 对于重转子a、中转子b、轻转子c的计算结果如图8、9、10所示. 为便于对比, 图上亦画出了无外加交叉刚度时之系统阻尼值.

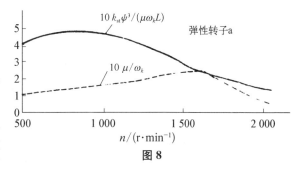

图 8

由图可见, 对于弹性转子, 系统阻尼值亦不能很好反映系统对付汽隙激振这类交叉刚度性的减稳因素的能力. 例如图 8 中, 转速为 1 500 r/min 时的 k_{st} 界限值约为 1 000 r/min 时之 62%, 亦即下降了约 38%;而相反, 系统阻尼值却增加了 46%, 达到了 146%.

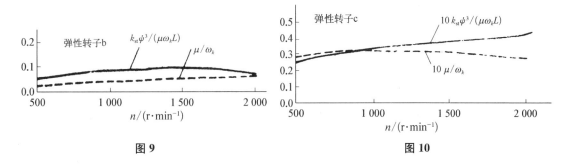

图 9　　　　　　　　　图 10

此外, 为了更明白无误起见, 还进行了这样的计算, 即:在同一转速下, 对转子施加一系列由小至大的 k_{st} 值, 来考察在此 k_{st} 作用下的系统阻尼值的变化. 因为施加的 k_{st} 越大, 系统越接近失稳状态, 其稳定裕度当然越低. 如果系统阻尼值能反映系统裕度的话, 则系统阻尼值应随 k_{st} 的增大而单调下降, 直到在 k_{st} 界限值下系统阻尼降为 0. 但实际计算结果(图 11)却不一定如此, 例如对于刚性转子 c, k_{st} 越大, U 确实越小, 但对于弹性转子 a, 则系统阻尼 U 随着 k_{st} 的增大起初反而上升, 只是 k_{st} 大到一定程度后系统阻尼 U 才下降. 这就更明显地表明了系统阻尼值不能很好反映系统抗交叉刚度的能力.

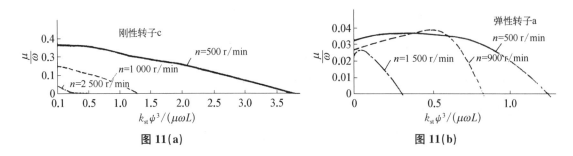

图 11(a)　　　　　　　　　图 11(b)

三、抗各向同性负阻尼的能力

系统阻尼既然不能如实代表抗交叉刚度的能力,那么它是否代表抗负阻尼减稳因素的能力呢？这就是本节分析的目的. 再者,各向同性负阻尼也是原理上最为简单的一种动力减稳因素. 而且,如果系统阻尼值不能跟负阻尼界限值一致的话,它跟其他的减稳因素就更不易一致了. 所以,将系统阻尼值与负阻尼界限值作对比,很有助于弄清系统阻尼值究竟能否代表稳定裕度.

当刚性转子受到各向同性阻尼 d_{st} 时,由运动方程可得

$$\left.\begin{array}{l}[m\nu^2+(d_{xx}+d_{st})\nu+k_{xx}]x_0+(d_{xy}\nu+k_{xy})y_0=0\\(d_{yx}\nu+k_{yx})x_0+[m\nu^2+(d_{yy}+d_{st})\nu+k_{yy}]y_0=0\end{array}\right\} \quad (20)$$

由存在非平凡解的条件可得关于 ν 或 λ 的四次方程:

$$A'_0\lambda^4+A'_1\lambda^3+A'_2\lambda^2+A'_3\lambda+A'_4=0$$

式中 $A'_i=A_i+B_i$. 各 A_i 见式(4). B_i 值则如下:

$$B_0=0 \quad B_1=2MD_{st} \quad B_2=D_{st}^2+D_{st}(D_{xx}+D_{yy})$$
$$B_3=D_{st}(K_{xx}+K_{yy}) \quad B_4=0$$
$$D_{st}=d_{st}\psi^3/(\mu L)$$

当 d_{st} 的负值($-d_{st}$)大到其界限值时,系统恰好失稳,ν 及 λ 为纯虚数,由此可算出负阻尼界限值. 计算结果如图 12、13、14 所示. 为了对比,图中亦示出相应的系统阻尼值. 由图可见,特别是由图 13 及 14 可见,二者的变化规律并不一致.

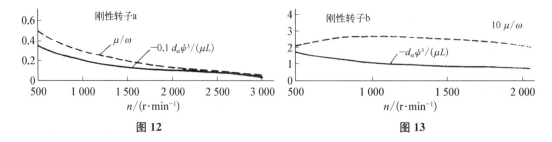

图 12　　　　　　　　　图 13

对于弹性转子,当转子与定子间加有各向同性阻尼时,由运动方程得

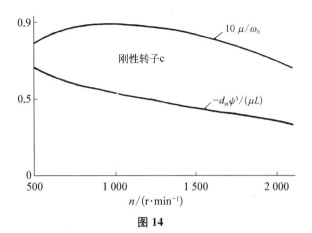

图 14

$$\left.\begin{array}{l}m\nu^2(x_0+\xi_0)+d_{st}\nu(x_0+\xi_0)+k\xi_0=0\\ m\nu^2(y_0+\eta_0)+d_{st}\nu(y_0+\eta_0)+k\eta_0=0\\ (d_{xx}\nu+k_{xx})x_0+(d_{xy}\nu+k_{xy})y_0-k\xi_0=0\\ (d_{yx}\nu+k_{yx})x_0+(d_{yy}\nu+k_{yy})y_0-k\eta_0=0\end{array}\right\} \quad (21)$$

由存在非平凡解的条件可得关于 ν 或 λ 的六次方程：

$$A'_0\lambda^6+A'_1\lambda^5+A'_2\lambda^4+A'_3\lambda^3+A'_4\lambda^2+A'_5\lambda+A'_6=0$$

式中 $A'_i=A_i+B_i$. 各 A_i 见式(10). 各 B_i 如下：

$$B_0=0$$
$$B_1=2MA_dD_{st}$$
$$B_2=A_dD_{st}^2+2A_gMD_{st}+2KM(D_{xx}+D_{yy})D_{st}$$
$$B_3=2K^2MD_{st}+2KA_dD_{st}+A_gD_{st}^2+2A_kMD_{st}+2KM(K_{xx}+K_{yy})D_{st}\\+K(D_{xx}+D_{yy})D_{st}^2$$
$$B_4=K^2D_{st}^2+2KA_gD_{st}+A_kD_{st}^2+K(K_{xx}+K_{yy})D_{st}^2+K^2(D_{xx}+D_{yy})D_{st}$$
$$B_5=2KA_kD_{st}+K^2(K_{xx}+K_{yy})D_{st}$$
$$B_6=0$$

按 ν 及 λ 为纯虚数的条件算出各向同性负阻尼($-d_{st}$)的界限值，如图 15、16、17 所示. 与相应的系统阻尼值对比，亦可看出二者不尽相似. 在重转子 a 和中转子 b 上，二者的变化规律还比较相似，但在轻转子 c 上，系统阻尼值从约 1 100 r/min 开始下降，而($-d_{st}$)的界限值则在 2 500 r/min 以前一直有所上升.

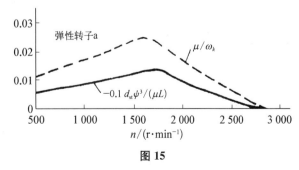

图 15

此外，还计算了刚性轻转子 c 加有一系列由小至大的负阻尼时的系统阻尼值(图 18). 由图可见,在 500,1 000,1 500 r/min 下,系统阻尼值并不随 $(-d_{st})$ 值之增大而单调下降. 这明确显示了系统阻尼值不能很好反映系统抗各向同性负阻尼的稳定裕度. 设想在 500 r/min 下受到 0.56 的无量纲负阻尼时,系统阻尼值很高,但这丝毫不意味着稳定裕度很大,因为当无量纲负阻尼略再增大到 0.65 时,系统迅即失稳(系统阻尼迅速下降到 0).

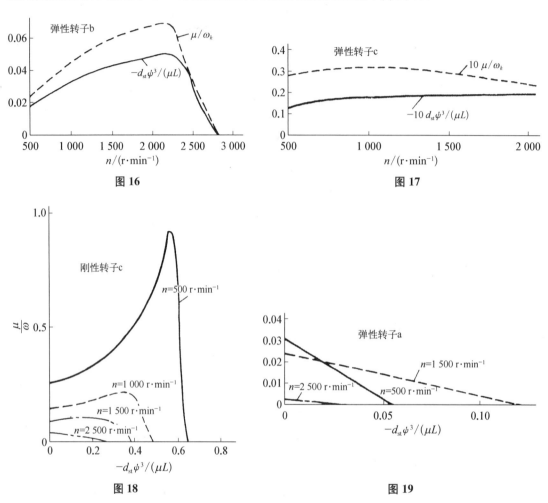

图 16　　图 17

图 18　　图 19

但也有某些情况下,当外加的 $(-d_{st})$ 由 0 逐渐增大到其界限值时,系统阻尼值几乎线性地单调下降到 0 的,如弹性重转子 a(图 19). 此种情况下,无外加负阻尼时系统阻尼值的大小与系统抗各向同性负阻尼的能力是近乎一致的.

四、结论

上述大多数计算结果表明,系统阻尼值不能代表系统抗恒定的减稳因素的能力,亦即不能代表稳定裕度,它只代表系统受扰后在自由振动下回复到其平衡位置上的迅速程度,而不宜引申转借来表达系统的稳定裕度. 这个结论具有普遍意义,并不仅限于圆柱轴承支撑的简单转子.

如果具体情况中该系统可能有的减稳因素明确的话,不如用相应减稳因素的界限值来表示系统的稳定裕度,更为直接和明确. 从计算方法的繁简程度来讲,计算某种减稳因素的界限值时,可以利用特征值为纯虚数这一特点,常可方便不少.

参 考 文 献

[1] Vogel, D. H. Die Stabilität gleitgelagerter Rotoren von Turbomaschinen unter besonderer Berücksichtigung einer Erregung durch Spaltströme. Diss. Technische Hochschule München,1969.

[2] Schirmer, G. F. Zur Stabilität der Schwingungen von Turbinenwellen. Diss. T. H. Darmstadt, 1969.

[3] Glienicke, J. 支承高速转子的滑动轴承(讲稿),西安交通大学,1979年9月.

[4] Glienicke, J. Theoretische und experimentelle Ermittlung der Systemdämpfung gleitgelagerter Rotoren und ihre Erhöhung durch eine aü cre Lagerdampfung. Fortschritt—Berichte der VDI Zeitsch-riften. Rethe И. Nr. 13. Dez. 1972.

[5] Kennedy, F. E. and H. S. Cheng. Computer-Aided Design of Bearing and Seals, 1976.

[6] 圆柱和椭圆轴承性能计算资料,西安交通大学科技参考资料78—010.

[7] 张直明. 流体动压轴承润滑理论,1979年2月.

The Correlation Between Systemdamping and Stability Margin of Rotor-Sliding Bearing Systems

Abstract: Symmetrical single-mass rotors supported on sliding bearings are taken to calculate systemdamping values and the ability of the systems against cross-coupling stiffness and negative damping. A comparison of the results shows that it is not always proper to estimate the stability margin of a system by its systemdamping. It is proposed to express the stability margin by means of the threshold value of the appropriate destabilizing factor whenever the kind of such factor is known.

计入弹性动变形的圆柱形径向滑动轴承动力特性的研究*

摘　要：本文用复数形式推导了滑动轴承中计入动变形时的扰动压力 Reynolds 方程及变形方程的联立方程组,以及动特性和稳定性计算公式.利用 SAP5 程序方便地取得了缩减的柔度矩阵.计算了典型情况下计入动变形时的轴承扰动压力分布、动特性和稳定性,并与只计入静变形时的结果作了比较和分析.可以看到,在一定条件下,这两种结果有显著差异,从而表明计入动变形的重要性.用锦纶轴承和青铜轴承进行了动特性测定实验,定性地证实了理论计算结果.

符 号 说 明

A——油膜区面积

B——阻尼无量纲系数：$B = b\psi^3/(\mu l)$,b——阻尼

d——轴承直径

c——半径间隙

l——轴承宽度

$\bar{E}$——弹性模量无量纲值：$\bar{E} = E\psi^3/(2\mu\omega_0)$,$E$——弹性模量

H——油膜厚度无量纲值：$H = h/c$,h——油膜厚度

H_δ——轴瓦内表面径向变形无量纲量：$H_\delta = h_\delta/c$,h_δ——径向变形

K——刚度无量纲系数：$K = k\psi^3/(\mu\omega_0 l)$,$k$——刚度

P——油膜压力无量纲值：$P = p\psi^2/(2\mu\omega_0)$,$p$——油膜压力

$\bar{Q}$——流量无量纲值：$\bar{Q} = 4Q/(r\omega_0 cl)$,$Q$——流量

r——轴承半径

S_0——承载量系数：$S_0 = W\psi^2/(\mu Ul)$,W——载荷

T——时间无量纲值：$T = t\omega_0$,t——时间

U——轴颈切向速度

γ_{st}——界限涡动比

ε_0——静平衡偏心率

λ——无量纲轴向坐标：$\lambda = z/(l/2)$,Z——由轴承宽度中央计的轴向坐标

θ_0——静平衡偏位角

μ——动力黏度

ν——泊松比

ϕ——由轴承孔顶计量的角坐标

φ——由间隙最大处计量的角坐标

ψ——间隙比

Ω——激振频率无量纲值：$\Omega = \omega/\omega_0$,ω——激振角频率

ω_0——工作角速

一、前言

近二十多年来,已有若干研究者对径向滑动轴承在弹性流体动力润滑工况下的静态特性作了研究[1,2].对其动力特性的研究,大多数仅限于计入静变形[3,4].[5]采用线变形这一较粗略的模型初步计算了动变形对动特性的影响,但对计入动变形后的稳定性问题未做广泛、深入的研究.

---

* 本文合作者：毛谦德、许汉平.原发表于《机械工程学报》,1986,22(1)：10 - 23.

本文采用频率法引入动变形,用复数形式推导了任意简谐小振动下的扰动压力Reynolds方程以及相应的弹性变形关系式.由其联立解即可得出计入动变形的扰动压力分布和动力特性.对轴瓦外表面有刚性支撑而不同弹性模量的圆柱形径向滑动轴承进行了较广泛的分析计算,由此得出了关于动变形对动特性和稳定性影响的一些值得注意的结果.采用弹性材料(锦纶)和较刚性的材料(铜)制成的轴承,以正弦激振法测得动特性,对理论结果作验证,获得了趋势上的一致.

全文以无量纲形式叙述.

二、理论推导

在非定常工况下,Reynolds 方程的无量纲形式为

$$\frac{\partial}{\partial \phi}\left(H^3 \frac{\partial P}{\partial \varphi}\right)+\left(\frac{d}{l}\right)^2 \frac{\partial}{\partial \lambda}\left(H^3 \frac{\partial P}{\partial \lambda}\right)=3 \frac{\partial H}{\partial \varphi}+6 \frac{\partial H}{\partial T} \tag{1}$$

图 1 示轴颈在轴承中的静平衡位置.轴颈在静平衡位置附近作简谐小扰动时,轴颈中心的位置参数可表达如下:

$$\varepsilon=\varepsilon_0+\Delta\varepsilon \mathrm{e}^{i\Omega T} ; \quad \theta=\theta_0+\Delta\theta \mathrm{e}^{i\Omega T} \tag{2}$$

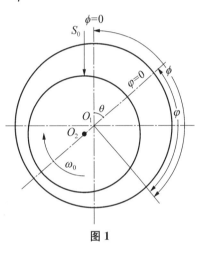

图 1

式中 $\Delta\varepsilon$ 和 $\Delta\theta$ 是偏心率和偏位角的复振幅.此时,油膜厚度的动态增量由两部分组成,其一为轴颈几何位置变动直接引起的增量

$$H_d=(\Delta\varepsilon\cos\varphi+\varepsilon_0\Delta\theta\sin\varphi)\mathrm{e}^{i\Omega T} \tag{3}$$

另一部分是由压力的动态增量引起的动态变形,其一般表达形式可写成:

$$H_\delta=\iint_A Q_\delta(\phi_1,\lambda_1)f(\varphi,\lambda,\varphi_1,\lambda_1)\mathrm{d}\varphi_1\mathrm{d}\lambda_1 \tag{4}$$

f 是点 (φ_1,λ_1) 对点 (φ,λ) 的影响函数;Q_δ 是点 (φ_1,λ_1) 上的油膜压力动态增量;A 是压力分布区.在简振微小振动假设下,Q_δ 亦可用简谐形式表示:

$$Q_\delta=Q\mathrm{e}^{i\Omega T} \tag{5}$$

由此可得动态下的油膜厚度为

$$H=H_0+H_d+H_\delta=H_0+(\Delta\varepsilon\cos\varphi+\varepsilon_0\Delta\theta\sin\varphi+H_\Delta)\mathrm{e}^{i\Omega T} \tag{6}$$

其中由油压扰动引起的膜厚度振幅部分 H_Δ 为

$$H_\Delta=\iint_A Q(\varphi_1,\lambda_1)f(\varphi,\lambda,\varphi_1,\lambda_1)\mathrm{d}\varphi_1\mathrm{d}\lambda_1$$

而 H_0 是包括了静变形的静态膜厚.式(6)与仅计入静变形的膜厚相比,增加了动变形 H_δ 一项,因而更为合理和完整,能如实反映动态下的全面情况.

动态下合成油膜压力为

$$P = P_0 + Q_\delta = P_0 + Q\mathrm{e}^{\mathrm{i}\Omega T} \tag{7}$$

将式(6)、(7)代入式(1),并整理得

$$\frac{\partial}{\partial \varphi}\left(H_0^3 \frac{\partial Q}{\partial \varphi}\right) + \left(\frac{d}{l}\right)^2 \frac{\partial}{\partial \lambda}\left(H_0^3 \frac{\partial Q}{\partial \lambda}\right) = -9\frac{M}{H_0}\frac{\partial H_0}{\partial \varphi}$$
$$-3H_0\left[\left(H_0\frac{\partial M}{\partial \varphi} - M\frac{\partial H_0}{\partial \varphi}\right)\frac{\partial P_0}{\partial \varphi} + \left(\frac{d}{l}\right)^2\left(H_0\frac{\partial M}{\partial \lambda} - \frac{\partial H_0}{\partial \lambda}M\right)\frac{\partial P_0}{\partial \lambda}\right] + 3\frac{\partial M}{\partial \varphi} + 6\mathrm{i}\Omega M \tag{8}$$

其中 $M = \Delta\varepsilon\cos\varphi + \varepsilon_0\Delta\theta\sin\varphi + H_\Delta$.

采用偏导数法计算动特性. 将式(8)对 $\Delta\varepsilon$ 取偏导以获得扰动压力 Q_ε 的 Reynolds 方程:

$$\mathrm{Rey}(Q_\varepsilon) = -9\frac{\cos\varphi + H_{\Delta\varepsilon}}{H_0}\frac{\partial H_0}{\partial \varphi} - 3H_0\left\{\left[H_0\left(-\sin\varphi + \frac{\partial H_{\Delta\varepsilon}}{\partial \varphi}\right) - (\cos\varphi + H_{\Delta\varepsilon})\frac{\partial H_0}{\partial \varphi}\right]\frac{\partial P_0}{\partial \varphi}\right.$$
$$\left.+ \left(\frac{d}{l}\right)^2\left[H_0\frac{\partial H_{\Delta\varepsilon}}{\partial \lambda} - (\cos\varphi + H_{\Delta\varepsilon})\frac{\partial H_0}{\partial \lambda}\right]\frac{\partial P_0}{\partial \lambda}\right\}$$
$$+ 3\left(-\sin\varphi + \frac{\partial H_{\Delta\varepsilon}}{\partial \varphi}\right) + 6\mathrm{i}\Omega(\cos\varphi + H_{\Delta\varepsilon}) \tag{9}$$

式中 $Q_\varepsilon = \frac{\partial Q}{\partial \varepsilon}$; $H_{\Delta\varepsilon} = \frac{\partial H_\Delta}{\partial \varepsilon}$; 算符 $\mathrm{Rey}(\ ) = \frac{\partial}{\partial \varphi}\left[H_0^3 \frac{\partial(\ )}{\partial \varphi}\right] + \left(\frac{d}{l}\right)^2 \frac{\partial}{\partial \lambda}\left[H_0^3 \frac{\partial(\ )}{\partial \lambda}\right]$.

式(6)对 $\Delta\varepsilon$ 取偏导:

$$H_{\Delta\varepsilon} = \iint_A Q_\varepsilon(\varphi_1, \lambda_1) f(\varphi, \lambda, \varphi_1, \lambda_1) \mathrm{d}\varphi_1 \mathrm{d}\lambda_1 \tag{10}$$

联立(9)、(10),反复迭代解出 Q_ε 分布,由此可积分得刚度系数 $K_{x\varepsilon}$, $K_{y\varepsilon}$ 和阻尼系数 $B_{x\varepsilon}$, $B_{y\varepsilon}$:

$$-\iint_A Q_\varepsilon \cos\phi \mathrm{d}\phi \mathrm{d}\lambda = K_{y\varepsilon} + \mathrm{i}\Omega B_{y\varepsilon}$$
$$-\iint_A Q_\varepsilon \sin\phi \mathrm{d}\phi \mathrm{d}\lambda = K_{x\varepsilon} + \mathrm{i}\Omega B_{x\varepsilon} \tag{11}$$

类似地,对 $\Delta\theta$ 取偏导数,并记 $Q_\theta = \frac{1}{\varepsilon_0}\frac{\partial Q}{\partial \theta}$, $H_{\Delta\theta} = \frac{1}{\varepsilon_0}\frac{\partial H_\Delta}{\partial \theta}$,则可解出扰动压力 Q_θ 分布,并积分得 $K_{x\theta}$, $K_{y\theta}$, $B_{x\theta}$ 和 $B_{y\theta}$:

$$-\iint_A Q_\theta \cos\phi \mathrm{d}\phi \mathrm{d}\lambda = K_{y\theta} + \mathrm{i}\Omega B_{y\theta}$$
$$-\iint_A Q_\theta \sin\phi \mathrm{d}\phi \mathrm{d}\lambda = K_{x\theta} + \mathrm{i}\Omega B_{x\theta} \tag{12}$$

如果希望用实数运算来代替上面的复数运算,只需以下列变换代入即可:

$$Q_\varepsilon = R + iS \qquad H_{\Delta\varepsilon} = U + i\Omega V$$
$$Q_\theta = R_1 + iS_1 \qquad H_{\Delta\theta} = U_1 + i\Omega V_1 \tag{13}$$

其中 R, R_1, S, S_1 均为实数. 每一复数方程均分解为两个实数方程, 实际上就得到了[5]中的形式. 将 R, S, U, V 联立求解, 以及 R_1, S_1, U_1, V_1 联立求解, 即可求出各动力特性.

将以上所得之动力特性进行坐标转换, 即可得直角坐标系中的动力特性系数 K_{ij} 和 $B_{ij}(i, j = x, y)$.

计入动变形后, 动力特性系数与激振频率有关, 故稳定性计算应与固定瓦轴承的常规方法略异. 先估取一 Ω 值, 求出动力特性系数后, 计算界限涡动比 γ_{st} 和相当刚度系数 K_{eq}:

$$K_{eq} = \frac{K_{xx}B_{yy} + K_{yy}B_{xx} - K_{xy}B_{yx} - K_{yx}B_{xy}}{B_{xx} + B_{yy}} \tag{14}$$

$$\gamma_{st} = \frac{(K_{eq} - K_{xx})(K_{eq} - K_{yy}) - K_{xy}K_{yx}}{B_{xx}B_{yy} - B_{xy}B_{yx}} \tag{15}$$

如算出的 $\gamma_{st} \neq \Omega$, 则修改 Ω 重新计算, 直至 $\gamma_{st} = \Omega$ 为止, 相应的 K_{ij}, $B_{ij}(i, j = x, y)$, K_{eq} 和 γ_{st} 即为失稳界限值, 可据以计算失稳转速.

三、计算方法

上列各 Reynolds 方程是用差分超松弛迭代法解算, 此过程中将膜厚分布作为已知, 解算压力或扰动压力分布. 由于轴向对称性, 取一半宽度计算, 网格划分为 9×40.

变形分布则利用 SAP5 程序中的非协调位移模型有限元法求出三维刚度矩阵, 换算成轴瓦内表面上压力分布与径向变形分布之间的缩减柔度矩阵, 将压力分布作为已知, 计算径向变形分布. 轴瓦单元划分为 $4 \times 22 \times 3$, 见图 2.

在两类方程间作交叉亚松弛迭代.

获得缩减的柔度矩阵的具体步骤如下:

(1) 用 SAP5 获得节点三维载荷与位移间的总刚度矩阵, 将其取出作为一个结果文件, 再处理成普通数据文件迭代计算时将其一次读入后存在公用块中;

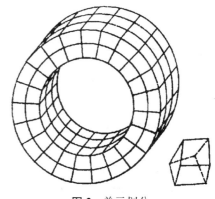

图 2 单元划分

(2) 用自编的专用子程序将其换算成缩减了的柔度矩阵, 即可反复使用于由压力分布求径向位移分布的每一次计算中.

由总刚度矩阵到缩减了的柔度矩阵的推导过程如下.

有关系式

$$[K]\{M\} = \begin{bmatrix} K_{11} & K_{12} \\ K_{21} & K_{22} \end{bmatrix} \begin{Bmatrix} M_1 \\ M_2 \end{Bmatrix} = \begin{Bmatrix} 0 \\ P_e \end{Bmatrix} \tag{16}$$

式中, $\{M_1\}$ 为除轴瓦内表面以外的各节点自由度所列成的矢量; $\{M_2\}$ 为轴瓦内表面自由度矢量; $\{M\}$ 为其总矢量; K 为总刚度矩阵; K_{11}, K_{12}, K_{21}, K_{22} 为相应的分块刚度矩阵; P_e 为内表面节点力. 此式可分解为两个子矩阵方程:

$$[K_{11}]\{M_1\} + [K_{12}]\{M_2\} = \{0\} \tag{17}$$

$$[K_{21}]\{M_1\} + [K_{22}]\{M_2\} = \{P_e\} \tag{18}$$

由此可得

$$\{M_2\} = [f]\{P_e\} \tag{19}$$

式中[f]为缩减到内表面上的柔度矩阵：

$$[f] = \{-[K_{21}][K_{11}]^{-1}[K_{12}] + [K_{22}]\}^{-1} \tag{20}$$

由于需要的只是表面正压力与表面径向位移之间的关系，[f]还可缩减. 如图3，作用在某 A_i 点上的径向力 P_{ri} 可分解为等效节点力 P_{et} 的三个分量：

$$P_{xi} = P_{ri}\sin\phi_i ; \quad P_{xi} = P_{ri}\cos\phi_i ; \quad P_{zi} = 0 \tag{21}$$

由 A_i 点的位移分量 (M_{2x}, M_{2y}, M_{2z}) 又可求得该点径向位移 $H_{\delta i}$：

$$H_{\delta i} = M_{2x}\sin\phi_i + M_{2y}\cos\phi_1 \tag{22}$$

M_{2z} 实际上与 $H_{\delta i}$ 是无关的.

综合以上关系，最后可得

$$\{H_\delta\} = [f]_r\{P_r\} \tag{23}$$

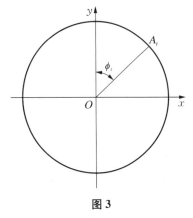

图 3

式中 $\{H_\delta\}$ 为纯由内表面各节点径向位移列成的矢量；$\{P_r\}$ 为这些节点上由油膜压力构成的径向节点力列成的矢量；$[f]_r$ 为缩减到内表面径向关系上的柔度矩阵. 用式(23)，就可从压力分布求得径向位移分布. 由于柔度矩阵的阶数远远低于总刚矩阵的阶数，因而计算时间可以大大缩短.

四、试验方法和装置

在倒置式试验台上测定了一个刚性较大的铜轴承和一个刚性较小的锦纶轴承的动特性，作为对比. 图4示试验装置简图. 电动机带动轴旋转. 轴承载荷由控制波纹管中气压加置. 用电磁激振器对轴承施加正弦交变力，迫使轴承相对于轴而振动. 由力的信号和位移信号可确定出油膜刚度和阻尼. 激振力信号由应变片输出，位移信号由电涡流式位移传感器输出，各信号由多通道磁带记录仪同时记录，在快速Fourier变换分析仪上将时域信号转换到频域，分析处理而定出动特性. 整套测试系统经过动态标定. 测试和标定详情.

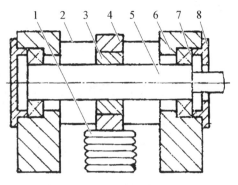

1—波纹管 2—铁链 3—试验轴承 4—轴承座
5—轴 6—基座 7—支撑轴承 8—端盖

图 4

五、结果

计算实例中的轴承参数取 $l/d=0.6$, $\varepsilon_0=0.1\sim0.95$, $\bar{E}=200$(相当于铜轴承)和20(相当于锦纶轴承),泊松比 $\nu=0.3$, $\Omega=0.5$ 及 1.0, 轴瓦外、内径之比 $D/d=1.6$.

静特性的结果与[4]一致,简述如下:

图 5 示 $\bar{E}=20$, $\varepsilon_0=0.8$ 的静态压力 p_0 分布. 与刚性轴承情况相比, p_0 随 $\bar{E}$ 的下降而下降,油压区增大,压力梯度降低,压力峰附近这种变化尤其明显.

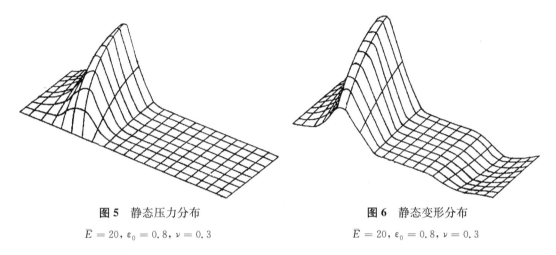

图 5 静态压力分布

$\bar{E}=20$, $\varepsilon_0=0.8$, $\nu=0.3$

图 6 静态变形分布

$\bar{E}=20$, $\varepsilon_0=0.8$, $\nu=0.3$

与图5相应的轴瓦内表面静态径向位移分布见图6. 位移分布与压力分布形状大致相似. $\bar{E}$ 越小则变形越大,最大变形发生在压力峰附近,在油膜区外会有较小的凸起. 变形从中截面向两端逐渐减小,最小油膜厚度发生在轴承两端.

图 7 示不同 $\bar{E}$ 下的轴心静平衡位置轨迹. 偏心不很大时,计入弹性变形的偏位角 θ_0 下降, $\bar{E}$ 越小下降越显著. 但偏心很大时则有相反趋势, $\bar{E}$ 越小 θ_0 越大.

图 8 示不同 $\bar{E}$ 下 S_0 数随 ε_0 变化的关系, $\bar{E}$ 越小则 S_0 数越小,这种变化随 ε_0 的增大而迅速增大.

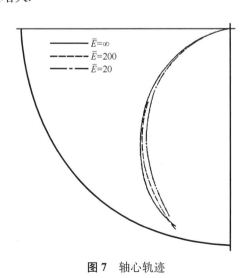

图 7 轴心轨迹

图 8 S_0 数随 ε_0 变化的曲线

图 9 示不同 $\bar{E}$ 下端泄随 ε_0 的变化关系,$\bar{E}$ 越小则端泄量越小,减小量随 ε_0 增大而增大.

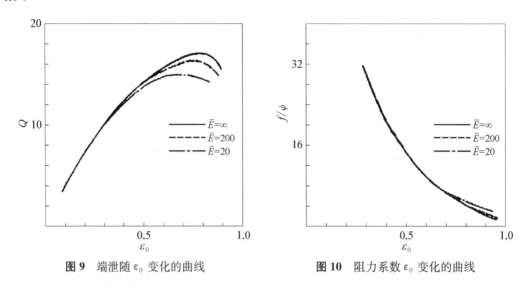

图 9 端泄随 ε_0 变化的曲线   图 10 阻力系数 ε_0 变化的曲线

图 10 示不同 $\bar{E}$ 下阻力系数随 ε_0 变化的关系,$\bar{E}$ 越小则阻力系数越大.

在动特性方面,主要结果如下:

表 1 是计入动变形和仅计入静变形的动特性系数比较. 计入动变形后,在 $\varepsilon_0 < 0.8$ 的范围内,K_{xx} 有所上升,到 $\varepsilon_0 = 0.8$ 时则下降了;K_{xy} 在计入动变形后变大了;当 $\varepsilon_0 \leqslant 0.2$ 时,计入动变形后 K_{yx} 略有增加,但 $\varepsilon_0 = 0.3$ 以后则下降了;K_{yy} 的变化规律与 K_{xx} 同. 刚度系数中,K_{xy} 变化较大些,其他三个系数不考虑动变形而产生的误差在 5% 左右. 计入动变形后,B_{xx} 和 B_{yx} 下降了,而且在较大偏心时不再随 ε_0 单调上升,而发生了下降,特别是 B_{yx},在 $\varepsilon_0 = 0.8$ 时降为负值. B_{xy} 和 B_{yy} 在偏心较小时稍有增大,但在偏心较大时则下降了. 计入动

表 1 计入动变形和仅计入静变形的动特性系数比较($E = 20$, $\Omega = 0.5$)

ε_0		K_{xx}	K_{xy}	K_{yx}	K_{yy}	B_{xx}	B_{xy}	B_{yx}	B_{yy}
0.1	*	0.243 3	−0.711 1	1.055 1	0.103 0	1.441 1	0.280 3	0.281 5	2.106 8
	,	0.262 1	−0.704 1	1.059 9	0.129 6	1.424 3	0.306 2	0.234 8	2.113 0
0.2	*	0.446 5	−0.681 8	1.225 4	0.304 8	1.447 2	0.485 2	0.475 3	2.416 5
	,	0.477 2	−0.667 7	1.227 1	0.339 4	1.408 3	0.517 0	0.405 8	2.406 8
0.3	*	0.669 6	−0.634 5	1.528 5	0.628 4	1.464 6	0.696 0	0.702 1	2.960 2
	,	0.702 4	−0.602 6	1.521 1	0.681 3	1.394 7	0.718 1	0.570 0	2.905 5
0.4	*	1.015 6	−0.689 1	2.066 4	1.119 1	1.777 4	1.107 7	1.122 4	3.944 0
	,	1.075 9	−0.610 8	2.046 8	1.228 9	1.618 5	1.121 8	0.834 6	3.798 8
0.5	*	1.508 5	−0.687 9	2.975 5	2.088 6	2.139 8	1.691 9	1.677 8	5.622 0
	,	1.589 1	−0.554 2	2.922 2	2.259 2	1.912 8	1.638 5	1.246 7	5.287 9
0.6	*	2.212 8	−0.467 4	4.611 2	4.233 8	2.418 0	2.361 9	2.375 3	8.563 6
	,	2.307 5	−0.161 0	4.336 0	4.541 9	1.916 8	2.076 6	1.302 8	7.377 6
0.7	*	3.500 5	0.044 2	7.846 8	9.287 3	2.964 4	3.501 6	3.434 7	14.260 5
	,	3.600 2	0.693 5	6.654 0	9.389 0	1.807 1	2.388 5	0.705 0	9.959 1
0.8	*	6.702 5	0.646 6	15.921 7	20.783 6	5.305 4	7.407 5	7.316 8	29.320 1
	,	6.292 4	2.485 1	10.188 0	18.803 2	1.528 2	2.877 5	−1.157 2	12.818 4

注:*—静变形;,—动变形.

变形后，B_{xy} 和 B_{yx} 彼此可有显著差异．从数值上看，计入动变形后阻尼系数变化大，即使在 $\varepsilon_0=0.2$ 这种小偏心下，不考虑动变形所引起的误差已值得注意．

按 $\gamma_{st}=\Omega=0.5$ 搜索了计入动变形时相应的 ε_0 值，并计算了该种 ε_0 值下只考虑静变形时的 γ_{st} 值，其对比见表 2．可见，动变形的计入使 γ_{st} 变小，这是有利于稳定性的．

表 2 $\bar{E}=20$，$\Omega=0.5$ 时计入动变形和只计入静变形的 γ_{st} 值比较

ε_0		γ_{st}
0.045 09	只计入静变形	0.500 5
	计入动变形	0.5
0.502 65	只计入静变形	0.502 7
	计入动变形	0.5

计入动变形后，动特性系数与激振频率有关．表 3 列示不同 $\bar{E}$ 下 $\Omega=0.5$ 和 1.0 时的刚度系数．可以看到，刚度系数随频率变化不大，即使在 $\varepsilon_0=0.9$ 这样大的偏心下，不考虑 Ω 而引起的刚度系数的相对误差也只有 5% 左右．

表 3 激振频率对刚度系数的影响

ε_0	Ω	$\bar{E}=200$				$\bar{E}=20$			
		K_{xx}	K_{xy}	K_{yx}	K_{yy}	K_{xx}	K_{xy}	K_{yx}	K_{yy}
0.1	0.5	0.244 6	−0.709 6	1.052 8	0.102 8	0.262 1	−0.704 1	1.059 9	0.126 9
	1.0	0.245 8	−0.710 6	1.054 9	0.101 2	0.289 3	−0.691 2	1.069 1	0.182 6
0.2	0.5	0.450 2	−0.677 6	1.221 9	0.305 1	0.472 2	−0.667 7	1.227 1	0.339 3
	1.0	0.454 8	−0.679 6	1.224 8	0.310 6	0.506 0	−0.641 9	1.247 7	0.414 7
0.3	0.5	0.674 6	−0.629 4	1.518 7	0.627 3	0.702 4	−0.602 6	1.521 1	0.681 3
	1.0	0.68	−0.626 0	1.523 5	0.637 5	0.745 8	−0.557 1	1.557 9	0.797 7
0.4	0.5	1.027 5	−0.679 0	2.055 5	1.128 4	1.075 9	−0.610 8	2.046 8	1.228 9
	1.0	1.038 5	−0.669 5	2.068 7	1.150 7	1.148 6	−0.510 6	2.120 4	1.445 5
0.5	0.5	1.536 1	−0.674 4	2.994 2	2.130 4	1.605 3	−0.520 2	2.914 8	2.311 2
	1.0	1.552 3	−0.651 9	3.013 6	2.176 5	1.718 3	−0.316 4	3.043 3	2.741 1
0.6	0.5	2.286 7	−0.442 9	4.688 6	4.440 9	2.305 7	−0.095 9	4.262 3	4.613
	1.0	2.314 2	−0.393 2	4.734 6	4.521 3	2.441 5	0.25	4.423 0	5.413 1
0.7	0.5	3.773 1	0.146 1	8.249 4	10.147 2				
	1.0	3.835 6	0.291 0	8.373 7	10.541 0				
0.8	0.5	6.865 4	2.694 5	16.951 5	29.180 1				
	1.0	7.004 9	3.187 6	17.282 7	30.822 8				
0.9	0.5	20.014 8	15.708 6	47.726 3	111.850 2				
	1.0	19.928 9	17.585 7	46.971 2	118.512 8				

表 4 是不同频率下的阻尼系数．频率对阻尼系数的影响比起对刚度系数的影响更小．

由以上可知．按工程计算的精度要求来看，实际上可不考虑频率影响．这说明，计入动变形后，方程中与频率无关的项仍起主要作用．对本例而言，这个结论是在 $\Omega=0.5\sim1.0$ 范围内得到的，超出这一范围，特别是 Ω 很大时，则需另作考虑．

表 4 激振频率对阻尼系数的影响

ε_0	Ω	$\bar{E}=200$				$\bar{E}=20$			
		B_{xx}	B_{xy}	B_{yx}	B_{yy}	B_{xx}	B_{xy}	B_{yx}	B_{yy}
0.1	0.5	1.439 2	0.280 9	0.281 1	2.099 5	1.424 3	0.366 2	0.234 8	2.113 0
	1.0	1.439 2	0.280 9	0.281 7	2.099 6	1.423 2	0.308 4	0.231 9	2.113 7
0.2	0.5	1.443 0	0.488 0	0.472 0	2.401 6	1.408 3	0.510 7	0.405 8	2.406 8
	1.0	1.441 8	0.488 8	0.470 3	2.406 1	1.405 5	0.512 0	0.403 3	2.406 5
0.3	0.5	1.462 0	0.706 1	0.696 8	2.933 8	1.394 7	0.718 1	0.570 0	2.905 5
	1.0	1.460 8	0.705 5	0.691 7	2.936 8	1.389 9	0.718 2	0.564 4	2.902 8
0.4	0.5	1.771 7	1.131 5	1.111 0	3.915 0	1.618 5	1.121 8	0.834 6	3.798 8
	1.0	1.768 3	1.129 5	1.102 7	3.917 7	1.609 6	1.119 7	0.821 6	3.774 6
0.5	0.5	2.140 3	1.737 1	1.679 8	5.621 9	1.802 0	1.619 6	1.037 3	5.189 9
	1.0	2.136 1	1.733 1	1.666 7	5.612 8	1.778 3	1.571 7	1.003 0	5.098 7
0.6	0.5	2.429 0	2.482 4	2.368 4	8.622 2	1.679 3	1.941 1	0.820 5	6.976 6
	1.0	2.426 1	2.479 4	2.368 2	8.616 7	1.652 4	1.820 7	0.786 1	6.721 0
0.7	0.5	3.024 8	3.909 0	3.424 3	14.595 1				
	1.0	3.017 9	3.904 0	3.416 9	14.578 4				
0.8	0.5	3.355 5	5.765 2	3.931 7	27.100 0				
	1.0	3.340 0	5.687 8	3.893 6	26.870 0				
0.9	0.5	2.737 7	7.097 4	−4.022 6	45.772 5				
	1.0	2.813 4	6.438 7	−3.723 5	43.515 1				

图 11 是计入动变形和只计静变形时相当刚度系数 K_{eq} 随 ε_0 变化的关系. 可见,计入动变形后 K_{eq} 稍有增加.

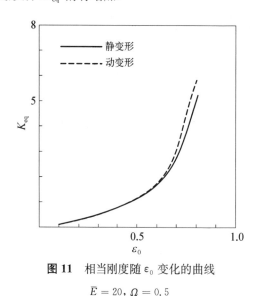

图 11 相当刚度随 ε_0 变化的曲线
$\bar{E}=20, \Omega=0.5$

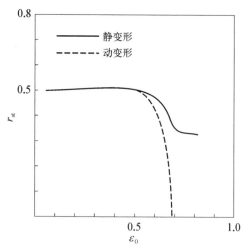

图 12 界限涡动比随 ε_0 变化的曲线
$\bar{E}=20, \Omega=0.5$

图 12 是计入动变形和只计静变形时界限涡动比 γ_{st} 随 ε_0 变化的关系. 可见,计入动变形后 γ_{st} 下降,恒稳情况提前到较小的 ε_0 下出现.

表 5 是临界质量无量纲值 K_{eq}/γ_{st} 随 ε_0 变化的关系. 计入动变形后, 在偏心较大时, 临界质量变大.

表 5 临界质量无量纲值 k_{eq}/γ_{st}^2 随 ε_0 变化的关系($\bar{E}=20, \Omega=0.5$)

ε_0		0.1	0.2	0.3	0.4	0.5	0.6	0.7	0.8
$\dfrac{K_{eq}}{\gamma_{st}^2}$	只计入静变形	0.632 3	1.267 7	2.011 7	3.005 4	4.552 0	8.196 2	24.701	48.426
	计入动变形	0.643 2	1.279 2	2.009 2	3.004 2	4.754 4	10.324	稳	稳

图 13～16 是不同 $\bar{E}$ 下刚度系数随 ε_0 变化的关系. 偏心不大时, $\bar{E}$ 越小则 K_{xx} 越大, 偏心较大时, $\bar{E}$ 越小则 K_{xx} 越小. $\bar{E}$ 越小则 K_{xy} 越大, 在大偏心下这种差别更明显一些. K_{yx} 和 K_{yy} 在小偏心下变化甚微, 在较大偏心下 $\bar{E}$ 越小则值越小.

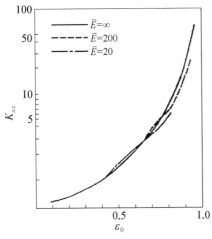

图 13 K_{xx} 随 ε_0 变化的曲线

$\Omega=0.5$

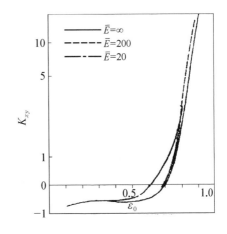

图 14 K_{xy} 随 ε_0 变化的曲线

$\Omega=0.5$

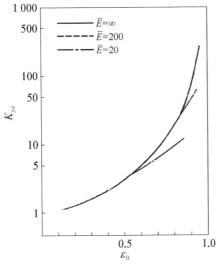

图 15 K_{yx} 随 ε_0 变化的曲线

$\Omega=0.5$

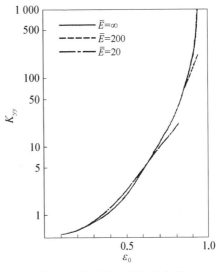

图 16 K_{yy} 随 ε_0 变化的曲线

$\Omega=0.5$

图 17～20 是不同 $\bar{E}$ 下阻尼系数随 ε_0 变化的关系. $\bar{E}$ 越小则 B_{xx} 和 B_{yx} 越小. 特别的是, 当偏心较大时, 它们会随 ε_0 增大而下降, B_{yx} 还会降为负值, $\bar{E}$ 越小则出现这种趋势越早. $\bar{E}$ 越小则 B_{xy} 和 B_{yy} 越小.

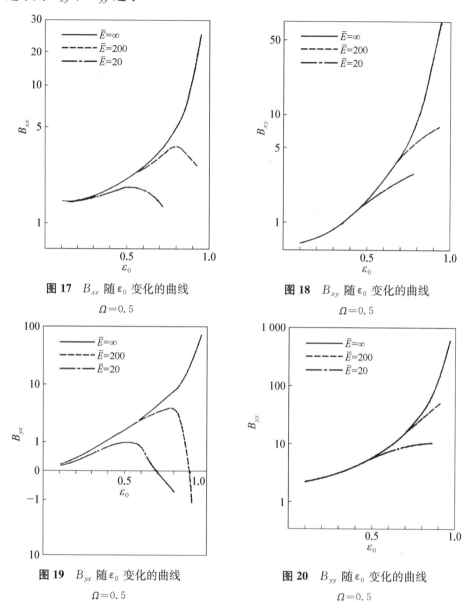

图 17　B_{xx} 随 ε_0 变化的曲线
$\Omega=0.5$

图 18　B_{xy} 随 ε_0 变化的曲线
$\Omega=0.5$

图 19　B_{yx} 随 ε_0 变化的曲线
$\Omega=0.5$

图 20　B_{yy} 随 ε_0 变化的曲线
$\Omega=0.5$

图 21 是不同 $\bar{E}$ 下相当刚度系数 K_{eq} 随 ε_0 变化的关系, $\bar{E}$ 越小则 K_{eq} 越大.

图 22 是不同 $\bar{E}$ 下界限涡动比 γ_{st} 随 ε_0 变化的关系, 在 ε_0 较小时变化甚少, ε_0 较大时 $\bar{E}$ 越小则 γ_{st} 越小, 且 $\gamma_{st}=0$ 的点前移, 即恒稳情况提前到较小的 ε_0 下发生.

图 23～26 是不同 $\bar{E}$ 下刚度系数试验结果与理论结果的比较. S_0 数较小时, $\bar{E}$ 越小则 K_{xx} 越大; S_0 数较大时, $\bar{E}$ 越小则 K_{xx} 越小. $\bar{E}$ 越小则 K_{xy} 越小. K_{yx} 的变化规律与 K_{xx} 同. K_{yy} 的理论值与试验值接近得较好; S_0 数较小时, $\bar{E}$ 越小则 K_{yy} 稍有增加; S_0 数大一点后, $\bar{E}$ 越小则 K_{yy} 越小.

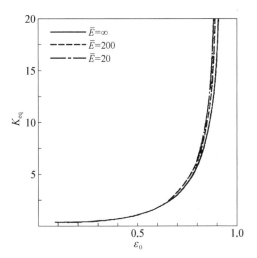

图 21 相当刚度 K_{eq} 随 ε_0 变化的曲线

$\Omega = 0.5$

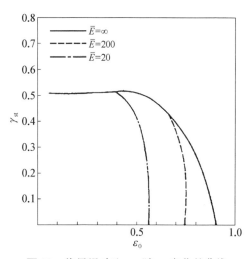

图 22 临界涡动比 γ_{st} 随 ε_0 变化的曲线

$\Omega = 0.5$

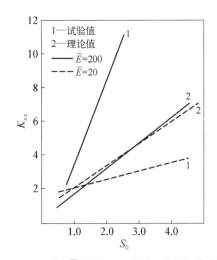

图 23 不同 $\bar{E}$ 值下 K_{xx} 随 S_0 变化的曲线

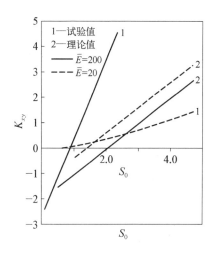

图 24 不同 $\bar{E}$ 值下 K_{xy} 随 S_0 变化的曲线

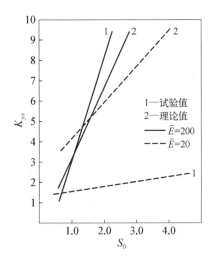

图 25 不同 $\bar{E}$ 值下 K_{yx} 随 S_0 变化的曲线

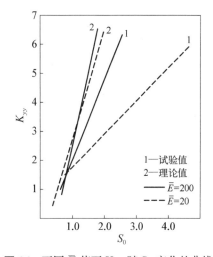

图 26 不同 $\bar{E}$ 值下 K_{yy} 随 S_0 变化的曲线

图 27～30 是不同 $\bar{E}$ 下阻尼系数试验结果与理论结果的比较. 对于软材料($\bar{E}=20$), B_{xx} 随 S_0 上升而下降, $\bar{E}$ 越小则 B_{xx} 越小. B_{xy} 的变化规律与 B_{xx} 同. $\bar{E}$ 越小则 B_{yx} 越小;对于软材料, B_{yx} 随 S_0 增大而减小,而且很快降为负值. $\bar{E}$ 越小则 B_{yy} 越小.

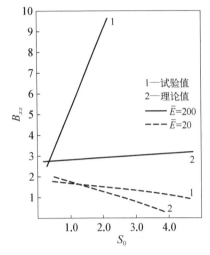

图 27　不同 $\bar{E}$ 值下 B_{xx} 随 S_0 变化的曲线

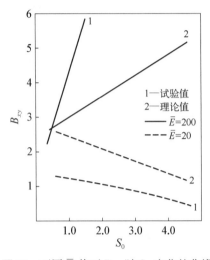

图 28　不同 $\bar{E}$ 值下 B_{xy} 随 S_0 变化的曲线

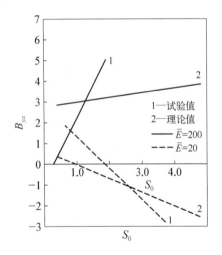

图 29　不同 $\bar{E}$ 值下 B_{yx} 随 S_0 变化的曲线

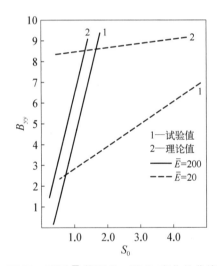

图 30　不同 $\bar{E}$ 值下 B_{yy} 随 S_0 变化的曲线

六、结论

在所计算的情况下:

1. 计入弹性变形后,静态压力分布降低,特别是压力峰附近的压力显著降低,承载量系数相应降低. 轴瓦内表面径向变形分布与压力分布大致相似. 最小油膜厚度发生在轴端.

2. 计入变形后,端泄量减小,阻力系数增大.

3. 计入动变形后,对刚度系数影响不大,其差别在 5～左右;但对阻尼系数影响显著,必须考虑. 阻尼系数的变化有如下特点:(i) 交叉阻尼不再彼此相等;(ii) B_{xx} 和 B_{yx} 在偏心较

大时不再随 ε_0 单调增大,而发生下降;(iii) B_{yx} 在大偏心下会出现负值.

4. 计入动变形后,动特性系数与频率有关,但在 $\Omega=0.5\sim1.0$ 范围内,这种影响微小,可以不计.

5. 在稳定性方面,计入动变形的影响是有利于稳定性.

致谢 在试验过程和 Reynolds 方程解算程序编制过程中,西安交通大学丘大谋副教授和虞烈讲师提出了有价值的意见,并给予不少帮助. 虞烈讲师还在分析和整理结果时给予帮助. 全部计算和试验是在西安交通大学润滑与轴承研究室完成的,该室其余工作人员亦给予了支持,在此一并致谢.

参 考 文 献

[1] Higginson, G. R. The theoretical effects of elastic deformation of the bearing liner on journal bearing performance. Proc. Symposium on Elastohydrddynamic Lubrication, Inst. Mech. Engrs., 1965, 180, Part 3B, pp. 31-38.

[2] Oh, K. P. and K. H. Hueber. Solution of the elastohydrodynamic finite journal bearing problem, Trans. ASME, J. Lub. Techn., Vol. 95, 1973, pp. 342-352.

[3] Jain, S. C. R. Sinhasen and D. V. Singh. A study of elastohybrodynamic lubrication of a centrally loaded 120° arc partial bearing in different flow reginmes, Proc. Inst. Mech. Engrs., Vol. 197, Part C, 1983, pp. 97-108.

[4] Mao Qiande, Han D.-C. and J. Glienicke. Stabilitaetseigenschaften von Gleitlagern bei Beruecksichtigung der Lagerelastizitaet. Konstruktion 35 (1983) H. 2, S. 45-52.

[5] Nilsson, L. R. K. The influence of bearing flexibility on the dynamic performance of radial oil film bearings. Proc. Fifth Leeds-Lyon Symposium on Tribology, 1978, Paper IX (i), pp. 311-319.

Innvestigation of Dynamic Performance of Cylindrical Journal Bearings in Consideration of Elastic Dynamic Deformation

Abstract: A system of simultaneous equations consisting of the Reynolds equations of pressure perturbations in consideration of dynamic deformation and the deformation equations of sliding bearings was derived in complex form. The formulae for the dynamic promperties and stability were also derived. A reduced flexibility matrix was conveniently obtained using SAP5 program. The pressure perturbations, bearing dynamic properties and stability in cosideration of dynamic deformation in typical cases were calculated, the results were analyzed and compared with those only consider the static deformation. It was noticed that the difference was significant under certain conditions, themrefore the importance of dynamic deformation became evident. Masurements of a nylon bearing and a bronze bearing were made and the results confirm the theoretical calculation qualitatively.

圆柱滚子轴承固有频率计算方法的改进*

摘　要：本文用"集中参数法",将滚子轴承视作若干集中质量(内圈连同并附的转子质量以及各个滚子)和若干弹性元件(各滚子与内圈和外圈接触变形弹性)在固定基础(外圈)上构成的振动系统,完整地考察了该系统的各阶振型和固有频率,获得了比滚动轴承行业在固有频率计算中常用方法(只能近似地获得两个固有频率)更为精确和全面的结果.这种计算方法可为分析滚子轴承振动和噪声问题提供一个方面的理论依据,较前有所改进.

此外,对于套圈的挠曲振动,除了截面中的弯矩外,还计入了剪切和拉伸,改进了原有的计算公式,更精确计算厚度较大的套圈及较高阶的固有频率.

本文对圆柱滚子轴承固有频率计算方法作了两点改进,进一步分析了滚子轴承的振动和噪声问题,其一是采用"集中参数法",将滚子轴承视为一个整体振动系统,考察该系统的各阶振型和固有频率;其二是对套圈的挠曲振动,除了截面中的弯矩外,还计入了剪切和拉伸.

一、轴承系统各阶固有频率的计算

1.1　概述

滚动轴承的振动和噪声特性,与轴承的各种振动形式和相应的固有频率有关,其中涉及的一个重要振动形式,是整个轴承作为一个以外圈为基础的系统的振动.以往的计算中(例如文献[1]),将全部滚动体折合为一个集中质量,从而简化为两个质量(内圈及并附的转子质量、滚动体集合)和两个弹性(各滚动体与内圈的接触弹性的折算集合,及与外圈的相应折算集合)所构成的简单振动系统.这样虽能使计算大为简化,但只能近似地获得两个固有频率,在完整性和精确性方面均有不足之处.为了对这样一种振动形式作更全面的考察,本文不采取并合各滚动体为一个质量的近似处理法,而将各滚动体视作独立的集中振动质量,使整个轴承的力学模型就如图1所示.这个模型虽有较多的自由度,但只需在数值计算上适当处理,就可在电子计算机上方便地求出各阶振型及相应的固有频率的准确数值.

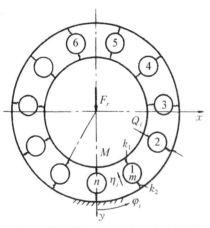

图1　轴承作为振动系统的力学模型

---

* 本文合作者：孙鋆强.原发表于《上海工业大学学报》,1986,7(3)：273 – 280.

1.2 圆柱滚子轴承中的滚子受力和接触刚度

滚子与套圈的接触处在载荷作用下发生变形,其接触刚度为

$$k_c = \mathrm{d}Q/\mathrm{d}\delta \tag{1}$$

式中,Q——滚子受力;δ——滚子与套圈由于接触变形而发生的趋近量.

趋近量可按下式计算[2]:

$$\delta = \frac{2Q}{\pi l}\left(\frac{1-\mu_1^2}{E_1} + \frac{1-\mu_2^2}{E_2}\right)\left(\ln\frac{l}{b} + 1.193\right) \tag{2}$$

式中,l——接触长度;μ_1 和 μ_2——滚子和套圈的泊松比;E_1 和 E_2——滚子和套圈的弹性模量;b——接触半宽:

$$b = \left[\frac{4Q}{\pi l\rho}\left(\frac{1-\mu_1^2}{E_1} + \frac{1-\mu_2^2}{E_2}\right)\right]^{1/2} \tag{3}$$

式中,ρ——综合主曲率.

由此可得:

$$k_c = \frac{\pi l}{a}\left[1.1444 + \ln\left(\frac{l^3\rho}{Qa}\right)\right]^{-1} \tag{4}$$

式中,$a = \dfrac{1-\mu_1^2}{E_1} + \dfrac{1-\mu_2^2}{E_2}$.

k_c 值随滚子受力 Q 而变.各滚子所受的力则由轴承载荷分配而来.设轴承径向载荷为 F_r,则当滚子数为 n 时,

$$F_r = \sum_{i=1}^{n}{}^{*} Q_i\cos\varphi_i \tag{5}$$

式中,$\sum^{*}$ 表示只在轴承承载区内求和;下标 i 表示第 i 个滚子.

由变形协调关系可知:

$$(u_r + \delta_1)/\cos\varphi_1 = (u_r + \delta_2)/\cos\varphi_2 = \cdots = \mathrm{const} \tag{6}$$

式中,u_r——径向游隙.

按式(2)、(3)、(5)、(6)可联立求出承载区范围及其中包含的滚子数(按 $Q_i \geqslant 0$ 的条件)以及各滚子所受的力 Q_i,然后由式(4)求出其内、外接触刚度.

1.3 运动方程组

滚动轴承的振动问题大多数可按小振幅的振动来处理,因此可近似地认为各处接触刚度是常数.内圈运动方程为

$$M\ddot{x} + \sum_{i=1}^{n}{}^{*} k_{1i}(x\sin\varphi_1 + y\cos\varphi_i + \eta_i)\sin\varphi_i = 0 \tag{7}$$

$$M\ddot{y} + \sum_{i=1}^{n}{}^{*} k_{1i}(x\sin\varphi_i + y\cos\varphi_i + \eta_i)\cos\varphi_i = 0 \tag{8}$$

式中，M——内圈连同并附的转子质量；x，y——内圈在振动过程中偏离静平衡位置的瞬时坐标；η_i——第 i 个滚子偏离静平衡位置的瞬时径向坐标；k_1——滚子与内圈的接触刚度.

滚子运动方程为（只对承载滚子列）

$$m\ddot{\eta}_i + k_{2i}\eta_i + k_{1i}(x\sin\varphi_i + y\cos\varphi_i + \eta_i) = 0, \quad i = 1, 2, \cdots \tag{9}$$

式中，m——滚子质量.

式(7)、(8)、(9)可写成矩阵形式：

$$[M][\ddot{Z}] + [K][Z] = 0 \tag{10}$$

式中，质量矩阵 $[M]$ 是对角矩阵，其对角元为 $(M, M, m, \cdots, m)$，行数和列数均为承载滚子数加2；$[Z]$——位移矢量，其分量为 $(x, y, \eta_1, \cdots, \eta_n)$，在 $\eta_1 \sim \eta_n$ 中只含受载滚子的 η_i；$[K]$——对称刚度矩阵，其上三角部分的形式如下（亦只含与受载滚子有关的元）：

$$\begin{pmatrix} \sum_{i=1}^{n}{}^{*} k_{1i}\sin^2\varphi_i & \sum_{i=1}^{n}{}^{*} k_{1i}\sin\varphi_i\cos\varphi_i & k_{11}\sin\varphi_1 & k_{12}\sin\varphi_2 & k_{13}\sin\varphi_3 & k_{14}\sin\varphi_4 & \cdots & k_{1n}\sin\varphi_n \\ & \sum_{i=1}^{n}{}^{*} k_{1i}\cos^2\varphi_i & k_{11}\cos\varphi_1 & k_{12}\cos\varphi_2 & k_{13}\cos\varphi_3 & k_{14}\cos\varphi_4 & \cdots & k_{1n}\cos\varphi_n \\ & & k_{11}+k_{21} & 0 & 0 & 0 & \cdots & 0 \\ & & & k_{12}+k_{22} & 0 & 0 & \cdots & 0 \\ & & & & k_{13}+k_{23} & 0 & \cdots & 0 \\ & & & & & \ddots & & \vdots \\ & & & & & & & k_{1n}+k_{2n} \end{pmatrix}$$

由于是无阻尼的自由振动，设固有频率以 ω 表示，则

$$[\ddot{Z}] = -\omega^2[Z] \tag{11}$$

于是式(10)成为

$$-\omega^2[M][Z] + [K][Z] = 0 \tag{12}$$

解出这一特征值问题，即得各阶振型的位移矢量及相应的固有频率值.

1.4 固有频率的计算[3]

先将式(12)的非标准型变换成标准型

$$\lambda[X] + [P][X] = 0$$

然后用迭代法求出一个特征值 λ_1 及特征矢量 $[X_1]$. 建立与 $[P]$ 同阶方阵 $[B_1]$，以矩阵 $[P]$ 中对应于 $[X_1]$ 中最大元素的那一行右乘 $[X_1]$ 即得 $[B_1]$. 将 $[P]$ 减去 $[B_1]$，舍去零值的一行

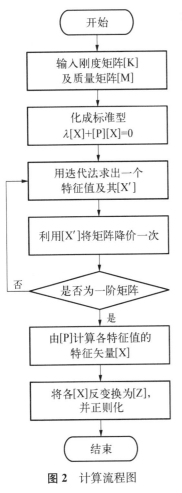

图 2 计算流程图

和该行序相同的一列后,得到降了一阶的矩阵$[P_1]$.再从$[P_1]$用迭代法求出特征值λ_2及特征矢量$[X_2']$,此λ_2亦是$[P]$的特征值,但$[X_2']$则不是$[P]$的特征矢量了.以此类推,可逐步求出全部特征值.然后从$[P]$依次求出每一个特征值所对应的特征矢量$[X]$.最后对$[X]$进行反变换,得到各个$[Z]$矢量,即所需的特征振型.计算流程如图 2 所示.

以上海滚动轴承厂生产的 2311 型短圆柱滚子轴承为例,设径向载荷F_r为 27 kg,径向游隙为$-1\ \mu m$(即有 1 μm 的径向预紧),用上法算出的全部 13 个固有频率为:

26 670、26 664、26 496、26 495、26 135、26 129、25 513、25 509、24 428、24 425、21 667、966、940 Hz.

图 3 是对于上述轴承考察径向游隙对实际上最重要的两个最低固有频率的影响.图中曲线的阶梯形突跳,是由于游隙(或预紧)的改变使承载滚子数增多或减少一个所致.

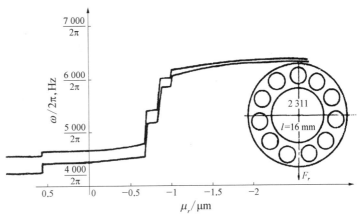

图 3 径向游隙与最低固有频率关系曲线

上述计算方法可比常用的简化方法得出更完整的固有频率谱,可用以细致考察结构参数对轴承系统频率特性的影响.

二、外套圈的平面挠曲振动频率计算

在外套圈的各种振型中,平面挠曲固有振动最为重要.由于过去所用方法[4]只考虑截面中的弯矩而略去剪切力和拉伸力的作用,致使对壁厚较大的套圈及高阶固有频率产生一定误差.为了改进计算精度,本文应用弹性力学[5]理论,全面考虑了弯矩、剪力、拉力的联合作用,计算了固有频率.

力学模型及计算公式的导出:图 4 示出套圈、坐标系统、及截面中的作用力和力矩.图中示出微元体$aabb$,对它可建立运动方程如下:

$$dM + Qr d\theta - J \frac{\partial^2 \psi}{\partial t^2} r d\theta = 0 \tag{13}$$

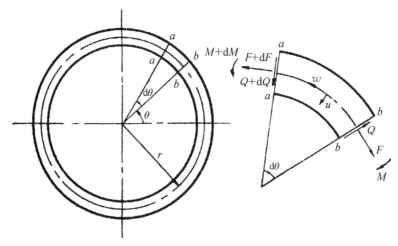

图 4 套圈截段受力图

$$dQ + Fd\theta - \rho \frac{\partial^2 u}{\partial t^2} r d\theta = 0 \tag{14}$$

$$dF - Qd\theta - \rho \frac{\partial^2 w}{\partial t^2} r d\theta = 0 \tag{15}$$

式中，u——微元体径向位移；w——切向位移；r——套圈平均半径；J——单位周长套圈的质惯矩；ρ——单位周长的质量；截面转角 ψ 为：

$$\psi = \frac{1}{r}\left(\frac{\partial u}{\partial \theta} + w\right) \tag{16}$$

考虑到：

$$M = \frac{EI}{r^2}\left(\frac{\partial^2 u}{\partial \theta^2} + \frac{\partial w}{\partial \theta}\right) \tag{17}$$

$$\frac{\partial w}{\partial \theta} = u \tag{18}$$

由式(13)~(15)可导出：

$$\frac{EI}{r^4}\left(\frac{\partial^6 w}{\partial \theta^6} + 2\frac{\partial^4 w}{\partial \theta^4} + \frac{\partial^2 w}{\partial \theta^2}\right) + \frac{\partial}{\partial t^2}\left[\rho\left(\frac{\partial^2 w}{\partial \theta^2} - w\right) - \frac{J}{r^2}\left(\frac{\partial^4 w}{\partial \theta^4} + 2\frac{\partial^2 w}{\partial \theta^2} + w\right)\right] = 0 \tag{19}$$

式中，I——截面对主形心轴线的主惯矩。

式(19)的解的形式可设为：

$$w(\theta, t) = w_1(\theta) e^{i\omega t} \tag{20}$$

代入式(19)，并整理得：

$$\frac{d^6 w_1}{d\theta^6} + \left(2 + \frac{r^2 J}{EI}\omega^2\right)\frac{d^4 w_1}{d\theta^4} + \left[1 + \omega^2\left(\frac{2Jr^2}{EI} - \frac{\rho r^4}{EI}\right)\right]\frac{d^2 w_1}{d\theta^2} + \left(\frac{r^2 J}{EI} + \frac{\rho r^4}{EI}\right)\omega^2 w_1 = 0 \tag{21}$$

设
$$w_1(\theta) = A\sin(n\theta + \varphi) \tag{22}$$

代入式(21)得：

$$-n^6 + \left(2 + \frac{r^2 J}{EI}\omega^2\right)n^4 - \left[1 + \omega^2\left(\frac{2Jr^2}{EI} - \frac{\rho r^4}{EI}\right)\right]n^2 + \left(\frac{r^2 J}{EI} + \frac{\rho r^4}{EI}\right)\omega^2 = 0 \tag{23}$$

即
$$\omega_n^2 = \frac{n^2(n^2-1)^2 EI}{\rho r^4(1+n^2) + r^2 J(n^2-1)^2} \tag{24}$$

由于 ρr^2 比 J 大得多，所以当 n 小时（即对于低阶振型），分母接近于 $\rho r^4(1+n^2)$，或式(24)就同常用公式一致了。但当厚度大，从而 J 大，以及对大的 n 值，式(24)的结果与略去剪力和拉力的常用公式差别就增大。图 5 例示两种方法对某个套圈计算结果的对比情况。

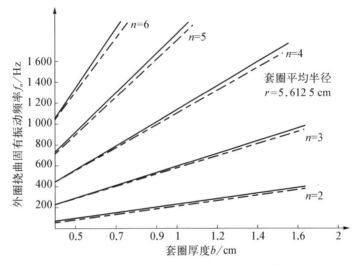

图 5 外圈挠曲固有频率与套圈厚度的关系

还用两种方法对 309 轴承的外圈作了计算，所得之若干阶固有频率如下表所列。表中还转引了由文献[6]中读出的实测值。证实了上述结论是正确的。

表 1 309 轴承外圈平面挠曲固有频率计算结果

计 算 方 法	平面挠曲固有频率/Hz					
	f_4[1)	f_5	f_6	f_7	f_8	f_9
按[4]计算	3 942	6 375	9 352	12 872	16 935	21 540
按式(24)计算	3 878	6 206	8 988	12 191	15 776	19 707
实测值,按[6]	≈3 400	≈6 200	≈10 000	—	≈14 000	≈18 000

1) 下标"4"为阶数，下同。

参 考 文 献

[1] Rippel, H. C., Tawresey J. S. An Analytical Study of Bearing Noise and Vibration Generation and Transmission. The Franklin Laboratories. AD 411438, 1961.

[2] Gupta, P. K. Mechanics of Rolling Instability in Rolling Bearing, 1978.

[3] Hatter, D. J. Matrix Computer Methods of Vibration Analysis. London, Butterworth & Co. Ltd., 1973: 180-194.

[4] Timoshenko, S. Vibration Problems in Engineering, 3rd ed., van Norstrand Co., 1954: 425.
[5] 钱伟长,叶开沅. 弹性力学. 北京：科学出版社,1980: 297 – 329.
[6] 张惠玲. 滚动轴承振动机理的研究. 上海交通大学硕士论文,1984.

Improvements to the Calculation of Natural Frequencies of Roller Bearings

Abstract: By the method of "lumped parameters", this article studies various vibration modes and natural frequencies of roller bearings considered as a vibration system composed of concentrated masses — the inner ring with the attached rotor mass and the roller masses, and elastic elements — the contact deformation elasticities between rollers and inner ring and outer ring, the outer ring being taken as the foundation of the system. The calculated results are more precise and comprehensive than those by the usual method, by which only two of the natural frequencies can be approximately calculated. The proposed method offers an improved theoretical basis for the analysis of bearing vibration and noise problems.

Besides bending moment in the cross section of the bearing ring, shear force and tensile force are also taken into considerations to calculate, the flexural vibration of the ring, so that more precise calculations can be made for rings with greater thickness and for higher-ordered vibration modes.

有限宽隧道孔轴承的动力特性和稳定性[*]

摘　要：本文利用有限元法,对在普通圆柱轴承上开设隧道孔后(简称隧道孔轴承)的压力分布、静特性、动特性和稳定性进行了理论计算.并用正弦激振法进行了动特性的实验测定.理论计算结果与实验测定结果一致.研究结果表明,参数选择适当的隧道孔轴承,较普通圆柱轴承更为稳定.由此提供了一种结构简单、基本上不降低静特性,且有效地提高稳定性的新颖工作原理.

0　引言

支承高速转子的滑动轴承,其动力稳定性是十分重要的.在工业上常用的各种轴承型式中,就动力稳定性的优劣而言,一般认为名次排列大致为:可倾瓦轴承、横错位轴承、三油叶轴承、椭圆轴承、圆柱轴承[1-7].此外,当圆柱轴承在实际工作中发生不稳定时,常采用中央加开周向油槽或截短轴承宽度等方法来应急补救.在轴承外加设弹性、阻尼支承,可获得很好的稳定性和制振性.上列各种提高稳定性的措施中,或者伴随静特性等的降低,或者使结构复杂化和影响可靠性的因素增多.寻找一种既能提高稳定性,又不使静特性显著下降,结构又不太复杂化的新结构或新原理,至今仍是有实用意义和学术价值的.[8]中用无限宽理论显示了开设隧道孔对提高稳定性的显著作用,是符合这一方向的工作.为了更加切合实际情况,本文用有限宽理论和实验验证来探讨隧道孔轴承的动特性、静特性和稳定性,以对这种型式作出更加确切的评价.

1　理论计算

1.1　基本关系

图 1 示出开有两个隧道孔的 180°轴承剖面图及几何关系,图 2 示轴工作展开形状.除隧道孔口以外,油膜中的压力分布满足 Reynolds 方程:

$$\nabla \cdot \left(\frac{h^3}{12\mu}\nabla P\right)=\frac{\omega}{2}\frac{\partial h}{\partial \varphi}+\dot{h} \tag{1}$$

式中油膜厚度分布 $h=c+e\cos\varphi$;膜厚增大率 $\dot{h}=V_x\sin\phi+V_y\cos\phi$;算符 $\nabla=\vec{i}_\varphi\frac{1}{R}\frac{\partial}{\partial \varphi}+\vec{i}_z\frac{\partial}{\partial z}$;坐标角 φ 由连心线 O_bO_j 上最大间隙处算起;角 ϕ 由 O_b 上方算起;$\varphi=\phi-\theta$;θ——偏位角;ω——轴颈转动角速度;V_x 和 V_y——轴颈中心变位速度;R——轴承半径;c——半

---

[*] 本文合作者:陈渭.原发表于《西安交通大学学报》,1986,20(4):53-62.

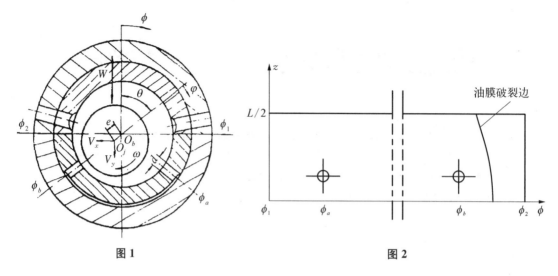

图1   图2

径间隙；e——轴颈相对于轴承的偏心距；z——由轴承宽度中央算起的轴向坐标；μ——油的动力黏度.

不计进油压力时,边界条件为:

$$\phi = \phi_1 \text{ 时}, P = 0 \tag{2}$$

$$\phi = \phi_2 \text{ 时}, P = 0 \text{ 或 } P = \frac{\partial P}{\partial \varphi} = 0 \tag{3}$$

$$z = \pm \frac{L}{2} \text{ 时}, P = 0 \tag{4}$$

式中 L——轴承宽度.

由于开设隧道孔,增加了下列关系.因孔径 d 比膜厚大得多,可忽略隧道孔中的流动阻力,认为每一隧道孔两出口周边 Γ_a 和 Γ_b 上压力相等:

$$P\mid_{\Gamma_a} = P\mid_{\Gamma_b} \tag{5}$$

隧道孔中流量连续:

$$\oint_{\Gamma_a}\left(\frac{hU_n}{2} - \frac{h^3}{12\mu}\frac{\partial P}{\partial n}\right)\mathrm{d}s + \oint_{\Gamma_b}\left(\frac{hU_n}{2} - \frac{h^3}{12\mu}\frac{\partial P}{\partial n}\right)\mathrm{d}s + f = 0 \tag{6}$$

式中 n——孔口周边 Γ_a 和 Γ_b 的外法线；U_n——轴颈圆周速度在 n 上的投影；$U_n = \omega R \vec{i}_\varphi \cdot \vec{i}_n$；$\phi_a$ 和 ϕ_b——隧道孔两出口中心的角位置；而 f 为:

$$f = \frac{\pi d^2}{4}[V_x(\sin\phi_a + \sin\phi_b) + V_y(\cos\phi_a + \cos\phi_b)] \tag{7}$$

在静平衡位置上计算时,$V_x = V_y = 0$. 解出压力分布(以 P_0 表示)后,按[1]中公式计算承载能力 F_x 和 F_y,修正 θ 角使 $|F_x/F_y| < 10^{-3}$ 以适应向下的外载荷. 无量纲承载力 $\bar{F} = \dfrac{F_y \psi^2}{\mu \omega R L}$；其中 $\psi = c/R$——相对间隙. 计算泄油量 q_L,无量纲泄油量 $Q_L = \dfrac{q_L}{2\psi\omega R^2 L}$；计算摩

擦阻力 F_i, 无量纲摩擦阻力 $\overline{F}_i = \dfrac{F_i \psi}{\mu \omega R L}$; 阻力系数 $f/\psi = \overline{F}_i / \overline{F}_y$.

计算油膜动特性时,压力对扰动参数 g(g 可为轴颈中心相对于静平衡位置的位移扰动 Δe 或 $e\Delta\theta$, 亦可为速度扰动 $\dot{e}$ 或 $e\Delta\dot\theta$)的偏导数 $P_g\left(=\dfrac{\partial P}{\partial g}\right)$ 的分布符合如下的扰动 Reynolds 方程[1]:

$$\nabla \cdot \left(\dfrac{h^3}{12\mu}\nabla P_g\right) = \dfrac{\omega}{2}\dfrac{\partial^2 h}{\partial\varphi\partial g} + \dfrac{\partial h}{\partial g} - \nabla \cdot \left(\dfrac{h^2}{4\mu}\dfrac{\partial h}{\partial g}\nabla P_0\right) \tag{8}$$

式中 P_0——静平衡工况($g=0$)下的压力分布.

扰动压力的边界条件为:

在全部油膜边界上, $\quad P_g = 0 \tag{9}$

由隧道孔流量连续条件:

$$\oint_{\Gamma_a}\left(\dfrac{U_n}{2}\dfrac{\partial h}{\partial g} - \dfrac{h^3}{12\mu}\dfrac{\partial P_g}{\partial n} - \dfrac{h^2}{4\mu}\dfrac{\partial h}{\partial g}\dfrac{\partial P_0}{\partial n}\right)\mathrm{d}s +$$
$$+ \oint_{\Gamma_b}\left(\dfrac{U_n}{2}\dfrac{\partial h}{\partial g} - \dfrac{h^3}{12\mu}\dfrac{\partial P_g}{\partial n} - \dfrac{h^2}{4\mu}\dfrac{\partial h}{\partial g}\dfrac{\partial P_0}{\partial n}\right)\mathrm{d}s + \dfrac{\partial f}{\partial g} = 0 \tag{10}$$

求出各 P_g (即 $P_e = \dfrac{\partial P}{\partial e}$, $P_\theta = \dfrac{\partial P}{e\partial\theta}$, $P_{\dot{e}} = \dfrac{\partial P}{\partial \dot{e}}$, $P_{\dot{\theta}} = \dfrac{\partial P}{e\partial\dot\theta}$) 后, 即算出了四个油膜刚度:

$$\begin{bmatrix} k_{ee} & k_{\theta e} \\ k_{e\theta} & k_{\theta\theta} \end{bmatrix} = -\iint_\Omega \begin{Bmatrix} P_e \\ P_\theta \end{Bmatrix} (\cos\varphi \quad \sin\varphi) R\,\mathrm{d}\varphi\mathrm{d}z \tag{11}$$

及四个油膜阻尼:

$$\begin{bmatrix} b_{ee} & b_{\theta e} \\ b_{e\theta} & b_{\theta\theta} \end{bmatrix} = -\iint_\Omega \begin{Bmatrix} P_{\dot{e}} \\ P_{\dot{\theta}} \end{Bmatrix} (\cos\varphi \quad \sin\varphi) R\,\mathrm{d}\varphi\mathrm{d}z \tag{12}$$

式中 Ω——油膜展开区.

取刚度相对单位 $\dfrac{\mu\omega L}{\psi^3}$, 阻尼相对单位 $\dfrac{\mu L}{\psi^3}$, 则得无量纲刚度 K_{ij} 及阻尼 B_{ij}:

$$K_{ij} = k_{ij}\psi^3/\mu\omega L, \quad B_{ij} = b_{ij}\psi^3/\mu L \quad i,j = e, \theta \tag{13}$$

刚度、阻尼在 (e, θ) 和 (x, y) 坐标系间的转化关系为:

$$\begin{bmatrix} K_{xx} & K_{xy} \\ K_{yx} & K_{yy} \end{bmatrix} = \begin{bmatrix} \sin\theta & \cos\theta \\ \cos\theta & -\sin\theta \end{bmatrix}\begin{bmatrix} K_{ee} & K_{e\theta} \\ K_{\theta e} & K_{\theta\theta} \end{bmatrix}\begin{bmatrix} \sin\theta & \cos\theta \\ \cos\theta & -\sin\theta \end{bmatrix} \tag{14}$$

用以支撑在旋转平面内各向同性的转子时, 油膜在失稳界限上的无量纲相当刚度 K_{eq}、界限涡动比 γ_{st}、刚性转子无量纲临界质量 M_{cr} 为[1]:

$$K_{\mathrm{eq}} = \dfrac{(K_{xx}B_{yy} + K_{yy}B_{xx} - K_{xy}B_{yx} - K_{yx}B_{xy})}{(B_{xx} + B_{yy})} \tag{15}$$

$$\gamma_{st}^2 = \frac{(K_{eq}-K_{xx})(K_{eq}-K_{yy})-K_{xy}K_{yx}}{B_{xx}B_{yy}-B_{xy}B_{yx}} \tag{16}$$

$$M_{cr}=\frac{K_{eq}}{\gamma_{st}^2} \tag{17}$$

如 $K_{eq}<0$ 即判为不稳;如 $K_{eq}>0$ 而 $\gamma_{st}^2<0$ 即判为恒稳[1];当 $K_{eq}>0$ 且 $\gamma_{st}^2>0$ 时,M_{cr} 越大则油膜稳定性越好.

1.2 有限元形式及计算过程

根据伽辽金变分原理,在函数空间 V(在油膜展开区 Ω 上一阶可微,且在区边界上为零)中求 P_0 分布,使下式对 V 中每一元素 v 都成立[9]:

$$\iint_\Omega \left(\frac{h^3}{12\mu}\nabla P_0 - \frac{h\vec{U}}{2}\right)\cdot\nabla v\,\mathrm{d}\Omega = 0 \tag{18}$$

式中 $\vec{U}=U\vec{i_\varphi}$.

将轴瓦工作面划分成单元(图3).用八节点双二次等参元作为计算单元.按式(18)构成刚度矩阵及载荷矢量.计入式(5),(6),解出 P_0.用数值积分法求出油膜压力的合力,检查合力方向是否与给定外载荷方向相反,根据方向误差修改偏位角 θ,重新计算压力分布,直至合力方向足够准确(相对误差小于 10^{-3} rad).至此,静平衡工况下的压力分布和偏位角已求得.计算各项静特性.用同样的有限元格式解算各扰动压力分布(式(8)、(9)、(10)),用数值积分计算油膜动特性.计算稳定性参数.

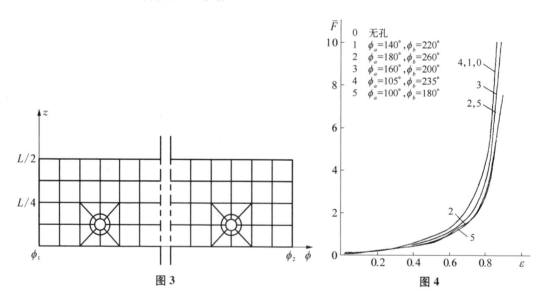

图3   图4

1.3 计算结果

计算结果是针对 $L/D=0.6$ 的 $180°$ 圆柱轴承开设各种隧道孔情况进行的:开一个隧道孔;开两个隧道孔(图2);开三个隧道孔;用隧道孔连通的两个轴向槽.同时亦计算了不开隧道孔的轴承性能以作对比.计算结果示于图4~图8.

图 4 示开两个隧道孔而孔口位置 ϕ_a 和 ϕ_b 不同时的影响. 由图可见, 当隧道孔口开设位置不在载荷作用线上或附近时, 对承载力影响甚小. 例如 $\phi_a=140°$, $\phi_b=220°$ 的轴承及 $\phi_a=105°$, $\phi_b=255°$ 的轴承的承载能力曲线几乎重合于不开隧道孔的轴承. 但如果隧道孔某个出口开在载荷作用线上或附近时, 会使承载力明显降低. 例如图中 $\phi_a=100°$ 而 $\phi_b=180°$, $\phi_a=180°$ 而 $\phi_b=260°$, 以及 $\phi_a=160°$ 而 $\phi_b=200°$ 这三种情况. 而且, ε 越大, 这种降低亦越严重, 其原因可以从压力分布(图5)上得到阐明.

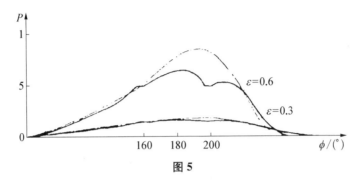

图 5

图 6 示与图 4 相应的各种情况的轴心静平衡位置轨迹. 可见变化不很大.

图 7 示上述各种情况的无量纲流量 Q_L 及阻力系数 f/ψ 曲线. 可见, 开隧道孔总是使 Q_L 降低, 但下降量不大. 孔口距载荷线较远时(如 $\phi_a=140°$ 而 $\phi_b=220°$), f/ψ 受隧道孔影响不大; 孔口开在载荷线上时, f/ψ 增大显著.

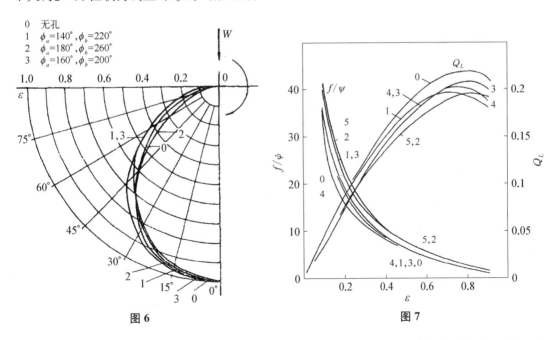

图 6 图 7

图 8 示开两个隧道孔时, 各种隧道孔口位置对八个无量纲油膜刚度和油膜阻尼的组合参数, 无量纲临界质量的影响. 经过搜索性计算可知, 在垂直向下的载荷下, 当隧道孔两出口位置对称于载荷时, $\phi_a=138°\sim148°$, 而 $\phi_b=222°\sim212°$ 这个范围内, 能使油膜稳定性显著提高(ε 不到 0.7 时即已恒稳). 孔口非对称开设时, 只要 ϕ_b 保持在 222°左右, ϕ_a 可从 112°~148°, 同样可在 ε 不到 0.7 时即为恒稳.

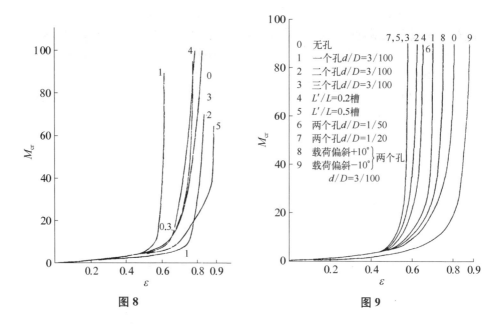

图 8 图 9

图 9 曲线 0,1,2,3 示出了隧道孔数目对稳定性的影响. 在 $\phi_a=140°$ 而 $\phi_b=220°$ 时,只开一个隧道孔还不足以使稳定性得到很大改善;开两个隧道孔时已使恒稳情况提前到了 $\varepsilon=0.7$ 以前;开三个隧道孔时进一步提前到了 $\varepsilon=0.6$ 以前,这已是一个不可低估的改进了. 图 9 曲线 4,5 示出在隧道孔口处开设一段轴向槽的效果. 与隧道孔数目的影响类似,轴向槽越长(本文中 $L'/L \leqslant 0.5$ 范围内),对稳定性影响越显著.

图 9 曲线 6,2,7 示出隧道孔口直径 d 的影响. 仍与隧道孔数目的影响相似,d 越大(本文只算到 $d/D=1/20$),则影响越大.

图 9 曲线 8,9 示出载荷方向偏斜对稳定性的影响. 看来这种轴承对于载荷方向在工作中要发生较大变化的工况不太相宜,根据某一载荷方向搜索得的最佳隧道孔位置,对其他载荷方向不一定相宜.

2 实验验证

试验轴承参数见表 1. 在西安交通大学轴承研究室的倒置式轴承动特性试验台上,用复合正弦激振法[10]测定了普通圆柱轴承和开隧道孔的轴承的动特性. 图 10 示实验测量框图.

表 1 试验轴承参数

	单位	轴承 1 圆柱轴承	轴承 2 ($\phi_a=115°,\phi_b=245°$)	轴承 3 ($\phi_a=180°,\phi_b=260°$)
$D_{外}$	mm	234	234	234
D	mm	100.23	100.22	100.218
L/D	—	0.6	0.6	0.6
d	mm	—	3	3
ψ	—	0.002 7	0.002 6	0.002 575

实验结果在趋势上很好地支持了理论计算. 根据实测的轴承油膜刚度和阻尼,计算了稳

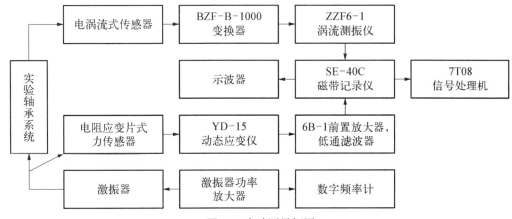

图 10 实验测量框图

定性参数,见表 2. 由表 2 可见,轴承 2 大大的改善了稳定性(在 ε＝0.5 以前就已是恒稳),轴承 3 则降低了稳定性,这都支持了理论结果.

表 2 按实测的动特性计算试验轴承的稳定性参数

	ε	K_{eq}	γ_{st}^2	M_{cr}	
轴承 1 圆柱轴承	0.5	1.619	0.174	9.305	
	0.6	2.821	0.022	128.2	
	0.7	5.760	−0.200	—	恒稳
轴承 2 $\phi_a=115°$ $\phi_b=245°$	0.5	1.722	−0.097	—	恒稳
	0.6	2.738	−0.355	—	恒稳
	0.7	5.385	−1.118	—	恒稳
轴承 3 $\phi_a=180°$ $\phi_b=260°$	0.5	0.217	0.184	1.179	
	0.6	0.955	0.203	4.704	
	0.7	2.527	0.179	14.12	

3 结论

1. 在普通圆柱轴承上开设隧道孔,可显著改变动特性.
2. 适当选择隧道孔口位置,能有效地提高动力稳定性,在较小的 ε 下即成为恒稳工作,而不降低承载能力及其他静特性. 因此,隧道孔轴承可以成为一种结构简单而性能好的轴承. 当圆柱轴承在运转中表现出不稳定时,加设适当参数的隧道孔可以作为改进措施.
3. 隧道孔口的连通面积应足够大(孔径、孔数、槽宽)才充分有效.
4. 隧道孔口应避免开在载荷方向上或其附近. 隧道孔轴承不宜用于载荷方向有变化的场合.

本文工作全过程中,得到了虞烈讲师的宝贵意见和细致帮助,实验部分还得到了丘大谋副教授的宝贵意见和大力帮助,以及王令安、杨建聪等同志的帮助,特此致谢.

参 考 文 献

[1] 张直明. 流体动压轴承润滑理论. 1979.

[2] D. F. Lietal. Stability and Transient Characteristics of Four Multilobe Journal Bearing Configuration. ASME J. of Lub. Techn, Vol. 102, No. 3,1980.

[3] Akkok, M. and G. M. Ettles. The Effect of Grooving and Bore Shape on the Stability of Journal Bearing. ASLE Trans., Vol. 23, No. 4, 1980.

[4] Gardner, D. R. et al. Stability of Profile Bore Bearing, Influence of Bearing Type Selection. Tribol. Int., Vol. 13, No. 5,1980.

[5] Malik, M. A Comparative Study of Some Two-Lobed Journal Bearing Configuration. ASLE Trans., Vol. 26, No. 1, 1983.

[6] Lanes, R. F. Experiments on Stability and Reponse of a Flexible Rotor in Three Types of Journal Bearing. ASLE Trans., Vol. 25, No. 3,1982.

[7] Akkok, m. et al. Effect of Load and Feed Pressure on Whirl in a Groove Journal Bearing. ASLE Trans., Vol. 23, No. 2, 1980.

[8] 张直明,吴西柳. 一种新型轴承——隧道孔轴承性能的初步分析. 西安交通大学科学技术报告,83—097 号.

[9] 李开泰,等. 有限元方法及其应用. 西安交大计算数学教研室,1981.

[10] 张优云. 在圆柱轴承上进行静特性和油膜刚度阻尼系数测定的研究. 西安交通大学科学技术报告,1981.

Dynamic Properties and Stability of Tunnel-Hole Bearings of Finite Width

Abstract: The pressure distribution, static and dynamic performances and stability parameters of cylindrical bearings with "tunnel holes" drilled on them are calculated theoretically by finite element method. Experimental measurements are also made by the method of sinusoidal excitation. The comparison between theoretical C results and experimental measurements is satisfactory. It is seen from the results that a tunnel-hole bearing with properly chosen geometry is more stable than a common cylindrical bearing. This suggests that a new working principle may be established to improve the stability of a bearing effectively without impairing its static performance.

EHL Analysis of Rib-Roller End Contact in Tapered Roller Bearings*

Abstract: Full numerical solutions of pressure and film thickness distributions are obtained by forward iterations for the EHL problem of elliptical contact between rib face and roller end in tapered roller bearings, with full consideration of the relevant geometric and kinematic peculiarities. It is found that the effect of elastic deformation is not negligible for the range of loads attendant with this problem. The effects of ratios of curvature in both principal planes and position of nominal point of contact on minimum film thickness and friction are studied and shown to be significant, and optimal values are deduced. Theory checks well with existing experimental data, and the validity of the theoretical treatment is evidenced.

Nomenclature

a, b = Semiaxes of the contact ellipse along Oz and Ox respectively

E = Modulus of elasticity

E' = Effective elastic modulus, $E' = 2 \Big/ \Big(\dfrac{1-\nu_1^2}{E_1} + \dfrac{1-\nu_2^2}{E_2} \Big)$

f = Friction force

F = Nondimensional friction force, $F = f/(aR_x E')$

G = Nondimensional materials parameter, $G = \alpha E'$

h = Film thickness

h_c = Central film thickness

h_{min} = Minimum film thickness

H = Nondimensional film thickness, $H = h/R_x$

K = Ellipticity parameter, $K = a/b$

p = Pressure

P = Nondimensional pressure, $P = p/E'$

r = Large end radius of roller

r_i, r_o = Inner and outer radii of rib

R = Cone distance of race on inner ring

R_i, R_o = Radii of curvature of lower and upper border lines of conjunction domain

R_x, R_z = Principal equivalent radii of curvature

R_{x1}, R_{z1} = Principal radii of curvature of rib face

R_{x2}, R_{z2} = Principal radii of curvature of roller end

$U_{1x(0)}$, $U_{2x(0)}$, $U_{1z(0)}$, $U_{2z(0)}$ = Speed components at nominal point of contact, where indices 1 and 2 relate to rib and roller respectively

X = Nondimensional coordinate in x-direction, $X = x/b$

Z = Nondimensional coordinate in z-direction, $Z = z/a$

W = Normal applied load

α = Pressure-viscosity coefficient

β = Half cone angle

σ = Half roller angle

δ = Elastic deformation

$\bar{\delta}$ = Nondimensional deformation, $\bar{\delta} = \delta/R_x$

ε = Distance from the nominal contact point to inner race

η = Absolute viscosity under local pressure

η_0 = Viscosity at atmospheric pressure

$\bar{\eta}$ = Nondimensional viscosity, $\bar{\eta} = \eta/\eta_0$

λ_0, λ_1, λ_2 = Nondimensional speed factors,

$\lambda_0 = (U_{1z(0)} + U_{2z(0)})/(U_{1x(0)} + U_{2x(0)})$

$\lambda_1 = (U_{2x(0)} - U_{1x(0)})/(U_{1x(0)} + U_{2x(0)})$

---

* In collaboration with X. QIU and Y. HONG. Reprinted from *STLE Tribology Transaction*, 1987, 31(4): 461–467.

$\lambda_2 = (U_{2x(0)} - U_{1x(0)})/(U_{1x(0)} + U_{2x(0)})$

ν_1, ν_2 = Poisson's ratios

ρ = Lubricant density

ρ_0 = Density at atmospheric pressure

$\bar{\rho}$ = Nondimensional density, $\bar{\rho} = \rho/\rho_0$

σ_0 = Nondimensional speed factor,

$\sigma_0 = \eta_0 b(U_{1x(0)} + U_{2x(0)})/E'R_x^2$

ψ = An angle formed by bearing axis and normal line of rib face at nominal contact point

Ω = Angular velocity of rotation of inner ring

Ω_0 = Factor, $\Omega_0 = (\Omega_{1y} - \Omega_{2y})(U_{1x(0)} - U_{2x(0)})$

Ω_1 = Factor, $\Omega_1 = (\Omega_{2y} - \Omega_{1y})(U_{1x(0)} - U_{2x(0)})$

Ω_{1x}, Ω_{1y} = Angular velocity components of rib along x- and y-axes

Ω_{2x}, Ω_{2y} = Angular velocity components of roller end along x- and y-axes

Introduction

It is well known that proper lubrication at the rib-roller end contact is indispensable for reliable operation of a tapered roller bearing. Many investigations of friction, film thickness and lubrication of rib-roller end contacts have been conducted[1-5].

Korrenn has carried out experimental work studying friction of rib-roller end contact[2]. Jamison et al. have studied influence of geometry and position of nominal contact point of rib-roller end contact on its lubrication[3]. They mentioned that it is necessary to make the nominal contact point located in the middle of the cone rib to favor lubrication. Dalmaz et al. have determined optimum geometric configuration of rib-roller end contact both theoretically using a hydrodynamic approach and experimentally on a self-built apparatus[4]. Recently, Gadallah and Dalmaz have further studied behavior of lubricated rib-roller end contact of a tapered roller bearing using a theoretical hydrodynamic approach and a special device to measure traction forces and film thicknesses[5]. In discussion of that paper, they pointed out that further theoretical work was needed, which should include surface deformations and a domain size close to the conjunction found in the rib-roller end contact.

Theoretical work on EHL for nominal point contact was started by Archard and Cowking, and led to the concept of a side-leakage[6]. Based on thirty four elastohydrodynamic solutions of nominal point contact computed by Hamrock and Dowson, approximate formulae for minimum and central film thicknesses were deduced[7]. Later developments include analysis of Wildhaber-Novikov gears by Evans and Snidle[8], study of effects of spin and variation of angle of lubricant entrainment by Mostophi and Gohar[9], and researches on effects of geometrical conditions resulting in lubricant entrainment directed along major axis of the contact ellipse, both theoretical and experimental, by Chittenden et al. [10-13]

The aim of this paper is to make a theoretical elastohydrodynamic study of rib-roller end contact of tapered roller bearings, with full consideration of their particular geometry and kinematic conditions. In obtaining full numerical solutions of pressure and film thickness distribution, the conjunction is to be taken as the bounding area of the film. Calculations are to be done under isothermal and steady state conditions, to study effects of ratios of curvature in both principal planes and position of nominal contact point on minimum film thickness and friction. Optimal values for specific working condition are to be deduced. The correspondence of theory with existing experimental data is to be surveyed.

Theory

Geometry and kinematics

The rib-roller end contact of a roller bearing is shown in Fig. 1. The EHL problem of the contact can be studied in the Cartesian coordinate frame $Oxyz$. Three kinds of mating surfaces are considered in this paper, viz., spherical rib face with spherical roller end, tapered rib face with spherical roller end, and tapered rib face with toroidal roller end. But all three forms can be adequately described by an equivalent ellipsoid near a plane as shown in Fig. 2. The equivalent radii in the principal planes are:

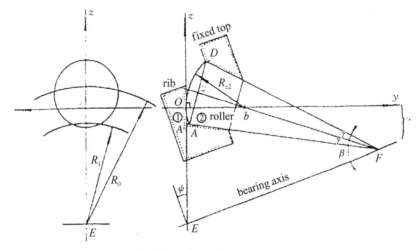

Fig. 1 Rib-roller end contact geometry

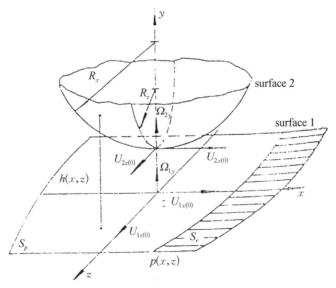

Fig. 2 EHD model

$$1/R_x = 1/R_{x2} - 1/R_{x1} \quad 1/R_z = 1/R_{z2} - 1/R_{z1} \tag{1}$$

The conjunction between rib face and roller end in a tapered roller bearing, which is taken as the integration domain, is bordered by four curves which may be approximated by elliptical or circular arcs. In this paper, the top and bottom borders are approximated by circular arcs, the radii of which are:

$$R_i = r_i/\cos\psi \quad R_0 = r_0/\cos\psi \tag{2}$$

The side borders are approximated by an ellipse:

$$\left[\frac{z - (R_{z2}^2 - r^2)^{1/2}\sin(\beta+\gamma-\psi)}{r\cos(\beta+\gamma+\psi)}\right]^2 + \frac{x^2}{r^2} = 1 \tag{3}$$

The velocity components of a point on the plane ($y=0$) are:

$$U_{1x} = U_{1x(0)} + z\Omega_{1y} \quad U_{1z} = U_{1z(0)} - x\Omega_{1y} \tag{4}$$

and those on the ellipsoid ($y=h$) are:

$$U_{2x} = U_{2x(0)} + z\Omega_{2y} \quad U_{2z} = U_{2z(0)} - x\Omega_{2y} \tag{5}$$

Note that for the rib-roller contact case, $U_{1y(0)} = U_{2y(0)} = 0$ and $\Omega_{1x} = \Omega_{2x} = 0$.

Equations

For the rib-roller end contact, under steady state conditions, the general Reynolds equation can be written as:

$$\frac{\partial}{\partial x}\left(\frac{\rho h^3}{\eta}\frac{\partial p}{\partial x}\right) + \frac{\partial}{\partial z}\left(\frac{\rho h^3}{\eta}\frac{\partial p}{\partial z}\right) = 6[(U_{1x(0)} + U_{2x(0)}) + (\Omega_{1y} + \Omega_{2y})z]\frac{\partial(\rho h)}{\partial x}$$
$$- 6[(U_{1z(0)} + U_{2z(0)}) - (\Omega_{1y} + \Omega_{2y})x]\frac{\partial(\rho h)}{\partial z} \tag{6}$$

The corresponding nondimensional form is:

$$\frac{\partial}{\partial X}\left(\frac{\bar{\rho}H^3}{\bar{\eta}}\frac{\partial P}{\partial X}\right) + \frac{1}{K^2}\frac{\partial}{\partial Z}\left(\frac{\bar{\rho}H^3}{\bar{\eta}}\frac{\partial P}{\partial Z}\right) = 6\sigma_0(1 - a\Omega_0 z)\frac{\partial(\bar{\rho}H)}{\partial X} + \frac{6\sigma_0}{K}(\lambda_0 - b\Omega_0 x)\frac{\partial(\bar{\rho}H)}{\partial Z}$$
$$\tag{7}$$

The retationship between lubricant density and pressure is taken as:

$$\bar{\rho} = \frac{0.6P}{1 + 1.7P} \tag{8}$$

The isorhermal viscosity-pressure dependence of lubricant is taken as:

$$\bar{\eta} = e^{CP} \tag{9}$$

The EHL film thickness H can be written as:

$$H = H_0 + \frac{b^2 X^2}{2R_x^2} + \frac{a^2 Z^2}{2R_x R_z} + \bar{\delta}(X, Z) \tag{10}$$

where H_0 corresponds to the nearest distance between rigid bodies with the same approach, and is a constant that is estimated mitially and corrected later.

The elastic deformation of an equivalent ellipsoid near a plane. $\delta(x, z)$ can be calculated by the method stated in [7]. With the contacting bodies assumed to behave as elastic half spaces:

$$\zeta = \frac{2a}{\pi R_x} \int_A \frac{P(X_k, Z_l) \mathrm{d}X_k \mathrm{d}Z_l}{[(X_l - X_k)^2 + K^2(Z_t - Z_l)^2]^{1/2}} \tag{11}$$

where A is the area over which the pressure p acts. The integration domain is divided into equal rectangular elements with the border curres approximated by line segments. The integration is first performed analytically for each elementary area, over which uniformly distributed unit pressure is assumed. Wherefrom a set of influence coefficients is obtamed, thereby allowing the deflections to be calculated by using the superposition principle:

$$\bar{\delta}_{ij} = \sum_{l=1}^{n} \sum_{k=1}^{m} P_{kl} D_{ij,kl} \tag{12}$$

where $D_{ij,kl}$ stands for the nondimensional influence coefficient.

The initial value of the constant H_0 in Eq. (10) has to be estimated, but thereafter corrected in the succeeding iteration steps, to attain balance between the normal applied load and the integrated normal resultant force which is calculated as:

$$W = abE' \int_A P \mathrm{d}X \mathrm{d}Z \tag{13}$$

In rib-roller end contact, the pressure which the lubricant experiences is not extreme. Therefore Newtonian behavior is assumed for the lubricant to calculate friction force. According to Reynolds boundary condition adopted by the authors, the boundary of film rupture Γ is found with $p = \mathrm{d}p/\mathrm{d}n = 0$, which divides the conjunction domain into a load carrying part S_p and a cavitated part S_r. Accordingly, the friction force can be calculated as:

on rib face

$$F_{1x} = -\frac{1}{2} \int_{S_p} H \frac{\partial P}{\partial X} \mathrm{d}X \mathrm{d}Z + \sigma_0 \int_{S_p} \frac{\bar{\eta}}{H} (\lambda_1 + a\Omega_1 Z) \mathrm{d}X \mathrm{d}Z + \sigma_0 \int_{S_r} \frac{\bar{\eta} H_\Gamma}{H^2} (\lambda_1 + a\Omega_1 Z) \mathrm{d}X \mathrm{d}Z$$

$$F_{1z} = -\frac{1}{2K} \int_{S_p} H \frac{\partial P}{\partial Z} \mathrm{d}X \mathrm{d}Z + \sigma_0 \int_{S_p} \frac{\bar{\eta}}{H} (\lambda_2 - b\Omega_1 X) \mathrm{d}X \mathrm{d}Z + \sigma_0 \int_{S_r} \frac{\bar{\eta} H_\Gamma}{H^2} (\lambda_2 - b\Omega_1 X) \mathrm{d}X \mathrm{d}Z$$

$$\tag{14}$$

on roller end

$$F_{2x} = -\frac{1}{2} \int_{S_p} H \frac{\partial P}{\partial X} \mathrm{d}X \mathrm{d}Z - \sigma_0 \int_{S_p} \frac{\bar{\eta}}{H} (\lambda_1 + a\Omega_1 Z) \mathrm{d}X \mathrm{d}Z - \sigma_0 \int_{S_r} \frac{\bar{\eta} H_\Gamma}{H^2} (\lambda_1 + a\Omega_1 Z) \mathrm{d}X \mathrm{d}Z$$

$$F_{2z} = -\frac{1}{2K}\int_{S_p} H\frac{\partial P}{\partial Z}dXdZ - \sigma_0\int_{S_p}\frac{\bar{\eta}}{H}(\lambda_2 - b\Omega_1 X)dXdZ - \sigma_0\int_{S_r}\frac{\bar{\eta}H_r}{H^2}(\lambda_2 - b\Omega_1 X)dXdZ$$

(15)

where H_r is the film thickness at the boundary of film rupture.

Method of solution

A finite difference method with Gauss-Seidel iterations is used to solve the quasilinear Reynolds equation simultaneously with the equations for elasticity, film thickness, and lubricant state. Reynolds boundary condition is incorporated.

1. Integration domain

For a chosen load and geometry, the elastostatic footprint dimensions are calculated as in [7]. When the ellipticity parameter K, the semimajor and semiminor axes of the contact ellipse are known, the computation zone is determined initially. Its size is such that it is about 4 times the semimajor axis of the contact ellipse in the zone upstream of the nominal point of contact, and about 1.25 times that factor in the zone downstream. In the transverse direction, it is about 1.6 times the semiminor axis. Then, if the computation zone is larger than the contact area determined by Eqs. (2) and (3), the superfluous parts must be cut off. The effective computation zone is thus fixed. A uniform mesh nodal structure is used. The nondimensional grid spacings of the coordinates X and Y are 0.05 and 0.2 respectively. Therefore, there are 105 and 16 spacings in the x- and y-directions, respectively.

2. Initial guess

An initial pressure distribution is set up by using a modified form of the Hertzian pressure distribution for dry contact[14]. The modification consists of an addition to the Hertzian curve in the inlet and a cutting-off at the exit. Thus it is possible to make the corresponding centre line film shape to exhibit characteristics typical of elastohydrodynamic contact at the very beginning of computation.

3. Convergence criterion

Overall convergence is measured by the sum of absolute values of pressure increments between successive iterations divided by the sum of pressures, and a value not more than 0.01 is thought to be acceptable. The balance between load and resultant force is accurate to 0.5 percent.

Results

The minimum film thickness h_{min}, the central film thickness h_c and the friction force f are of interest for design purposes. Before calculation, we have to analyze the relationship of geometric parameters which are not independent of each other. The distance from nominal contact point to inner race ε can be written as (15):

$$\varepsilon = R_{x2}\left(\sin(\psi-\beta) - \frac{1}{\cos(\psi-\beta)\cot\gamma + \sin\gamma}\right) + \frac{R}{\cot\gamma + \tan(\psi-\beta)} \tag{16}$$

If γ is small, it can be rewritten as:

$$R_{x2} \approx \frac{R\sin\gamma - \varepsilon}{\sin\gamma - \sin(\psi-\beta)} \tag{17}$$

From Fig. 1, we can write

$$R_{x1} = R_{x2} + [R\cos\gamma + (R_{x2}^2 - R^2\sin^2\gamma)^{1/2}]\frac{\sin(\beta+\gamma)}{\sin\psi} \tag{18}$$

It can be seen that there are two degrees of designing freedom for roller bearings of spherical rib face with spherical roller end, or of tapered rib face with spherical roller end. But there are three degrees of designing freedom for bearings of tapered rib face with toroidal roller end, that it to say, R_{z1} can also be chosen freely, independent of R_x.

All the results presented below are for the bearing number 7518, in which $r = 9.739$ mm, $\gamma = 2°$, $\beta = 11.6422°$, $r_i = 57.486$ mm and $r_o = 60.5$ mm. After choosing ε and ratio R_{x2}/R_{x1} initially, the values of ψ, R_{x1} and R_{x2} can be calculated by solving Eqs. (19) and (20). A mineral oil of viscosity $\eta_0 = 0.02831$ Pa · s and $\alpha = 1.9895 \times 10^{-8}$ Pa^{-1} at 25°C is used, and the angular speed of rotation of the inner ring is $\Omega = 1000$ r/min. The contacting bodies are of elasticity modulus $E = 2.307 \times 10^{11}$ Pa and Poisson's ratio $\nu = 0.3$. The position of nominal point of contact is held fixed near middle of rib face.

Fig. 3 shows a typical result for a roller bearing of tapered rib face with spherical roller end. The distributions of pressure and film thickness in the direction of motion close to the midplane of the contact are shown in Fig. 3(a). No pressure spike has been detected. This might be attributed to the moderate load in rib-roller end contact of tapered roller bearings under average load conditions. Contour plots of film thickness and of nondimensional pressure are shown in Figs. 3(b) and 3(c). No well defined side lobe appears in this case.

The influence of ratio R_{x2}/R_{x1} on the value of film thickness and friction coefficients are shown in Fig. 4 for a roller bearing of spherical rib face with spherical roller end. Fig. 4(a) gives a comparison between elastohydrodynamic lubrication results and hydrodynamic results. It is evident that the effects of elastic deformation are not negligible for the range of loads attendant with problem. The optimal value of ratio R_{x2}/R_{x1} is about 0.9.

Fig. 5 shows the influence of ratio R_{x2}/R_{x1} on film thickness and friction coefficient for tapered rib face with spherical roller end. The optimal value of ratio R_{x2}/R_{x1} is about 0.8.

The variations of both the central and minimum film thicknesses with varied ratio R_z/R_x are shown in Fig. 6 for tapered roller bearings of tapered rib face with toroidal roller

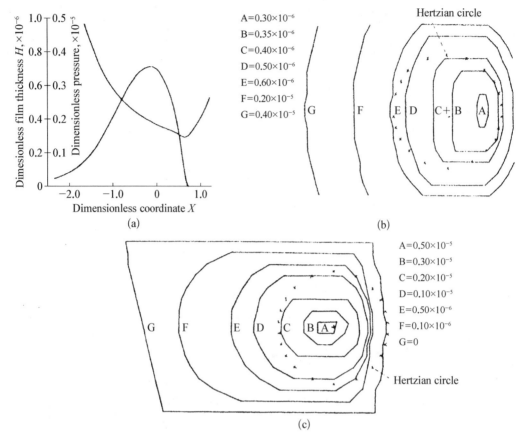

Fig. 3 A typical result for a roller bearing of tapered rib face with spherical roller end. $W = 160$ N, $R_{x1} = 282.294$ mm, $R_{x2} = 254.0646$ mm, $\varepsilon = 2.735$ mm, $\psi = 12.0622°$.

(a) Variation of dimensionless pressure and film thickness on x-axis. The value of z is held fixed near axial center of contact.
(b) Contour plots of dimensionless film thickness.
(c) Contour plots of dimensionless pressure.

end, with its ratio of R_{x2} to R_{x1} fixed at 0.97. It can be seen that the minimum film thickness predicted by EHL theory reaches its maximum at a value of R_z/R_x of about 0.55, while HL theory predicts an optimal value of R_z/R_x much higher than 1. The side lobe appearing in the film thickness distribution obtained by EHL theory causing the film thickness under the lobe to be less than that in the midplane can be used to explain this fundamental difference of predictions between EHL and HL theories.

The effects of position of nominal point of contact on minimum film thickness are shown in Fig. 7. When the position of point of contact lies in the central region of rib face, the film thickness reaches its highest value.

Numerical calculations are also performed for the conditions under which the experiments of [5] were done. Fig. 8 compares the obtained theoretical results by elastohydrodynamic theory with the experimental results stated in [5]. It can be seen that the EHL results correspond well with the experimental results.

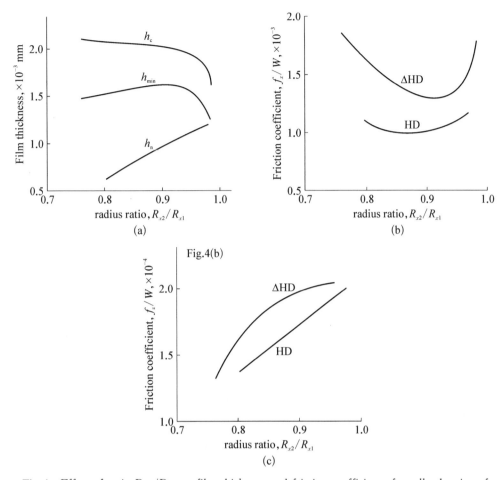

Fig. 4 Effect of ratio R_{x2}/R_{x1} on film thickness and friction coefficient of a roller bearing of spherical rib face with spherical roller end. $W=550$ N, $\varepsilon=2.735$ mm.

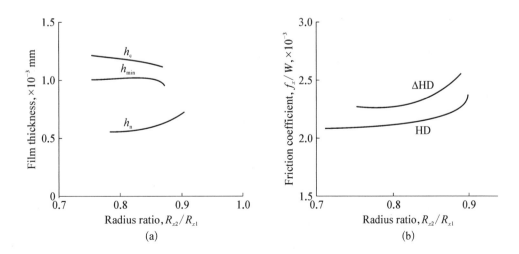

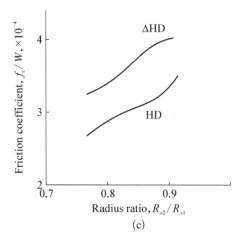

(c)

Fig. 5 Effect of ratio R_{x2}/R_{x1} on film thickness and friction coeficient of a roller bearing of tapered rib face with spherical roller end. $W=160$ N, $\varepsilon=2.735$ mm.

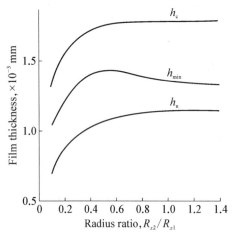

Fig. 6 Effect of ratio R_z/R_x on film thickness of a tapered roller bearing of tapered rib face with toroidal roller end. $W=550$ N, $\varepsilon=2.735$ mm, $R_{z1}=279.955$ mm, $R_{x2}=271.556$ mm, $\psi=12.1642°$.

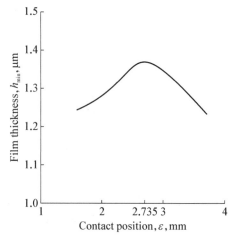

Fig. 7 Effect of position of nominal point of contact on minimum film thickness of a roller bearing of spherical rib face with spherical roller end. $\varepsilon=2.735$ mm, which means the position of point of contact lies in the central region of rib face. $W=550$ N, $R_{x2}/R_{x1}=0.97$.

Conclusions

1. Full numerical solutions of pressure and film thickness distributions are obtained by forward iterations for the FHL problem of elliptical contact between rib face and roller end in tapered roller bearings. It is found that the effect of elastic deformation is not negligible for the range of loads attendant with this problem.

2. The EHL theory checks well with the experimental data stated in [5], which verifies the validity of the theoretical treatment.

3. Deductions of optimal values of ratio R_{x2}/R_{x1} are done by EHL theory for spherical rib face with spherical roller end and for tapered rib face with spherical roller end under the conditions considered in this paper, which result in values of 0.9 and 0.8 respectively.

4. Deductions of optimal values of R_z/R_x for tapered rib face with toroidal roller end under the conditions considered in this paper by EHL and HL theories differ from each other fundamentally.

5. It is recommended that the nominal contact point be located in the central region of the rib face to favour lubrication.

6. Further study of this problem is needed to include such effects as thermal aspects.

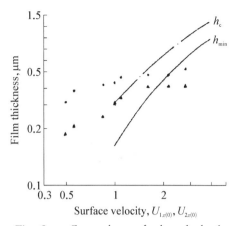

Fig. 8 Comparlson of the obtalned theoretlcal results with existing experimental data for a tapered roller bearing.
● Experlmental results h_c (from [5])
▲ Experinsental results h_{min} (from [5])
— · — Theoretlcal results h_c
— Theoretical results h_{min}
$E' = 0.4517 \times 10^{11}$ Pa, $\alpha = 1.57 \times 10^{-8}$ Pa^{-1}, $\eta_0 = 0.026$ Pa · s, $W = 80$ N, $R_{x1} = 185.283$ mm, $R_{x2} = 163.262$ mm

References

[1] Frase, D. and Brockmüler, U., "Correct Bearing Lubrication Increases Reliability of Differentials," Bull Bearing Journal, 199, pp 1 – 8(1982).

[2] Korrenn. H. "Gleitreibung und Grenzbelastung an den Bordflachen von Kegelrollerlagern, Einfluss von Drehzahl," Fortschritt Berichte VDI Zeitschrift, I, 11, pp 1 – 121(1967).

[3] Jamison, W. E., Kauzlarich, J. J., and Mochel, E. V., "Geometric Effects on the Rib-Roller Contact in Tapered Roller Bearings," ASLE Trans., 20, pp 78 – 88(1976).

[4] Dalmaz, G., Tessier, J. E., and Dudragne, G., "Friction Improvement in Cycloidal Motion Contacts: Rib-Roller End Contact in Tapered Roller Bearing." Proc. of the 7th Leeds Lyon Symp, on Trib., pp 175 – 185(1980).

[5] Gadallah, N. and Dalmaz, G., "Hydrodynamic Lubrication of the Rib-Roller End Contact of a Tapered Roller Bearing," ASME J. of Trib., 106, pp 265 – 274(1984).

[6] Archard, J. F. and Cowking, E. W., "Elastohydrodynamic Lubrication of Point Contact," Proc. Instn. Mech. Engrs., 180, Part 3B, pp 47 – 56(1966).

[7] Harnrock, B. I. and Dowson, D., Ball Bearing Lubrication. Wiley, New York, 1981.

[8] Evans, H. P. and Snidle, R. W., "Analysis of Elastohydrodynamic Lubrication of Elliptical Contacts with Rolling Along the Major Axis," Proc. Instn. Mech. Engrs., 197C, pp 209 – 211(1983).

[9] Mostophi, A. and Gohar, R., "Oil Film Thickness and Pressure Distribution in Elastohydrodynamic Point Contacts," Jour. Mech. Engng. Sci., 24, 4, pp 173 – 182(1982).

[10] Chittenden, R. J., Dowson, D., Dunn, J. F., and Taylor, C. M., "A Theoretical Analysis of the Isothermal Elastohydrodynamic Lubrication of Concentrated Contacts, Part 1. Direction of Entrainment Coincident with the Major Axis of the Hertzian Contact Ellipse," Proc. R. Soc. Lond., Ser. A, 397, pp 245 – 269(1985).

[11] Chittenden, R. J., Dowson, D., Dunn, J. F., and Taylor, C. M., "A Theoretical Analysis of the Isothermal Elastohydrodynamic Lubrication of Concentrated Contacts. Part II. General Case with Lubricant Entrainment Along Either Principal Axis of the Hertzian Contact Ellipse or at Some Intermediate Angle," Proc. R. Soc. Lond., Ser. A, 397, pp 271 – 294(1985).

[12] Chittenden, R. J., Dowson, D., and Taylor, C. M., "Elastohydrodynamic Film Thickness in Concentrated Contacts. Part I: Experimental Investigation for Lubricant Entrainment Aligned with the Major Axis of the Contact Ellipse," Proc. Instn. Mech. Engrs., 200, C3, pp 207-218(1986).

[13] Chittenden, R. J., Dowson, D., and Taylor, C. M., "Elastohydrodynamic Film Thickness in Concentrated Contact. Part II: Correlation of Expetimental Results with Elastohydrodynamic Theory," Proc. Instn. Mech. Engrs., 200, C3, pp 219-226(1986).

[14] Evans, H. P. and Snidle, R. W., "Towards a Refined Solution of the Isothermal Point Contact EHD Problem." Fundamentals of Tribology. MIT Press, pp 1103-1128(1978).

[15] Jamison, W. E., Kauzlarich, J. J., and Mochel, E. V., "Geometric Effects on the Rib-Roller Contact in Tapered Roller Bearings," Trans. ASIE. 20, pp 79-88(1976).

圆锥滚子轴承挡边接触副弹性流体动力润滑分析*

摘　要：本文对具有摆线相对运动的圆锥滚子轴承挡边接触副进行了流体动力润滑和弹性流体动力润滑分析，考察了这种润滑问题中弹性变形的影响程度，并计算了几何形状和参数对接触面承载力、摩擦力、摩擦力矩的影响，由此归纳出最佳几何参数. 理论计算结果与现有实验数据基本吻合.

一、引言

圆锥轴承内、外圈跑道的锥角不相等，因此它们对滚子的联合作用构成基本上沿滚子轴向而指向大端的力. 内圈上的大挡边就是为了承受这个力以使滚子保持在所需位置上之用. 挡边与滚子大端面构成了具有复杂相对运动——摆线运动的摩擦副.

挡边接触副的几何形状和参数影响着此处润滑条件. 如设计不当，将难以形成有足够厚度的油膜，以致挡边与滚子端面直接接触摩擦，引起温度过分升高，使轴承损坏. 这种损坏情况，在圆锥滚子轴承的实际应用中屡见不鲜. 因此，选用合理的几何形状和参数以改善挡边接触副的润滑性能，是提高这种轴承承载能力和寿命的关键之一.

为了使摩擦副中能形成承载油膜，并为了避免不正常的边缘接触，一般取滚子端面的曲率半径小于挡边工作面的曲率半径，所以构成了名义上的点接触副. 由于接触区润滑膜内压力相当高，引起了不容忽视的表面弹性变形和润滑油压粘效应，因此必须用点接触的弹性流体动力润滑理论分析研究这一润滑问题. 迄今已有若干国外学者在挡边接触副的润滑问题及改进其油膜承载力方面做过有成效的工作[1-7]，但尚未见到在这种具体的几何条件和运动条件下直接用弹流基本方程来严格解算这一问题的著作.

本文针对挡边接触副所特有的几何条件和运动条件，用等温全膜弹流润滑理论进行分析计算，求出膜厚分布、压力分布、摩擦力和摩擦力矩等数值. 从计算结果中找出有利于增大膜厚和减小摩擦的几何参数. 同时亦平行地进行全膜流体动力润滑理论计算，以期在两种计算结果的比较中弄清弹性变形对这种润滑问题的影响程度.

二、物理模型

挡边与滚子端面的重叠区域可看成是两个圆和一个椭圆所围成的区域，如图 1 所示.

---

* 本文合作者：洪跃、裘新君. 原发表于《上海工业大学学报》，1987，8(5)：498-507.

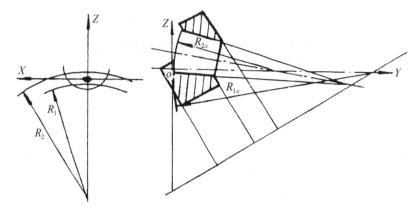

图 1 接触区几何参数及坐标系

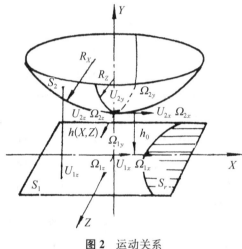

图 2 运动关系

2.1 区域内各点的速度

接触区内各点的速度为(图 2)：

$$\begin{aligned} U_{1x} &= U_{1x}^{(o)} + z\Omega_{1y} \\ U_{1z} &= U_{1z}^{(o)} - x\Omega_{1y} \\ U_{2x} &= U_{2x}^{(o)} + z\Omega_{2y} \\ U_{2z} &= U_{2z}^{(o)} - x\Omega_{2y} \end{aligned} \quad (1)$$

式中 U_{1x}、U_{2x}、U_{1z} 和 U_{2z} 依次表示挡边 1 和滚子 2 沿 x 方向的速度及二者沿 z 方向的速度；上标 (o) 表示名义接触点处的；Ω_{1y} 和 Ω_{2y} 依次表示二者沿 y 轴的角速度分量.

2.2 膜厚分布

膜厚分布如下式所示：

$$h = h_0 + \frac{x^2}{2R_{2x}} + \frac{z^2}{2R_{2z}} - \frac{x^2}{2R_{1x}} - \frac{z^2}{2R_{1z}} + \delta(x,z) \quad (2)$$

式中 h_0 为假想无变形时的最小油膜厚度，它发生在名义接触点上；x 和 z 表示以名义接触点为原点的坐标，依次沿挡边圆周切线方向及垂直方向；R_{1x} 和 R_{2x} 为挡边和滚子端面在包含 z 轴的主平面中的曲率半径；R_{1z} 和 R_{2z} 为二者在包含 z 轴的主平面中的曲率半径；$\delta(x,z)$ 为各点的弹性变形所致的膜厚增量. 上式亦可写成：

$$h = h_0 + \frac{x^2}{2R_x} + \frac{z^2}{2R_z} + \delta(x,z) \quad (3)$$

式中 R_x 和 R_z 是两主平面中的综合曲率半径.

球面挡边与球形滚子端面接触时，$R_{1z} = R_{1x}$ 且 $R_{2z} = R_{2x}$，故

$$R_x = \frac{R_{1x}R_{2x}}{R_{1x} - R_{2x}} \quad R_z = R_x \quad (4)$$

锥曲挡边与球形滚子端面接触时，$R_{1z}=\infty$，$R_{2z}=R_{2x}$，故

$$R_x=\frac{R_{1x}R_{2x}}{R_{1x}-R_{2x}} \quad R_z=R_{2x} \tag{5}$$

锥面挡边与弧环形滚子端面接触时，$R_{1z}=\infty$，故

$$R_x=\frac{R_{1x}R_{2x}}{R_{1x}-R_{2x}} \quad R_z=R_{2z} \tag{6}$$

可见，因为末一种型面配对的 R_x 和 R_z 是互相独立的，所以在设计上可有更多的适应性，是上列三种配对中最能满足优化要求的型面.

三、弹流润滑的基本方程和解算

3.1 基本方程及边界条件

在本文所考察问题的运动情况下，描述油膜内压力分布的 Reynolds 方程为：

$$\begin{aligned}&\frac{\partial}{\partial x}\left(\frac{\rho h^3}{\eta}\frac{\partial p}{\partial x}\right)+\frac{\partial}{\partial z}\left(\frac{\rho h^3}{\eta}\frac{\partial p}{\partial z}\right)\\&=6[(U_{1x}^{(o)}+U_{2x}^{(o)})+z(\Omega_{1y}+\Omega_{2y})]\frac{\partial(\rho h)}{\partial x}\\&+6[(U_{1z}^{(o)}+U_{2z}^{(o)})-x(\Omega_{1y}+\Omega_{2y})]\frac{\partial(\rho h)}{\partial z}\end{aligned} \tag{7}$$

滚子和挡边接触变形之和可按 Boussinesq 公式计算：

$$\delta=\frac{2}{\pi E'}\int_A\frac{p(x_1,z_1)}{\sqrt{(x_1-x)^2+(z_1-z)^2}}dx_1dz_1 \tag{8}$$

接触变形使膜厚分布发生改变，如式(3)所示.

考虑到油的黏度在高压下会增大，计算中应计入压粘效应，本文用 Barus 公式表示：

$$\eta=\eta_0 e^{ap} \tag{9}$$

以上各式中，$E'=2/[(1-\nu_1^2)/E_1+(1-\nu_2^2)/E_2]$——当量弹性模量；$A$——$x$-$z$ 平面中的油膜展区；a——压粘系数；η_0——大气压下的油黏度；ν_1 和 ν_2——泊松比；E_1 和 E_2——弹性模量.

压力边界条件定为：在油膜展区的上游边和两侧边上，表压力为零；油膜破裂边界用雷诺边界条件定出，即：破裂边界上 $p=\partial p/\partial n=0$.

3.2 解算过程

用顺解法求解上述联立方程组，其计算框图如图 3 所示.先估取一个压力分布及 h_0 初值：由式(8)计算接触区内各点变形；由式(7)及雷诺边界条件计算膜厚分布；由式(9)计算黏度分布；以所得的膜厚分布和黏度分布代入式(7)，解出新的压力分布.按下式比较前后二次（即第 $k-1$ 和第 k 次）压力分布，以判定是否已相对收敛到足够精度：

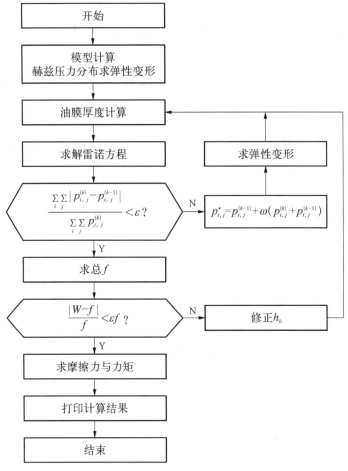

图 3 迭代过程框图

$$\frac{\sum_{i}\sum_{j}|p_{i,j}^{(k)}-p_{i,j}^{(k-1)}|}{\sum_{i}\sum_{j}p_{j,j}^{(k)}}<\varepsilon=0.01 \tag{10}$$

如不符合收敛准则,则按下式取定下一次迭代的压力分布初值:

$$p_{i,j}^{*}=p_{i,j}^{(k-1)}+\omega(p_{i,j}^{(k)}-p_{i,j}^{(k-1)}) \tag{11}$$

式中 ω 为亚松弛因子,其值小于 1.

压力分布收敛后,用数值积分求出总承载力,按下式检验它是否足够接近给定负荷:

$$\frac{|W-F|}{F}<\varepsilon_F \tag{12}$$

式中 W 为给定的负荷值;F 为算出的总承载力;ε_F 取为 0.000 5. 如不符合上述条件,则改变 h_0 值使其逐步逼近.

由于挡边副不属于重载工况,适当选取初始压力分布和 h_0 值时,上述顺解程序一般均能收敛.

为了考察弹性变形对挡边润滑问题的影响程度,平行地作了不计弹性变形的流体动力润滑理论计算,以作对比.

3.3 摩擦力计算

因为挡边接触副的最大压力通常小于 0.3 GPa[5],在正常负载下工作时则压力在 0.07~0.15 GPa 之间,所以可假定牛顿流体模型适用于一般机油,按此来计算摩擦力.

挡边上的摩擦力:

$$F_{1x} = \iint_{A_p} \left(-\frac{h}{2}\frac{\partial p}{\partial x} + \eta \frac{U_{2x}-U_{1x}}{h}\right) dx dz + h_\Gamma \iint_{A_r} \eta \frac{U_{2x}-U_{1x}}{h^2} dx dz \tag{13}$$

滚子上的摩擦力:

$$F_{2x} = \iint_{A_p} \left(-\frac{h}{2}\frac{\partial p}{\partial x} - \eta \frac{U_{2x}-U_{1x}}{h}\right) dx dz - h_\Gamma \iint_{A_r} \eta \frac{U_{2x}-U_{1x}}{h^2} dx dz \tag{14}$$

挡边上的摩擦力矩:

$$\begin{aligned} M = & \iint_{A_p} \left(-\frac{h}{2}\frac{\partial p}{\partial x} + \eta \frac{U_{2x}-U_{1x}}{h}\right)(z+R)\cos\psi\, dx dz \\ & + h_\Gamma \iint_{A_r} \eta \frac{U_{2x}-U_{1x}}{h^2}(z+R)\cos\psi\, dx dz \\ & - \iint_{A_p} \left(-\frac{h}{2}\frac{\partial p}{\partial z} + \eta \frac{U_{2z}-U_{1z}}{h}\right) x\cos\psi\, dx dz \\ & - h_\Gamma \iint_{A_r} \eta \frac{U_{2z}-U_{1z}}{h^2} x\cos\psi\, dx dz \end{aligned} \tag{15}$$

上列式中,A_p 和 A_r 依次为油膜展区中的完整膜部分和破裂膜部分;h_Γ 为破裂边界上相应的油膜厚度;R 为轴承回转中心线到挡边接触点的距离;ψ 为轴承回转中心线与挡边接触点法线的夹角.

四、计算结果与分析

现以上海滚动轴承厂生产的 7518 轴承为例,作了球形滚子端面对锥面挡边和球面挡边接触副的弹流润滑计算.计算参数为:

    挡边内侧半径:57.486 mm
    挡边外侧半径:60.500 mm
    滚子大端半径:9.739 mm
    滚子锥角:2°
    内圈滚道与回转轴线的夹角:11.642 2°
    润滑油动力黏度:0.028 314 0 Pa·s
    润滑油压粘指数:$0.198 950 \times 10^{-1}$ mm²/N
    弹性模量:$0.232 967 \times 10^6$ N/mm²
    轴承内圈转速:1 000 r/min

4.1 弹流润滑的压力分布和膜厚分布

球形滚子端面对球面挡边,曲率半径比为 0.76,负荷为 550 N 时的接触区压力分布和膜厚分布如图 4 所示.

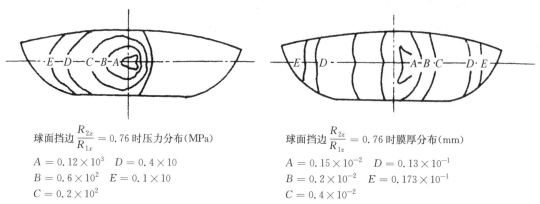

球面挡边 $\dfrac{R_{2x}}{R_{1x}} = 0.76$ 时压力分布(MPa)
$A = 0.12 \times 10^3 \quad D = 0.4 \times 10$
$B = 0.6 \times 10^2 \quad E = 0.1 \times 10$
$C = 0.2 \times 10^2$

球面挡边 $\dfrac{R_{2x}}{R_{1x}} = 0.76$ 时膜厚分布(mm)
$A = 0.15 \times 10^{-2} \quad D = 0.13 \times 10^{-1}$
$B = 0.2 \times 10^{-2} \quad E = 0.173 \times 10^{-1}$
$C = 0.4 \times 10^{-2}$

图 4 压力分布和膜厚分布(曲率半径比 0.76 的球面挡边)

球形滚子端面对锥面挡边,曲率半径比为 0.8,负荷为 160 N 时的压力分布和膜厚分布如图 5 所示.

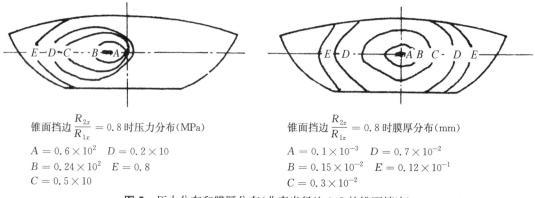

锥面挡边 $\dfrac{R_{2x}}{R_{1x}} = 0.8$ 时压力分布(MPa)
$A = 0.6 \times 10^2 \quad D = 0.2 \times 10$
$B = 0.24 \times 10^2 \quad E = 0.8$
$C = 0.5 \times 10$

锥面挡边 $\dfrac{R_{2x}}{R_{1x}} = 0.8$ 时膜厚分布(mm)
$A = 0.1 \times 10^{-3} \quad D = 0.7 \times 10^{-2}$
$B = 0.15 \times 10^{-2} \quad E = 0.12 \times 10^{-1}$
$C = 0.3 \times 10^{-2}$

图 5 压力分布和膜厚分布(曲率半径比 0.8 的锥面挡边)

由图可知,计算结果中未出现二次压力峰,这亦表明此种挡边接触副尚不属于重载情况[8].

4.2 曲率半径比对挡边承载力的影响

在给定的负载下,计算不同几何参数下的膜厚.比较膜厚大小,即可评价挡边副的承载能力.

图 6 表示膜厚与曲率半径比的关系.其中不计弹性变形而按流体动力润滑计算的膜厚比弹流润滑计算的膜厚小得多.可见挡边副中弹性变形的作用是不容忽略的.

计算结果还表明,球形滚子端面与球面挡边相配的承载能力,显著大于球形滚子端面与锥面挡边相配的承载能力.不过这一结论只适用于光洁度相同的这两种型面比较.考虑到锥面挡边比球面挡边容易加工到较好的光洁度,前者实际上的承载能力不一定逊于后者.在实际评定承载能力高低时,应综合考虑理论计算结果和实际粗糙度.

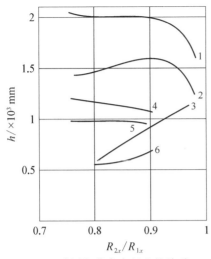

图 6 膜厚与曲率半径比的关系

1—球面挡边,中心膜厚
2—同上,最小膜厚
3—同上,不计弹性变形
4—锥面挡边,中心膜厚
5—同上,最小膜厚
6—同上,不计弹性变形

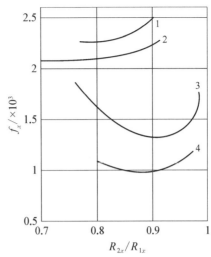

图 7 摩擦系数与曲率半径的关系

1—在弹流润滑时锥面挡边上 X 方向的摩擦系数
2—在动力润滑时锥面挡边上 X 方向的摩擦系数
3—在弹流润滑时球面挡边上 X 方向的摩擦系数
4—在动力润滑时球面上 X 方向的摩擦系数

从能获得尽可能大的最小油膜厚度看,球形滚子端面对球面挡边的最佳曲率半径比为 0.9 左右;对锥形挡边的最佳曲率半径比为 0.8 左右,而在 0.75~0.85 的范围内对轴承性能无明显影响.这与[2]的实验结果基本吻合.

至于弧环形滚子端面,尚有待具体计算.但是,如果取 $R_{2x}/R_{1x}=0.9$ 且 $R_{2x}=R_x$(见式 6),则其接触区几何全同于优化的球面挡边与球形滚子端相配的情况,其理论上的承载能力当亦与后者同.由于弧环形滚子接触副有更多的设计自由度,故可单独改变 R_{2x} 值,即在此 R_{2x} 值附近(即大于 R_x 或小于 R_x)进行搜索,可望达到进一步提高承载能力的目的,至少有相同承载能力.

4.3 曲率半径比对摩擦力和摩擦力矩的影响

图 7 表示与球形滚子端面相配时,锥面挡边和球面挡边上沿 X 方向的摩擦系数与曲率半径比的关系.图 8 表示其摩擦力矩.

由图可见,均与球形滚子端面相配时,球面挡边的摩擦系数在理论上要比锥面挡边的低些.但这当然只适用在光洁度相同的两种型面比较时.

还可看到,在膜厚最大的最佳曲率半径处,摩擦系数和摩擦力矩也是较低的.

图中亦示出了流体动力润滑计算所得的摩擦系数,它比弹流计算结果偏小.作者认为这主要是由于未计入压粘效应所致.

4.4 与 Dalmaz 实验结果[4]的对比

针对[4]的实验条件,用本文的方法作了计算.图 9 对比了计算结果和[4]的实验结果.由图 9 可见,本文计算结果与实验数据在趋势上是一致的,在一段范围的数量亦接近.

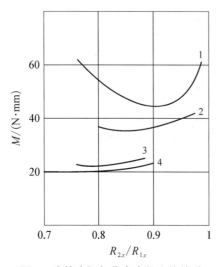

图 8 摩擦力矩与曲率半径比的关系

1—弹流润滑时球面挡边上摩擦力矩($f=550$ N)
2—动力润滑时球面挡边上摩擦力矩($f=550$ N)
3—弹流润滑时锥面挡边上摩擦力矩($f=160$ N)
4—动力润滑时锥面挡边上摩擦力矩($f=160$ N)

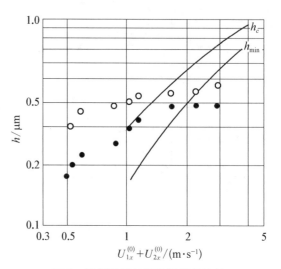

图 9 计算结果与实验结果的比较

$E' = 0.451\,7 \times 10^{11}$ Pa, $a = 1.57 \times 10^{-8}$ Pa^{-1},
$\eta_0 = 0.026$ Pa·s, $W = 80$ N, $R_{x1} = 185.283$ mm,
$R_{x2} = 163.262$ mm

h_c—中心膜厚; ○ 实验值[4]; — 计算值
$h_{\min}$—最小膜厚; ● 实验值[4]; — 计算值

五、结论

1. 本文用弹性流体运力润滑理论,计入挡边接触副的具体几何条件和运动条件,计算了 7518 轴承在特定工况下的油膜压力和厚度分布及摩擦阻力. 与现有实验结果的对比表明本文方法可行.

2. 挡边接触副润滑问题中,弹性变形的作用不容忽视. 略去弹性变形的结果将使算出的膜厚(或承载能力)和摩擦系数偏低,速度越低则偏差越大.

3. 与球形滚子端面相配时,锥面挡边接触副的最佳曲率半径比为 0.8 左右,且在 0.75~0.85 范围内对膜厚和摩擦影响不显著;球面挡边接触副的最佳曲率半径比则为 0.9 左右.

4. 与球形滚子端面相配时,球面挡边可比锥面挡边在理论上达到最高的承载能力;弧环形滚子端面与锥形挡边相配时,由于有更多的设计自由度,在理论上可望达到最高的承载能力. 但上述结论是在同样的光洁度下来比较各种型面配对而得到的. 在实际选取型面时,应充分考虑加工粗糙度对承载能力的影响,而锥边挡边比球面挡边更易加工到优良的光洁度.

参 考 文 献

[1] Frase, D., Brockmuler, U. Correct bearing lubrication increases reliability of differentials. Ball Bear. J., 199, 1982: 1-8.

[2] Korrenn, H. Gleitreibung und Grenzbelastung an den Bordflachen von Kegelrollenlagern-Einfluß von Drehzahl, Belastung, Schmierstoff und Gestaltung der Gleitflachen nach Versuch und Berechnung. Fortschritt-Berichte VDI Zeitschrift, 1967,1(11): 1-121.

[3] Jamison, W. E., Kauzlarich, J. J., Mocher, E. V. Geometric effects on the rib-roller end contact in a tapered roller bearing. ASLE Trans., 1976, 20: 78-88.

[4] Dalmaz, G., Tessier, J. F., Dudraghe, G. Friction improvement in cycloidal motion contacts: rib-roller end contact in a tapered roller bearing. Proceedings of the 7th Leeds-Lyon symposium on tribology, Science and Technology Press, 1980: 175-185.

[5] Gadallah, N., Dalmaz, G. Hydrodynamic lubricatication of the rib-roller end contact of a tapered roller bearing. Trans. ASME, J Lubr Technol, 1984, 116: 265-272.

[6] Wallin, E., Low-energy bearing arrangement state of the art, requirements of the 80's. Ball Bear. J., 1982,212: 1-8.

[7] Swingler, C. L. Regimes of fluid film lubrication at the rib-roller contact in a tapered roller bearing. AGARD-CP-323, 1982.

[8] Hamrock, B. J., Dowson, D. Ball Bear. Lubr. John Wiley & Sons Inc.,1981.

Elastohydrodynamic Lubrication Analysis of Rib-roller End Contact of Tapered Roller Bearing

Abstract: This paper makes hydrodynamic and elastohydrodynamic lubrication analyses of rib-roller end contact of tapered roller bearings with cycloidal relative motion, and makes a study of the effect of elastic deformation in such cases. Influence of geometric form and parameters on load carrying capacity, frictional force and frictional moment are calculated, and optimal geometric parameters are deduced. The result of comparison with existing experimental data is satisfactory.

粘弹性轴瓦对滑动轴承动力特性的影响*

摘　要：本文提出了一种以粘弹性材料作轴瓦，从而提高轴承稳定性和减振性的新设想，相应地在理论上构成了滑动轴承小扰动粘弹性流体动力润滑问题. 联立求解了在微小简谐扰动下的复数形式的粘弹性变形方程和流体动力润滑方程，在变形计算中利用缩减的柔度矩阵求解三维弹性变形. 用超松弛迭代法求解润滑方程和扰动润滑方程，解扰动压力的复振幅分布得油膜-支承面在一定扰动频率下的复合动力特性，经迭代使界限涡动频率与扰动频率达成一致，由此求出稳定性基本参数. 此外亦计算了同步涡动下的动特性作为确定油膜减振性的依据，由此获得了共振图谱. 结果表明，当用粘弹性材料轴承支撑转子系统时，可使其稳定性和减振性明显提高，这种理论上预示了一条构成轴承优异动力特性的新途径.

随着近代科学技术的发展和研究工作的逐步深入，对滑动轴承支承的高速旋转机械，提出了越来越高的稳定性和减振性要求. 一般的轴承形式，已难于完满地解决问题. 有效措施之一是在轴承外或轴瓦外加入弹性阻尼支承[1,2]，例如挤压膜式阻尼器[3,4]，在参数匹配时能大大提高轴承-转子系统的稳定性. 但因其整体结构复杂，弹性零件可能产生疲劳破坏等原因，在应用上还受到一定限制. 近10余年来在美国等国获得较大发展的波箔式气体动压轴承具有良好的动力特性，但多用于小型轻载场合[5]. 如果将轴承外弹性阻尼支承、轴瓦外弹性阻尼支承、波箔式轴承的发展历史和它们的共性联系起来，就可以看到一种"弹性阻尼和摩擦支承面愈益密切结合"的趋势. 由此可以合乎逻辑地设想：将适当的弹性、阻尼值与摩擦支承面直接结合成"粘弹性材料支承面"，将可发挥膜外阻尼对油膜动力特性的改善作用，以最简单的结构来实现这种有效的增稳的原理. 文献[6]和[7]分别建立了圆柱轴承在简谐小扰动下的弹流润滑(EHL)和粘弹流(VEHL)方程，并计算了轴瓦材料的弹性、阻尼值对轴承动特性和稳定性的影响. 文献[8]按粘弹性力学研究了橡皮推力轴承的膜厚变化. 本文针对具体的粘弹性阻尼材料，进一步考察这种"粘弹性轴承"对转子系统稳定性和减振性的改善作用.

一、理论基础

在非定常工况下，雷诺方程的无量纲形式为：

$$\frac{\partial}{\partial \phi}\left(H^3 \frac{\partial p}{\partial \phi}\right)+\left(\frac{d}{l}\right)^2 \frac{\partial}{\partial \lambda}\left(H^3 \frac{\partial p}{\partial \lambda}\right)=3 \frac{\partial H}{\partial \phi}+6 \frac{\partial H}{\partial T} \tag{1}$$

图1示出圆柱轴承静平衡位置. 对轴承施加微小简谐扰动时，轴颈中心瞬时位置如下：

$$\begin{aligned}\varepsilon_j &= \varepsilon + \varepsilon_a e^{i\omega T} \\ \theta_j &= \theta + \theta_a e^{i\omega T}\end{aligned} \tag{2}$$

---

* 本文合作者：蒋晓飞. 原发表于《润滑与密封》，1988(6)：23-28.

其中 ε_a、θ_a 是偏心率和偏位角的振幅,它们可以是相互独立的复数值. 此时油膜压力和油膜厚度亦由静态和动态增量构成:

$$P = P_0 + Q e^{i\omega T} \tag{3}$$

$$H = H_0 + H_j e^{i\omega T} + \delta_v e^{i\omega T} \tag{4}$$

式中,H_0、P_0——静态膜厚、压力分布;Q——动态压力增量之复振幅;ω——相对扰动频率,$\omega = \omega_有/\Omega$,$\omega_有$——扰动频率,Ω——轴承角速度;δ_v——粘弹性动变形之复振幅.

$$H_j = \varepsilon_a \cos\varphi + \varepsilon \cdot \theta_a \sin\varphi \tag{5}$$

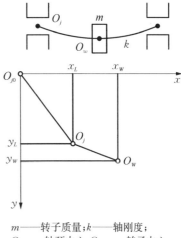

m——转子质量;k——轴刚度;
O_j——轴颈中心;O_w——转子中心;
O_{j0}——轴颈中心静位置

图 1 轴承-转子系统

是轴颈几何位置变动引起的膜厚增量. 将式(3)~(5)代入式(1),则得小扰动下计入粘弹性动变形的扰动压力润滑方程:

$$\frac{\partial}{\partial \phi}\left(H_0^3 \frac{\partial Q}{\partial \phi}\right) + \left(\frac{d}{l}\right)^2 \frac{\partial}{\partial \lambda}\left(H_0^3 \frac{\partial Q}{\partial \lambda}\right) = -9\frac{M}{H_0}\frac{\partial H_0}{\partial \phi} - 3H_0\left[\left(H_0 \frac{\partial M}{\partial \phi} - M\frac{\partial H_0}{\partial \phi}\right)\right.$$
$$\left.\frac{\partial P_0}{\partial \phi} + \left(\frac{d}{l}\right)^2 \frac{\partial}{\partial \lambda}\left(H_0 \frac{\partial M}{\partial \lambda} - M\frac{\partial H_0}{\partial \lambda}\right)\frac{\partial P_0}{\partial \lambda}\right] + 3\frac{\partial M}{\partial \phi} + 6i\omega M \tag{6}$$

其中

$$M = \varepsilon_a \cos\varphi + \varepsilon \cdot \theta a \sin\varphi + \delta_v$$

粘弹性材料是一种兼有某些黏性液体和弹性固体特性的材料. 黏性液体在受力时损耗能量而不贮存能量;相反,弹性固体则能贮存能量而不损耗能量. 粘弹性材料兼有两者之特性,当它发生动态应力和应变时,既能将部分能量作为位能贮存起来,又能将另一部分能量转化为热能而耗掉. 前者表现为弹性,后者表现为阻尼. 粘弹性材料的恰当使用,可起到极佳的减振和降噪作用.

粘弹性材料的复模量为

$$E^* = E' + iE'' \tag{7}$$

式中 E'、E'' 依次称为储能模量和损耗模量;后者对前者之比通常称为损耗因子 $\eta = E''/E'$.

在三维弹性变形的有限元计算中,刚度矩阵中的元素仅与形函数和泊松比 ν 有关,而弹性模量 E 总可提到矩阵外面来:

$$E[K]\{\delta_e\} = \{F\} \tag{8}$$

这里,$[K]$、$\{\delta_e\}$、$\{F\}$ 分别是无量纲总纲矩阵、位移矢量和载荷列阵.

由稳态简谐运动下的弹性-粘弹性对应原理[9],相应的粘弹性位移为:

$$\{\delta_v\} = \frac{1}{E^*}[K]^{-1}\{F\} \tag{9}$$

将式(7)代入(9)得: $\{\delta_v\} = (f' - if'')\{\delta_e\}$

式中 $(f' - \mathrm{i}f'')$ 可称为相对复柔度:

$$f^* = f' - \mathrm{i}f'' = 1/(E^*/E) \tag{10}$$

其中　$f' = (E'/E)/[(E'/E)^2 + (E''/E)^2]$——相对复柔度的实部;

$f'' = (E''/E)/[(E'/E)^2 + (E''/E)^2]$——相对复柔度的虚部.

油膜动特性系数采用偏导数法计算[6].计算时先联立解出某一相对扰动频率下的压力和变形的复振幅分布,再从扰动压力复振幅分布积分得油膜刚度系数和阻尼系数.

稳定界限状态的油膜无量纲相当刚度 K_{eq} 和界限涡动比 γ_{st} 依次为[10]:

$$K_{eq} = (K_{xx}B_{yy} + K_{yy}B_{xx} - K_{xy}B_{yx} - K_{yx}B_{xy})/(B_{xx} + B_{yy}) \tag{12}$$

$$\gamma_{st}^2 = [(K_{eq} - K_{xx})(K_{eq} - K_{yy}) - K_{xy}K_{yx}]/(B_{xx}B_{yy} - B_{xy}B_{yx}) \tag{13}$$

计入动变形效应后,动特性系数和界限涡动比与扰动频率有关,在稳定界限状态应保证界限涡动比 γ_{st} 等于计算动特性时所依据的相对扰动频率 ω,这可通过迭代达到.

对一圆柱轴承支承的对称单质量弹性转子(图 2),不计圆盘质量不平衡影响,失稳角速度计算公式的无量纲形式为:

$$\frac{\Omega_{st}}{\omega_k} = -\frac{So_k}{2k_{eq} \cdot f/c} + \sqrt{\left(\frac{So_k}{2k_{eq} \cdot f/c}\right)^2 + \frac{1}{\gamma_{st}^2}} \tag{14}$$

且稳定界限状态有

$$So_k = So\frac{\Omega_{st}}{\omega_k} \tag{15}$$

故　$$\frac{\Omega_{st}}{\omega_k} = 1/[\gamma_{st}\sqrt{1 + So/(k_{eq} \cdot f/c)}] \tag{16}$$

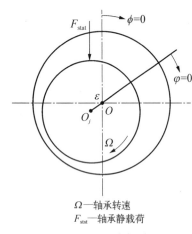

Ω——轴承转速
F_{stat}——轴承静载荷

图 2　不平衡振动

式中,Ω_{st}/ω_k——相对失稳角速度,$\omega_k = \sqrt{k/m}$——转子刚支时的固有频率;f/c——轴的相对柔度,$f = mg/k$——转子静挠度,c——轴承半径间隙;So_k——以刚支临界转速为参考速度的承载量系数,$So_k = \dfrac{\psi^2 W/2}{\mu\omega_k rl}$;$So$——承载量系数,$So = \dfrac{\psi^2 W/2}{\mu\Omega rl}$.

在稳定状态下工作的转子,由于残余不平衡,仍会作受迫振动,某一瞬时的端视图如图 3 所示.一般情况下,圆盘几何中心 O_W 在 x-y 座标系下作椭圆涡动.设定一参考承载量系数 So_k,由相对转速 Ω/ω_k 下的 So 可确定轴承在相对扰动频率 $\omega = 1$ 下的 8 个动特性系数,则可进一步求得椭圆涡动的长半轴(相对最大振幅)Q_W/e 与 Ω/ω_k 的对应关系[11],取定一系列的 So,并以 f/c 为参变量,则可得一族 $Q_W/e \sim \Omega/\omega_k$ 的关系曲线.

当 Q_W/e 最大时,相应的 Ω/ω_k 就代表了临界转速.在临界状态下的 Q_W/e 称为共振放大倍数,它越低表

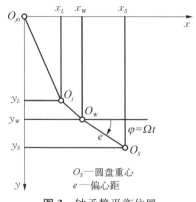

O_s——圆盘重心
e——偏心距

图 3　轴承静平衡位置

明轴承的"减振性"越好.

对于每一种参考承载量系数 So_k,都有一个 $Q_W/e \sim \Omega/\omega_k$ 曲线图,如只要知道转子最大振幅的话将每一个最大的 Q_W/e 汇集到一张图中,就可得到 $Q_{W_{max}}/e \sim So_k$ 的关系,同时可得相应的临界转速 $\Omega_{cr}/\omega_k \sim So_z$ 的关系,它们常称为"共振图谱".

二、数值计算

对 360°圆柱轴承的油膜区均分成 16×80 个网格,用有限差分法将雷诺方程离散成线性代数方程组,经超松弛迭代可解得压力或扰动压力分布.此过程中油膜厚度分布作为已知,并取雷诺边界条件.压力和变形用亚松弛迭代耦合求解.

对轴瓦的弹性变形计算,由于圆柱轴承的轴对称形状,并假设沿圆周方向均匀支撑,先用少量计算建立缩减柔度矩阵的一个列矢量,再利用叠加原理和柔度矩阵的自动生成来计算三维弹性变形是非常合适的,它使变形的求解工作量降低到最小限度.计算关系可表达为:

$$\delta_0(\phi, \lambda) = \iint_A p(\phi', \lambda') \cdot f(\phi', \lambda', \phi, \lambda) d\phi' d\lambda' \tag{17}$$

其中,$f(\phi', \lambda', \phi, \lambda)$——点 (ϕ', λ') 对点 (ϕ, λ) 的柔度影响系数,所有节点的柔度影响系数即组成柔度矩阵.

许多工程材料对于机械能的内耗能力十分微弱,而高分子聚合物则有较高的内耗,见表1[12].

表 1 几种常用结构材料的阻尼性能

材 料	损耗因子	材 料	损耗因子
金 属	0.0001~0.001	混凝土、砖	0.001~0.1
玻 璃	0.001~0.005	木材、软木	0.01~0.2
粒状介质	0.01~0.05	橡胶、塑料	0.001~10.0

现有的阻尼材料有:阻尼粘弹材料,包括橡胶类与塑料类;阻尼合金,包括基体为铁基或其他有色金属的金属;复合材料,包括层压材料及混合材料.

我国现今已研制出几种粘弹性高阻尼材料,其中有些具有损耗因子大、模量高、温度范围宽等特点,可望构成轴承材料或附加材料.表 2 例示[13]中所述 3103 号材料的储能模量、损耗因子与频率的关系(温度 60℃时).

表 2 3103 粘弹性阻尼材料性能

f(Hz)	2	10	20	30	40	50	60	70	80	90	100	150	200	600	1000
$E'(\times 10^{10}$ 达因$/cm^2)$	0.6	0.6	0.6	0.62	0.63	0.64	0.66	0.69	0.71	0.73	0.77	0.81	0.9	1.5	1.8
η	0.145	0.31	0.44	0.51	0.60	0.67	0.71	0.74	0.79	0.82	0.83	0.89	0.92	0.8	0.7

三、结果分析

计算实例中的轴承参数取宽径比 $l/d = 0.6$,轴瓦外径、内径之比 $D/d = 1.6$.本文比较

了无量纲复模量 $E'=20.2;\eta=1$ 和 $E=200$（铜轴承）的失稳转速及共振放大倍数. 还对 3103 粘弹性材料作了计算, 计算时取 $l/d=0.6$, 轴承直径 $d=50$ mm, 转子重量 $W=50$ kg, 间隙 $\psi=1.5‰$, 20 号透平油, 60℃ 时其动力黏度 $\mu=11.8\times10^{-3}$ Pa·s.

图 4 是以 So_k 为横坐标的圆柱轴承支撑的失稳角速度. 由图可知, 与刚性轴承相比, 轴瓦带有粘弹性时, 失稳角速度增大, 尤其是 $E'=2$, $\eta=1$ 和 3102 号材料, 轴承稳定性明显提高. 例如当 $f/c=2.0$, $So_k=1.6$ 时, 相对材料 1 的失稳角速度, 材料 2 提高了 4.7%, 材料 3 提高了 12.4%, 而材料 4 则提高了 38.2.

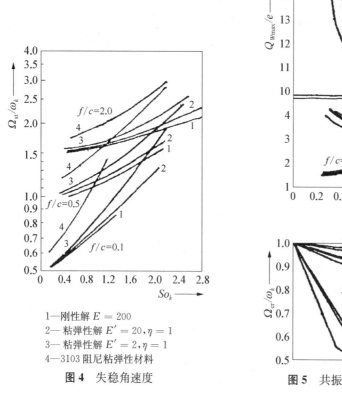

1—刚性解 $E=200$
2—粘弹性解 $E'=20, \eta=1$
3—粘弹性解 $E'=2, \eta=1$
4—3103 阻尼粘弹性材料

图 4 失稳角速度

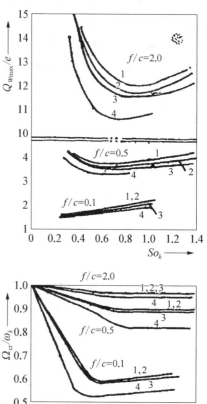

图 5 共振放大倍数、临界转速

图 5 是共振放大倍数 $Q_{W\max}/e$、临界转速 Ω_{cr}/ω_k 与参考承载量系数 So_k 的关系曲线. 当材料增加阻尼效应后, 共振放大倍数大大降低, 说明轴瓦材料阻尼对抑制轴的共振是很有效的. 又以 $f/c=2.0$, $So_k=0.8$ 为例, 相对材料 1 的共振放大倍数, 材料 2 降低了 2.9%, 材料 3 降低了 4.6%, 而材料 4 则降低了 9.5%. 考虑轴瓦粘弹性变形后, 临界转速 Ω_{cr}/ω_k 较刚性轴承降低了.

（在本文工作中, 得到了轴承研究室易子夏等同志的热心帮助, 在此深表感谢.）

符 号 说 明

B_{ij} —— 油膜阻尼系数, $B_{ij}=b_{ij}\psi^3/(\mu l)$. ψ: 间隙比, μ: 动力黏度, l: 轴承宽度

E —— 杨氏弹性模量, $E=E_{有}\psi^3/(2\mu\Omega)$. Ω: 轴承角速度

E^* —— 粘弹性复模量, 相对单位同上

f/c —— 轴的相对柔度, $f=mg/k$. m: 转盘质量, k: 轴刚度

H —— 油膜厚度, $H=h/c$. c: 半径间隙

K_{ij}—— 油膜刚性度系数，$K_{ij} = k_{ij}\psi^3/(\mu\Omega l)$

P—— 油膜压力，$p = P\psi^2/(2\mu\Omega)$

So—— 承载量系数，$So = W\psi^2/(2\mu\Omega rl)$，$W$：载荷

Ω_{st}/ω_k—— 相对失稳角速度，$\omega_k = \sqrt{k/m}$

T—— 时间，$T = t\Omega$

λ—— 纵向坐标，$\lambda = z/(l/2)$

ω—— 相对扰动频率，$\omega = \omega_有/\Omega$，$\omega_有$：扰动频率

参 考 文 献

[1] Glienicke, J., Stanski, U. External bearing damping — means of preventing dangerous shaft vibrations on gas turbines and exhaust turbo-changers, Proc. 11th CIMAC, 1975：287 - 311.

[2] Zhang, Z. M. System damping factor of an elastic rotor supported on tilting-pad bearings with elastic and damped pivots, proc. ISME Inter. Tribology Conf., July 1985：541 - 546.

[3] Nikolajsent, J. L., Holmes, R. Investigation of squeeze-film isolators for the vibration control of a flexible rotor, J. Mech. Engng. Sci., 1979,21(4)：247 - 252.

[4] Holmes, R., Dogan, R., Investigation of a rotor bearing assembley incor-porating a squeeze-film damper bearing, J. Mech. Engng. Sci., 1982,24(3)：129 - 137.

[5] Heshmat, H., Walowit, J. A. and Pinkus, O., Analysis of gas lubrication foil journal bearings, Trans. ASME, J. Lub. Tech., 1983,105(4)：638 - 646.

[6] 张直明,毛谦德,许汉平. 计入弹性动变形的圆柱径向滑动轴承动力特性研究. 机械工程学报,1986(1)：10 - 23.

[7] 蒋晓飞,张直明. 轴瓦粘弹性动变形对圆柱径向滑动轴动特性和稳定性的研究. 上海工业大学学报,1987(5)：471 - 482.

[8] Hori, Y., Kato, T. and Narumiya, H. Rubber surface squeeze film. Trans. ASME, J. Lub. Tech., 1982,103：398 - 405.

[9] Chrstensen, R. M. Theory of viscoelasticity, An Introduction. 2nd ed. Academic Press Inc., New York,1982.

[10] 张直明,等. 滑动轴承的流体动力润滑理论. 高教出版社出版,1986：94 - 95.

[11] 张直明,秦学明. 转子动力学. 上海工业大学研究生教材,1986：9 - 11.

[12] 北京橡胶工业研究所. 阻尼材料研究小结.

[13] 张浩勤. 粘弹性高阻尼材料性能研究和应用. 机械强度,1986(4)：63 - 67.

Effects of Visco-Elastic Bush on Dynamic characteristic of Sliding Bearing

Abstract：This paper puts forward a new idea of improving bearing stability and buffering with the bush made of visco-elastic material, and correspondingly forms perturbation visco-elastic hydrodynamic lubrication theory for sliding bearing. Results show that when rotor system is supported by visco-elastic material bearings, its stability and buffering are obviously improved, and the theory predicts a new way to form excellent dynamic characteristic of bearings.

小间隙浮动理论研究*

提　要：本文分析了磁记录中头盘间气膜的静态特性和动态特性，并给出 IBM3370 型磁头浮动块在 5.25 英寸硬盘上的静、动特性的数值结果. 作者用 Galerkin 法导出了动态压力方程的等价变分方程，并推导了有限元离散方程及动态特性的求解方法. 考虑了分子平均自由程效应后的静态和动态计算结果与现有的实验数据完全一致.

符 号 表

a：磁盘偏摆振幅，μm；$A:=a/h_0$

b：磁头浮动导轨宽度，mm；$B:=b/l$

C_{ij}：气膜阻尼系数，
　　　$i=j=1$，平行模式，$kgf \cdot s/mm$
　　　$i=j=2$，俯仰模式，$kgf \cdot mm \cdot s/rad$

h_0：最小气膜厚度，μm；h：各点气膜厚度，μm

$H:=h/h_0$，H_0：无量纲静平衡气膜厚度

H_d：无量纲气膜厚度增量

I_θ：绕俯仰轴的浮动块转动惯量，$kgf \cdot s^2 \cdot mm$

k_{ij}：气膜刚度系数，$i=j=1$，平行模式，kgf/mm
　　　$i=j=2$，俯仰模式，$kgf \cdot mm/rad$

k, k_θ：加载弹簧平行刚度，俯仰刚度，kgf/mm，
　　　$kgf \cdot mm/rad$

l, l_1：浮动块长度，斜楔长度，mm

m：浮动块质量，$kgf \cdot s^2/mm$

$M:=\lambda_a/h_\varphi$，克努森数（Knudsen Number）

P_a, P：大气压力、各点气膜压力，N/m^2

s：经拉普拉斯变换的时间变量，t：时间，s

U_x：磁记录表面速度，沿导轨方向的分量，m/s

W：浮动块承载力，g

$X:=x/l$，$Y:=y/l$，无量纲直角坐标

$X_G:=x_G/l$，无量纲载荷作用中心

δ_1：斜楔高度，μm

$Z:=z/h_0$，$D=D \cdot l/h_0$，浮动块运动坐标

λ_a：大气压下空气分子平均自由程，0.064 μm

$\Lambda_x:=6\mu U_x l/(P_a h)$，气体压缩数

μ：动力黏度，$N \cdot s/m^2$

$\sigma:=12\mu\omega_0 l^2/(P_a h_0^2)$，挤压数

$\tau:=\omega_0 t$

$\Omega:=\omega/\omega_0$

一、引言

磁头浮动机构是电子计算机磁盘存储器的关键部件，为加大存储容量，减小磁盘存储器的尺寸，必须加大磁记录密度，因而引起磁头和磁盘材料、结构和加工工艺的不断发展，并使关键部件的几何参数不断发生变化. 磁头与磁盘间的气膜厚度目前已降到亚微米级，最小的已达 $0.3 \mu m$，克努森数达 0.2，故在界面上的滑流不可忽视. 同时，磁头对磁盘的跟踪性能也是需要研究的根本问题.

据国内外业已发表的文献看，对于静态特性计算，采用的方法有两种：有限差分法[1,2,9]及有限元法[4,5]. 在动态特性计算中，主要采用频率响应法[6,10,11]及非线性轨迹法[7,8]，其中离散方法使用较多的是差分法[6-8,10]. 而用有限元法进行动态特性计算，使数值计算的收敛性、适用范围得到改善，是目前国内外学者[1]的一个研究方向.

---

* 本文合作者：胡近. 原发表于《电子计算机外部设备》，1989(1)：7-14.

本文主要致力于对磁头浮动块气膜建立一套既可用于静态分析,也可用于频率响应分析的有限元方法,以构成比较有效的分析、计算手段. 在此基础上,分析计算了小偏摆下的磁头动态响应.

二、基本方程及有限元离散

1. 基本方程

磁头浮动块的几何图形及坐标关系如图1所示.

等温条件下,考虑分子平均自由程的修正雷诺方程为

$$\nabla \cdot \left[PH^3 \left(1 + \frac{6M}{PH}\right) \nabla P - \Lambda PH \right] - \sigma \frac{\partial PH}{\partial \tau} = 0 \quad (1)$$

磁头浮动块的运动方程是

$$m \frac{d^2 z}{dt^2} + kz = \iint \Delta P \, dx \, dy \quad (2)$$

$$I_\theta \frac{d^2 \theta}{dt^2} + k_\theta \theta = \iint \Delta P (x_G - x) \, dx \, dy \quad (3)$$

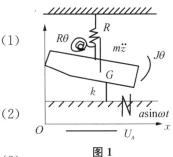

图1

由磁头浮动块垂直位移 Z 及俯仰运动 θ 引起的气膜厚度的变动量 h_d 由下式决定:

$$h_d = z + \theta(x - x_G) - a \sin \omega t \quad (4)$$

令 P_0、H_0 分别为静态平衡压力、气膜厚度,P_d、H_d 为相应的变动项. 设 $|P_d| \ll P_0$,$|H_d| \ll H_0$,$P = P_0 + P_d$,$H = H_0 + H_d$,代入式(1),略去二次及更高次项,可得

$$\nabla \cdot \left[(P_0 H_0^3 + 6M H_0^2) \nabla P_0 - \Lambda P_0 H_0 \right] = 0 \quad (5)$$

边界条件: 在导轨四周 $P_0 = 1$;对称线上 $\frac{\partial P}{\partial n} = q$(流量)

$$\nabla \cdot g(G_t) - j \Omega \sigma f(G_t) = 0 \quad (t = 1, 2) \quad (6)$$

边界条件: 导轨四周,$G_t = 0$;对称线上 $g(G_t) \cdot n = 0$

其中,

$$g(G_t) = \left[P_0 H_0^3 \left(1 + \frac{6M}{P_0 H_0}\right) \right] \nabla G_t + H_0^3 \nabla P_0 \cdot G_t$$

$$+ \left[3 P_0 H_0^2 \left(1 + \frac{4M}{P_0 H_0}\right) \cdot F_t \nabla P_0 - \Lambda f(G_t) \right] \quad (7)$$

$$f(G_t) = H_0 G_t + P_0 F_t \quad (t = 1, 2) \quad (8)$$

$F_1 = 1$,平行模式;$F_2 = X_G - x$,俯仰模式.

为研究头盘间气膜的变化规律,必须求解方程(5)、(6).

2. 有限元离散

对于静态压力方程的离散,用文献[4]的方法,可得到

$$\sum_e [M_{r\theta}]_e \bar{P}_0^{(n+1)} = \sum_e [M_{r\theta}]_e \bar{P}_0^{(n)} - \sum_e [D_r]_e \tag{9}$$

而对于动态压力方程的离散,则是本文须较详细叙述的内容. 观察式(6)可知,G_t 为复数, $G_t = G_{tr} + \mathrm{j}G_{ti}(\mathrm{j} = \sqrt{-1}$,虚数),代入式(6),按实部、虚部分开,得到两个等式,再用 δG_{tr}、δG_{ti} 分别乘这两个等式,并相加,在积分区域求积分,运用格林公式则有下列等价变分方程:

$$\delta \psi(G_t) = \iint [g(G_t) \cdot \nabla \delta G_t + \Omega \sigma f(G_{tr}) \delta G_{tr} - \Omega \sigma f(G_{ti}) \delta G_{ti}] \mathrm{d}\sigma = 0 \tag{10}$$

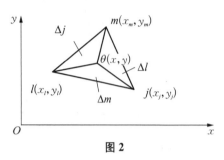

图 2

同静态方程离散方法一样,也采用图 2 所示的三角形单元.

$[L]_e = (L_i, L_j, L_m)^T$,单元面积坐标

$$L_k = a_k + b_k X + c_k Y \ (k = i, j, m)$$

$$\begin{bmatrix} a_i & b_i & c_i \\ a_j & b_j & c_j \\ a_m & b_m & c_m \end{bmatrix} = \begin{bmatrix} 1 & 1 & 1 \\ X_i & X_j & X_m \\ Y_i & Y_j & Y_m \end{bmatrix} \tag{11}$$

记:

$$\begin{aligned} &[G_{tr}]_e = [L]_e^T [\bar{G}_{tr}], \ [G_{ti}]_e = [L]_e^T [\bar{G}_{ti}] \\ &[\bar{G}_{tr}] = (G_{trl} \quad G_{trj} \quad G_{trm}), \ [\bar{G}_{ti}] = (G_{til} \quad G_{tij} \quad G_{tim}) \end{aligned} \tag{12}$$

将式(12)代入到式(10)中,经整理可得

$$\begin{aligned} &\sum_e [R_{rs}]_e [\bar{G}_{tr}] + \sum_e [S_{rs}]_e [\bar{G}_{ti}] + \sum_e [TR_{t,r}]_e = 0 \\ &\sum_e [R_{rs}]_e [\bar{G}_{tr}] - \sum_e [S_{rs}]_e [\bar{G}_{ti}] - \sum_e [TI_{t,r}]_e = 0 \end{aligned} \tag{13}$$

其中,下标 r、s 分别取单元节点号 i、j、m;$t = 1, 2$;e:单元.

$$\left.\begin{aligned} &[R_{rs}]_e = A_{rs}\bar{P}\bar{H}_3 + B_r\bar{H}_3\bar{L}(S) + 6MA_{rs}\bar{H}_2 - C_r\bar{H}_1\bar{L}(S) \\ &[S_{rs}]_e = \Omega\sigma\bar{H}\bar{L}_2(r, s) \\ &[TR_{1,r}]_e = B_r[3P\bar{H}_2 + 4M(\sum_u H_u) \cdot \Delta e] - C_r\bar{P}_1 \\ &[TI_{1,r}]_e = \Omega\sigma\left[(\sum_u P_u) + P_r\right] \cdot \Delta e/12 \\ &[TR_{2,r}]_e = X_G \cdot [TR_{1,r}]_e - B_r\{\sum_u [(\sum_v \bar{H}_2\bar{L}_2(u, v) \cdot X_v) \cdot P_u + 12M \cdot \\ &\qquad \overline{H_1 L}(u) \cdot X_u]\} + (\sum_u C_u P_u)[\sum_u (\sum_v \bar{L}_2(u, v) \cdot X_v) \cdot P_u] \\ &[TI_{2,r}]_e = X_G \cdot [TI_{1,r}]_0 - \Omega\sigma[\sum_u \overline{PL}_2(r, u) \cdot X_u] \end{aligned}\right\} \tag{14}$$

$$\overline{H_3L}(S) = \left[\left(\sum_u H_u\right)\left(\sum_u H_u^2\right) + H_s\left(\sum_u H_u\right)^2 + 2H_s^3\right] \cdot \Delta e/60$$

$$\overline{H_2L}(S) = \left[\left(\sum_u H_u\right)^2 + 2H_s\left(H_s + \sum_u H_u\right)\right] \cdot \Delta e/60$$

$$\overline{H_1L}(S) = \left[\left(\sum_u H_u\right)^2 + H_s\right] \cdot \Delta e/12$$

$$\overline{PH_3} = \sum_u \overline{H_3L}(u) P_u$$

$$\overline{PH_2} = \sum_u \overline{H_2L}(u) P_u$$

$$\overline{PH_1} = \sum_u H_1L(u) P_u$$

$$\overline{P}_1 = \left(\sum_u P_u\right) \cdot \Delta e/3 \qquad (15)$$

$$\overline{H}_2 = \left[\left(\sum_u H_u^2\right) + \left(\sum_u H_u\right)^2\right] \cdot \Delta e/12$$

$$\overline{H_2L_2}(r,s) = \begin{cases} \left[\left(\sum_u H_u\right)^2 + 4H_r\left(3H_r + \sum_u H_u\right)\right] \cdot \Delta e/180 & (r=s) \\ \left[\left(\sum_u H_u\right)^2 + (H_r + H_s)^2 + H_r^2 + H_s^2\right] \cdot \Delta e/180 & (r \neq s) \end{cases}$$

$$\overline{HL_2}(r,s) = \begin{cases} 2\left[\left(\sum_u H_u\right) + 2H_r\right] \cdot \Delta e/60 & (r=s) \\ \left[\left(\sum_u H_u\right) + H_r + H_s\right] \cdot \Delta e/60 & (r \neq s) \end{cases}$$

$$\overline{PL_2}(r,s) = \begin{cases} 2\left[\left(\sum_u P_u\right) + 2P_u\right] \cdot \Delta e/60 & (r=s) \\ \left[\left(\sum_u P_u\right) + P_r + P_s\right] \cdot \Delta e/60 & (r \neq s) \end{cases}$$

$$\overline{L}_2(r,s) = \begin{cases} \Delta e/6 & (r=s) \\ \Delta e/12 & (r \neq s) \end{cases}$$

其中,下标 r, s 分别取 i、j、m;Δe:单元面积;$\sum_u$ 对三角形单元节点求和.

3. 静动特性计算公式

静态承载力 $$W = \iint (P-1) \mathrm{d}\sigma \qquad (16)$$

压力中心 $$X_G = \left(\iint (P-1) \cdot X \mathrm{d}\sigma\right)/W \qquad (17)$$

动态响应 $$\left|\frac{\Delta h}{a}\right| = \{[\mathrm{Re}(\bar{Z}) + \mathrm{Re}(\bar{\theta})(X_G - X) - 1]^2 + [\mathrm{Im}(Z) + \mathrm{Im}(\bar{\theta}) \cdot (X_G - X)]^2\}^{1/2} \qquad (18)$$

气膜刚度与阻尼 $$k_{1t} + \mathrm{j}\omega C_{1t} = -\frac{P_a l^{1+t}}{h_0} \iint G_t \mathrm{d}\sigma$$

$$k_{2t} + \mathrm{j}\omega C_{2t} = -\frac{P_a l^{2+t}}{h_0} \iint G_t \mathrm{d}\sigma \qquad (19)$$

三、计算实例与结果分析

本文全部计算程序是在上海工业大学计算中心 IBM-4361 机上调试完成的. 式(9)的求解采用了全带宽带状转换存储与高斯列主元消去法以及其他的节省内存和计算时间的方法. 式(13)的求解则相应地运用了复数的运算. 整个程序内存量小,计算速度快.

1. 与差分法解[6]的计算结果及实验数据对比

实验用磁头和参数如图 3 所示. 用有限元计算的结果与用差分法计算的结果及实验数据的对比见图 4.

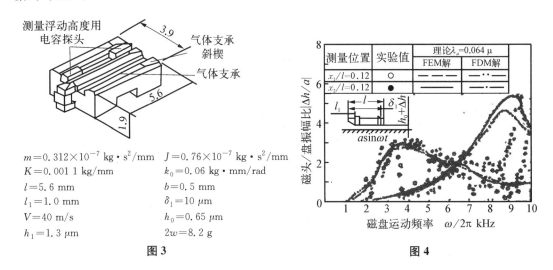

图 3   图 4

经比较发现,考虑了分子平均自由程效应后,有限元计算的共振振幅及共振频率与实验结果相符,证实了本计算方法的正确性.

2. IBM3370 型磁头的静动特性计算结果

结构图、几何参数及工况参数如图 5 所示[8].

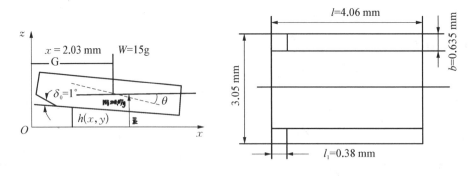

图 5 IBM 3370 型磁头结构及尺寸

对 5.25 英寸盘,当 $V=14.36$ m/s, $W=15$ g, $h_0=0.3089$ μm, $\theta_0=0.4125\times10^{-4}$ rad 时,静态压力分布如图 6 所示. 最小气膜厚度及姿态角随速度变化关系见图 7. 压力分布的特点,在前楔角末端及浮动块末端均有两个压力峰值,前峰与后峰的比值与姿态角及压力中心有关,对导轨几何中心线 $X-X$,压力呈近似对称分布.

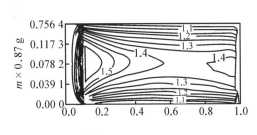

图 6 静态压力分布图

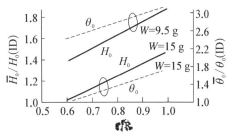

$H_0(\mathrm{ID})=0.308\,9\,\mu m, R=63.5\,mm, \theta_0(\mathrm{ID})=4125E-4$

图 7 最小气膜厚度及姿态角随速度变化

分子平均自由程对压力分布的影响必须考虑. 由实验证实[2], $h_0 \doteq 0.31\,\mu m$, 本例计算结果,在考虑分子平均自由程的影响,亦即 $\lambda_a = 0.064\,\mu m$ 时, $h_0 = 0.368\,9$; 若不考虑时, 即 $\lambda_a = 0.0$ 时, 则 $h_0 = 0.48\,\mu m$. 因此, 当 $h_0 < 1\,\mu m$ 时, 在计算中必须采用修正雷诺方程.

图8讨论了分子平均自由程对动态响应的影响.可以发现,由于 w 的不同,导致了支承气膜的刚度与阻尼的不同,使考虑了分子平均自由程效应的共振幅值、共振频率小于不考虑其影响的情形.

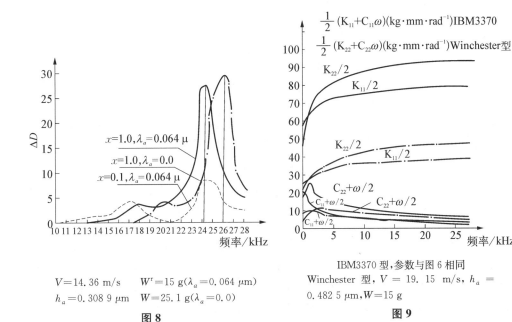

$V = 14.36\,m/s$ $W^r = 15\,g\,(\lambda_a = 0.064\,\mu m)$
$h_a = 0.308\,9\,\mu m$ $W = 25.1\,g\,(\lambda_a = 0.0)$

图 8

IBM3370 型,参数与图6相同
Winchester 型, $V = 19.15\,m/s$, $h_a = 0.482\,5\,\mu m$, $W = 15\,g$

图 9

图9给出了气膜刚度与阻尼的频率特性.我们根据计算结果可以发现,IBM3370 型磁头较 3350 型磁头[10]具有较高的共振频率,因而其动力性态较好,这是因为其体积小、质量小、气膜间隙小及支撑系统性能好的综合结果.

四、气膜对小偏摆的动态响应

据文献[13],盘面振动基本上是属于周期性的,少数盘片含有一定量的准周期性振动,但处理方法相同.

盘面振动可用 Fourier 级数来表示：

$$X(t) = \frac{a_0}{2} + \sum_{n=1}^{\infty} (a_n \cos n\omega_0 t + b_n \sin n\omega_0 t), \quad n = 1, 2, \cdots \quad (20)$$

式中
$$a_0 = \frac{2}{T} \int_{-T/2}^{T/2} X(t) \mathrm{d}t, \quad a_n = \frac{2}{T} \int_{-T/2}^{T/2} X(t) \cos n\omega_0 t \, \mathrm{d}t$$

$$b_n = \frac{2}{T} \int_{-T/2}^{T/2} X(t) \sin n\omega_0 t \, \mathrm{d}t, \quad \omega_0 = \frac{2\pi}{T}, \quad T\text{——周期(s)}$$

由试验结果[13]，磁盘偏摆的最大幅值出现在基频 ω_0 附近，对于振动的频率范围，经统计分析，频率超过 600 Hz，振动幅值仅在 0.3 μm 以内，磁盘的小偏摆幅值与频率的数量关系可用一定的拟合公式来表示. 为了从理论上对磁头浮动性能进行分析，对小偏摆下的频率特性响应，可以利用前面的线性理论来计算. 从已知的实验数据及经验公式出发，将各阶小偏摆分量求出，然后乘以头/盘响应系数，即可得到实际的磁头浮动对于小偏摆跟踪响应.

计算结果如下表所示：

F/Hz	40.0	80.0	120.0	160.0	200.0	240.0	280.0
HDA	0.000 00	0.000 02	0.000 05	0.000 08	0.000 13	0.000 19	0.000 25
A/μm	8.074 09	4.037 50	2.691 66	2.018 75	1.615 00	1.345 83	1.153 57
OH/μm	0.000 03	0.000 08	0.000 12	0.000 17	0.000 21	0.000 25	0.000 29
F/Hz	320.0	360.0	400.0	600.0	800.0	1 000.0	2 000.0
HDA	0.000 33	0.000 40	0.000 49	0.000 92	0.001 22	0.001 24	0.004 43
A/μm	1.009 37	0.897 22	0.607 50	0.538 33	0.403 75	0.323 00	0.161 50
DH/μm	0.000 32	0.000 36	0.000 39	0.000 49	0.000 49	0.000 39	0.000 71
F/Hz	3 000.0	4 000.0	5 000.0	6 000.0	7 000.0	8 000.0	9 000.0
HDA	0.010 64	0.017 57	0.029 17	0.041 54	0.055 66	0.071 77	0.089 05
A/μm	0.107 67	0.080 75	0.064 60	0.053 83	0.046 14	0.040 37	0.035 89
DH/μm	0.001 14	0.001 41	0.001 88	0.002 23	0.002 56	0.002 89	0.003 19

其中，F——频率(Hz)，A——各种频率下的偏摆幅值(μm)，HDA——磁头/盘振幅比 ($|\Delta h/a|$)，DH——磁头振动响应(μm). 计算中的主要参数为，磁头系 IBM3370 型，V = 16.75 m/s，h_0 = 0.316 μm，θ_0 = 0.495E－4 rad，转速 2 400 r/min，基频 40 Hz. 若改变转速，或改变偏摆频率结构，则磁头响应的动力特性将有很大变化，因此对磁盘偏摆的频率结构研究是需要进行的工作之一.

五、结论与展望

用有限元法进行磁头静动特性计算，适用范围大，收敛性好. 磁头的几何参数对磁头静动特性影响较大. 在亚微米浮动高度下，分子平均自由程的效应必须考虑. IBM3370 型磁头不仅浮动高度小，且动态特性具有气膜刚度大、共振频率高之优点，是我国磁头设计的一个发展方向. 因此，本文的工作以及进一步的研究工作可以使我国的磁头独立设计能力得到加强，为发展我国的计算机外部设备奠定良好的理论及设计基础. 对于理论分析计算与进一步验证及用于设计指导,可望在今后的工作中得以完成和实现.

本工作得到上海工业大学姚俊源副教授、王汝霖副教授、王保良老师及电子部外部设备所秦如镜高工等的大力支持和帮助,在此表示衷心的感谢.

参 考 文 献

[1] Gastelli, V. and Parrics, J. Review of Numerical Methods in Gas Bearing Film Analysis. Trans. of ASME, J. of Lubr. Tech. , Series F, 1968, Oct: 777 – 792.

[2] White, J. W. and Nigam, A. A Factorial Implicit Scheme for the Numerical Solution of the Reynolds Equation at very Low Spacing. Trans. of ASME, J. of Lubr. Tech. , 1980, 102: 80 – 85.

[3] Burgdofer, A. The Influence of the Molecular Mean Free Path on the Performance of Hydrodynamic Gas Lubricated Bearings. Trans. of ASME, J. of Basic Engineering, 1959, 81: 94 – 100.

[4] Mitsuya, Y. Molecular Mean Free Path Effects in Gas Lubricated Slider Bearings. Bulletin of JSME, 1979, 22(167): 863 – 870.

[5] Mitsuya, Y. Molecular Mean Free Path Effects in Gas Lubricated Slider Bearings (2nd Report, Expermental Studies). Bulletin of JSME, 1981, 24(187): 236 – 242.

[6] Ono, K. Dynamic Characteristics of Air-Lubricated Slider Bearing under Submicron Spacing Conditions. Bulletin of ISME, 1979, 22(173): 1672 – 1677.

[7] White, J. W. Flying Characteristics of the "Zero-Load" Slider Bearings. Trans. of ASME, J. of Lubr. Tech. 1983, 105: 484 – 490.

[8] White, J. W. Flying Characteristics of the 3370-type Slider on $5\frac{1}{4}$-inch Disk. Part 1: Static Analysis. Part 2: Dynamic Analysis. ASLE SP – 16: 72 – 84.

[9] 傅仙罗,王春海. 浮动磁头气动力静态特性. 计算机学报,1985(6): 462 – 469.

[10] 杨勃,杨俊. 5.25英寸温盘磁头浮动特性的研究. 电子计算机外部设备,1986(2): 22 – 26.

[11] 华伟,等. 磁头浮动块动态浮动性能的有限元研究. 西北电讯工程学院学报,1987,14(2): 16 – 30.

[12] 周恒,刘延柱. 气体动压轴承的原理及计算. 国防工业出版社,1981.

[13] 倪裴铭. 磁盘盘面动态特性的测试与分析. 外部设备,1984(1): 33 – 46.

[14] 施文康,等. 磁盘机中头/盘系统对各种不同干扰的动态响应. 电子计算机外部设备,1986(4): 12 – 19.

关于高速内圆磨头主轴-滑动轴承系统稳定性的讨论*

摘 要：本文用线性稳定性理论分析、计算了工件作为主轴附加支撑对系统稳定性的改善作用. 结果表明：这种作用不可忽略，在一定条件下可达到很强烈的程度，在实际计算主轴系统稳定性时必须计入. 附加支撑的方位不同时，所起增稳作用程度亦不同，存在最佳方位.

一、前言

笔者不久前参加了一个高速内圆磨头的鉴定会. 被鉴定的磨头主轴在工作速度下空转时有十个微米左右的振幅，但在磨削时却比较平稳，磨出的工件在光洁度和圆度方面都较好. 这一现象引起了与会者的兴趣，但尚需对这种现象的机理作出合理的阐明. 本文即拟对此作一动力学分析.

磨削工况与空转情况相比，从两个方面改变了主轴-滑动轴承系统的动力性态. 其一为砂轮与工件间的压力和磨削力改变了轴承载荷，影响了轴承的静态工作点（偏心比 ε 和偏位角 θ）从而改变了轴承动特性（刚度和阻尼）. 这个方面是易被认识到和理解的，亦不难对此作定性定量的分析计算，本文不拟对它作进一步讨论.

另一个方面则较易被忽略，迄今亦未见有关的详尽分析文章. 这就是：磨削工况改变了转子的支承条件，从而影响稳定性. 本文将专就这一因素作主要是定性的分析讨论.

图 1 示最简单形式的高速内圆磨头主轴简图. 在空转时，它是一个具有两个支承的转子系统. 但在磨削工件时，工件约束了砂轮在接触法线方向的运动自由，构成了转子的附加支承. 不过这个附加支承不像轴承那样有两维约束，而是可视为一维约束支承. 因此，磨削工作时，转子系统的简化力学模型，即如图 2 所示. 在包含磨削点的轴向平面 zx 中，有三个支承，其中两个为滑动轴承，第三个则由工件对转子的支撑作用构成. 附加支承的刚度主要由磨削

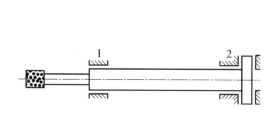

图 1　磨头轴系简图

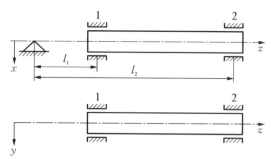

图 2　转子-轴承系统力学模型

---

* 原发表于《上海工业大学学报》，1989，10(2)：95 - 100.

接触部位的接触刚度和工件刚度所构成. 在被鉴定的高速内圆磨头中,其刚度显然比轴承(柔性空气轴承)刚度大得多,可近似视为刚性支点. 在与之垂直的另一轴向平面 yz 中,则仍可视为原来的两支承轴系.

由于本文目的仅在定性地显示并讨论支承条件变化对稳定性的影响,为了简便起见,在下文的分析计算中再假设：转子与轴承相比可视作刚性转子,且质心位于两轴承间之中点上；轴承的刚度和阻尼暂按圆柱轴承的短轴承理论结果计算. 当然,对于其他结构情况作定量分析时,应当用具体的参数条件为依据,不过问题的复杂程度和规模将大得多. 而上述简化假设并不违逆本文主旨,却能以最简洁的方式清楚表达基本的分析逻辑和现象规律.

二、基本关系

转子在 zx 平面中的运动方程为：

$$\frac{m}{2}(\ddot{x}_1+\ddot{x}_2)\frac{l_1+l_2}{2}+(b_{xx}\dot{x}_1+k_{xx}x_1+b_{xy}\dot{y}_1+k_{xy}y_1)l_1$$
$$+(b_{xx}\dot{x}_2+k_{xx}x_2+b_{xy}\dot{y}_2+k_{xy}y_2)l_2=0 \quad (1)$$

$$x_1 l_2 = x_2 l_1 \quad (2)$$

在 yz 平面中,

$$\frac{m}{2}(\ddot{y}_1+\ddot{y}_2)+b_{yx}(\dot{x}_1+\dot{x}_2)+k_{yx}(x_1+x_2)+b_{yy}(\dot{y}_1+\dot{y}_2)+k_{yy}(y_1+y_2)=0 \quad (3)$$

$$b_{yx}\dot{x}_1+k_{yx}x_1+b_{yy}\dot{y}_1+k_{yy}y_1=b_{yx}\dot{x}_2+k_{yx}x_2+b_{yy}\dot{y}_2+k_{yy}y_2 \quad (4)$$

上列式中,m——转子质量；$x_i, y_i (i=1,2)$——转子在轴承 1 和 2 处沿 x 和 y 方向的瞬位移；l_1 和 l_2——砂轮中点到轴承 1 和 2 中点的距离；$b_{ij}, k_{ij} (i,j=x,y)$——轴承的四个阻尼系数和四个刚度系数.

齐次方程组(1)~(4)的解的形式可取为：

$$r = r_a e^{i\gamma} \quad (5)$$

式中 r 代表 x_1, y_1, x_2, y_2；r_a——其复振幅.

将式(5)代入方程组,得非平凡解存在的条件为：

$$\begin{vmatrix} \left[\frac{m}{2}\frac{l_1+l_2}{2}\gamma^2+(b_{xx}\gamma+k_{xx})l_1\right] & \left[\frac{m}{2}\frac{l_1+l_2}{2}\gamma^2+(b_{xx}\gamma+k_{xx})l_2\right] & (b_{xy}\gamma+k_{xy})l_1 & (b_{xy}\gamma+k_{xy})l_2 \\ l_2 & -l_1 & 0 & 0 \\ b_{yx}\gamma+k_{yx} & b_{yx}\gamma+k_{yx} & \frac{m}{2}\gamma^2+b_{yy}\gamma+k_{yy} & \frac{m}{2}\gamma^2+b_{yy}\gamma+k_{yy} \\ b_{yx}\gamma+k_{yx} & -b_{yx}\gamma-k_{yx} & b_{yy}\gamma+k_{yy} & -b_{yy}\gamma-k_{yy} \end{vmatrix}=0$$
(6)

经展开为特征多项式方程并化成无量纲式,得：

$$a_0\bar{\gamma}^5+a_1\bar{\gamma}^4+a_2\bar{\gamma}^3+a_3\bar{\gamma}^2+a_4\bar{\gamma}+a_5=0 \tag{7}$$

式中,$\bar{\gamma}=\gamma/\omega$——无量纲特征数;$\omega$——转动角速度;系数 $a_0\sim a_5$ 为:

$$\begin{aligned}
a_0 &= M^2 B_{xx} \\
a_1 &= M^2 K_{xx}+M(\beta B_2+B_{xx}B_1) \\
a_2 &= M(\beta G+B_{xx}K_1+B_1 K_{xx})+(\beta+1)B_{xx}B_2 \\
a_3 &= M(\beta K_2+K_{xx}K_1)+(\beta+1)(B_{xx}G+B_2 K_{xx}) \\
a_4 &= (\beta+1)(K_{xx}G+B_{xx}K_2) \\
a_5 &= (\beta+1)K_{xx}K_2
\end{aligned} \tag{8}$$

式中,$M=m\psi^3\omega/(2\mu l)$——无量纲转子质量;ψ——轴承相对间隙;μ——润滑剂动力黏度;l——轴承长度;$\beta=\left(\dfrac{l_2-l_1}{l_2+l_1}\right)^2$;$B_1=B_{xx}+B_{yy}$;$K_1=K_{xx}+K_{yy}$;$B_2=B_{xx}B_{yy}-B_{xy}B_{yx}$;$K_2=K_{xx}K_{yy}-K_{xy}K_{yx}$;$G=B_{xx}K_{yy}+B_{yy}K_{xx}-B_{xy}K_{yx}-B_{yx}K_{xy}$;$B_{ij}=\dfrac{b_{ij}\psi^3}{\mu l}$,$K_{ij}=\dfrac{k_{ij}\psi^3}{\mu\omega l}(i,j=x,y)$——轴承的无量纲阻尼和刚度系数.

按短轴承理论结果[1],可推导得(x,y)坐标系中的轴承无量纲动特性系数为(设轴承所受径向载荷沿 y 方向):

$$\begin{aligned}
K_{xx} &= \left(\frac{l}{d}\right)^2 \frac{4\varepsilon[\pi^2(2-\varepsilon^2)+16\varepsilon^2]}{(1-\varepsilon^2)^2[\pi^2(1-\varepsilon^2)+16\varepsilon^2]} \\
K_{xy} &= \left(\frac{l}{d}\right)^2 \frac{\pi[16\varepsilon^4-\pi^2(1-\varepsilon^2)^2]}{(1-\varepsilon^2)^{5/2}[\pi^2(1-\varepsilon^2)+16\varepsilon^2]} \\
K_{yx} &= \left(\frac{l}{d}\right)^2 \frac{\pi[\pi^2(1-\varepsilon^2)(1+2\varepsilon)+32\varepsilon^2(1+\varepsilon^2)]}{(1-\varepsilon^2)^{5/2}[\pi^2(1-\varepsilon^2)+16\varepsilon^2]} \\
K_{yy} &= \left(\frac{l}{d}\right)^2 \frac{4\varepsilon[\pi^2(1-\varepsilon^2)(1+2\varepsilon)+32\varepsilon^2(1+\varepsilon^2)]}{(1-\varepsilon^2)^3[\pi^2(1-\varepsilon^2)+16\varepsilon^2]} \\
B_{xx} &= \left(\frac{l}{d}\right)^2 \frac{2\pi[\pi^2(1+2\varepsilon^2)-16\varepsilon^2]}{(1-\varepsilon^2)^{3/2}[\pi^2(1-\varepsilon^2)+16\varepsilon^2]} \\
B_{xy} &= B_{yx} = \left(\frac{l}{d}\right)^2 \frac{8\varepsilon[\pi^2(1+2\varepsilon^2)-16\varepsilon^2]}{(1-\varepsilon^2)^2[\pi^2(1-\varepsilon^2)+16\varepsilon^2]} \\
B_{yy} &= \left(\frac{l}{d}\right)^2 \frac{2\pi[\pi^2(1-\varepsilon^2)^2+48\varepsilon^2]}{(1-\varepsilon^2)^{5/2}[\pi^2(1-\varepsilon^2)+16\varepsilon^2]}
\end{aligned} \tag{9}$$

式中,d——轴承直径.

根据轴承在工作时的偏心比 ε,即可由上列式算出无量纲动特性系数.

按 Routh-Hurwitz 判稳准则,当 $a_0>0$ 时,要求:

$$a_i>0 \quad (i=1\sim 5) \tag{10}$$

且

$$(a_1 a_2-a_0 a_3)(a_3 a_4-a_2 a_5)-(a_1 a_4-a_0 a_5)^2>0 \tag{11}$$

由此可定出相应的无量纲质量临界值 M_{st}. $M<M_{st}$ 时系统稳定.显然,M_{st} 越高,系

的稳定性越好.

当 $\beta \to 0$ 亦即 $l_2/l_1 \to 1$ 时,工件所起的附加支点作用对系统稳定性的影响趋近于 0. 此时系统蜕化为普通两支承的对称刚性转子情况. 相应的特征方程为:

$$a'_0 \bar{\gamma}^4 + a'_1 \bar{\gamma}^3 + a'_2 \bar{\gamma}^2 + a'_3 \bar{\gamma} + a'_4 = 0 \tag{12}$$

式中系数 $a'_0 \sim a'_4$ 为:

$$a'_0 = M^2;\ a'_1 = MB_1;\ a'_2 = MK_1 + B_2;\ a'_3 = G;\ a'_4 = K_2 \tag{13}$$

稳定性要求:当 $a'_0 > 0$ 时,

$$a'_i > 0 \quad (i = 1 \sim 4) \tag{14}$$

$$a_3(a_1 a_2 - a_0 a_3) - a_1^2 a_4 > 0 \tag{15}$$

由此得无量纲质量临界值为:

$$M_{\text{st}} = \frac{B_1 B_2 G}{G^2 + B_1 K_2 - B_1 K_1 G} \tag{16}$$

当 $\beta = 0$ 时,由式(11)定出的 M_{st} 应当就等于由式(16)定出的值. 这可作为检验计算程序正确性的一个方面.

在短轴承理论结果中,各 B_{ij} 和 K_{ij} 均正比于 $(l/d)^2$,因此 M_{st} 亦正比于 $(l/d)^2$. 由无量纲式(7)~(11)可知,比值 $M_{\text{st}}/(l/d)^2$ 取决于 ε 和 β 这两个无量纲量,前者表征了轴承的静态工作点从而决定着轴承无量纲动特性,后者表征了附加支点位置的效应程度.

三、计算和分析

在 $\varepsilon = 0.1 \sim 0.9$,$\beta = 0 \sim 1$ 的广泛参数范围内作了计算,其结果如图 3 所示. 图中表达 $M_{\text{st}} \sim \varepsilon$ 关系,并以 β 为参变量.

由图可知,附加支承提高了系统的稳定性. β 愈大(即 l_2/l_1 愈大),提高的程度愈大. 例如在 $\varepsilon = 0.2$ 时,$\beta = 0.3$ 的 M_{st} 相对于 $\beta = 0$ 的提高为 134.3%;$\beta = 0.6$ 的为 173.4%;$\beta = 1$ 达到了 237.0%. 在 $\varepsilon = 0.4$ 时,这种效应更显著,与 $\beta = 0$ 的 M_{st} 相比,$\beta = 0.3$ 的提高为 139.0%;$\beta = 0.6$ 的为 189.3%;$\beta = 1$ 的达到了 301.1%.

因为轴承动特性是各向异性的(除 $\varepsilon = 0$ 的情况外),所以附加支点的作用方向不同时,其对系统稳定性的影响亦理应不同. 这从式(8)中当 x 方向附加支承时出现 B_{xx} 和 K_{xx} 即可看出. 为了在数值结果上显示这一点,又假设磨削接触处位于 yz 轴向平面内,按相似于上述的原理,亦作了计算. 其结果如图 4 所示. 由图可知,特别当 ε 较大时,附加支承的增稳作用比位于 zx 平面时强烈得

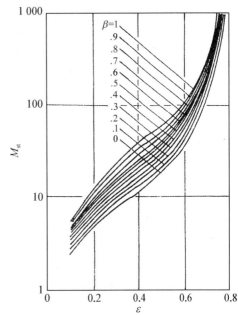

图 3 在 x 方向有一维附加支承的计算结果

多.在 $\varepsilon=0.2$ 时,情况与在 zx 平面时差不多(略有所逊).在 $\varepsilon=0.4$ 时,$\beta=0.3$ 的 M_{st} 提高到 139.2%;$\beta=0.6$ 的到 196.2%;$\beta=1$ 的提高到 390.6%.$\varepsilon=0.6$ 时就更显著了;$\beta=0.3$ 的到 193.1%;$\beta=0.6$ 的到 979.9%,已接近绝对稳定;$\beta=1$ 时达 ∞,已进入绝对稳定区.

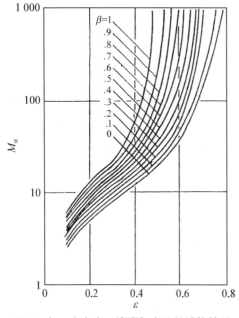

图 4　在 y 方向有一维附加支承的计算结果　　图 5　在 ε 方向有一维附加支承的计算结果

由上可见,在不同的 ε 下,存在最佳的附加支点方位(相对于轴承载荷方向而言).因为轴承在 ε 方向上的刚度和阻尼都接近于最大,所以不难设想,如果在 ε 方向上设附加支承(即在 ε 方向上磨削接触工件),其增稳效果可能接近于最佳.其物理原因是:这个方向的附加支承将最大限度地增强系统的各向异性,从而获得最大增稳效果.为了显示这一点,又针对 ε 方向上设附加支承的情况作了计算,其结果如图 5 所示.相应的 M_{st} 提高到:$\varepsilon=0.2$ 时略高于上述两种支承方向,这是因为 ε 小时轴承动特性的各向异性不强烈;在 $\varepsilon=0.4$ 时,$\beta=0.3$ 的到 178.4%;$\beta=0.6$ 的到 2649.8%,已接近绝对稳定;$\beta=1$ 的到 1897.6%,亦接近绝对稳定;在 $\varepsilon=0.6$ 时,$\beta=0.3,0.6$ 和 1 时均已进入绝对稳定区.这些结果都再现了上述的设想.

四、结论

滑动轴承支承的高速内圆磨头在磨削工件时,除砂轮与工件的压力和磨削力改变了轴承载荷从而影响转子-轴承系统稳定性外,工件所起的附加支承作用亦会显著改变系统稳定性,而且存在最佳支承方位.在最佳方位上附加一维支承,将极有效地起增稳作用.这一知识或许对这类磨头的设计者会有参考价值.

在对砂轮主轴-滑动轴承系统作稳定性分析时,不仅要计算其空转状态下的稳定性态(例如 M_{st} 值),更要计算其磨削工况下的性态,此时应计入附加支承的作用.

需再次说明的是:本文只是为说明作用机理而作的一种主要是定性的分析.如需针对具体条件作更确切的分析计算,应当计入更切实的参数条件,包括转子的弹性和质量分布、轴承的更准确的动特性值以及附加支承的刚度,或许还有阻尼.

参 考 文 献

[1] 张直明. 滑动轴承的流体动力润滑理论. 北京：高等教育出版社,1986：78 - 79.

A Study on Stability of Spindle-Bearing System of High Speed Inned Grinder

Abstract: The beneficial effect of a work piece as an additional support to the main spindle on the stability of the rotor-bearing system is analyzed and calculated. The results show that this effect is unnegligible, and may become very strong under certain circumstances. Therefore it must be taken during practical calculation of stability of the spindle system. The effectiveness of an additional support on stabilization varies with its dierction of action. There exists an optimal direction.

浮环轴承支承的转子系统的
静态特性和动态稳定性研究*

摘　要：本文研究了浮环轴承支承的转子系统的静态特性和动态稳定性,用能量方程计算油膜的黏度变化.用数值法将运动方程的特征行列式展开为多项式.理论计算结果与现有实验结果基本一致.对计算结果作了一些分析讨论.

符 号 说 明

b_{ij}	动态阻尼系数	T_{E_o}	相对温度因子 $T_{E_o} = \rho C_V \psi_1^2/(\mu_0 \omega)$
c	半径间隙	W	轴承载荷
D	轴承直径	a	环轴转速比 $a = \omega_2/\omega_1$
F_c	剪切流摩擦力	β	内外层间隙比 $\beta = c_2/c_1$
F_P	压力流摩擦力	δ	内外层半径比 $\delta = R_2/R_1$
H	无量纲油膜厚度	μ	润滑油动力黏度
k_{ij}	动态刚度系数	ω	角速度
k_s	轴刚度	ε	轴承偏心率
L	轴承宽度	ψ	半径间隙比 $\psi = c/R$
P	油膜压力	Π	$\Pi = \bar{P}H^{3/2}$
R	轴承半径	ϕ	相对柔度系数 $\phi = W/(k_s C_1)$
S	承载量系数 $S = W\psi^2/(\mu\omega LR)$		
T	油膜温度		

一、前言

在以往的理论研究中,对于浮环轴承的油膜黏度均作了一些简化的假设,Li 和 Rohde[1],Orcutt 和 Ng[2]等人假定油膜黏度为常数且内外层黏度相等,陆红涛等[3]假定内外层黏度比为1.1.以上这些文献的计算结果与实验值均有误差,高速时尤为显著.周晓光和葛中民[4]用热平衡方法分别计算内外层油膜黏度,Nikolajsen[5]用短轴承理论结合能量方程计算内外层油膜黏度,误差有所减小.但由于假设条件与实际情况仍有一定差距,所以计算结果仍有误差.本文采用有限长轴承理论和雷诺边界条件代替短轴承理论,用能量方程、粘温关系和雷诺方程联立求解计算,计算结果与现有实验结果进行了比较,获得了较好的一致性.

国内外对浮环轴承-转子系统的动态稳定性已进行了大量的研究[5,6].本文在考虑油膜黏度变化的基础上,采用较为简便和准确的计算方法对这一系统的稳定性进行了分析计算,并作了一些机理性的探讨.

另外,本文还讨论了浮环的起浮性能,以线图的形式给出计算结果,并作了一些讨论.

---

* 本文合作者：屠尉立.原发表于《上海工业大学学报》,1989,10(4)：302-309.

二、静特性计算

在高速轴承中一般认为油流带走的热量远大于传导散热,而且轴承向周围介质散热的规律颇为复杂.为简便起见,假设系统为绝热流动,即全部热量均由油吸收而从两侧泄出带走.同时根据一般轴承的计算结果和实测结果,温度沿轴向变化相对很小,因此可以认为油温仅沿周向变化.

假定润滑油不可压缩、层流状态、忽略润滑油的重力和惯性力的影响,在静态平衡状态时,轴承中的油膜压力分布可用如下以 Π 函数表示的无量纲雷诺方程来描述.

$$\frac{\partial^2 \Pi}{\partial \varphi^2} - \frac{1}{\bar{\mu}}\frac{\partial \bar{\mu}}{\partial \varphi}\frac{\partial \Pi}{\partial \varphi} - \frac{3}{4}\left[\frac{2}{H}\frac{\partial^2 H}{\partial \varphi^2} + \left(\frac{1}{H}\frac{\partial H}{\partial \varphi}\right)^2 - \frac{2}{H\bar{\mu}}\frac{\partial H}{\partial \varphi}\frac{\partial \bar{\mu}}{\partial \varphi}\right]\Pi$$
$$+ \left(\frac{D}{L}\right)^2 \frac{\partial^2 H}{\partial \lambda^2} = 3\bar{\mu}H^{-3/2}\frac{\partial H}{\partial \varphi} \tag{1}$$

润滑油的粘温关系用美国材料试验学会(ASTM)的线图拟合而成的近似公式表示.

$$\lg\lg(\gamma + 0.6) = A + B\lg T \tag{2}$$

式中,γ 为润滑油运动黏度(cm/s);

T 为绝对温度(K),A,B 为常数.

假定润滑油定压比热和密度不变,能量方程可用如下一维积分形式表示

$$\bar{T} = \int_0^\varphi \frac{\frac{\bar{\mu}}{H} + \frac{H^3}{3\bar{\mu}}\left(\frac{\partial \bar{P}}{\partial \varphi}\big|_{\lambda=0}\right)^2}{H - \frac{H^3}{3\bar{\mu}}\frac{\partial \bar{P}}{\partial \varphi}\big|_{\lambda=0}} d\varphi + \bar{T}_o \tag{3}$$

式中,$\bar{T} = \frac{\rho C_r \psi^2}{\mu_0 \omega}T$;$\bar{T}_0$ 为无量纲进油温度;ρ 为润滑油密度;C_r 为滑润油定容压热.

采用雷诺边界条件,用数值计算法联立求解方程(1)~(3).解出 $\bar{P}$,$\bar{\mu}$ 和 $\bar{T}$ 的分布,就可以求出轴承的静态特性.

浮环轴承正常工作时,根据平衡条件,作用在浮环内外层上的油膜力应大小相等,方向相反,而作用在浮环内外层表面上的摩擦力矩之和应等于零.即:

$$\frac{S_1 \mu_{01}(\omega_1 + \omega_2)L_1 R_1}{\psi_1^2} = \frac{S_2 \mu_{02}\omega_2 L_2 R_2}{\psi_2^2} \tag{4}$$

$$\frac{\mu_{01}L_1 R_1^3}{c_1}\left[(\omega_1 - \omega_2)\bar{F}_{C1} - (\omega_1 + \omega_2)\bar{F}_{P2}\right] = \frac{\mu_{02}L_2 R_2^3 \omega_2}{c_2}\left[\bar{F}_{C2} + \bar{F}_{P2}\right] \tag{5}$$

同时,内层油膜力应与外载荷平衡,即:

$$W_1 = \frac{S_1 \mu_{01}(\omega_1 + \omega_2)L_1 R_1}{\psi_1^2} \qquad (6)$$

式中,下标 1 表示内层油膜,2 表示外层油膜.

方程(4)~(6)表示了 ε_1,ε_2 和 ω_2 之间的关系. 在给定轴承参数,润滑油黏度,工作参数和载荷的条件下,联立求解上述三个方程,就可求出内、外层油膜偏心率 ε_1 和 ε_2,以及浮环转速 ω_2.

如把整个系统视为一个当量圆柱轴承系统,则这个当量轴承系统的承载量系数 So 为:

$$So = \frac{W_1 \psi_1^2}{\mu_{01} \omega_1 L_1 R_1} = (1+a)S_1 \qquad (7)$$

方程(4)~(6)是一个关于 ε_1,ε_2 和 a 的非线性方程组,求解时包含雷诺方程的求解. 本文采用拟牛顿法求解这一非线性方程组. 这一方法具有超线性收敛速度,不需要求导数值,对初值要求不高等特点. 经计算实践证明,这是一种十分行之有效的计算方法.

浮环轴承正常工作的基础是浮环的稳定运转. 在轴启动运转时,浮环并不转动. 在轴速达到一定数值,油膜作用在浮环内表面上的摩擦力矩大于浮环外表面与轴承座之间的摩擦力矩时,浮环才开始转动. 设 ω_s 为浮环起动时轴的角速度,则得:

$$\frac{\mu_{01} R_1^3 L_1 \omega_s}{c_1}[\overline{F}_{C1} - \overline{F}_{P1}] = W_1 f_2 R_2 \qquad (8)$$

式中,f_2 是浮环与轴承座之间的摩擦系数.

式(8)可转化成如下的无量纲形式:

$$So_s \cdot f_2 \cdot \frac{R_2}{R_1} \cdot \frac{R_1}{c_1} = \overline{F}_{C1} - \overline{F}_{P1} \qquad (9)$$

式中,So_s 是浮环起浮时的当量轴承承载量系数.

$$So_s = \frac{W_1 \psi_1^2}{\mu_{01} \omega_s L_1 R_1} \qquad (10)$$

当轴承静特性参数确定后,利用式(9)和(10),可求得 ω_s.

三、稳定性计算

虽然实际的浮环轴承-转子系统往往不是单质量系统. 但为了简便起见,以探讨基本规律为目的的理论分析均以单质量对称系统作为研究对象,并不失其一般性.

转子系统如图 1 所示. 假定系统在平衡位置附近作小扰动振动,以 (X, Y)、(x_1, y_1) 和 (x_2, y_2) 分别代表转子、轴颈和浮环的位移,则根据线性理论,系统的运动方程如下:

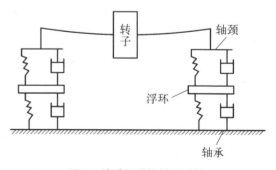

图 1 单质量弹性转子系统

$$\begin{aligned}
&m_1\ddot{X} + k_s(X - x_1) = 0 \\
&m_1\ddot{Y} + k_s(Y - y_1) = 0 \\
&k_s(X - x_1) = k_{xx1}(\dot{x}_1 - \dot{x}_2) + k_{xy1}(\dot{y}_1 - \dot{y}_2) + b_{xx1}(\dot{x}_1 - \dot{x}_2) + b_{xy1}(\dot{y}_1 - \dot{y}_2) \\
&k_s(Y - y_1) = k_{yx1}(\dot{x}_1 - \dot{x}_2) + k_{yy1}(\dot{y}_1 - \dot{y}_2) + b_{yx1}(\dot{x}_1 - \dot{x}_2) + b_{yy1}(\dot{y}_1 - \dot{y}_2) \\
&m_2\ddot{x}_2 + k_{xx2}x_2 + k_{xy2}y_2 + b_{xx2}\dot{x}_2 + b_{xy2}\dot{y}_2 + k_{xx1}(x_2 - x_1) + k_{xy1}(y_2 - y_1) \\
&\quad + b_{xx1}(\dot{x}_2 - \dot{x}_1) + b_{xy1}(\dot{y}_2 - \dot{y}_1) = 0 \\
&m_2\ddot{y}_2 + k_{yx2}x_2 + k_{yy2}y_2 + b_{yx2}\dot{x}_2 + b_{yy2}\dot{y}_2 + k_{yx1}(x_2 - x_1) + k_{yy1}(y_2 - y_1) \\
&\quad + b_{yx1}(\dot{x}_2 - \dot{x}_1) + b_{yy1}(\dot{y}_2 - \dot{y}_1) = 0
\end{aligned} \quad (11)$$

方程组(11)的求解是很复杂的,牵涉到阶数较高的特征行列式展开和高阶多项式判稳等数学问题. Tanaka[6]采用解析法展开,计算烦琐,而且容易出错,阶数很高时无法采用. 本文采用[7]中介绍的特征行列式数值展开法和[8]中介绍的交叉除法判稳,迅速而准确地解决了这一问题.

四、计算结果及讨论

4.1 静特性计算结果

图2表明,油膜黏度随着转速升高而降低. 转速越高,下降越多. 由图3可以看出,假设粘度为常数时,转速比 a 随着轴颈转速 N_1 的增大基本不变. 而考虑黏度变化时,a 随 N_1 的上升而下降. 由于内层有效旋转速度(N_1+N_2)大于外层有效旋转速度 N_2,而且随着 N_1 的上升,二者差距越来越大. 所以油膜黏度的下降内层大于外层(图2),使得由于黏度下降而造成的内层摩擦力的下降大于同样原因造成的外层摩擦力的下降. 所以,随着 N_1 的上升,环轴转速比 a 不断下降. 这一计算结果同国内外许多文献的实验结果是一致的.

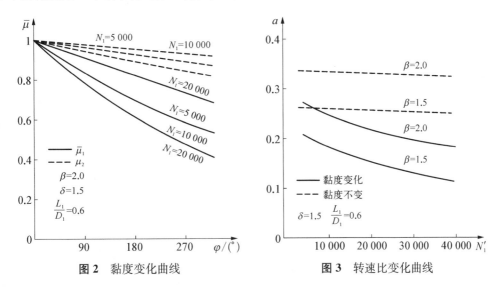

图2 黏度变化曲线　　　图3 转速比变化曲线

图4是计算与实验结果对比. 以本文方法并采用[3]的系统实验参数进行了计算,其结果与[3]的计算和实验结果进行了比较. 由图可见,本文计算结果与实验值的变化趋势是一

致的,与[3]的计算结果相比,误差有明显改善,这是由于采用了能量方程确定内外层油膜黏度之故,而在高转速和大间隙比时,计算精度有更明显的提高.

图 5 是浮环起浮转速与外层摩擦系数之间的关系.从图中可以看出,在一定参数条件下,So_s 与 f_2 基本上成反比关系.由于其他参数不变时,So_s 与浮环起浮时的轴速 ω_s 也成反比关系,所以 ω_s 与 f_2 基本上成正比关系,摩擦系数小的材料配对将使浮环易于起浮.同时,So_s 中的其他参数对 ω_s 也有影响,减小 ψ_1 值,减小轴承比压和提高润滑油黏度均能提高浮环的起浮性能.从图中还可以看到,内外层半径比 δ 对浮环起浮性能也有影响,δ 值小则 So_s 值大,有利于浮环起浮.这是由于外层半径的减小使外层摩擦力矩下降,从而减小浮环时的阻力矩.

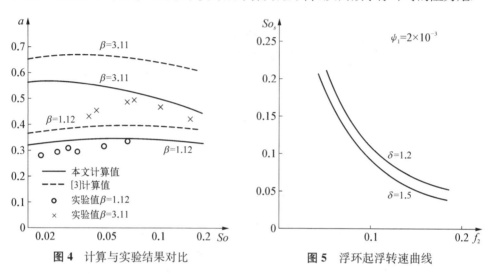

图 4　计算与实验结果对比　　　　图 5　浮环起浮转速曲线

4.2　稳定性计算结果

为了能直接查知系统的失稳转速,本文的稳定图谱横坐标采用以刚支临界转速为参考角速度的承载量系数 So_k.

$$So_k = So \cdot \frac{\omega_1}{\omega_k} = \frac{W_1 \psi_1^2}{\mu_{01} \omega_k L_1 R_1} \tag{12}$$

式中,
$$\omega_k = \sqrt{\frac{k_s}{m_1}}$$

图 6 是不同参数条件下的稳定图谱.从图中可以看出,稳定边界存在突变,这是稳定模态发生转换的结果.根据[2]的实验结果,内外层油膜以各自的涡动频率发生涡动,其幅值也不同,也就是说系统中存在二种涡动模式.但是,内外层油膜是相互影响的,尤其是外层油膜的阻尼作用能胜过内层油膜的激振作用,从而大大增强内层油膜的稳定性.失稳是整个系统的失稳,并由其中某一层油膜起主导作用.在 So_k 值较大的区域内,内层油膜起主导作用,当转速达到某一临界值时,外层油膜的阻尼作用已无法抑制内层油膜的激振作用,从而导致整个系统失稳.在 So_k 值较小的区域内,外层油膜起主导作用,当转速达到某一临界值时,外层油膜不但不起抑振作用,反而产生一种激振作用,从而导致系统的失稳.两个区域边界的交点就形成了系统稳定边界的突变点.

从图 6(a)中可以看到,β 值越小,稳定边界发生突变时的 So_k 值越大.图中 $\beta=1.0$ 的稳

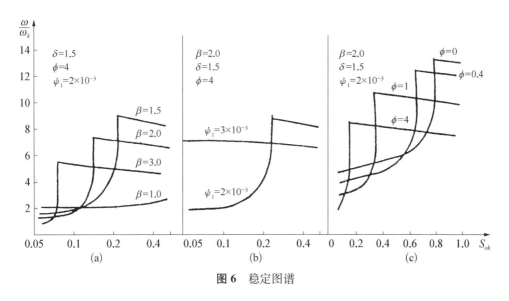

图 6 稳定图谱

定边界突变点已超出图中的 S_{ok} 值范围. 在内层间隙不变的条件下, β 值小就意味着外层偏心率 ε_2 小, 因此可以认为系统的稳定模态转换与外层油膜的偏心率有关. ε_2 值越小, 模态转换处的 S_{ok} 值就越大. 这一结论可以从图 6(b) 中得到进一步证实. 在图 6(b) 中, 由于 β 和 δ 值均保持不变, ψ_1 值增加就意味着 c_2 的增加, 也就是 ε_2 增加. 当 ψ_1 值增至 3×10^{-3} 时, 系统稳定边界在图示 S_{ok} 值范围内未发生突变, 稳定边界一直由内层油膜起主导作用. 从图 6(c) 中可以发现, 系统稳定边界的转换与轴的刚度有关, 刚度越大, 稳定边界发生突变时的 S_{ok} 值也就越大. 这同[6]的计算结果是一致的.

从图 6(a) 中也可以看到, 当系统稳定性由内层油膜起主导作用时, β 值越小, 系统越稳定. 这是由于外层油膜的阻尼作用不同而引起的. 当 β 值小时, 外层间隙小, 外层油膜的阻尼作用大, 所以系统稳定性较好. β 值对系统稳定性影响很大, 这说明外层油膜的阻尼作用对系统稳定性影响很大. 图 6(c) 的结果进一步论证了上述分析, 在系统稳定性由内层油膜起主导作用时, ψ_1 值较大, 系统稳定性较差. 虽然此时内层偏心率上升了, 但由于外层间隙增大, 外层油膜阻尼减弱, 从而使系统稳定性下降.

为了进一步验证本文计算方法的准确性, 以本文方法并采用[6]的实验参数进行了计算, 并与实验值进行了比较. [6]中的实验装置是一个单质量弹性对称系统, 主要参数为: 轴颈 $D_1 = 25$ mm, 浮环直径 $D_2 = 33$ mm, 轴承宽度 $L = 13.15$ mm.

计算结果如表 1 所示. 由表可见计算结果与实验值基本相符. 只是编号 9 的误差较大, 但[6]中的误差更大.

表 1 理论结果与实验结果比较

编号	c_1 /mm	β	ψ	P_m /(kg·cm^{-2})	失稳转速/(r·min^{-1})		
					[6]理论解	本文理论解	实验值
1	0.019	3.84	11.2	0.59	5 830	5 469	5 300
2	0.019	3.63	11.2	0.59	6 000	5 781	5 400
3	0.020	3.10	10.6	0.59	6 420	6 406	5 600
4	0.020	2.50	10.6	0.59	7 150	7 031	6 200
5	0.018	2.28	11.8	0.59	7 610	7 656	7 100

(续表)

编号	ε_1/mm	β	ψ	P_m/(kg·cm^{-2})	失稳转速/(r·min^{-1}) [6]理论解	本文理论解	实验值
6	0.031	2.33	6.86	0.59	7 090	6 719	6 900
7	0.041	1.77	5.20	0.59	7 950	7 656	7 500
8	0.053	1.98	4.0	0.59	7 190	7 719	7 600
9	0.031	2.33	12.0	1.00	5 740	6 406	8 600

五、结论

1. 在高速浮环轴承中,润滑油黏度变化较大,内外层油膜黏度差随着转速的升高而不断增大.因此,假设黏度不变或内外层黏度比为常数将会产生较大误差.

2. 浮环的起浮性能同浮环与轴承间的摩擦系数,内层间隙,轴承比压和内外层半径比等参数有关.

3. 浮环轴承-转子系统中存在两种涡动模态,内外层油膜均可能作为主导因素而使系统失稳.两种失稳模态的转换与外层偏心率和轴的刚度有关.

4. 外层油膜的阻尼作用对系统稳定性有很大影响,而外层油膜阻尼作用的大小取决于外层间隙.

参 考 文 献

[1] Li, C. H, Rohde, S. M. On the steady state and dynamic performance charactcristic of floating ring dearings. Trans. ASME. J. Lubr. Tech., 1981(3): 389-397.

[2] Orcutt. F. K, Ng. G. W. Steady-state and dynamic properties of the floating ring journal bearing. Trans. ASME. SERIES F, 1986(2): 243-253.

[3] 陆红涛,等. 浮环轴承的静特性研究. 上海交通大学,1980.10.

[4] 周晓光,葛中民. 浮环轴承的静态特性研究. 润滑与密封,1985(5): 14-20.

[5] Nikolajsen, I. L. The effects of variable viscosity on the stability of plain journal bearings and floating-ring journal bearings. Trans. ASME. J. Lubr. Tech., 1973(4): 447-456.

[6] Tanaka. M., Hori. Y. Stability charactcistics of floating bush bearings. Trans. ASME. J. Lubr. Tech., 1972(3): 248-259.

[7] 张直明. 滑动轴承的流体动力润滑理论. 高等教育出版社,1986.

[8] 张直明. 计入支点弹性和阻尼时可值瓦轴承支撑的转子系统的动力稳定性. 机械工程学报,1983(2): 9-21.

[9] 屠尉立,张直明. 轴向进油对浮环轴承特性和稳定性的影响. 上海工业大学学报,1988(2): 144-151.

The Static Charactistics and Dynamic Stability of Rotor Systems Supported by Floating Ring Bearings

Abstract: This paper deals with the static characteristics and dynamic stability of rotor systems supported by floating ring bearings. The effects of variable viscosity are determined by means of energy equation. The characteristic determinant of the equations of the motion is developed into a polynomial by numerical method. The theoretical results are consistent with the existing experiments. Some discussions are given to the results.

复套式转子-轴承系统的固有复频率计算[*]

摘　要：本文介绍一种计算复套式转子-滑动轴承系统固有复频率的方法.用正向和反向隐式传递矩阵法建立特征行列式,用有限复平面轨迹显示及三角元折半法搜索特征根,形成了相应的有效方法和保证计算精度的程序.用锤击模型试验验证了计算,获得满意的符合.

一、前言

现代某些超大功率齿轮箱中,为了增大输入、输出轴的扭转柔度以减小冲击和振动的传递,同时又不加长轴以节省传动装置的轴向地位,采用了复套式转子(图1).其中细而长的挠性传动轴穿过空心的齿轮轴,内外轴在右端相连,内轴的左端则与外界相连传递扭矩.外轴有两个滑动轴承,内轴左端有一个滑动轴承.这样就使转子有较大的扭转柔度,套轴又有足够的弯曲刚度来支承齿轮以确保良好啮合,整体又构紧凑.

现有的传递矩阵法转子动力学程序[1],多数不能直接用来计算这种转子-轴承系统的固有复频率.为此,作者拟订了相应的计算方法和程序.并在简单的模型上作了实验验证.

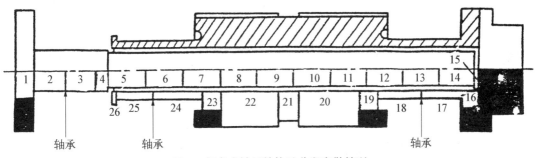

图1　复套式转子结构及分段离散情况

二、计算方法

2.1　特征行列式的建立

将转子离散化为若干段等直径的分段,并将其质量集中置于该分段两端节点上.每个节点上,有八个分量构成的状态矢量：

$$R_j = (x, \varphi, M, S, y, \Psi, N, Q)_j^T \tag{1}$$

式中(x, φ, M, S)和(y, Ψ, N, Q)分别为水平和垂直平面内的(位移、转角、弯矩、切力).

---

* 本文合作者：王云根.原发表于《上海工业大学学报》,1990,11(1)：27-35.

在内轴左端面,弯矩和切力均为零,可记为:

$$\boldsymbol{R}_0 = \boldsymbol{B}_0 \cdot (x_0, \varphi_0, y_0, \Psi_0)^T = \boldsymbol{B}_0 \cdot \boldsymbol{R}_0^* \quad (2)$$

式中

$$\boldsymbol{B}_0 = \begin{pmatrix} 1 & 0 & 0 & 0 \\ 0 & 1 & 0 & 0 \\ 0 & 0 & 0 & 0 \\ 0 & 0 & 0 & 0 \\ 0 & 0 & 1 & 0 \\ 0 & 0 & 0 & 1 \\ 0 & 0 & 0 & 0 \\ 0 & 0 & 0 & 0 \end{pmatrix} \quad (3)$$

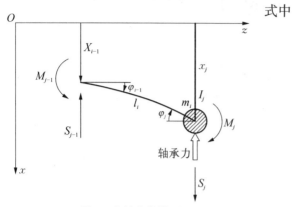

图 2 内轴分段模型

$$\boldsymbol{R}_0^* = (x_0, \varphi_0, y_0, \Psi_0)^T$$

传递过程从内轴左端开始,逐个分段用该段左端状态量 $\boldsymbol{R}_{j-1}$ 表达右端状态量 $\boldsymbol{R}_j$. 到内轴右端后,进入外轴反向传递,直到其左端为止.对于每一分段,有:

$$\boldsymbol{R}_j = \boldsymbol{A}_j \cdot \boldsymbol{R}_{j-1} \quad (j = 1 \sim n) \quad (4)$$

式中 n 为转子分段数.内轴分段(图2)的传递矩阵为:

$$\boldsymbol{A}_j = \begin{pmatrix} 1 & l_j & b_j & -c_j & 0 & 0 & 0 & 0 \\ 0 & 1 & a_j & -b_j & 0 & 0 & 0 & 0 \\ 0 & I_j\gamma^2 & 1+a_jI_j\gamma^2 & -l_j-b_jI_j\gamma^2 & 0 & 0 & 0 & 0 \\ F_j & l_jF_j & b_jF_j & 1-c_jF_j & G_j & l_jG_j & b_jG_j & -c_jG_j \\ 0 & 0 & 0 & 0 & 1 & l_j & b_j & -c_i \\ 0 & 0 & 0 & 0 & 0 & 1 & a_j & -b_j \\ 0 & 0 & 0 & 0 & 0 & I_j\gamma^2 & 1+a_jI_j\gamma^2 & -l_j-b_jI_j\gamma^2 \\ H_j & l_jH_j & b_jH_j & -c_jH_j & K_j & l_jK_j & b_jK_j & 1-c_jK_j \end{pmatrix}$$

(5)

其中 γ 为复特征值,即复频率; l_j 为分段长; I_j 为轴质惯矩; $a_j = l_j/(EJ_j)$; $b_j = l_j^2/(2EJ_j)$; $c_j = l_j^3/(6EJ_j)$; E 为弹性模量; J_j 为截面轴惯矩; m_j 为集中质量(包括轴段的及附加的); $F_j = m_j\gamma^2 + b_{xx}\gamma + k_{xx}$; $G_j = b_{xy}\gamma + k_{xy}$; $H_j = b_{yx}\gamma + k_{yx}$; $K_j = m_j\gamma^2 + b_{yy}\gamma + k_{yy}$; $b_{ij}(i,j=x,y)$ 为轴承阻尼系数; $k_{ij}(i,j=x,y)$ 为轴承刚度系数.

在内轴的末一分段右端(图3),计入内、外轴的联接部分结构后,有外悬的集中质量(外悬距 e),且发生传递方向的改变.其传递矩阵为:

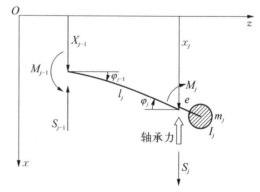

图 3 内轴末分段模型

$$\boldsymbol{A}_j = \begin{bmatrix} 1 & l_j & b_j & -c_j \\ 0 & 1 & a_j & -b_j \\ m_je\gamma^2 & (I_j+m_je^2+m_jl_je)\gamma^2 & 1+(a_jI_j+b_jm_je)\gamma^2 & -l_j-(b_jI_j+c_jm_je)\gamma^2 \\ F_j & l_jF_j+m_je\gamma^2 & b_jF_j+a_jm_je\gamma^2 & 1-c_jF_j-b_jm_je\gamma^2 \\ 0 & 0 & 0 & 0 \\ 0 & 0 & 0 & 0 \\ 0 & 0 & 0 & 0 \\ H_j & l_jH_j & b_jH_j & -c_jH_j \\ 0 & 0 & 0 & 0 \\ 0 & 0 & 0 & 0 \\ 0 & 0 & 0 & 0 \\ G_j & l_jG_j & b_jG_j & -c_jG_j \\ 1 & l_j & b_j & -c_j \\ 0 & 1 & a_j & -b_j \\ m_je\gamma^2 & (I_j+m_je^2+m_jl_je)\gamma^2 & 1+(a_jI_j+b_jm_je)\gamma^2 & -l_j-(b_jI_j+c_jm_je)\gamma^2 \\ K_j & l_jk_j+m_je\gamma^2 & b_jK_j+a_jm_je\gamma^2 & 1-c_jK_j-b_jm_je\gamma^2 \end{bmatrix}$$

(6)

外轴轴段(图 4)则为由右向左传递,其传递矩阵为:

$$\boldsymbol{A}_j = \begin{bmatrix} 1 & -l_j & -b_j & -c_j & 0 & 0 & 0 & 0 \\ 0 & 1 & a_j & b_j & 0 & 0 & 0 & 0 \\ 0 & I_j\gamma^2 & 1+a_jI_j\gamma^2 & l_j+b_jI_j\gamma^2 & 0 & 0 & 0 & 0 \\ F_j & -l_jF_j & -b_jF_j & 1-c_jF_j & G_j & -l_jG_j & -b_jG_j & -c_jG_j \\ 0 & 0 & 0 & 0 & 1 & -l_j & -b_j & -c_j \\ 0 & 0 & 0 & 0 & 0 & 1 & a_j & b_j \\ 0 & 0 & 0 & 0 & 0 & I_j\gamma^2 & 1+a_jI_j\gamma^2 & l_j+b_jI_j\gamma^2 \\ H_j & -l_jH_j & -b_jH_j & -c_jH_j & K_j & -l_jK_j & -b_jK_j & 1-c_jK_j \end{bmatrix}$$

(7)

这样,任一节点的状态量均可由 $\boldsymbol{R}_0^*$ 来表达:

$$\boldsymbol{R}_1 = \boldsymbol{A}_1 \cdot \boldsymbol{B}_0 \cdot \boldsymbol{R}_0^* = \boldsymbol{B}_1 \cdot \boldsymbol{R}_0^*;$$
$$\boldsymbol{R}_2 = \boldsymbol{A}_2 \cdot \boldsymbol{B}_1 \cdot \boldsymbol{R}_0^* = \boldsymbol{B}_2 \cdot \boldsymbol{R}_0^*; \cdots;$$
$$\boldsymbol{R}_n = \boldsymbol{A}_n \cdot \boldsymbol{B}_{n-1} \cdot \boldsymbol{R}_0^* = \boldsymbol{B}_n \cdot \boldsymbol{R}_0^*.$$

每次运算均为 8 行 8 列的矩阵 $\boldsymbol{A}_j$ 与 8 行 4 列的矩阵 $\boldsymbol{B}_{j-1}$ 相乘. 矩阵的每个元素都是 γ 的多项式. 随着传递过程的进行, $\boldsymbol{B}_j$ 矩阵元素的多项式阶数便越来越高.

在外轴左端,弯矩和切力均为零. 由此,可

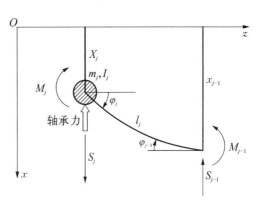

图 4 外轴分段模型

选出 $\boldsymbol{B}_n$ 矩阵中与 M_n，S_n，N_n，Q_n 相应的第三、四、七、八行，构成 4 行 4 列的特征行列式：

$$D = \begin{pmatrix} b_{31} & b_{32} & b_{33} & b_{34} \\ b_{41} & b_{42} & b_{43} & b_{44} \\ b_{71} & b_{72} & b_{73} & b_{74} \\ b_{81} & b_{82} & b_{83} & b_{84} \end{pmatrix} \tag{8}$$

式中 b_{ij} 表示 $\boldsymbol{B}_n$ 矩阵中第 i 行第 j 列的元素.

由 $D=0$ 的条件，确定 γ 的各个复根，就得到了系统的各个复固有频率.

2.2 复特征值的寻求

以 u 和 v 分别表示 γ 的实部和虚部：$\gamma = u + iv$. 以转动角速度 ω 为参考量，则其相对值为 $U = u/\omega$，$V = U/\omega$. 我们对 V 值的感兴趣范围，通常在 $0 \sim 3$ 之间，因为共振和倍频共振相应于 $V=1$ 和 2，而油膜稳定性问题则一般总在 $V<1$ 的区域内. U 的感兴趣范围取在 $-2 \sim 2$ 之间似亦足够了，因 $U < -2$ 时系统阻尼和对数衰减率极大，这种复模态不致引起麻烦，而 $U > 2$ 则意味着系统早已丧失稳定性，如果我们从必定是稳定的低角速 ω 开始计算，逐步提高 ω 算到开始失去稳定为止，一般可避免发生 $U > 2$ 的极少见情况.

在 $U + iV$ 复平面的这个限定区域内，以 $\Delta U (= 0.1$ 或 $0.2)$ 和 ΔV（同 ΔU）等分为网格，计算每一网点上的 D 复值. 将 $\mathrm{Re}(D)$ 和 $\mathrm{Im}(D)$ 的符号分布情况打印出来，就得到了 $\mathrm{Re}(D) = 0$ 和 $\mathrm{Im}(D) = 0$ 的粗略轨迹（图 5）. 这两种轨迹的交点上，$\mathrm{Re}(D) = \mathrm{Im}(D) = 0$，相应的 $U + iV$ 即为一复根.

为了以足够精度算出 U 和 V 的根值（我们取为 0.000 1），将含有两种轨迹交点的小区域选出，然后用三角元折半法逼近到规定精度. 根据每一小单元四角网点上 $\mathrm{Re}(D)$ 符号和 $\mathrm{Im}(D)$ 符号是否均存在＋号及－号，即可认为该小单元是否应属"可疑"，对此即应进一步分割搜索. 这种判断法偶尔亦会误判，即漏过了含交点的小单元而抓出了其邻近单元，这是由于网点分布不很密之故. 进一步的搜索，在错抓出的小单元中将得不到根. 好在已打印了 $U + iV$ 的复平面中 $\mathrm{Re}(D)$ 和 $\mathrm{Im}(D)$ 的轨迹图，不难判断出漏检，并在旁邻区域内将交点搜出.

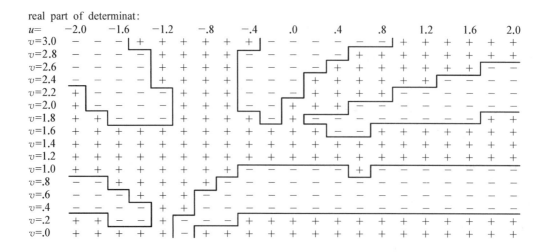

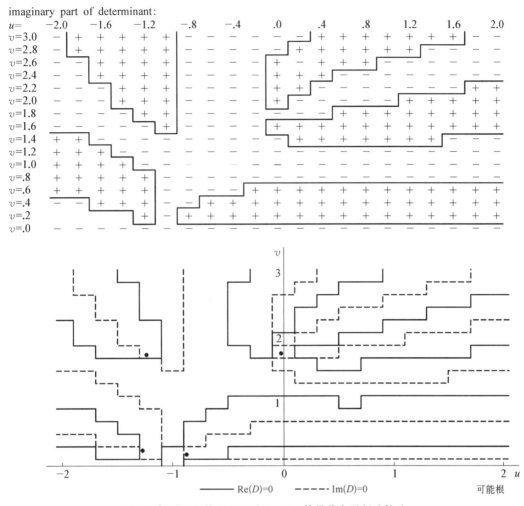

图 5 复平面上的 Re(D) 和 Im(D) 符号分布及粗略轨迹

三角元折半搜索,如下进行(图 6). 先将抓出的小矩形单元分成两个三角形. 根据每个三角形的三个角点上 Re(D) 和 Im(D) 是否均存在 + 号及 − 号,判定其是否可疑. 将可疑三角形的斜边中点与直角顶点相连,即划分成两个三角形补充计算斜边中点上的 D 复值,又可按上述法分别判断这两个分三角形是否可疑. 以此类推,不断进行到分三角形的边长小于规定精度,取可疑三角形中点作为复根值.

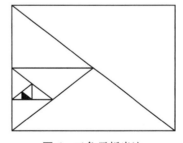

图 6 三角元折半法

三角元折半搜索法的优点,是每次折半细分只需补充计算一个 D 复值.

2.3 对称转子-轴承系统的节约处理

本文的计算方法和程序,亦适用于普通非复套式的转子系统. 对于对称的普通转子系统,可以在中央截开,只取右半部计算. 这样可减少分段,大大降低 D 所含的 γ 阶数,减少计算时间而提高精度.

此时应分别按"柱型"模态(左、右对称)和"锥型"模态(左、右反对称)作不同计算.

对于柱型计算,中央截分节点上 φ, S, Ψ, Q 均为零,此时将 R_0^* 改为 $(x, M, y, N)^T$,相应的 B_0 按式(9).对于锥型计算,中央截分节点上 x, M, y, N 均为零,此时将 R_0^* 改为 $(\varphi, S, \Psi, Q)^T$,相应的 B_0 按式(10).

$$B_0 = \begin{pmatrix} 1 & 0 & 0 & 0 \\ 0 & 0 & 0 & 0 \\ 0 & 1 & 0 & 0 \\ 0 & 0 & 0 & 0 \\ 0 & 0 & 1 & 0 \\ 0 & 0 & 0 & 0 \\ 0 & 0 & 0 & 1 \\ 0 & 0 & 0 & 0 \end{pmatrix} \tag{9}$$

$$B_0 = \begin{pmatrix} 0 & 0 & 0 & 0 \\ 1 & 0 & 0 & 0 \\ 0 & 0 & 0 & 0 \\ 0 & 1 & 0 & 0 \\ 0 & 0 & 0 & 0 \\ 0 & 0 & 1 & 0 \\ 0 & 0 & 0 & 0 \\ 0 & 0 & 0 & 1 \end{pmatrix} \tag{10}$$

其余计算均照旧.

2.4 计算程序

为了保证计算精度,基本上采用双精度运算.过程中,将建立的特征行列式 D 记存文件中,备补充计算时调用.

经在 PC—XT 型微机上多次运算(分段数不大于30),本程序功能稳定可靠,切合实用.

三、模型试验验证

试验模型如图7所示.支承在三个同样刚度的弹性支座上.支座只有垂直方向的弹性,且可调节其弹性大小.支座的刚度均用静态标定.为简单起见,系统中未设阻尼.

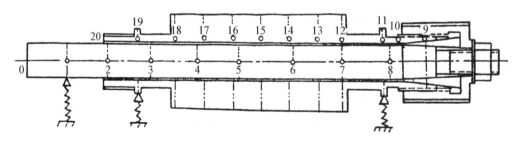

图7 复套式模型及其离散化分段情况

用锤击法激起该系统的各阶固有振动,测试系统如图 8 所示.

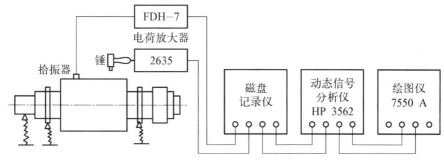

图 8 模型测试系统

图 9 示一组典型的实验结果,图(a)与(b)是敲击点与拾振点互换以作对比. 这组结果表明:在 0~1 000 Hz 范围内有四个较明显的固有频率:20 Hz,42.5~45 Hz,172.5~175 Hz,950~951.2 Hz. 各组实验结果如表 1 所示.

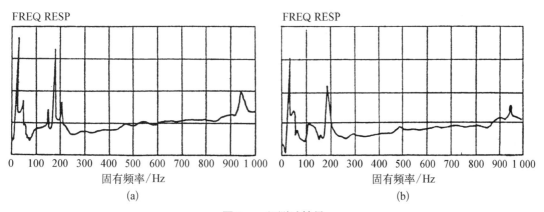

图 9 一组测试结果

为了排除支座可能带来的一些不确定因素,还将转子模型用绳子悬吊起来进行锤击测试,以获得它在自由-自由情况下的固有频率. 所得数据可更单纯地验证正、反传递矩阵法的可用性. 其结果亦列于表 1.

表 1 模型系统的固有频率测试结果和计算结果

支承刚度 k /(N·m^{-1})	测得的固有频率/Hz	算出的固有频率/Hz 根据 k	设 k 与 5×10^5 N/m 串连	支承刚度 k /(N·m^{-1})	测得的固有频率/Hz	算出的固有频率/Hz 根据 k	设 k 与 5×10^5 N/m 串连
0 (即自由-自由)	197.5	199.4		117 000	20.0	22.99	20.69
	993.5	1 082			44.0	46.58	42.01
46 900	15.0	14.54	13.90		174.0	182.9	181.4
	29.65	28.36			951.0	996.8	996.7
	172.6	178.2	177.9	168 000	23.7	27.47	23.77
	935.0	996.3	996.3		43.7	55.41	48.13
86 900	17.5	19.78	18.26		183.7	186.3	183.5
	35.1	40.20	37.15		909.0	997.1	996.9
	176.3	180.9	180	496 000	35.0	47.23	33.47
	945.0	996.6	996.5		71.2	92.65	67.03

(续表)

支承刚度 k /(N·m^{-1})	测得的固有频率/Hz	算出的固有频率/Hz		支承刚度 k /(N·m^{-1})	测得的固有频率/Hz	算出的固有频率/Hz	
		根据 k	设 k 与 5×10^5 N/m 串连			根据 k	设 k 与 5×10^5 N/m 串连
	182.0	207.8	191.7		88.10	175.0	84.45
	987.0	999.3	997.7		180.0	299.2	202.0
2.15×10^6	54.95	98.15	42.74		954.3	1 011	998.7

从测试结果可知,各组数据中的末两个固有频率总是大致等于自由-自由情况下测得的频率值(由于模型支座与转子模型直接连结,支座的一部分质量随该处转子振动,使在支座上测得的这两个固有频率反而略低于自由-自由情况下的频率.至于支承在轴承油膜上的转子,则将有不同情况).各组数据中头上两个频率则随支座刚度强烈变化.前者可称为"转子主导"型,后者可称为"轴承主导"型.前者数值由于受支座刚度影响小,更能用来验证正、反传递法的计算结果.后者数值则很大程度上依赖于支承动刚度的准确性.

用本文所述方法,对模型系统作了计算.共分 20 段,如图 7 所示.在相邻段直径阶差很大的地方,利用 45°影响锥的概念,适当计入了粗段端部弯曲刚度的减弱效应.支座与转子直接连结的托架质量作为转子该处的附加质量纳入计算.轴段参数见表 2.计算结果如表 1 所列.

表 2 复套式模型的分段参数

段号	外径/mm	内径/mm	长度/mm	附加质量/kg	轴质惯矩/(kg·m^2)
1	30	0	42.5	0.221	0
2	30	0	40	0	0
3	30	0	40	0	0
4	30	0	40	0	0
5	30	0	40	0	0
6	30	0	40	0	0
7	30	0	40	0	0
8	30	0	40	0	0
9	30	0	50	1.37*	0
10	49	35	26	0	0
11	47.25	35	13	0.357	0
12	50	35	41.5	0.332 9	0
13	85	35	23.667	0	0
14	85	35	23.667	0	0
15	85	35	23.667	0	0
16	85	35	23.667	0	0
17	85	35	23.667	0	0
18	85	35	23.667	0.332 9	0
19	50	35	41.5	0.357	0
20	50	35	32.5	0	0

注:* 表示附加质量离传递方向反逆节点的外悬距 $e=-4.6$ mm.

对比测试结果与计算结果.可以看到:在自由-自由情况下,二者满意地符合.支承在弹性支座上时,"转子主导"型的测试和计算结果亦能满意地符合.这证实了正、反传递矩阵法的可用性.

至于"轴承主导"型,则在支座刚度小时,亦能很好符合. 但支座刚度大时,却有较大偏差. 仔细观察可知:测得的第一阶固有频率并不大致随支座刚度的平方根而增减,而是作缓慢得多的变化. 这说明:模型支座在小振动时实际动刚度并不完全符合它在静态大变形情况下标定的值,而是不同程度地更小. 考虑到支座上V形槽与轴颈接触处可能不太清洁以及不太密贴或有夹尘,接触处可能有微小变形. 各接合面间可能有微小相对运动等因素(这些附加影响因素在实际轴承油膜中并不存在或完全不同),对于试验模型转子与支承静刚度间设想为串连有另一附加弹性环节. 如果取该附加串连刚度为 5×10^5 N/m,则计算结果的"轴承主导"型频率亦与实测值满意地符合,如表1所示.

四、结论

用正、反传递矩阵法建立复套式转子-滑动轴承系统的弯曲固有振动特征行列式;用复平面上显示轨迹及三角元折半搜索法计算复频率;对于对称系统作节约处理. 由此建立了一套适用于复套式、普通式及对称式转子-滑动轴承系统的固有复频率计算方法和程序. 实例运算(分段数不大于30)表明它稳定可靠.

用简单复套式模型系统的锤击试验验证了这套计算法是切实可用的.

需指出的是:在内、外轴接合处,本文是作为完善的整体来处理的. 对于不是非常紧固的内、外连接情况,这样处理显然会带来误差,需要今后加以更完善的处理.

本工作由南京高速齿轮箱厂资助完成. 该厂并提供了模型系统以供测试验证. 该厂李钊刚工程师对本工作给予了热情支持和关心,测试工作并得到了上海工业大学沈沛涛和奚风丰的帮助. 特此一并致谢.

参 考 文 献

[1] Glienicke. J. Dynamik glietgelagerter Wellen- Programme zur Berechnung der selbst- und unwuchterregten Schwingungen allgemeiner gleitgelagerter (und walzgelägerter) Rotoren mit Zusatzeinflüssen. Forschunugsheft. Heft(61)1978.

Calculation of Complex Frequencies of Complex Rotor-Bearing Systems

Abstract: This paper presents a method of calculation of the complex natural frequencies of complex rotor-bearing systems. Forward and backward implicit transfer matrix method is used to establish the characterisic determinant, and loci display on finite coplex plane and triangle halving are used to search out the characteristic roots, wherefore, an effective and accurate program is formed. The calculation is verified by model impulse experiment, and satisfactory correlation is obtained.

负压型磁头的设计分析[*]

提 要: 负压型磁头作为实现高密度贮存的一项重要技术受到高度重视. 本文采用有限元法对两种类型的负压型磁头(FTO 型及 HS 型)进行数值计算及分析,得到了稳态时的气膜压力分布、气膜厚度、姿态角等特性值,探讨了若干因素对稳态性能的影响. 根据数值计算结果,提出了影响负压型磁头性能的关键因素——负压槽深度的最佳取值准则,并对比了这两种类型磁头的性能. 计算实践还表明,有限元法由于有便于适应不同网格划分及台阶型膜厚突变之长处,是分析计算负压型磁头的有力工具.

一、引言

磁头是磁盘存储器中的关键部件之一. 为得到更高的存储密度并减小磁头与磁盘介质间的接触与磨损,研究者不断地探索新型的磁头. 因为普通的双轨式磁头(以 IBM3370 为例)其浮动间隙约为 $0.31\sim0.33\ \mu m$,但其加载力却超过了 Winchester 型的加载力. 如要进一步减小头盘间浮动间隙,须增大加载力或减小支承导轨的面积,前者不利于接触起停,且头盘跟踪性能也无法保证,后者则将使支承刚度降低. 这种矛盾推动了研究者寻找新的出路.

负压型磁头最早报道于 1974 年的美国专利[1],它具有加载力小、头盘间隙小的特点. 80 年代中,发表了对这种磁头的分析研究[2],其他一些结构类型的负压型磁头也有所分析研究[3,4]. 同时,这种磁头的生产技术也逐步地由试验室研制转入生产开发研究[5]. 国内学者也开始关心负压型磁头的研究工作[6],但还未见有关研究报道.

本文采用有限元法[7]对负压型磁头用考虑边界滑移影响的 Reynolds 方程来进行数值求解,计算其稳态工况下的各项性能参数,并与普通型进行比较,以期掌握负压型磁头的一些特点和规律,为我国的磁头部件设计研制及生产奠定基础.

二、磁头形状及参数简介

普通双轨型磁头也叫正压型磁头,形状如图 1 所示,负压型磁头如图 2 所示,这两种磁头均属于 H 型磁头. 图 2 中(a)为 FTO 型负压磁头,(b)为 HS 型负压磁头. 从结构上看,负压磁头是在普通的双轨型基础上加一个台阶,以便在

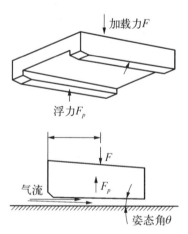

图 1 正压型磁头形状

---

[*] 本文合作者:胡近. 原发表于《电子计算机外部设备》,1990(2):38-42,28.

一部分区域内产生吸力(即负压),这样,加上较小的弹簧力就构成了有效加载力,可以与正压部分产生的浮力进行平衡. 对于 H 型磁头,可以通过恰当地选择设计参数(即负压槽的深度及位置)而获得合适大小的吸力及姿态角,使其性能优于正压型磁头.

图 3、4 为本文所考察的 H 型磁头的两种结构形式,取自参考文献[5].

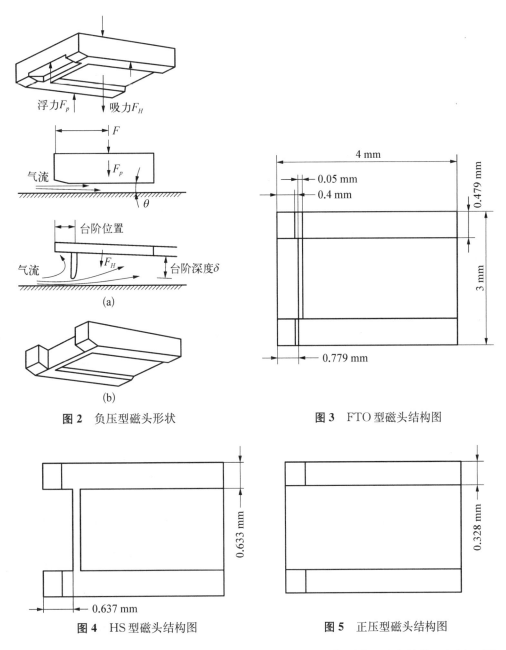

图 2 负压型磁头形状

图 3 FTO 型磁头结构图

图 4 HS 型磁头结构图

图 5 正压型磁头结构图

图 3 为 FTO 型(快速启停)磁头结构,图 4 为 HS 型(高刚度)磁头结构. 为便于同正压型磁头比较,平行地考察了图 5 所示的相同外形尺寸的正压型磁头. 加载力的作用位置均在磁头对称线的中点,但加载力的大小是不同的. 对于负压型磁头而言,从理论上说,加载力可调整到任意值. 实算中则可设定飞行高度 h_0,通过调整几何尺寸、姿态角来达到预选加载力的大小及位置.

三、基本方程及数值方法

按照等温润滑理论,采用考虑边界滑移的修正 Reynolds 方程

$$\frac{\partial}{\partial x}\left[PH^3\left(1+\frac{6M}{PH}\right)\frac{\partial P}{\partial x}\right]+\frac{\partial}{\partial y}\left[PH^3\left(1+\frac{6M}{PH}\right)\frac{\partial P}{\partial y}\right]=\Lambda\frac{\partial}{\partial x}(PH) \tag{1}$$

式中,P 为无量纲气膜压力,H 为无量纲气膜厚度,M 为克努森数,Λ 为气体压缩数.这些无量纲值的定义如下:$P=p/p_a$,p_a 为大气压力,$H=h/h_0$,h_0 为最小气膜厚度,$M=\lambda_a/h_0$,λ_a 为空气分子平均自由程,$\Lambda=6\mu UL/(p_a h_0^2)$,$\mu$ 为空气动力黏度,U 为相对速度,L 为磁头浮动块长度.

利用文献[7]的有限元法对方程(1)进行离散,并用 Newton-Raphson 法拟线性化,得到关于气膜压力的线性方程组,用 Gauss 消去法求解.

静态特性计算方法亦与[7]中的相同,只是在求合力时需要区分浮力与吸力.

需要指出的是,就我们所知,从业已发表的负压型磁头的文献看,数值方法都是用的有限差分法,因此,在离散过程中对台阶前后的润滑方程要用流量平衡式导出.有限元法在这方面具有优越性,不必在气膜厚度突变处加入流量平衡式.并且,有限元法便于适应不同网格划分,使用上灵活方便.

四、数值计算结果

负压型磁头的几何尺寸具有对称性,故取其一半为离散求解区域,如图 6 所示.图 7 给出了相应区域上的压力分布,其特点是在负压槽附近沿气流方向有一个压力差,产生负压区,负压区的合力即为吸力.

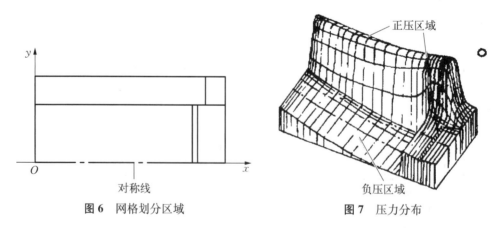

图 6 网格划分区域　　图 7 压力分布

图 8 给出了 HS 型磁头压力分布随速度变化的情况.其中 $V=27$ m/s 时,$h_0=0.23\ \mu$m,姿态角 $\theta=0.65\times 10^{-4}$ rad.最大正压产生在浮动导轨的中线,而最大负压则发生在磁头几何对称中线上.在相同的负压槽深度下,速度高则产生较小的负压,因而其平衡状态时正压也较小.最小飞行高度 h_0 及姿态角 θ_0 随速度变化的关系对于三种不同类型磁头的比较见图 9,图 10.

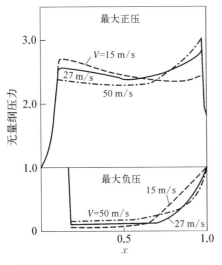

图 8 HS 型磁头速度对压力分布的影响($\delta=4\ \mu m$)

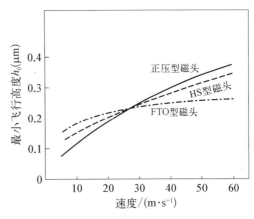

图 9 三种磁头的最小飞行高度 h_0 随速度变化的关系

由图 9 看出,两种负压型磁头最小飞行高度 h_0 随速度的变化幅值均小于正压型磁头,同时 FTO 型磁头由于负压槽前部仍产生一部分浮力,故其最小飞行高度要比 HS 型的小. 同时,还可以从图 11 所示的有效加载力(弹簧加载力与吸力之和)与速度的关系中加以说明. 对于 HS 型磁头来说,吸力随速度增加而迅速增大到一定值,然后随速度增加而逐渐减小. 对于 FTO 型磁头,其吸力只是随速度的增大而逐步增加,因此,当速度较高时,其最小飞行高度仍变化不大. 所以,负压型磁头与正压型磁头不同之处就在于最小飞行高度小,且随速度变化小,这正是我们所需要的特性.

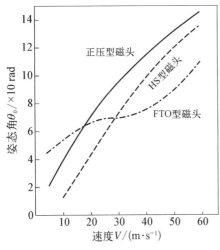

图 10 三种磁头的姿态角 θ_0 随速度变化的关系

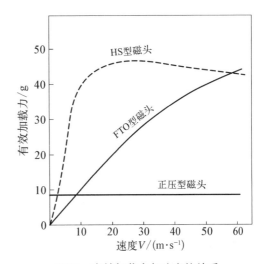

图 11 有效加载力与速度的关系

对于设计者来说,怎样才能选择一个较理想的负压槽深度值呢?图 12 给出了 $V=27\ m/s$ 时三种不同槽深的压力分布. 图 13 给出了三种速度值下吸力随槽深的变化关

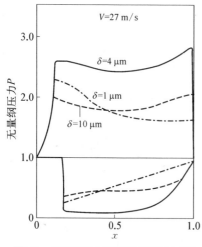

图 12 HS 型磁头不同负压槽(台阶)深度对压力分布的影响

系. 从图中可以看出,吸力值并不是槽深的单调函数,而是具有服从一定规律的特性. 由图 13 可知,对于不同的速度,吸力值随槽深的变化均有一个极值. 不过,单纯从最大吸力来考虑取定槽深是不全面的,因为,它不仅影响吸力大小,还影响磁头的平衡状态,尤其是在平衡状态附近的性能. 图 14 表明了台阶深度对平衡状态的影响. 对最小飞行高度来说,当速度不变($V=27$ m/s),对应于不同的负压槽深度,h_0 有一个极小值. 综合考虑上述因素,并考虑到槽深的加工公差,可在一定的数值范围内选取槽深. 反之,如果确定了飞行高度值及其允许变化的范围,就可分析计算出相应的负压槽深度值,并使磁头工作在比较稳定的工作状态.

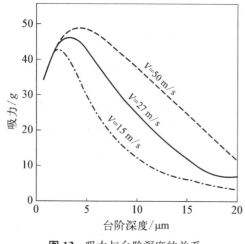

图 13 吸力与台阶深度的关系

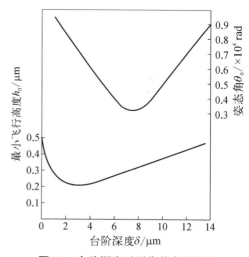

图 14 台阶深度对平衡状态的影响

五、结论

(1) 负压型磁头具有较小的加载力,有利于改善磁头接触启停性能,盘片不易划伤,磁头、磁盘的磨损可减轻.

(2) FTO 型磁头由于吸力随速度增大而增大,其飞行高度随速度的变化较小,因此,FTO 型磁头能较好地满足减小磁盘内、外道飞行高度差的要求,并且较 HS 型更易于在低速时起浮,具有较好的不易磨损性.

(3) 与正压型磁头不同的是,随着负压槽深度的变化,负压磁头的姿态角及最小膜厚在一定的范围内变化,且变化较小,这为设计与制造提供了较好的条件.

(4) 有限元法便于适应不同的网格划分及台阶型突变,对分析计算不同的磁头类型及几何尺寸提供了较大的优越性,对于负压型磁头更是有力的分析计算工具.

本文对于负压型磁头的分析只是给出了部分研究结果,作者将进一步对负压型磁头动态特性进行分析.

本工作得到上海市高教局基金的资助(87B-13),在此特表谢意.

参 考 文 献

[1] Garnier, M. F., Tang, T., and White, J. W. Magnetic Head Slider Assembly. U. S. Patent 3,885,625, Dec. 17, 1974.

[2] White, J. W. Flying Characteristics of the "Zero-Load" Slider Bearing. Trans. ASME, Vol. 105, July 1984: 484-490.

[3] Kogure, K., et al. Design of Negative Pressure Slider for Magnetic Recording Disks. Trans ASME, Vol. 105, July 1983: 496-502.

[4] Seiji Yoneoka, et al. Design Considerations for Negative Pressure Head Slider. Fujitsu Sci Tech. J., Vol. 21, No. 1, March, 1985: 40-49.

[5] Soneoka S., et al. Fast Take-Off Negative Pressure Slider. IEEE Trans, on Mag, Vol Mag-23, No. 5, Sept., 1987: 3464-3466.

[6] 华伟. 负压型磁头. 电子计算机外部设备, 1988(2): 63-68.

[7] 胡近, 张直明. 小间隙浮动理论研究. 电子计算机外部设备, 1989(1): 7-14.

Analysis of Cylindrical Journal Bearing With Viscoelastic Bush*

Abstract: Stationarily loaded cylindrical journal bearings with viscoelastic deformable bearing bush is analyzed in this paper. FEM is used to calculate the deformation of bush induced by the pressure of the oil film, which is in its turn influenced by this deformation causing an increment of the oil film thickness. The simultaneous solution of distributions of oil film pressure and thickness is used to obtain the static performances of the bearing. In the analysis of dynamic behavior of the system under small harmonic vibrations, the effect of linear viscoelastic dynamic deformation of bush is taken into consideration. The frequency method is used to obtain the distributions of the derivatives of complex amplitudes of dynamic pressure increment in the oil film with respect to amplitudes of variation of journal eccentricity and attitude angle, and the simultaneously coexisting distributions of derivatives of dynamic deformation amplitudes with respect to the same, wherefrom the linear dynamic coefficients of the viscoelastically supported oil film are obtained. Unbalance response properties and stability threshold speeds of the rotor-bearing system are calculated. Numerical results enable conclusions to be made that suitable viscoelastic property of the bush will effectively improve the dynamic behavior, both in unbalance response and stability.

Nomenclature

B_{ij} = nondimensional damping coefficients, $B_{ij} = b_{ij}\psi^3/(\mu l)(i, j = x, y)$

b_{ij} = damping coefficients

c = radial clearance

d = bearing diameter

D = outer diameter of bearing bush

$\bar{E}$ = nondimensional elasticity modulus, $\bar{E} = E\psi^3/(2\mu\Omega)$

E = Young's modulus

E^*, E' and E'' = nondimensional viscoelastic complex modulus, nondimensional storage modulus, and nondimensional loss modulus respectively, with the same relative unit as that of $\bar{E}$

f = static flexibility of rotor, $f = mg/k$

f^* = relative complex flexibility

g = gravitational acceleration

H = nondimensional film thickness, $H = h/c$

h = film thickness

K_{ij} = nondimensional stiffness coefficients, $K_{ij} = k_{ij}\psi^3/(\mu\Omega l)(i, j = x, y)$

k_{ij} = stiffness coefficients

k = static stiffness of rotor

l = bearing width

m = mass of rotor

P = nondimensional pressure, $P = p\psi^2/(2\mu\Omega)$

p = oil film pressure

r = bearing radius

So = Sommerfeld number, $So = \psi^2 W/(2\mu\Omega r l)$

T = nondimensional time, $T = t\Omega$

t = time

x = circumferential coordinate, $x = r\varphi$

---

* In collaboration with X. Jiang. Reprinted from *Transactions of the ASME, Journal of Tribology*, 1990, 112: 442 – 446.

W = load
Z = nondimensional axial coordinate, $Z = z/(l/2)$
z = axial coordinate measured from middle plane of bearing width
β = amplifying factor at resonance
μ = dynamic viscosity of lubricant
ϕ = angular coordinate measured from steady state film force direction
φ = angular coordinate measured from maximum clearance, $\varphi = \phi - \theta$
ψ = clearance ratio, $\psi = c/(d/2)$
Ω = angular velocity of rotor
Ω_{st} = stability threshold speed
ω = frequency of whirling or oscillation
$\bar{\omega}$ = nondimensional whirling frequency, $\bar{\omega} = \omega/\Omega$
ω_k = natural frequency of rotor with rigid supports
ε = steady-state eccentricity ratio
θ = steady-state attitude angle

1 Introduction

It is necessary for high speed rotating machines supported on fluid film bearings to have good stability and vibration suppressing ability. Ordinary bearings offer limited possibility in this respect. A well-known effective method is introducing elastic-damping supports outside of bearing or shell (Glienicke and Stanski, 1975, Holmes and Dogan, 1982, Nikolajsent and Holmes, 1979, Zhang, 1985), the stability of the rotor-bearing system can be significantly improved by proper elasticity and damping parameters. But the application is accompanied by complicated structure or possibility of fatigue failure of the elastic suspension. Gas lubricated foil bearings have excellent dynamic characteristics. However, they are mostly used in miniature light-load cases (Heshmat et al., 1983). Yet from the above, we can clearly see the general principle underlying them, viz., the dynamic behavior of the bearing system can be effectively improved by combining elasticity and damping property with the supporting structure in series with the lubricating film, either a little distant from the later as in external bearing dampers, or more adjacent to it as in corrugated foil bearings. With this view in mind, it is logical to assume that if the bearing bush itself is given appropriate elasticity and damping properties, resulting in the most intimate combination of the lubricating film with the additional elastic-damping mechanism, a significant improvement of dynamic behavior could be expected for a very simple structure. The aim of this paper is to explore theoretically the possibility brought forth by such an idea. The fundamental methodology adopted in this paper is based on linear theory of simple harmonic vibration, which is justified by onset of threshold instability, or unbalance-induced synchronous whirl when its amplitude is small.

2 Analysis

For an isoviscous incompressible lubricant, the pressure field under isothermal laminar flow condition in the clearance space of journal bearing is governed by nonstationary Reynolds equation:

$$\partial(h^3\partial p/\partial x)/\partial x + \partial(h^3\partial p/\partial z)/\partial z = 6\mu\Omega r\partial h/\partial x + 12\mu\partial h/\partial t \qquad (1)$$

or its nondimensional form:

$$\partial(H^3\partial P/\partial\varphi)/\partial\varphi + (d/l)^2\partial(H^3\partial P/\partial Z)/\partial Z = 3\partial H/\partial\varphi + 6\partial H/\partial T \qquad (1a)$$

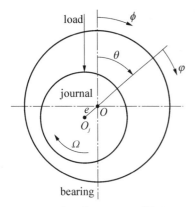

Fig. 1 Steady-state equilibrium position of journal within bearing

Figure 1 shows the steady state equilibrium position of a journal within a cylindrical bearing. Assuming small arbitrary harmonic whirling of the journal within the bearing with frequency ω, we can express the instantaneous position of the journal center as

$$\varepsilon_i = \varepsilon + \varepsilon_a e^{i\omega t} \quad \theta_i = \theta + \theta_a e^{i\omega t} \qquad (2)$$

where ε and θ express the eccentricity ratio and attitude angle of the journal center at its steady-state equilibrium position, and the mutually independent parameters ε_a and θ_a express the complex amplitudes of eccentricity ratio and attitude angle, respectively. The corresponding film pressure and film thickness are composed of their steady state parts and dynamic increments. Neglecting errors of higher orders, they can be expressed in nondimensional form as

$$P = P_0 + Qe^{i\bar{\omega}T} \qquad (3)$$

$$H = H_0 + H_j e^{i\bar{\omega}T} + \delta_\nu e^{i\bar{\omega}T} \qquad (4)$$

where H_0 is the distribution of nondimensional steady state film thickness including nondimensional steady-state deformation δ_s of the bush, $H_0 = 1 + \varepsilon\cos\varphi + \delta_s$; P_0 is the distribution of steady-state pressure; Q is the nondimensional complex amplitude of dynamic pressure increment; $\bar{\omega}$ is the relative whirl frequency of the journal, $\bar{\omega} = \omega/\Omega$, while Ω is the angular rotational speed of shaft; δ_ν is the nondimensional complex amplitude of bush viscoelastic deformation; H_j is the distribution of nondimensional amplitude of dynamic increment of film thickness due to small dynamic displacement of journal, $H_j = \varepsilon_a\cos\varphi + \varepsilon\theta_a\sin\varphi$. Substituting equations (3)–(4) into (1a), we get the governing equation of pressure perturbation (Zhang et al., 1986a) taking into account bush viscoelastic deformation under small harmonic oscillation of journal in nondimensional form as

$$\begin{aligned}&\partial(H_0^3\partial Q/\partial\varphi)/\partial\varphi + (d/l)^2\partial(H_0^3\partial Q/\partial Z)/\partial Z\\&= -9(M/H_0)\partial H_0/\partial\varphi - 3H_0[(H_0\partial M/\partial\varphi - M\partial H_0/\partial\varphi)\partial P_0/\partial\varphi +\\&(d/l)^2\partial(H_0\partial M/\partial Z - M\partial H_0/\partial Z)\partial P_0/\partial Z] + 3\partial M/\partial\varphi + 6i\bar{\omega}M\end{aligned} \qquad (5)$$

where the variable M stands for $\varepsilon_a\cos\varphi + \varepsilon\theta_a\sin\varphi + \delta_\nu$.

Steady-state harmonic viscoelasticity problems can be solved in the same manner as that for the corresponding elasticity problems. According to the elastic-viscoelastic correspondence principle of isothermal, linear viscoelasticity theory (Christensen, 1982), steady-state harmonic elastic solutions can be converted to corresponding viscoelastic solutions by replacing the elastic moduli with the complex viscoelastic moduli.

To be consistent with elastic problems, the relations of complex moduli are given by the following nondimensional equations:

$$\lambda^*/G^* = 2\nu^*/(1-2\nu^*) \quad E^* = 2(1+\nu^*)G^* \tag{6}$$

where λ^* and G^* are complex Lamé constants of the viscoelastic body, E^* is complex modulus and ν^* is complex Poisson's ratio. All the variables are expressed in their nondimensional forms. If Poisson's ratio is assumed constant, that is $\nu^* = \nu = $ const, it is apparent from (6) that the viscoelastic solutions can be achieved from the corresponding elastic solutions simply through replacing the elastic modulus $\overline{E}$ by the complex modulus E^*.

The complex modulus E^* is commonly decomposed into a real part E' and an imaginary part E'' which are called the storage and the loss moduli, respectively.

$$E^* = E' + iE'' \tag{7}$$

Assume strain to be a harmonic function of time

$$\varepsilon(t) = \varepsilon_0 e^{i\omega t} \tag{8}$$

where ε_0 is the amplitude and ω the frequency of vibration. The corresponding stress is also a harmonic function with the same frequency:

$$\sigma(t) = |E^*| \varepsilon_0 e^{i(\omega t + \alpha)} \tag{9}$$

where $\alpha = \tan^{-1}(\eta)$, with $\eta = E''/E'$, and $|E^*|$ is the magnitude of E^*. α is sometimes referred to as loss angle of the viscoelastic material, and η is called the loss factor. The larger α and η are, the higher the damping effect is. For a given material, E^*, E', E'', α, and η are usually dependent of the exciting frequency.

In the calculation of the three-dimensional elastic deformation by the finite element method, the elastic modulus can be separated from the stiffness matrix $[K]$ so that the elements of the later are only relative to shape function and Poisson's ratio:

$$\overline{E}[K]\{\delta_e\} = \{F\} \tag{11}$$

where $[K]$, $\{\delta_e\}$, and $\{F\}$ are nondimensional global stiffness matrix, displacement vector and load vector, respectively. The steady-state deformation δ_s is calculated by this equation with nodal loads calculated from the steady-state pressure distribution. When calculating the viscoelastic displacement vector caused by the dynamic increment of film pressure, we write

$$\{\delta_v\} = [K]^{-1}\{F\}/E^* \tag{12}$$

Substituting (7) into (12), we get

$$\{\delta_v\} = (f' - \mathrm{i}f'')\{\delta_e\} \tag{13}$$

where f' and f'' are the real and imaginary parts of the relative complex flexibility respectively. $(f' - \mathrm{i}f'')$ may be called the complex relative flexibility, and the following relationships hold

$$f^* = f' - \mathrm{i}f'' = 1/(E^*/\overline{E}) \tag{14}$$

$$f' = (E'/\overline{E})/[(E'/\overline{E})^2 + (E''/\overline{E})^2] \tag{15}$$

$$f'' = (E''/\overline{E})/[(E'/\overline{E})^2 + (E''/\overline{E})^2] \tag{16}$$

The oil film dynamic coefficients for a given value of frequency ratio $\bar{\omega}$ are calculated by the partial differential method. Derivatives of complex amplitudes of pressure and deformation distributions with respect to whirl amplitude ε_a or θ_a are first solved simultaneously, then the film stiffness coefficients and damping coefficients are obtained by integrations of the derived distributions in the same way as when film resultant forces are calculated. It is evident that the dynamic coefficients thus obtained are dependent on the assumed whirl ratio $\bar{\omega}$. A value of unity is given to $\bar{\omega}$ when the dynamic coefficients to be used in calculating synchronous vibrations are aimed at.

For stability calculations, the equivalent stiffness coefficient K_{eq} and the whirl ratio γ_{st} at stability threshold are calculated as follows (Zhang et al., 1986b):

$$K_{eq} = (K_{xx}B_{yy} + K_{yy}B_{xx} - K_{xy}B_{yx} - K_{yx}B_{xy})/(B_{xx} + B_{yy}) \tag{17}$$

$$\gamma_{st}^2 = [(K_{eq} - K_{xx})(K_{eq} - K_{yy}) - K_{xy}K_{yx}]/(B_{xx}B_{yy} - B_{xy}B_{yx}) \tag{18}$$

The values of K_{eq} and γ_{st} thus obtained are also dependent on the assumed value of relative frequency $\bar{\omega}$. At stability threshold, the whirl ratio γ_{st} should keep in coincidence with the exciting frquency $\bar{\omega}$. This is arrived at by an iterative procedure.

For the cylindrical whirling of a symmetrical single mass rotor supported on a pair of identical journal bearings, the threshold speed can be calculated by the following nondimensional equation (Zhang et al., 1986b):

$$\Omega_{st}/\omega_k = 1/[\gamma_{st}\sqrt{1 + So_{st}/(k_{eq} \cdot f/c)}] \tag{19}$$

where Ω_{st}/ω_k is the relative threshold speed, $\omega_k = \sqrt{k/m}$ is the natural frequency of the rotor when rigidly supported, f/c represents the relative flexibility of the rotor, $f = mg/k$ denotes the static deflection of the rotor caused by its own weight, c is the radial clearance of the bearing, $So_{st} = \psi^2 W/(2\mu\Omega_{st}rl)$ is the Sommerfeld number of the bearing under threshold speed.

From a multitude of calculated results of this sort, a "stability map" can be deduced, showing the relationship between Ω_{st}/ω_k and So_k, with f/c as a parameter. A higher curve means a better stability. In the above, $So_k = \psi^2 W/(2\mu\omega_k rl)$ is the Sommerfeld number with reference to ω_k.

Response of the rotor system to residual unbalance is also influenced by the viscoelastic property of the bearing bush. The major semiaxis of the elliptical whirling orbit of the rotor center under unit unbalance can be calculated from the corresponding nonhomogeneous equations of motion, using the dynamic coefficients of the bearing calculated under unit whirl ratio. A "resonance map" can be deduced from a multitude of such calculations, showing the resonant amplifying factor and the relative critical speed Ω_{cr}/ψ_k in relationship with So_k, with f/c as a parameter. An improvement of vibration suppressing ability is linked with a lowering of the resonant amplifying factor.

3 Methods of Solution

In the calculations, the oil film domain is discretized, non-dimensional Reynolds equation and its perturbated forms are replaced by their corresponding finite difference equations, and solved by successive over-relaxation (SOR) method. Underrelaxation iterations between Reynolds equations and deformation calculations are performed to obtain the simultaneous solutions of nondimensional oil film pressure (or the derivatives of its complex amplitudes) and bearing deformation (or the derivatives of its complex amplitudes).

The finite element method is used for calculating the elastic deformation of the bearing. For cylindrical journal bearings, where the bearing structure and boundary condition are axially symmetrical, the computational and storage amount can be significantly decreased in the following manner.

Figure 2 shows the oil film domain discretized (16 * 80), the position of a node is denoted by (i, j), P_{kl} is the average pressure on the subdomain A_{kl}. Let $[\Delta]$ represent the radial displacement matrix of bearing surface composed of nodal displacements δ_{ij}, and $[F]_{kl}$ the flexibility coefficient matrix expressing radial deformations at all nodes caused by unit pressure at region A_{kl}.

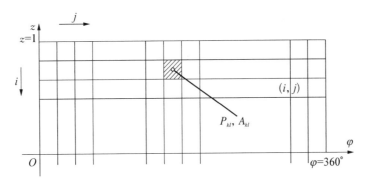

Fig. 2 Discretization of the oil film domain

$$[\Delta] = [\delta_{ij}] \quad i = 1, 2, \cdots, 17 \quad k = 1, 2, \cdots, 16 \tag{20}$$
$$[F]_{kl} = [f_{ij}]_{kl} \quad j = 1, 2, \cdots, 81 \quad l = 1, 2, \cdots, 80$$

From principle of superposition,

$$[\Delta] = \sum_{k=1}^{16} \sum_{l=1}^{80} \overline{P}_{kl} \cdot [F]_{kl} \cdot \Delta A \tag{21}$$

or

$$\delta_{ij} = \sum_{k=1}^{16} \sum_{l=1}^{80} \overline{P}_{kl} \cdot [f_{ij}]_{kl} \cdot \Delta A \tag{22}$$

where $(f_{ij})_{kl}$ is the element (i, j) of the flexibility matrix (k, l), $\Delta A = \Delta\varphi \cdot \Delta Z$, $\Delta\varphi$ and ΔZ are step widths in circumferential and axial directions, respectively.

Because of the axial symmetry of cylindrical bearings, we need not obtain 16 * 80 flexibility matrixes for all the small areas A_{kl}, but only 16 matrices for a single column in the axial direction. The matrices of all other columns can be formed automatically through corresponding transpositions of these 16 matrices.

4 Results and Discussion

Calculations have been done for a cylindrical bearing with $l/d = 0.6$, $D/d = 1.6$, nondimensional complex modulus $E' = 20$ or 2, loss factor $\eta = 0.5$ or 1.0 and $E = 200$, $\eta = 0$ (corresponding approximately to bronze) for comparison, and Poisson's ratio $\nu = 0.3$. The bearing bush is assumed to be uniformly and rigidly supported on its outer surface.

Figures 3 and 4 represent the comparisons between various stiffness coefficients. Figures 5 and 6 show those between damping coefficients. It can be seen that significant increases in damping coefficients are caused by increases in η and So within the calculated range. The two cross damping coefficients differ from each other when dynamic elastic or viscoelastic deformation is taken into account.

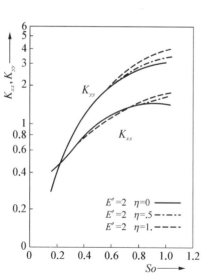

Fig. 3 Stiffness coefficients K_{xx}, K_{yy}

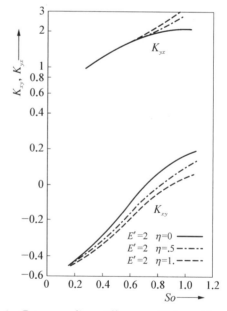

Fig. 4 Cross coupling stiffness coefficients K_{xy}, K_{yx}

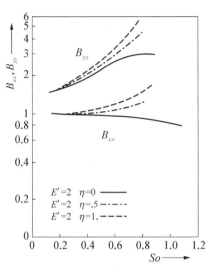

Fig. 5 Damping coefficients B_{xx}, B_{yy}

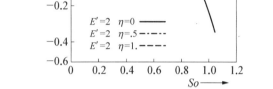

Fig. 6 Cross coupling damping coefficients B_{xy}, B_{yx}

It has been noticed from our numerous results of calculations with η value ranging from 0 up to 2 that, for a definite non-dimensional storage modulus E', the optimal value of loss factor η is approximately 1, since an η value of unity is always accompanied by the highest stability threshold speed and the lowest amplifying factor at resonance. Accordingly, the following discussions are limited to the case with unity η value.

Figure 7 shows stability threshold speeds of a flexible rotor supported on journal bearings made of materials of different E' modulus, with unity loss factor. It can be seen that the threshold speed increases significantly for viscoelastic bush as compared to rigid bush. For example, for $f/c = 2.0$,

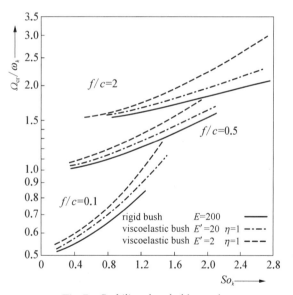

Fig. 7 Stability threshold speeds

$So_k = 1.6$, when compared to the rigid bush (no. 1), the threshold speed for the second material (no. 2) increases by 4.7 percent, while for the third one increases by 12.4 percent.

A comparison of the results of β and Ω_{cr}/ω_k is given in Fig. 8. With viscoelastic bush, β decreases significantly, which implies that it can be effectively used to suppress the resonance amplitude of rotors. For another example, with $f/c = 2.0$, $So_k = 0.8$, compared to the rigid bush, the resonant amplifying factor of the second material reduces by 2.9

percent, while that of the third one by 4.6 percent.

5 Conclusions

The present study exhibits that if a rotor system is supported on bearings with bush made from suitable viscoelastic materials, its stability and vibration suppressing ability can be significantly improved, thus the theory confirms the suggested concept of improving journal bearing stability and vibration suppressing quality by using bearing bushes made of viscoelastic materials.

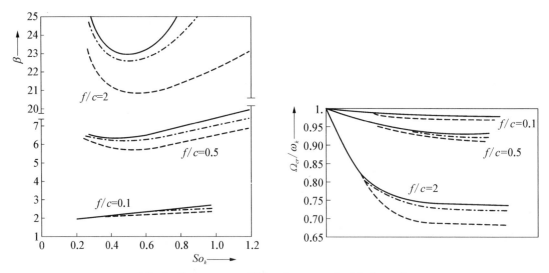

Fig. 8 Resonant amplifying factors and critical speeds

References

[1] Christensen, R. M., 1982, Theory of Viscoelasticity, an Introduction, 2nd ed., Academic Press Inc., New York.

[2] Glienicke, J., and Stanski, U., 1975, "External Bearing Damping—Means of Preventing Dangerous Shaft Vibrations on Gas Turbines and Exhaust Turbo-Chargers," Proc. 11th CIMAC, pp. 287-311.

[3] Heshmat, H., Walowit, J. A., and Pinkus, O., 1983, "Analysis of Gas Lubricated Foil Journal Bearings," ASME Journal of Lubrication Technology, Vol. 105, pp. 638-646.

[4] Holmes, R., and Dogan, R., 1982, "Investigation of a Rotor Bearing Assembly Incorporating a Squeeze-Film Damper Bearing," Journal of Mechanical Engineering Science, Vol. 24, pp. 129-137.

[5] Nikolajsent, J. L., and Holmes, R., 1979, "Investigation of Squeeze-Film Isolators for the Vibration Control of a Flexible Rotor," Journal of Mechanical Engineering Science, Vol. 21, pp. 247-252.

[6] Zhang, Z., 1985, "System Damping Factor of an Elastic Rotor Supported on Tilting-Pad Bearings with Elastic and Damped Pivots," Proceedings of JSME International Tribology Conference, Tokyo, pp. 541-546.

[7] Zhang, Z., Mao, Q., and Xu, H., 1986a, "The Effect of Dynamic Deformation on Dynamic Properties and Stability of Cylindrical Journal Bearings," Proceedings of 13th Leeds-Lyon Symposium on Tribology, Paper XI (iv), pp. 363-366.

[8] Zhang, Z., Zhang, Y., Chen, Z., Xie, Y., Qui, D., and Zhu, J., Theory of Hydrodynamic Lubrication of Sliding Bearings (in Chinese), Zhang, Z., ed., Chinese High Education Publication Corp., Beijing, pp. 94-95.

负压型磁头的静动态特性*

摘　要：本文论述了磁盘存贮器中负压磁头的静动特性. 作者采用小扰动法及有限元法,对考虑边界滑移的修正雷诺方程及两自由度的磁头运动方程进行求解,得到相应的静态特性和稳态特性. 通过所得结果的分析,兼顾静态及动态性能,确定了最佳的台阶深度值. 比较结果还表明负压型磁头的动态性能远比正压型的好.

符 号 说 明

a：磁盘偏摆振幅,μm, $A = a/h$

b：磁头浮动导轨宽度,mm, $B = b/l$

C_{ij}：气膜阻尼系数,$i = j = 1$,平行模式,kN·s/min；$i = j = 2$,俯仰模式,kN·mm/rad

h_0：最小气膜厚度,μm

h：各点气膜厚度,μm

$H = h/h_0$

H_0：无量纲静平衡气膜厚度

H_d：无量纲气膜厚度增量

I_θ：绕俯仰轴的浮动块转动惯量,g·mm^2

K_{ij}：气膜刚度系数,$i = j = 1$,平行模式,kN/mm, $i = j = 2$,俯仰模式,kN·mm/rad

k, k_θ：加载弹簧平行刚度、俯仰刚度,kN/mm；kN·mm/rad

l, l_1：浮动块长度、斜契长度,mm

m：浮动块质量,g

$M = \lambda_a/h$, 克努森数(Knudsen Number)

P_a, P：大气压力、各点气膜压力,N/mm^2

S：经拉普拉斯变换的时间变量

U：磁记录表面相对速度,m/s

W：浮动块承载力,mN

$X = x/l, Y = y/l$,无量纲直角坐标

$X_G = x_G/l$,无量纲载荷作用中心

δ_t：斜楔高度,μm

δ：负压型磁头台阶深度,μm

$Z = z/h, \theta = \theta \cdot l/h_0$,浮动块运动坐标

λ_a：大气压下空气分子平均自由程,$0.064\ \mu m$

A_x：$6\mu Ul/(P_a \cdot h_0^2)$,气体压缩数

μ：气体动力黏度,N·s/m

σ：$12\mu\omega_0 l(P_a \cdot h_0^2)$,抗压数

τ：$\omega_0 t$, $\Omega = \omega/\omega_0$

一、引言

在磁盘存贮器中,为提高记录密度,必须减小磁头飞行高度. 目前,飞行高度小于 $0.3\ \mu m$ 的磁头已付诸使用. 由于设计参数本身的限制,普通的双轨型(即正压型)磁头无法同时满足轻载(以减小磨损)与高支承刚度的要求,也不可能改变其飞行高度随速度变化较大的缺点.

负压型磁头能较好地满足上述要求. 80 年代以来,国外已发表了有关文章[1—3]. 作者在文献[4]中,已分析了负压型磁头的稳态特性,较文献[1—3]不同的是,在离散方法中,采用了有限元法. 这个方法可适应不同网格划分及台阶型膜厚的突变. 本文主要就负压型磁头动态特性,采用有限元法及小扰动法[5],进行数值求解,得出了负压磁头动态特性的数值解及相应的正压型磁头的数值解.

---

* 本文合作者：胡近. 原发表于《上海工业大学学报》,1990,11(5)：456 - 462.

二、磁头形状及参数

普通双轨型磁头也叫做正压型磁头,如图 1(a)所示.负压型磁头如图 1(b)所示,从结构上看,是在普通的双轨型基础上,磁头前部横向加一个台阶,以形成吸力(即负压),这样加上弹簧力就构成了有效加载力,可与正压部分产生的浮力进行平衡.对于这类负压型磁头,可以通过适当选择参数,即台阶的深度与位置,而获得合适的吸力大小与姿态角,使其工作性能优于正压型磁头.图 2 为本文所考察的快速起浮负压型(FTO-Fast take-off)及正压型磁头的结构参数.加载力的作用位置均在磁头对称线中点.正压型磁头的加载力为 0.078 N.而对负压型磁头而言,加载力可以调节到任意较小的值,实际计算中可根据设定的飞行高度 h_0 改变几何尺寸及姿态角来满足预选的加载力大小.

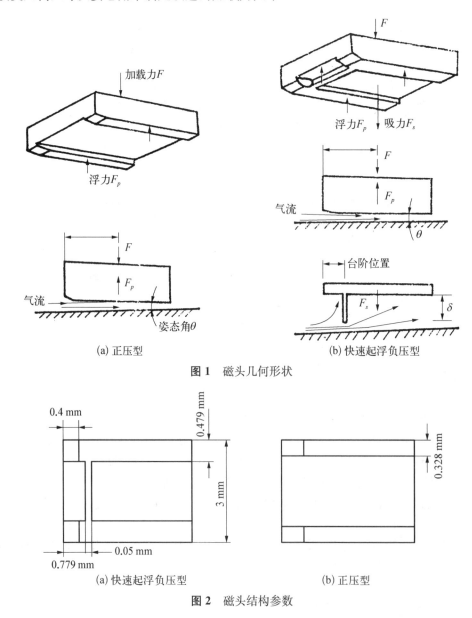

(a) 正压型 (b) 快速起浮负压型

图 1 磁头几何形状

(a) 快速起浮负压型 (b) 正压型

图 2 磁头结构参数

三、基本方程及数值方法

磁头浮动块的物理模型及座标关系如图 3 所示. 等温条件下,考虑分子平均自由程的修正雷诺方程为

$$\nabla \cdot \left[PH^3 \left(1 + \frac{6M}{PH}\right) \nabla P - \Delta PH \right] - \sigma \frac{\partial PH}{\partial \tau} = 0 \quad (1)$$

磁头浮动块的运动方程是

$$m \frac{d^2 z}{dt^2} + kz = \iint \Delta p \, dx \, dy$$
$$I_\theta \frac{d^2 \theta}{dt^2} + k_\theta \theta = \iint \Delta p (x_G - x) \, dx \, dy \quad (2)$$

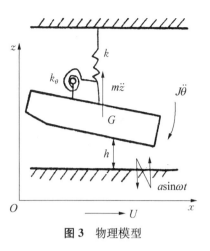

图 3 物理模型

由磁头浮动块垂直位移 z 及俯仰运动 θ 引起的气膜厚度的变动量 h_d 由下式决定

$$h_d = z + \theta(x - x_G) - a\sin\omega t \quad (3)$$

采用小扰动法[5],对方程(1)扰动,得到关于扰动压力的方程,然后用有限元法进行离散及数值求解.

负压型磁头的几何形状具有对称性,故取其一半为离散求解区域,如图 4(a)所示.

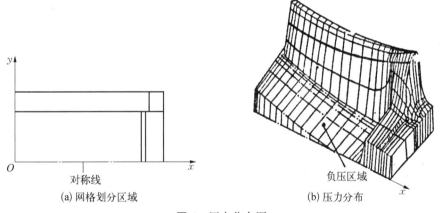

(a) 网格划分区域　　　　　　(b) 压力分布

图 4　压力分布图

四、数值结果分析

文献[4]给出了负压型磁头的稳态特性,包括气膜压力分布,气膜厚度随速度的变化关系,以及影响吸力大小的原因,这里给出其典型的压力分布,如图 4(b)所示. 其特点是在台阶附近沿气流流动方向有一个压力突降,产生负压区,负压区的合力即为吸力.

对正弦小偏摆,磁头飞行高度的响应,仍是负压型磁头的重要特性. 为此,作者对不同激振频率下的磁头动态响应及气膜刚度、阻尼进行了大量计算,并相应计算了具有相同弹簧常数及磁头质量与转动惯量的正压型磁头的性能. 参数均取自[2,3],其中,质量 $m = 0.2$ g,加载弹簧俯仰刚度 $k_\theta = 6.2 \times 10^{-3}$ kN·mm/rad,垂直刚度 $k = 1.53 \times 10^{-5}$ kN/mm.

图 5 给出了快速起浮负压型与正压型磁头动态特性的比较关系. 在计算中已假定正压型磁头及负压型磁头的支撑参数及质惯量相同. 负压型磁头较正压型有较高的共振频率,这可由图 5(b)作出解释,负压磁头的气膜刚度大大超过正压型气膜厚度,在所计算的工况下($V=27$ m/s)为 3~6 倍,因此,负压型磁头具有极轻的载荷及较大的刚度值,稳定性能大大提高. 同时,负压型磁头由于阻尼系数的减小,所以其共振点处的幅值较大,但在低于共振频率的激振频率下,负压型磁头的响应幅值却低于正压型的. 因此,负压型的动态特性远比正压型好. 至于负压型磁头的阻尼值小,这是由于当气体压缩效应增大时,阻尼系数减小,这种现象在重载支承的情形中也是较普遍的.

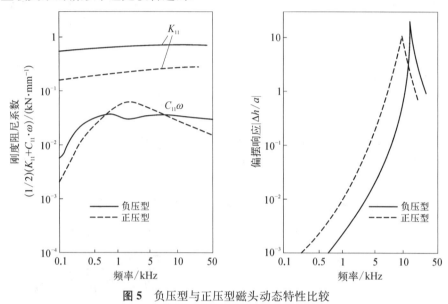

图 5 负压型与正压型磁头动态特性比较

图 6 比较了三种不同台阶深度下的负压型磁头的动态特性. 从中可以看出,台阶深度 δ 对于动态特性的影响是很大的,我们可以利用图 7 来分析.

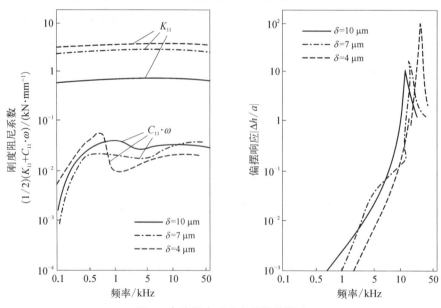

图 6 台阶深度对动态特性的影响

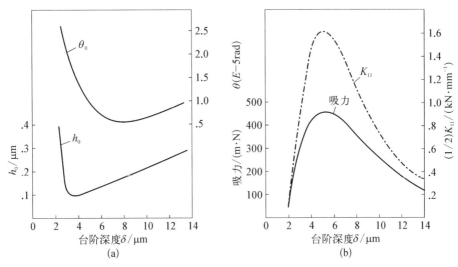

图 7 台阶深度的影响

三种不同的 δ 值中，$\delta = 4\ \mu m$，由于产生了最大的吸力，h_0 最小，所以刚度值最大. 由此可知，设计者可以通过适当选定 δ 值，同时获得最大吸力、最小 h_0 及 h_0 随 δ 变化最小的理想设计状况. 图 5、图 6 的部分典型值比较见表 1.

表 1 不同磁头的特性值 （比较频率 600 Hz，单位：kN/mm）

磁头类型	负压型(FTO 型)			正压型
	$\delta = 4\ \mu m$	$\delta = 7\ \mu m$	$\delta = 10\ \mu m$	
共振频率/kHz	26	19	16	11
共振幅值 $|\Delta h/a|$	107	26.4	22	11.7
刚度系数 (K_{11})	1.49	1.37	75	0.24
吸力/mN	426.3	404.7	245	0

负压型磁头除了本身结构特点外，还具有与正压型磁头相同结构的几何形状. 因此，设计中除正确选择台阶深度、位置外，还要注意磁头其他几何尺寸的影响，如浮动导轨宽度、斜楔高度、斜楔长度等. 为此，作者对各种影响磁头性能的参数，以正压型磁头为例，作了分析计算. 图 8 给出了磁头几何参数变化对浮动性能的影响曲线.

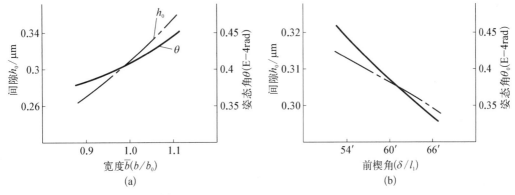

正压型(3370 磁头)，$l = 4.05$ mm，$B = 3.05$ mm，$l_1 = .35$ mm，$b_0 = .635$ mm
楔角 $(\delta/l_1) = 1°$，$V = 14.36$ m/s，$h_0 = .308\,9\ \mu m$，$\theta_0 = .412\,5E{-}4$ rad

图 8 几何尺寸对浮动性能的影响

五、结论

1. 根据修正雷诺方程,用扰动法和有限元法建立了负压型磁头气膜动态特性的理论及计算方法.

2. 负压型磁头能在较小的弹簧加载力下,提供很大的气膜支承刚度,可以提高飞行高度的稳定性和抗干扰性,并提供了较宽的装配公差.

3. 负压型磁头不仅具有较好的稳态性能及更宽的可调设计参数范围,而且其动态性能也远比正压型的好.

4. 为获得静态及动态性能俱佳的负压型磁头,可以在一系列台阶深度下进行计算,确定产生最大吸力及最小气膜厚度的最佳取值. 从制造角度看,具有较小的 δ 值的负压台阶易于加工并允许更大的尺寸公差.

本课题研究工作得到上海市高教局基金资助(87B—13),特表谢意.

参 考 文 献

[1] Kogure, K., Fukui, S., Mitsuya, Y. et al. Design of negative pressure slider for magnetic recording disks. Trans. ASME, 105 1983: 496-502.
[2] White, J. W. Flying characteristics of the zero-load slider bearings. Trans ASME, 106 1984: 484-490.
[3] Yoneoka, S., Yamada, T., Aruga, K. et al. Fast take-off negative pressure slider. IEEE Trans. on Magn., Mag. 23(5). 1987: 3464-3466.
[4] 胡近,张直明. 负压型磁头的设计分析. 电子计算机外部设备,1990(2):48-52.
[5] 胡近,张直明. 小间隙浮动理论研究. 电子计算机外部设备,1989(1):7-14.

Static and Dynamic Characteristics of Negative Pressure Magnetic Head

Abstract: States the static and dynamic characteristics of a negative pressure slider for magnetic recording disks. By virtue of modified Reynolds equation with inclusion of slip flow effects, the static and dynamic characteristics are calculated by the perturbation method and finite element method. Using the calculated results, the optimum reverse step depth is determined in consideration of both static and dynamic characteristics. Comparison also shows that the dynamic flying characteristics of the negative pressure slider is far better than those of the comparative positive pressure slider.

径向滑动轴承轴瓦反转的现象和机理分析*

摘 要：综述了我国若干高速旋转机械径向轴承实际发生过的一些"轴瓦反转"现象，讨论了这种现象产生的原因，定性地分析并提出了两种蠕动机理：由正旋的"局部松脱"和正旋的"局部滑移"累积效应造成轴瓦整体反转，同时提出最简单条件下相应的临界力学条件，还以示范性模型佐证了基本观点.

一、引言

近年来，我国接连发生了数起大型汽轮发电机组灾难性的毁机事故. 在事故现场，都可看到一些支持轴承的轴瓦已发生了不同程度的反转，即发生了逆于轴旋转方向的某个角度转动，小到几度，大到几十度以至进油口被错位堵塞. 实际上，除了严重事故外，在试车、试验等场合，以及在其他一些中小型旋转机械中，也发生过轴瓦反转现象. 这种现象曾引起国内同行的关注，但迄今未获得令人信服的论述. 由于轴瓦反转现象与机组的事故和机组可靠性密切相关，它既可以是事故造成的后果，又有可能成为引起事故的原因，所以必须弄清其发生的机理和条件. 这样不仅可使我们加深对有关事物的认识，还有助于改进设计以提高机组可靠性. 为此，作者在进行广泛调查研究的基础上，对轴瓦反转的现象及其发生的条件进行了定性分析，还从动态力学观点提出两种可能引起轴瓦反转的机理及其发生的基本条件，并制作了简单的示范性模型再现了反转现象，以佐证基本观点.

二、轴瓦反转现象的实例与原因分析

作者曾调查访问了一些电厂. 调查结果表明，轴瓦反转已在我国多次发生，有的还造成了重大事故. 现将有关资料摘录如下. 为慎重起见，厂名从略.

（1）1972 年 9 月 18 日上午，××热电厂 125 机组因副励磁机烧坏失励而引起系统振荡的大事故，全厂出力从 43.75×10^4 kW 减至 2.2×10^4 kW. 事故原因是：5 号机副励磁机组第七号轴瓦未设定位销，运行中轴瓦发生反转，油眼被阻塞，轴瓦合金因缺油而烧坏，使轴位下降，以致副励磁机转子碰摩定子甚至产生火花，造成 5 号机失励运行. 事故后曾对此机组同类轴承进行全面检查，发现第六号轴承的轴瓦也有反转现象，油眼约被阻塞 1/2. 后在这两个轴承内加装了尺寸足够大的轴瓦定位销，才防止了同类事故发生.

（2）1972 年 7 月 1 日 9：35，××电厂并网运行不到 4 小时就发现 2 号机组的合金温度由 79℃升到最大值(150℃)，其振幅达 0.2 mm，被迫紧急停机. 在降速过程中发现 4 号轴承回油槽无回油，判断是由于轴瓦转动错位使油口受堵所致. 后对 4 号轴承解体检查，证实轴

---

* 本文合作者：朱礼进、易子夏. 原发表于《上海工业大学学报》，1991，12(2)：122-128.

瓦已由于瓦顶定位销未装好而发生反转,上、下瓦面均有较大面积的磨损.

(3) 60 年代××第二热电厂一台 $5×10^4$ kW 机组的 2 号轴承(三油楔轴承),在运行中反转了约 7°左右,以致进油口被堵,下瓦合金由于缺油而被烧化并且大部分被挤出,甚至被旋转的轴颈带到上瓦范围内而与上瓦合金相粘贴.水平中分面处的锁定件未能挡住轴瓦反转,反而被强大的反转力挤推变形而完全丧失原状.只是在将锁定件充分加大后,才消除了轴瓦反转.

(4) ××电厂的 $10×10^4$ kW 机组一个三油楔轴承和另一电厂的又一台 $5×10^4$ kW 机组的 2 号轴承(三油楔),也都发生过轴瓦反转.

对"轴瓦反转"问题已引起了科技人员的重视.例如哈尔滨汽轮机厂就曾在 70 年代将其列为十大技术难题之一,高洪元工程师等曾在试验中证实了轴瓦确会在运行中发生反转,并且可能形成很大的反转力,以致剪断尺寸不够大的定位件.东方汽轮机厂金志伟工程师等也曾专门在圆柱轴承支撑的转子上再现了轴瓦反转现象.他们发现,当轴瓦与轴箱配合较松并且转速到达第一临界转速 6 000 r/min 的 2.5 倍左右时(即 14 000~17 000 r/min 时),产生较强的轴颈涡动,其原因估计是油膜振荡),可明显观察到轴瓦反转.

综上所述,可以认为:① 轴瓦反转与轴颈涡动密切相关.东方汽轮机厂试验中显然是在油膜振荡时发生明显的轴瓦反转,虽然缺乏数据,但圆柱轴承支持的转子在 2 倍一阶临界转速以上的强烈振动大都是油膜振荡,而油膜振荡时轴颈有最强烈的正向涡动(正进动).当然,在油膜振荡以外的转子振动(例如不平衡振动)情况下,轴颈也是有涡动的,而且在滑动轴承支持时通常是正向涡动.② 当轴瓦与外座面配合松时容易发生反转.③ 如果发生反转,定位件会受到很大的力.

上海发电设备成套设计研究所杨光海高工和本文作者曾分别与美国 Beno Sternlicht 和日本 Yukio Hori 等润滑理论和轴承专家讨论过轴瓦反转问题.前者认为轴瓦反转确示了油膜振荡;后者则认为油膜振荡和其他共振都会引起反转,并认为反转是一种蠕动的结果.

其实,早在 50 年代苏联在有关汽轮机的书籍中就叙述过轴瓦反转问题,虽未涉及机理,但已提出了只有加紧配合才能防止反转的意见.

本文作者之一早在 1979 年就曾提出轴瓦反转是由轴颈正向涡动(正进动)所致,并在配合不够紧时发生.其机理大致可以从少齿差传动和波导传动中得到启示,亦可由此说明为何此时定位销会受到很大的剪力.只有从轴颈正向涡动(正进动)时轴瓦内面所受的正向旋转径向力,才能找到轴瓦反转的真正原因.

三、对"轴瓦反转"现象产生原因的讨论

目前有下列两种解释:① 轴颈在静平衡位置时,其中心偏离轴承中心,通过轴颈中心的载荷绕轴承中心的力矩方向反于旋转方向(图1),这个力矩引起了轴瓦反转;② 过大的横振动使转子碰擦定子,造成转子的反进动,从而引起轴瓦反转.

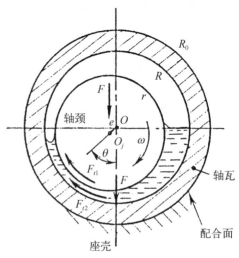

图 1 轴承的定常工作状态
r—轴颈半径 R—轴承半径
R_0—轴瓦配合面半径

第一种看法中,只考虑到轴颈 O_j 偏心 e 后轴颈上的载荷 F 绕轴承中心 O 有一个反于轴旋转方向的力矩.这一点虽然是不错的,但它没有考虑到这个力矩并未直接作用到轴瓦上而是作用到油膜上,并且除径向力外同时必定还有正向摩擦力 F_{t1}(矩)的作用.轴瓦所受的,是油膜作用于它的径向力 F(通过轴瓦中心)和摩擦力 F_{t2}(矩),且后者是正向的.由于油膜在定常工况下总是处于平衡中(非定常工况下,也是如此,因惯性项通常可以忽略),所以轴颈对油膜的作用的总和必定等于油膜对轴瓦作用的总和.就力矩而言,必有[1]

$$F_{t1}r - Fe\sin\theta = F_{t2}R \approx F_{t2}r$$

由此可知,在定常工况下,无论从轴颈对油膜的作用看,还是从油膜对轴瓦的作用看,绕轴承中心 O 的力矩作用只能是一个正向的力矩,它的大小等于 $F_{t2}r$ 或即 $F_{t1}r - Fe\sin\theta$.这个力矩的大小一般不超过轴瓦外表面的摩擦力矩,因后者的数量级至少是轴瓦与座壳配合面上的摩擦系数 f 与轴承力 F 和配合面半径 R_0 的乘积,即 fFR_0,而大多数情况下 f 比 F_{t2}/F 大得多.如果这个力矩引起轴瓦转动,也只能是正转而不见反转.因此这种看法的考虑不够全面,只看到了较小项 $Fe\sin\theta$ 的方向,而未看到较大项 $F_{t1}r$ 的正向作用.

第二种看法,简单地认为轴颈的反进动就会引起轴瓦的反转.其实,这二者并无必然的关系.从大量实际情况和后面的分析可知,轴瓦松转方向在绝大多数情况下是反于轴颈旋转方向.有不少机组的轴瓦反转,是发生在显然不存在转子与定子碰擦的所谓"动、静摩擦"的工况下,特别是在小型模拟试验台上试验时更是如此.这些情况下都是没有轴颈反进动却发生了轴瓦反转的.因此,一则有许多反转现象是发生在显然没有转子与定子碰擦的情况下;二则即使有了"动、静摩擦"所造成的轴颈反进动,也没有任何理由认为会引起轴瓦反转.

这两种看法都认为:既有"反转"现象,就一定有个宏观上的"反向旋转"因素.殊不知这个现象是个动态问题,涉及内在的应力和应变分布及其变化,其彻底解决要依靠大量的非线性动态力学分析及细致的实验工作.正因为如此,本文亦只能从反转现象的基本机理和关键因素出发,进行一些定性的分析.

四、正旋径向力引起轴瓦反转的机理和力学条件

根据大量事实、试验和示范性简单实验(见后),作者认为轴瓦反转是由于轴颈在轴承内的正向涡动所带来的正旋径向力所引起的.其力学机理可能有以下两种:

第一种为"局部松脱"机理.为了简单起见,此处以轴承静载荷为零,轴承只受大小不变的正旋径向油膜力的情况来说明,并且假设为平面问题.图 2(a)示圆柱形轴瓦安装在外壳中并受到均匀支撑的简单情况,在配合面上设有均布的装配压力 p_0(如果是动配合则无 p_0).图 2(b)中,轴瓦内的压力分布 p_a 表示轴颈正向涡动所引起的油膜旋转压力.二者复合时,配合面上的压力分布即如图 2(c)或图 2(d)所示.当油膜压力较小时,瓦外配合面上的压力分布 p 变为非均布的,一部分部位压力增大,另一部分部位压力减小,但最小处仍有压力 p_{min}(暂不计剪应力).此时配合面无一处发生局部的松脱.随着油膜压力的正向旋转,瓦外配合面上的压力分布亦作正向旋转.由于两配合面处保持相互压紧,在这个旋转压力场的运转过程中,配合面上任何局部都不发生相对运动.力场旋转一周,全部情况复原.这样周而复始,轴瓦不发生松转.但如果瓦内压较大,以致瓦外配合面中某个局部压力下降为0[如图 2(d)],则这个局部处发生了轴瓦与外壳的微小脱开——"松脱".图 2(e)是松脱局部的夸大

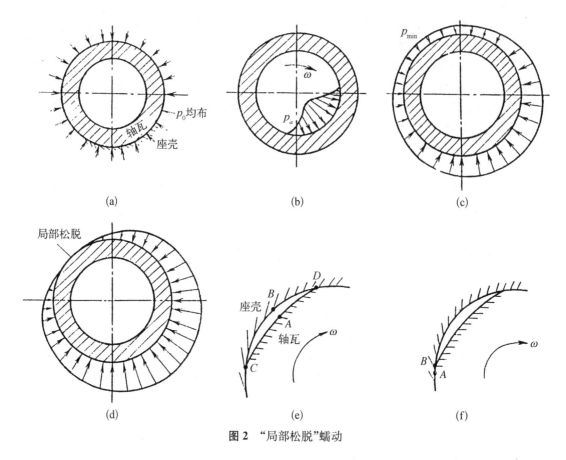

图 2 "局部松脱"蠕动

示意图,图中轴瓦上的 A 点和外壳上的 B 点在松开前原是相重合的. 当瓦内压力场随着轴颈的正向涡动而正旋时,松脱部位亦正向变位. 一个微小时间间隔后,发生过松脱的局部处,轴瓦与外壳又重新贴合. 但此时轴瓦上的 A 点不再与外壳上的 B 点相重合,而是落到了 B 点的后面(反于旋转方向)一个极微小的距离上,如图 2(f). 这个极微小的局部反向错位,正是轴瓦发生整体反转的基因. 为什么在重新贴合时 A 点会落到 B 点反向处呢? 由图 2(e) 可知,局部松开时,轴瓦上一段弧长 $\overparen{CA}$ 必定略小于外壳上对应的弧长 $\overparen{DB}$. 由于弧长 $\overparen{CA}$ 在由松开状态逐渐与外壳恢复贴合(此时 AD 部分并不同时发生贴合,而是保持松脱状态)的过程中保持小于弧长 $\overparen{DB}$,所以重新贴合后就发生了上述的局部错位. 当这样的松脱部位沿旋转方向绕移一周后,在整周配合上都留下了局部微小错位,于是构成了轴瓦整体的极微小反转. 高速旋转的轴会引起每分钟上千次乃至上万次的上述循环,在一段时间后就有可能发生明显的轴瓦反转.

在所假设的简单情况下,发生"松脱"性反转的力学条件是:$p_{\min}<0$,即在正旋力的作用下,轴瓦与座壳配合面上的最小压力消失.

第二种为"局部滑移"机理. 当轴瓦内表面受到油膜的不均布压力,轴瓦与座壳的配合面上不仅发生正应力分布的变化,而且发生不均布的剪应力 τ[图 3(a)]. 如果剪应力处处小于该处的摩擦系数 f 乘正压力 p,则剪应力不会引起配合面上的局部滑移. 油膜压力旋转 360° 后,一切复原,轴瓦与座壳间没有整体的错位. 但如果某个局部上的剪应力 τ 大于该处的局部摩擦力 f_p,此处就会发生局部滑移[图 3(b)]. 当油膜压力旋转时,滑移区亦随之旋转. 由于应力场和滑移分布不可能精确地对称于轴心,当油膜压力作 360°旋转后,轴瓦与座壳的整体相对位置会产生微小的错位.这种错位效应随着成千上万次旋转而积累,造成明显的轴瓦相对于机壳的整体转动.这种整体转动的机理比前一种复杂,其方向也不像前一种那样明显易判.简单的模型试验证实,这种整体转动既可能是正方向的,也可能是反方向的,随条件而异.但在可与轴承工况相比拟的情况下,则由这种"局部滑移"将发生轴瓦的整体反转.

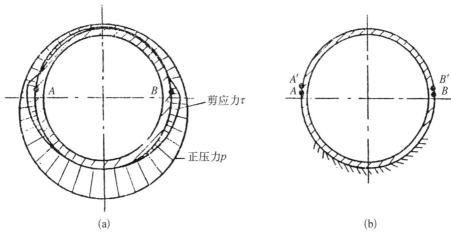

图 3 "局部滑移"蠕动

由上可见,在上述简单情况下,发生"局部滑移"性反转的力学条件是:$|\tau| > fp$.

当然,在实际情况下,问题要复杂得多,因为配合面的装配压力本来就不均匀,油膜力又是静载与动载的复合,而且应力、应变场沿轴向亦不是均匀的.因此在油膜力旋转的每一个角位置上,应力、应变的分布都是不同的.不过就发生轴瓦反转的根本机理来说,仍不外乎由于正向旋转的"局部松脱"或"局部滑移"的积累效应这两种基本原因,只不过数量关系要复杂得多.

众所周知,如果要用定位销之类挡住这类蠕动性的相对转动,定位件将受到很大的剪应力,在极端的情况下,其数量级可接近整个蠕动面(即轴瓦与轴承座的配合面)上摩擦力之总和.这就当然会使尺寸不够大的定位件被剪断或损坏了.

五、简单的模型实验

图 4 所示为模型装置的示意图.有机玻璃筒 1 代替轴承座,橡胶套筒 2 代替轴瓦,不太紧地装在 1 内,压力套 3 对橡胶套筒内表面施加压力,代替油膜对轴瓦的压力.压紧力的大小是可以调节的.当旋转手柄 6 时,压力套对橡胶套施加了径向旋转压力.压力足够大时,可观察到橡胶套筒在缓慢地转动.在径向压力不变的情况下,橡胶套筒与玻璃筒之间的配合较紧时,橡胶套筒转动就较慢些,当配合较松时,转动就较快,说明轴瓦的转动与轴瓦和轴承座之间的配合力大

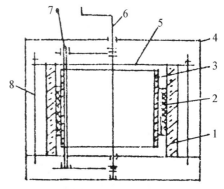

1—玻璃筒(内径 ϕ85 mm)
2—橡胶筒(内径 ϕ78 mm)
3—加压套筒(外径 ϕ76 mm 或 ϕ16 mm)
4—箱体;5—压板;6—摇手柄;
7—销轴;8—螺杆.

图 4 示范实验装置

小有关;在配合力一定的情况下,加大径向压力,橡胶套筒转动变快,径向压力变小时,转动变慢,这说明轴瓦的转动与径向力的大小有关;在径向压力及配合压力不变的情况下,采用半径较大的压力套,从而径向压力分布较宽时[如图 5(a)、(b)],橡胶套筒呈现反转,当采用半径较小的压力套,从而径向压力分布较集中时[如图 6(a)、(b)],橡胶套筒呈现正转.由此可知轴瓦的转向还同径向力分布情况有关.

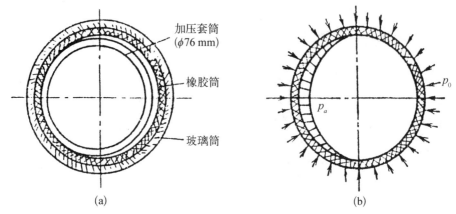

图 5 压力分布范围较大的情况

在原理相似的另一套筒模型上,以弹性不大的有机玻璃套筒代替橡胶套筒,亦即 1 和 2 均用弹性不大的材料制成.此时,尽管压力套的半径很小而压力分布集中,内套筒 2 总是发生反转.

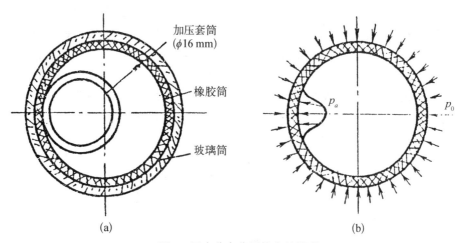

图 6 压力分布范围较小的情况

这些简单实验,证实了轴瓦内正向旋转的压力确实能引起轴瓦反转.虽还有待进一步的数量规律分析计算,但弹性不大的内套筒在同样材料外套筒内所发生的反转,不难估计是由"局部松脱"性的蠕动所造成.弹性较大的内套筒在加压范围较大时的反转,则既可能由"局部松脱"亦可能由"局部滑移"性的蠕动所造成;在加压范围较小时的正转,则只能由"局部滑移"来解释了.

实际的轴承中,因为瓦内受到的是油膜作用而来的压力,其作用不会十分集中,所以与上述简单实验相对照地看来,亦不难理解为何总是发生轴瓦反转而看不到轴瓦正转的记述了.

参 考 文 献

[1] 张直明. 滑动轴承的流体动力润滑理论. 北京: 高等教育出版社, 1986: 38.

An Investigation on the Mechanism of Counter-rotation of Journal Bearing Bush

Abstract: A number of cases of "counter rotation of bearing bush" found in domestic high-speed rotating, machinery is recited. The relevent causal discussion and a qualitative analysis are made. Two kinds of local creep motion are suggested for explaining the counter-rotation of the bush, namely, the accumulative effect of progressive "local loosening", and that of progressive "local slip". Corresponding mechanical criteria of the occurance of counter-rotation under the simplest conditions are proposed. The basic stand-point is justified by experiments done on a demonstrative model.

计入弹性动变形的单块径向轴承可倾瓦动特性*

摘　要：本文论述了弹性变形对可倾瓦轴承动特性和转子系统动力性态的影响. 用小简谐扰动原理导出了径向轴承单块可倾瓦的动态弹流方程组及动特性计算公式. 用 SOR（Successive Over-Relaxtion）法解算油膜静态压力和扰动压力差分方程组，用基于平面曲梁模型的柔度矩阵计算瓦块的静态变形和扰动变形分布，并以亚松弛迭代的总体顺解法获得上列各量的联立解. 从而求得可变形径向可倾瓦在不同涡动比下的动特性. 归纳了弹性变形对可倾瓦动特性的影响规律，并作了机理上的阐释. 为转子-可倾瓦轴承系统动力性态的预测，提供了较深入合理的分析基础和基本数据.

符 号 说 明

B_{rr}——可倾瓦径向阻尼系数 b_{rr} 的无量纲值：$B_{rr} = b_{rr}\psi^3/(2\eta l)$

B_{ij}^*——"固定"瓦阻尼系数无量纲值（$i,j = t,r$）

c——瓦弧半径与轴颈半径之差，即瓦间隙

d——轴颈直径

$f(\varphi, \varphi_1)$——影响系数，即 φ_1 处作用单位径向力时，瓦在 φ 处的径向位移

H——膜厚 h 的无量纲值：$H = h/c$；H_0——其静态分量；δH——其变形所致增量；$(\delta H)_0$——δH 的静态分量；H_d——其动变形复幅

K_δ——瓦弹性变形系数：$K_\delta = 0.75(1 - \nu^2) \cdot \eta\omega/(T_1^3 E \psi^3 \alpha_J)$，式中 ν——泊松比，T_1——瓦厚度对瓦中性直径之比，E——弹性模量，$\alpha_J = 1 + \dfrac{3}{5}T_1^2 + \dfrac{3}{7}T_1^4 + \cdots$

K_{rr}——可倾瓦径向刚度系数 K_{rr} 的无量纲值：$K_{rr} = K_{rr}\psi^3/(2\eta\omega l)$

K_{ij}^*——"固定"瓦刚度系数无量纲值（$i,j = t,r$）

l——轴颈长度，即轴瓦宽度

P——油膜压力 p 的无量纲值：$P = p\psi^2/(\eta\omega)$；P_0——其静态分量；Q——其扰动复幅

T——时间 t 的无量纲值：$T = t \cdot \omega$

Z——轴向坐标 z 的无量纲值：$Z = z/(l/2)$

$\varepsilon_{to}, \varepsilon_{ro}$——轴颈相对于瓦弧中心的切向、径向静态偏心比；$\Delta\varepsilon_t, \Delta\varepsilon_r$——此二偏心比的扰动幅值

η——动力黏度

Ω——扰动频率 ω_1 的相对值，即涡动比：$\Omega = \omega_1/\omega$

ω——轴的角速度

ψ——相对间隙：$\psi c/(d/2)$

一、前言

近年来，轴承研究者愈益关注可倾瓦变形对转子-轴承动力性态的影响问题. 例如文献[1]根据现代透平机械中未计入可倾瓦变形效应而算出的临界转速与实测结果的显著差别，明确指出必须计入瓦变形对轴承特性的影响. 与此相应，研究计入瓦变形而计算轴承动特性的文章亦逐年有增. 在综合评价文献[2～6]所代表的研究现况的基础上，根据模型合理性和

---

* 本文合作者：吴西柳、郑志祥. 原发表于《上海工业大学学报》，1992，13(3)：189 - 196.

计算效率两方面的考虑,文献[7]选择平面曲梁模型作为计算可倾瓦变形的依据.本文即在[7]的基础上,提出径向可倾瓦静态弹流及小简谐动态弹流的基本方程组及其解算方法和动特性计算公式,在径向可倾瓦常用参数的全部范围内进行计算,归纳瓦变形对动特性的效应及作机理上的阐释,并为"组装"确定轴承动特性值提供单块可倾瓦的基础数据.因此,本文是[7]的后续,也是下一步工作的先篇,即最终确定可倾瓦径向轴承计入静动变形效应的特性,以及相关的转子-可倾瓦轴承动力性态问题的解决.至于支点变形效应的计入,不难在本文的基础上,用串联复刚度的计算及对瓦径向静偏心比的修正来解决,今后将以另文论述.

二、基本关系

瓦与轴颈间的油膜内的流动和压力分布取决于轴颈与瓦的相对位置和相对运动,而与瓦的绝对倾角无关.因此可先在固结于瓦的相对坐标系中进行动态分析(即所谓"固定"瓦分析,"固定瓦"在概念上相当于已作静态偏转而不作振动性倾摆的轴瓦),然后换算得可倾瓦的动特性[8].

当轴颈在静平衡位置邻域内作小扰动时(图1),油膜内的压力分布和间隙分布均可视为其静态分量和扰动分量之和,且后者数量级小于前者.对于转子-轴承系统的不平衡响应问题和稳定界限问题,可认为各扰动分量均作简谐变化.以复数形式表示时,压力和膜厚分布依次为:

$$P(\varphi, Z, T) = P_0(\varphi, Z) + Q(\varphi, Z) \cdot e^{i\Omega T} \tag{1}$$

$$H(\varphi, T) = H_0(\varphi) + M(\varphi) \cdot e^{i\Omega T} \tag{2}$$

后一式的扰动量中,既包含轴颈扰动直接引起的膜厚扰动,又包含扰动压力引起的扰动

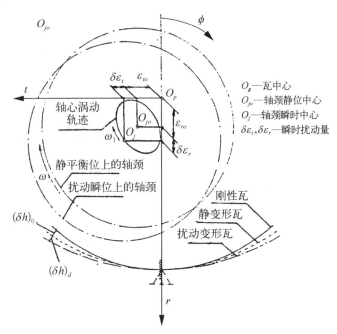

图1 在固结于瓦的相对坐标系中的几何和运动

变形:

$$M(\varphi) \cdot e^{i\Omega T} = (\Delta\varepsilon_t \sin\varphi + \Delta\varepsilon_r \cos\varphi + H_d) \cdot e^{i\Omega T} \tag{3}$$

式中 $\Delta\varepsilon_t$ 和 $\Delta\varepsilon_r$ 依次为轴颈简谐小扰动的 r 和 t 方面振幅,二者之适当得数组合即可表达轴颈之任意简谐小扰动.

压力分布和膜厚分布是互相影响的,其关系由非定常雷诺方程、变形方程和几何关系表达:

$$\frac{\partial}{\partial\varphi}\left(H^3 \frac{\partial P}{\partial\varphi}\right) + \left(\frac{d}{l}\right)^2 \frac{\partial}{\partial Z}\left(H^3 \frac{\partial P}{\partial Z}\right) = 6\frac{\partial H}{\partial\varphi} + 12\frac{\partial H}{\partial T} \tag{4}$$

$$\delta H = (\delta H)_0 + H_d e^{i\Omega T} = \iint_A P(\varphi_1, Z, T) f(\varphi, \varphi_1) d\varphi_1 dZ \tag{5}$$

$$H = 1 + (\varepsilon_{to} + \Delta\varepsilon_t e^{i\Omega T})\sin\varphi + (\varepsilon_{ro} + \Delta\varepsilon_r e^{i\Omega T})\cos\varphi \tag{6}$$

式中 A ——油膜区.

将式(1)~(3)代入式(4)~(6),将定常分量与时变分量分开,并略去高阶小项,可得定常(静态)方程组:

$$\frac{\partial}{\partial\varphi}\left(H_0^3 \frac{\partial P_0}{\partial\varphi}\right) + \left(\frac{d}{l}\right)^2 \frac{\partial}{\partial Z}\left(H_0^3 \frac{\partial P_0}{\partial Z}\right) = 6\frac{dH_0}{d\varphi} \tag{7}$$

$$(\delta H)_0 = \iint_A P_0 f d\varphi dZ \tag{8}$$

$$H_0 = 1 + \varepsilon_{to}\sin\varphi + \varepsilon_{ro}\cos\varphi + (\delta H)_0 \tag{9}$$

及非定常(动态)分量的复幅方程组:

$$\frac{\partial}{\partial\varphi}\left(H_0^3 \frac{\partial Q}{\partial\varphi}\right) + \left(\frac{d}{l}\right)^2 \frac{\partial}{\partial Z}\left(H_0^3 \frac{\partial Q}{\partial Z}\right)$$
$$= -18\frac{M}{H_0}\frac{dH_0}{d\varphi} + 6\frac{dM}{d\varphi} + 12i\Omega M - 3H_0\left(H_0\frac{dM}{d\varphi} - M\frac{dH_0}{d\varphi}\right)\frac{\partial P_0}{\partial\varphi} \tag{10}$$

$$H_d = \iint_A Q f d\varphi_1 dZ \tag{11}$$

$$M = \Delta\varepsilon_t \sin\varphi + \Delta\varepsilon_r \cos\varphi + H_d \tag{12}$$

式(7)~(9)连同 P_0 的边界条件(瓦周边 $P_0=0$,开扩间隙内油膜破裂处的雷诺边界条件),构成其静弹流问题,见[7].

动态方程组(10)~(12)含有两个独立的轴颈扰动幅 $\Delta\varepsilon_t$ 和 $\Delta\varepsilon_r$. 为分别考察其微分效应,将方程组分别对 $\Delta\varepsilon_t$ 和 $\Delta\varepsilon_r$ 求偏导,得复幅偏导数的下列方程组:

$$\frac{\partial}{\partial\varphi}\left(H_0^3 \frac{\partial Q_s}{\partial\varphi}\right) + \left(\frac{d}{l}\right)^2 \frac{\partial}{\partial Z}\left(H_0^3 \frac{\partial Q_s}{\partial Z}\right) = -18\frac{M_s}{H_0}\frac{dH_0}{d\varphi} + 6\frac{dM_s}{d\varphi}$$
$$+ 12i\Omega M_s - 3H_0\left(H_0\frac{\partial M_s}{\partial\varphi} - M_s\frac{dH_0}{d\varphi}\right)\frac{\partial P_0}{\partial\varphi} \tag{13}$$

$$(Hd)_s = \iint_A Q_s f \mathrm{d}\varphi_1 \mathrm{d}Z \tag{14}$$

$$M_s = \frac{\partial(\Delta\varepsilon_t \sin\varphi + \Delta\varepsilon_r \cos\varphi)}{\partial s} + (Hd)_s \tag{15}$$

式中 $Q_s = \dfrac{\partial Q}{\partial s}$, $M_s = \dfrac{\partial M}{\partial s}$, $(s = \Delta\varepsilon_t, \Delta\varepsilon_r)$.

式(13)~(15)连同压力复幅导数 Q_s 的边界条件(全部油膜周边上 $Q_s = 0$),构成小简谐扰动下可倾瓦动态弹流问题. 由此联立解出 Q_s, M_s 和 $(Hd)_s$, 即可按下列式计算"固定"瓦的8个动特性系数:

$$K_{ts}^* + i\Omega B_{ts}^* = -\iint_A Q_s \sin\varphi \mathrm{d}\varphi \mathrm{d}Z \tag{16}$$

$$(s = \Delta\varepsilon_t, \Delta\varepsilon_r)$$

$$K_{rs}^* + i\Omega B_{rs}^* = -\iint_A Q_s \cos\varphi \mathrm{d}\varphi \mathrm{d}Z \tag{17}$$

略去瓦块质量时,一块可倾瓦对于轴颈只有 r 方向的刚度和阻尼,它们可由"固定"瓦动特性换算而得[8]:

$$K_{rr} = K_{rr}^* - \frac{K_{tt}^*(K_{rt}^* K_{tr}^* - \Omega^2 B_{rt}^* B_{tr}^*) + \Omega^2 B_{tt}^*(K_{rt}^* B_{tr}^* + K_{tr}^* B_{rt}^*)}{K_{tt}^{*2} + \Omega^2 B_{tt}^{*2}} \tag{18}$$

$$B_{rr} = B_{rr}^* - \frac{B_{tt}^*(\Omega_2 B_{rt}^* B_{tr}^* - K_{rt}^* K_{tr}^*) + K_{tt}^*(K_{rt}^* B_{tr}^* + K_{tr}^* B_{rt}^*)}{K_{tt}^{*2} + \Omega^2 B_{tt}^{*2}} \tag{19}$$

三、计算方法

上述动态弹流方程组的解算方法与静态方程组的解法相似[7]:用 Vogelpohl 变换以 $H_0^{3/2} Q_s$ 代替 Q_s,用五点格式差分方程组代替式(13),用逐点超松弛(SOR)法解算 Q_s 的离散分布;用基于平面应变曲梁模型的柔度矩阵代替式(14)中的影响系数 $f(\varphi, \varphi_1)$,该矩阵与 Q_s 构成的节点载荷矢量相乘即得到 $(Hd)_s$ 的离散分布;每次迭代计算中按式(15)计入 $(Hd)_s$ 对 M_s 的修正时,均采用亚松弛因子(0.03~0.4);将新的 M_s 代入式(13)重新求 Q_s,如此顺序迭代,直至 Q_s 和 $(Hd)_s$ 同时收敛;按式(16)和(17)用数值积分计算"固定瓦"动特性系数;按式(18)和(19)换算得可倾瓦动特性系数. 上述计算过程,需在取定相对扰动频率 Ω 后进行. 在不同的 Ω 值下,所得之动特性值亦不同,构成可倾瓦的动特性频率谱. 因[7]中已有较详细叙述,本文对解算方法不再展开叙述. 可指出的仅是:动态问题的总体亚松弛因子一般需取得与静态问题时的相同或稍低,以保证有效的收敛.

关于选择平面应变曲梁模型作为变形计算的依据的理由,在[7]中已作说明. 事实上,除[7]以外,现有文献中计入可倾瓦弹性动变形的计算法[2,4,5],亦均依据曲梁模型,显见其在兼顾合理性和易算性方面的特点是被广泛认可的. 本文和[7]之特点则在于免除了不必要的简略或缺陷([2]只考虑曲梁的平均曲率变化;[4]只考虑曲率对截面惯矩的影响;[5]的变形公式忽略了光滑条件). 显然本文的变形计算有较高的合理性. 大量运算表明,本文的方法有很好的收敛性和计算性,无须再作进一步的简化处理.

本文的一切计算中,对于压力、扰动压力、变形、扰动变形的相接两次迭代值的相对误差以及切向合力对径向合力的比值,均取 0.001 为限值来设立相对收敛准则.

为了考验本文方法的正确性,将它在 $K_\delta = 0$ 时的计算结果与用[9]的程序计算的结果进行了比较. 表 1 示其一例. 除 $\varepsilon_{ro} = 0.9$ 时因对分格数和相对收敛精度很敏感因而有一定差别外,在 $\varepsilon_{ro} < 0.9$ 的范围内均相差不大. 这从一个侧面证实了本文计算的正确性.

表 1 刚性瓦计算结果的考核(瓦张角 75°,长径比 0.5,偏支系数 0.6,瓦"固定")

	ε_{ro}	K_{tt}	K_{tr}	K_{rt}	K_{rr}	B_{tt}	$B_{tr} \approx B_{rt}$	B_{rr}
本文	−0.45	0.001 53	−0.001 40	0.053 33	0.060 80	0.006 15	−0.004 70	0.102 10
[9]		0.001 55	−0.001 39	0.053 45	0.061 06	0.006 17	−0.004 68	0.102 32
本文	−0.3	0.004 86	−0.000 50	0.073 23	0.100 68	0.008 01	−0.003 77	0.139 06
[9]		0.004 85	−0.000 53	0.073 46	0.101 06	0.008 04	−0.003 74	0.139 40
本文	−0.1	0.011 93	0.001 65	0.118 89	0.188 67	0.012 40	−0.001 93	0.223 70
[9]		0.011 86	0.001 53	0.119 11	0.188 98	0.012 45	−0.001 94	0.224 39
本文	0.1	0.027 85	0.007 87	0.212 20	0.378 88	0.021 20	0.003 97	0.395 87
[9]		0.027 36	0.006 91	0.211 71	0.375 44	0.021 29	0.003 50	0.396 81
本文	0.3	0.068 31	0.023 44	0.429 23	0.841 05	0.041 66	0.022 60	0.802 98
[9]		0.068 76	0.023 94	0.431 77	0.847 88	0.041 86	0.023 16	0.806 22
本文	0.5	0.210 79	0.093 60	1.102 1	2.393 2	0.103 32	0.112 24	2.053 2
[9]		0.210 03	0.090 25	1.101 6	2.388 3	0.103 58	0.111 45	2.059 0
本文	0.7	0.992 59	0.498 07	4.372 2	10.474	0.410 29	0.776 72	8.303 5
[9]		0.972 28	0.484 99	4.375 7	10.466 79	0.411 12	0.775 16	8.327 2
本文	0.9	12.050	10.722	58.942	200.29	4.240 1	14.640	120.31
[9]		12.858	8.246 6	62.906	191.78	5.310 6	18.262	132.76

四、计算结果和讨论

用上述方法,在广泛参数范围内计算了单块径向可倾瓦的静、动特性. 计算范围为:瓦张角 60°,75°,100°;偏支系数 0.5,0.6,0.65;长径比 0.5,0.6,0.8;径向偏心比≤0.9 的各种数值(间距≯0.1);瓦弹性变形系数 $K\delta = 0, 20, 40, 60, 80, (100)$. 以下仅以瓦张角 75°(四瓦轴承常用参数)长径比 0.5,偏支系数 0.5 的计算结果为例讨论. 在其他参数下,变形影响的大体趋势与之相似.

图 2 例示"固定瓦"4 个刚度系数($\Omega = 1$)的计算结果. 图 3 例示"固定瓦"4 个阻尼系数($\Omega = 1$)的计算结果. 图 4 例示可倾瓦的径向刚度和阻尼系数($\Omega = 1$)的计算结果. 图 5 例示可倾瓦动特性系数与相对频率 Ω 的关系.

由图可见,在相同 ε_{ro} 下,瓦变形效应一般是使刚度和阻尼降低. ε_{ro} 愈大,瓦变形的这种效应亦愈强. 而且阻尼的降低程度更严重. 这与实际情况中观察到的可倾瓦变形效应是一致的[1,2]. 因此,如不计瓦的弹性变形效应,计算得的转子-轴承系统临界转速就可能偏高,过临界转速的共振放大倍数则偏低. 这种预测上的误差,不利于实际机组的运行安全性. 因此轴瓦的变形效应应当尽可能予以计入.

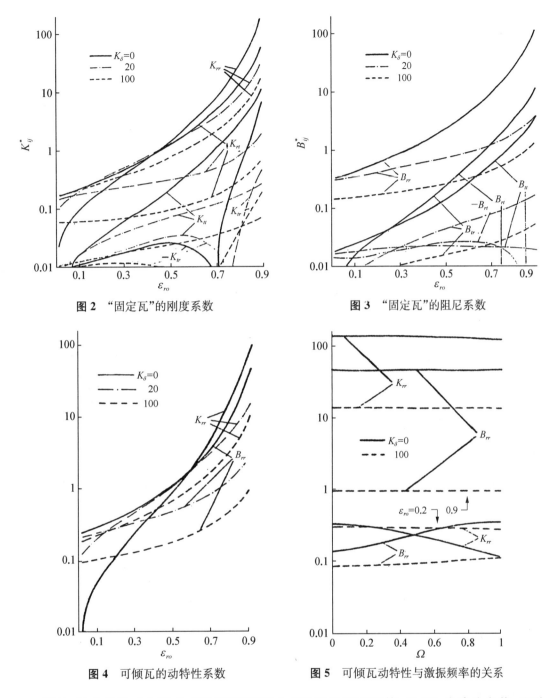

图 2 "固定瓦"的刚度系数

图 3 "固定瓦"的阻尼系数

图 4 可倾瓦的动特性系数

图 5 可倾瓦动特性与激振频率的关系

另外,计入弹性动变形时,"固定瓦"的两个交叉阻尼不再相等,且 B_{rt}^* 常降为负值.这种现象与普通固定瓦轴承的动变形效应是一致的[10].

[7]中曾指出,当 ε_{ro} 小时,瓦弹性变形的效应可能不仅不使瓦的承载力降低,反而由于 ε_{to} 增大而导致承载力加大.这是可倾瓦显著区别于普通固定瓦的特点之一.相似地,由图亦可看到,当 ε_{ro} 很小时,瓦弹性变形效应也可能使"固定瓦"的刚度系数略为增大,其原因亦应归于 ε_{to} 的增大.但"固定瓦"的阻尼则总随变形而降低.至于由"固定瓦"刚度和阻尼的综合效应构成的可倾瓦径向刚度和阻尼,则在全部计算范围内总是随瓦的弹性变形而降低.

计入动变形效应后,"固定瓦"动特性随相对扰动频率(即涡动比)而异,但其依赖关系并不强烈. 至于由相对坐标转换到绝对坐标而获得的可倾瓦动特性,当偏心比 ε_{ro} 和瓦柔度系数 K_δ 较小时,随 Ω 的变化较大,否则变化较小.

五、结论

(1) 计入径向可倾瓦的静动弹性变形后,单瓦油膜刚度降低,油膜阻尼降低更多. 这将影响转子-轴承的动力性态. 因此应尽可能计入瓦静动弹性变形对油膜动特性的影响.

(2) 与现有计入可倾瓦动变形效应的文章相比,本文的平面应变曲梁模型有较高的合理性.

(3) 大量运算表明本文方法有很好的收敛性和计算性.

(4) 在瓦张角、偏支系数、长径比、变形系数、径向偏心比和涡动比等参数的广泛范围(基本上可覆盖全部实用范围)内取得了计入(及不计)弹性变形效应的单块径向可倾瓦的油膜静动特性. 为改进可倾瓦支撑的转子动力性态预测计算,建立了比较合理的单瓦性能数据集.

(5) 瓦的 ε_{ro} 值愈大,瓦弹性变形效应愈强. 在 ε_{ro} 很小的个别情况下,瓦弹性变形有可能反而使油膜刚度略为增大,这可以从 ε_{to} 的增大上找到原因,并且不代表一般趋势.

本文的方法是建立在简谐小振动的假设上,所得轴承动特性本质上只适用于预测临界转速、共振放大倍数、不平衡响应、稳定界限分析、减稳因素界限值确定等问题. 对于对数衰减率(或系统阻尼值)的计算问题,则需对本文方法作适当修改和扩展,非本文内容所及.

参 考 文 献

[1] Caruso, W. J., Gans, B. E., Catlow, W. G. Application of recent rotor dynamics develompments to mechanical drive turbines//Proceedings of the 11th turbomachinery symposium, USA, 1983: 1-17.

[2] Lund, J. W., Pederson, L. B. The influence of the pad flexibility on the dynamic coefficients of a tilting pad journal bearing. Trans. ASME, J. Trib., 1987,109(Jan): 65-70.

[3] 李小江,朱均. 热弹变形对大型可倾瓦轴承性能影响研究. 第四届全国摩擦学学术会议论文,兰州,1985.

[4] 陈祥华,张直明. 计入弹性动变形的可倾瓦径向轴承动力特性研究. 上海工业大学学报,1989(4): 310-316.

[5] Nilsson, L. R. K. The influence of bearing flexibility on the dynamic performance of radial oil film bearings// Proceedings of the 5th Leeds-Lyon Symposium on Tribology, UK, 1969: 311-319.

[6] Brugier, D., Pascal, M. T. Influence of elastic deformations of turbo-generator tilting pad bearings on the static behavior and on the dynamic coefficients in different designs. Trans. ASME, J. Trib, 1989,111(Apr): 364-371.

[7] Zhen, Z., Zhang, Z., Wu, X. Effect of pad deformation on static performance of journal bearing tilting pad. In: Proceedings of the international conference on hydrodynamic bearing-rotor system dynamics, PR China, 1990: 60-66.

[8] 张直明. 滑动轴承的流体动力润滑理论. 高等教育出版社,1986: 106.

[9] Glienicke, J., Han, D.-C. Turbinenlager-Grundprogramm. Universitat Karlsruhe, Bundesrepublik Deutshland, 1981: 91.

[10] Zhang, Z., Mao, Q., Xu, H. The effect of dynamic deformation on dynamic properties and stability of cylindrical journal bearings//Proceedings of 13th Leeds-Lyon Symposium on Tribology, UK, Sept. 1986: 363-366.

Dynamic Properties of Single Journal Bearing Tilting Pads with Consideration of Pad Elastic Deformation

Abstract: This paper investigates the effects of pad elastic deformation on journal bearing tilting pad dynamic properties and rotor dynamic behavior. Small harmonic oscillations are assumed for deriving the dynamic EHL equations of individual pads and the formulae of dynamic property calculation. The finite difference equations for steady state pressure and perturbated pressure distributions of oil film are solved by SOR method. The steady state deformation and perturbated deformation of pad are calculated with a flexibility matrix based on plan strain curved beam. Global forward iterations with under-relaxation are performed to get the simultaneous solutions of the above values. The dynamic properties of deformable tilting pads under different whirl ratios are thus obtained. The general tendency of the effects of pad dynamic deformation on pad dynamic properties are summarized, and explanations are given for the relevent mechanisms. Comprehensive analytical basis and fundamental data are available for better prediction of dynamic behavior of rotor-tilting pad bearing systems.

油叶型轴瓦性能数据库研究*

摘 要：对现有的压力分布解算方法（SOR，PIJ，ADI，Castelli 法，Lund 法，Pan 法，Kato 法和 MGM）进行了大范围的系统实算比较，从精度和速度两方面选出最适合建库用的基本解算法。分析了单块轴瓦各项决定性几何相似参数的最经济而又确保高精度的分档方法及相应的插值方法，使一个长径比的轴瓦的静动特性数据能贮存于一张高密度软盘上。使用了快速读取法。将单瓦数据库支持的通用轴承程序与联邦德国 80 年代透平轴承基本程序作了系统的实算比较。结果表明，在能实现对比程序等温层流全部功能的前提下，仅需约 1/20（可倾瓦轴承）至约 1/100（固定瓦轴承）的时间即可获得相同结果。

符 号 说 明

H—— 无量纲油膜厚度；$H = 1+\varepsilon\cos\varphi$

ε—— 偏心率；$\varepsilon = e/C_{\min}$

$C_{\min}$—— 轴瓦最小半径间隙；$C_{\min} = C - \sqrt{\delta_1^2 + \delta^2}$

δ_1—— 错位距

δ—— 预偏心距

C—— 轴瓦间隙；$C = R-r$

r—— 轴颈半径

$\psi_{\min}$—— 轴承最小相对间隙；$\psi_{\min} = C_{\min}/R$

$\bar{F}_\theta, \bar{F}_\varepsilon$—— θ, ε 方向无量纲油膜力；$\bar{F}_i = \dfrac{\psi_{\min}}{\mu\Omega d l}F_i$

P—— 无量纲油膜压力；$P = \dfrac{p\psi^2}{\mu\Omega}$

$\bar{Q}_1, \bar{Q}_2, \bar{Q}_3$—— 无量纲进、出、侧泄油量；$\bar{Q}_i = \dfrac{Q_i}{r^3\psi_{\min}\Omega}$

$\bar{F}_t$—— 无量纲油膜阻力；$\bar{F}_t = \dfrac{F_t\psi_{\min}}{\mu\Omega d l}$

$\bar{K}_{ij}$—— 无量纲线性化油膜刚度系数；$\bar{K}_{ij} = \dfrac{\psi_{\min}^2}{2\mu\Omega l}k_{ij}$

$\bar{B}_{ij}$—— 无量纲线性化油膜阻尼系数；$\bar{B}_{ij} = \dfrac{\psi_{\min}^3}{2\mu l}b_{ij}$

φ—— 由最大间隙处计量的角坐标

Ω—— 轴颈角速度

l—— 轴承宽度

一、前言

流体动力润滑滑动轴承的设计和分析，通常需要对若干块单块轴瓦多次地解算其油膜压力分布，计算各项静、动特性，并合并成为整个轴承的特性。众多的滑动轴承研究者在各种工作中将不少时间消耗在重复进行基本解算方法和计算程序的研究上。现有的一些通用轴承程序仍未避免对单块轴瓦作多次重复的运算，使运算时间较长，使用不够方便。现有手册和专用轴承数据库所能提供的轴承数据，仅包括了几种常用类型的轴承，不能满足更广的优化设计的要求。

建立滑动轴承中应用最多的油叶类单块轴瓦的性能数据库，配以快速检索取用和精确的插值方法，可以在很大程度上免除轴承研究、设计人员这种重复烦琐的工作，能使轴承的

---

* 本文合作者：李志刚。原发表于《上海工业大学学报》，1992，13(3)：213-219。

许多设计、计算工作很快地完成,并可使原来因工作量太大或工作周期太长而难以实现的工作成为可行.

二、建数据库用的基本解算法的选用

用差分法解算油叶类轴瓦油膜的雷诺方程,可有足够精度,且有运算速度快之优点.目前世界上用于雷诺差分方程组的有效算法很多,如张直明的特征矢量法(PIJ)[3]、Castelli 的矩阵法[4]、Lund 的矩阵法[5]、交叉方向隐式法(Alternative Direction Implicit,ADI)[6]、多重网格法(Multi Grid Method,MGM)[8],以及最常用的 Successive Over-Relaxtion Method (SOR 法)等.此外,Pan 以流量平衡为依据的局部偏微分方程法(Local Partial Differetial Equation Method,LPDEM)[9],及 Kato 的一单元法(One Element Method,OEM,属于迦辽金法)[7]也是近年来出现的有效算法.

Castelli 法计算过程中数值稳定,计算结果精确,计算速度快.其计算过程中大部分时间是化在若干个矩阵 L_j 的求逆上.矩阵的阶数等于轴向半宽分格数 n(图1),需求逆的矩阵的个数等于周向分格数 m 减1.可见 n 值的选取对计算时间有重大影响.作者根据大量实算经验,取 $n=8$,这样既可保证精度,又可最大限度地节约时间.由于压力沿周向的变化较剧,取 $m=60$ 以保证数据库的高精度.固然,如果把此

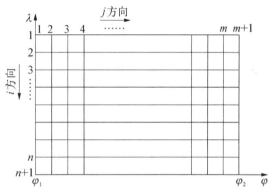

图1 差分法解算轴瓦压力分布网格划分

方法用在计算整个轴承特性上,则为了寻求偏位角 θ,需多次在不同的 H 分布下解算各块瓦的静态压力分布,而每一次均需重建 $m-1$ 个 L_j,计算速度并不显得优越.但对建单块轴瓦数据库,由于是在给定的 H 分布下一次建立 L_j 并求逆后,连续地解算静态压力和数次扰动压力,故整个计算过程中速度相当快.作者对此作过广泛参数范围内的实算比较.比较内容是:各种解算法连续计算一块轴瓦在 0.1～0.95 的 10 种 ε 值下的静动特性参数所需 cpu.轴瓦的参数为 $L/d=0.6$;起始角 $\varphi_1=0°$;终止角 $\varphi_2=210°$;周向分为 60 个格,轴向半宽分为 8 格(即 $n×m=8×60$).各种迭代法的松弛因子或加速因子均选用其在此种计算工况下的最佳值:SOR 法中选用 1.82;LPDEM 中选用 1.65;MGM 中选用 1.34;ADI 法中 $R_k=4\sin^2\left[\dfrac{\pi(2k-1)}{4n}\right]$.各种迭代法计算时需赋初值的地方均赋于零.收敛精度均取残量 $\delta=10^{-4}$ 为判据.

比较结果如表1所列(cpu 单位为 0.01 s).

表1 各种解算法连续计算单块轴瓦在 10 种 ε 下静动特性所需 cpu

计算方法	LPDEM	CASTELLI	PIJ	MGM	ADI	OEM	SOR
cpu	17.82	21.6	30.94	32.77	40.77	41.51	72.88

从比较结果上看,LPDEM 和 CASTELLI 法两种计算方法速度明显优于其他方法,但 LPDEM 法在动特性计算的精度上有所欠缺[2].从速度和精度两个方面综合考虑,选用

CASTELLI法作为建库用的基本解算法较为合适.

三、单块轴瓦性能数据库的建立

建立单块轴瓦数据库主要需考虑以下三个方面的要求：① 数据库的结构尽可能小；② 数据库的数据有足够的计算精度；③ 使用方便，计算省时.

为了达到这三点要求，作者在研究过程中主要作了下列工作：

3.1 单块轴瓦各项决定性几何相似参数分析和采用

油膜二维无量纲雷诺方程为：

$$\frac{\partial}{\partial \varphi}\left(H^3 \frac{\partial P}{\partial \varphi}\right)+\left(\frac{d}{l}\right)^2 \frac{\partial}{\partial \lambda}\left(H^3 \frac{\partial P}{\partial \lambda}\right)=6\left(\frac{\partial H}{\partial \varphi}+2 \frac{\partial H}{\partial T}\right) \tag{1}$$

其边界条件为：

进油边：$\varphi=\varphi_1, P=0$,

出油边：$\varphi=\varphi_2, P=0$,

或 $\varphi=\varphi_p, P=0, \frac{\partial P}{\partial \varphi}=0$

轴向：$\lambda=\pm 1, P=0$

(φ_p 为油膜自然破裂边界的角坐标)

单块轴瓦对轴承的油膜作用力 $\bar{F}_{xi}$ 和 F_{yi}（图2）为：

$$\begin{matrix} \bar{F}_{xi} \\ F_{yi} \end{matrix} = -\int_{-1}^{1}\int_{\varphi_1}^{\varphi_2} P \begin{Bmatrix} \sin(\varphi+\theta) \\ \cos(\varphi+\theta) \end{Bmatrix} d\varphi d\lambda \tag{2}$$

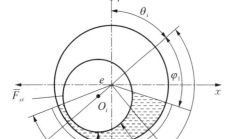

图2 轴瓦油膜力对轴颈的作用

由此可知，单块轴瓦的几何相似参数为：$\varepsilon, \varphi_1, \varphi_2$, l/d. 其中 θ 只是由于轴瓦相对于轴承整体坐标系 $(x-y)$ 的角位置而出现的，并不影响油膜的动力相似性. 在轴瓦的相对坐标系 $(\varepsilon-\theta)$ 中积分计算油膜力 $\bar{F}_{\varepsilon i}$ 和 $\bar{F}_{\theta i}$ 时，则不出现 θ_i.

$$\begin{matrix} \bar{F}_{\varepsilon i} \\ \bar{F}_{\theta i} \end{matrix} = -\int_{-1}^{1}\int_{\varphi_1}^{\varphi_2} P \begin{Bmatrix} \sin\varphi \\ \cos\varphi \end{Bmatrix} d\varphi d\lambda \tag{3}$$

显然有，

$$\begin{Bmatrix} \bar{F}_X \\ \bar{F}_Y \end{Bmatrix}_i = \begin{bmatrix} \sin\theta & \cos\theta \\ \cos\theta & -\sin\theta \end{bmatrix} \begin{Bmatrix} \bar{F}_\varepsilon \\ \bar{F}_\theta \end{Bmatrix} \tag{4}$$

从缩小数据库的结构，节省磁盘空间出发，可选用 $\varepsilon, \varphi_1, \varphi_2, l/d$ 这四个几何相似数为单块瓦数据库的几何参变量. 数据库中每一个纪录行中存放的轴瓦性能数据为：$\bar{F}_\theta, \bar{F}_\varepsilon$, $\bar{F}_t, \bar{Q}_1, \bar{Q}_2, \bar{Q}_3, K_{ij}, B_{ij}(i, j=\varepsilon, \theta)$，而 θ_i 角的影响在用组装法求取轴承性能参数时，由

式(4)完成$(\varepsilon - \theta)$坐标系向$(x - y)$坐标系的转换.

3.2 几何参变量的离散化和插值

存放在数据库内的数据是离散型的,各离散点之间需进行插值计算.对单瓦的四个几何参变量如何离散才能保证插值计算的精度,是数据库技术的一个关键,如果仅从插值精度方面考虑,离散节点的间距当然是越小越好,但如果$\Delta\varepsilon$,$\Delta\varphi_1$,$\Delta\varphi_2$,$\Delta l/d$取得过小,会使节点增加,数据库结构庞大.经过一系列计算验证,四个几何参变量中,$\Delta\varepsilon$的大小是至关重要的.ε越大,插值精度对$\Delta\varepsilon$的敏感性愈强烈,因此$\Delta\varepsilon$应愈小.而其他几何参变量没有类似的特点.所以作者采用了分段设置$\Delta\varepsilon$的办法,以同时保证插值精度和数据库的小结构化问题.目前一个长径比的轴瓦特性数据能存放在一张高密度的软盘上,而插值精度仍很高.据验证:当$\varepsilon \leqslant 0.95$,$\bar{F}_\varepsilon$和$\bar{F}_\theta$的相对误差$\delta < 0.5\%$,当$0.95 < \varepsilon < 0.993$时,$\delta < 3\%$.

插值公式为四维三节点拉格朗日插值公式,采用这个插值公式是基于这样的考虑:如采用二节点的线性插值,每一次插值时读取的节点数虽少些,但要保证较高插值精度,势必要减少插值节点间距,不利于数据库的小结构化.如采用高于三节点以上的插值公式,$\Delta\varepsilon$,$\Delta\varphi_1$,$\Delta\varphi_2$和$\Delta l/d$比目前的方法的取值可以有所增大,但需读取过多的节点数,造成速度减慢,这也是不可取的.从以上两个方面综合考虑,选用了三节点的抛物线插值公式.

3.3 快速读取法

使用单瓦数据库省时的关键在于检索速度的快慢.单瓦数据库实质上是一个(由FORTRAN77语言支持下的)大型直接存取数据文件.数据的存入规律决定了它们在数据库内的排列顺序也存在着某种规律.在调用数据时,只需按其在库内存放的规律,计算出该数据在库内的纪录行号NUM(i,j,k),即可用FORTRAN77所支持的直接存放文件的Read语句从数据文件中按号读出该行纪录.这样可使检索速度非常快,从而使数据库省时的特点得到充分发挥.

四、计算速度与精度的比较

4.1 计算精度比较

将调用数据库内单块轴瓦性能参数(经插值计算的)组装计算轴承性能的通用计算程序HIEF的计算结果与调用子程序从头计算单瓦性能数据的通用计算程序DBADI的计算结果进行比较.DBADI所调用的基本算法子程度仍为CASTELLI法,计算时网格划分$n \times m = 8 \times 60$,均与建库时相同.两个计算程序除了在单瓦数据获得的方法上的区别外,其他均完全一样.这样处理的目的是为了防止由基本解算法不同,网格划分不同或组装方式不同造成的误差在上述两种计算结果的比较中出现.

图3为$2 \times 150°$剖分式圆轴承计算结果的比较.

图4为$3 \times 100°$三油叶轴承计算结果的比较.预负荷系数$m=0.5$.

图5为$5 \times 60°$均布可倾瓦轴承计算结果比较,$m=0$,$\beta_2 = 30°$(β_2为轴瓦的始边到支点的角度).

从上述图中可清楚地看出,两种方法计算结果的误差非常小,两种曲线基本重合.从而

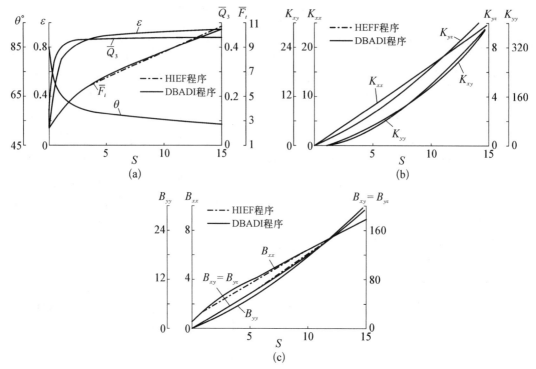

图 3 2×150°剖分式圆轴承计算结果比较($l/d=0.6$)
(a) 静特性系数 (b) 刚度系数 (c) 阻尼系数

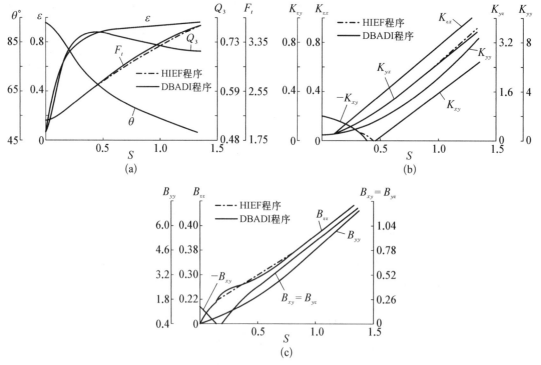

图 4 3×100°三油叶轴承计算结果比较($m=0.5; l/d=0/6$)
(a) 静特性系数 (b) 动特性系数 (c) 阻尼系数

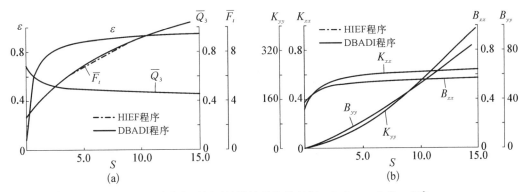

图 5 5×60°均布可倾瓦计算结果比较($l/d=0.6; m=0, \beta_2=30°$)
(a) 静特性系数　(b) 动特性系数

证实了用单瓦轴承数据库取代从头计算单瓦数据的办法,在计算精度上是完全可行的.

4.2 计算速度的比较

计算速度的比较是在 HIEF 和 80 年代初的联邦德国 ALPE 轴承通用计算程序之间进行的. 计算工况为:层流,等粘. 比较是在 IBM4361 机上进行的. 计算了五种轴承:2×150°剖分式圆轴承;2×150°椭圆轴承($m=0.5$);3×100°三油叶轴承($m=0.5$);5×60°可倾瓦轴承($m=0.5$ 和 0 两种). 每种轴承均连续计算 19 种偏心率($\varepsilon=0.05\sim0.95, \Delta\varepsilon=0.05$)的轴承全部静,动特性系数. 结果表明:对于固定瓦轴承,HIEF 比 ALPE 快 71~101 倍,对于可倾瓦轴承则快近 20 倍.

五、结论

油叶型单块轴瓦性能数据库因其数据精确,使用方便,高效实用等特点,为径向滑动轴承的研究和设计提供了一种较好的计算手段. 目前建立的数据库,已可解决常规和非常规设计的油叶类轴承和可倾瓦轴承在层流、等粘、不计进油压力的条件下的性能计算. 为了扩大应用范围,还将研究计入若干附加因素的方法.

参 考 文 献

[1] 张直明,等. 滑动轴承的流体动力润滑理论. 北京:高等教育出版社,1986.
[2] 李志刚. 滑动轴承数据库研究. 上海工业大学硕士学位论文,1990.
[3] 张直明. PIJ 过程(特征矢量法),西安交通大学(大型计算程序),1979.
[4] Castelli, V., Shapiro, W. Improved method for solution of the general imcompressible fluid film lubrication problem. ASME Trans.,J. Lubr. Tech., 1967,89:211-218.
[5] Lund, J. W. The pivoted, spherical cap slider bearing. ASME Trans., J. Lubr. Tech., 1982, 104(Apr.):216-219.
[6] Han Dong-Chul. Statische und Dynamische Eigenchuften Von Gleitagern bei Hohen Umfangsge Schwindigkeiten und bei Verkantung. Dissertation,Karlsruhe,1979.
[7] Kato, T. A fast method for calculating dynamic coefficients of finite width journal bearing with quasi Reynolds boundary condition. J. Trib., 1988,110:387-393.
[8] Hackbusch, W. 多重网格法. 林群,等译. 北京:科学出版社,1984.

[9] Pan, C. H. T. A new numerical technique for the analysis of lubricating films, part 1: imcompressive, isoviscous lubricant//Proc. 13th Leeds-Lyon Symposium on Tribology, Leeds, Sept. 1986: 8 - 12.

Investigation on Database of Properties of Bearing Bush of Lobe Type

Abstract: The fundamental algorithm most suitable for building bush property database is chosen from a number of existing effective methods-SOR, PIJ, ADI, Castelli's method, Lund's method, Pan's method, Kato's method and MGM, based on a comparision in accuracy and speed of calculation shown in a series of actual calculation in a wide range of parameters. Analysis of effects of various geometric similarity parameters is made, which enables selecting the most economic discretization and corresponding method of interpolation of these parameters, while maintaining high accuracy. This results in the possibility of storing all the static and dynamic coefficients of bearing bush of a given length-to-diameter ratio in a single floppy disk of high density. Quick access is effected. Examplar calculations are made to compare a general purpose bearing program supported by this database with a West German turbin bearing program of the nineteen eighties. The results are: to get the same numerical results, the present technique requires only about 1/20 (for tilting pad bearings) to about 1/100 (fixed bush bearings) of computational time, while capable of carrying out all the funtions of the German program in isothermal, laminar regime.

计入弹性变形的可倾瓦轴承和
转子系统的动力性态*

摘　要：本文在对径向可倾瓦轴承单块瓦的静态和动态弹流分析基础上,将计入瓦弹性变形效应的可倾瓦特性用"组装"法获得整个轴承的静、动特性,并分析计算了它对转子系统主要动力性态的影响,得到了一些有实用意义的结论.

一、引言

可倾瓦轴承中的变形对轴承和转子系统动力性态的影响已引起广泛注意[1-5]. 主要的变形环节为支点、瓦及壳体. 引起变形的原因主要是温度变化及弹性变形. 对于径向可倾瓦轴承,温度变形受边界条件的影响很大,在各种情况下颇不一致,不易精确预测,至今尚无公认的完善预测方法. 一般只宜根据具体条件下的经验,通过对轴承间隙和预负荷系数的校正而粗略计入其影响,[2]即采取此种现实处理法. 从分析的观点看,温度变形是一种定常的或缓变的过程,它主要是通过对轴承静态参数的影响而间接影响到轴承动特性和转子动力性态. 本文认同[2]的方法,假设轴承的间隙和预负荷系数已计入了温度变形,对之不再进行讨论.

弹性变形主要发生于支点和瓦,壳体的弹性变形通常要小得多. 就轴承动特性和转子动力性态而言,不仅要考虑由定常载荷和油膜压力引起的支点和轴瓦静变形,而且要考虑与轴颈扰动相伴随的力和压力的动态增量引起的变形动态增量(动变形)[5]. 瓦的静、动变形及其效应已于[5-7]中分析处理. 如所周知,当略去瓦惯性时,可倾瓦在简谐小振动下的动特性可归结为一个径向刚度和一个径向阻尼[8]. 支点的弹性动变形效应,亦主要表现为一个径向刚度和一个径向阻尼. 因此,支点和瓦动变形的复合效应,可以用该方向上两个复刚度的串连关系来计算. 对于给定的轴颈(相对于轴承中心)的偏心位置,支点的静变形效应,在于改变了瓦的位置,亦不难在由轴颈位置计算各瓦偏心时予以表达. 至于支点的刚度,一般不像轴瓦那样涉及变形分布的分析计算,而常可根据支点与瓦的接触几何和材料性能直接用赫茨公式算出,或再根据实际经验略作修正[1].

综上可见,从分析观点看来,问题的关键在于单块可倾瓦的弹性静变形和动变形,特别是因为它的计算与润滑方程的解算紧密偶合,以致必须用弹流分析专门处理. 本文即在[5-7]已对实用广泛参数范围内的径向可倾瓦计入弹性变形而建成的单瓦特性数据库基础上,用工程上对可倾瓦轴承常用的"组装"法计算整个轴承计入静、动变形后的动特性;在"组装"过程中用校正瓦偏心的方法计入支点静变形,用串连复刚度的方法计入支点动变形. 然后将轴承动特性与转子动力方程相结合,获得可倾瓦轴承弹性变形对转子系统动力性态的

---

* 本文合作者：吴西柳、郑志祥. 原发表于《上海工业大学学报》,1992,13(4)：303-310.

影响规律.

二、支点弹性变形的计入方法

依靠单块可倾瓦的静、动弹流计算,已建立了给定参数(长径比 l/d、瓦张角 β、偏支系数 γ)下的单瓦性能数据库,其内容为各种瓦径向偏心率 ε_{ro} 和瓦变形因子 K_δ 下的单瓦无量纲静特性[承载量系数 $\bar{F}_r = F_r \psi^2/(\eta \omega dl)$、无量纲最小膜厚 $H_{\min} = h_{\min}/c$、阻力系数 $\bar{F}_t = F_t \psi/(\eta \omega dl)$、侧泄量系数 $\bar{Q}_s = 2Q_s/(\omega rcl)$ 及动特性频谱(小简谐扰动时各种涡动比 Ω 下的径向刚度系数 $K_{rr} = k_{rr} \psi^3/(2\eta \omega l)$ 和阻尼系数 $B_{rr} = b_{rr} \psi^3/(2\eta l)$,即:$\varepsilon_{ro}$,$K_\delta \sim \bar{F}_r$,$H_{\min}$,$\bar{F}_t$,$\bar{Q}_s$,$K_{rr}$,$B_{rr}$ 关系数表].

支点静变形可如下计入. 由图 1 可知,当轴心在某瓦的径向(r 方向)有绝对位移 e'_r 时,由于支点受力变形 δ_p,使瓦沿 r 方向移动了距离 δ_p,所以轴心相对于移动了的瓦心 O' 的实际径向偏心为

$$e_{ro} = e'_r - \delta_p \tag{1}$$

换言之,要使轴心相对于瓦达到某个实际径向偏心 e_{ro},轴心沿 r 方向的绝对位移要加大一个量 δ_p:

$$e'_r = e_{ro} + \delta_p \tag{2}$$

且

$$\delta_p = F_r/k_p \tag{3}$$

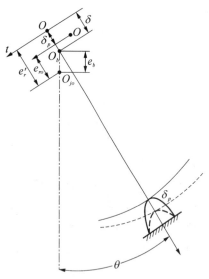

图 1 支点的静变形效应

式中 k_p 为支点刚度.

以瓦间隙 c 为相对单位时,(2)的无量纲形式为:

$$\varepsilon'_r = \varepsilon_{ro} + \alpha_p \bar{F}_r \tag{4}$$

式中 α_p 为支点柔度系数,$\alpha_p = 2\eta \omega l/(k_p \psi^3)$.

因为 $\bar{F}_r$ 视 ε_{ro} 而定,故当根据支点刚度算出支点柔度系数后,即可对每一 ε_{ro} 算出相应的 ε'_r 值,从而直接将轴心对一瓦的绝对径向偏心率 ε'_r 与该瓦的无量纲特性值相联系起来. 由于对各瓦的 ε'_r 可直接由对轴承的偏心比 ε_b 算出,这样对于用"组装"法确定轴承特性是十分方便的.

支点动变形则用串连法计入. 瓦在扰动频率 ω_1 下的径向刚度 k_{rr} 和阻尼 b_{rr} 可视为复刚度 k'_{rr}:

$$k'_{rr} = k_{rr} + i\omega_1 b_{rr} \tag{5}$$

瓦与支点的串连复刚度 k'_s 则为:

$$k'_s = \frac{1}{\dfrac{1}{k'_{rr}} + \dfrac{1}{k_p}} \tag{6}$$

或将其表达为串连刚度 k_s 和阻尼 $b_s(k'_s = k_s + i\omega_1 b_s)$，则：

$$k_s = k_p \frac{k_{rr}(k_p + k_{rr}) + (\omega_1 b_{rr})^2}{(k_p + k_{rr})^2 + (\omega_1 b_{rr})^2}; \quad b_s = \frac{k_p^2 b_{rr}}{(k_p + k_{rr})^2 + (\omega_1 b_{rr})^2} \tag{7}$$

或以无量纲形式表达为：

$$K_s = \frac{1}{\alpha_p} \frac{K_{rr}(K_{rr} + 1/\alpha_p) + (\Omega B_{rr})^2}{(K_{rr} + 1/\alpha_p)^2 + (\Omega B_{rr})^2}; \quad B_s = \frac{B_{rr}/\alpha_p^2}{(K_{rr} + 1/\alpha_p)^2 + (\Omega B_{rr})^2} \tag{8}$$

式中 $K_s = k_s \psi^3/(2\eta\omega l); B_s = b_s \psi^3/(2\eta l)$.

当 α_p 算出后，K_s 和 B_s 亦可直接依决于 ε'_r.

用 ε'_r，$K_\delta \sim \overline{F}_r$，$H_{\min}$，$\overline{F}_t$，$\overline{Q}_s$，$K_s$，$B_s$ 的关系取代 ε_{ro}，$K_\delta \sim \overline{F}_r$，$H_{\min}$，$\overline{F}_t$，$\overline{Q}_s$，$K_{rr}$，$B_{rr}$ 的关系，就计入了支点的静、动变形效应. 据此即可用一般可倾瓦轴承的"组装"法获得计入支点变形的轴承特性.

三、"组装"轴承特性

对于各瓦支点的布置对称于轴承载荷 F 的情况（图 2），轴颈相对于轴承中心的静态偏心 e_b 总是沿载荷 F 方向. 根据各瓦的预偏心 δ，瓦 k 的支点的位置角 θ_k，以及偏心距 e_b，即可算出瓦 k 的 e'_{rk} 值（图 1）：

$$e'_{rk} = \delta + e_b \cos\theta_k \tag{9}$$

或即

$$\varepsilon'_{rk} = m + \varepsilon_b(1-m)\cos\theta_k \tag{10}$$

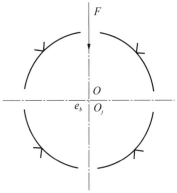

图 2 支点布置对称于轴承载荷

式中 $\varepsilon'_r = e'_r/c; \varepsilon_b = e_b/c; m = \delta/c$. 如前所述，由 ε'_{rk} 值即可从单块可倾瓦数据库中插值求得瓦 k 的各项无量纲特性值，由此再合成整个轴承的无量纲特性值：

$$So_b = \frac{F\psi_{\min}^2}{\eta\omega dl} = \sum_{k=1}^n \overline{F}_{r,k} \cos\theta_k (1-m)^2 \tag{11}$$

$$H_{\min,b} = h_{\min,b}/c_{\min} = \min(H_{\min,k})/(1-m) \tag{12}$$

$$\overline{F}_{t,b} = \frac{\overline{F}_{t,b}\psi_{\min}}{\eta\omega dl} = \sum_{k=1}^n \overline{F}_{t,k}(1-m) \tag{13}$$

$$\overline{Q}_{s,b} = \frac{2Q_{s,b}}{\omega r c_{\min} l} = \sum_{k=1}^n \overline{Q}_{s,k}/(1-m) \tag{14}$$

对每一 Ω，
$$K_{ij,b} = \frac{k_{ij,b}\psi_{\min}^3}{2\eta\omega l} = \sum_{k=1}^n K_{rr,k} U_i U_j (1-m)^3 \quad (i,j = x,y) \tag{15}$$

$$B_{ij,b} = \frac{b_{ij,b}\psi_{\min}^3}{2\eta l} = \sum_{k=1}^n B_{rr,k} U_i U_j (1-m)^3$$

$$(i, j = x, y) \tag{16}$$

其中 $U_x = \sin\theta_k; U_y = \cos\theta_k$.

当给定瓦数、支点布置和预负荷系数 m 后,即可根据单块可倾瓦的数据库"组装"得一系列 ε_b 下的上述无量纲轴承特性,而构成该轴承的数据文件. 由于 $So_b \sim \varepsilon_b$ 是一一对应关系,所以轴承的数据文件亦即表达了 $So_b \sim H_{\min,b}$, $\bar{F}_{t,b}$, $\bar{Q}_{s,b}$, $K_{ij,b}(\Omega)$, $B_{ij,b}(\Omega)$ $(i, j = x, y)$ 的关系,可直接用于设计计算和转子动力学计算.

四、转子系统的动力性态

以图3所示的单质量对称转子和一对相同轴承构成的系统为分析对象. 对于由残余不平衡引起的同步振动,轴承的作用可以用涡动比为1的小简谐扰动下的动力特性来表达. 至于其他各种涡动比的小简谐扰动下的轴承动力特性,可以用来对转子系统在稳定界限上的性态作分析. 例如可以计算转子系统在工作转速下所能抗受的减稳因素的界限值,借以评价系统的稳定裕度. 本文仅以正旋交叉刚度和负阻尼[11]作为减稳因素之例,来进行此种分析.

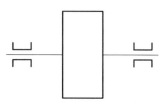

图3 转子-轴承系统模型

关于不平衡响应的计算方法和原理,许多文献(例如[10])中都可查到,此处不再赘述. 本文用通常的方法,考察了瓦和支点的弹性变形对临界转速、共振放大倍数和不平衡响应的影响.

关于抗减稳因素的计算,系以齐次运动方程为依据. 在交叉刚度界限值 $k_{\rm st}$ 下,有:

$$m_r \ddot{x}_r + k_r(x_r - x_j) - k_{\rm st} y_r = 0 \tag{17}$$

$$m_r \ddot{y}_r + k_r(y_r - y_j) + k_{\rm st} x_r = 0 \tag{18}$$

$$k_r(x_r - x_j) = 2(k_{xx,b} x_j + b_{xx,b} \dot{x}_j + k_{xy,b} y_j + b_{xy,b} \dot{y}_j) \tag{19}$$

$$k_r(y_r - y_j) = 2(k_{yx,b} x_j + b_{yx,b} \dot{x}_j + k_{yy,b} y_j + b_{yy,b} \dot{y}_j) \tag{20}$$

在稳定界限上,可令

$$[x_r, y_r, x_j, y_j]^{\rm T} = [\hat{x}_r, \hat{y}_r, \hat{x}_j, \hat{y}_j]^{\rm T} e^{i\omega_1 t} \tag{21}$$

采用对称布置的可倾瓦轴承时,$k_{xy,b} = k_{yx,b} = b_{xy,b} = b_{yx,b} = 0$. 由方程(17)~(21)存在非平凡解的条件可得交叉刚度的无量纲界限值 $K_{\rm st}$ 和无量纲界限涡动频率 $\bar{\omega}_1$ 为:

$$K_{\rm st} = 2\alpha \left[\frac{\bar{\omega}_1^2 \alpha^2 \beta_{xx} B_{yy}}{(\beta_x^2 + \bar{\omega}_1^2 B_{xx}^2)(\beta_y^2 + \bar{\omega}_1^2 B_{yy}^2)} \right.$$
$$\left. - \left(\frac{\bar{\omega} K_{xx}\beta_x + \bar{\omega}_1^2 B_{xx}^2}{\beta_x^2 + \bar{\omega}_1^2 B_{xx}^2} - \bar{\omega}_1^2 \right) \left(\frac{\bar{\omega} K_{yy}\beta_y + \bar{\omega}_1^2 B_{yy}^2}{\beta_y^2 + \bar{\omega}_1^2 B_{yy}^2} - \bar{\omega}_1^2 \right) \right]^{\frac{1}{2}} \tag{22}$$

$$\bar{\omega}_1 = \left(\frac{-b + \sqrt{b^2 - 4ac}}{2a} \right)^{\frac{1}{2}} \tag{23}$$

式中 $\alpha = \dfrac{2So_k}{f/c_{\min}}; \beta_x = \alpha + \bar{\omega}K_{xx}; \beta_y = \alpha + \bar{\omega}K_{yy};$

$a = B_{xx}B_{yy}(B_{xx}+B_{yy}); b = B_{xx}\beta_y^2 + B_{yy}\beta_x^2 - a; c = -\bar{\omega}(K_{xx}B_{yy}\beta_x + K_{yy}B_{xx}\beta_y).$

上述计算需用选代法进行,即:先设取一涡动比 Ω 值,查取相应的轴承动特性系数,由(23)计算 $\bar{\omega}_1$,再计算与此相应的涡动比 $\Omega' = \bar{\omega}_1/\bar{\omega}$,当 Ω' 与 Ω 的相对误差超过 0.001 时,将 Ω' 作为新的涡动比估取值 Ω,重新计算,直至收敛. 此时的轴承动特性值,即用以计算 K_{st} 值.

在负阻尼界限值 b_{st} 下,运动方程为:

$$m_r\ddot{x}_r + k_r(x_r - x_j) - b_{st}\dot{x}_r = 0 \tag{24}$$

$$m_r\ddot{y}_r + k_r(y_r - y_j) - b_{st}\dot{y}_r = 0 \tag{25}$$

$$k_r(x_r - x_j) = 2(k_{xx,b}x_j + b_{xx,b}\dot{x}_j) \tag{26}$$

$$k_r(y_r - y_j) = 2(k_{yy,b}y_j + b_{yy,b}\dot{y}_j) \tag{27}$$

用类似于上的方法,可得负阻尼的无量纲界限值 B_{st} 为:

$$B_{st} = \min(B_{st,x}, B_{st,y}) \tag{28}$$

而

$$B_{st,i} = \dfrac{\alpha^2 B_{ii}}{\beta_i^2 + \bar{\omega}_{1,i}^2 B_{ii}^2} \quad (i = x, y) \tag{29}$$

无量纲界限涡动频率 $\bar{\omega}_{1,x}$ 或 $\bar{\omega}_{1,y}$ 为:

$$\bar{\omega}_{1,i} = \left(\dfrac{-b_i + \sqrt{b_i^2 - 4a_ic_i}}{2a_i}\right)^{\frac{1}{2}} \tag{30}$$

式中 $a_i = B_{ii}^2; b_i = \beta_i^2 - a; c_i = -\bar{\omega}K_{ii}\beta_i (i = x, y).$

亦需先选代至涡动比收敛,然后计算 B_{st}.

五、计算结果及讨论

以图 2 的四瓦轴承为例进行了计算. 瓦参数为:长径比 0.5,瓦张角 75°,偏支系数 0.6,预负荷系数 0.5. 为了检验计算方法的正确性,将不计瓦和支点变形的轴承性能计算结果与 [10] 中的数值作了对比,如图 4. 可见二者符合良好.

由图 4 可见,当瓦有弹性变形时,在同样承载量系数 So 下,最小膜厚随瓦变形系数 K_δ 增大而急速减低. 轴承刚度在轻载时(So 小)降低,重载时(So 大)增大. 轴承阻尼则随瓦变形系数的增大而显著下降.

支点弹性变形对轴承性能的影响,如图 5 所示. 它对最小膜厚影响不很大,但使轴承刚度和阻尼显著下降. 瓦变形和支点变形的同时计入,则兼有上述两种效应.

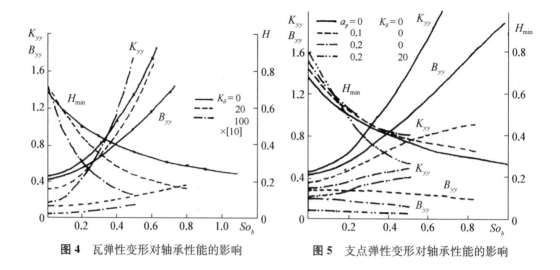

图 4　瓦弹性变形对轴承性能的影响　　　　　图 5　支点弹性变形对轴承性能的影响

轴承特性的这种变化,必须影响到转子系统动力性态.不难预计,轴承刚度的变化将使临界转速相应地有所变化,轴承阻尼显著下降则将严重增高共振振幅.计算结果亦无不如此.此处仅以图 6 所示数条转速-振幅曲线例示瓦变形的这种影响.图 7 集合了众多这类计算所得之共振振幅,更清楚表明瓦变形之不良影响.一些高速旋转机械中采用可倾瓦轴承时,有时其共振振幅很大而成为困难问题,宜分析是否与弹性变形有关.

图 6 中亦表示了各种转速下转子系统抗受减稳因素的能力.由图可见,弹性变形不利于系统稳定性.它虽不会直接导致失稳,但严重影响稳定裕度.K_{st} 和 B_{st} 的变化趋势大致相同.

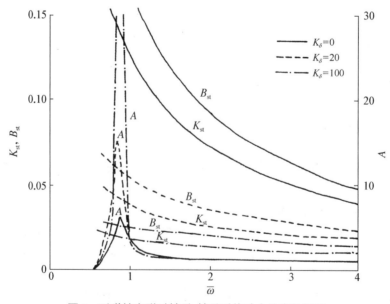

图 6　瓦弹性变形对转子-轴承系统动力性态的影响

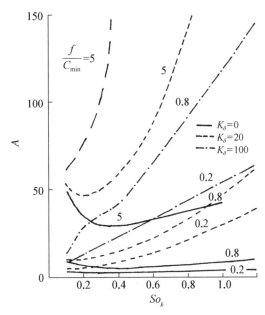

图 7 瓦变形对共振振幅的影响

六、结论

1. 用基于平面应变曲梁模型的单块可变形可倾瓦数据库方法,可以在较[2]更合理的基础上有效地实现可变形可倾瓦轴承及其转子系统的计算.

2. 本文算例表明,瓦的弹性变形使轴承最小膜厚降低,使轴承刚度有所变化,并使轴承阻尼显著降低;支点弹性变形则使轴承刚度和阻尼均下降.

3. 瓦和支点的弹性变形使转子系统的临界转速有所变化.

4. 瓦和支点的弹性变形使转子系统的共振振幅急剧增高.

5. 瓦和支点的弹性变形使转子系统的稳定裕度(以抗受正旋交叉刚度或负阻尼的界限值为标志)降低.

6. 由于瓦和支点弹性变形造成的影响均属不良,有时甚为严重,因此设计时应注意保证瓦和支点刚度. 对于变形系数较大的轴承必须计入弹性变形的影响.

7. 本文将支点视为线性弹性元件,主要在于考察其影响趋势. 如需计入实际支点刚度的非线性,应适当修改本文对支点静变形的处理方法. 支点动变形仍可按线性元件处理,因为动态分析是以小振动为前提.

参 考 文 献

[1] Caruso, W. J., Gans, B. E., Catlow, W. G. Application of recent rotor dynamics developments to mechanical drive turbines. Proceedings of the 11th Turbomachinery Symposium, 1983: 1 – 17.

[2] Lund, J. W., Pederson, L. B. The influence of the pad flexibility on the dynamic coefficients of a tilting pad journal bearing. Trans. ASME, J. Trib., 1987,109(Jan.): 65 – 70.

[3] Nilsson, L. R. K. The influence of bearing flexibility on the dynamic performance of radial oil film bearings. Proceedings of the 5th Leeds-Lyon Symposium on Tribology, 1969: 311 – 319.

[4] Brugier, D., Pascal, M. T. Influence of elastic deformations of turbo-generator tilting pad bearings on the static behavior and on the dynamic coefficients in different designs. Trans. ASME, J. Trib., 1989, 111 (Apr.): 364-371.

[5] Zhen, Z., Zhang, Z., Wu, X. Effect of pad deformation on static performance of journal bearing tilting pad. Proceedings of the International Conference on Hydrodynamic Bearing-Rotor System Dynamics, Xi'an, 1990: 60-66.

[6] 郑志祥,吴西柳,张直明. 用曲梁模型对单块可倾瓦进行弹流研究. 上海工业大学学报,1991(3): 213-221.

[7] 张直明,吴西柳,郑志祥. 计入弹性动变形的径向轴承可倾瓦特性. 上海工业大学学报,1992(3): 189-196.

[8] 张直明,张言羊,谢友柏,等. 滑动轴承的流体动力润滑理论. 高等教育出版社,1986: 106.

[9] Zhang, Z., Mao, Q., Xu, H. The effect of dynamic deformation on dynamic properties and stability of cylindrical journal bearings. Proceedings of 13th Leeds-Lyon Symposium on Tribology, Sept. 1986: 363-366.

[10] Glienicke, J., Leonhard, M. Stabilitätsprobleme bei der Lagerung schnellaufender Wellen, Bericht, Universitaet Karlsruhe, Juni 1981: 136.

[11] 张直明,虞烈. 滑动轴承-转子系统的系统阻尼值与稳定裕度的相互关系. 上海工业大学学报,1985(4): 11-20.

Dynamic Behavior of Tilting Pad Bearings and Rotor Systems with Consideration of Pad Elastic Deformation

Abstract: Based on steady state and dynamic EHL analysis of individual pads in tilting pad journal bearings, an "assembly" method is used to get the steady state and dynamic properties of the complete bearings with consideration of the effect of pad elastic deformation. The influence of pad deformation on the main dynamic behaviors of rotor systems supported on tilting pad bearings is analyzed. Results of practical meaning are obtained.

Non-newtonian Elastohydrodynamic Lubrication Analysis of Rib-roller End Contact in Tapered Roller Bearings *

Abstract: A complete numerical solution has been obtained for elastohydrodynamically lubricated contacts of rib-roller end in tapered roller bearings taking into account the effects of non-Newtonian behaviour of lubricants. A limiting shear stress model of lubricant shear rheology proposed by Bair & Winer is adopted to investigate its influence on oil film thickness, pressure distribution and traction forces. The optimal values of the ratio of curvature are deduced with full consideration of the special geometric and kinematic aspects. The theoretical non-Newtonian film thickness agrees well with the existing experimental data.

Nomenclature

a, b = semiminor and semimajor axes of contact ellipse respectively, m

E' = effective elastic modulus, $E' = 2/((1 - \nu_1^2)/E_1 + (1 - \nu_2^2)/E_2)$, N/m^2

f_{1x} = friction coefficient on rib face

F_{1x} = dimensionless traction force acting on rib face, $F_{1x} = F_{1x}/(abE')$

$F(\tau_e)$ = rheology function

h_c, $h_{\min}$ = central and minimum film thickness, m

H = dimensionless film thickness, $H = h/R_x$

H_0 = dimensionless constant used in calculation of H

J_0, J_1, J_2 = integrals involving rheology function

k = ellipticity parameter, $k = a/b$

M_1 = traction torque on rib, Nm

p = pressure, N/m^2

P = dimensionless pressure, $P = p/E'$

Q = dimensionless reduced pressure

r = large end radius of roller, m

r_i, r_o = inner and outer radii of rib, m

R_x, R_z = equivalent radii of curvature in x- and z-direction, m

R_{x1}, R_{x2} = radii of curvature in x-direction of rib and roller end respectively, m

$u_{1(0)}$, $u_{2(0)}$ = velocities in moving direction of rib and roller, respectively, m/s

$w_{1(0)}$, $w_{2(0)}$ = velocities in transverse direction of rib and roller respectively, m/s

W = dimensionless normal load on rib-roller end contact, $W = w/(abE')$

X = dimensionless coordinate in x-direction, $X = x/b$

Y = dimensionless coordinate in direction of film, $Y = y/b$

y = variable transformation, $y = y/h$

Z = dimensionless coordinate in z-direction, $Z = z/a$

β = half cone angle, 11.642 2°

β_r = sliding-rolling ratio, $\beta_r = (u_{1(0)} - u_{2(0)})/(u_{2(0)} + u_{2(0)})$

β_s = spinning-rolling ratio, $\beta_s = (\Omega_{1y} - \Omega_{2y}) \cdot \sqrt{ab}/(u_{1(0)} + u_{2(0)})$

γ = half roller angle, 2.0°

σ = variable, $\sigma = b\eta_0/(E'Rx)$, s/m

$\bar{\delta}$ = dimensionless elastic deformation, $\bar{\delta} = \delta/R_x$

ε = distance from the nominal contact point to inner-race, m

$\bar{\eta}$ = dimensionless absolute viscosity, $\bar{\eta} = \eta/\eta_0$

---

* In collaboration with JIANG Xiaofei and WONG Pat Lam Patrick. Reprinted from *ME Research Bulletin*, 1993, 1(1): 31–46.

η_0 = ambient viscosity of lubricant, Pa·s

$\bar{\rho}$ = dimensionless density, $\bar{\rho} = \rho/\rho_0$

ρ_0 = ambient density of lubricant, kg/m^3

Ω_{1y}, Ω_{2y} = angular velocities along y-direction of rib and roller respectively, rad/s

τ_L = limiting shear stress, N/m^2

$\bar{\tau}_{YX}$, $\bar{\tau}_{YZ}$ = dimensionless shear stresses, $\bar{\tau}_{YX} = \tau_{yx}/E'$, $\bar{\tau}_{YZ} = \tau_{yz}/E'$

$\bar{\tau}_{YX}|_{Y=0}$, $\bar{\tau}_{YZ}|_{Y=0}$ = dimensionless shear stresses on rib face

1 Introduction

Industrial problems related to the elastohydrodynamic lubrication continue to arise despite widespread research efforts accompanied by numerous theoretical and experimental investigations over the last few decades. For example, the elastohydrodynamic lubrication of rib-roller end contact (Fig. 1) in tapered roller bearings has received more attentions. Gadallah and Dalmaz[1] applied the hydrodynamic theory to this contact and established an analogous apparatus made of a flat glass disk and a steel toric specimen to measure film thicknesses and traction forces. However, they pointed out that further theoretical work which includes the consideration of surface deformations is needed. Zhang, Qiu and Hong[2] have made the first attempt to achieve a numerical solution of elastohydrodynamic lubrication of this type of contact by considering its specific geometry and kinematic conditions. A domain size close to the conjunction found in the rib-roller end contact was used. It was found that the effect of elastic deformation is not negligible for the range of loads attendant with this problem. For the sake of simplicity, Newtonian behaviour of the liquid lubricant was assumed. Subsequent numerical analyses were carried out under isothermal and steady state conditions. As a result, although it was relatively accurate in predicting the film thickness to the same order of magnitude as in experimental data, the increase of film thickness

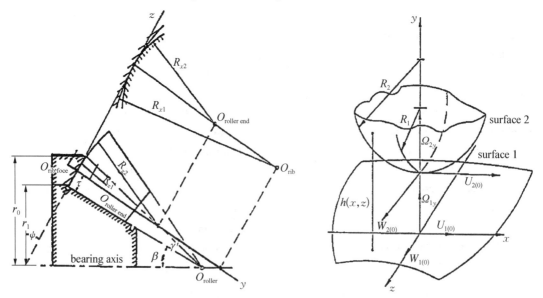

Fig. 1 Rib-roller end contact geometry

against running speed was found to be much greater than that found in the experiments. It is possible that this can be explained by the non-Newtonian and thermal effects.

In the present analysis, a full numerical solution to the elastohydrodynamic problem in elliptical contacts at the rib-roller end contact of tapered roller bearings is obtained through incorporating a limiting shear stress model of lubricant rheology. The generalized two-dimensional Reynolds equation is modified to include the non-Newtonian characteristics. Inferences on oil film thicknesses and traction forces are thus obtained. The Reynolds and film thickness equations, with viscosity and density as functions of pressure, are solved simultaneously by using the straight-forward iteration method to estimate film thicknesses, pressure distributions, friction coefficients and traction torques. The effects of ratios of curvature in both principal planes on the minimum film thickness, the central film thickness, friction coefficients and traction torques are also studied. Further, it is shown that optimal values for the geometric layout of the bearing under specific working conditions can be determined accurately. Two main types of conjunction in tapered roller bearings are considered in this paper: spherical rib face with spherical roller end contact and tapered rib face with spherical roller end contact. Calculations are carried out as assuming smooth surfaces, isothermal and steady-state conditions.

2 The Governing Equations

2.1 Reynolds Equation

The differential equation governing the pressure distribution in a lubricant film is known as the Reynolds equation. The general Reynolds equation for a non-Newtonian fluid can be derived from the reduced form of the Navier-Stokes and continuity equations.

The equilibrium equation for an infinitesimal fluid element, the Navier-Stokes equations by making certain assumptions, are given by

$$\frac{\partial p}{\partial x} = \frac{\partial \tau_{yx}}{\partial y} \tag{1a}$$

$$\frac{\partial p}{\partial y} = 0 \tag{1b}$$

$$\frac{\partial p}{\partial z} = \frac{\partial \tau_{yz}}{\partial y} \tag{1c}$$

The shear rheological model proposed by Bair & Winer[3] is employed in this analysis and can be written as

$$\frac{\partial u}{\partial y} = \frac{\tau_{yx}}{\tau_e} \frac{\tau_L}{\eta} \ln \frac{1}{1 - \tau_e/\tau_L} \tag{2a}$$

$$\frac{\partial w}{\partial y} = \frac{\tau_{yz}}{\tau_e} \frac{\tau_L}{\eta} \ln \frac{1}{1 - \tau_e/\tau_L} \tag{2b}$$

where

$$\tau_e = \sqrt{\tau_{yx}^2 + \tau_{yz}^2} \tag{2c}$$

By defining a rheology function $F(\tau_e)$ as

$$F(\tau_e) = \frac{\ln[1/(1-\tau_e/\tau_L)]}{\tau_e/\tau_L} \tag{3}$$

Equation (2a) and (2b) can be written as

$$\frac{\partial u}{\partial y} = \frac{\tau_{yx}}{\eta} F(\tau_e) \tag{4a}$$

$$\frac{\partial w}{\partial y} = \frac{\tau_{yz}}{\eta} F(\tau_e) \tag{4b}$$

It is noted that $F(\tau_e)$ equals to unity for Newtonian fluids.

Substitution of equation (4a) into (1a) gives

$$\frac{\partial p}{\partial x} = \frac{\partial}{\partial y}\left[\frac{\eta}{F(\tau_e)} \frac{\partial u}{\partial y}\right] \tag{5}$$

Integrating Eq. (5) with respect to y and rearranging yields

$$\frac{\partial u}{\partial y} = \frac{F(\tau_e)}{\eta} \frac{\partial p}{\partial x} y + A \frac{F(\tau_e)}{\eta} \tag{6}$$

Assuming $\partial p/\partial x$ is not a function of y. Integrating again from zero to y, yields

$$u = \int_0^y \frac{F(\tau_e)}{\eta} \frac{\partial p}{\partial x} y\, dy + A \int_0^y \frac{F(\tau_e)}{\eta} dy + B \tag{7}$$

The velocity boundary conditions are

$$u_1(x, z) = u_{1(0)} + z\Omega_{1y} \quad \text{at} \quad y = 0 \tag{8a}$$

$$u_2(x, z) = u_{2(0)} + z\Omega_{2y} \quad \text{at} \quad y = h \tag{8b}$$

Applying the boundary conditions, Eq. (7) can be written as

$$u = \frac{1}{\eta} \frac{\partial p}{\partial x} \int_0^y F(\tau_e) y\, dy + \frac{u_2 - u_1 - \frac{1}{\eta} \frac{\partial p}{\partial x} \int_0^h F(\tau_e) y\, dy}{\int_0^h F(\tau_e) dy} \int_0^y F(\tau_e) dy + u_1 \tag{9}$$

where η is assumed not to be a function of y. Similarly, integrating Eq. (1b) twice with respect to y and applying the velocity boundary conditions

$$w_1(x, z) = w_{1(0)} - x\Omega_{1y} \quad \text{at} \quad y = 0 \tag{10a}$$

$$w_2(x, z) = w_{2(0)} - x\Omega_{2y} \quad \text{at} \quad y = h \tag{10b}$$

yields

$$w = \frac{1}{\eta}\frac{\partial p}{\partial z}\int_0^y F(\tau_e)y\,dy + \frac{w_2 - w_1 - \frac{1}{\eta}\frac{\partial p}{\partial z}\int_0^h F(\tau_e)y\,dy}{\int_0^h F(\tau_e)\,dy}\int_0^y F(\tau_e)\,dy + w_1 \quad (11)$$

The continuity equation for a steady state condition is

$$\frac{\partial(\rho u)}{\partial x} + \frac{\partial(\rho v)}{\partial y} + \frac{\partial(\rho w)}{\partial z} = 0 \quad (12)$$

Assuming $\rho \neq \rho(y)$ and integrating Eq. (12) with respect to y from zero to h yields

$$\int_0^h \frac{\partial(\rho u)}{\partial x}dy + \int_0^h \frac{\partial(\rho w)}{\partial z}dy + \rho[v(h) - v(0)] = 0 \quad (13)$$

In Eq. (13), $v(0) = 0$ and $v(h) = u_2 \partial h/\partial x + w_2 \partial h/\partial z$

Using the Leibnitz's rule,

$$\frac{\partial}{\partial x}\left[\int_0^h \rho u\,dy\right] + \frac{\partial}{\partial z}\left[\int_0^h \rho w\,dy\right] = 0 \quad (14)$$

Substituting Eq. (9) and Eq. (11) into (14) and rearranging it gives

$$\frac{\partial}{\partial x}\left\{\frac{\rho}{\eta}\frac{\partial p}{\partial x}\left[\int_0^h\int_0^y F(\tau_e)y\,dy\,dy - \frac{\int_0^h F(\tau_e)y\,dy}{\int_0^h F(\tau_e)\,dy}\int_0^h\int_0^y F(\tau_e)\,dy\,dy\right]\right\}$$

$$+ \frac{\partial}{\partial z}\left\{\frac{\rho}{\eta}\frac{\partial p}{\partial z}\left[\int_0^h\int_0^y F(\tau_e)y\,dy\,dy - \frac{\int_0^h F(\tau_e)y\,dy}{\int_0^h F(\tau_e)\,dy}\int_0^h\int_0^y F(\tau_e)\,dy\,dy\right]\right\}$$

$$+ (u_{2(0)} + z\Omega_{2y} - u_{1(0)} - z\Omega_{1y})\frac{\partial}{\partial x}\left[\rho\frac{\int_0^h\int_0^y F(\tau_e)\,dy\,dy}{\int_0^h F(\tau_e)\,dy}\right] + \frac{\partial}{\partial x}[(u_{1(0)} + z\Omega_{1y})\rho h]$$

$$+ (w_{2(0)} - x\Omega_{2y} - w_{1(0)} + x\Omega_{1y})\frac{\partial}{\partial z}\left[\rho\frac{\int_0^h\int_0^y F(\tau_e)\,dy\,dy}{\int_0^h F(\tau_e)\,dy}\right] + \frac{\partial}{\partial z}[(w_{1(0)} - x\Omega_{1y})\rho h] = 0$$

$$(15)$$

For the simple Newtonian fluid, $F(\tau_e) = 1$, equation (15) can be reduced to

$$\frac{\partial}{\partial x}\left(\frac{\rho h^3}{6\eta}\frac{\partial p}{\partial x}\right) + \frac{\partial}{\partial z}\left(\frac{\rho h^3}{6\eta}\frac{\partial p}{\partial z}\right) = [u_{1(0)} + u_{2(0)} + z(\Omega_{1y} + \Omega_{2y})]\partial\frac{(\rho h)}{\partial x}$$

$$+ [w_{1(0)} + w_{2(0)} - x(\Omega_{1y} + \Omega_{2y})]\partial\frac{(\rho h)}{\partial z} \quad (16)$$

This is identical to the equation derived by Zhang, et al[2].

Defining the dimensionless parameter group

$X = x/b$, $Y = y/b$, $Z = z/a$, $H = h/R_x$, $k = a/b$, $P = p/E'$, $\bar{\rho} = \rho/\rho_0$ and $\bar{\eta} = \eta/\eta_0$

For the rib-roller end contact, under steady-state conditions, the general non-Newtonian Reynolds equation in dimensionless form can be written as

$$\frac{\partial}{\partial X}\left[\frac{\bar{\rho}}{\bar{\eta}}H^3\left(\frac{J_1^2}{J_0}-J_2\right)\frac{\partial P}{\partial X}\right]+\frac{1}{K^2}\frac{\partial}{\partial Z}\left[\frac{\bar{\rho}}{\bar{\eta}}H^3\left(\frac{J_1^2}{J_0}-J_2\right)\frac{\partial P}{\partial Z}\right]$$

$$+\sigma[(u_{2(0)}-u_{1(0)})+aZ(\Omega_{2y}-\Omega_{1y})]\frac{\partial}{\partial X}\left[\bar{\rho}H\left(1-\frac{J_1}{J_0}\right)\right]$$

$$+\frac{\sigma}{K}[(w_{2(0)}-w_{1(0)})+bX(\Omega_{1y}-\Omega_{2y})]\frac{\partial}{\partial Z}\left[\bar{\rho}H\left(1-\frac{J_1}{J_0}\right)\right]$$

$$+\sigma(u_{1(0)}+aZ\Omega_{1y})\frac{\partial}{\partial X}(\bar{\rho}H)+\frac{\sigma}{K}(w_{1(0)}-bX\Omega_{1y})\frac{\partial}{\partial Z}(\bar{\rho}H)=0 \quad (17)$$

where

$$J_0=\int_0^1 F(\tau_e)\mathrm{d}\dot{y}^* \quad (18a)$$

$$J_1=\int_0^1 F(\tau_e)\dot{y}^*\mathrm{d}\dot{y}^* \quad (18b)$$

$$J_2=\int_0^1 F(\tau_e)\dot{y}^{*2}\mathrm{d}\dot{y}^* \quad (18c)$$

and

$$\sigma=b\eta_0/(E'R_x^2) \quad (18d)$$

2.2 Film Thickness

The film thickness in dimensionless form may be expressed by

$$H=H_0+\frac{b^2X^2}{2R_x^2}+\frac{a^2Z^2}{2R_xR_z}+\bar{\delta}(X,Z) \quad (19)$$

where H_0 corresponds to the central film thickness of undeformed bodies. H_0 is a constant that is initially estimated and corrected later in the succeeding iterations.

The surface elastic deformation is often determined by employing a semi-infinite body. According to Boussinesq's solution[4] of two-dimensional deformation field, the normal displacement at any point (x, z) can be calculated by the following equation

$$\delta(\bar{x},\bar{z})=\frac{2}{\pi E'}\iint_A\frac{p(x,z)}{\sqrt{(x-\bar{x})^2+(z-\bar{z})^2}}\mathrm{d}x\mathrm{d}z \quad (20)$$

where the area A includes the full region of pressure generation. However, integration of the right-hand side of Eq. (20) causes two problems. The first problem arises from the singularity at $x=x$ and $z=z$ point. The second is the considerably large amount of work in the numerical integration. This is due to the fact that the integration must be carried out for the whole area in order to evaluate the deformation at one point, and for each iteration the deformation of every node on the finite difference grid must be evaluated. Presently, in order to overcome the above mentioned difficulties, the following approach is adopted by many researchers. For grid elements with singularity in integration, a polynomial function is used to approximate the practical pressure distribution so that the analytical solution for the integration on that element can be obtained. A constant pressure field in the region surrounding the singularity was assumed by Hamrock and Dowson[5]. The analytical expressions for the elastic deformation were then obtained. The deformation of every node could thus be expressed as a linear combination of the nodal pressure and the deformation matrix.

$$\delta_{kl} = \frac{2}{\pi E'} \sum^i \sum^j D_{ij}^{kl} p_{ij} \Delta x \Delta z \tag{21}$$

where D_{ij}^{kl} stands for the dimensionless deformation matrix.

2.3 Load Capacity

Once the pressure distribution for the appropriate cavitation boundary conditions has been determined numerically, we can express the load capacity in dimensionless form as

$$\overline{W} = \iint_A P(X, Z) dX dZ \tag{22}$$

2.4 Lubricant Density and Viscosity

The variation of density with pressure can be expressed by the empirical formula proposed by Dowson and Higginson[6].

$$\bar{\rho} = 1 + \frac{0.6e - 9E'P}{1 + 1.7e - 9E'P} \tag{23}$$

The Roelands equation[7] which represents the viscosity-pressure relationship reads as

$$\bar{\eta} = \exp(\ln\eta_0 + 9.67)[-1 + (1 + 5.1e - 9E'P)^{0.68}] \tag{24}$$

2.5 Shear Stress and Traction

Integrating the equilibrium Eqs. (1a) and (1c), the dimensionless film shear stresses are given by

$$\bar{\tau}_{YX} = Y \frac{\partial P}{\partial X} + \bar{\tau}_{YX} |_{Y=0} \tag{25a}$$

$$\bar{\tau}_{YZ} = \frac{Y}{k} \frac{\partial P}{\partial Z} + \bar{\tau}_{YZ} |_{Y=0} \tag{25b}$$

Applying the velocity conditions (Eq. (8) and Eq. (10)), the surface shear stresses on the rib face ($y=0$), in the X- and Z-direction become

$$\bar{\tau}_{YX} |_{Y=0} = \frac{R_x \sigma}{b}(u_2 - u_1) \frac{\bar{\eta}}{H J_0} - \frac{R_x}{b} H \frac{\partial P}{\partial X} \frac{J_1}{J_0} \tag{26a}$$

$$\bar{\tau}_{YZ}\mid_{Y=0} = \frac{R_x\sigma}{b}(w_2-w_1)\frac{\bar{\eta}}{HJ_0} - \frac{R_x}{Kb}H\frac{\partial P}{\partial Z}\frac{J_1}{J_0} \tag{26a}$$

The friction coefficient in rolling direction on rib face is solved by

$$f_{1x} = \frac{\bar{F}_{1x}}{\bar{W}} = \frac{\iint_\Gamma \bar{\tau}_{YX}\mid^*_{Y=0} dXdZ}{\iint_A P(X,Z)dXdZ} \tag{27}$$

The traction torque acting on rib with the rotating axis of inner ring is

$$M_1 = abE'\left[\iint_\Gamma \bar{\tau}_{YX}\mid^*_{Y=0}(aZ+e+R_i)\cos\psi dXdZ - \iint_\Gamma \bar{\tau}_{YZ}\mid^*_{Y=0}bX\cos\psi dXdZ\right] \tag{28}$$

where Γ is the entire computing area including the cavitation zone.

2.6 Geometry and Kinematics

The rib-roller end contact is a specific lubrication problem because of its particular geometry and kinematic conditions. The conjunction between rib and roller ends in a tapered roller bearing is bordered by four curves which may be approximated by elliptical or circular arcs (Fig. 2). However, for the sake of simplicity, a rectangular zone is chosen as the computing area. The domain size in which the pressure can be built up is small. Therefore, the fully flooded condition cannot be maintained. Starvation and side leakage have significant effects on film thicknesses. Another unusual geometric feature is that the ratio of the equivalent curvature radii R_z/R_x is less than unity, where R_x is the radius of curvature in the rolling direction and R_z is that in the perpendicular direction.

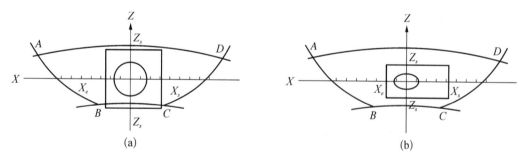

Fig. 2 Computing zone for
(a) spherical rib-spherical roller end (b) tapered rib-spherical roller end

The velocity components which affect the performance of the contact include not only the rolling speed, but also the sliding speed and the spinning of each contact body. The sliding speed and spinning are especially important at high speed conditions.

3 Numerical Solution Techniques

In our calculations, the oil film domain is discretized. The dimensionless Reynolds equation (17) is replaced by its corresponding finite difference equation, and solved by the successive over-relaxation method. Under-relaxation iteration between the Reynolds equation and the deformation calculation are performed to obtain the simultaneous solution of dimensionless oil film pressure distributions, shear stress distributions and film shapes. Although the conjunction of the rib-roller end contact is more precisely represented by circular and elliptical curves, a rectangular shape is assumed for the sake of computational simplicity. The inlet distance should be carefully determined to take account of lubrication starvation. By considering the geometries of the two contact surfaces, the dimensionless inlet position X_s is found to be -3.6 for the tapered rib/spherical roller end contact with a load of 160 N, and -2.0 for the spherical rib/spherical roller end contact with a load of 550 N, respectively. In the transverse direction, dimensionless size Z_s is about 1.6. Nevertheless, the starting positions X_s and Z_s will decrease when the applied load or the ratio of curvature of two mating surfaces R_{x2}/R_{x1} increases. A uniform mesh configuration is set up for the whole computing zone. The dimensionless grid increments in directions X and Z are 0.05 and 0.2 respectively.

It is convenient, for a numerical solution, to rewrite the Reynolds equation (17) in terms of a reduced pressure which exists in a general definition

$$Q = \frac{1}{E'} \int_0^{E'P} \frac{d\omega}{\bar{\eta}(\omega)} \tag{29}$$

For the variation of viscosity with pressure, the Roelands equation is initially chosen. When substituted into the reduced pressure transformation (Eq. (29)), it is noted that the finished form is not a simple algebraic expression. It is therefore simpler to use the Power law viscosity model to approximate the Roelands equation. The error involve with this approach is less than 4% over the typical pressure range of 0.05 to 0.7 GPa[8]. The resulting expression for Q is

$$Q = \frac{1}{E'} \int_0^{E'P} \frac{d\omega}{(1+c\omega)^K} = \frac{1-(1+cE'P)^{1-K}}{E'c(K-1)} \tag{30}$$

where constants C and K are equal to 9.310 1e-10 Pa and 22.743 1, respectively.

For computational purposes, a further substitution is made in the Reynolds equation. It is

$$\Phi = QH^{3/2} \tag{31}$$

The use of the parameter Φ instead of the reduced pressure Q can produce a more gentle curve because localized high values of the first and second derivatives of Q are often present. A further benefit is that the substitution eliminates terms containing the products of H and Q or H and Φ. This can speed up the computation processes[9].

At the edges of the rectangular computation zone, the pressure is zero, and therefore, Φ is also zero. This condition, commonly known as the Reynolds boundary condition, can be satisfied by simply resetting Φ to zero whenever it occurs as a negative value.

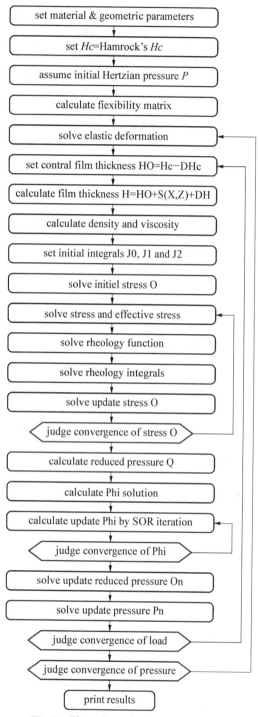

Fig. 3 Flow chart of computer program

The Reynolds equation (17) contains a set of integrals J_0, J_1 and J_2 which are integrated by the rheology function $F(\tau_e)$ (Eq. (18)). The function $F(\tau_e)$ is composed of the shear stresses τ_{yx} and τ_{yz} (Eqs. (3) and (2c)) which include those evaluated on the rib face (Eq. (26)). However, the rib face shear stresses $\tau_{yx}|_{y=0}$ and $\tau_{yz}|_{y=0}$ are related in turn to the integrals J_0 and J_1. Hence, an under-relaxation iteration of the shear stresses on the rib face is required. This process is continued until convergence is reached. The flow chart is shown in Figure 3.

4 Results and Discussion

Non-Newtonian EHL solutions have been determinated for two basic contact types stated previously, viz., the spherical rib face with spherical roller end contact; and the tapered rib face with spherical roller end contact. The geometric and kinematic parameters in the analysis are chosen to be identical to those in [2], in order to check the validity of the Newtonian solutions. The effective elastic modulus $E' = 2.329\,67\text{e}11$ Pa and a mineral oil with the ambient viscosity $\eta_0 = 0.028\,3$ Pa·s at 25℃ are used. The rotational speed of the bearing is 1 000 r/min. With the rib-roller in Fig. 1 having the following dimensions: $r = 0.009\,739$ m, $r_i = 0.057\,486$ m, $r_o = 0.060\,5$ m, $\gamma = 2°$ and $\beta = 11.642\,2°$. The nominal contact height ε is $0.002\,735$ m as mentioned in [2] in order to facilitate the necessity to place the contact at the central region of rib to enhance lubrication. A numerical approach has been achieved at moderate and high loads. The minimum film thickness $h_{\min}$, the central film thickness h_c, the friction coefficient f_{1x} and the traction torque M_1 are presented.

Fig. 4 depicts the influence of ratio R_{2x}/R_{1x} on the film thickness and the friction coefficient/torque for a tapered roller bearing of the spherical rib face/spherical roller

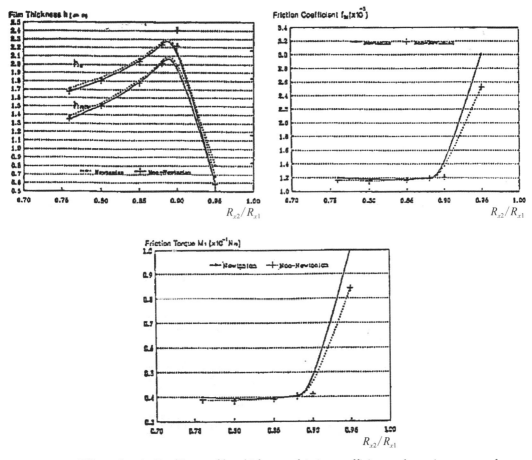

Fig. 4 Effect of ratio R_{2x}/R_{1x} on film thickness, friction coefficient and traction torque of spherical rib-spherical roller end contact with a load of 550 N

end. The optimal value of this ratio R_{2x}/R_{1x} is about 0.9. Fig. 5 represents the influence of ratio R_{2x}/R_{1x} on the above characteristic performances for the tapered rib face/spherical roller end. A range of ratios, R_{2x}/R_{1x}, from 0.7 to 0.8 has been detected to favour lubrication. The sensitivity of R_{2x}/R_{1x} for tapered rib-spherical roller end contact is weaker than spherical rib-spherical roller end contact due to its better flood lubrication. The effect of non-Newtonian behaviour of lubricant on film thickness and traction has also been investigated. A nonlinear shear stress-strain rate relationship (Eqs. (2a) and (2b)) proposed by Bair & Winer[3] with the limiting shear stress assumed to be $\tau_L = 7e5 + 0.05E'P$ Pa is incorporated into Reynolds equation to arrive at the theoretical results. The difference between the film thickness obtained by using Newtonian and non-Newtonian approaches is not clear under this light load and low speed conditions.

Numerical calculations are performed for the cases under which the experiment of [1] were done. Fig. 6 illustrates the comparison of the theoretical results with the experimental data for $W = 80$ N, $E' = 9.183\,145\mathrm{e}10$ Pa, $\eta_0 = 0.026$ Pa·s, $R_{x1} = 0.185\,283$ m, $R_{x2} = 0.163\,262$ m, $R_{z1} = \infty$, and $R_{z2} = R_{x2}$. The Newtonian EHL calculated and the interferometrically measured film thickness appears to be well correlated

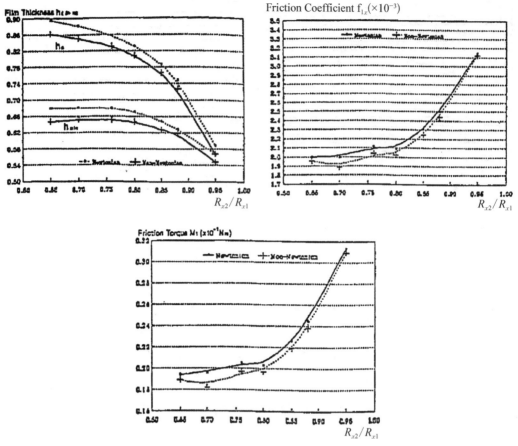

Fig. 5 Effect of ratio R_{2x}/R_{1x} on film thickness, friction coeffieient and traction torque of tapered rib-spherical roller end contact with a load of 160 N

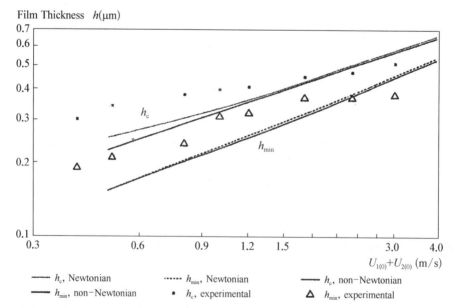

Fig. 6 A comparison of numerical solution with experimental data at the $W=80$ N, $\beta_r=0.185\,5$ and $\beta_s=0.083\,5$

at low speeds. However, the predicted film thicknesses are larger than the experimental values at high speeds. This discrepancy could be reduced through introducing a non-Newtonian model of lubricant. It is obvious that the non-Newtonian EHL approach for the heavier load of 320 N can be improved significantly (see Fig. 7). Good agreement between numerical and experimental results has been achieved.

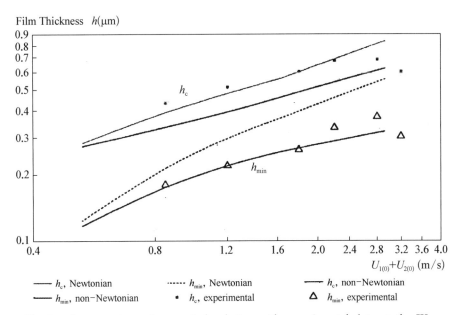

Fig. 7 A comparison of numerical solution with experimental data at the $W = 320$ N, $\beta_r = 0.185\ 5$ and $\beta_s = 0.132\ 62$

5 Conclusion

A numerical procedure has been developed for examining the two-dimensional non-Newtonian fluid flow in the rib-roller end contact of tapered roller bearings. The effect of nonlinear rheological behaviour of a liquid lubricant is not negligible under the conditions of heavy load or high speed.

The non-Newtonian EHL numerical solution agrees well with the experimental data stated in [1], which verifies the validity of the analysis treatment.

Deductions of optimal values of ratio R_{x2}/R_{x1} are carried out for the spherical rib face/spherical roller end contact and the tapered rib face/spherical roller end contact under the conditions considered in this paper. They are in values of 0.9 and about 0.75 respectively. It is found that in the rib-roller end contact of the tapered roller bearings, neither the compressibility of the lubricant nor the pressure-viscosity relationship (Roelands equation) affected the pressure and film thickness distributions significantly.

Further research into incorporating aspects such as thermal and surface roughness effects will be worth exploring.

Acknowledgment

The authors would like to express their appreciation to the City Polytechnic of Hong Kong for its financial support of this research project (Research project No. 700150).

References

[1] Gadallah, N. and Dalmaz, G., 1984, "Hydrodynamic Lubrication of the Rib-Roller End Contact of a Tapered Roller Bearings," Journal of Tribology, Trans. ASME, Vol. 106, pp. 265-274.

[2] Zhang, Z., Qiu, X. and Hong, Y., 1988, "EHL Analysis of Rib-Roller End Contact in Tapered Roller Bearings," STLE Trans., Vol. 31, No. 4, pp. 461-467.

[3] Bair, S. and Winer, W. O., 1979, "A Rheology Model for EHD Contacts Based on Primary Laboratory Data," J. Lub. Tech., Trans. ASME, Vol. 101, pp. 258-265.

[4] Timoshenko, S. P. and Goodier, J. N., 1982, "Theory of Elasticity" 3rd Edition, McGraw-Hill, New York.

[5] Hamrock, B. J. and Dowson, D., 1981, "Ball Bearing Lubrication, the Elastohydrodynamics of Elliptical Contacts," John Wiley and Sons, New York.

[6] Dowson, D. and Higginson, G. R., 1977, "Elastohydrodynamic Lubrication" SI Edition, Pergamon, Oxford.

[7] Roelands, C. J. A., 1966, "Correlational Aspects of the Viscosity-Temperature-Pressure Relationship of Lubricating Oils," V. R. B., Groningen, Netherlands.

[8] Zhu, D. and Wen, S., 1984, "A Full Numerical Solution for the Thermoelastohydrodynamic Problem of Elliptical Contacts," J. Lub. Tech., Trans. ASME, Vol. 106, pp. 246-254.

[9] Gohar, R., 1988, "Elastohydrodynamics" Ellis Horwood Ltd., New York, p. 127.

油叶型轴承非线性油膜力数据库

摘　要：本文比较了获得油叶型轴承非线性油膜力的现有方法,提出采用油叶型轴承非线性油膜力数据库来获得非线性油膜力,介绍了数据库的结构,叙述了对动态雷诺方程的处理技巧以适合于建库.从精度和速度上显示了数据库的杰出优越性,证实数据库方法是对转子-轴承系统进行精确非线性动力分析的极其有效的方法.

主 要 符 号

H：无量纲油膜厚度,$H = 1 + \varepsilon\cos\Phi$　　　　ε：偏心率,$\varepsilon = e/c$
c：轴承半径间隙,$c = R - r(\text{m})$　　　　　　R：轴承半径(m)
r：转子半径(m)　　　　　　　　　　　　　D, d：轴承直径(m)
L、l：轴承长度(m)　　　　　　　　　　　　Ψ：相对间隙 $\Psi = c/R$
m：转子上的集中质量(kg)　　　　　　　　　k：转子刚度(N/m)
η：动力黏度(Pa·s)　　　　　　　　　　　　ω：转子角速度
$'$：代表对无量纲时间 ωt 的求导　　　　　　　P：无量纲油膜压力 $P = p\Psi^2/(2\mu\omega)$

前言

越来越多的现象和事实表明,用线性理论来研究分析转子-轴承系统在一定情况下存在不全面性,以该理论为基础的设计、计算不能保证转子-轴承系统在地震、工况突变、大激励下安全工作,也不能解释很多重要现象(如：1.转子能稳定工作在线性极限转速之上；2.转子工作在线性极限转速之下,若有较大的激励,仍会产生油膜振荡现象).因此,国际上越来越多的研究者投入对转子轴承系统的非线性分析的研究,而首当其冲的是轴承非线性油膜力的获得.关于这一方面的资料参见文献[1,2].

目前国内外获得轴承非线性油膜力的方法大致有两类,一类是直接应用数值方法解雷诺方程获得油膜力,如文献[1]中采用的方法.这类方法虽可达到较高精度,但由于对转子轴承系统进行非线性分析需相当多次求解雷诺方程,虽目前计算机计算速度很快,但所耗计算工作量仍非常巨大.目前只能用在一些简单的非线性分析计算,而且[1]中为了提高计算效率采用了简化的边界条件,在一定程度上降低了一些精度.

另一类方法可采用各种近似方法来获得油叶型轴承的非线性油膜力.主要有：1.图线法,如[3]中所述的 Field Mapping of Bearing Characteristics Method,该方法利用静态时的力和方向角与偏心 e 的图线以及阻尼与 e 的关系而获得油膜力；2.采用短轴承和长轴承的近似理论来获得非线性油膜力,如文献[3].3.相似于油膜线性模型,建立油膜的非线性模型,如文献[4-6].以上几种近似方法由于经过了较多的近似处理,虽计算速度快,但精度较

---

* 本文合作者：王文.原发表于《上海工业大学学报》,1993,14(4)：299-305.

低,不能胜任较精确的转子-轴承系统的非线性动力分析.

以上两类获得非线性油膜力的方法都存在局限性,这给转子-轴承系统的非线性分析带来很大的阻力,我们在吸取两类方法各自优点的基础上,参照单块油叶型轴瓦静、动特性数据库[8]的成功经验,提出了以单块瓦为基础建立非线性油膜力数据库,从而在基本不牺牲精度条件下,得到比第一类方法速度高得多的方法. 又因其可拼装成各种常规或非常规设计的油叶型轴承,而油叶型轴承是径向轴承最多的型式,所以有极大的方便性和实用性.

1 非线性油膜力数据库

建立非线性油膜力数据库首先遇到的是动态雷诺方程中扰动项的数值范围如何选择,动态雷诺方程为:

$$\frac{\partial}{\partial \phi}\left(H^3 \frac{\partial P}{\partial \phi}\right)+\left(\frac{d}{l}\right)^2 H^3 \frac{\partial^2 P}{\partial \lambda^2}=6 \frac{\partial H}{\partial \phi}+12(\varepsilon' \cos\phi + \varepsilon\theta' \sin\phi) \tag{1}$$

$$\frac{\partial}{\partial \phi}\left(H^3 \frac{\partial P}{\partial \phi}\right)+\left(\frac{d}{l}\right)^2 H^3 \frac{\partial^2 P}{\partial \lambda^2}=-6\varepsilon(1-2\theta')\sin\phi + 12\varepsilon' \cos\phi \tag{2}$$

令:

$$RS = -6\varepsilon(1-2\theta')\sin\phi + 12\varepsilon' \cos\phi$$

式(1)中的 ε'、θ' 不像其他参数如 ε 有一定的范围,它的范围可在 $(-\infty, +\infty)$ 间,这给建库带来了极大的阻碍. 通过对动态雷诺方程的分析,作者们根据"旋转效应"和"挤压效应"的相对比重采用了如下的处理技巧:

(a) $\varepsilon(1-2\theta')=0$, $\varepsilon'=0$ 时,$FS=0.0$ $FC=0.0$

(b) $\varepsilon|1-2\theta'| \geqslant 2|\varepsilon'|$ 时,

令: $P_1 = \dfrac{P}{\varepsilon|1-2\theta'|}$, 则 $RS = -6 \dfrac{\varepsilon(1-2\theta')}{\varepsilon|1-2\theta'|}\sin\phi + 6 \dfrac{2\varepsilon'}{\varepsilon|1-2\theta'|}\cos\phi$

令: $q_1 = \dfrac{2\varepsilon'}{\varepsilon|1-2\theta'|}$, 即有: $\dfrac{\partial}{\partial \phi}\left(H^3 \dfrac{\partial P_1}{\partial \phi}\right)+\left(\dfrac{d}{l}\right)^2 H^3 \dfrac{\partial^2 P_1}{\partial \lambda^2} = RS$

(a) $\varepsilon(1-2\theta')>0$ 时,$RS = -6\sin\phi + 6q_1\cos\phi$

(b) $\varepsilon(1-2\theta')<0$ 时,$RS = 6\sin\phi + 6q_1\cos\phi$

(c) $\varepsilon|1-2\theta'| \leqslant 2|\varepsilon'|$ 时,

令: $P_2 = \dfrac{P}{2|\varepsilon'|}$, $RS = -6\dfrac{\varepsilon(1-2\theta')}{2|\varepsilon'|}\sin\phi + 6\dfrac{2\varepsilon'}{2|\varepsilon'|}\cos\phi$

令: $q_2 \dfrac{\varepsilon(1-2\theta')}{2|\varepsilon'|}$, $\dfrac{\partial}{\partial \phi}\left(H^3 \dfrac{\partial P_2}{\partial \phi}\right)+\left(\dfrac{d}{l}\right)^2 H^3 \dfrac{\partial^2 P_2}{\partial \lambda^2} = RS$

① $\varepsilon'>0$ 时,$RS = -6q_2\sin\phi + 6\cos\phi$

② $\varepsilon'<0$ 时,$RS = 6q_2\sin\phi + 6\cos\phi$

综上,可得出下面的处理情况:

$$RS(1) = -6\sin\phi + 6q_1\cos\phi \quad [\varepsilon \mid 1-2\theta' \mid \geqslant 2\mid\varepsilon'\mid, \varepsilon(1-2\theta') > 0] \quad (3)$$

$$RS(2) = 6\sin\phi + 6q_1\cos\phi \quad [\varepsilon \mid 1-2\theta' \mid \geqslant 2\mid\varepsilon'\mid, \varepsilon(1-2\theta') < 0] \quad (4)$$

$$RS(3) = -6q_2\sin\phi + 6\cos\phi \quad [\varepsilon \mid 1-2\theta' \mid \leqslant 2\mid\varepsilon'\mid, \varepsilon' > 0] \quad (5)$$

$$RS(4) = 6q_2\sin\phi + 6\cos\phi \quad [\varepsilon \mid 1-2\theta' \mid \leqslant 2\mid\varepsilon'\mid, \varepsilon' < 0] \quad (6)$$

$$FS = 0, \ FC = 0 \quad [\varepsilon(1-2\theta') = 0, \varepsilon' = 0] \quad (7)$$

通过上面对动态雷诺方程的处理,解决了 ε'、θ' 范围难以确定的矛盾,使方程参数的变化在 $q_1, q_2 = -1 \sim +1$ 范围内,从而使非线性油膜力数据库的建立成为可能. 这也许是一些学者即使想到了用数据库的方法,但仍觉无从下手的障碍所在,如[4]中也曾提起数据库方法,但仍要采用近似非线性模型. 以上处理方法解决了建立非线性油膜力库的关键问题.

非线性油膜力库的建立考虑到精度、存储量和方便使用的性能,采用以单瓦为基础,分别建立各种长径比和瓦张角下的单瓦库,在长径比和瓦张角确定后的单瓦库由如下三个变量决定:瓦起始角($TH=0°\sim360°$,定义见图1)、ε 和 Q(包括雷诺方程四种处理情况下的 q_1 和 q_2),通过对非线性油膜力与这三个变量之间关系的摸索,对三个变量分别取 40, 25, 20 个分散点来建库,这样一种长径比和瓦张角下的单瓦非线性油膜力库的存储量为 0.64 M,一张 1.2 M 软盘即可放下,比较方便.

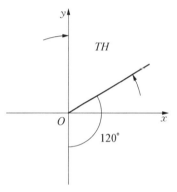

图 1　瓦起始角 TH 定义

由于编写了可建立各种长径比和瓦张角下的建库程序,并且采用了效能较好的算法,建一个单瓦库不需多长时间,因此新的长径比和瓦张角的单瓦库极易建立. 我们对单瓦库的文件名按如下规则定义:如长径比为 0.5,瓦张角为 150°的单瓦库文件名为 050.150,相应的,长径比为 1.5,瓦张角为 360°,则文件名为 150.360. 采用上面的命名方法,可使单瓦库的编排明了,已建立的单瓦库可留待以后再次应用. 建库时解算雷诺方程的算法采用 Castelli 法(参见[8]),由于动载下的油膜破裂条件尚无公认的表达形式,此处仍沿用雷诺边界条件.

在建好单瓦库的基础上,对 ε、TH 和 Q 分别采用两点、三点、三点插值,根据各种实用轴承的结构,编写了各个相应的拼装子程,已编好的拼装子程有: 1. 360°圆轴承,2. 剖分式圆轴承,3. 椭圆轴承,4. 错位椭圆轴承,5. 三油叶轴承,6. 四油叶. 其他结构的轴承的拼装子程根据其结构特性,结合单瓦检索子程,是极易编写的. 非线性油膜力的获得,只要调用相应的拼装子程和单瓦库即可.

2　非线性油膜力库的精度和速度

在 TH、ε 和 Q 三个变量方向上对数据库方法确定的油膜力与直接计算值进行了比较. 图 2 中 $L/D=0.5$, $\varepsilon=0.5$, 瓦张角为 120°, $Q=0.5$[(3)式下的 Q],变化瓦起始角 TH 值,步长取 1°, FS 为无量纲切向力, FC 为无量纲径向力. 从图 2 的计算曲线上可以看到两种算法的计算值基本重合. 在 ε 和 Q 方向上也进行了类似的比较,都证实数据库计算方法的精度很高,类似图 2 的曲线就不一一列出.

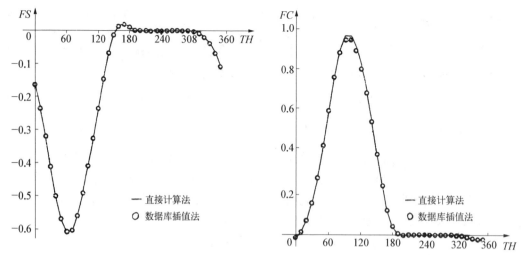

图 2 数据库和直接计算方法在 TH 方向精度比较($FS^\#$ 无量纲切向力;$FC^\#$ 无量纲径向力)

下面举两个数据库与直接计算方法的综合比较算例,即对单质量弹性转子作一些非线性动力分析,图 3 为坐标定义.

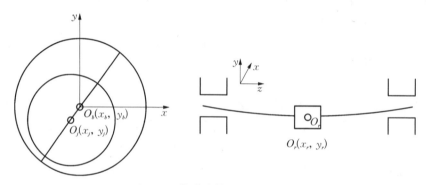

图 3 单质量弹性转子示图

(1) 单质量弹转子突加不平衡响应算例:

$$L=0.05 \text{ m}, D=0.1 \text{ m}, \eta=5.6\times 10^{-3} \text{ Pa·s} \quad c=5.0\times 10^{-4} \text{ m}$$
$$m=15.605 \text{ kg}, k=3.0575\times 10^5 \text{ N/m}, \omega_n=140, \omega=210$$

不平衡量 $AA=m\delta/mc=0.2$,初始位置在静平衡位置.

采用四阶龙格-库塔变步长法进行时间积分,图 4 为数据库计算所得轴心轨迹,图 5 为

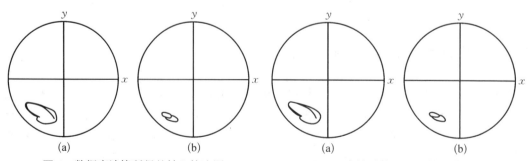

图 4 数据库计算所得的轴心轨迹图 **图 5** 直接计算所得的轴心轨迹图
(a) $t=0\sim 0.1$ s;(b) $t=1.9\sim 2.0$ s (a) $t=0\sim 0.1$ s;(b) $t=1.9\sim 2.0$ s

直接计算所得轴心轨迹,两图中(a)为 $t=0\sim 0.1$ s 里的轴心轨迹,(b) 为 $t=1.9\sim 2.0$ s 里的轴心轨迹. 比较两图,可看出两者差不多一样.

(2) 产生油膜振荡的算例:

$$L=0.05 \text{ m}, D=0.1 \text{ m}, \eta=5.6\times 10^{-3} \text{ Pa}\cdot\text{s}, c=5.9\times 10^{-4} \text{ m}$$
$$m=62.42 \text{ kg}, k=1.223\times 10^{6} \text{ N/m}, \omega_n=140, \omega=448$$

轴心初始位置在静平衡位置,给轴承座以如下激励:

$$\overline{X}_b=\begin{cases}A\sin(\omega_e t) & \text{当 } 0\leqslant t\leqslant 2\pi/\omega_e,\ \omega_e=0.8\omega_n \\ 0 & \text{当 } t>2\pi/\omega_e,\ \ \ \ \ \ \ A=0.3g\end{cases}; \overline{Y}_b=0$$

通过时间积分,可得如图 6 的轨迹图,(b) 为数据库的结果,(a) 为直接计算结果,比较(a)、(b)两图,两者可谓完全一样.

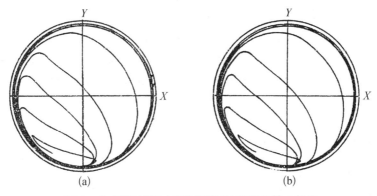

图 6 由直接计算(a)和数据库(b)得出的轴心轨迹

上面油膜振荡算例的参数取自文献[1],另外作者还应用数据库分析了[1]中部分算例,并获得了相似结论,进一步证实了数据库的可靠. 通过上面的一系列的精度比较,可以看到数据库的精度完全足够,在精度上并不比直接计算方法差多少,能应用于实际计算.

我们来看一下数据库方法和直接计算的速度比较. 首先以如下算例为例:360°圆轴承,$L/D=0.5$,$Q=0.5$[式(3)中的 Q],$TH=0°$,ε 从 $0.02\sim 0.96$,步长为 0.001. 两种方法的计算时间比较如表 1 所示.

表 1 对 ε 插值时的速度比较

机　　种	Castelli	数据库	倍　　数
COMPAQ 386	14 min	5 s	168

另外数据库方法又特适合于计算轴心扰动轨迹,在检索子程中加入一定的处理,可使扰动轨迹的计算速度大大提高,下面表 2 为图 6 算例下两种算法的速度比较,各自进行 0.05 s 的轴心轨迹计算所用的时间.

表 2 计算动轨迹时的速度比较

机　　种	Castelli	数据库	倍　　数
COMPAQ 386	31 min	5 s	372

通过上面的精度和速度比较,可以看到数据库方法的突出的优越性,这一点在多跨转子-轴承系统非线性分析中将更为明显. Kato[7]对双跨转子-轴承系统进行非线性动力分析时,由于采用直接计算方法获得非线性油膜力,感到即使采用简化的雷诺边界条件仍将耗去大量的计算时间,故稍牺牲精度而采用半 Sommerfeld 条件,以求缩短计算油膜力的时间. 两相对照,本文的数据库技术,是对多跨转子轴承-系统进行非线性动力分析的极其有效的方法.

3 结束语

我们采用数据库不仅很快地再现了文献[1]中的部分算例,而且已用来进行了单质量弹性转子突加不平衡响应的分析和各种结构轴承动力性态的分析,并将用于多跨转子-轴承系统的非线性动力分析. 无论用于何处,数据库都显示了精度高、速度快的优点. 当然数据库也有其待发展的方面,如考虑紊流影响,选择更能反映动态下油膜的边界条件. 但无论怎样,数据库方法的思路仍将是获得非线性油膜力的精确、有效的快速方法,它可解决转子-轴承系统非线性分析中的主要障碍,开拓了对转子-轴承系统进行精确非线性分析的成功之路.

致谢 李志刚先生曾在建单瓦库的方法和结构上,以及减少数据库读取数据时间等方面,提出了宝贵的建议和帮助,有效地支持了本项工作的成功,特致谢意.

参 考 文 献

[1] Hori, Y. Anti-earthquake considerations in rotordynamics. Proceedings of the IMechE 4th Intl. Conf. on Vibration in Rotating Machinery, Edinburgh, England, Sept., C318/88, 1988: 1-8.

[2] Hori, Y., Kato, T. Earthquake-induced instability of a rotor supported by oil film bearings, J. Vib. Acoust., Trans. ASME, 112, 1990: 160-165.

[3] Shapiro, W. Fluid-film bearing response to dynamic loading. In: Anderson, W. J. ed. Bearing Design—Problems, Technology and Future, New York, 1980.

[4] Jakeman, R. W. Non-linear oil film response model for the dynamically misaligned sterntube bearing, Trib. Int., 22(1), 1989: 3-10.

[5] Parszzewski, Z. A., Carter, B. M. Theoretical and experimertal determination of the non-linear characteristics of a journal bearing. Proc. 13th Leeds-Lyon Symp. on Tribol, 1986: 579-581.

[6] Hashish, W. A. Improved mathematical models and dynamic analysis of light rotor-bearing systems unbalance and stochastic excitation, PhD Dissertation, Concordia University, Montreal, Canada. 1981.

[7] Kato, T., Koguchi, K., Hori, Y. Seismic response of a linearly stable multirotor system. to be published.

[8] 李志刚. 滑动轴承数据库研究. 硕士论文,上海工业大学,1990.

[9] 张直明. 滑动轴承流体动力润滑理论. 北京:高等教育出版社,1986.

Nonlinear Oil Film Force Database

Abstract: Methods to get value of nonlinear oil film force have been reviewed in short and database method to get accurate value of nonlinear oil film force in very short time, which is a difficult problem in nonlinear analysis of rotor-bearing system, has been put forward. Methods of how to set up the nonlinear oil film force database including the method dealing with the dynamic Reynolds equation, together with the process of determining the parameters and assembly of single-lobe database were presented in detail. Precision and calculating speed have been compared between direct calculation and database method. The results showed that database method is much more efficient in performing accurate nonlinear dynamic analysis on rotor-bearing system.

应用高级紊流模式的紊流润滑理论分析*

摘　要：分析了常用紊流润滑理论基础,并在充分考察润滑流场的边界条件及内部结构的基础上,采用复合型高级紊流模式理论,即在近壁区采用低紊流雷诺数的 k-ε 模式,而在紊流核心区采用代数雷诺应力模式,对复杂流场的紊流润滑进行了分析. 同时在计入惯性效应的情况下,推导出了一种适用于高压密封和重载轴承等复杂流场的紊流雷诺润滑方程,并运用该理论对 Couette 型紊流流场进行了考核性分析计算. 计算结果与实验数据十分吻合,验证了模式的有效性.

前言

随着机器向高速、大功率发展,越来越多的动力润滑轴承处于紊流工况下工作. 为了满足工程发展所需,人们对紊流润滑理论作了大量的研究,提出了不少有效的紊流润滑理论,并取得了很多成功的应用经验. 但由于历史发展的原因,现有的一些常用紊流润滑理论主要是针对高速轻载工况提出的,满足不了高压密封对和高速重载轴承之类复杂紊流流场的应用要求. 鉴于目前的这种情况,有必要对紊流润滑的应用理论和计算方法做更进一步的深入研究,以期建立一套较完善、实用的紊流润滑理论和方法.

1　目前常用的紊流润滑理论分析

由于紊流方程组的不封闭性,在研究紊流的历史发展过程中,形成了各种各样的紊流模式理论. 最早提出的主要是零方程模式理论,随后有一方程理论、两方程理论和多方程理论等. 随着紊流模式理论研究的不断深入,紊流理论在工程应用中也得到了不断发展. 自从 Constantinescu[1] 首先将 Prandtl 混合长度理论引入润滑领域以来,各种不同的模式理论也相继被用于解决紊流润滑问题,其中最主要的还有 Elrod,Ng 和 Pan[2,3] 的壁面律理论,Hirs 的整体流动理论等等.

在 Constantinescu 理论中[1],引用 Prandtl 混合长度理论来表示雷诺应力,将雷诺应力代入不计惯性项的运动方程,并与连续方程联立求解,最先推导得到紊流雷诺润滑方程,对紊流润滑领域作出了开创性的贡献. 但他在进行雷诺润滑方程推导时,只考虑了黏性底层和完全紊流区,而没有考虑过渡区的存在. 此外,他在引入雷诺应力表达式时将周向流动与轴向流动人为地解耦,即没有考虑两向流动间的相互影响,而是将空间流分别看成是两种独立的平面流. 这种处理除近似的适用于高速轻载轴承等类以 Couette 型为主的流动外,对其他复杂流场显然是不太合适的.

第二种常用的紊流润滑理论是由 Elrod,Ng 和 Pan[2,3] 提出的. 他们引用了 Reichardt 的

---

*本文合作者：张运清. 原发表于《上海工业大学学报》,1994,15(1)：55-60.

涡粘公式,并将公式中的壁面剪应力换成局部总剪应力,试图以此考虑空间流的影响,通过联立求解不计惯性的运动方程和连续方程得到了非线性的 Elrod 和 Ng[2]润滑方程,但由于求解复杂,不便于应用,Ng 和 Pan[3]采用了小扰动简单 Couette 流线性假设,得到了目前广泛采用的线性化润滑方程. Ng 和 Pan 在推导润滑方程时,采用小扰动简单 Couette 流假设虽说使整个问题的处理简单化,但同时也使该理论的适用范围得到限制,即线性化润滑方程也只适用于以 Couette 流为主的流场. 另外,Elrod,Ng 和 Pan 采用的 Reichardt 涡粘系数经验公式是在"常应力区"条件下,通过对实验数据的拟合而得到的. 而实际润滑流场中由于压力梯度存在,壁面间的剪应力为非均匀分布,故经验公式的应用前提条件不满足,只有在高速轻载的场合,由于 Couette 流效应远大于压力流效应,故两壁面间的总剪应力变化不大,可以近似看作"常应力区". 所以说由经验公式推导得到的润滑方程基本上只能满足高速轻载轴承的分析所需,但对于高压密封和高速重载轴承之类的复杂流场,由于流场中存在较大的周向和轴向压力梯度,两壁面间的剪应力远非"常应力",而且不宜再作小扰动简单 Couette 流假定. 故 Elord-Ng-Pan 理论不宜于分析复杂流场的润滑问题.

Hirs 理论[4]也是一种应用较广的紊流润滑理论,但由于整个理论是基于经验公式之上的,当应用场合不同于经验公式的条件时,就必然存在不可避免的误差. 因而该理论可以说完全是经验性的,仅就这一点而言,Hirs 理论的理论基础不及半经验的 Constantinescu 理论和 Elrod-Ng-Pan 理论. 此外,Hirs 理论本身还存在着严重的理论缺陷[5].

由于零方程类模式理论存在着一定的不足,故随着紊流模式理论的发展,越来越多的高级模式理论不断地引入到润滑领域,并较成功地解决了一些实际紊流润滑问题. 但由于润滑流场的复杂性,可以说很难用一种模式理论来完整地描述整个润滑流场. 因此本文采用了较合理的两种高级紊流模式理论,即低紊流雷诺数的 k-ε 理论和代数雷诺应力理论推导出了一种合理的紊流润滑雷诺方程,并将惯性项的影响直接引入雷诺润滑方程.

2 紊流润滑雷诺方程的推导

2.1 紊流模式

在现有的紊流模式理论中,两方程 k-ε 模式理论和代数雷诺应力模式理论由于他们都引入了最能反映紊流内部机理的紊动能传输方程和能量耗散方程,使得这两种模式理论较其他的模式方程具有更坚实、完善的理论基础,尤其是代数雷诺应力模式理论,它直接计入了固壁对紊流脉动参数的阻滞效应. 由于润滑流场中有两个固壁面存在,所以这种固壁效应的影响就更显得重要. 因而应该说代数雷诺应力模式理论是目前研究润滑流场最合理的紊流模式,但由于该理论在进行模式化处理简化时,忽略了黏性效应的影响,这使得该理论不适用于紊流近壁区即由黏性底层和过渡区组成的区域,而只适用于紊流核心区. 因此本文采用一种复合型的紊流模式理论,即在近壁区采用低紊流雷诺数的 k-ε 方程,而在紊流核心区采用代数雷诺应力模式,通过这两种模式理论的联立使用,既直接考虑了固壁效应的影响,又考虑了近壁区内黏性效应的影响,从而使得这种复合型高级紊流模式理论具有远较目前常用的紊流润滑理论更先进合理、考虑因素更全面的特点. 在复合型紊流模式理论中,近壁区和紊流核心区按紊流剪应力占总剪应力的 95% 的分界条件来划分,同时保证在分界面上剪应力连续.

1. 低紊流雷诺数的 $k-\varepsilon$ 模式方程

低紊流雷诺数的 $k-\varepsilon$ 两方程模式理论是由 Jones 和 Launder[6] 提出的,后经过 Hassid 和 Poreh[7] 对该理论进行了改进. 在此采用 Hassid 和 Poreh 提出的公式[不计 Dk/Dt 和 $D\varepsilon/Dt$ 项].

$$\begin{cases} \dfrac{\partial}{\partial y}\left[\left(\nu+\dfrac{\nu_t}{\sigma_k}\right)\cdot\dfrac{\partial k}{\partial y}\right]+\nu_t\left[\left(\dfrac{\partial u}{\partial y}\right)^2+\left(\dfrac{\partial w}{\partial y}\right)^2\right]-\varepsilon-\dfrac{2\cdot\nu\cdot k}{y^2}=0 \\ \dfrac{\partial}{\partial y}\left[\left(\nu+\dfrac{\nu_t}{\sigma_\varepsilon}\right)\cdot\dfrac{\partial \varepsilon}{\partial y}\right]+C_{\varepsilon1}\cdot\nu_t\left[\left(\dfrac{\partial u}{\partial y}\right)^2+\left(\dfrac{\partial w}{\partial y}\right)^2\right]\dfrac{\varepsilon}{k} \\ \quad -C_{\varepsilon2}[1-0.3\exp(-R_k^2)]\dfrac{\varepsilon^2}{k}-2\nu\left(\dfrac{\partial \varepsilon^{1/2}}{\partial y}\right)^2=0 \end{cases} \quad (1)$$

其中:$\nu_t=C_m[1-0.98\exp(-A_m\cdot R_k)](k^2/\varepsilon)$,式中:$C_m=0.09$,$\sigma_k=1.0$,$\sigma_\varepsilon=1.3$,$C_{\varepsilon1}=1.45$,$C_{\varepsilon2}=2.0$,$R_k=k^2/\nu\varepsilon$,为紊流雷诺数,对牛顿流体:$A_m=0.0015$.

2. 代数雷诺应力模式理论

由 Ljuboja 和 Rodi[8] 提出. 他们在模式处理时直接计入了固壁效应的影响,但忽了黏性效应的影响.

$$\begin{cases} \dfrac{\partial}{\partial y}\left(\dfrac{\nu_t}{\sigma_k}\cdot\dfrac{\partial k}{\partial y}\right)+P_d-\varepsilon=0 \\ \dfrac{\partial}{\partial y}\left(\dfrac{\nu_t}{\sigma_\varepsilon}\cdot\dfrac{\partial \varepsilon}{\partial y}\right)+C_{\varepsilon1}\cdot P_d\cdot\dfrac{\varepsilon}{k}-C_{\varepsilon2}\cdot\dfrac{\varepsilon^2}{k}=0 \end{cases} \quad (2)$$

其中:$\nu_t=0.09G_\mu(k^2/\varepsilon)$;$P_d=\nu_t\left[\left(\dfrac{\partial u}{\partial y}\right)^2+\left(\dfrac{\partial w}{\partial y}\right)^2\right]$;

$$G_\mu=\dfrac{1+\dfrac{3}{2}\dfrac{C_2\cdot C_2'}{1-C_2}\cdot f}{1+\dfrac{3}{2}\dfrac{C_1'}{C_1}f}\cdot\dfrac{1-2\dfrac{C_2\cdot C_2'\cdot P_d/\varepsilon}{C_1-1+C_2\cdot P_d/\varepsilon}\cdot f}{1+2\dfrac{C_1'}{C_1-1+P_d/\varepsilon}\cdot f};\quad f=\dfrac{k^{3/2}}{C_w\cdot y\cdot\varepsilon}$$

$C_w=3.72$; $C_1=1.8$; $C_2=0.6$; $C_1'=0.6$; $C_2'=0.3$;
$C_{\varepsilon1}=1.44$; $C_{\varepsilon2}=1.92$; $\sigma_k=1.0$; $\sigma_\varepsilon=1.3$.

2.2 运动方程一般形式

对稳态流场而言:

$$\begin{cases} u\dfrac{\partial u}{\partial x}+v\dfrac{\partial u}{\partial y}+w\dfrac{\partial u}{\partial z}=-\dfrac{1}{\rho}\dfrac{\partial p}{\partial x}+\dfrac{1}{\rho}\dfrac{\partial \tau_{xy}}{\partial y} \\ u\dfrac{\partial w}{\partial x}+v\dfrac{\partial w}{\partial y}+w\dfrac{\partial w}{\partial z}=-\dfrac{1}{\rho}\dfrac{\partial p}{\partial z}+\dfrac{1}{\rho}\dfrac{\partial \tau_{zy}}{\partial y} \end{cases} \quad (3)$$

其中:$\tau_{xy}=\mu\dfrac{\partial u}{\partial y}-\rho\overline{u'v'}$; $\tau_{zy}=\mu\dfrac{\partial w}{\partial y}-\rho\overline{w'v'}$;

$-\overline{u'v'}=\nu_t\dfrac{\partial u}{\partial y}$; $-\overline{w'v'}=\nu_t\dfrac{\partial w}{\partial y}$; ν_t——涡粘系数

2.3 用修正平均法对惯性项进行处理

现有的紊流润滑理论一般不计惯性项的影响,但随着机器向高速重载发展,惯性效应的影响日显重要,不宜忽略. 通常考虑惯性的方法大概有:(a) 小参数扰动法;(b) 流函数法;(c) Constantinescu 平均法;(d) 直接解法;(e) 修正平均法. 本文在此采用修正平均法的思想,但采用不同的处理方法,将惯性项直接引入封闭的解析滑润方程.

令:

$$\begin{cases} \bar{f}(x,z) = \dfrac{1}{h}\int_0^h \left[u\dfrac{\partial u}{\partial x} + v\dfrac{\partial u}{\partial y} + w\dfrac{\partial u}{\partial z} \right] dy = \dfrac{1}{h}\left[\dfrac{\partial I_{xx}}{\partial x} + \dfrac{\partial I_{xz}}{\partial z} \right] \\ \bar{g}(x,z) = \dfrac{1}{h}\int_0^h \left[u\dfrac{\partial w}{\partial x} + v\dfrac{\partial w}{\partial y} + w\dfrac{\partial w}{\partial z} \right] dy = \dfrac{1}{h}\left[\dfrac{\partial I_{xz}}{\partial x} + \dfrac{\partial I_{zz}}{\partial z} \right] \end{cases} \quad (4)$$

其中:

$$I_{xx} = \int_0^h u^2 dy, \quad I_{xz} = \int_0^h uw\, dy, \quad I_{zz} = \int_0^h w^2 dy$$

令等效压力梯度为:

$$\begin{cases} \dfrac{\partial p'}{\partial x} = \dfrac{\partial p}{\partial x} + \rho \bar{f} \\ \dfrac{\partial p'}{\partial z} = \dfrac{\partial p}{\partial z} + \rho \bar{g} \end{cases} \quad (5)$$

将式(4)和式(5)代入运动方程(3)得

$$\begin{cases} \dfrac{\partial p'}{\partial x} = \dfrac{\partial \tau_{xy}}{\partial y} \\ \dfrac{\partial p'}{\partial z} = \dfrac{\partial \tau_{zy}}{\partial y} \end{cases} \quad (6)$$

从式(6)可见,采用修正平均法的思想对惯性项进行近似处理后得到的运动方程的形式完全类似于不计惯性项时的运动方程,只是将压力梯度 $(\partial p/\partial x)$ 和 $(\partial p/\partial z)$ 换成了式(6)中的等效压力梯度 $(\partial p'/\partial x)$ 和 $(\partial p'/\partial z)$. 因而不考虑惯性项的运动方程的求解过程及有关结果完全适用于式(6)的求解,也就是说,在对惯性项进行修正平均处理后,考虑惯性的润滑问题的求解就完全可以方便地借用我们已经非常习惯处理的无惯性项的润滑问题的有关方法,这是本文采用修正平均法考虑惯性项的目的之一.

2.4 雷诺方程的推导

联立求解运动方程(6)和连续方程,并计入边界条件得到含有惯性项的雷诺润滑方程:

$$\dfrac{\partial}{\partial x}\left(G\dfrac{\partial p'}{\partial x}\right) + \dfrac{\partial}{\partial z}\left(G \cdot \dfrac{\partial p'}{\partial z}\right) = U_j \cdot \dfrac{\partial}{\partial x}(Fh)$$

或

$$\dfrac{\partial}{\partial x}\left[G\left(\dfrac{\partial p}{\partial x} + p\bar{f}\right)\right] + \dfrac{\partial}{\partial z}\left[G\left(\dfrac{\partial p}{\partial z} + p\bar{g}\right)\right] = U_j \cdot \dfrac{\partial}{\partial x}(Fh) \quad (7)$$

其中:

$$G = \dfrac{A_2}{A_1} \cdot B_1 - B_2, \quad F = 1 - \dfrac{B_1}{A_1 \cdot h},$$

$$A_1 = \int_0^h \frac{\mathrm{d}y}{\nu + \nu_t}, \quad A_2 = \int_0^h \frac{y\mathrm{d}y}{\rho(\nu + \nu_t)}, \quad B_1 = \int_0^h \int_0^y \frac{\mathrm{d}y\mathrm{d}y}{\nu + \nu_t}, \quad B_2 = \int_0^h \int_0^y \frac{y\mathrm{d}y\mathrm{d}y}{\rho(\nu + \nu_t)}$$

3 计算结果分析

利用本文提出的理论对两平板间的速度场(如图 1 所示)进行了分析计算,其结果如图 2 所示. 曲线(1)表示平面 Couette 紊流($Re = 5\,408, B_x = B_z = 0$)的速度分布,并与 Leutheusser 的相应实验结果[9]进行了比较(该结果在图中以"·"表示). 从图中可见计算结果与实验非常相符. 这足以说明本文采用的紊流分析方法是合理的、可信的. 曲线 3 和曲线 4 分别表示了有轴向压力梯度 B_z 存在时,速度 $\overline{U}$ 和 $\overline{W}$ 的分布规律. 分别比较曲线 1 和曲线 3 以及曲线 2 和曲线 4 可以发现,由于轴向压力梯度 B_z 的存在,轴向流动 $\overline{W}$ 对 $\overline{U}$ 值的影响是不容忽略的,尤其是对有较大 B_x 和 B_z 存在的复杂流场更是如此. 而目前常用的紊流润滑理论中,由于采用线性化解耦,忽略了两向流动之间的相互影响,这显然是不合理的. 因此对高压密封和高速重载轴承之类复杂流场采用本文提出的复合型模式理论进行分析是合理的.

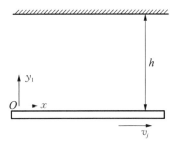

图 1 润滑流场

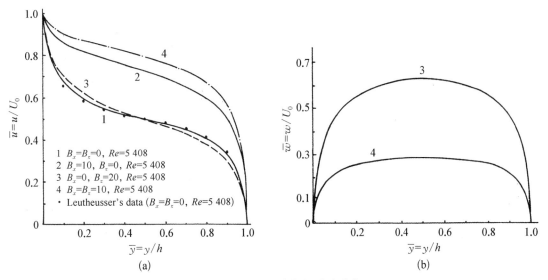

图 2 两平板间速度场的速度分布

4 结论

(1) 文章详细分析了常用紊流润滑理论中引用的模式的理论基础及适用范围.

(2) 通过对润滑流场的特点分析,认为采用复合型的紊流模式对复杂润滑流场进行分析是较完善、合理的.

(3) 文章推导了适用于复杂流场的紊流雷诺润滑方程,并将惯性项的影响直接引入雷诺润滑方程中.

(4) 对平面 Couette 流场进行了分析计算,结果与实验非常吻合.同时分析了轴向流动对周向流的影响,认为对复杂流场不宜采用目前常用的紊流润滑理论,而建议采用本文提出的分析计算方法.

参 考 文 献

[1] Constantinescu, V. N. On turbulent lubrication. Proc. Inst. Mech. Eng., 173, 1959: 881-889.
[2] Elrod, H. G., Ng, C. W. A theory for turbulent films and its application to bearings, Trans. ASME, Ser. F. 89, 1967: 346-362.
[3] Ng, C. W., Pan, C. H. T. A linearized turbulent lubrication theory. Trans, ASME, Ser. F, 95, 1973: 137-146.
[4] 张直明.关于紊流理论中的 Hirs 法的理论基础.西安交通大学学报,1978(4):25,26.
[5] Jones, W. P., Launder, B. E. The calculation of low Reynolds number phenomena with a two-equation model of turbulence, Int. J. Heat mass transfer, 1973,16: 119.
[6] Hassid, S., Poreh, A. M. Turbulent energy dissipation model for flows with drag reduction. ASME, J. Fluids Eng.,1978,100: 107.
[7] Ljuboja, M., Rodi, W. Calculation of turbulent wall jets with an algcbraic Reynolds stress model. ASME, J. Fluids Eng., 1980,102: 350.
[8] Loutheusser, H. J., Agdin, E. M. Plane Couette flow between smooth and rough walls, Experiments in Fluids, 1991(11): 302-312.

A New Reynolds Lubrication Equation Using Advanced Turbulent Models

Abstract: The paper analyzes the widely used turbulent lubrication theories as well as boundary conditions and inner structures of the lubricating fluid field in details. Using the most appropriate compound turbulent model, e. g. low Reynolds number $k-\varepsilon$ model for turbulent core, a new Reynolds turbulent lubrication equation is derived, which is appropriate to analyze complex fluid fields and to consider the effects of fluid inertia. The velocity prediction of this paper for plane Couette flow agrees very well with the experimental results obtained by Leutheusser.

应用高级紊流模式对雷诺润滑方程的解算方法*

摘　要：提出了一种解复杂流场雷诺润滑方程的工程实用计算方法，即在紊流参数数据库支持下雷诺润滑方程的实用求解方法。同时介绍了紊流参数数据库建立的理论，叙述了计算方法的求解步骤，并运用该方法对周隙高压密封的特性进行了分析计算。计算结果与实验数据十分吻合，证明了工程计算方法的简便实用性和有效性。

前言

自从 Constantinescu 首先将 Prandtl 混合长度模式理论引入润滑领域以来，各种不同的模式也相继被用于解决工程紊流润滑问题。但由于历史发展的原因，目前常用的紊流润滑理论主要是为高速轻载轴承而开发的，难以满足像高压周隙密封和高速重载轴承之类复杂紊流流场的特性分析。作者曾分析了常用紊流润滑理论基础，并在考察润滑流场的边界条件及内部结构的基础上，提出了一种在理论上较合理的复合型高级紊流模式理论，即在近壁区采用低紊流雷诺数的 k-ε 理论，而在紊流核心区采用代数雷诺应力理论，并对复杂流场的紊流润滑问题进行了分析，在计入平均化惯性项的情况下推导出一种适用于复杂流场的紊流雷诺型润滑方程，并验证了该理论的有效性和可靠性。详见[1]。但是正像其他的高级紊流模式理论一样，它是由非常复杂的、具有强烈非线性且相互耦合的方程组形式给出的。这就给整个紊流场的求解带来了很大的不便，往往计算非常复杂费时，很难求解。这一缺点极大地限制了高级紊流模式理论在工程中的应用。鉴于这种情况，作者提出了一种方便实用的工程计算方法，克服了高级紊流模式理论所具有的计算复杂、费时的缺点，从而使高级紊流模式理论用于工程实际成为可能。

1　应用复合型高级紊流模式的雷诺型润滑方程

作者采用复合型高级紊流模式理论作为紊流平均运动基本方程组的封闭方程，并在计入润滑边界条件的情况下，得到了含有均效惯性项的雷诺型润滑方程[1]如下：

$$\frac{\partial}{\partial x}\left[G\left(\frac{\partial p}{\partial x}+\rho\bar{f}\right)\right]+\frac{\partial}{\partial z}\left[G\left(\frac{\partial p}{\partial z}+\rho\bar{g}\right)\right]=\frac{\partial}{\partial x}(UFh) \tag{1}$$

式中：

$$\begin{cases} G=\dfrac{A_2}{A_1}\cdot B_1-B_2 \\ F=1-\dfrac{B_1}{A_1 h} \end{cases} \quad \begin{cases} A_1=\displaystyle\int_0^h \dfrac{\mathrm{d}y}{\nu+\nu_t} \\ A_2=\displaystyle\int_0^h \dfrac{y\,\mathrm{d}y}{\rho(\nu+\nu_t)} \end{cases} \quad \begin{cases} B_1=\displaystyle\int_0^h\int_0^y \dfrac{\mathrm{d}y\,\mathrm{d}y}{\nu+\nu_t} \\ B_2=\displaystyle\int_0^h\int_0^y \dfrac{y\,\mathrm{d}y\,\mathrm{d}y}{\rho(\nu+\nu_t)} \end{cases}$$

---

* 本文合作者：张运清。原发表于《上海工业大学学报》，1994，15(4)：337-342。

$$\begin{cases} \bar{f} = \dfrac{1}{h}\left[\dfrac{\partial I_{xx}}{\partial x} + \dfrac{\partial I_{xz}}{\partial z}\right] \\ \bar{g} = \dfrac{1}{h}\left[\dfrac{\partial I_{xz}}{\partial x} + \dfrac{\partial I_{zz}}{\partial z}\right] \end{cases} \quad \begin{cases} I_{xx} = \displaystyle\int_0^h u^2 \mathrm{d}y \\ I_{xx} = \displaystyle\int_0^h uw \mathrm{d}y \\ I_{zz} = \displaystyle\int_0^h w^2 \mathrm{d}y \end{cases}$$

p——油膜压力；U——表面相对移动速度；ρ——密度；$\bar{f}$，$\bar{g}$——平均惯性项；ν——运动黏度；h——膜度；u，w——x，z方向的紊流平均运动速度；ν_t——紊流黏度，其值由复合型紊流模式方程确定.

显然，当我们对雷诺方程(1)采用常规 SOR 法离散求解时，由于每一点的 ν_t 值随该点的压力值变化，故在迭代求解压力 p 的过程中，需要由复合型紊流模式方程多次计算各点的 ν_t 值，然后再求得各点的其他系数 G、$\bar{f}$、$\bar{g}$ 和 F，而每计算一点的 ν_t 值都需要求解非常复杂的非线性方程组，从而导致雷诺方程的求解非常困难，往往要费很长的时间，这给工程应用带来了极大的不便. 但倘若我们事先将与 ν_t 有关的参数建成一个紊流参数数据库，那么在求解雷诺方程(1)时，只需每次按一定的规律从数据库中调出相应的参数值，避免在迭代求解过程中多次耗费冗长的计算时间来求解这些参数值，从而大大简化雷诺方程的求解，这也是本文提出的工程实用计算方法的本质所在.

2 紊流参数数据库

通过对润滑流场的边界条件及内部结构的分析，我们采用复合型紊流模式来分析润滑流场特性，即在近壁区采用低紊流雷诺数的 k-ε 方程，而在紊流核心区采用代数雷诺应力模式，两区的分界条件取为分界面上紊流雷诺应力占总剪应力的 95%，并保证分界面上剪应力连续.

1. 低紊流雷诺数的 k-ε 模式方程[2]

$$\frac{\partial}{\partial y}\left[\left(\nu + \frac{\nu_t}{\sigma_k}\right)\cdot\frac{\partial k}{\partial y}\right] + \nu_t\left[\left(\frac{\partial u}{\partial y}\right)^2 + \left(\frac{\partial w}{\partial y}\right)^2\right] - \varepsilon - \frac{2\nu k}{y^2} = 0 \quad (2)$$

$$\frac{\partial}{\partial y}\left[\left(\nu + \frac{\nu_t}{\sigma_\varepsilon}\right)\frac{\partial\varepsilon}{\partial y}\right] + C_{\varepsilon 1}\cdot\nu_t\left[\left(\frac{\partial u}{\partial y}\right)^2 + \left(\frac{\partial w}{\partial y}\right)^2\right]\frac{\varepsilon}{k} -$$

$$C_{\varepsilon 2}[1 - 0.3\exp(-R_k^2)]\frac{\varepsilon^2}{k} - 2\nu\left(\frac{\partial\varepsilon^{\frac{1}{2}}}{\partial y}\right)^2 = 0$$

其中：

$$\nu_t = C_m[1 - 0.98\exp(-A_m\cdot R_k)]\frac{k^2}{\varepsilon} \quad R_k = \frac{k^2}{\nu\cdot\varepsilon}(\text{紊流雷诺数})$$

$$C_m = 0.09, \sigma_k = 1.0, \sigma_\varepsilon = 1.3, C_{\varepsilon 1} = 1.45, C_{\varepsilon 2} = 2.0,$$

对牛顿流体：$A_m = 0.0015$

2. 代数雷诺应力模式理论[3]

$$\begin{cases} \dfrac{\partial}{\partial y}\left(\dfrac{\nu_t}{\sigma_k}\cdot\dfrac{\partial k}{\partial y}\right) + P_d - \varepsilon = 0 \\ \dfrac{\partial}{\partial y}\left(\dfrac{\nu_t}{\sigma_\varepsilon}\cdot\dfrac{\partial \varepsilon}{\partial y}\right) + C_{\varepsilon 1}\cdot P_d\cdot\dfrac{\varepsilon}{k} - C_{\varepsilon 2}\cdot\dfrac{\varepsilon^2}{k} = 0 \end{cases} \quad (3)$$

式中：

$$\nu_t = 0.09 G_\mu \cdot \frac{k^2}{\varepsilon}, \quad P_d = \nu_t \left[\left(\frac{\partial u}{\partial y}\right)^2 + \left(\frac{\partial w}{\partial y}\right)^2 \right]$$

$$G_\mu = \frac{1 + \frac{3}{2} \frac{C_2 C_2'}{1 - C_2} f}{1 + \frac{3}{2} \frac{C_1'}{C_1} f} \cdot \frac{1 - 2 \frac{C_2 C_2' \cdot P_d/\varepsilon}{C_1 - 1 + C_2 \cdot P_d/\varepsilon} f}{1 + 2 \frac{C_1'}{C_1 - 1 + P_d/\varepsilon} f}, \quad f = \frac{k^{\frac{3}{2}}}{C_w y \varepsilon}$$

$C_w = 3.72$, $C_1 = 0.6$, $C_2 = 0.6$, $C_2' = 0.3$, $C_1' = 0.6$, $C_{\varepsilon 1} = 1.44$, $C_{\varepsilon 2} = 1.92$, $\sigma_k = 1.0$, $\sigma_\varepsilon = 1.3$.

由于模式(2)和(3)都是非常复杂、具有强烈非线性且相互耦合的方程，因此在数值求解模式方程(2)和(3)时应特别注意这类方程的特点，需作一些相应的数学处理，否则难以求解得到所需的结果.

引进三个无量纲量：

$$B_x' = -\frac{h^2}{\mu U} \cdot \frac{\partial p}{\partial x}, \quad B_z' = -\frac{h^2}{\mu U} \cdot \frac{\partial p}{\partial z}, \quad R_h = \frac{Uh}{\nu}$$

式中：B_x'、B_z'——无量纲等效压力梯度；R_h——局部雷诺数.

模式方程(2)和(3)具体表达了紊流工况参数（以 B_x'、B_z' 和 R_h 的值表征）与相应的 $\nu_t(y)$ 之关系，可由此解出后者，并进一步求得紊流参数 A_1、A_2、B_1、B_2、I_{xx} I_{xz} 和 I_{zz} 及 $\bar{G}$、$\bar{F}$、$\bar{f}$ 和 $\bar{g}$ 的值. 也就是说，上述紊流特性参数依决于三个无量纲数 B_x'、B_z' 和 R_h 的函数关系可通过求解模式方程而定出，依据这种函数关系我们即可建立起以无量纲数 B_x'、B_z' 和 R_h 为依变量的紊流参数数据库，该数据库是一个由 FORTRAN77 语言支持下的大型直接存取数据文件，数据是按某种规律存入的，其在库内的排列顺序也存在着一定的规律，在调用时只需按其在库内的存放规律算出该数据在库内的记录号即可直接读出，检索速度非常快.

从目前的工业水平及发展趋势看，数据库中无量纲参数的取值范围取为：$800 \leqslant R_h \leqslant 30\,000$，$-75 \leqslant B_x' \leqslant 75$，$-370 \leqslant B_z' \leqslant 370$. 在目前的工业应用中一般不会超出该范围，在特别情况下，若计算时超出该范围，可依据基本紊流参数随 B_x'、B_z' 和 R_h 在上、下限处的变化规律作外推插值处理.

3 雷诺型润滑方程的计算方法

对径向轴承，将方程(1)变形并无量纲化得到：

$$\frac{\partial}{\partial \varphi}\left[H^3 \cdot \bar{G} \cdot \frac{\partial p}{\partial \varphi}\right] + \left(\frac{d}{l}\right)^2 \cdot \frac{\partial}{\partial \lambda}\left[H^3 \cdot \bar{G} \cdot \frac{\partial p}{\partial \lambda}\right] = \\ = \frac{\partial}{\partial \varphi}(\bar{F} \cdot H) - \frac{H^2 C^2 U}{\nu \cdot r}\left[\frac{\partial}{\partial \varphi}(\overline{Gf}) + \frac{d}{l} \cdot \frac{\partial}{\partial \lambda}(\overline{G}\bar{g}')\right] \tag{4}$$

式中：

$$\begin{cases}\overline{f}=\dfrac{U^2}{r\cdot H}\cdot\overline{f}\\ \overline{g}=\dfrac{U^2}{r\cdot H}\overline{g}'\end{cases}\quad\begin{cases}\overline{f}=\dfrac{\partial}{\partial\varphi}(\overline{I}_{xx}\cdot H)+\dfrac{d}{l}\dfrac{\partial}{\partial\lambda}(\overline{I}_{xz}\cdot H)\\ \overline{g}'=\dfrac{\partial}{\partial\varphi}(\overline{I}_{xz}\cdot H)+\dfrac{d}{l}\cdot\dfrac{\partial}{\partial\lambda}(\overline{I}_{zz}\cdot H)\end{cases}\quad(5)$$

$$\begin{cases}\overline{G}=\dfrac{\overline{A}_2}{A_1}\overline{B}_1-\overline{B}_2\\ \overline{F}=1-\dfrac{\overline{B}_1}{A_1}\end{cases}\quad P=\dfrac{p}{p_0}=\dfrac{c^2p}{\mu Ur},\ H=\dfrac{h}{c}$$

等效压力梯度 B'_x 和 B'_z 的计算公式：

$$\begin{cases}B'_x=-\dfrac{h^2}{\mu U}\cdot\dfrac{\partial p'}{\partial x}=-\dfrac{h^2}{\mu U}\left(\dfrac{\partial p}{\partial x}+\rho\overline{f}\right)\\ B'_z=-\dfrac{h^2}{\mu U}\cdot\dfrac{\partial p'}{\partial z}=-\dfrac{h^2}{\mu U}\left(\dfrac{\partial p}{\partial z}+\rho\overline{g}\right)\end{cases}\quad(6)$$

即：

$$\begin{cases}B'_x=-H^2\cdot\dfrac{\partial p}{\partial\varphi}-\dfrac{HUc^2}{\nu\cdot r}\overline{f}\\ B'_z=-H^2\cdot\dfrac{d}{l}\cdot\dfrac{\partial p}{\partial\lambda}-\dfrac{HUc^2}{\nu\cdot r}\overline{g}\end{cases}$$

采用松弛迭代法对雷诺方程(4)的离散格式方程整理得：

$$P_{ij}^{(m)}=\beta\left[\dfrac{A_{ij}\cdot P_{i+1,j}^{(m-1)}+B_{ij}\cdot P_{i-1,j}^{(m)}+C_{ij}\cdot P_{i,j+1}^{(m-1)}+D_{ij}\cdot P_{i,j-1}^{(m)}-F_{ij}}{E_{ij}}-P_{ij}^{(m-1)}\right]\\ +P_{ij}^{(m-1)}\quad(7)$$

式中：

$$\begin{cases}A_{ij}=H^3_{i+\frac{1}{2},j}\cdot\overline{G}_{i+\frac{1}{2},j}\\ B_{ij}=H^3_{i-\frac{1}{2},j}\cdot\overline{G}_{i-\frac{1}{2},j}\\ C_{ij}=\left(\dfrac{d}{l}\cdot\dfrac{\Delta\varphi}{\Delta\lambda}\right)^2\cdot H^3_{i,j+\frac{1}{2}}\cdot\overline{G}_{i,j+\frac{1}{2}}\\ D_{ij}=\left(\dfrac{d}{l}\cdot\dfrac{\Delta\varphi}{\Delta\lambda}\right)^2\cdot H^3_{i,j-\frac{1}{2}}\cdot\overline{G}_{i,j-\frac{1}{2}}\\ E_{ij}=A_{ij}+B_{ij}+C_{ij}+D_{ij}\\ F_{ij}=\Delta\varphi\left[(H\overline{F})_{i+\frac{1}{2},j}-(H\overline{F})_{i-\frac{1}{2},j}\right]-\dfrac{H^2C^2U}{\nu\cdot\Gamma}\Delta\varphi\left\{(\overline{G}\overline{f})_{i+\frac{1}{2},j}-(\overline{G}\overline{f})_{i-\frac{1}{2},j}\right.\\ \qquad\left.+\dfrac{d}{l}\cdot\dfrac{\Delta\varphi}{\Delta\lambda}\left[(\overline{G}\overline{g}')_{i+\frac{1}{2},j}-(\overline{G}\overline{g}')_{i-\frac{1}{2},j}\right]\right\}\end{cases}\quad(8)$$

求解步骤：

(1) 给各内节点的压力 P_{ij} 赋初值，同时对各点的惯性项 $\overline{f}'_{ij}$ 和 $\overline{g}'_{ij}$ 赋初值；

(2) 由公式(6)求得各相应点的等效压力梯度 B'_{xij} 和 B'_{zij} 值,并求出各点的雷诺数 R_{hij};

(3) 由已求得的各点的 B'_{xij}、B'_{zij} 和 R_{hij} 值,以紊流参数数据库中查取并插值得到各点的紊流参数 $\overline{A}_{1ij}$、$\overline{A}_{2ij}$、$\overline{B}_{1ij}$、$\overline{B}_{2ij}$、$\overline{I}_{xxij}$、$\overline{I}_{xzij}$ 和 $\overline{I}_{zzij}$ 值,并求得各点的 $\overline{G}_{ij}$ 和 $\overline{F}_{ij}$ 值;

(4) 由公式(5)计算各点的惯性项 $\overline{f}'^{(m)}_{ij}$ 和 $\overline{g}'^{(m)}_{ij}$ 的值,并与假定初值作误差判断。若收敛,则求得在初值压力条件下各点的真实惯性项,当然同时也得到了在初始压力值条件下各点的紊流参数值;若不收敛,则采用松弛迭代法求得新一轮迭代各点的惯性项 $\overline{f}'_{ij}$ 和 $\overline{g}'_{ij}$ 的新初值,然后返回到(2)重新执行,如此反复,直到收敛;

(5) 利用式(8)求得各点的系数 A_{ij}、B_{ij}、C_{ij}、D_{ij}、E_{ij} 和 F_{ij};

(6) 由公式(7)求各点的压力 $P^{(m)}_{ij}$ 值;并采用误差判据判断压力的收敛性。若收敛,则求得各点的最终压力值;反之,则返回到(2)重新执行,直至收敛。在压力迭代过程中采用雷诺边界条件;

(7) 在迭代求得了各点的压力值后,即可进行静、动特性的计算。

4 周隙高压密封的静特性计算

图1为一周隙高速高压密封,由于其轴向存在较大的压差,使得周隙内不仅存在周向的流动,而且沿轴向还存在强烈的流动。对于这种复杂紊流流场的分析,应用目前常用的紊流润滑理论显然是不合适的,完全应当应用复合型紊流模式理论。我们采用本文提出的计算方法非常方便地对图示周隙高压密封的静特性进行了计算,计算时间不长,在486/25 微机上大约只需 0.5 h,并将计算结果与 Hori 等人的实验结果[4]进行了考核,如表1所示.

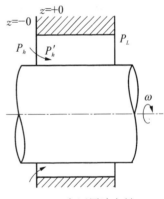

图 1 高压周隙密封

实验参数为:

$P_d = P_h - P_L = 0.7$ MPa, $N = 4\,080$ r/min.

$c = 0.175$ mm, $D = 70$ mm, $L = 35$ mm,

$R_a = \dfrac{Q}{\pi D \nu}$ —— 轴向雷诺数;$F_m = \dfrac{F}{DL}$;

Q —— 轴向流量.

表 1 高压密封特性分析

偏心率 静特性	$\varepsilon=0.2$		$\varepsilon=0.8$	
	$F_m/\times10^{-4}$ Pa	$R_a/\times10^{-2}$	$F_m/\times10^{-4}$ Pa	$R_a/\times10^{-2}$
实验值	2.5	17.3	11.5	18.5
计算值	2.314	16.89	9.73	17.74

从表1可看出,计算结果与实验数据十分吻合。这不但说明,对于复杂流场采用作者提出的模式理论是合理的,而且也证明了本文提出的工程实用计算方法既具有足够的精度又具有方便、快速的特点.

5 小结

鉴于高级紊流模式理论在实际应用中的复杂性,本文提出了一种简便实用的工程计算方法,即将应用中非常复杂、计算费时的紊流流场参数事先求出放入数据库中,建立一个以 B'_x、B'_z 和 R_h 为变量,以无量纲紊流参数 $\overline{A}_1$、$\overline{A}_2$、$\overline{B}_1$、$\overline{B}_2$、$\overline{I}_{xx}$、$\overline{I}_{xz}$ 和 $\overline{I}_{zz}$ 为依变量的数据库,在该数据库的支持下,采用松弛迭代法求解高级紊流模式的雷诺润滑方程. 本方法既具有坚实的理论基础,又避免了繁杂的基础性紊流润滑流场的解算. 通过实例分析计算说明,本文提出的工程实用计算方法是分析复杂流场紊流场润滑特性的一种方便、有效的方法,在工程上将具有推广价值.

参 考 文 献

[1] 张运清,张直明. 应用高级紊流润滑模式的紊流润滑理论分析. 上海工业大学学报,1994,15(1): 55-60.
[2] Hassid, S., Poreh, A. M. Turbulent energy dissipation model for flows with drag reduction. ASME, J. Fluids Eng., 1978,100: 107.
[3] Ljuboja, M., Rodi, W. Calculation of turbulent wall jets with an algebraic Reynolds stress model. ASME, J. Fluids Eng.,1980,102: 350.
[4] Kaneko, S., Hori, Y. Static and dynamic characteristics of annular plain seals. IMechE., 1984.

An Engineering Method for the Solution of Reynolds Lubrication Equation

Abstract: The paper presents an engineering method for the solution of Reynolds lubricaton equation with the aid of a comprehensive database of characteristics of complicated turbulent lubricating film. The method of forming a group of dimensionless turbulent parameters is introduced, and the numerical calculating procedure of the engineering method is given in details. The prediction using the engineering method for static characteristics of annular plain seals with great axial pressure gradient agrees very well with the experimental results obtained hy Hori.

高速重载轴承紊流润滑特性分析*

摘　要：文章应用作者提出的复合型紊流润滑理论对高速重载轴承的紊流润滑特性进行了分析，并与现今常用的 Ng-Pan 紊流润滑理论的计算结果进行了对比，分析了两种方法的异同及产生差异的主要原因，表明了对高速重载轴承等类复杂紊流润滑膜不宜再采用目前常用的零方程类模式理论，而更加宜于采用本文提出的复合型紊流润滑理论.

前言

由于目前常用的紊流润滑理论大都属于零方程类模式理论，主要是以高速轻载轴承为对象的，而现在随着机器向高速、大功率发展，越来越多的轴承处于高速中重载工况下工作. 显然，对这类工况下工作的轴承再来用现今常用的紊流润滑理论进行分析不太合适的. 为了满足工程发展所需，作者通过理论分析并经与多项实验结果校核[1,2]提出了一种适用于复杂流场的紊流润滑分析方法——复合型紊流润滑理论. 它不仅可用于计算流场时均速分布[1]和高速周隙密封性能[2]，亦可用于非轻载的高速滑动轴承计算，并能获得比零方程类紊流润滑理论更合适的结果.

1　高速重载轴承紊流特性分析

1.1　应用复合型高级紊流模式的雷诺型润滑方程[1]

无量纲雷诺型润滑方程为：

$$\frac{\partial}{\partial \varphi}\left[H^3 \bar{G} \cdot \frac{\partial P'}{\partial \varphi}\right] + \left(\frac{d}{l}\right)^2 \cdot \frac{\partial}{\partial \lambda}\left(H^3 \bar{G} \cdot \frac{\partial P'}{\partial \lambda}\right) = \frac{\partial}{\partial \varphi}(\bar{F}H)$$
$$+ K'\left[\cos\varphi + \Psi \cdot \frac{\partial}{\partial \varphi}(\bar{F}H\sin\varphi)\right] + K\theta'\left[\sin\varphi - \Psi \cdot \frac{\partial}{\partial \varphi}(\bar{F}H\cos\varphi)\right] \quad (1)$$

式中：P'——等效压力；$\bar{G}$、$\bar{F}$——无量纲紊流参数；Ψ——相对间隙；K——偏心率；H——无量纲膜厚；K'、θ'——无量纲扰动速度.

在紊流参数数据库支持下，采用松弛迭代法求解雷诺方程(1)即可求得流场压力 P 值[2].

1.2　高速重载轴承静特性计算

1.2.1　油膜力 F_ξ 和 F_η 的计算

对有限宽轴承，如图 1 所示.

---

* 本文合作者：张运清、易子夏、孙美丽. 原发表于《上海工业大学学报》，1994，15(6)：514-518.

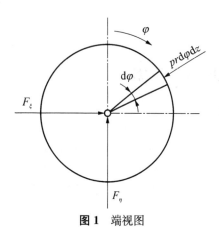

图 1 端视图

$$\begin{cases} F_\xi = -\int_{-l/2}^{l/2} \int_\theta^{2\pi+\theta} pr\sin\varphi\mathrm{d}\varphi\mathrm{d}z \\ F_\eta = -\int_{-l/2}^{l/2} \int_\theta^{2\pi+\theta} pr\cos\varphi\mathrm{d}\varphi\mathrm{d}z \end{cases} \quad (2)$$

无量纲化得：

$$\begin{cases} \overline{F}_\xi = \dfrac{F_\xi \cdot \Psi^2}{\mu U_j l} = -\dfrac{1}{2}\int_\theta^{2\pi+\theta}\int_{-1}^{1}(P\cdot\mathrm{d}\lambda)\sin\varphi\mathrm{d}\varphi \\ \overline{F}_\eta = \dfrac{F_\eta \cdot \Psi^2}{\mu U_j l} = -\dfrac{1}{2}\int_\theta^{2\pi+\theta}\int_{-1}^{1}(P\cdot\mathrm{d}\lambda)\cos\varphi\mathrm{d}\varphi \end{cases}$$

1.2.2 侧泄流量 Q 的计算

$$Q = \int_\theta^{2\pi+\theta}(q_z|_{\lambda=+1} - q_z|_{\lambda=-1})r\mathrm{d}\varphi$$

而

$$q_z = \int_0^h w\mathrm{d}y = U_j h b'_z \overline{G}$$

故

$$Q = \int_{\varphi_{pb}}^{\varphi_{pe}}[(B'_z U_j h \overline{G})|_{\lambda=+1} - (B'_z U_j h \overline{G})|_{\lambda=-1}]r\mathrm{d}\varphi$$

即

$$\overline{Q} = \dfrac{Q}{\Psi U l d} = \dfrac{1}{4}\cdot\dfrac{d}{l}\int_{\varphi_{pb}}^{\varphi_{pe}}[(B'_z\overline{G})|_{\lambda=+1} - (B'_z\overline{G})|_{\lambda=-1}]\mathrm{d}\varphi \quad (3)$$

式中：φ_{pb}、φ_{pe}——油膜起始角和破裂角；$B'_z = -H^2\cdot\dfrac{d}{l}\cdot\dfrac{\partial P}{\partial\lambda} - \dfrac{H\Omega C^2}{\nu}\overline{g}'$——等效压力梯度；$\Omega$——角速度；$\nu$——运动黏度；$C$——半径间隙；$\overline{g}'$——平均惯性项；$P$——无量纲压力；$\Psi$——相对间隙.

1.3 高速重载轴承动特性计算

润滑油膜对于轴颈在其中振动时的力学反应称为油膜的动特性. 当轴颈在静平衡位置附近作扰动时，油膜作用在轴颈上的力将发生变化，这种变化是非线性的. 但当扰动量小时，可将油膜力的动态增量在静平衡位置作线性化处理，由此得到四个刚度系数 k_{ij} 和四个阻尼系数 $C_{ij}(i=x,y;j=x,t)$ 即：

$$\begin{cases} \Delta F_x = k_{xx}\cdot\Delta x + k_{xy}\cdot\Delta y + C_{xx}\cdot\Delta\dot{x} + C_{xy}\cdot\Delta\dot{y} \\ \Delta F_y = k_{yx}\cdot\Delta x + k_{yy}\cdot\Delta y + C_{yx}\cdot\Delta\dot{x} + C_{yy}\cdot\Delta\dot{y} \end{cases} \quad (4)$$

式中：

$$\begin{cases} k_{xx} = \left(\dfrac{\partial F_x}{\partial x}\right)_0, \; k_{xy} = \left(\dfrac{\partial F_x}{\partial y}\right)_0, \; k_{yx} = \left(\dfrac{\partial F_y}{\partial x}\right)_0, \; k_{yy} = \left(\dfrac{\partial F_y}{\partial y}\right)_0 \\ C_{xx} = \left(\dfrac{\partial F_x}{\partial\dot{x}}\right)_0, \; C_{xy} = \left(\dfrac{\partial F_x}{\partial\dot{y}}\right)_0, \; C_{yx} = \left(\dfrac{\partial F_y}{\partial\dot{x}}\right)_0, \; C_{yy} = \left(\dfrac{\partial F_y}{\partial\dot{y}}\right)_0 \end{cases} \quad (5)$$

按动特性系数定义式(5)，采用在图 2 中静平衡位置 $O_j(\theta,e)$ 附加一个微小位移或微小速度扰动的方法来计算八个动态特性系数(静平衡位置的油膜力为 F_{xO} 和 F_{yO}).

1.3.1 刚度系数 k_{ij} 的求解

在静平衡位置 $O_j(\theta, e)$ 给定一个沿 x 方向的扰动 Δx,则有:

$$\begin{cases} \Delta e = \Delta x \cdot \sin\theta \\ e\Delta\theta = \Delta x \cdot \cos\theta \end{cases}$$

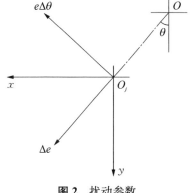

图 2 扰动参数

由雷诺方程(1)求得相应扰动位置 $(\theta+\Delta\theta, e+\Delta e)$ 的载荷 F_{xx} 和 F_{yx},从而有:

$$k_{xx} \doteq \frac{F_{xx} - F_{xo}}{\Delta x}, \quad k_{yx} \doteq \frac{F_{yx} - F_{yo}}{\Delta x}$$

同理,在静平衡位置 $O_j(\theta, e)$ 给定一个沿 y 方向的扰动 Δy,则有:

$$\begin{cases} \Delta e = \Delta y \cdot \cos\theta \\ e\Delta\theta = -\Delta y \cdot \sin\theta \end{cases}$$

由雷诺方程(1)即可求得相应扰动位置的载荷 F_{xy} 和 F_{yy},从而得:

$$k_{xy} \doteq \frac{F_{xy} - F_{xo}}{\Delta y}, \quad k_{yy} \doteq \frac{F_{yy} - F_{yo}}{\Delta y}$$

1.3.2 阻尼系数 C_{ij} 的求解

同样,分别通过在静平衡位置 $O_j(\theta, e)$ 处沿 x 方向给定一个扰动速度 $\Delta \dot{x}$ 和沿 y 方向给定一个扰动速度 $\Delta \dot{y}$ 求得相应的油膜力,从而求得相应的阻尼系数,即:

$$\begin{cases} C_{xx} \doteq \dfrac{F_{xx} - F_{xo}}{\Delta \dot{x}} & C_{yx} \doteq \dfrac{F_{yx} - F_{yo}}{\Delta \dot{x}} \\ C_{xx} \doteq \dfrac{F_{xy} - F_{xo}}{\Delta \dot{y}} & C_{yy} \doteq \dfrac{F_{yy} - F_{yo}}{\Delta \dot{y}} \end{cases}$$

2 算例

分别采用作者提出的复合型紊流润滑理论和线性化的 Ng-Pan 紊流理论对一个剖分式径向圆柱轴承进行了动、静特性分析计算.

圆柱轴承的参数为:

$d=100$ mm,$l=50$ mm,$c=0.1$ mm,$n=50\,000$ r/min,双侧进油,进油压力为零,圆柱轴承由上下两块瓦组成,瓦张角为 $150°$.

计算结果如图3～图10所示,图中:

$$SO = \frac{F\Psi^2}{2\mu U l} \text{——索麦菲尔德数}; \quad \overline{Q} = \frac{Q}{\Psi U l d} \text{——无量纲侧泄流量};$$

$$K_{ij} = \frac{k_{ij}\Psi^3}{\mu\Omega l} \text{——无量纲油膜刚度}; \quad C_{ij} = \frac{C_{ij}\Psi^3}{\mu l} \text{——无量纲油膜阻尼}.$$

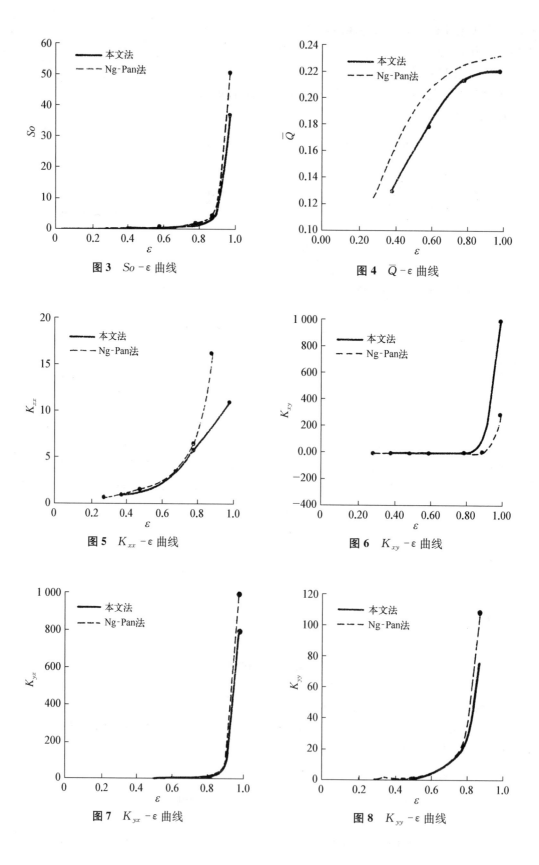

图 3　So-ε 曲线

图 4　$\bar{Q}$-ε 曲线

图 5　K_{xx}-ε 曲线

图 6　K_{xy}-ε 曲线

图 7　K_{yx}-ε 曲线

图 8　K_{yy}-ε 曲线

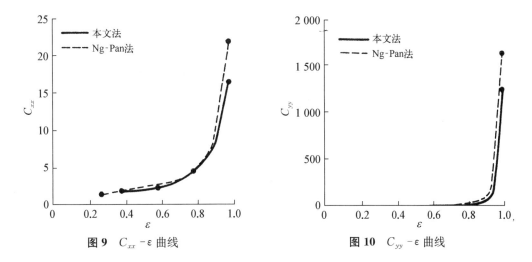

图 9 C_{xx}-ε 曲线　　　　图 10 C_{yy}-ε 曲线

由图示计算结果可见,当偏心率 ε 较小即高速轻载时,用本文提出的分析方法得到的结果(实曲线)与 Ng-Pan 方法计算值(虚曲线)相近;当偏心率 ε 较大即高速重载时,两种方法的计算结果有较大的差别.出现这种情况主要原因是,Ng-Pan 理论采用了 Reichardt 涡粘公式,该经验公式是在"常应力区"的条件下,通过对实验数据的拟合而得到的,而实际润滑流场中由于有压力梯度存在,沿膜厚的剪应力分布为非均匀的,尤其是在 B_x 和 B_z 较大的高速重载场合.尽管 Ng-Pan 理论将经验公式中的壁面剪应力 τ_w 以局部剪应力 τ 替代,试图在某种程度上考虑 B_x 和 B_z 的影响,但由于作了线性化处理,故使得实际公式中远未充分地计入压力梯度或 B_x,B_z 对紊流性态的影响.所以,当偏心率 ε 较小即高速轻载时,由于 Couette 流效应远大于压力流效应即 B_x 和 B_z 影响较小,壁面间的总剪应力变化不大,可以近似看作"常应力区",这时运用 Ng-Pan 理论分析计算的结果当然应与本文提出的方法的结果相一致;而当流场为高速重载时,B_x 和 B_z 对流场紊流性态的影响较大,此时如仍采用以 Couette 附近小摄动为基础的紊流润滑理论,就会有较大偏差,而如采用充分考虑了 B_x 和 B_z 影响的复合型紊流润滑理论,当然就合理多了.

参 考 文 献

[1] 张运清,张直明.应用高级紊流模式的紊流润滑理论分析.上海工业大学学报,1994,15(1):55-60.
[2] 张运清,张直明.应用高级紊流模式的雷诺型润滑方程的解算方法.上海工业大学学报,1994,15(4):337-342.

The Analysis of Turbulent Characteristics of Heavily Loaded Bearings with High Rotation

Abstract: With the compound turbulent lubrication theory proposed by the authors, the characteristics of turbulent lubrication of heavily loaded bearings with high rotation are analyzed in detail. The predictions obtained by this theory are compared with the results of Ng-Pan's theory, and the conclusion is got that the compound turbulent lubrication theory is more appropriate to dealing with the complicated turbulent fields than the widely used turbulent lubrication theory.

Thermal Non-Newtonian EHL Analysis of Rib-Roller End Contact in Tapered Roller Bearings*

Abstract: An EHL approach to the rib-roller end contact in tapered roller bearings has been achieved by taking into account the non-Newtonian behavior of lubricants and thermal effects and with full consideration of the peculiar geometrical and kinematic conditions. Two kinds of geometrical configurations of rib and roller end were investigated: tapered rib/spherical roller end and spherical rib/spherical roller end. Optimal ratios of curvature radius of roller end to rib face were deduced. The film thickness, friction torque, lubricant temperature, and surface temperature at various speeds and loads were calculated.

Nomenclature

a, b = semiminor and semimajor axes of contact ellipse respectively, m

B = factor considering temperature effect on τ_L, 585 K

c_f, c_1, c_2 = heat capacities of lubricant and solids, J/(kg·K)

E' = effective elastic modulus, $E' = 2/((1-\nu_1^2)/E_1 + (1-\nu_2^2)/E_2)$, N/m²

$F(\tau_e)$ = rheology function

h_c, h_{min} = central and minimum film thickness, m

H = dimensionless film thickness, $H = h/R_x$

H_0 = dimensionless constant used in calculation of H

J_0, J_1, J_2 = integrals of the rheology function $F(\tau_e)$

K = ellipticity parameter, $K = a/b$

k_f, k_1, k_2 = thermal conductivity of lubricant and solids, W/(m·K)

m = limiting-shear-stress proportionality constant, 0.047

M_1 = friction torque acting on rib around bearing axis, N·m

N_Ω = rotational speed of bearing, r/min

p = pressure, N/m²

p_h = maximum Hertzian pressure, N/m²

P = dimensionless pressure, $P = p/E'$

r = large end radius of roller, 0.009 739 m

r_i, r_o = inner and outer radii of rib, 0.057 486 and 0.060 5 m

R_i, R_o = curvature radii of lower and upper border lines of conjunction domain, respectively, $R_i = r_i/\cos\psi$; $R_o = r_o/\cos\psi$

R_x, R_z = equivalent radii of curvature in x- and z-direction, m

R_{x1}, R_{x2} = radii of curvature in x-direction of rib and of roller end respectively, m

t = temperature, K

T = dimensionless temperature, $T = t/t_0$

t_0 = inlet temperature, K

t_1, t_2 = temperature of solids, K

t_m = mean lubricant temperature, K

t_c = centerline lubricant temperature, K

$u_{1(0)}$, $u_{2(0)}$ = velocities in moving direction of rib and of roller at nominal contact point, respectively, m/s

u_s = average entraining velocity, $u_s = 0.5(u_{1(0)} +$

---

* In collaboration with Xiaofei Jiang, Pat Lam Wong. Reprinted from *Transactions of the ASME*, *Journal of Tribology*, 1995, 117(4): 646–654.

$u_{2(0)}$), m/s

U_1, W_2 = dimensionless velocity components, $U_1 = u_1/u_s$, $W_2 = w_2/u_s$

$w_{1(0)}$, $w_{2(0)}$ = velocities in transverse direction of rib and of roller at nominal contact point, respectively, m/s

W = dimensionless normal applied load on rib-roller end contact, $W = w/(abE')$

X = dimensionless coordinate in x-direction, $X = x/b$

Y = dimensionless coordinate in direction of film thickness, $Y = y/b$

$\mathring{y}$ = variable transition, $\mathring{y} = y/h$

Z = dimensionless coordinate in z-direction, $Z = z/a$

β = half cone angle, 11.642 2°

β_r = sliding-rolling ratio, $\beta_r = (u_{1(0)} - u_{2(0)})/(u_{1(0)} + u_{2(0)})$

β_s = spinning-rolling ratio, $\beta_s = (\Omega_{1y} - \Omega_{2y})R_x/(u_{1(0)} + u_{2(0)})$

β_t = thermal expansivity, K^{-1}

γ = half roller angle, 2.0°

γ_t = temperature-viscosity coefficient, K^{-1}

σ = constant, $\sigma = b\eta_0 u_s/(E'Rx)$

ε = distance from the nominal contact point to inner-race, 0.002 735 m

$\bar{\eta}$ = dimensionless absolute viscosity, $\bar{\eta} = \eta/\eta_0$

η_0 = ambient viscosity of lubricant, Pa·s

$\bar{\rho}$ = dimensionless density, $\bar{\rho} = \rho/\rho_0$

ρ_0 = ambient density of lubricant, kg/m³

ρ_1, ρ_2 = density of solids, kg/m³

Ω_{1y}, Ω_{2y} = angular velocities along y-direction of rib and of roller respectively, rad/s

τ = shear stress, N/m²

τ_e = equivalent shear stress, N/m²

τ_L = limiting shear stress, N/m²

τ_{L0} = initial shear strength, 2.3 MPa

$\bar{\tau}_{YX}$, $\bar{\tau}_{YZ}$ = dimensionless shear stresses, $\bar{\tau}_{YX} = \tau_{yx}/E'$, $\bar{\tau}_{YZ} = \tau_{yz}/E'$

$\bar{\tau}_{YX}|_{Y=0}$, $\bar{\tau}_{YZ}|_{Y=0}$ = dimensionless shear stresses on rib face

ψ = angle formed by bearing axis and normal line of rib at nominal contact point

1 Introduction

The elastohydrodynamic lubrication of line and point contacts have received extensive attention over the last few decades. Some important issues, which are often encountered in industry, such as non-Newtonian behavior of lubricants, transient problems, thermal effects, and topography effects of contact surfaces, can now be wholly or partly incorporated into the classic EHL theory. As a result, some complex and difficult practical problems can now be further understood. The rib-roller end contact in tapered roller bearings presents a special EHL problem. The successful lubrication of this contact is of vital importance to the bearing performance. This problem has received special treatments, due to lack of existing solutions with full consideration of the relevant peculiar geometrical and kinematic situation. The study of rib-roller end contact in tapered roller bearings started with Korrenn (1970) who carried out an experiment to observe the frictional behavior of rib-roller end contacts as illustrated in Fig. 1. Jamison et al. (1976) studied the influences of bearing geometry, roller skewing and manufacturing irregularities on the bearing performances. Significant progress was achieved in the theoretical analysis and experimental investigation by Dalmaz et al. (1980, 1984). However, their calculations were only based on the hydrodynamic lubrication theory and no surface deformation was considered. It was not until

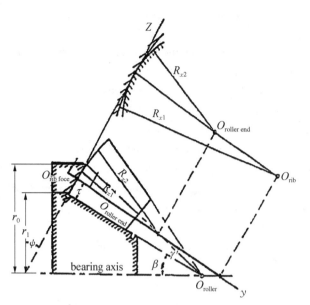

Fig. 1 Geometry of rib-roller end contact

1988 that a complete numerical solution of the elastohydrodynamic lubrication of this end contact was published by Zhang et al. (1988). A domain size close to the conjunction found in the rib-roller end contact was used. However, the numerical analyses were carried out under Newtonian fluid and isothermal conditions. Their results, though qualitatively valid for selecting the optimum geometry to ensure as thick a film as possible, yet predicted an over steep growth of film thickness with speed as compared to Dalmaz's experimental data (1984). Evidently, non-Newtonian rheology and thermal effects have to be considered to improve the analysis quantitatively. A non-Newtonian EHL solution has been obtained by the authors (1993). Recently, a torque model for the prediction of each torque component due to raceway rolling resistances and frictional forces of rib-roller end contact in each roller of the tapered roller bearings was presented by Zhou and Hoeprich (1991). The film thickness, friction, and scuffing failure of rib-roller end contacts in cylindrical roller bearings was investigated by Aramaki et al. (1992).

This paper presents an elastohydrodynamic lubrication analysis of rib-roller end contact in tapered roller bearings including both non-Newtonian and thermal effects. The influences of geometrical parameters of rib face and roller end, lubricated with a limiting shear stress type lubricant, and with consideration of the energy equation were studied. Optimal ratios of curvature radius of roller end to rib face were deduced. The rise of fluid film and surface temperatures, film thicknesses and friction torques were numerically solved at various bearing speeds and loads.

2 Reynolds Equation

For the rib-roller end contact, under steady-state and smooth surface conditions, the nondimensional Reynolds equation can be written as follows:

$$\frac{\partial}{\partial X}\left[\frac{\bar{\rho}}{\bar{\eta}}H^3\left(\frac{J_1^2}{J_0}-J_2\right)\frac{\partial P}{\partial X}\right]+\frac{1}{K^2}\frac{\partial}{\partial Z}\left[\frac{\bar{\rho}}{\bar{\eta}}H^3\left(\frac{J_1^2}{J_0}-J_2\right)\frac{\partial P}{\partial Z}\right]$$
$$+\sigma(U_2-U_1)\frac{\partial}{\partial X}\left[\bar{\rho}H\left(1-\frac{J_1}{J_0}\right)\right]$$
$$+\frac{\sigma}{K}(W_2-W_1)\frac{\partial}{\partial Z}\left[\bar{\rho}H\left(1-\frac{J_1}{J_0}\right)\right]$$

$$+\sigma U_1 \frac{\partial}{\partial X}(\bar{\rho}H) + \frac{\sigma W_1}{K}\frac{\partial}{\partial Z}(\bar{\rho}H) = 0 \tag{1}$$

where the integrals of the rheology function $F(\tau_e)$ across the film thickness are given by:

$$J_0 = \int_0^1 F(\tau_e)\mathrm{d}\mathring{y}; \quad J_1 = \int_0^1 F(\tau_e)\mathring{y}\mathrm{d}\mathring{y};$$

$$J_2 = \int_0^1 F(\tau_e)\mathring{y}^2\mathrm{d}\mathring{y}. \tag{2}$$

and the equivalent shear stress τ_e in two dimensions expressed by Johnson and Tevaarwerk (1977) is:

$$\tau_e = \sqrt{\tau_{yx}^2 + \tau_{yz}^2} \tag{3}$$

The non-Newtonian model proposed by Bair and Winer (1979) is employed in this analysis and the corresponding rheology function is defined as:

$$F(\tau_e) = \frac{\ln[1/(1-\tau_e/\tau_L)]}{\tau_e/\tau_L} \tag{4}$$

where τ_L is the limiting shear stress and is a function of pressure and temperature. In view of the fact that the variation of τ_L with both pressure and temperature is only available for very few lubricants and for very limited temperature ranges, the following expression used by Wang and Zhang (1987) is employed here:

$$\tau_L = (\tau_{L0} + mp)\exp\left[B\left(\frac{1}{t_m} - \frac{1}{t_0}\right)\right] \tag{5}$$

3 Film Thickness and Load Equations

The oil film shape in dimensionless form is expressed by:

$$H = H_0 + \frac{b^2 X^2}{2R_x^2} + \frac{a^2 Z^2}{2R_x R_z} + \frac{2a}{\pi R_x}\iint_A \frac{P(\bar{X},\bar{Z})\mathrm{d}\bar{X}\mathrm{d}\bar{Z}}{\sqrt{(X-\bar{X})^2 + (Z-\bar{Z})^2}} \tag{6}$$

The load capacity in dimensionless form is given by:

$$\overline{W} = \iint_A P(X,Z)\mathrm{d}X\mathrm{d}Z \tag{7}$$

4 Density-Pressure-Temperature Relationship

The variation of density with pressure and temperature can be expressed by the empirical formula proposed by Dowson and Higginson (1977):

$$\bar{\rho} = \left[1 + \frac{0.6e-9E'P}{1+1.7e-9E'P}\right][1 - \beta_t t_0(T-1)] \tag{8}$$

223

5 Viscosity-Pressure-Temperature Relationship

The Roelands equation which represents the viscosity-pressure-temperature relationship reads as (Roelands, 1966):

$$\bar{\eta} = \exp\{(\ln\eta_0 + 9.67)[-1 + (1 + 5.1e - 9E'P)^{0.67}] + \gamma_t t_0(1-T)\} \tag{9}$$

6 Energy Equation

The temperature distribution within the lubricant film in the contact region can be obtained by solving the energy equation. The three-dimensional energy equation is written as:

$$\rho_f c_f \left(u\frac{\partial t}{\partial x} + w\frac{\partial t}{\partial z}\right) - k_f \frac{\partial^2 t}{\partial y^2} = \eta_f\left[\left(\frac{\partial u}{\partial y}\right)^2 + \left(\frac{\partial w}{\partial y}\right)^2\right] - \frac{t}{\rho_f}\frac{\partial \rho_f}{\partial t}\left(u\frac{\partial p}{\partial x} + w\frac{\partial p}{\partial z}\right) \tag{10}$$

Because of the complexity in solving the above equation in three dimensions, a simplification was adopted to reduce the energy equation into a two-dimensional one by assuming a parabolic temperature profile across the film thickness:

$$t = t_1 - (3t_1 + t_2)\left(\frac{y}{h}\right) + 2(t_1 + t_2)\left(\frac{y}{h}\right)^2 + 4t_c\left[\left(\frac{y}{h}\right) - \left(\frac{y}{h}\right)^2\right] \tag{11}$$

where t_c is the centerline ($y = h/2$) lubricant temperature and t_1 and t_2 are the surface temperatures of the rib face and roller end, respectively. For the sake of simplicity, the lubricant viscosity in the Reynolds equation is assumed to be dependent on the mean temperature across the film, it is therefore more convenient to re-express the above equation in terms of the mean lubricant temperature t_m.

$$t_m(x, z) = \frac{1}{h}\int_0^h t(x, z, y)\,\mathrm{d}y = (t_1 + t_2 + 4t_c)/6 \tag{12}$$

Substituting Eq. (12) into Eq. (11) yields:

$$t = t_1 - (3t_1 + t_2)\left(\frac{y}{h}\right) + 2(t_1 + t_2)\left(\frac{y}{h}\right)^2 + (6t_m - t_1 - t_2)\left[\left(\frac{y}{h}\right) - \left(\frac{y}{h}\right)^2\right] \tag{13}$$

Surface temperatures of the rib face and roller end are not known a priori and must be obtained as part of the solution of the energy equation. The boundary conditions for the energy equation are:

$$t_1 = t_0 + \frac{k_f}{\sqrt{\pi\rho_1 c_1 u_1 k_1}}\int_{x_s}^{x}\frac{\left(\frac{\partial t}{\partial y}\right)\big|_{y=0}}{\sqrt{x-\xi}}\mathrm{d}\xi \tag{14}$$

$$t_2 = t_0 + \frac{k_f}{\sqrt{\pi\rho_2 c_2 u_2 k_2}}\int_{x_s}^{x}\frac{\left(-\frac{\partial t}{\partial y}\right)\big|_{y=h}}{\sqrt{x-\xi}}\mathrm{d}\xi \tag{15}$$

At the inlet location, the temperature of the lubricant and the two solid surfaces were taken as t_0.

By substituting the velocity distributions in x- and z-directions, temperature profile (Eq. (13)), density-pressure-temperature relation (Eq. (8)), and viscosity-pressure-temperature relation (Eq. (9)) into the three-dimensional energy equation (Eq. (10)), and integrating it along the y direction from 0 to h, a two-dimensional temperature distribution in terms of the mean film temperature across the film thickness and the surface temperatures of rib and roller end is then obtained.

7 Shear Stress and Friction Torque

The shear stress on the rib surface ($Y=0$), in x- and z-directions were, respectively, determined by:

$$\bar{\tau}_{YX}|_{Y=0} = \frac{R_x \sigma}{b}(U_2 - U_1)\frac{\bar{\eta}}{HJ_0} - \frac{R_x}{b}H\frac{\partial P}{\partial X}\frac{J_1}{J_0} \tag{16}$$

$$\bar{\tau}_{YZ}|_{Y=0} = \frac{R_x \sigma}{b}(W_2 - W_1)\frac{\bar{\eta}}{HJ_0} - \frac{R_x}{Kb}H\frac{\partial P}{\partial Z}\frac{J_1}{J_0} \tag{17}$$

The friction torque acting on the rib with the bearing axis is given by:

$$M_1 = abE'\left[\iint_\Gamma \bar{\tau}_{YX}|_{Y=0}(aZ+\varepsilon+R_i)\cos\psi \mathrm{d}X\mathrm{d}Z - \iint_\Gamma \bar{\tau}_{YZ}|_{Y=0}bX\cos\psi \mathrm{d}X\mathrm{d}Z\right] \tag{18}$$

where Γ is the entire computation area including the cavitation zone.

8 Computation Scheme

For a given load and geometry, the domain size between rib and roller end contacts is determined by the small area ABCD surrounded by four curves (Fig. 2). The longitudinal dimension of the area is strongly nonuniform, being particularly short at the lower region and rather long at the upper region. The transverse dimension of this area is also nonuniform and limited by the rib width.

Therefore, the flooded conditions of the oil film deviates significantly from one when the conjunction area is unlimited in dimension, especially when the ratio of the radii of

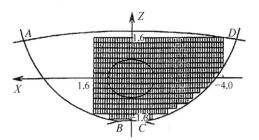

Fig. 2 Computation zone of rib-roller end contact

curvature is nearly unity and/or the applied load is large so that the Hertzian ellipse is large. The starvation effect in the lower region and side leakage effect will reduce the film thickness significantly. The computation zone should take full consideration of this geometrical peculiarity. This is one of the main reasons why the EHL of rib-roller end contacts should be specially treated, rather than approximated by simply adopting any existing solution based on fully flooded or uniformly starved condition.

In the present work, the computation zone is firstly defined as a rectangle with 5.6 times of semimajor axis of the contact ellipse in the entraining direction and 3.2 times of semiminor axis in the transverse direction. Those area of the rectangle being outside ABCD must be trimmed and the remaining area inside ABCD is the effective computation zone.

The Reynolds equation, Eq. (1), contains a set of integrals J_0, J_1, and J_2 (Eq. (2)) which are different weighted integrations from the rheology function $F(\tau_e)$ across the film thickness. The value of $F(\tau_e)$ is dependent on the distributions of the shear stresses τ_{yx} and τ_{yz} (Eqs. (4) and (3)). The latter can be obtained from their boundary values (Eqs. (16) and (17)) and the local pressure gradients solved from Eq. (1), and are therefore dependent on the integrals J_0 and J_1 in their turn. An under-relaxation iteration is employed to get the simultaneous solutions of the pressure and shear stress.

To solve the thermal non-Newtonian lubrication problem, under-relaxation iterations among the Reynolds, elasticity, and energy equations were employed to obtain simultaneous solutions of the pressure, film thickness, and temperature distributions. The computation procedure used by Jiang et al. (1993) was extended to allow solutions for a thermal analysis. The initial value of central film thickness was calculated from the isothermal film thickness formula of Chittenden et al. (1985). This formula is more suitable to the tapered rib/spherical roller end contact in which the direction of lubricant entrainment coincides with the major axis of the Hertzian contact ellipse. All thermal non-Newtonian EHL results reported in this work started from corresponding distributions obtained by isothermal analysis. Then, the thermal effects were introduced and iterated to achieve the full solution.

Table 1 Lubricant and solid properties

Lubricant:	
Inlet temperature (t_0), K	298
Thermal conductivity (k_f), W/(m·K)	0.14
Heat capacity (c_f), J/(kg·K)	2 000
Ambient viscosity (η_0), Pa·s	0.028 3
Ambient density (ρ_0), kg/m³	800
Temperature-viscosity coefficient (γ_t), K^{-1}	0.042
Thermal expansivity (β_t), K^{-1}	0.000 64
Initial shear strength (τ_{L0}), MPa	2.3
Limiting-shear-stress proportionality constant (m)	0.047
Factor considering temperature effect on τ_L (B), K	585
Solid:	
Thermal conductivity (k_1, k_2), W/(m·K)	45
Heat capacity (c_1, c_2), J/(kg·K)	460
Density (ρ_1, ρ_2), kg/m³	7 850

The properties of lubricant and solids are given in Table 1. The lubricant viscosity and pressure-viscosity coefficient used by Zhang et al. (1988) are for mineral oil 20 at 25℃. The other lubricant properties such as thermal conductivity, heat capacity and limiting shear stress are not readily available for this oil. Therefore, only the representative values are adopted.

9 Results and Discussion

Because of the difference between the inner and outer raceway contact angles, there is a force component, which balances the forces between outer ring/roller and inner ring/roller, and drives the tapered roller against the rib. Depending on the magnitude of axial loads to be supported, the bearing may have a small or steep contact angle. The greater the inclination of the rollers, the greater the axial load-carrying capacity. The load acting on the rib is normally less than ten percent of the applied load in the roller-race-way. In addition, the mating surfaces in end contacts can be designed to make a conforming contact which results in a thicker film. Therefore, the pressure generated in rib-roller end contacts is low when compared with that in heavily loaded, concentrated contact EHD problems.

The spinning motion of each contact body as well as the rolling and sliding motions were taken into consideration in present analysis.

Some basic parameters of the rib and roller end are listed below:

the apex length = 279.045 mm
large end radius of roller, r = 9.739 mm
inner radius of rib, r_i = 57.486 mm
outer radius of rib, r_o = 60.5 mm
height of nominal contact point, ε = 2.735 mm
half cone angle, β = 11.642 2°
half roller angle, γ = 2.0°

Figure 3 shows the dimensionless pressure distribution and the contour plot of film thickness for the tapered rib/spherical roller end contact with curvature radius ration of $R_{x2}/R_{x1}=0.85$, a load of $W=550$ N, and a bearing speed of $N_a=2,000$ r/min. Two essential phenomena in elastohydrodynamic lubrication, the pressure spike and the side lobes of film thickness have been detected in this case. Due to the small domain size of end contacts and the occurrence of starvation and strong side leakage, the pressure distribution is sharp and the film thickness is reduced. The spinning velocity of contact bodies in tapered roller bearings has little effects on the distributions of pressure and film thickness. The shapes of pressure and film thickness within the central region of significance remain nearly symmetrical about x-axis. The maximum pressure is usually located at the contact center and a concentrated pressure spike occurs near the exit zone. The formula of Chittenden et al. (1985) predicts the central film thickness more accurately than that of Hamrock and Dowson (1981) under values of ellipticity $K<1$.

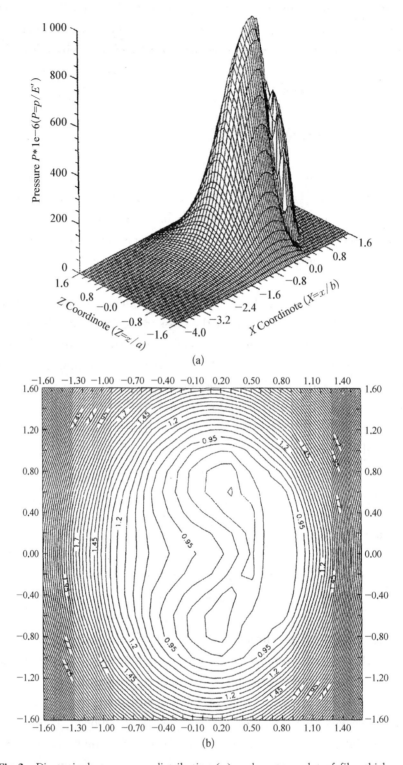

Fig. 3 Dimensionless pressure distribution (a) and contour plot of film thickness (b) for the tapered rib/spherical roller end contact with $R_{x2}/R_{x1}=0.85$, $W=550$ N, $N_\Omega=2,000$ r/min

Non-Newtonian thermal EHL solutions have been obtained for the two contact types: (i) tapered rib with spherical roller end and (ii) spherical rib with spherical roller end. The same geometrical and kinematic parameters of bearing analysis used by Zhang et al. (1988) were employed in the present work. Effects of ratios of radius of roller end to rib face were investigated at moderate speed ($N_\varOmega = 1,000$ r/min) and load ($W = 160$ N) conditions. The minimum film thickness, $h_{\min}$, central film thickness, h_c and friction torque, M_1, are depicted in Fig. 4. With the aim of pursuing the thickest film and the least friction torque, a range of ratio R_{x2}/R_{x1} from 0.6 to 0.8 was detected to favor lubrication for tapered rib/spherical roller end. If the ratio surpasses this range, the film thickness will decrease. In the extreme case where the ratio equals to unity, the fluid film will breakdown because of lack of wedge action. For the spherical rib/spherical roller end contact, there is a clearer indication of an optimal ratio R_{x2}/R_{x1} at a value of 0.9. This result confirms the previous work (Zhang et al., 1988) made by one of the authors and strongly supports the rules of modern bearing design (Jamison et al., 1976).

By comparing Fig. 4(a) and (b), it is noted that the film thickness of a tapered rib/spherical roller end contact is obviously smaller than that of a comparable spherical rib/spherical roller end contact. In other words, the maximum load-carrying capacity of the latter is greater than that of the former. This can be explained by their different geometrical characteristics. In the case of spherical rib/spherical roller end, the Hertzian contact zone is a circle instead of an ellipse of $K<1$ as in the case of tapered rib/spherical roller end. This results in a lower maximum-Hertz-pressure and less side leakage, and hence a higher minimum-film-thickness. The outcome was obtained, however, under the assumptions that the surfaces are ideally smooth and the nominal contact point is always located at the middle of rib. Practically the concave spherical rib has a higher demands on manufacturing. The accuracy of surface profile and finish of spherical rib in mass production are usually not good enough to ensure the predicted film thickness. In addition, it was revealed from Figs. 4(a) and (b) that the effect of R_{x2}/R_{x1} on film thicknesses for tapered rib/spherical roller end contact is less sensitive than for spherical rib/spherical roller end contact. A possible explanation is due to its smaller effects of side starvation. The isothermal Newtonian solution of the film thickness are also presented in Fig. 4. The thermal non-Newtonian film thickness is smaller than that of isothermal Newtonian. This can evidently be attributed to the reduction of viscosity in the thermal non-Newtonian analysis. However, the influences of thermal and non-Newtonian behavior of the lubricant to the results are not pronounced at this very low pressures (maximum Hertzian pressure $p_h \leqslant 0.17$ GPa) and very low speed (dimensionless velocity $U^* = \eta_0 u_s/(E'R_x) \leqslant 4e-13$). The ratios of sliding-rolling spinning-rolling are also low, being 0.185 3 and 71.78, respectively.

The geometrical influences on film thicknesses and traction torques were also calculated at higher load and speed conditions. Figure 5 shows the thermal non-Newtonian solution for both configurations at $W = 320$ N and $N_\varOmega = 5,000$ r/min. It is noted that the

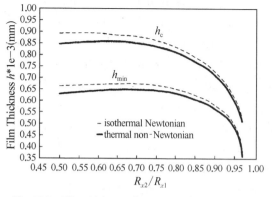

Fig. 4(a) Film thickness of tapered rib/spherical roller end contact

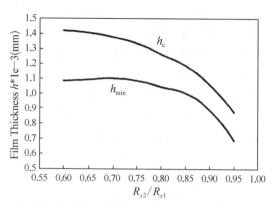

Fig. 5(a) Film thickness of tapered rib/spherical roller end contact

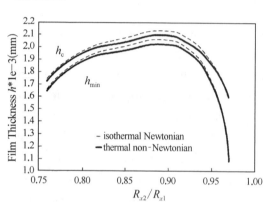

Fig. 4(b) Film thickness of spherical rib/spherical roller end contact

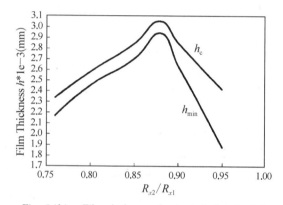

Fig. 5(b) Film thickness of spherical rib/spherical roller end contact

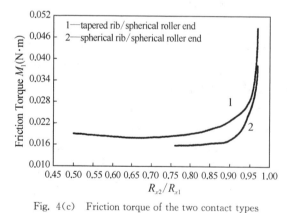

Fig. 4(c) Friction torque of the two contact types

Fig. 4 Effect of ratio R_{x2}/R_{x1} on film thicknesses and friction torques for $W = 160$ N and $N_\Omega = 1,000$ r/min

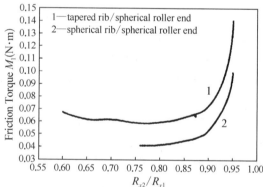

Fig. 5(c) Friction torque of the two contact types

Fig. 5 Effect of ratio R_{x2}/R_{x1} on film thicknesses and friction torques for $W = 320$ N and $N_\Omega = 5,000$ r/min

variation of h_{min}, h_c, and M_1 with R_{x2}/R_{x1} are quite similar to those in Fig. 4. The only difference is that the optimal value of R_{x2}/R_{x1} for spherical rib/spherical roller end moves from 0.9 to 0.88 as shown in Fig. 5(b). Therefore, it can be concluded that a range of 0.6 to 0.8 is optimum for the ratio of curvature radius of tapered rib/spherical roller end and 0.9 for that of spherical rib/spherical roller end. Thermal non-Newtonian analyses were carried out by varying the applied load acting on rib while keeping the other parameters constant, viz., $R_{x2}/R_{x1} = 0.8$, $N_\Omega = 1,000$ r/min, and $E' = 233$ GPa. The resulting film thicknesses and friction torques on dependence of loads are shown in Fig. 6. For the tapered rib/spherical roller end contact, an increase in load produces a small decrease in minimum film thickness. However, the film thickness decreases remarkably with an increase of loads in the case of spherical rib/spherical roller end contact. This considerable reduction in the film thickness is due to the considerable effect of side leakage. For spherical rib/spherical roller end contact, the Hertzian ellipse becomes a circle. The ratio of rib width to the semiminor axis of Hertzian ellipse has a smaller value. Consequently, the side leakage shows a more significant effect on film thicknesses for spherical rib in comparison with tapered rib. In our calculations in Figs. 4, 5, and 6, the ratio of rib width to the semiminor axis of Hertzian contact ellipse varies from 2.0 to 0.6. It is obvious that the minimum film thickness is considerably reduced when the Hertz contact width exceeds the rib width. For the case of $W = 550$ N, the minimum film thickness moves to the side lobes. The friction torque against the rib face increases in a nearly linear relation with the load. It was also found again that the torque is smaller and at a lesser increasing-rate for spherical rib/spherical roller end contact than for tapered rib/spherical roller end contact. The central film thickness for tapered rib/spherical roller end contact is depicted in Fig. 6 as a function of load. It increases slightly with an increase of load instead of a slight decrease if one has Hamrock and Dowson's film thickness formula in mind. It looks as if the result is somewhat different from the usual. In fact, this is one of noteworthy characteristics in rib-roller end contacts. Because of the low local pressure, small elastic deformations around the edge make the film thicknesses here very small, resulting in a strong restriction to the side flow from the central region. Due to the limited width of contact area in the related case, the pressure in the central region is higher with a steep pressure drop to both sides to keep in balance with the applied load. That leads to a

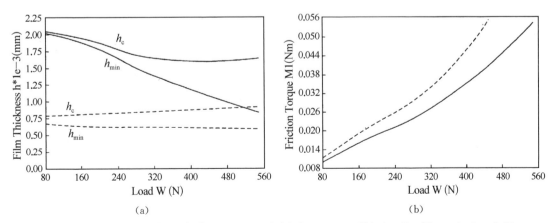

Fig. 6 Effect of load on film thickness (a) and friction torque (b) for $R_{x2}/R_{x1} = 0.8$ and $N_\Omega = 1,000$ r/min. ---- tapered rib/spherical roller end contact; —— spherical rib/spherical roller end contact

larger deformation at the central region and a reduction of the total contact width. The rate of increase of the deformation exceeds that of the mutual approach in this case. So, a slight increase of central film thickness is detected. This bears some similarity to the phenomenon found at the end-crowned roller/inclined rib in cylindrical roller bearings (Aramaki et al., 1992).

The variation of the minimum film thickness and friction torque for relatively high speed is plotted in Fig. 7. There is a considerable change in the film thickness of both types of conjunctions as the speed is increased. This clearly illustrates the dominant effect of speed on the film thickness in EHL contacts. Thermal effects cause a remarkable reduction in both film thickness and traction. The minimum film thickness of thermal non-Newtonian solution is 11.7 percent at 7,500 r/min and 20.6 percent at 20,000 r/min smaller than that of isothermal Newtonian solution. The figure depicts a pronounced reduction of traction as the thermal effects were introduced. The friction torque is almost decreased by half (48.2 percent at 7,500 r/min and 51.0 percent at 20,000 r/min). The ratios of sliding-rolling and spinning-rolling are the same as the case of Figs. 4 and 5. It is obvious that the thermal influence is very important and cannot be ignored in high speed applications of tapered roller bearings. The non-Newtonian effect seems to be not so important so far as the film thickness is concerned. However, a non-Newtonian rheological model is still necessary for a realistic estimate of the traction torque in high speed tapered roller bearings. Figure 8(a) elucidates the distribution of centerline ($y=h/2$) lubricant temperature of the tapered rib/spherical roller end contact in tapered roller bearings. Having solved the mean lubricant temperature from the energy equation and solid temperatures from boundary Eqs. (14) and (15), the centerline lubricant temperature is given by the expression $t_c=(6t_m-t_1-t_2)/4$. For a given set of operating conditions: load $W=550$ N, ratio of radii of roller end to rib $R_{x2}/R_{x1}=0.85$, bearing speed $N_\Omega=2,000$ r/min, sliding-rolling ratio $\beta_r=0.1853$, and spinning-rolling ratio $\beta_s=71.78$ values, the maximum temperature is 7.09℃ above the ambient and occurs near the contact

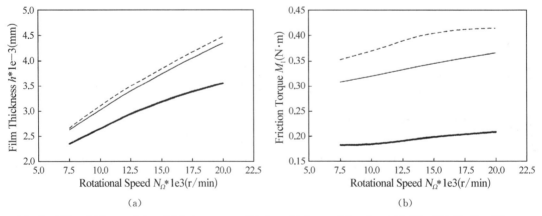

Fig. 7 Film thickness (a) and friction torque (b) for a high speed bearing with $W=550$ N, $R_{x2}/R_{x1}=0.7$ and tapered rib/spherical roller end contact. ---- isothermal Newtonian solution; ——— isothermal non-Newtonian solution; ——— thermal non-Newtonian solution.

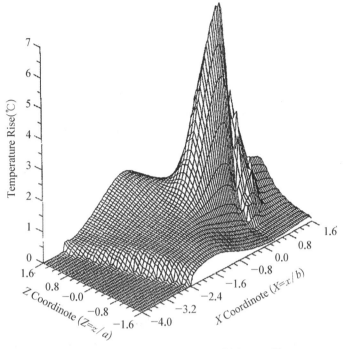

(a) Temperature rise at centerline of lubricant film

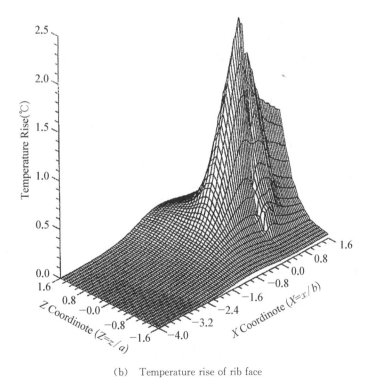

(b) Temperature rise of rib face

Fig. 8 Temperature distributions of tapered rib/spherical roller end contact for $W = 550$ N, $R_{x2}/R_{x1} = 0.85$, $N_\Omega = 2,000$ r/min, $\beta_r = 0.185\ 3$, and $\beta_s = 71.78$

center. The positions of the maximum temperature rise and maximum pressure are usually closely located. This indicates the primary effect of pressure on the temperature rise for the sliding and spinning cases. Downstream of the maximum values, the temperature drops rapidly and fluctuates near the exit. This is caused by the tremendous drop of pressure after the pressure spike. As stated previously, the rib-roller end contact has its particular geometrical and kinematic conditions causing non-uniform starvation. A part of the inlet region (flat grids illustrated in Fig. 8) is actually beyond the computation zone. The temperature distribution is not symmetrical about the x-axis. Oil starvation reduces the film thicknesses and, therefore, the amount of convection dissipation. Hence, the temperature on the side of greater starvation increases somewhat faster. Another source for unsymmetry arises from the spinning motion of rib and roller. The spinning decreases the slip in x-direction in the half plane ($z<0$) and thus the temperature rise is alleviated. On the other half plane ($z>0$), the situation is contrary. Moreover, the spinning causes a velocity difference in z-direction which generates an additional viscous shear work, and thus the temperature rise in both inlet and outlet zones is slightly increased.

Figure 8(b) depicts the temperature distribution of rib face at the same operating conditions as in Fig. 8(a). The maximum temperature is 2.26℃ above the ambient and occurs between the contact center and the exit. The shear heating in z-direction caused by spinning were not included in the analysis because the calculation of temperature rise on solid surfaces is based on Jaeger's approach to a moving-heat-source problem (Carsiaw and Jaeger, 1959) in which only the heat flow in the oil entraining direction is considered. The magnitude of the maximum spin speed differences, $(\Omega_{1y}-\Omega_{2y})_x$, and the maximum rolling speed differences, $(u_{1(0)}-u_{2(0)})$, of the two surfaces are of the same order. It would be expected to resolve the exact temperature boundary conditions. However, numerical results show that the spinning velocities, which are not very large in tapered roller bearings, do not affect much the maximum film temperature rise. This is because the temperature rise mostly occurs near the center of the contact zone. The contribution on velocity components by spinning are small near the contact center. Hence, the influence of spinning on the maximum temperature rise is small. Since the maximum temperature rise is important to designers, the present treatment to solid surface temperatures seems acceptable. The temperature distribution of roller end face is very much similar with that of rib face. The maximum temperature is 2.62℃ above the ambient which is slightly higher than that of rib face due to the slower roller velocity of roller. It is also observed that the compression cooling has less significant effect on surface temperatures than on lubricant temperatures.

Figure 9 depicts the temperature distributions for the same contact configuration and load as in Fig. 8, but with $R_{x2}/R_{x1}=0.7$ and $N_\Omega=7\ 500$ r/min conditions. The figure indicates that as the ratio of curvature radius is decreased and the speed is increased, the heat generation becomes pronounced. The maximum temperatures are 40.25, 4.40, and

5.26℃ above the ambient in the film centerline, rib surface, and roller end surface, respectively.

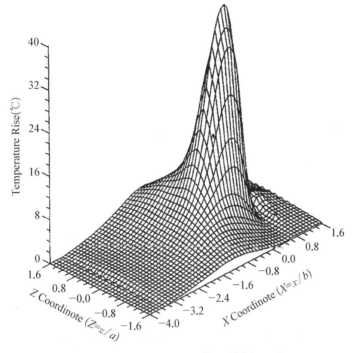

(a) Temperature rise at centerline of lubricant film

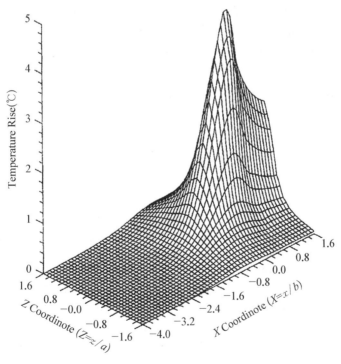

(b) Temperature rise of rib face

Fig. 9 Temperature distributions of tapered rib/spherical roller end contact for $W=550$ N, $R_{x2}/R_{x1}=0.7$, $N_\Omega=7,500$ r/min, $\beta_r=0.1853$, and $\beta_s=71.78$

10 Conclusions

A numerical procedure has been developed to aid in understanding the two-dimensional thermal non-Newtonian fluid flow in the rib-roller end contact of tapered roller bearings.

Optimal R_{x2}/R_{x1} ratios are calculated for tapered rib/spherical roller end contact and spherical rib/spherical roller end contact under various load and speed conditions. These optimal ratios are 0.6 to 0.8, and 0.9, respectively, which support the findings of Zhang (1988). The spherical rib/spherical roller end contact has a higher axial load-carrying capacity and a lower friction loss.

For the tapered rib/spherical roller end contact, an increased load produces a small decrease in the minimum film thickness. However, the minimum film thickness decreases remarkably with increased load for spherical rib/spherical roller end contact when the semiminor axis of the Hertzian contact ellipse exceeds the rib width.

The non-Newtonian rheological model should be used for an accurate prediction of the traction torque in high speed tapered roller bearings.

The spinning velocity causes a slight temperature rise in the inlet and exit zones, and also causes an asymmetrical temperature distribution in the rolling direction; however, it has little effect on the maximum temperature rise. Numerical results also show spinning motions of the rib and roller have a negligible effect on the film shape and pressure distribution for realistic spin velocities encountered in tapered roller bearings.

For high speed applications, the effects of temperature on the minimum film thickness and friction torque are significant and cannot be neglected.

Acknowledgment

The authors would like to express their appreciations to the City University of Hong Kong for the financial support of this project.

References

[1] Aramaki, H., Cheng, H. S., and Zhu, D., 1992, "Film Thickness, Friction, and Scuffing Failure of Rib/Roller End Contacts in Cylindrical Roller Bearings," ASME Journal of Tribology, Vol. 114, pp. 311–316.

[2] Bair, S., and Winer, W. O., 1979, "A Rheology Model for EHD Contacts Based on Primary Laboratory Data," ASME Journal of Lubrication Technology, Vol. 101, pp. 258–265.

[3] Carslaw, H. S., and Jaeger, J. C., 1959, Conduction of Heat in Solids, Oxford University Press.

[4] Chittenden, R. J., Dowson, D., Dunn, J. F., and Taylor, C. M., 1985, "A Theoretical Analysis of the Isothermal Elastohydrodynamic Lubrication of Concentrated Contacts, I. Direction of Lubricant Entrainment Coincident with the Major Axis of the Hertzian Contact Ellipse," Proc. R. Soc. (London), Series A 397, pp. 245–269.

[5] Dalmaz, G., Tessier, J. F., and Dudragne, G., 1980, "Friction Improvement in Cycloidal Motion Contact: Rib-Roller End Contact in Tapered Roller Bearings," Proc. 7th Leeds-Lyon Symp. on Tribology, VII (iii), pp. 175.

[6] Dowson, D., and Higginson, G. R., 1977, Elastohydrodynamic Lubrication, SI Ed., Pergamon, Oxford.

[7] Gadallah, N., and Dalmaz, G., 1984, "Hydrodynamic Lubrication of the Rib-Roller End Contact of a Tapered Roller Bearing," ASME Journal of Tribology, Vol. 106, pp. 265−274.

[8] Hamrock, B. J., and Dowson, D., 1981, Ball Bearing Lubrication, the Elastohydrodynamics of Elliptical Contacts, Wiley, New York.

[9] Jamison, W. E., Kauzlarich, J. J., and Mochel, E. V., 1977, "Geometric Effects on the Rib-Roller Contact in Tapered Roller Bearings," ASLE Trans., Vol. 20, pp. 79−88.

[10] Jiang, X. F., Wong, P. L., Qiu, X. J., and Zhang, Z. M., 1993, "Effects of Non-Newtonian Model and New Viscosity-Pressure Law on EHD Lubrication of Rib-Roller End Contact in Tapered Roller Bearings," Proc. of the International Symposium on Tribology'93, Tsinghua University, Beijing, China, pp. 147−154.

[11] Johnson, K. L., and Tevaarwerk, J. L., 1977, "Shear Behavior Elastohydrodynamic Oil Films," Proc. R. Soc. (London), Series A 356, pp. 215−236.

[12] Korrenn, H., 1970, "Gleitreibung und Grenzbelastung an den Bord-flachen von Kegelrollenlagern," Fortschritt Berichte, V. D. I. Zeitschrift, Ser. 1, Vol. 11.

[13] Roelands, C. J. A., 1966, "Correlational Aspects of the Viscosity-Temperature-Pressure Relationship of Lubricating Oils," V. R. B., Groningen, Netherlands.

[14] Wang, S. H., and Zhang, H. H., 1987, "Combined Effects of Thermal and Non-Newtonian Character of Lubricant on Pressure, Film Profile, Temperature Rise, and Shear Stress in E. H. L.," ASME Journal of Tribology, Vol. 109, pp. 666−670.

[15] Zhang, Z., Qiu, X., and Hong, Y., 1988, "EHL Analysis of Rib-Roller End Contact in Tapered Roller Bearings," STLE Trans., Vol. 31, pp. 461−467.

[16] Zhou, R. S., and Hoeprich, M. R., 1991, "Torque of Tapered Roller Bearings," ASME Journal of Tribology, Vol. 113, pp. 590−597.

一种紊流润滑理论分析新方法
——复合型紊流模式理论*

摘　要：在对主要用于高速轻载工况的常用紊流理论进行简要分析和充分考察润滑流场的边界条件及内部结构的基础上，采用理论上比现有紊流润滑理论更为合理的复合型紊流模式理论，即在近壁区采用低紊流雷诺数的 $k-\varepsilon$ 模式，而在紊流核心区采用代数雷诺应力模式，对复杂流场的紊流润滑进行了分析，同时在计入惯性效应的情况下，推导出了一种适用于高压密封和高速重载轴承等复杂流场的紊流雷诺润滑方程. 利用这种复合型紊流模式理论对 Couette 型紊流流场进行了分析计算，计算结果与实验数据十分吻合，验证了模型的有效性，可以应用于高压密封和高速重载轴承之类有复杂流场的紊流润滑分析.

1　前言

随着机器向高速和大功率化发展，主要用于高速轻载工况的经典紊流润滑理论已不能满足日益增多的复杂紊流流场（如高压密封和高速重载滑动轴承等）的实用要求. 早在本世纪 50 年代，Constantinescu[1] 就以 Prandtl 混合长度理论为依据建立了紊流润滑理论，但其中未计入惯性项，而且只考虑了黏性底层和完全紊流区而忽视了过渡区的存在，同时周向流和轴向流也都被人为地解偶，未计入两者之间的相互影响. 显然，这种处理只能够近似地适用于以单向 Couette 流为主的高速轻载滑动轴承. 60 年代相继问世的且应用相当广泛的以 Reichardt 壁面定律为依据的 Elrod-Ng[2] 和 Ng-Pan[3] 紊流润滑理论，同样也都未计入惯性项，而且后者还作了线性近似处理以简化计算. 但是，由于 Reichardt 式是在单向常应力条件下得出的，尽管经过他们的人为"局部化"处理，终究还是难以解决高压密封和重载轴承之类有复杂流场的紊流润滑问题. 基于经验建立起来的 Hirs 紊流润滑理论[4]，其适用范围更具有局限性，这是因为它还存在着严重的理论缺陷[5].

以上所述均为零方程模式理论，都不敷实际应用所需，因而一些新的模式理论如一方程和两方程的模式理论等都陆续被引入紊流润滑领域. 但是，由于复杂的紊流润滑流场含有一个方向的剪切流和两个方向的压力流，而且三者可属同一数量级，在膜厚方向上又有附壁区、核心区和过渡区之分，故很难用单一模式来完善地描述整个润滑流场.

因此，本文采用两种较为先进的紊流模式理论的复合，即以低紊流雷诺数 $k-\varepsilon$ 理论描述过渡区，而以代数雷诺应力理论描述核心区，推导了一种更为合理的紊流雷诺润滑方程，并将平均惯性效应直接纳入雷诺方程.

---

* 本文合作者：张运清. 原发表于《摩擦学学报》，1995，15(3)：271 - 275.

2 基本方程

2.1 紊流模式

在现有的紊流模式中,由于两方程 k-ε 模式理论和代数雷诺应力模式理论引入了能反映紊流内部机理的紊动能传输方程和能量耗散方程,因而比其他模式方程具有更坚实、更完善的理论基础,尤其代数雷诺应力模式理论还直接计入了固体壁面对紊流脉动参数的阻滞效应. 由于润滑流场中有两个固体壁面存在,固壁效应就更为重要. 由此可见,代数雷诺应力模式理论是目前研究润滑流场比较合理的紊流模式. 但是,这种理论由于在进行模式处理简化时略去了黏性效应的影响,致使其不适用于近壁区即由黏性底层和过渡区组成的区域,而只适用于紊流核心区.

本文采用一种复合型的紊流模式理论,即在近壁区用低紊流雷诺数的 k-ε 方程,在紊流核心区用代数雷诺应力模式,两区的分界条件按紊流雷诺应力占总应力的 95% 来确定.

低紊流雷诺数的 k-ε 模式理论是由 Jones 等[6]提出的,后经 Hassid 等[7]作了改进. 本文在此采用后者(不计 Dk/Dt 和 $D\varepsilon/Dt$ 项):

$$\left.\begin{aligned}&\frac{\partial}{\partial y}\left[\left(\nu+\frac{\nu_t}{\sigma_k}\right)\frac{\partial k}{\partial y}\right]+\nu_t\left[\left(\frac{\partial u}{\partial y}\right)^2+\left(\frac{\partial w}{\partial y}\right)^2\right]-\varepsilon-\frac{2\nu\cdot k}{y^2}=0,\\ &\frac{\partial}{\partial y}\left[\left(\nu+\frac{\nu_t}{\sigma_\varepsilon}\right)\frac{\partial\varepsilon}{\partial y}\right]+C_{\varepsilon 1}\cdot\nu_t\left[\left(\frac{\partial u}{\partial y}\right)^2+\left(\frac{\partial w}{\partial y}\right)^2\right]\frac{\varepsilon}{k}\\ &\quad-C_{\varepsilon 2}[1-0.3\exp(-R_k^2)]\frac{\varepsilon^2}{k}-2\nu\left(\frac{\partial\varepsilon^{1/2}}{\partial y}\right)^2=0.\end{aligned}\right\} \quad (1)$$

式中 $\nu_t=C_m[1-0.98\exp(-A_m\cdot R_k)]\dfrac{k^2}{\varepsilon}$,

$R_k=\dfrac{k^2}{\nu\cdot\varepsilon}$,

$C_m=0.90$, $\sigma_k=1.00$, $\sigma_\varepsilon=1.30$, $C_{\varepsilon 1}=1.45$, $C_{\varepsilon 2}=2.00$.

对牛顿流体,$A_m=0.0015$.

代数雷诺应力模式理论[8]:

$$\left.\begin{aligned}&\frac{\partial}{\partial y}\left[\frac{\nu_t}{\sigma_k}\frac{\partial k}{\partial y}\right]+P_d-\varepsilon=0,\\ &\frac{\partial}{\partial y}\left[\frac{\nu_t}{\sigma_\varepsilon}\frac{\partial\varepsilon}{\partial y}\right]+C_{\varepsilon 1}\cdot P_d\frac{\varepsilon}{k}-C_{\varepsilon 2}\frac{\varepsilon^2}{k}=0.\end{aligned}\right\} \quad (2)$$

式中 $\nu_t=0.09G_\mu\dfrac{k^2}{\varepsilon}$,

$P_d=\nu_t\left[\left(\dfrac{\partial u}{\partial y}\right)^2+\left(\dfrac{\partial w}{\partial y}\right)^2\right]$,

$G_\mu=\dfrac{1+\dfrac{3}{2}\dfrac{C_2\cdot C_2}{1-C_2}f}{1+\dfrac{3}{2}\dfrac{C_1}{C_1}f}\cdot\dfrac{1-2\dfrac{C_2\cdot C_2\cdot P_d/\varepsilon}{C_1-1+C_2\cdot P_d/\varepsilon}f}{1+2\dfrac{C_1}{C_1+P_d/\varepsilon-1}f}$,

$$f=\frac{k^{3/2}}{C_w \cdot y \cdot \varepsilon},$$

$C_w=3.72$, $C_1=1.80$, $C_2=0.60$, $C_1=0.60$, $C_2=0.30$, $C_{\varepsilon1}=1.44$, $C_{\varepsilon2}=1.92$, $\sigma_k=1.00$, $\sigma_2=1.30$.

2.2 运动方程的一般形式

对定常、稳态流场而言，有：

$$\left.\begin{array}{l} u\dfrac{\partial u}{\partial x}+v\dfrac{\partial u}{\partial y}+w\dfrac{\partial u}{\partial z}=-\dfrac{1}{\rho}\dfrac{\partial P}{\partial x}+\dfrac{1}{\rho}\dfrac{\partial \tau_{xy}}{\partial y}, \\ u\dfrac{\partial w}{\partial x}+v\dfrac{\partial w}{\partial y}+w\dfrac{\partial w}{\partial z}=-\dfrac{1}{\rho}\dfrac{\partial P}{\partial z}+\dfrac{1}{\rho}\dfrac{\partial \tau_{xy}}{\partial y}. \end{array}\right\} \tag{3}$$

式中 $\tau_{xy}=\mu\dfrac{\partial u}{\partial y}-\rho\cdot\overline{u'v'}$，$\tau_{zy}=\mu\dfrac{\partial w}{\partial y}-\rho\cdot\overline{w'v'}$，$-\overline{u'v'}=\nu_t\dfrac{\partial u}{\partial y}$，$-\overline{w'v'}=\nu_t\dfrac{\partial w}{\partial y}$.

ν_t——紊流扩散系数.

2.3 用修正平均法对惯性项进行处理

现有的紊流润滑理论一般不计惯性项的影响，但随着机器向高速重载化发展，惯性效应不可忽略. 通常，考虑惯性效应的方法大致有：小参数摄动法、流函数法、Constantinescu 平均法、直接解法和修正平均法. 本文采用的是修正平均法，但处理方法却与之不同，将惯性项直接引入封闭的解析润滑方程. 即取：

$$\left.\begin{array}{l} \bar{f}(x,z)=\dfrac{1}{h}\int_0^h\left[u\dfrac{\partial u}{\partial x}+v\dfrac{\partial u}{\partial y}+w\dfrac{\partial u}{\partial z}\right]dy=\dfrac{1}{h}\left[\dfrac{\partial I_{xx}}{\partial x}+\dfrac{\partial I_{xz}}{\partial z}\right], \\ \bar{g}(x,z)=\dfrac{1}{h}\int_0^h\left[u\dfrac{\partial w}{\partial x}+v\dfrac{\partial w}{\partial y}+w\dfrac{\partial w}{\partial z}\right]dy=\dfrac{1}{h}\left[\dfrac{\partial I_{xz}}{\partial x}+\dfrac{\partial I_{zz}}{\partial z}\right]. \end{array}\right\} \tag{4}$$

式中 $I_{xx}=\int_0^h u^2 dy$，$I_{xz}=\int_0^h u\cdot w\, dy$，$I_{zz}=\int_0^h w^2 dy$.

令等效压力梯度为：

$$\dfrac{\partial P'}{\partial x}=\dfrac{\partial P}{\partial x}+\rho\cdot\bar{f},\quad \dfrac{\partial P'}{\partial z}=\dfrac{\partial P}{\partial z}+\rho\cdot\bar{g}. \tag{5}$$

将式(4)和式(5)代入运动方程(3)，得：

$$\dfrac{\partial P'}{\partial x}=\dfrac{\partial \tau_{xy}}{\partial y},\quad \dfrac{\partial P'}{\partial z}=\dfrac{\partial \tau_{zy}}{\partial y}. \tag{6}$$

由式(6)可见，利用修正平均法对惯性项近似处理后得到的运动方程与不计惯性项时的运动方程相似，只不过将压力梯度 $\partial P/\partial x$ 和 $\partial P/\partial z$ 换成了等效压力梯度 $\partial P'/\partial x$ 和 $\partial P'/\partial z$. 因此，不考虑惯性项的运动方程的求解过程及有关结果，可以引用于对惯性项进行修正平均处理后式(6)的求解，这是本文采用修正平均法考虑惯性效应的目的之一.

2.4 雷诺型润滑方程的推导

联立求解运动方程(6)和连续方程，并计入有关边界条件得含惯性项的雷诺方程：

$$\frac{\partial}{\partial x}\left(G\frac{\partial P'}{\partial x}\right)=\frac{\partial}{\partial x}\left(G\frac{\partial P'}{\partial z}\right)=U\frac{\partial}{\partial x}(F\cdot h) \tag{7}$$

或

$$\frac{\partial}{\partial x}\left[G\left(\frac{\partial P}{\partial x}+\rho\cdot\bar{f}\right)\right]+\frac{\partial}{\partial z}\left[G\left(\frac{\partial P}{\partial z}+\rho\cdot\bar{g}\right)\right]=U\frac{\partial}{\partial x}(F\cdot h).$$

式中 $G=\dfrac{A_2}{A_2}B_1-B_2$, $A_1=\int_0^h\dfrac{\mathrm{d}y}{\nu+\nu_t}$, $B_1=\int_0^h\int_0^y\dfrac{\mathrm{d}y\mathrm{d}y}{\nu+\nu_t}$,

$F=1-\dfrac{B_1}{A_1\cdot h}$, $A_2=\int_0^h\dfrac{y\mathrm{d}y}{\rho(\nu+\nu_t)}$, $B_2=\int_0^h\int_0^y\dfrac{y\mathrm{d}y\mathrm{d}y}{\rho(\nu+\nu_t)}$.

3 计算结果与分析

利用本文提出的复合型紊流模式理论对图1所示两平板之间的速度场进行了分析计算,所得结果如图2所示. 曲线(1)表示平面Couette紊流($Re=5\,408$, $B_x=B_z=0$)的速度分布,并与Leutheusser等[9]的相应实验结果进行了比较. 由图2可以看出,计算结果与实验结果吻合得很好,表明本文采用的紊流分析法是合理可靠的. 曲线(3)和(4)分别表示有轴向压力梯度 B_z 时速度 u 和 w 的分布规律. 分别比较曲线(1)与(3)及曲线(2)与(4)可以发现,轴向压力梯度所导致的轴向流动对周向速度分布的影响不容忽视,尤其在有大的压力流动时更是如此. 但是,在目前常用的紊流润滑理论中,由

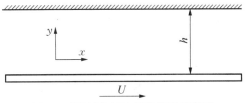

图1 两平板之间流场的简明示意图

于采用线性化解耦而忽略了两方向流动之间的相互影响,这显然有其局限性. 因此,对高压密封和高速重载轴承之类有复杂流动的场合,采用本文提出的复合型模式进行分析更为合理.

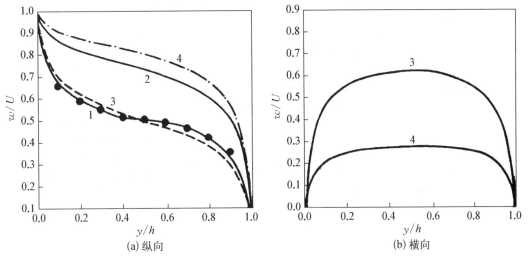

图2 纵横向速度分布($Re=5\,408$)

1. $B_x=B_z=0$; 2. $B_x=10$, $B_z=0$; 3. $B_x=0$, $B_z=20$; 4. $B_x=10$, $B_z=20$.

注:●表示Leutheusser等[9]所获得的实验结果, $B_x=B_z=0$.

4 结论

a. 通过对紊流润滑膜各个分区流动特点的分析,提出了一种复合型的紊流模式理论.利用这种理论能够比较充分地表达紊流润滑膜的流动特点.

b. 推导出了适用于复杂流场的紊流雷诺方程.并且将惯性项的影响直接引入其中.

c. 利用复合型紊流模式理论对平面 Couette 流场进行了分析计算,结果与实验值非常吻合,表明利用本文提出的方法对复杂流场进行分析是合理可靠的.

参 考 文 献

[1] Constantinescu V N. On turbulent lubrication. Proc. Inst. Mech. Eng. ,1959,173: 881 - 889.

[2] Elrod H G. , Ng C W. A theory for turbulent films and its application to bearings. Trans. ASME(F),1967,89: 346 - 362.

[3] Ng C W. , Pan C H T. A linearized turbulent lubrication theory. Trans. ASME(D), 1965,87: 675 - 688.

[4] Hirs G G. A bulk flow theory for turbulence in lubricant films. Trans. ASME(F),1973,95: 137 - 146.

[5] 张直明.关于紊流润滑理论中的 Hirs 法的理论基础.西安交通大学学报,1978(4):25.

[6] Jones W P, Launder B E. The calculation of low Reynolds number phenomena with a two-equation model of turbulence. Int. J. Heat and Mass Transfer, 1973,16: 1119.

[7] Hassid S, Poreh A M. Turbulent energy dissipation model for flows with drag reduction. J. Fluids Eng. (ASME), 1978, 100: 107.

[8] Ljuboja M, Rodi W. Calculation of turbulent wall jets with an algebraic Reynolds stress model. J. Fluids Eng. (ASME), 1980, 102: 350.

[9] Leutheusser H J, Aydin E M. Plane Couette flow between smooth and rough walls. Experiments in Fluids, 1991, 11: 302 - 312.

A New Method of Theoretical Analysis of Turbulent Lubrication Using a Combined Model of Turbulence

Abstract: Theoretical bases of the conventional turbulent lubrication theories used for high speed and light load condition are briefly analyzed. A theoretically sounder combined model of turbulent lubrication is formed, based on consideration of boundary conditions and internal structure of the lubrication flow field. The low Reynolds turbulence number $k - \varepsilon$ model is used for the near-wall region, and the algebraic Reynolds stress model used for the turbulence kernel. Complex turbulent lubrication cases are analyzed by this model. A turbulent lubrication equation of the Reynolds type applicable to high pressure annular seals and heavily loaded journal bearings is derived with consideration of inertia effect. Calculational results of turbulent Couette flow obtained by applying this combined turbulent lubrication model compare well with experimental data, and validity of the theory is confirmed. It can be used for turbulent lubrication analysis under high pressure annular seals and high speed-heavily loaded journal bearings.

核电水泵在热冲击下瞬态
热效应的研究和分析（Ⅰ）*

——热弹性问题微分控制方程的建立和分析

提 要：本章简要推导了基于连续介质力学理论而建立的热传导方程和热弹性运动方程；进而采用量纲分析法并结合所研究的核电水泵材料及所受热载的特点，着重对方程中的耦合项和动力项进行了定性讨论；最后，还以一半空间的热冲击问题作为算例，着重研究了在不同温度梯度下动态热应力的变化行为. 本文的结论为采用基于非耦合和拟静态处理而形成的三维时变有限元法对核电水泵进行瞬态热效应的数值分析提供了理论前提.

在恶劣的工况下，核电停堆冷却泵中的水温在 10 s 内会上升 135℃，在水泵有关生产和设计部门此现象通常被称为"热冲击"(Thermal Shock). 在热冲击作用下，核电水泵机组处于强瞬变温度场、热应力场和热变形场之中，水泵中一些关键部位如：密封处以及其他薄弱环节或受强应力处，即可能发生损毁现象；此外，由于整个水泵机组结构的热变形从而使得泵内动件和静件间原有的间隙发生变化，从而有可能导致水泵中转子的动力性态恶化. 水泵的有关设计部门在多年的实践中已确认此问题为必需予以重视和亟待解决的关键性技术难题，考核核电水泵机组的热冲击适应性能力亦已纳入有关的工程技术规范书中. 由于问题的复杂性，有关此现象的理论研究还缺乏有效的探索，迄今为止，水泵机组热冲击适应能力的考核手段国内还尚停留在代价昂贵的实物实验阶段；经联机检索亦未见国外有相关的研究文献发表.

1 微分方程的建立和分析**

考察核电水泵机组结构在热冲击载荷作用下的强度和热变形问题是属热弹性力学研究范畴，弹性力学意义上的所谓"热冲击"现象是指：在非定常的急剧冷却或急剧加热的情形下，热弹性体上产生剧烈的温度变化，并相应地产生较大的非定常热应力(见文[1—3]). 由于它出现于短促的时间间隔内，因而赋予此种热应力带有冲击的特征，且由此产生的高速变形还会对弹性体的温度场分布产生影响. 热冲击问题的这些特征体现在热传导方程中的耦合项和热弹性运动方程中的动力项. 换言之，热冲击问题的严密求解原则上必须计入这些项的影响. 然而，数学上处理这类方程是很困难的，这些困难一定程度上制约了热冲击问题的研究在实际工程领域中的开展. 研究表明：耦合项、动力项对问题求解的影响，主要取决于热弹性材料的特性和热冲击载荷的强度，对于一般的金属材料，在热冲击载荷强度不太大的情形下，热冲击问题可转化成一般的非定常问题(此种情形可称为"伪热冲击"问题)，即可采用数学上较易处理的非耦合热传导方程和拟静态的热弹性运动方程对问题进行求解，因此，

---

* 本文合作者：曹清、吴益敏. 原发表于《上海大学学报(自然科学版)》，1995,1(6)：633-639.
** 为使推导过程简洁，在建立微分控制方程时，采用了爱因斯坦标记法.

在研究热冲击问题之前,依据材料特性和热载特点对问题的性质进行定位显得很有必要. 如是,则可使得某些"伪热冲击"问题得到有依据的简化,从而避免烦琐的数学处理使问题的求解变为可能且结论同样有效.

1.1 热传导方程的建立

1.1.1 基本方程的推导

对于任何一种固体材料,假定其求解域,则热传导方程可经由以下步骤导出. 由熵不等式:

$$T^{ij}\gamma_{ij} - \rho\left(\frac{\mathrm{d}\Psi}{\mathrm{d}t} + \Phi\frac{\mathrm{d}T}{\mathrm{d}t}\right) - \frac{Q}{T}T_{ij} \geqslant 0 \tag{1}$$

及自由能密度 ψ 的时间导数表达式:

$$\frac{\mathrm{d}\Psi}{\mathrm{d}t} = \frac{\partial\Psi}{\partial t} + \frac{\partial\Psi}{\partial T}\dot{T} \tag{2}$$

可得以下各式:

$$T^{ij} - \rho\frac{\partial\Psi}{\partial\gamma} = 0 \tag{3}$$

$$\frac{\partial\Psi}{\partial T} + \varphi = 0 \tag{4}$$

$$-\frac{Q^i}{T}T_j = 0 \tag{5}$$

将式(4)代入熵密度变化率表达式:

$$\rho\frac{\mathrm{d}\varphi}{\mathrm{d}t} = \rho\frac{\partial\varphi}{\partial\gamma_{ij}} + \rho\frac{\partial\varphi}{\partial\theta}\dot{T} = \rho\frac{\gamma}{T} - \frac{1}{T}\frac{\partial q_i}{\partial x_i} \tag{6}$$

有下式成立:

$$-\rho\frac{\partial^2\Psi}{\partial\gamma_{ij}\partial\theta}\gamma_{ij} - \rho\frac{\partial^2\Psi}{\partial\theta^2}\dot{T} = \rho\frac{\gamma}{T} - \frac{1}{T}\frac{\partial q_i}{\partial x_i} \tag{7}$$

引入热力系数 $\beta_{ij} = -\dfrac{\partial^2\Psi}{\partial\gamma_{ij}\partial\theta}$ 和常应变比热系数 $C_r = -\dfrac{\partial^2\Psi}{\partial\theta^2}T$,有:

$$\rho\beta_{ij}\gamma_{ij} + \frac{1}{T}C_r = \rho\frac{r}{T} - \frac{1}{T}\frac{\partial q_i}{\partial x_i} \tag{8}$$

将 Fourier 定律:

$$q_i i_i = -k_{ij}\frac{\partial T}{\partial x_j}i_i \tag{9}$$

代入式(8),在小变情形下有:

$$\frac{\partial}{\partial x_i}\left(k_{ij}\frac{\partial T}{\partial x_i}\right) = C_r\dot{T} + T\beta_{ij}\dot{\gamma}_{ij} - \rho_0 r \tag{10}$$

式(10)即是各向异性体在线性热弹性问题中的热传导方程,对均质各向同性体,导热系数和热力系数分别以常数 K 和 β 表示,则相应的热传导方程可由式(10)简化而得:

$$k\Delta T = C_r \dot{T} + T\beta_{ij}\dot{\gamma}_{ij} - \rho_0 \gamma \tag{11}$$

式中:$T\beta\dot{\gamma}_{ij}$ 称为热传导方程中的耦合项.

为了方便求解,通常假定温度变化 θ 较之 T_0 是很小的,这样耦合项即从非线性项近似转化为线性项 $T_0\beta\dot{\gamma}_{ij}$,由于 $\dot{\gamma}_{ij} = \dot{u}_{ij}$,则最终式(10)可简化为:

$$k\Delta T = C_r \dot{T} + T_0 \beta \dot{u}_{ij} - \rho_0 r \tag{12}$$

1.1.2 初始条件和边界条件

方程(12)称为泛定方程,需要给出定解条件,即微分方程的边界条件的初始条件.假定 S 为求解域 V 的边界,并有 $S = S_1 \cup S_2 \cup S_3$,则在 S 上满足:

1. 已知边界温度(第一类边界条件 Direchlet 条件)

$$T|_{S_1} = T_0 \qquad \text{在边界 } S_1 \text{ 上} \tag{13}$$

2. 已知边界热流输(第二类边界条件 Neumann 条件)

$$k_n \frac{\partial T}{\partial n}\bigg|_{S_2} = q_0^S \qquad \text{在边界 } S_2 \text{ 上} \tag{14}$$

3. 已知边界的对流或辐射条件(第三类边界条件 Cauchy 条件)

$$k_n \frac{\partial T}{\partial n}\bigg|_{S_c} = h(T_c - T^S) = q^S c \tag{15}$$

$$k_n \frac{\partial T}{\partial n}\bigg|_{S_r} = h(T_r - T^S) = q^S r \tag{16}$$

4. 初始条件

时刻 $t = 0$ 时,V 内的温度分布

$$T = T(x, y, z, 0) \tag{17}$$

1.2 耦合效应的分析

由于热传导方程中耦合项的存在(参见式(12)),温度场 $T(x, y, z, t)$ 不能独立由热传导方程解出,而必须与热弹性运动方程耦合求解.在推导热传导方程的过程中,并未涉及热弹性运动方程,因此,严格地说:只要问题是非定常的,热传导方程中总有耦合项的存在.耦合项的出现,加深了人们对能量转换的认识,但同时也增加了求解热弹性问题的困难.实践表明(见文[1,3]):在很多情形下,略去耦合效应对问题解的影响并不大,有的甚至微乎其微.换言之,耦合效应对问题解的影响是有条件的.考察一略去内热源 $\rho_0 r$ 的热传导方程(本问题所涉及的核电水泵亦无内热源的影响):

$$k\Delta T = C_r \dot{T} + T_0 \beta \dot{u}_{ij} \tag{18}$$

将线热膨胀系数 α 代入,则式(18)可转化为:

$$k\Delta T = C_r \dot{T}\left[1 + \frac{T_0 \beta \alpha}{C_r}\left(\frac{\dot{u}_{ij}}{\alpha \dot{T}}\right)\right] \tag{19}$$

将 $C_r = \rho C_p$ 和 $\beta = (3\lambda + 2\mu)\alpha$ 代入上式,并以 V_e 表示弹性波在物体中的传播速度 $\left(V_e = \sqrt{\dfrac{(\lambda + 2\mu)}{\rho_0}}\right)$,则式(19)又可转化为:

$$k\Delta T = C_r \dot{T}\left[1 + \frac{(3\lambda + 2\mu)^2 \alpha^2 T_0}{\rho^2 C_p V_e^2} \frac{\lambda + 2\mu}{3\lambda + 2\mu}\left(\frac{\dot{u}_{ij}}{\alpha \dot{T}}\right)\right] \tag{20}$$

引入无量纲系数 $\delta = \dfrac{(3\lambda + 2\mu)^2 \alpha^2 T_0}{\rho^2 C_p V_e^2}$,则

$$k\Delta T = C_r \dot{T}\left[1 + \delta \frac{\lambda + 2\mu}{3\lambda + 2\mu}\left(\frac{\dot{u}_{ij}}{\alpha \dot{T}}\right)\right] \tag{21}$$

由式(21)可见,耦合项的影响取决于 δ 值的大小和热冲击的高速变形,在热冲击载荷不太大的情形下(即在 $\ddot{u}_{ij}$ 的数值较小的情况下(参考方程(25)),$\dot{u}_{ij}$ 与 $\alpha \dot{T}$ 的比值可视为同一量级),耦合项的影响主要取决于反映热弹性材料特性的系数 δ 上,当 δ 远小于零时,略去耦合项的影响是合理的和有依据的.

本问题的热弹性材料特性系数列于表1,据这些数据换算出的 δ 在基准温度为时的值为 0.007 7,因此,可以得出结论:在本问题的研究中可以略去耦合效应的影响.

表1 核电停堆冷却泵材料特性系数表*

线膨胀系数 α	16.5×10^{-6}	Lame 弹性常数 μ	72.2 GPa
质量密度 ρ_0	7.85 g/cm³	单位质量常应变比热 C_p	502 J/(kg·℃)
泊桑比 γ	0.33	弹性模量 E	192.08 GPa
Lame 弹性常数 λ	140.17 GPa	弹性波速度 V_e	6.028×10^3 m/s
导热系数 K	16.7 W/(m·℃)		

* 以上数据由上海水泵厂提供.

1.3 热弹性运动方程的建立

1.3.1 基本方程的推导

在小变形下均质、各向同性体,应力、应变和温度关系的热弹性本构方程可写成:

$$\sigma_{ij} = \lambda u_{k,k}\delta_{ij} + \mu(u_{i,j} + u_{j,i}) - \beta\theta\delta_{ij} \tag{22}$$

由此得:

$$\frac{\partial \sigma_{ij}}{\partial x_i} = \lambda u_{k,ki} + \mu(u_{i,ji} + u_{j,ii}) - \beta\theta_j = (\lambda + \mu)u_{i,jj} - \beta\theta_j \tag{23}$$

将式(23)代入动力平衡方程

$$\rho_0 \frac{dV_j}{dt} = \rho_0 f_j + \frac{\partial \sigma_{kj}}{\partial x_k} \quad i, j = 1, 2, 3 \tag{24}$$

忽略 $\dfrac{\partial}{\partial x_i}$ 与 $\dfrac{\partial}{\partial x^t}$ 之间的差异,并以 $\ddot{u}_j$ 表示 $\dfrac{D^2 u_j}{Dt^2}$,经整理可得均质各向同性体的热弹性运动方程:

$$\Delta u_i + \frac{1}{1-2\gamma}e_i + \frac{\rho_0}{\mu}f_j - \frac{2(1+\gamma)}{1-2\gamma}\alpha\theta_i = \frac{\rho_0}{\mu}\ddot{u}_i \tag{25}$$

方程中的右端 $\frac{\rho_0}{\mu}\ddot{u}_i$ 称为动力项(亦称惯性项).

1.3.2 边界条件

弹性问题中边界条件有两类,假定 $S \in \mathbf{R}^2$ 为求解域 $V \in \mathbf{R}^3$ 的边界,并且 $S=S_f+S_u$,则在 S 上应满足

在 S_f 边界上

$$F_i = \sigma_{ij}l_j \quad (i,j=1,2,3) \tag{26}$$

在 S_u 边界上

$$U_i = U_i(x,y,z,t) \quad (i=1,2,3) \tag{27}$$

1.4 动力效应的分析

热弹性问题研究中的热变形和热应力,都是由于作用于热弹性体上的温度场的变化而引发的,其变化的特性取决于温度场的性态. 研究表明:热弹性体上的温度场的急剧变化而引起的热变形和热应力需计入动力效应(即方程(25)中的 $(\rho_0/\mu)\ddot{u}_i$)此时热变形和热应力的研究便转化为热弹性波的研究.

本文研究中所涉及的换热边界条件是边界面上的强迫对流换热,亦即泵结构内的温度变化是由施加在边界面上的温度变化而产生的. 从某种意义上来说,也就是边界面上温度变化的响应. 其响应速度的快慢取决于热弹性材料的特性. 在热学中通常是以热量传播速度 v_q 来描述此效应的

$$v_q = \sqrt{k_d/t_0} \tag{28}$$

式中:k_d 为热扩散系数,$k_d=K/\rho_0 C_p$;t_0 为松弛时间,对于一般金属材料 t_0 的数量级大约在 10^{-10} s.

由于热量传播速度通常是一有限的值,因此,对于热弹性体内热量的传递,热弹性材料总是存在着阻尼的作用,这也意味着换热边界面上温度的变化与物体内部的温度变化是完全不同的,且变化的强度随距换热边界的距离的增大而逐渐趋缓. 正是由于此原因,对于由于边界面上温度的骤然变化(非周期性)而引起的动态热应力、热变形,其冲击的特征大都在距换热边界面相当近的区域内发生(见文[4]),一般都在的 k_d/V_e 量级上. 对于本学位论文所研究的核电水泵中的金属材料,其弹性波的传播速率(大约在 6.028×10^3 m/s 量级,参考表 1)远远大于温度波的扩散率(大约在 2.04×10^2 m/s 量级,由式(28)估算得),因此,可以认为一旦温度场形成,那么相应的热应力场也几乎在同时刻形成. 因为热量传播率很低,故若不计表层热冲击效应的影响,则对于该核电水泵,由瞬态温度场而引发的热应力、热变形场完全可以用拟静态的求解方法来计算(即忽略 $(\rho_0/\mu)\ddot{u}_i$ 项).

很多研究表明:既使在热冲击效应最为明显的表层,其由于热冲击而造成的动态热应力、热变形的冲击特征在初始的瞬间 τ_0' 最为显著,且随着时间的增加而迅速趋于拟静态

解.这个初始瞬间 τ_0' 与边界热交换发生变化的瞬间 τ_0 在同一量级上(见文[1—8]).对于一般的金属材料的量级大约在 k_v/V_e^2 s 阶.这也意味着若换热边界面上的温度变化在 k_v/V_e^2 内有较大的变化的话,研究计入动力效应的动态热应力、热变形才有意义.反之,若在 k_d/V_e^2 时间内换热边界温度变化十分缓慢的话,则完全可以用拟静态的求解方法来求解瞬态热应力、热变形.

经换算本问题的 τ_0 大约是 10^{-13} s 阶,而本问题的对流介质温度的变化情况是在 10 s 内发生从 0℃ 到 135℃ 的线性变化.不难看出其在 10^{-13} s 内的温度变化是相当平缓的,因此,本问题可视为"伪"热冲击问题来研究,即在处理热弹性运动方程时可略去动力项的影响.

2 算例

为验证以上对动力效应机理性的描述,本文比造核电水泵换热工况及核电水泵的材料特性,对一边界面上温度在 τ_0 内由 T_0 线性地增长到 T_A 的半空间体的热冲击问题进行了较为细致的分析和研究(限于篇幅在此仅对一些结论进行讨论).

图 1 为非定常温度场的计算结果及不同 τ_0 值时动态热应力的变化曲线.从图 1 可以看出随着据换热边界面距离的增大弹性体内的温度响应逐渐趋缓,若计入"热松弛"的影响则还可发现弹性体内各点的温度改变滞后于边界条件的改变.图 2 表明了在 $l=1$(l 是距半空间表面的无量纲长度)处对不同的 τ_0 无量纲热应力作为无量纲时间 τ 的函数的过程曲线.可以看出,动态热应力的峰值随着 τ_0 的增加迅速减小.对于本文所研究核电水泵而言,根据其材料特性系数(参考表 1)可推算出无量纲时间 τ_0 于实际时间 t 间的关系式为 $t=1.177\times10^{-13}\tau_0$,当 τ_0 等于 1 和 2 时,t 的值分别为 1.177×10^{-13} s 和 2.354×10^{-13} s,从图 2 可以看出:在这样极其短促的时间内动态热应力的峰值可减少近 50%.因此可以得出结论:对于实际可能遇到的情况,一般由动力学效应而产生的应力提高是无足轻重的.这也是目前绝大多数的工程实际问题不考虑热交换产生的热弹性波,而采用拟静态处理来求得非定常温度场引起的热应力和热变形的主要缘由.

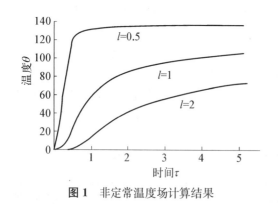

图 1 非定常温度场计算结果   图 2 动态热应力 σ 随 τ 及 τ_0 的变化曲线

3 结论

通过以上分析和讨论,可得出以下结论:1. 由于该核电水泵的边界热交换发生的变化是在 10 s 内完成的且温度上升的幅度在 135℃,故相对于足以引起明显动态热应力 τ_0 的是足够大以至于可以完全忽略动态热应力(即方程(25)中的动力项)的影响;2. 此外,将忽略动力项的热弹性运动方程(参考方程(25))进行时间微分后可看出: $\ddot{u}_{ij}$ 与 $\alpha \dot{T}$ 基本是在同一数量级上,则热传导方程中的耦合项可否略去的主要依据在于反映热弹性材料特性的系数 δ 上,根据上海水泵厂所提供的泵材料的有关参数,可推算出 δ 的值为 0.007 7,因此在本问题的研究中可以略去耦合项的影响. 3. 综上所述,可推断该核电水泵所承受的此种热载强度热冲击过程是一"伪"热冲击过程,在此种情形下热传导方程和热弹性运动方程的解完全可采用非耦合和拟静态的方法来处理.

参 考 文 献

[1] 王洪纲. 热弹性力学概论. 北京:清华大学出版社,1989.
[2] H. 帕尔内库斯. 非定常热应力. 何善育,等译. 北京:科学出版社,1965.
[3] 竹内洋一朗. 郭廷玮,等译. 热应力. 北京:科学出版社,1977.
[4] 姜任秋,等. 冲击加热半无限体热应力问题的理论分析. 工程热物理学报,1993,14(4): 409-433.
[5] Takeuti Y. On an Axisymmetric Coupled Thermal Stress Problem in a Finite Circular Cylinder. Journal of Applied Mechanics. 1983, 50: 116-122.
[6] Toshiaki Hata. Thermal Shock in a Hollow Sphere Caused by Rapid Uniformed Heating. Journal of Applied Mechanics,1991, 58: 64-69.
[7] Naobumi SUMI. Yashimoto ITO. Dynamic Thermal Stress in Composite Hollow Cylinders and Spheres Caused by Sudden Heating. JSME International Journal Series I, 1993, 35(3).
[8] Takeuti Y. Some Comsideration on Thermal Shock Problems in a Plate. Journal of Applied Mechanics,1981,48: 113-118.

The Research and Analysis on the Effect of the Transient Heating of the Nuclear Power Station Pump under Thermal Shock
Part I: The Foundation And Analysis of the Differential Control Equation for the Thermoelasticity Problem

Abstract: In this paper, a heat conduction equation and a dynamic thermoelastic equation are briefly deduced and established on the basis of continuous medium mechanics theory. On top of that, an qualitative discussion is emphatically centered around the couple term and the dynamic term of the equations by means of the dimensional analysis and by considering the combination of the characteristics of the materials of and thermal load effected on the nuclear power station pump under study. Finally, a half-space thermal shock problem is used as a computational example in the highlighted research on the varying behavior of the dynamic thermal stress on the temperature slope. The conclusion of the paper provides reliable justification for applying the numerical method (three-dimensional finite element method developed from non-coupled and quasi-static treatment) on the analysis of the transient heating effect in the nuclear power station pump.

核电水泵在热冲击下瞬态热效应的研究和分析（Ⅱ）*

——三维时变有限元分析和计算

提 要：本文首先推导了非耦合热传导方程和拟静态热弹性运动方程的有限元求解公式；然后，根据核电水泵的结构及其性能参数，估算出了水泵不同区域处的对流换热系数；最后，在进行了大量的试算工作和有关数值考核计算的基础上，采用三维时变温度场、热应力和热变形场有限元分析方法，详细考察了核电停堆冷却泵在受热载冲击下各瞬时的瞬态温度场及热应力、热变形场的分布状况和变化行为。本文的研究工作初步形成了对核电水泵机组在热冲击下的瞬态热效应分析的研究方法；同时亦为最终替代代价昂贵的实物实验提供了坚实的理论基础和关键的计算手段。

0 引言

旨在详细了解核电水泵机组结构在 13.5℃/s 温度梯度载荷冲击下的温度分布状况及热应力、热变形变化行为的瞬态热效应的理论研究和数值模拟，是考核其热冲击适应能力的一项重要的研究内容。一旦有了这些理论分析和计算的结果，就可以准确预测此种强度的热载荷对核电水泵机组结构强、刚度的影响，并为其进一步的抗热冲击设计提供可资借鉴的数据和手段；此外，这些理论和计算的结果亦是接续进行的水泵机组中轴系动力性态研究中的重要的原始数据。

由于有限元法可以求解复杂、多样的边界条件和变化的物性系数情况下的热弹性问题[1]，这里所指的边界条件是广义的，即包括几何形状、温度和热量在边界面上的传输、位移约束的分布等等。因此，本文采用了三维时变有限元法对核电水泵的瞬态热效应进行了数值模拟。

1 有限元方程的推导

1.1 瞬态温度场有限元求解阵列的建立

对略去耦合项的瞬态热传导方程，其相应的泛函为

$$\Pi = \int_v \frac{1}{2} k \left\{ \left(\frac{\partial T}{\partial x_1}\right)^2 + \left(\frac{\partial T}{\partial x_2}\right)^2 + \left(\frac{\partial T}{\partial x_3}\right)^2 \right\} dv - \int_v T\left(\bar{q}_B - C_\gamma \frac{\partial T}{\partial t}\right) dv - \int_{s_2} T^s q^s ds_2 \\ - \int_{s_c} h\left[T_s T^s - \frac{1}{2}(T^s)^2\right] ds_c - \int_{s_r} k\left\{T_r T^s - \frac{1}{2}(T^s)^2\right\} ds_r \tag{1}$$

---

* 本文合作者：曹清、吴益敏。原发表于《上海大学学报（自然科学版）》，1996,2(1)：25-31.

假定空间域 $V \in R^3$ 被 M 个具有 n_e 个结点的单元所离散,V 内共有 N 个结点,在每个单元内各点的温度用单元的结点温度来表达,即

$$T = [N]\{T\}^e \tag{2}$$

$\delta \Pi = 0$ 的矩阵形式为

$$\int_v \delta T C_\gamma \frac{\partial T}{\partial t} \mathrm{d}v = \int_{s_c} h(T_e - T^s) \delta T^s \mathrm{d}s_c + \int_{s_r} k(T_r - T^s) \delta T^s \mathrm{d}s_r$$
$$+ \int_{s_2} q^s \delta T \mathrm{d}s_2 + \int_v \bar{q}^B \delta T \mathrm{d}v - \int_v \{\delta T'\}^T [k] \{T''\} \mathrm{d}v \tag{3}$$

上式经整理并组集后(考虑到对流和辐射项的表现形式在上式中是完全一致的,故在以下的方程推导中忽略了辐射项),可写成:

$$\left(\sum_e [K]^e + \sum_e [H]^e\right)\{T\} + \left(\sum_e [C]^e\right)\{\dot{T}\} = \left(\sum_e \{R_{\bar{q}}\}^e + \sum_e \{R_{q_s}\}^e + \sum_e \{R_h\}^e\right) \tag{4}$$

或表示成:

$$[C]\{\dot{T}\} + [K]\{T\} = \{R\} \tag{5}$$

式中各项表达式分别为

单元对热传导矩阵的贡献:

$$[K]^e = \int_{v_e} [B]^T \{K\} [B] \mathrm{d}v \tag{6}$$

单元热交换边界对热传导矩阵的修正:

$$[H]^e = \int_{s_c} h[N]^T [N] \mathrm{d}s \tag{7}$$

单元对热容矩阵的贡献:

$$[C]^e = \int_{v_e} \rho C_\gamma [N]^T [N] \mathrm{d}v \tag{8}$$

单元热源产生的温度载荷:

$$[R_{\bar{q}}]^e = \int_{v_e} \rho \bar{q} [N]^T \mathrm{d}v \tag{9}$$

单元给定热流边界产生的温度载荷:

$$[R_{q_s}]^e = \int_{s_z} q^s [N]^T \mathrm{d}s \tag{10}$$

单元对流换热边界产生的温度载荷:

$$[R_h]^e = \int_{s_c} h T_e [N]^T \mathrm{d}s \tag{11}$$

这样,包含空间域和时间域的偏微分问题在空间域内被离散为有 N 个节点常微分初值问题. 至此,空间域的离散就完成了.

1.2 时间域离散格式

方程(3)本质上是一抛物型方程的积分表达形式. 为求解此类方程,在时域上通常采用差分格式离散,即将一阶常微分方程组进一步化成代数联立方程组,然后进行求解. 本文选用了下列差分格式:

$$\frac{\{T_{n+1}\}-\{T_n\}}{\Delta t}=\frac{\partial}{\partial t}\{T_n\}+\alpha\left(\frac{\partial}{\partial t}\{T_{n+1}\}-\frac{\partial}{\partial t}\{T_n\}\right) \tag{12}$$

式中:α 为积分因子,可根据温度或热源变化的缓急而取不同的值. 将方程(3—4)代入上式,并将含$\{T_{n+1}\}$的项移至等号左端,经整理后有:

$$[([C]/\Delta t)+\alpha[K]]\{T_{n+1}\}=[([C]/\Delta t)-(1-\alpha)[K]]\{T_n\} \\ +[\alpha\{R_{n+1}\}+(1-\alpha)\{R_n\}] \tag{13}$$

一旦给定了初值$\{T_0\}$就可用上述递推格式求出时域内任一时刻 t_n 时空间域的温度分布.

1.3 热变形场、热应力场有限元求解阵列的建立

将应力、应变与温度间的关系式写成矩阵形式:

$$\{\sigma\}=[D](\{\varepsilon\}-\{\varepsilon_r\}) \tag{14}$$

忽略了动力项,因此可对 V 域内每个单元应用虚功原理,可得:

$$\int_{v_e}\{\delta\varepsilon\}^T[D](\{\varepsilon\}-\{\varepsilon_T\})\mathrm{d}v=(\{\varepsilon u\}^e)^T(\{F_P\}^e+\{F_G\}^e+\{F_Q\}^e) \tag{15}$$

式中$\{F_P\}^e$、$\{F_G\}^e$、$\{F_Q\}^e$ 分别为单元节点力、体积力、分布力载荷向量,并且有 $\{\delta\varepsilon\}^e=[B]\{\delta u\}^e$,代入上式后有:

$$(\{\delta u\}^e)^T\left(\int_{v_e}[B]^T[D][B]\mathrm{d}v\{u\}^e-\int_{v_e}[B]^T[D]\{\varepsilon_T\}\mathrm{d}v\right) \\ =(\{\delta u\}^e)(\{F_P\}^e+\{F_G\}^e+\{F_Q\}^e) \tag{16}$$

因$\{\delta u\}^e$为虚位移,将热载荷向量移至右端,有:

$$[K]^e\{u\}^e=(\{F_P\}^e+\{F_G\}^2+\{F_Q\}^e+\{F_T\}^e) \tag{17}$$

经组集后有$[K]\{u\}=\{F\}$,其中热载荷向量为

$$\{F_T\}^e=\int_{v_e}[B]^T[D]\{\varepsilon_T\}\mathrm{d}v=\int_{v_e}[B]\{\sigma_T\}\mathrm{d}v \tag{18}$$

2 计算模型构造

2.1 有限单元的划分

有限单元的划分工作是有限元分析的一项基础性工作. 是根据对分析对象的温度和应

力的初步估计将计算模型作恰当划分,既保证计算分析具有足够的精度,又使整个计算需要的内、外存空间能在普通微机上实现,以便在有关水泵设计和生产部门推广使用.经过大量的试算工作,本文确立了该水泵机组的有限元计算几何模型(见图1、图2).在建模的同时本文也注意到了温度场模型的基础性及其对应力场模型的延续性[2,3].

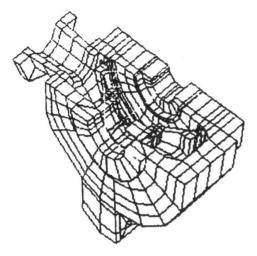

图1 核电停堆冷却泵静部件有限元几何模型构造图

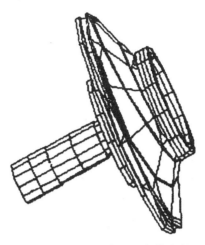

图2 核电停堆冷却泵动部件有限元几何模型构造图

对于核电停堆冷却泵中的静部件(含泵体、泵盖、导流体等部件),由于对称性,计算采用了实际模型的1/2.共选用了572个八结点的等参三维元作为模型的基本组成单元,共计1 165个结点.此外,还根据所分析对象的实际换热工况,生成了六组共计764个四结点对流单元.

对于核电停堆冷却泵中的动部件(含叶轮、转轴等部件),共选用了588个八结点的等参三维元作为模型的基本组成单元,共计1 023个节点.此外,还根据所分析对象的实际换热工况,生成了两组共计688个四结点对流单元.

2.2 边界条件的确立

2.2.1 换热边界条件的确立

换热边界条件主要有二,其一是泵内介质与泵内与之相接触的构件间的强迫对流换热.本文根据该核电水泵的结构及其有关的性能参数采用基于一元设计理论而发展的方法来推算介质在水泵中的流速分布,然后参照了传热学中[4,5]有关的准则关联式和图表估算了水泵中不同区域处的对流换热系数,并将这些参数代入相应的对流边界单元组中.其二是泵体外表面与周围环境间的自然对流换热.本文采用了文[3]所推荐使用的自然对流换热系数.

此外,在对称面处采用了热流强度为零的处理.

2.2.2 约束边界条件的确立

在对称面处采用了垂直于该截面方向的位移为零的处理;由于机架部分的刚度很大且还有冷却循环装置,因而其受热冲击热效应的影响是很小的(上海水泵厂曾对该水泵的机架内部件在受热冲击的过程中进行过温度实测,发现最大也不超过28℃).因此,将水泵与机架结合部位的相关节点处视为刚性的近似处理(即限制6个方向的自由度)是基本符合实际工况的.

表1 核电水泵各区域中平均对流换热系数最终估算值(W/m²·℃)

水泵介质入口段	20 857.16	导流体流通	35 687.27
水泵介质出口段	30 634.81	泵体与导流体间型腔	25 172.53
叶轮流道	34 558.12	导流体与叶轮和泵盖间的型腔	36 194.29

3 计算结果的分析和讨论

本文完整地计算出了热冲击过程中核电水泵机组结构的温度分布状况和热应力、热变形变化行为.限于篇幅,在此仅列出基于计算结果所整理得的最大热应力、热变形时间历程变化曲线.图3是核电水泵静部件最大热应力和最大Von Mises应力的时间历程变化曲线;图4是核电水泵动部件最大热应力和最大Von Mises应力的时间历程变化曲线;图5和图6分别是核电水泵中叶轮前后密封环最大径向热变形时间历程变化曲线.

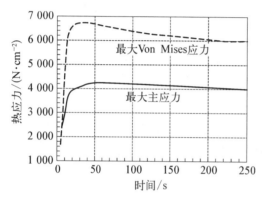

图3 水泵静部件最大主应力和最大 Von Mises 应力时间历程变化曲线

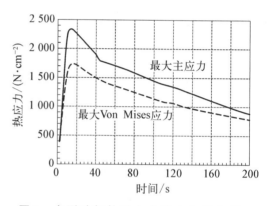

图4 水泵动部件最大主应力和最大 Von Mises 应力时间历程变化曲线

4 结论

图3和图4分别给出了该核电水泵在承受热冲击的过程中,静部件和动部件中的最大正应力和最大Von Mises应力的时间历程变化曲线.从图中可以看出在此种热冲击载荷作用下,水泵结构中的应力开始时以较大的速度增大,随着水泵内介质温度趋于稳恒和水泵结构的温度差逐渐降低,水泵结构中的应力也相应减小;由于水泵动部件的结构厚度相对于静部件的结构厚度要薄得多、且受热载较为均匀,因此,水泵动部件的热应力的降低速度要比水泵静部件快得多;基于同样的原因,水泵动部件中的热应力也要比水泵静部件的热应力小得多.

从图中可知:在温度梯度为13.5℃/s的热载荷的冲击下,水泵结构中的最大正应力为42.91 Mpa,最大 Von Mises 应力为67.65 Mpa;均远远小于水泵材料的许用应力(水泵材料中最小的抗拉强度为550 Mpa,最小屈服强度为200 Mpa).

从图5和图6中可以发现,叶轮前、后二密封环处的最大径向热变形值分别为

0.188 mm 和 0.188 1 mm，且由于结构本身具有 45°反对称性，因此叶轮密封处的热变形基本上是均匀向外膨胀. 相对于泵体密封环和叶轮密封环的径向间隙值 0.25～0.38 mm，即便不计泵体密封环的向外膨胀的热变形量，叶轮密封处单纯的热变形不足以引起叶轮与泵体间的碰擦.

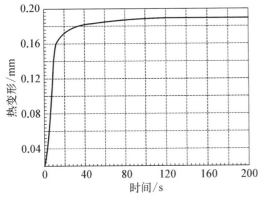 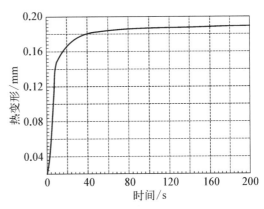

图 5　叶轮前密封环最大径向变形时间变化历程图　　图 6　叶轮后密封环最大径向变形时间变化历程图

综上可以得出以下结论：纯考虑热冲击载荷作用时，(1) 由于按第一强度理论(最大拉应力理论)进行校核得出的安全数值为 12.8；按第四强度理论(歪形能理论)进行校核所得出的安全系数值为 8.1，均符合 PC 工程技术规范要求，因此，在强度上该核电停堆冷却泵是完全能承受此种强度的热载荷冲击. (2) 在热冲击载荷作用下，水泵动、静件间的最小间隙≥0.062～0.192 mm，因此，此种强度的热冲击载荷不会使该核电停堆冷却泵的动静件发生碰撞、干摩擦等破坏性的影响. (3) 虽然在强、刚度方面该核电停堆冷却泵均能承受此种强度的热载荷冲击，但从严格意义上来说这些结论仅仅还是停留在静态考察的观点上. 从动力学的观点来看，密封是该水泵转子-支承系统中的一个很关键的环节，周隙密封间隙的变化会导致密封动力特性的变化，周隙进而会影响水泵转子的动力性态. 因此，在此种强度的热载荷冲击下，该核电停堆冷却泵的运行是否安全、可靠，还需进一步考察热冲击过程中水泵轴系的动力性态及其变化行为.

参 考 文 献

[1] Bathe Klaus-Jurgen, Wilson Edward L. Numerical Method in Finite Element Analysis. Prentice-Hall, Inc Englewood Cliffs, New Jersey, 1976.
[2] 欧阳华江. 关于热传导方程的有限元算法. 工程力学学报,1994,14(4)：58-61.
[3] 纪铮,钟万勰. 关于离散热传导物理模型的探讨. 计算结构力学及应用,1994,11(4)：408-413.
[4] Thomas Lindon C. Heat Tranfer. Prentice-Hall, Inc. A Simon & Schuster Company Englewood Cliffs, New Jersey 07632,1992.
[5] 阿巴兹 VS,拉森 PS. 顾传保,等译. 对流换热. 北京：高等教育出版社,1992.

The Research and Analysis on the Transient Heating Effect in the Pump of Nuclear Power Station under Thermal Shock（Ⅱ）
——Three Dimensional Finite Element Analysis and Computation

Abstract：Formulations of the finite element methods for non-coupled heat conduction equations and quasi-static thermoelastic equations are derived in this paper. And estimations of the convection heat transfer coefficients in different parts of the nuclear pump under research are worked out. Then on the basis of a great deal of tryout computational efforts and relevant numerical examining examples, the finite element method is adopted to solve the three-dimensional transient temperature field the thermal stress field and the thermal deformation field of the nuclear pump. The distribution status and varying features of the temperature fild, the thermal stress field and the thermal deformation field at various transient moments are given. Through computing the primary analytical research methods for studying the transient heating effects in the pump of nuclear power station under thermal shock are obtained.

大自由度的转子-滑动轴承系统非线性动力学分析*

提　要：本文以研究多跨转子-滑动轴承系统在大激励下的动力性态为目标,提出了一套新的关于大自由度的转子-轴承-基础系统的非线性动力学计算方法.其中引入块三对角矩阵追赶法求解上述问题构成的大型矩阵；提出中心差分法＋Houbolt法的预估计-校正时间积分法,以解决二阶隐式非线性方程的时间积分问题；系统方程在物理坐标 x 和 y 上的"解耦",使计算速度和计算精度得以进一步的提高.本文介绍了系统初始位移值的计算方法、计入基础参振的计算方法和非线性油膜力数据库的应用.上述方法为大自由度的多跨转子-轴承-基础系统的非线性动力学性态的研究提供了一种高效、高精度的计算工具.

符 号 说 明

M_j, C_j, U_j, V_j, W_j——转子第 j 个离散节点上的质量矩阵、阻尼矩阵和刚度分矩阵；

X_j, F_j——转子第 j 个离散节点上的位移矢量和力矢量；

M, C, K——转子离散系统方程的总质量矩阵、总阻尼矩阵和总刚度矩阵；

X_t, F_t——转子离散系统方程在 t 时间的位移矢量和力矢量；

$\widetilde{M}$——转子离散系统方程时间积分的等效质量矩阵；

R_t——系统方程在 t 时刻广义力矢量；

Ω——转子工作角速度；

E——材料杨氏模量；

$J_{dj}(J_{pj})$——转子离散系统第 j 个节点的半径转动惯量(极转动惯量)；

l_j, I_j——第 j 个轴段的长度和质惯性矩；

$f_{xk}(x, y, \dot{x}, \dot{y})$, $f_{yk}(x, y, \dot{x}, \dot{y})$——第 k 个轴承非线性膜力在 x 和 y 坐标上的分量；

α_j, β_j, γ_j——第 j 个轴段的刚度系数, $\alpha_j = \dfrac{6EI_j}{l_j^2}$；$\beta_j = \dfrac{12EI_j}{l_j^3}$；$\gamma_j = \dfrac{2EI_j}{l_j}$

　　高速旋转机械如大型汽轮发电机的转子轴承系统对动力性态有很高的要求,以确保安全工作.由于滑动轴承的油膜力具有强烈的非线性特性,目前常用的线性分析方法在某种程度上存在着很大的不足,对许多实际现象无法作出合理的解释.如：(1)某些转子在线性失稳转速之上,轴心可涡动于一个小的轨迹而不发散；(2)转速低于线性失稳转速的转子在受到较大的外加激励时,如地震、叶片脱落等,可能会产生油膜振荡而致失稳.以线性理论为基础的计算结果的可靠程度由此经常受到怀疑.国内外越来越多的研究者开始把注意力转向对该系统的非线性动力学分析,并由此发展了一些用于转子轴承系统非线性动力学分析的计算方法.其中较为突出的有 M. L. Adams[1] 的多轴承支承的柔性转子在模态坐标下的非线性动力学分析方法；Joseph Padovan 等[2] 的有限元分析法；H. D. Nelson[3] 的模态综合法；A. Selva Kumer 和 T. S. Sanber[4],Songyuan LU[5] 的 DT-TMM；R. Rsubbiah 等的有限元传递矩阵组合法[6] 等.然而,已有的各种算法在使用上各有不便.如：转换到模态坐标分析计算,需精确地计算系统方程的特征值和特征向量,有一定的难度,高阶模态的

---

* 本文合作者：李志刚.原发表于《上海大学学报(自然科学版)》,1995,1(6)：640-651.

截断也存在着计算精度和计算经验的问题. 如直接采用有限元法,当系统的自由度较大时,即使在压缩自由度后仍会对计算机容量有较高的要求,计算速度也较难提高. 由于必须同时迭代满足边界的约束条件和转子的位移和轴承的非线性油膜力的耦合关系,DT-TMM 在计算速度上和计算精度上较难完善. 此外,转子-轴承系统构成的动力学方程在时域上是隐式的,二阶时间积分的预估计-校正方尚未见文献加以叙述. 对计算中大量用到的非线性油膜力获得方法也需改进. 由于上述原因,目前对转子-轴承系统的非线性动力学计算较多的局限于 Jeffcote 转子模型,即便是多跨转子模型,也是用较少的节点来描述.

本文以研究滑动轴承支承的多跨转子在大激励下的动力性态为目标,提出了一套新的关于大自由度的转子-轴承-基础系统的非线性动力学计算方法. 即:首次将块三对角矩阵追赶法用于上述问题构成的大型矩阵的求解中;提出中心差分法加 Houbolt 法的预估计-校正时间积分法,以解决二阶隐式非线性方程的时间积分问题;建立了轴瓦的非线性油膜力值的数据库[7,8],并应用于多跨转子-轴承-基础系统非线性动力学计算上,在计算速度和计算精度上同时获得了令人满意的效果;提出了较为精确的系统各离散节点位移初始值的计算方法和计入基础参振的计算方法. 由于涉及内容较多,本文将分两部分给出. 第一部分介绍计算方法的基本思想和主要计算公式. 第二部分给出某些非线性计算的结果及与线性分析结果的对比、实验验证等.

1 系统力学模型及广义动力学方程

转子-轴承-基础系统的力学模型如(图1)所示. 系统广义动力学方程的建立方法可参照文献[10],也可用有限元法[2,9]. 这两种方法所建立的广义动力学方程对我们要求解的问题而言,在矩阵结构上的相同的. 本文采用前一种方法. 第 j 节点的动力学方程可为:

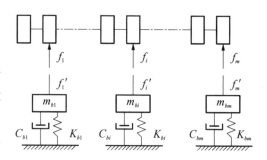

图1 转子-轴承-基础系统的力学模型

$$M_j \ddot{X}_{jt} + C_j \dot{X}_{jt} + W_j X_{j-i_t} + U_j X_{jt} + V_j X_{j+i_t} = F_{jt} \tag{1}$$

其中:

$$M_j = \begin{bmatrix} m & & & \\ & m & & \\ & & J_x & \\ & & & J_y \end{bmatrix}_j ; \quad X_j = \begin{bmatrix} X \\ Y \\ \varphi \\ \Psi \end{bmatrix}_j ; \quad V_j = \begin{bmatrix} -\beta_{j+1} & 0 & \alpha_{j+1} & 0 \\ 0 & -\beta_{j+1} & 0 & \alpha_{j+1} \\ -\alpha_{j+1} & 0 & \gamma_{j+1} & 0 \\ 0 & -\alpha_{j+1} & 0 & \gamma_{j+1} \end{bmatrix} ;$$

$$U_j = \begin{bmatrix} \beta_{j+1}+\beta_j & 0 & \alpha_{j+1}+\alpha_j & 0 \\ 0 & \beta_{j+1}+\beta_j & 0 & \alpha_{j+1}+\beta_j \\ \alpha_{j+1}-\alpha_j & 0 & 2(\gamma_{j+1}+\gamma_j) & 0 \\ 0 & \alpha_{j+1}-\alpha_j & 0 & 2(\gamma_{j+1}+\gamma_j) \end{bmatrix} ; \quad F_j = \begin{bmatrix} -f_x - p_{cx} \\ -f_y - p_{cy} \\ -M_{fy} \\ -M_{fx} \end{bmatrix}_j ;$$

$$W_j = \begin{bmatrix} -\beta_j & 0 & -\alpha_j & 0 \\ 0 & -\beta_j & 0 & -\alpha_j \\ \alpha_j & 0 & \gamma_j & 0 \\ 0 & \alpha_j & 0 & \gamma_j \end{bmatrix}; \quad C_j = \begin{bmatrix} 0 & 0 & 0 & 0 \\ 0 & 0 & 0 & 0 \\ 0 & 0 & 0 & -J_p\Omega \\ 0 & 0 & J_p\Omega & 0 \end{bmatrix}_j$$

系统的广义动力学方程可组装成：

$$M\ddot{X}_t + C\dot{X}_t + KX_t = F_t \tag{2}$$

其中系统的质量矩阵 M 和阻尼矩阵 C 分别是由节点的块质量矩阵 M_j 和块阻尼矩阵 C_j 构成的块对角矩阵，而系统的刚度矩阵 K 是由节点的刚度分矩阵 U_j, V_j, W_j 构成的块三对角矩阵．即：

$$K = \begin{bmatrix} U_1 & V_1 & & & & & \\ W_2 & U_2 & V_2 & & & & \\ & \ddots & \ddots & \ddots & & & \\ & & W_j & U_j & V_j & & \\ & & & \ddots & \ddots & \ddots & \\ & & & & W_{n-1} & U_{n-1} & V_{n-1} \\ & & & & & W_n & U_n \end{bmatrix} \tag{3}$$

2 中心差分法＋Houbolt 法的预估计-校正时间积分法

式(2)这样的方程在时域上的求解，目前常用的方法有两种：(1) 降阶转换为一阶的状态方程，然后用四阶 Rung-Kutte-Gill 法求解[11]．(2) 直接用二阶的时间积分求解，如 Houbolt 法、Wilson θ 法和 Newmark β 法[12]．

式(2)是一个 $(4n \times 4n)$ 的大型矩阵．如采用第一种方法处理，则状态矢量的系数矩阵为 $(8n \times 8n)$，且不再具有块三对角矩阵的形式．当转子离散节点数 n 较大时，矩阵的求逆受计算机内存的限制，计算速度也较慢．第二种处理方法中，目前所采用的时间积分法均为单一的隐式时间积分法，迭代过程中合理初值的预估计处理尚未见文献报道．

实际上，二阶问题的时间积分法有显式和隐式两大类[13]，其形式分别为：

显式： $$\widetilde{M}X_{t+\Delta t} = R_t \tag{4a}$$

隐式： $$\widetilde{M}X_{t+\Delta t} = R_{t+\Delta t} \tag{4b}$$

显式时间积分法中主要有中心差分法．隐式的有 Houbolt 法、Wilson θ 法和 Nemnark β 法．后两种方法在推导过程中加有线性加速度的条件，故对其应用范围有所限制．文献[13] 对这几种时间积分法的计算稳定性问题和计算精度问题作了较为详尽的分析．从中可得知：中心差分法属条件稳定，其条件为：$\Delta t \leqslant \Delta t_{cr} = T_n/\pi$，Houbolt 法属无条件稳定；Wilson θ 法和 Newmark β 法则是在给定参数下无条件稳定．

图 2(a) 和图 2(b) 分别表示了上述各算法积分计算余统振动一周期后振幅的相对误差 AT 和周期的相对误差 PE/T．对中心差分法有 $PE/T = 0$，对 Newmark β 法有 $AT = 0$．

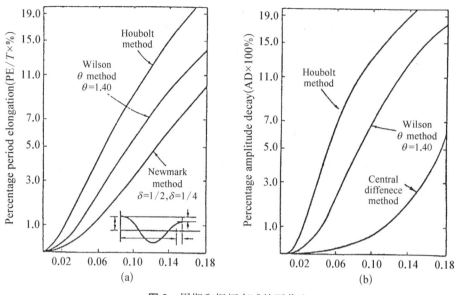

图 2 周期和振幅衰减的百分比

由计算中数值稳定性决定,中心差分法的计算步长有所限制. T_n 是系统的最小振动周期. 由于转子是由连续介质构成,从理论上讲 $T_n \longrightarrow 0$. 即使是离散化的转子模型,当轴段划分过小时, T_n 是相当小的. 从这个意义上讲,这种计算方法不宜单独用在对式(4b)问题的求解上. 然而,转子系统的高阶模态在振动中不易激发,实际计算中可以忽略不计,对步长的限制条件可修改为: $\Delta t \leqslant \Delta t_{cr} = T_p / \pi$ (T_p 为高阶模态缩减后的系统最小振动周期). 从(图2)来看,中心差分法的计算精度无疑是最高的. 从两个方面综合考虑,把中心差分法作为二阶隐式问题时间积分初值的预估计方法是十分合适的. 其在每个 Δt 上预估的初值和迭代矫正后的"精确值"相比有很高的精确度,可以有效地迭代次数,提高计算速度.

尽管 Newmark β 法和 Wilson θ 法计算精度上优于 Houbolt 法,但由于受线性化的加速度的限制,应用范围和收敛性比不上 Houbolt 法. 由此作者认为,二阶隐式式问题的预估计—矫正时间积分法由中心差分法和 Houbolt 法组合应当优于现有其他方法. 这已被作者系统的计算考察所证实. 预估计—矫正的计算步骤如下:

 Ⅰ 由中心差分法求得 $X'_{t+\Delta t}$, $\dot{X}'_{t+\Delta t}$

 Ⅱ 求得 $R'_{t+\Delta t}$

 Ⅲ 由 Houbolt 求得 $\widetilde{X}_{t+\Delta t}$, $\dot{\widetilde{X}}_{t+\Delta t}$

 Ⅳ 迭代收敛判据 $\delta = | X'_{t+\Delta t} - \widetilde{X}_{t+\Delta t} | \leqslant \delta^*$. ($\delta^*$ 为预定的收敛判据)

 Ⅴ 如果 $\delta \leqslant \delta^*$,松弛法修正 $X'_{t+\Delta t}$, $\dot{X}'_{t+\Delta t}$,重复 Ⅱ～Ⅳ;当 $\delta \leqslant \delta^*$,令 $X_{t+\Delta t} = \widetilde{X}_{t+\Delta t}$, 进入一下个 Δt 的计算步.

3 块三对角矩阵的追赶法

当用式(4b)求解 $\widetilde{X}_{t+\Delta t}$ 时, $\widetilde{M}$ 为块三角阵,对 $\widetilde{M}$ 的求逆无法直接转化为对块矩阵的求逆问题,如直接求 $\widetilde{M}^{-1}$,当 n 较大时,即使计算机的内存允许,计算速度和计算精度也将受到较大的影响.本文介绍一种块三对角矩阵追赶法,将上述结构的矩阵的求逆转化为 n 个块矩阵的求逆.

块三角矩阵形式如式(3),问题的求解可简记为:

$$KX=F \tag{5}$$

设:

$$K=\begin{bmatrix} O_1 & & & & & & \\ P_2 & O_2 & & & & & \\ & \ddots & \ddots & & & & \\ & & P_j & O_j & & & \\ & & & \ddots & \ddots & & \\ & & & & P_{n-1} & O_{n-1} & \\ & & & & & P_n & O_n \end{bmatrix} \begin{bmatrix} I_m & Q_1 & & & & & \\ & I_m & Q_2 & & & & \\ & & \ddots & \ddots & & & \\ & & & I_m & Q_1 & & \\ & & & & \ddots & \ddots & \\ & & & & & I_m & Q_{n-1} \\ & & & & & & I_m \end{bmatrix} \tag{6}$$

其中,O_j,P_j,Q_j 是 $(m\times m)$ 的待定块矩阵,由矩阵乘法可得:

$$\begin{cases} U_1=O_1 \\ V_1=O_1Q_1 \end{cases} \tag{7}$$

$$\begin{cases} W_i=P_i \\ U_i=P_iQ_{i-1}+O_i \\ V_i=O_iQ_i \end{cases} \tag{8}$$

从式(8)中可得

$$\begin{cases} O_i=U_i-P_iQ_{i-1} \\ Q_i=O_i^{-1}V_i \end{cases} \quad (i=2,3,\cdots,n) \tag{9}$$

求解(5)的问题等效为求解:

$$\begin{cases} LY=F \\ SX=Y \end{cases} \tag{10}$$

计算过程为:

Ⅰ 递推求解 $\{Q_i\}$,

$$\begin{cases} Q_1=U_1^{-1}V_1 \\ Q_i=[U_i-W_iQ_{i-1}]^{-1}V_i \end{cases} \quad (i=2,3,\cdots,n) \tag{11}$$

Ⅱ 求解 $LY=F$

$$\begin{cases} Y_1=U^{-1}F_1 \\ Y_i=[U_i-W_iQ_{i-1}]^{-1}[F_i-W_iY_{i-1}] \end{cases} \quad (i=2,3,\cdots,n) \tag{12}$$

Ⅲ 求解 $SX=Y$

$$\begin{cases} X_n = Y_n \\ X_i = Y_i - Q_i X_{i+1} \end{cases} \quad (i = n-1, \cdots, 2, 1) \tag{13}$$

用上述方法求解由块矩阵构成的三对角矩阵的先决条件是 U_1^{-1} 和 $[U_i - W_i Q_{i-1}]^{-1}$ 存在. 当 K^{-1} 存在的条件满足时, U_1^{-1} 和 $[U_i - W_i Q_{i-1}]^{-1}$ 的存在是完全合理的[17]. 式(11)为求 n 个 $(m \times m)$ 的块矩阵的逆阵, 工作量比原来的求 $(nm \times nm)$ 的矩阵 K 的逆阵要少得多, 而且基本上不受 n 的大小的限制. 式(12)和(13)的计算过程中的工作量仅为 $3n-2$ 个 $(m \times m)$ 的块矩阵和 $(m \times 1)$ 的列阵的乘法. 如采用直接计算的方法, K^{-1} 和 F 的相乘需计算 nn 个 $(m \times m)$ 块矩阵和 $(m \times 1)$ 列阵的乘法, 另外, 当式(11)的求解完成后, 在以后的计算中, 当仅有 F 的变化时, 只需进行式(12)和(13)的回代计算. 这对需大量且反得求解式(4b)和(5)的计算问题, 其优点是不言而喻的.

块三对角矩阵的追赶法求解具有式(2)结构的大型稀疏矩阵的极为有效的方法, (2)式结构的大型稀疏矩阵在工程应用和科学研究中又是常见的. 文献[17]、[18]和[19]曾介绍此种方法. V. Castelli 用在解润滑油膜压力分布的"column method"的计算公式和块三对角矩阵的追赶法公式完全相同. 不过推导过程不同, 其定义也仅基于润滑油膜压力分布的列矢量概念. 将块三对角矩阵的追赶法用在转子动力学问题的求解上, 还未见文献记载.

4 系统方程的"解耦"简化

由于在式(4b)所涉及的问题中, 广义力矢量 $\vec{R}_{t+\Delta}$ 是由 $X_{t+\Delta}$ 决定的未知矢量, 计算是一个预估计-矫正的过程. 我们可以把使 x 和 y 耦合的项移到方程的右方归入广义力矢量项, 使系统方程在等号左边在 x 和 y 两个坐标上解耦为如下形式:

$$M\ddot{X}_t + KX_t = F_{xt} \tag{14a}$$

$$M\ddot{Y}_t + KY_t = F_{yt} \tag{14b}$$

其中 M 和 K 是 (2×2) 块矩阵的构成的对角阵和三对角阵, (4b)的求解已被简化为对由 (2×2) 的块矩阵构成的三对角施行"追赶法"的问题. (2×2) 的矩阵的求逆可直接用公式计算, 不仅计算工作量降为未经简化前的 1/4, 计算精度还有一个质的飞跃. 式(11)~(13)的求解过程中的数值误差仅剩下计算机字长的截断误差, 对整个计算过程中的变量采用双精度型时, 截断误差非常之小, 以至与离散误差或预估计-矫正过程中的迭代误差相比成为微不足道.

5 系统各离散节点位移初始值的确定方法

在对多跨转子-轴承系统进行非线性动力学分析的仿真计算时, 系统各离散节点的位移妆始值的确定是一个十分重要的问题. 对单跨转子而言, 外载荷一旦确定, 各轴承的载荷不随转速而变化. 而在多跨转子-轴承系统中, 情况就较为复杂. 各轴承的载荷分配在这里是一个静不定问题. 在转速和外载荷给定的情况下, 各轴承的安装标高的确定须引入 $(m-2)$ 个定解条件, 同时还必须计入在这些定解条件下轴颈在各轴承中浮起量和分配到各轴承载荷之间的关系.由于轴颈在轴承中位移和轴承所承受的载荷呈非线性关系,这个静不定问题没有封闭解,只能进行迭代计算.且这个静不定问题是二维的,即因各轴承中 e_{xi} 的不同,转子的挠曲线是空间的,各轴承中 $F_{xi} \neq 0$. 这样,在 X 方向也存在位移和载荷分配的静不定问题,且与 Y 方向的静不定问题互相耦合,构成了复杂的固流耦合的二维静不定问题.此外,$(m-2)$ 个定解条件通常是在某个转速下给出的.即使满足这些定解条件的解可以求得,当计算条件(如计算转速)变化后,轴颈在各轴承中的浮起量的变化通常是不同的,各轴承中分配到的载荷将由此而发生变化,须重新反复求解上述固流耦合的二维静不定问题.由上述特点决定,整个计算将有相当难度,计算工作量也较为庞大.过去许多研究者在涉及这个问题时,为了求解方便和减少计算量,大都把 X 方向的静不定问题忽略不计.这样的简化计算,对线性分析来讲,计算精度尚可接受,但在非线性分析的仿真计算时就显得过于粗糙.因为位移初始值的误差过大就相当于使被计算的系统在计算开始的瞬间受到较大的冲击,这种冲击响应要经过较多的计算周期才能得以衰减,严重的甚至会使计算出错中断.为了防止这种现象在计算中出现和使整个计算过程有较合理的和较为精确的计算初值,本文作者研究设计了一套求解上述固流耦合的二维静不定问题的精度高而速度快的计算方法.现将此套方法介绍如下:

5.1 建立系统凝聚柔度矩阵 A 和凝聚的刚度矩阵 K

定义: $$\widetilde{Y} = AP \tag{15}$$

其中:$A = \{\alpha_{ji}\}$;$P = [p_1, p_2, \cdots, p_{m-3}, p_{m-2}]^T$;

$\widetilde{Y} = [\widetilde{y}_1, \widetilde{y}_2, \cdots, \widetilde{y}_{m-3}, \widetilde{y}_{m-2}]^T$

α_{ji} 为单独在支承 i 处施加单位力在支承 j 处引起的挠度(图 3).建立凝聚柔度矩阵 A 的方法可以有很多种,为了提高计算精度和计算速度,本文采用三对角块矩阵追赶法求解的方法.当 $t=0$ 时有 $\ddot{X} = \dot{X} = 0$ 和 $\ddot{Y} = \dot{Y} = 0$,系统的运动方程(14)可演化为如下形式:

$$KX_0 = F_{x0} \tag{16a}$$
$$KY_0 = F_{y0} \tag{16b}$$

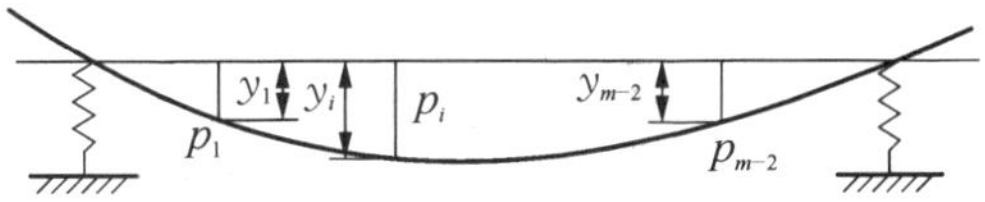

图 3 转子内节点载荷和挠度

以 Y 方向为例,式(16b)中 Y_0 是一组比值,K 的逆阵不存在,故无法直接施用三对角块矩阵追赶法求解.对此可选某一个(或几个)支承处的位移值为参考点,代入式(16b),采用有限元计算中常用的方法对矩阵(16b)进行修改.

在式(16b)中的力矢量 F'_{y0} 中有 n 个元素,其中作为基准点的元素为已知值.轴承的个数为 m 个,其中 $m-2$ 轴承为内支承点.现取基准点的位移值为零值,分别逐次取这 $m-2$ 个内支承点 i 上的载荷为单位力($i=1, 2, \cdots, m-2$),而力矢量中的其他元素为零元素,共需求解(16b)式 $m-2$ 次,以分别求得 $m-2$ 组的每个内支承点上的位移值 $Y_j(j=1, 2, \cdots, m-2)$,即为所需求得的 $\alpha_{ji}(j=1, 2, \cdots, m-2)$. 由于三对角块矩阵追赶法的特点,$n$ 个(2×2)块矩阵求逆只要在 $j=1$ 时求解一次,在后面 $j=2, 3, \cdots, m-2$ 的求解 α_{ji} 的过程中,只需进行回代计算的块矩阵的乘法运算,计算速度非常快.

凝聚柔度矩阵 A 建立后即可求逆得到凝聚的刚度矩阵 $\widetilde{K}$.

5.2 协调方程计入轴颈浮起量

转子相对两个作为基准的轴承的轴颈中心的挠度曲线可成为"相对挠度曲线"(图4),设各支承截面上的相对挠度为 x_i 和 $y_i (i=1, 2, \cdots, m-2)$,组成 X 和 Y. Y 可看成由两部分组成,即由自重引起的挠度 Y_0 和支承力引起的挠度 Y_1. X 则只有支承力引起的挠度 X_1. 由式(15)得:

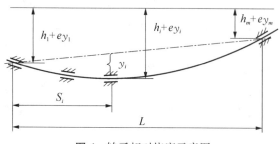

图4 转子相对挠度示意图

$$\begin{cases} X_1 = -AF_x \\ Y_1 = -AF_y \end{cases} \quad (17)$$

则相对挠度 X 和 Y 为:

$$\begin{cases} X = -AF_x \\ Y = Y_0 - AF_y \end{cases} \quad (18)$$

从图4可得,第 i 个轴承处的相对挠度 $y_i'(x_i')$ 可用下列公式求得:

$$y_i' = \begin{cases} h_i + e_{y_i} - (h_1 + e_{y_1}) - \dfrac{(h_m + e_{y_m}) - (h_1 + e_{y_1})}{L} S_i & (h_m + e_{y_m}) \leqslant (h_1 + e_{y_1}) \\ h_i + e_{y_i} - (h_m + e_{y_m}) - \dfrac{(h_1 + e_{y_1}) - (h_m + e_{y_m})}{L} (L - S_i) & (h_1 + e_{y_1}) \leqslant (h_m + e_{y_m}) \end{cases} \quad (19)$$

由(19)式可得 X' 和 Y':

$$X' = [x_1', x_2', \cdots, x_{m-3}', x_{m-2}']^T \quad (20a)$$

$$Y' = [y_1', y_2', \cdots, y_{m-3}', y_{m-2}']^T \quad (20b)$$

协调 Y 和 Y'(X 和 X'),迭代计算使之一致. 计算路线和方法如下:

(1) 先设各 e_{x_i} 和 $e_{y_i} (i=1, 2, \cdots, m-1)$,迭代计算满足由定解条件求得的 F_{x_i} 和 F_{y_i} 时的 e_{x_i} 和 e_{y_i},以及该 e_{x_i} 和 e_{y_i} 下的轴承的4个刚度系数 $K_{lj}(l, j = x, y)$.

(2) 由(18)式求得 X 和 Y.

(3) 由(19)式求得 X' 和 Y'.

(4) 求出位移误差 ΔX 和 ΔY,并合并成 ΔS.

$$\begin{cases} \Delta X = X - X' \\ \Delta Y = Y - Y' \end{cases} \quad (21)$$

$$\Delta S = [\Delta x_1, \Delta y_1, \Delta x_2, \Delta y_2, \cdots, \Delta x_{m-2}, \Delta y_{m-2}]^T \quad (22)$$

(5) 设由 Δe_{x_i} 和 Δe_{y_i} 得校正 ΔX 和 ΔY 的力矢量的增量为 ΔF_{x_i} 和 ΔF_{y_i},有下列成立:

$$\begin{bmatrix} k_{xx_1} & k_{xy_1} & & & & & & \\ k_{yx_1} & k_{yy_1} & & & & & & \\ & & k_{xx_2} & k_{xy_2} & & & & \\ & & k_{yx_2} & k_{yy_2} & & & & \\ & & & & \ddots & \ddots & & \\ & & & & & & k_{xx_{m-2}} & k_{xy_{m-2}} \\ & & & & & & k_{yx_{m-2}} & k_{yy_{m-2}} \end{bmatrix} \begin{bmatrix} \Delta e_{x_1} \\ \Delta e_{y_1} \\ \Delta e_{x_2} \\ \Delta e_{y_2} \\ \vdots \\ \Delta e_{x_{m-2}} \\ \Delta e_{y_{m-2}} \end{bmatrix} = \begin{bmatrix} \Delta F_{x_1} \\ \Delta F_{y_1} \\ \Delta F_{x_2} \\ \Delta F_{y_2} \\ \vdots \\ \Delta F_{x_{m-2}} \\ \Delta F_{y_{m-2}} \end{bmatrix}$$

简计为:

$$G \Delta \varepsilon = \Delta F \tag{23}$$

(6) 由于 $\Delta \varepsilon$ 的出现,实际需修正的位移误差为 $\Delta S'$

$$\Delta S' = \Delta S - \Delta \varepsilon$$

将系统的凝聚刚度矩阵 $\vec{K}$ 扩展为二维的刚度矩阵 $\overline{K}$,可得:

$$\overline{K} \Delta S' = \Delta F \tag{24}$$

(7) 联立式(23)和(24)得:

$$\overline{K} [\Delta S - \Delta \varepsilon] = G \Delta \varepsilon \tag{25}$$

即可得到所需求解的 $\Delta \varepsilon$.

$$\Delta \varepsilon = [G + \overline{K}]^{-1} \overline{K} \Delta S \tag{26}$$

(8) 由于修正 $\Delta S'$ 造成中间各支承上 ΔF 的出现,为使系统静力平衡,在作为基准参考点的第 0 和 $m-1$ 个支承上出现 ΔF_{x_0}、ΔF_{y_0} 和 $\Delta F_{x_{m-1}}$、$\Delta F_{y_{m-1}}$,引起 Δe_{x_0}、Δe_{y_0} 和 $\Delta e_{x_{m-1}}$、$\Delta e_{y_{m-1}}$ 的出现. 这样,原来由式(26)求得到的 $\Delta \varepsilon$ 仅是一次近似修正,为了高精度地计算各轴承的载荷分配和轴颈在轴承中的浮起量,计算将是一个 $(2°\sim 8°)$ 迭代求解的过程.

采用上述计算方法亚松弛迭代计算,一般仅需 3~5 次迭代即可在二维方向上同时满足精度要求,得到各轴承在计算转速下的静平衡位置 e_{x_i}、e_{y_i} 和载荷 F_{x_i}、F_{y_i} ($i=0, 1, \cdots, m-1$),计算速度很快. 当上述过程完成后,即可调用块三对角矩阵追赶法子程序精确快速计算系统各节点的位移值.

6 非线性油膜力数据库的应用

非线性动力学分析的仿真计算是在时间域上的逐点积分计算. 因为系统的运动方程(2)是隐式方程,在时域上的每一离散节点上必须对轴颈在轴承处的位移和轴承的油膜力进行反复迭代计算,轴承非线性油膜力值的精度和求取的速度将直接影响系统非线性动力学分析计算的精度和效率. 目前国内外的研究者在涉及这个问题时采用的方法主要有两种:(1) 直接用数值方示解 Reynolds 方程以获得非线性油膜力值;(2) 采用近似计算方法获得非线性油膜力值,如采用短轴承理论或无限宽轴承理论、图线法等.

用数值计算方示获得非线性油膜力可以有较高的计算精度,但须耗费大量计算机时. 尽管现在计算机技术发展日新月异,直接数值计算的方法因其庞大的计算工作量目前还只能用在一些简单的非线性分析计算上. 近似计算的方法虽有较高的计算速度,计算精度上却不尽人意. 作者在研究建立线性化的油叶型轴承单瓦性能数据库[14,15]的基础上指导研究建立了非线性油膜力数据库[7,8],并已由作者成功地应用于多跨转子-轴承系统非线性动力学分析的仿真计算中.

非线性油膜力数据库的数据是事先通过数值计算所得,其应用仅是检索调用插值计算的过程,无须再进行繁重的数值计算. 故由数据库获得非线性油膜力数据的方法同时具有了直接数值计算时计算精度高和检索调用时快速高效的特点. 数据库技术在多跨转子-轴承系统非线性动力学分析的仿真计算中的应用使大规模而又高精度的计算成为可能.

7 计入基础参振的方法

计入基础弹性变形和轴承座及基础参振的力学模型见(图1). 基础的运动方程如下:

$$M_B \ddot{X}_{B_t} + C_B \dot{X}_{B_t} + K_B X_{B_t} = F'_t \tag{27}$$

其中: M_B, C_B 和 K_B 分别是由块矩阵 M_{b_t}, C_{b_t} 和 K_{b_t} 构成的块对角矩阵.

$$M_{b_t} = \begin{bmatrix} m_{b_t} & 0 \\ 0 & m_{b_t} \end{bmatrix}; \quad C_{b_t} = \begin{bmatrix} C_{xx_t} & C_{xy_t} \\ C_{yx_t} & C_{yy_t} \end{bmatrix}; \quad K_{b_t} = \begin{bmatrix} Kb_{xx_t} & Kb_{xy_t} \\ Kb_{yx_t} & Kb_{yy_t} \end{bmatrix};$$

$$X_B = [x_{b_1}, y_{b_1}, x_{b_2}, y_{b_2}, \cdots, x_{b_m}, y_{b_m}]^T$$

$$F' = [f'_{x_1}, f'_{y_1}, f'_{x_2}, f'_{y_2}, \cdots, f'_{x_m}, f'_{y_m}]^T$$

对式(27)和式(2)在时域上联立求解,与式(2)不同的是,式(27)中的刚度矩阵 K_B 是由块矩阵构成的对角阵. 求广义质量矩阵的逆阵 $\widetilde{M}_B^{-1}$ 可以直接化为块矩阵求逆而不必再施用块三对角矩阵追赶法.

轴颈对轴承的相对位移 Y 由式(28)和图5示出.

$$Y = Y_Z - Y_B \tag{28}$$

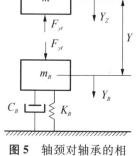

图5 轴颈对轴承的相对位移 Y

8 结论

本文以一系列分析技术建立了一套新的大自由度的转子-轴承-基础系统的非线性动力学计算方法. 提出中心差分法+Houbolt法的预估计-校正时间积分法,以解决二阶隐式非线性方程的时间积分问题;引入块三对角矩阵追赶法法成功地用于上述问题构成的大型矩阵的求解中;系统方程在物理坐标 x 和 y 上的"解耦"简化,使计算速度和计算精度得以进一步的提高. 大自由度的转子-轴承-基础系统的非线性动力学计算时系统各离散节点的位移初始值的计算方法;计入基础参振的计算方法和非线性油膜力数据库在多跨转子-轴承-基础系统的非线性动力学计算中的应用. 后续文章中将叙述的大量实算表明,该方法的建立为大自由度的多跨转子-轴承-基础系统的非线性动力学性态的研究提供了一种高效而又精确的计算工具.

参 考 文 献

[1] Adams M L. Non-linear Dynamics of Flexble Multi-Bearing Rotor. Journal of sound and Vibration, 1980, 71(1): 129-144.

[2] Joseph Padovan, et al. Nonlinear Transient Finite Element Analysis of Rotor-Bearing-Stator Systems. Computers & structure, 1984, 18: 629-639.

[3] Nelson H D, et al. Nonlinear Analysys of Rotor-Bearing Systems Using Component Mode Systhesis. Journal of Engineering for Power, Transaction of the ASME. 105: 606-614.

[4] Selva Kunar A, Sankar T S. A New Transfer Matrix Method for Response Analysis of Large Dynamic Systems. Computers & Structures, 1986, 23(4): 545-552.

[5] Songyuan Lu. Atransfer Matrix Mehtod for Nonliear Vibration Analysis of Rotor-Bearing systems. DE 37, Vibration Analysis and Computational, ASME 1991.

[6] Subbiah R, et al. Transient Dynamic Analysis of Rotor Using the Combined Methodologies of Finite Elements and Transfer Matrix. Journal of Applied Mechanics, Transction of the ASME, 1988, 55: 448-452.

[7] Zhang Z, Li Z, Wang W. Calculation of Lobe Type Hydrodymanic Bearings Aided by Database of Properties of Bush Segment Proceedings of International Conference on Tribology 93, Beijing, China, 1993.

[8] 王文. 非线性油膜力数据库. 硕士学位论文. 上海：上海工业大学, 1991.

[9] 钟一谔, 等. 转子动力学. 北京：清华大学出版社, 1987.

[10] 虞烈. 轴承-转子系统的稳定性与振动控制研究. 西安：西安交通大学博士论文, 1987.

[11] Hori Y, Kato T. Earthquack-Induced Instability of a Rotor Supported hy Oil Film Bearing. Journal of Vibration and Acoustics, Transaction of the ASME, 1990, 112: 160-165.

[12] Subbiah R, Rieger N F. On the transient Analysis of Rotor-Bearing Syitems. Journal of Vibration, Acoustic, Stress, and Reliability in Design, Transaction of the ASME, 1988, 110: 515-520.

[13] Bathe & Wilson. Numeerical Methods in Finite Element Analysis, John Wlley Publishers.

[14] 李志刚, 滑动轴承数据库研究. 硕士学位论文, 上海：上海工业大学, 1991.

[15] Zhigang Li & Zhiming Zhang. Calculation of Lobe Type Hydrodymanic Bearings Aided by Database of Properties of Single Bush Segment. Proceedings of ASIA—PACIFIC VIBRATION CONFERENCE93, KITAKYUSHU, JAPAN.

[16] 张直明. 滑动轴承流体力润滑理论. 高等教育出版社, 1986.

[17] Golub G H, Van Loan C F. Matrix Computations. The Johns Hopkins University Press, 1983.

[18] George J A. On Block Elimination for Sparse Linear Systems, SIAM J. Num. Anal. 1974, 11: 585-603.

[19] Varah J M. On the Solution of Block-Tridiagonal Systems Arising from Certain Finite Difference Equations. Math Comp., 1972, 26: 859-868.

Non-Linear Dynamic Analysis of Multi-span Rotor-Journal Bearing System

Abstract: A new calculation method of non-linear dynamics of rotor-bearing-fundation system of high degree of freedom is proposed in this paper, with the aim of studing the dynamic behavior of mult-span rotor supported by journal bearings under large excitations. Triadiagonal block matrix algorithm is successfully applied to solve the large matrix involved; central difference method is supplemented to Houbolt method to form a time integration method with estimation and correction, to solve the time integration of second order implicit non-linear equation; simplication of system equation by decoupling between the physical coordinates x and y further improves the calculation accuracy and speed. Also introduced in the paper are the method of calculation of the initial displacement of the system; the method of consideration of fundation vibration; and the application of datadase of the system; the method of consideration of fundation vibration of rotor-bearing systems. The program realized with these techniques provides a calculation tool of high efficiency and accuracy for the study of non-linear dynamic behavior of multi-span rotor-bearing-fundation system with high degree of freedom.

大自由度的转子-滑动轴承系统非线性动力学分析(Ⅱ)*

提　要：本文以作者提出的关于大自由度的转子-轴承-基础系统的非线性动力学计算方法,详细地考察了转子轴承系统非线性的稳定性问题.其中主要包括：不平衡质量大小和分布对转子轴承系统稳定性的影响；某些转子轴承系统,当其转速高于线性失稳转速时,转子轴心可涡动于一个小的轨迹而不发散的现象之原因的探索；多跨转子轴承系统在不同类别大激励下的动力性态研究等.仿真计算的结果与对比性的实验结果较为吻合.

1　不平衡质量大小和分布对转子轴承系统稳定性的影响

转子轴承系统线性失稳转速是通过求解系统动力学方程的齐次线性方程组得到,因此无法计入实际存在的转子不平衡质量造成的同步强迫振动的影响.许多实验和研究表明,不平衡质量的大小对系统稳定性有明显的作用[1].受计算速度和计算规模的限制,目前考察不平衡质量对系统稳定性的影响大多局限于刚性转子或单质量弹性转子轴承系统,关于柔性转子上不平衡质量的分布对系统稳定性的影响的研究尚未见文献报道.

作者考察了滑动轴承支承转子上不平衡质量大小及分布对系统稳定性的影响[2].大量计算结果除支持了文献[1]中关于刚性转子不平衡质量大小对系统稳定性影响的结论外,更重要的还发现：对柔性转子而言,不平衡质量的分布是一个重要的影响因素,其程度可能要超过不平衡质量大小的因素.计算表明：强迫振动的挠曲线与转子的一阶振型接近时,不平衡质量的存在和增加对提高系统的稳定性无益,与转子的二阶以上的振型接近时,不平衡质量的存在和适量增加能有效地提高系统的稳定性.造成这种现象的原因是,转子-轴承系统的失稳转速通常是和系统的一阶固有振动有关,当强迫振动的挠曲线与转子的一阶固有振型接近时,容易激起系统的大幅度一阶振动,而强迫振动所引起的油膜特性的增稳效应则是有限的,不足以抑制激发的油膜涡动.当强迫振动的挠曲线与转子二阶以上振型接近时,一阶振动的成分在系统振动中不占主导地位,与一阶固有振动有关失稳现象不易发生或即便发生也容易被抑制.一般来讲,实际转子的一阶动平衡是做得较好的,残余的不平衡质量往往对二阶或二阶以上的振动作用才较明显.这种分布的残余不平衡质量对转子轴承系统的稳定性提高会起有益的作用.

图 1 是单跨三圆盘转子试验台的 3 个圆盘分别加不同偏心质量的模拟升速峰-峰值曲线.转子几何尺寸见图 2.仿真计算时采用的轴承类型和参数为：$2 \times 170°$椭圆轴承,椭圆度 $m=0.1$,$\psi_{\min}=0.0027$,$\psi=0.003$,$1/d \approx 0.6$,润滑剂黏度 $\mu=0.03(\mathrm{Pa \cdot s})$,和试验条件基本一致.所不同的是：试验轴承是下瓦 $180°$、上瓦 $150°$圆轴承,$\psi=3‰$,上瓦双侧进油,进油

---

* 本文合作者：李志刚.原发表于《上海大学学报(自然科学版)》,1996,2(1)：50-59.

压力为 0.6 kgf/cm². 考虑到进油压力会使轻载轴承的进油口附近轴承间隙,呈缓慢开扩楔形区域建立局部的完整油膜区,当轴颈在轴承中涡动时,轴承的非线性特性会使这局部的完整油膜区的作用加大,起到增大轴承的椭圆度的作用. 作者采用以较小的椭圆度的椭圆轴承来等效计入进油压力效应的圆轴承.

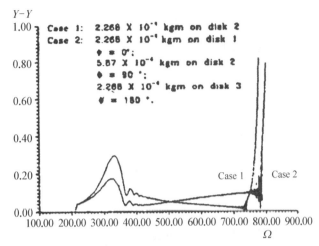

图 1　单跨三圆盘转子试验台上各圆盘分别加不同的偏心质量的模拟升速图

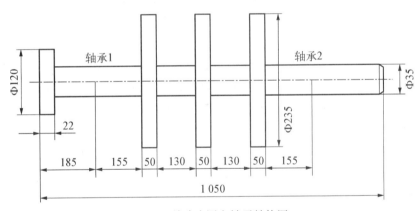

图 2　单跨多圆盘转子结构图

从图上得知：激起二阶模态的偏心质量分布明显地提高了系统的失稳转速,而且使系统的失稳相对于激起一阶模态的偏心质量分布的失稳过程而言有一个渐进的过程. 仿真计算所得的两种不同偏心质量分布下的升速时的峰-峰值曲线的形状以及系统的临界转速、失稳转速与作者的实验曲线和数据较为吻合.

2　滑动轴承油膜力的非线性特性对系统稳定性的影响

由于滑动轴承的油膜力具有强烈的非线性特性,许多实际现象无法用线性理论为基础的分析方法作出合理的解释. 某些情况下,转速高于线性失稳转速,转子轴心可涡动于一个小的轨迹而不继续发散的现象就是其中之一. 对石横引进机组的小轴系(两跨三支承)转子轴承系统的稳定性问题的争议就是一个较为典型的例证. 作者对该系统的稳定性问题进行了非线性仿真计算,并和相同计算条件下的线性计算的结果进行了对比,对这两种方法在计算结果上的差异进行了探索性的分析.计算中采用的转子和轴承的参数由上海发电设备成套研究所提供.

2.1 线性分析计算结果

计算参数:1#和2#轴承为$2\times 155°$椭圆轴承,椭圆度$m=0.2875,\psi_{\min}=1.32‰,\psi=1.85‰,l/d\approx 0.8$;3#轴承为$360°$圆轴承,$\psi=1.31‰,l/d=0.446;\mu=0.018$ Pa·s.

线性分析计算结果:$n_1=860$ r/min;$n_{st}=1900$ r/min.

从计算结果来看,该系统的失稳转速较低.为了分析其原因,作者曾分别采用不同的轴承类型和参数进行了对比性计算.

对比性计算之一:3#轴承改用$4\times 80°$可倾瓦轴承,其他轴承的类型和参数不变.计算结果:$n_1=860$ r/min;$n_{st}=2000$ r/min.

对比性计算之二:1#和2#轴承的椭圆度修正为$m=0.575,\psi_{\min}=0.78625‰,\psi=1.85‰$,而其他轴承的类型和参数不变.计算结果:$n_1'=908$ r/min;$n_1=1001$ r/min;$n_{st}=3000$ r/min.

从上述结果得知:该系统中1#和2#轴承(电机轴承)动力性态能对整个系统的稳定性的影响较大,3#轴承(励磁机轴承)的影响较小.值得注意的是,1#和2#轴承的椭圆度仅从0.2875改为0.575,系统的线性失稳转速提高了1100 r/min(约为57.89%).

2.2 非线性仿真法模拟升速过程的计算结果

计算中采用的轴承类型和参数与线性分析相同.计算起始转速$n_s=700$ r/mm;$\Delta t=T_n/256$;在本文中定义:$T_n=2\pi/\omega_s$.每个Δt中角速度的增量为$\Delta\Omega=0.01$.设第一跨的中心节点上有质量偏心质距0.14(kg·m),相位角$\varphi_1=0°$;第二跨的中心节点上有质量偏心质距0.05(kg·m),相位角$\varphi_2=180°$.模拟升速过程各轴承轴颈水平方向振动峰-峰值曲线由图3给出.$\omega_1\approx 98\sim 105$ s^{-1},$\omega_2\approx 280\sim 300$ s^{-1},$\Omega_{st}'\approx 350$ s^{-1},从图中可以得出:

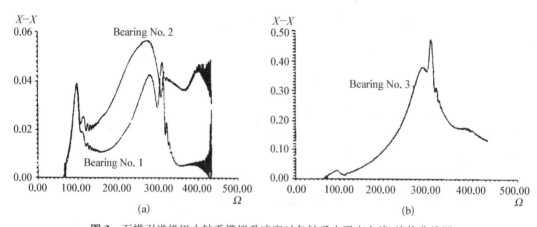

图3 石横引进机组小轴系模拟升速度时各轴承水平方向峰-峰值曲线图

(1)系统的失稳并非在所有的轴承中同时开始的,而是在某一个轴承处或某一跨先出现不衰减的低速涡动,再逐渐传递到其他的轴承,使其他的轴承处也依次出现不衰减的低速涡动,并导致整个系统的失稳.如计算中得出:1#轴承在$\Omega_{st}'\approx 350$ s^{-1};2#轴承在$\Omega_{st}'\approx$

370 s^{-1}；3#轴承在 $\Omega'_{st} \approx 410 \text{ s}^{-1}$ 时，相继出现此种涡动.

(2) 与线性方法计算的线性失稳转速相比，该系统在上述计算条件和参数下用非线性的仿真计算方法模拟升速过程得出的失稳转速 Ω'_{st} 较高，与实际情况较为吻合.

2.3 考察突加不平衡响应对系统失稳转速的影响

升速过程中失稳转速有滞后现象，这是许多实验以及实际机组运行过程中被多次证明的，故模拟升速计算得到的失稳转速 Ω'_{st} 还不能简单的被定为是系统的安全可靠的设计依据. 对系统进行激振(模拟锤击)或突加不平衡响应(模拟叶片脱落)计算所得的系统失稳转速 Ω_{st} 往往要小于模拟升速所得的失稳转速 Ω'_{st}. 图4是系统突加不平衡响应后各轴承中轴颈在 X 方向振动曲线图. 图中横坐标 r 代表旋转次数. 突加不平衡量质偏心距和位置如上所述.

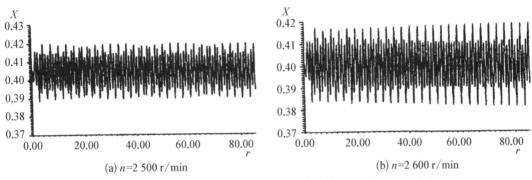

图4 石横引进机组小轴系突加不平衡质量时1#轴承的响应曲线

计算表明，突加不平衡质量时，系统出现的失稳转速 Ω_{st} 约在 2 500～2 600 r/min 之间. $n = 2 600$ r/min 时，1#和2#轴承上低速涡动成分逐渐增加，振动逐渐发散，而 $n = 2 500$ r/min 时，1#和2#轴承上低速涡动成分逐渐减少，振动逐渐收敛. Ω_{st} 明显地小于 Ω'_{st}，又明显地大于线性失稳转速(注意：后一点特征并非在所有的转子轴承系统的稳定性问题中都存在). 以作者之见，激振所得的 Ω_{st} 应作为该系统安全可靠的失稳阈速较为妥切.

对石横机组小轴系转子轴承系统非线性动力学分析计算得到失稳转速较高，转子可以稳定运转于线性失稳转速之上. 分析这种现象的原因，主要可以归纳为以下几点：

(1) 椭圆度较小的椭圆轴承，接近线性失稳转速的静平衡位置较特殊(ε 较小，$\theta \approx 90°$)，静平衡位置上轴颈对上下两块轴瓦的 θ_i 和 $\varepsilon_i (i = 1, 2)$ 与不平衡涡动轨迹上的轴颈对上下两块轴瓦的 θ'_i 和 ε'_i 有一定的差别，实际涡动轨迹上的轴颈对上下两块轴瓦的 ε'_i 较 ε_i 要大，上瓦的实际承载区会由 θ'_i 与 θ_i 的不同而相对显得大些，所有这些在效果上如同加大了轴承的椭圆度，该系统的稳定性受1#和2#轴承的椭圆度变化的敏感程度已有前面的计算得以证实.

(2) 静平衡位置上求得的各轴瓦的承载区范围没有考虑轴颈涡动的挤压效应，线性化的刚度系数和阻尼系数是在静态的承载区内求得. 非线性油膜力由于计入了每个瞬时轴颈涡动的具体挤压效应，其动态的承载区的边界处理较为合理，动态的承载区比静态的承载区相对来讲要大些，在上瓦尤为明显. 由此会增加轴承在水平方向的抗振能力，提高系统的稳定性.

3 多跨转子轴承系统在大激励下的动力性态

转速低于线性失稳转速的转子轴承系统在受到外界激励的情况下(如:锤击、地震、叶片脱落等现象)会进入失稳状态.这种用线性理论无法解释的现象引起了许多研究者的关注和探索[5].作者以一个模拟 600 MW 汽轮发电机组试验台为计算模型图 5,考察了多跨转子轴承系统在大激励下的动力性态.计算中转子划分为 267 个节点,11 个支承轴承均为 $2\times150°$ 椭圆轴承,椭圆度 $m=0.5, \psi_{min}=0.002, l/d=0.6, \mu=0.018(Pa \cdot s)$.

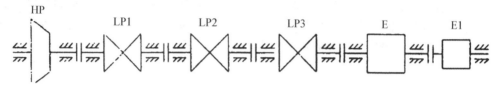

图 5 模拟 600 MW 机组试验台计算模型

图 6 是模拟升速过程中各轴承(为省略起见,每一跨中取一个轴承为例)中轴颈在垂直方向振动的峰-峰值曲线.从图中得知,$\omega_1 \approx 285\ s^{-1}, \Omega'_{st} \approx 780\ s^{-1}$.振动的薄弱环节是电机转子.失稳先从电机转子这跨开始,逐渐传递到其他转子,从而带动整个系统失稳.

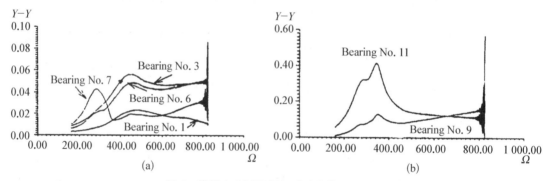

图 6 模拟 600 MW 机组试验台模拟升速图

图 7 模拟机组在转速为 6 000 r/min 时受到"地震波"[3]的冲击,激振频率为系统的一阶临界转速 ω_1,冲击力的幅值 $A=0.2$.此时系统 $\Omega/\omega \approx 2.204\ 6$,远小于 Ω_{st}/ω_1.在转速为 6 000 r/min 时机组的每一跨的中央突加不平衡质量的冲击(模拟叶片脱落现象),各轴承中轴颈涡动轨迹如图 8 所示.对这两种不同类型的外加激励,系统反映出的动力形态是不同的.

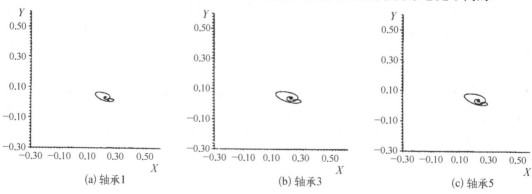

(a) 轴承1   (b) 轴承3   (c) 轴承5

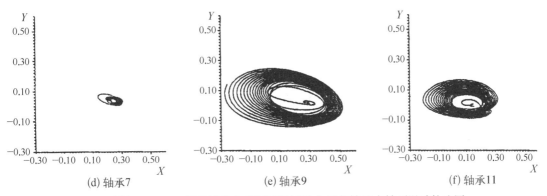

图 7 模拟 600 MW 机组试验台受"地震波"冲击后各轴承中轴颈涡动轨迹图

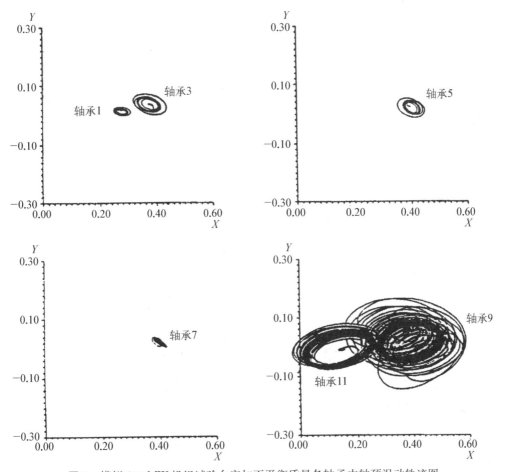

图 8 模拟 600 MW 机组试验台突加不平衡质量各轴承中轴颈涡动轨迹图

激振频率接近系统的一阶临界转速 ω_1 的外加激励极易激起系统的一阶固有振动,当激振强度达到一定幅值,会导致系统的失稳,尽管此时 Ω/ω_1 远小于 Ω_{st}/ω_1. 激振频率为工频的激励(如不平衡质量的强迫激振)在 Ω/ω_1 远小于 Ω_{st}/ω_1 情况下并不能引起系统的失稳,只是在冲击的瞬态会激发系统的各阶固有振动,过后异步涡动成分会逐渐衰减(图 9). 当然,如果此时 Ω/ω_1 接近于 Ω_{st}/ω_1,突加一定的不平衡质量的冲击可能会激起失稳.

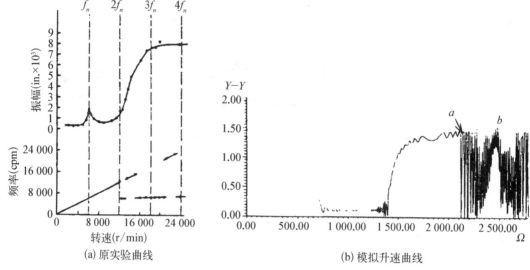

(a) 原实验曲线　　　　　(b) 模拟升速曲线

图 9　O. Pinkus 实验及其仿真

系统对上述两种激励的响应的另一个不同之处是:"地震波"的冲击是瞬时的和随机的,冲击消失之后,系统中某些固有频率与激振力的频率接近的环节(如该计算模型中的电机转子和励磁机转子)固有振动被激起. 其他固有频率远离激振频率的环节不受外加激励所激发,其振动只是受被激振的环节振动的传递影响而产生些较低频率的"随波逐流"的涡动. 同步激振的响应与上述情况不同,系统中受激振的环节均按同一频率在振动.

从图 8 中还可以得知,多跨转子轴承系统中轴颈在各轴承中水平方向的静平衡位置是不同的. 如只考虑垂直方向的安装杨度和变形问题会造成较大的计算误差[4,5].

4　实验验证

有关滑动轴承-转子系统稳定性研究的实验已有许多文献记载. Ocsar Pinkus 实验[6]是早期的和较为有代表性的. 本文作者用非线性仿真方法模拟了该实验. 另外,作者对本文 1.3 的算例作了对比性实验,仿真计算结果与实验结果较为吻合.

4.1　Ocsar Pinkus 实验的仿真

计算所用的参数根据[6]中记述而取定或推定:转子 $D=63.5$ mm;$L=1\,099.2$ mm;轴承 $d=50.8$ mm;$l=50.8$ mm;半联轴器直径 $D_1=76.2$ mm;厚度 $l_2=25.4$ mm;其与轴承处的过渡段的直径 $d_1=38.1$ mm;长度 $l_1=50.8$ mm. 且设:轴材密度 $\rho=7\,550$ kg/m^3;弹性模量 $E=2.1\times10^{11}$. $2\times160°$椭圆轴承;轴承半径间隙 $C=0.063\,5$(mm);润滑剂黏度 $\mu=0.018$ Pa·s;椭圆度 $m=0.1$. 设:在转子的 $0.25L$ 处有质量偏心质距 2.75×10^{-4} kg·m,相位角 $\varphi_1=0°$;在转子的 $0.5L$ 处有质量偏心质距 4.39×10^{-4} kg·m,相位角 $\varphi_2=120°$;在转子的 $0.75L$ 处有质量偏心质距 2.75×10^{-4} kg·m,相位角 $\varphi_3=240°$.

图 9 是模拟升速轴颈水平方向振动的峰-峰值曲线与原实验所得的曲线图的对比. 图

10 是图 9 上 a 和 c 处的轴颈涡动轨迹图. 系统一阶临界转速 $n_1 = 6\,200$ r/min, 失稳转速 $n_{st} = 11\,936$ r/min, 二阶临界转速 $n_2 = 23\,230$ r/min, 与实验数据很接近. 实验中得出的一个重要的现象也被清楚的仿真出来. 即: 转子进入失稳之后是进行低频涡动, 但在过二阶临界转速时恢复了同步涡动 (仿真计算得出此时振动中以同步涡动成分为主), 过了二阶临界转速之后又进入低频涡动状态.

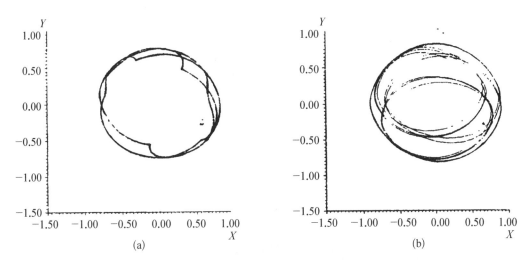

图 10 模拟 O. Pinkus 实验的升速曲线上 a 点和 c 点的轴颈涡动轨迹图

本算例清楚显示, 只有用计入不平衡的非线性仿真计算, 才可能得出如实反映转子系统在升速过程中的全部动力特性.

4.2 不平衡质量的分布对转子轴承系统稳定性影响的实验

试验台的转子参数 (图 2)、轴承参数如前所述, 采用 22♯ 透平油, 进油温度约为 35～40℃. 试验中不平衡质量的加载分两种情况:

情况 1: 中间圆盘加 1.4 g 不平衡质量;

情况 2: 1♯ 盘上加 1.1 g 不平衡质量, 相位角为 $\varphi_1 = 0°$; 3♯ 盘上加 0.9 g 不平衡质量, 相位角为 $\varphi_3 = 180°$.

试验中用 Bently NEVADA 108 数据采集仪采集数据, 由 Bently 的 ADRE3 软件完成数据的传输和分析; 同时采用小野 CF‐350 双通道 FFT 分析仪监控振动的频谱和 ARON Bs‐601 20 MHz 示波器监控涡动轨迹.

图 11 是 1♯ 轴承垂直方向振动的瀑布图, 从这些图中可以得出如下结论:

(1) 中心盘加不平衡质量的情况下系统的失稳转速较低 ($n_{st} \approx 6\,000$ r/min), 1♯ 和 3♯ 盘反相加不平衡质量的情况下系统的失稳转速较高 ($n_{st} \approx 6\,780$ r/min).

(2) 中心盘加不平衡质量的情况下系统的失稳较突然, 一旦出现半频涡动即产生油膜震荡. 而后一种情况下从出现半频涡动到产生油膜震荡有一个过程 (瀑布图上连续出现几个不大的半频峰值), 和仿真计算的结果是一致的. 产生这种现象的原因可归结为后一种不平衡质量分布情对系统有增稳的作用, 转子强迫振动的挠曲线与二阶振型接近使得一阶固有振动较难自激激发.

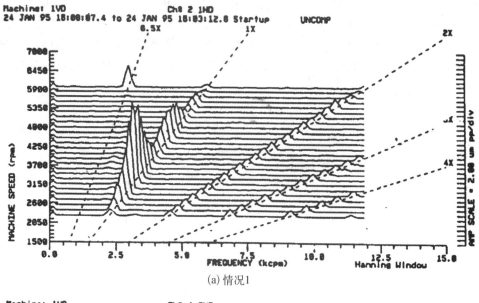

(a) 情况1

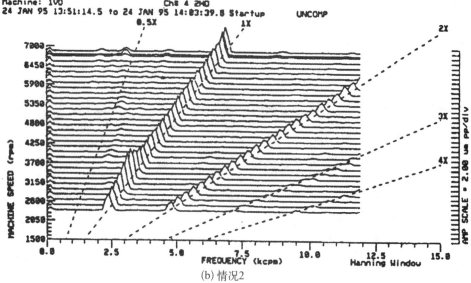

(b) 情况2

图 11　单跨三圆盘转子试验台实验在垂直方向振动的瀑布图

5　结论

本文以作者提出的关于大自由度的转子-轴承-基础系统的非线性动力学计算方法,详细地考察了若干有关转子轴承系统非线性的稳定性问题.仿真计算的结果与对比性的实验结果较为吻合,能相当全面地如实反映转子-轴承系统的动力性态,充分证明了该套计算方法的可靠性和实用性.为用计算机仿真的方法来模拟多跨转子轴承系统的实际运行状态从而进一步研究此运行状态,提供了一整套有效的计算手段.本文中有关转子轴承系统非线性的稳定性问题的探索和分析,可供广大研究者参考.

参 考 文 献

[1] 袁小阳. 轴系的稳定裕度、非线性振动和动力性能优化. 西安交通大学博士论文, 1994.
[2] 李志刚, 张直明. 不平衡质量大小和分布对转子轴承系统稳定性的影响. 全国第四届转子动力学学术讨论会论文集, 西安, 1995: 52-59.
[3] 李志刚. 多跨转子-轴承系统非线性动力学分析. 上海大学博士论文, 1995.
[4] 李志刚, 张直明. 大自由度的转子-滑动轴承系统非线性动力学分析. 上海大学学报, 1995, 1(6): 640-651.
[5] Hori Y, Kato T. Earthquack-Induced Instability of a Rotor Supported by Oil Film Bearing. Journal of Vibration and Acoustics, Transaction of the ASME, 1990, 112: 160-165.
[6] Ocsar Pinkus, Beno Sternlicht. Theory of Hydrodynamic Lubrication. McGraw-Hill Book Company, Inc, 1961: 450-451.

Non-linear Dynamic Analysis of Multi-span Rotor-Journal Bearing System (II)

Abstract: Non-linear stability problems of rotor-bearing systems are studied in detail in this paper, using the calculation method of non-linear dynamics of rotor-bearing-fundation system with high degree of freedom proposed by the authors. The problems treated consist of: the effects of unbalance magnitude and distribution on the stability of rotor system; investigation of the cause of non-divergence of the rotor orbit for certain rotors running at a speed higher than the linear stability threshold; study of the dynamic behavior of multi-span rotor systems under different kinds of large excitations. Correspondence between the simulated and the experimental results is satisfactory.

Partial EHL Analysis of Rib-roller End Contact in Tapered Roller Bearings*

Abstract: A partial EHL analysis was performed for the tapered rib/spherical roller end contact in tapered roller bearings. The average Reynolds equation, the elasticity equation and the pressure-viscosity relation were solved simultaneously. The effects of the surface roughness as well as the peculiar geometrical and kinematics parameters of the rib-roller end contact on the friction torque and film thickness were investigated. The optimal ratios of radius of curvature of roller end to rib face were deduced, which confirm the previous finding with the theory of smooth surfaces. The significant range of surface roughness, and the optimal surface roughness for the roller big end were obtained. It was found that asperity contacts extend into the outlet zone. The results are significant for the design of rib faces and roller ends. The theoretical treatment is validated by its good correlation with the existing experimental data for smooth surface contact.

Notation

a, b	semiminor and semimajor axes of contact ellipse respectively (m)	Ψ	angle formed by bearing axis and normal line of rib at nominal contact point
K	ellipticity parameter, $K = a/b$	ε	distance from the nominal contact point to inner race (m)
h_T	nominal film thickness (m)		
h_c	nominal central film thickness (m)	X	nondimensional coordinate in x, $X = x/b$
h_m	nominal minimum film thickness (m)		
h_{sm}	nominal minimum film thickness for smooth surface contact (m)	Z	nondimensional coordinate in z, $Z = z/a$
h_0	$h_c - \delta(0, 0)$ (m)	Y	nondimensional coordinate in film thickness direction, $Y = y/b$
δ	elastic deformation (m)		
Δ_1, Δ_2	the roughness of contact surface (m)	H_T	nondimensional nominal film thickness, h_T/R_x
σ_1, σ_2	surface roughness standard deviation (m)	n	rotation speed (nps)
σ	composite roughness standard deviation (m)	u_s	$0.5(u_{1(0)} + u_{2(0)})$, average entraining velocity (m/s)
r_i, r_o	inner and outer radii of rib (m)	U_1, U_2	dimensionless speed, u_1/u_s and u_2/u_s
r	large end radius of roller (m)	W_1, W_2	dimensionless speed, w_1/u_s and w_2/u_s
R_x, R_z	equivalent radii of curvature in x and z (m)	E_1, E_2	Young's moduli (N/m^2)
		μ_1, μ_2	Poisson's ratios
R_{x1}, R_{x2}	radii of curvature in x of rib and of roller end respectively (m)	E	effective elastic modulus
		p	pressure (N/m^2)
β_1	half cone angle	$\bar{p}$	mean hydrodynamic pressure (N/m^2)

---

* In collaboration with W. Wang and P. L. Wong. Reprinted from *Tribology International*, 1996, 29(4): 313–321.

$\bar{P}$	nondimensional mean pressure, p/E	η_0	ambient viscosity of lubricant
$\bar{P}_a$	nondimensional asperity contact pressure	τ_e	equivalent shear stress
λ	h_T/σ	$F(\tau_e)$	rheological function, for Newtonian, $=1$
$\bar{\sigma}$	σ/R_x	M	friction torque acting on rib around bearing axis (N·m)
N	number of asperities per unit area		
β	mean radius of curvature of asperity	$M_{r,a}$	asperity contact friction torque
W_T	total dimensionless load, $w_T/(ER_x^2)$	$M_{r,h}$	hydrodynamic friction torque
W_h	hydrodynamic load	α	half roller angle
W_a	asperity contact load	h_T	average gap
τ	shear stress	q_x, q_z	flow in x and z direction
$\bar{\eta}$	dimensionless viscosity, η/η_0	$\bar{q}_x, \bar{q}_z$	average flow

1 Introduction

Tapered roller bearings are widely used in automobiles and industrial machinery because they have the ability to carry not only heavy radial loads but also axial and combined loads. Due to the conical shape of the rollers, the forces acting on a roller from the inner and outer races differ slightly in direction, their resultant pushes the roller towards its bigger end and friction is thus induced between the rib surface and roller end. For heavy load or high speed conditions, lubrication problems in rib/roller end contact can be serious. Hence, many investigations of friction, film thickness and lubrication of rib/roller end contacts have been conducted[1-6].

Korrenn[1] started the study of the friction of rib/roller end contacts by carrying out experiments. The effects of geometric configuration and position of nominal contact point to favour lubrication were found by Jamison et al.[2] Dalmaz et al.[3,4] determined the optimal geometric configuration of rib/roller end contact by using an optical simulator rig and simulated the results with a hydrodynamic approach. It was pointed out that the elastic deformation of surfaces has to be considered in the theoretical study. With the development of elastohydrodynamic lubrication (EHL) theory, Zhang et al.[5] started an EHL analysis of the rib/roller end contact with the inclusion of special geometry and kinematics conditions, assuming Newtonian lubricant and isothermal conditions. Recently, non-Newtonian and thermal effects were included in an EHL study by Jiang et al.[6] It was found that under the investigated working conditions the non-Newtonian effect is very minimal and the thermal effect is somehow greater but only noticeable at high speed.

However, all previous research, both experimental and theoretical, was conducted with the assumption of perfectly smooth surfaces. In practice, the tapered roller bearings often operate with minimum film thickness of the order of a few tenths of a micron. Furthermore, the surface roughnesses of roller ends and rib surface are poorer than that of rollers and raceway surfaces. It has been pointed out by Kleckner and Pirvics[7] that the dimensions of the asperities of roller ends and the rib surface are of the same order of magnitude as the built up oil film. Figure 1 shows a typical surface roughness measurement of the roller end of a tapered roller bearing. It can be seen that the magnitude of roughness is of the same order of film thickness as those which were obtained experimentally[4] and theoretically[6]. In order to simulate as closely as possible the true physical situation, the surface roughness was thus included in the present analysis. Only one particular type of rib/roller end contact, tapered rib to spherical roller end, was selected in the study of roughness effects. This was not only because it is the most popular configuration but also it has the smallest film thickness. Hence, roughness effects are conspicuous with this configuration.

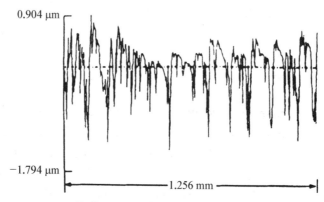

Fig. 1 Roller end surface roughness $\sigma=0.418\ \mu m$

During the last two decades, a number of partial EHL analyses have been carried out taking into account the surface roughness effect on the film thickness. One of the major advances is to use a stochastic method to describe lubrication between rough surfaces. Utilizing the stochastic approach, Patir and Cheng[8] put forward an average flow model for determining the effects of three-dimensional roughness on partial hydrodynamic lubrication. The effects of surface roughness are expressed by some flow factors which can be obtained through numerical flow simulation and the model was confirmed by References [9–11]. Hence, the average flow model was adopted in this work.

2 Theory

2.1 Geometry and kinematics

Figure 2 shows the rib/roller end contact geometry. It can be described by an equivalent sphere on a plane geometry as shown in Fig. 3. For a tapered rib to spherical end contact, the nominal film shape can be given by:

$$h_T = h_0 + \frac{x^2}{2R_x^2} + \frac{z^2}{2R_x R_z} + \delta(x, z) \tag{1}$$

where $\delta(x, z)$ is the elastic deformation.

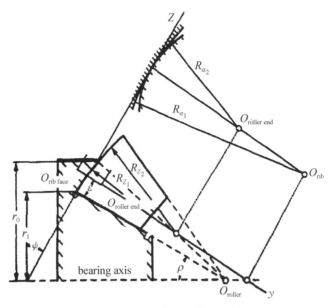

Fig. 2 Geometry of rib-roller end contact

When under operation, rolling, sliding and spinning exist simultaneously, as illustrated in Fig. 3. The velocities of the surfaces in the two principal directions are:

$$\begin{aligned}
u_1(z) &= u_{1(0)} + z \cdot \Omega_{1(0)} \\
u_2(z) &= u_{2(0)} + z \cdot \Omega_{2(0)} \\
w_1(x) &= w_{1(0)} - x \cdot \Omega_{1(0)} \\
w_2(x) &= w_{2(0)} - x \cdot \Omega_{2(0)}
\end{aligned} \quad (2)$$

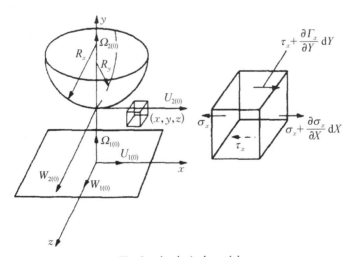

Fig. 3 Analytical model

2.2 Governing equations

Considering the force balance of a fluid element in the x-direction (Fig. 3), it can be shown that:

$$\frac{\partial \tau_x}{\partial y} = \frac{\partial p}{\partial x}, \quad \frac{\partial u}{\partial y} = \frac{\tau_x}{\eta} F(\tau_e) \tag{3}$$

By integration:

$$u = \frac{1}{\eta} \frac{\partial p}{\partial x} \int_0^y F(\tau_e) y \, dy + \frac{u_2 - u_1 - \frac{1}{\eta} \frac{\partial p}{\partial x} \int_0^h F(\tau_e) y \, dy}{\int_0^h F(\tau_e) \, dy} \int_0^y F(\tau_e) \, dy + u_1 \tag{4}$$

Hence, the flow rate per unit width of cross-section can be found as:

$$q_x = \int_0^h u \, dy = \frac{1}{\eta} \frac{\partial p}{\partial x} \left[\int_0^h \int_0^y F(\tau_e) y \, dy \, dy - \frac{\int_0^h F(\tau_e) y \, dy}{\int_0^h F(\tau_e) \, dy} \int_0^h \int_0^y F(\tau_e) \, dy \, dy \right]$$

$$+ \frac{u_2 - u_1}{\int_0^h F(\tau_e) \, dy} \int_0^h \int_0^y F(\tau_e) \, dy \, dy + \int_0^h u_1 \, dy \tag{5}$$

With a control volume (dx, dz, h) selected as shown in Fig. 4, the average flow in the x direction is expressed as:

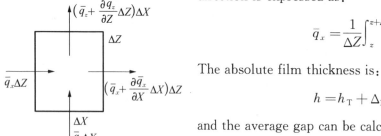

Fig. 4 Control volume

$$\bar{q}_x = \frac{1}{\Delta Z} \int_z^{z+\Delta z} q_x \, dz \tag{6}$$

The absolute film thickness is:

$$h = h_T + \Delta_1 + \Delta_2 \tag{7}$$

and the average gap can be calculated as:

$$\bar{h}_T = \int_{-h_T}^{\infty} (h_T + \Delta) f(\Delta) \, d\Delta \tag{8}$$

where $\Delta = \Delta_1 + \Delta_2$. For a Gaussian surface[11], the function f is:

$$f(\Delta) = \frac{1}{\sigma \sqrt{2\pi}} \cdot \exp\left(-\frac{\Delta^2}{2\sigma^2}\right) \tag{9}$$

Equation (5) can be written in terms of pressure flow factors ϕ_x, ϕ_y and shear flow factors ϕ_{sx}, ϕ_{sz}, their definition being given by Patir and Cheng[8]. The mean flow can then be expressed as:

$$\bar{q}_x = \phi_x \frac{1}{\eta} \frac{\partial \bar{p}}{\partial x} \left[\int_0^{h_T} \int_0^y F(\tau_e) y \, dy \, dy - \frac{\int_0^{h_T} F(\tau_e) y \, dy}{\int_0^{h_T} F(\tau_e) \, dy} \int_0^{h_T} \int_0^y F(\tau_e) \, dy \, dy \right]$$

$$+ \frac{u_2 - u_1}{\int_0^{\bar{h}_T} F(\tau_e) \, dy} \int_0^{\bar{h}_T} \int_0^y F(\tau_e) \, dy \, dy + \int_0^{\bar{h}_T} u_1 \, dy + \frac{u_2 - u_1}{2} \sigma \phi_{sx} \tag{10}$$

Similarly:

$$\bar{q}_z = \phi_z \frac{1}{\eta} \frac{\partial \bar{p}}{\partial z} \Big[\int_0^{\bar{h}_T} \int_0^y F(\tau_e) y \, dy \, dy - \frac{\int_0^{\bar{h}_T} F(\tau_e) y \, dy}{\int_0^{\bar{h}_T} F(\tau_e) \, dy} \int_0^{\bar{h}_T} \int_0^y F(\tau_e) \, dy \, dy \Big]$$

$$+ \frac{w_2 - w_1}{\int_0^{\bar{h}_T} F(\tau_e) \, dy} \int_0^{\bar{h}_T} \int_0^y F(\tau_e) \, dy \, dy + \int_0^{\bar{h}_T} w_1 \, dy + \frac{w_2 - w_1}{2} \sigma \phi_{sz} \quad (11)$$

Considering the mean flow balance, we have:

$$\frac{\partial (\rho \bar{q}_x)}{\partial x} + \frac{\partial (\rho \bar{q}_z)}{\partial z} = -\frac{\partial \bar{h}_T}{\partial t} \quad (12)$$

Substituting Equations (10) and (11) into (12), the general average Reynolds equation can be obtained and its dimensionless form is:

$$\frac{\partial}{\partial X} \Big[\phi_x \frac{\bar{\rho}}{\bar{\eta}} H_T^3 \Big(\frac{J_1^2}{J_0} - J_2 \Big) \frac{\partial \bar{P}}{\partial X} \Big] + \frac{1}{K^2} \frac{\partial}{\partial Z} \Big[\frac{\phi_z \bar{\rho}}{\bar{\eta}} H_T^3 \Big(\frac{J_1^2}{J_0} - J_2 \Big) \frac{\partial \bar{P}}{\partial Z} \Big]$$

$$= -\theta (U_2 - U_1) \frac{\partial}{\partial X} \Big[\bar{\rho} \frac{H_T^2}{\bar{H}_T} \Big(1 - \frac{J_1}{J_0} \Big) \Big] - \frac{\theta}{K} (W_2 - W_1) \frac{\partial}{\partial Z} \Big[\bar{\rho} \frac{H_T^2}{\bar{H}_T} \Big(1 - \frac{J_1}{J_0} \Big) \Big]$$

$$- U_1 \frac{\partial}{\partial X} (\bar{\rho} \bar{H}_T) - \frac{\theta}{K} W_1 \frac{\partial}{\partial Z} (\bar{\rho} \bar{H}_T) - \frac{U_2 - U_1}{2} \sigma \frac{\partial (\bar{\rho} \varphi_{sx})}{\partial X} - \frac{W_2 - W_1}{2} \sigma \frac{\partial (\bar{\rho} \sigma_{sz})}{\partial Z} \quad (13)$$

where

$$\theta = \frac{b \eta_0 u_s}{(E' R_x^2)}$$

$$J_0 = \int_0^1 F(\tau_e) \, d\bar{y}, \quad J_1 = \int_0^1 F(\tau_e) \bar{y} \, d\bar{y},$$

$$J_2 = \int_0^1 F(\tau_e) \bar{y}^2 \, d\bar{y} \quad (14)$$

For a Newtonian lubricant, $F(\tau_e) = 1$. Hence, J_0, J_1 and J_2 are equal to 1, 1/2, 1/3 respectively. Substitution into Equation (13), yields the same average Reynolds equation as derived by Zhu and Cheng[11].

2.3 Elastic deformation

In dimensionless form, the surface elastic deformation is expressed as:

$$\delta(X, Z) = \frac{2}{\pi} \iint_\Gamma \frac{\bar{P}(X, Z) + \bar{P}_a(X, Z)}{\sqrt{(X-X)^2 + (Z-Z)^2}} \, dX \, dZ \quad (15)$$

where Γ is the calculation zone. The mean asperity contact pressure is calculated by the method given by Greenwood and Tripp[12] as:

$$\bar{P}_a = K' F_{2.5}(\lambda) \quad (16)$$

where

$$K' = \frac{8}{15}\sqrt{2\pi}(N\beta\sigma)\sqrt{\sigma/\beta} \tag{17}$$

and

$$F_{2.5} = \frac{1}{\sqrt{2\pi}}\int_{\lambda}^{\infty}(s-\lambda)^2 \exp\left(-\frac{s^2}{2}\right)ds \tag{18}$$

The film thickness is given as:

$$H_T = H_0 + \frac{b^2 X^2}{2R_x^2} + \frac{a^2 Z^2}{2R_x R_z} + \delta(X, Z) \tag{19}$$

2.4 Viscosity

Among many isothermal viscosity-pressure formulae that have been proposed, the most accurate one for moderate pressure ranges is that of Roelands[13] and hence it is adopted here:

$$\bar{\eta} = \exp\{(\ln\eta_0 + 9.67)[-1 + (1 + 5.1e-9 \cdot E' \cdot \bar{P})]^{0.67}\} \tag{20}$$

2.5 Load-carrying capacity and traction torque

The total partial EHL load is the sum of the two components: hydrodynamic and asperity contact loads and can be given by:

$$\begin{aligned} W_T &= \iint_\Gamma [\bar{P}(X,Y) + \bar{P}_a(X,Y)]dXdY \\ &= \iint_\Gamma \bar{P}(X,Y)dXdY + \iint_\Gamma \bar{P}_a(X,Y)dXdY \\ &= W_h + W_a \end{aligned} \tag{21}$$

Similarly, the frictional force on the rib face of the inner-ring, F_{rib} also consists of two parts:

$$F_{rib} = F_{r,a} + F_{r,EHL} \tag{22}$$

The first part can be calculated as:

$$F_{r,a} = \mu_a W_a \tag{23}$$

The asperity contact friction coefficient μ_a is equal to 0.2, as suggested by Aihara[14] and Zhou and Hoeprich[15] for metal surfaces. The second part is given by:

$$F_{r,EHL} = abE' \iint_\Gamma \bar{\tau}_x \Big|_{Y=0} dXdZ \tag{24}$$

where

$$\bar{\tau}_x \Big|_{Y=0} = \frac{R_x \theta}{b}(U_2 - U_1)\frac{\bar{\eta}}{H_T J_0} - \frac{R_x}{b}H_T \frac{\partial P}{\partial X}\frac{J_1}{J_0} \tag{25}$$

The total traction torque of the rib to the bearing rotational axis is:

$$M = M_{r,a} + M_{r,EHL} \tag{26}$$

As derived by Aihara[14], the friction torque induced by the asperity contact can be expressed as:

$$M_{r,a} = \frac{\varepsilon}{D_a} \cdot F_{r,a} \cdot r_i \tag{27}$$

and similarly:

$$M_{r,EHL} = \frac{\varepsilon}{D_a} \cdot abE' \Big[\iint_\Gamma \bar{\tau}_X \Big|_{Y=0} (a \cdot Z + \varepsilon + R_i) \cos\Psi \, dX \, dZ \\ - \iint_\Gamma \bar{\tau}_Z \Big|_{Y=0} \cdot b \cdot X \cos\Psi \, dX \, dZ \Big] \tag{28}$$

where

$$\bar{\tau}_Z \Big|_{Y=0} = \frac{R_x \theta}{b}(W_2 - W_1) \frac{\bar{\eta}}{H_T J_0} - \frac{1}{K} \frac{R_x}{b} H_T \frac{\partial P}{\partial Y} \frac{J_1}{J_0} \tag{29}$$

$R_i = r_i/\cos\Psi$ and D_a is the average diameter of the roller.

2.6 Numerical method

For rib/roller end contact in tapered roller bearings, the induced pressure is generally smaller than 0.3 GPa. It is in fact not a heavily loaded situation. Hence, the average Reynolds equation can be solved by the successive over-relaxation (SOR) method and with good convergence. Further, the simultaneous system including the elasticity equation can be solved by the under-relaxation iterative method. The numerical procedure is similar to the one used by Zhu and Cheng[11].

3 Results and discussion

In the present work, a true configuration of an actual tapered roller bearing was used. The selected one was the Chinese made 7518E tapered roller bearing. Its rib/roller end contact is of the tapered rib to spherical end type. According to the result of Jiang et al.[6], this type of configuration comes with the smallest film thickness of all the others. The roughness effects are thus expected to be pronounced. The relevant material and geometrical parameters of the bearing and parameters of a typical bearing lubricant are listed in Table 1. The value of $N\beta\sigma$ is taken as 0.04 as recommended by Zhu and Cheng[11]. For most engineering surfaces, the value of σ/β is in the range of 0.0001 to 0.01[11]. In the present analysis, the value of σ/β was chosen to be 0.001 as a reference.

The measured roughness of the rib face is much better than that of the roller big end. This may be due to the fact that grinding the tapered rib surface is much more straightforward. Generally speaking, the roughness of the roller end is in the range of

0. 1 to 0. 45 μm. Table 2 shows two sets of results calculated for two different values of combined surface roughness, each of which were calculated for three cases: (i) two surfaces having the same roughness; (ii) rough rib face to smooth roller end; (iii) smooth rib face to rough roller end. It can be seen that the effects on film thickness and frictional torque are dominated by the combined roughess σ. The domination of roughness on different surfaces induces only small differences in the results. Hence, for the sake of convenience, the roughnesses of the two contact surfaces were set to be the same in this work. This makes the flow factors ϕ_{sx} and ϕ_{sz} equal to 0 and the governing equations (10) and (11) are thus simplified.

Table 1 Main parameters in the calculation

$E_1 = E_2 = 2.12 \times 10^{11}$ n/m^2, $\mu_{11} = \mu_2 = 0.3$, $\eta = 2.83 \times 10^{-2}$ Pa·s
$r = 0.009\ 739$ m, $r_o = 0.060\ 5$ m, $r_i = 0.057\ 486$ m, $\alpha = 2°$
$\beta_1 = 11.64°$, $\varepsilon = 0.002\ 735$ m, $N \cdot \beta \cdot \sigma = 0.04$, $\sigma/\beta = 0.001$
$W = 80$ N, $n = 1\ 000$ r/min

Table 2 Effects of different roughness combinations: $R_{x2}/R_{x1} = 0.8$, $n = 1\ 000$ r/min, $W = 80$ N

	$\sigma = 0.353\ 6$ μm			$\sigma = 0.424\ 3$ μm		
	$\sigma_1 = \sigma$ $\sigma_2 = 0$	$\sigma_1 = \sigma_2$ $= 0.25$ μm	$\sigma_1 = 0$ $\sigma_2 = \sigma$	$\sigma_1 = \sigma$ $\sigma_2 = 0$	$\sigma_1 = \sigma_2$ $= 0.3$ μm	$\sigma_1 = 0$ $\sigma_2 = \sigma$
h_m/μm	0.828 1	0.824 21	0.812 4	0.850 4	0.840 8	0.833 2
h_c/μm	0.931 5	0.926 58	0.917 7	0.953 1	0.947 76	0.940 3
M/(N·mm)	1.03	1.076	1.096	2.5	2.553	2.617

The validity of the theoretical treatment has been checked by comparing the numerical results calculated with a very small value of σ to the existing experimental data. Numerical calculations were performed under the same conditions for the experiments as carried out by Gadallah and Dalmaz[4]. Figure 5 shows that the numerical results correlate well with the experimental data.

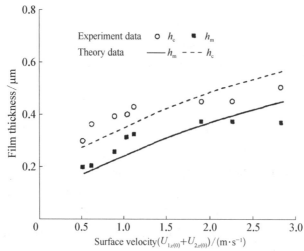

Fig. 5 Comparison between theoretical results for smooth contact and Dalmaz experimental data[4]

In the present analysis, the orientation of roughness is assumed to be isotropic to simplify the consideration of spinning motion in the contact. As results of a typical partial EHL analysis, Fig. 6 shows the mean EHL pressure, mean roughness contact pressure and nominal film thickness distribution along the centre axis of the contact ellipse. The effects of roughness are thicker nominal film thickness and smaller mean EHL pressure. In fact, this is an opposite trend to the results obtained by Zhu and Cheng[11]. This is attributable to the difference in the values of the ellipticity parameter K. In Reference[11], the value of K is about 4 while for a typical rib/roller end elliptical contact, K is nearly 0.35. Thus the contact ellipticity parameters have an influence on the effects of the roughness. This was also observed by Zhu and Cheng[11]. The small hump at the right bottom of Fig 6 shows the asperity contact pressure. It locates in the minimum film thickness area. As expected, the maximum asperity contact pressure occurs at the minimum film thickness point. It it noted that the asperity contacts happen not only in the lubricated contact region but also at the outlet where the lubrication film ruptures.

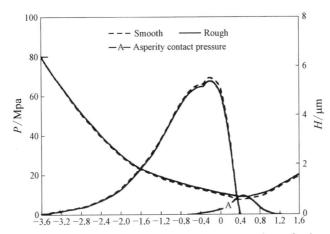

Fig. 6 Typical PEHL results for rib-roller end contact at moderate load rotation speed

The roughness effects on the film thickness and traction torque at moderate speed and load are shown in Fig 7. When $h_{sm}/\sigma > 3$, the roughness effects are small.

While with $h_{sm}/\sigma < 3$, the roughness effects become significant and the total traction torque M, which is the sum of hydrodynamic and asperity contact traction torque, increases rapidly. When $h_{sm}/\sigma < 2$, the total traction is dominated by the asperity contact traction. This leads to the ineffectiveness of lubrication in the contact as a result of the increase in the number of asperity contacts. Hence, a suitable value of h_{sm}/σ must be larger than 2. However, a large value of h_{sm}/σ which means a very smooth surface is not necessary. Figure 7 shows a decreasing trend in film thickness with the increase in h_{sm}/σ values. Therefore, the surface roughness of the roller big end is recommended to be in the range of 0.1 to 0.3 μm for the investigated case, which is equivalent to 5 to 2 in h_{sm}/σ.

Figure 8 shows the effects of the ratios of radius of curvature of roller end to rib

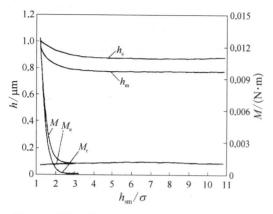

Fig. 7 The effects of h_{sm}/σ on friction torque and film thickness

Fig. 8 Effects of curvature radius ratio on friction torque and film thickness

face. It is noted that the curves shown in Fig 8 demonstrate the same trends as those in the case of smooth surface contact described by Zhang et al.[5] and Jiang et al.[6]. This confirms their finding that the optimal ratios of curvature radius of spherical roller end to tapered rib face is in the range of 0.6 to 0.8. For values less than 0.6, the radius of the roller big end becomes too small so that it is difficult to control the location of the nominal contact point at the middle of the rib face. Even worse, the contact point may be located in the chamfer or out of rib due to an inaccuracy in manufacturing[2].

Figure 9 shows the effects of rotation speed on the film thickness and friction torque. Slower speed results in a drop of film thickness and the friction torque is thus increased. For the high speed range, even though the film thickness becomes larger, the total friction torque still goes up slightly due to the increase in the hydrodynamic friction. Hence the optimum speed in the investigated case ought to be 1 500 r/min.

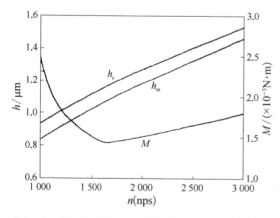
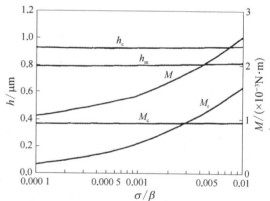

Fig. 9 Speed effects on friction torque and film thickness

Fig. 10 Effects of σ/β on friction torque and film thickness

σ/β was taken to be 0.001 in all of the above calculations. In fact, its value is governed by the manufacturing technique employed and is likely to vary from 0.000 1 to 0.01[11]. The effects of σ/β on the film thickness and friction torque are shown in Fig 10. It is seen that when σ/β is larger, the curvature of asperities is smaller for a constant value of surface roughness and thus the asperity contact pressure is higher. This results in a higher friction torque. Its effect on the film thickness is negligible.

4 Conclusions

A partial EHL analysis of the tapered rib/spherical roller end contact in tapered roller bearings has been completed. By adopting Patir-Cheng's average flow model and Greenwood and Tripp's asperity contact model, the effects of surface roughness on film thickness and friction torque were investigated. Based on the numerical results, the following conclusions can be drawn:

(1) By choosing a very small value for σ, results of the present analysis for rough surface contact correlate well with the experimental data of smooth surfaces.

(2) The optimal values of the ratio R_{2x}/R_{1x} for tapered rib/spherical roller contact were deduced to be in the range of 0.6 to 0.8 which confirms findings with the theory of smooth surfaces.

(3) The effects of surface roughness are significant when $h_{sm}/\sigma < 3$, and become dominant when $h_{sm}/\sigma < 2$. The increase in the amount of asperity contacts leads to ineffectiveness of lubrication such that the friction torque increases drastically. The roughness effects can only be totally neglected when $h_{sm}/\sigma > 5$.

(4) The optimal surface roughness for the roller big end is in the range of 0.1 μm to 0.3 μm.

(5) The effect of σ/β on the film thickness is almost negligible but not the friction torque which becomes larger with the increase in σ/β. Hence, appropriate manufacturing techniques should be selected for the finishing of the roller big end surface such that larger curvature of asperities, β, can be achieved.

Acknowledgements

The authors thank the City University of Hong Kong for financial support for the project. Thanks are given to Prof. X. Jiang for his helpful advice.

References

[1] Korrenn H. Gleitreibung und Grenzbelastung an den Bordflachen von Kegelrollenlagern. Fortschritt Berichte V. D. I. Zeitschrift, 1967, Ser. 1, 11.

[2] Jamison W. E., Kauzlarich J. J. and Mochel E. V. Geometric effect on the rib-roller contact in tapered roller bearings. ASLE Trans., 1976, 20, 79-88.

[3] Dalmaz G., Tessier J. F. and Dudragne G. Friction improvement in cycloidal motion contact: rib-roller end contact in tapered roler bearings. Proc. 7th Leeds-Lyon Symp. on Tribology, VII(iii), 1980, 175.

[4] Gadallah N. and Dalmaz G. Hydrodynamic lubrication of the rib-roller end contact of a tapered roller bearing. Trans. ASME J. Tribol, 1984, 106, 265-274.

[5] Zhang Z. , Qiu X. and Hong Y. EHL analysis of rib-roller end contact in tapered roller bearings. STLE Trans. , 1988. 31,461-467.

[6] Jiang X. F. , Wong P. L. and Zhang Z. Thermal non-Newtonian EHL analysis of rib-roller end contact in tapered roller-bearings. ASME/STLE annual Conference, Hawaii, USA, October 1994.

[7] Kleckner R. J. and Pirvics J. Spherical roller bearing analysis. ASME Trans. J. Lubric. Technol. , 1982, 104, 99.

[8] Patir N. and Cheng H. S. An average flow model for determining effects of three-dimensional roughness on partial hydrodynamic lubrication. ASME Trans. J. Lubric. Technol. , 1978, 100, 12-17.

[9] Elrod H. G. A general theory for laminar lubrication with Reynolds roughness. ASME Trans. J. Lubr. Technol. , 1979, 101, 8-14.

[10] Tripp J. H. Surface roughness effects in hydrodynamic lubrication: The flow factor method. ASME Trans. J. Lubr. Technol. 1983, 105, 458-465.

[11] Zhu D. and Cheng H. S. Effect of surface roughness on the point contact EHL. ASME Trans. J. Tribol. , 1988, 110, 32-37.

[12] Greenwood J. A. and Tripp J. H. The contact of two nominally flat rough surfaces. Proc. Inst. Mech. Eng. Part 1, 1970, 185, 48, 625-633.

[13] Roelands C. J. A. Correlational aspects of the viscosity-temperature-pressure relationship of lubricating oils. V. R. B. Groningen, Netherlands, 1966.

[14] Aihara S. A new running torque formula for tapered roller bearing under axial load. ASME Trans. J. Tribol. , 1987, 109, 471-478.

[15] Zhou R. S. and Hoeprich M. R. Torque of tapered roller bearing. ASME Trans. J. Tribol. 1991, 113, 590-597.

不平衡质量的大小和分布对柔性转子-轴承系统稳定性的影响*

摘　要：本文详细考察了滑动轴承支承的转子不平衡质量的大小及分布对系统稳定性的影响.大量计算结果除支持了有关文献中关于刚性转子不平衡质量的大小对系统稳定性影响的结论外,更重要的还发现：对柔性转子而言,不平衡质量的分布也是一个重要的影响因素,其程度可能要超过不平衡质量的大小的因素.计算结果表明：当转子不平衡量所引起的强迫振动的挠曲线与转子的一阶振型相吻合或接近时,对转子的稳定性是不利的.而当转子强迫振动的挠曲线与转子的二阶或三阶振型相吻合或接近时,增加不平衡量会提高系统的稳定性.

0　前言

高速旋转机械的转子轴承系统的稳定性分析,目前通常采用以线性理论为基础的计算方法.线性失稳转速是通过求解独立于系统动力学方程中之非齐次项的齐次线性方程组得到,故无法计入转子实际存在的不平衡质量造成的同步强迫振动的影响.许多实验和学者的研究[1]表明,不平衡质量的大小对系统稳定性有明显的作用.因受计算速度和计算规模的限制,目前用仿真方法考察不平衡质量对系统稳定性的影响大多局限于刚性转子或单质量弹性转子-轴承系统,对柔性转子上不平衡质量的分布对系统稳定性的影响的研究尚未见文献记载.

本文作者以研究滑动轴承支承的多跨转子在大激励下的动力性态为目标,提出了一套新的关于大自由度的转子-轴承-基础系统非线性动力学计算方法[2].以该方法实现的软件可高效而又精确对具有的 300 个离散节点和 12 个轴承支承的转子-轴承-基础系统进行非线性动力学性态计算分析.本文用这套方法考察了滑动轴承支承转子不平衡质量的大小和分布对系统稳定性的影响.受文章篇幅所限,本文对计算所采用的方法仅作一简单的介绍.

1　非线性仿真计算方法简介

转子-轴承-基础系统的力学模型如图 1 所示.系统的广义动力学方程可用下式表示：

$$M\ddot{X}_t + C\dot{X}_t + KX_t = F_t \qquad (1)$$

式中　M——转子离散系统方程的总质量矩阵；
　　　C——转子离散系统方程的总阻尼系数矩阵；

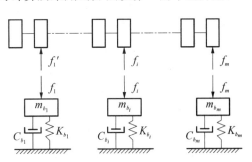

图 1　转子-轴承-基础系统力学模型

---

* 本文合作者：李志刚.原发表于《振动与冲击》,1997,16(1)：15－19,94.

K——转子离散系统方程的总刚度矩阵;

X_t——转子离散系统方程 t 时间的位移矢量;

F_t——转子离散系统方程 t 时间的力矢量;

f_i——非线性油膜力,$f_i = f_i(x, y, \dot{x}, \dot{y})$.

将式(1)在时间域上离散和积分. 对式(1)的时间积分法有显式和隐式两大类,其形式分别为:

显式:
$$\widetilde{M} X_{t+\Delta t} = R_t \tag{2a}$$

隐式:
$$\widetilde{M} X_{t+\Delta t} = R_{t+\Delta t} \tag{2b}$$

显式时间积分法中主要有中心差分法,隐式的时间积分法有 Houbolt 法、Willson θ 法和 Newmark β 法[3],作者详细地分析对比了上述各种方法的特点,首次提出了由中心差分法和 Houbolt 法组合成的预估计-矫正时间积分法. 计算公式分别给出如下:

中心差分法:
$$\widetilde{M} = \left[M + \frac{\Delta t}{2} C \right]$$
$$R_t = \Delta t^2 F_t - [\Delta t^2 - 2M] X_t - \left[M - \frac{\Delta t}{2} C \right] X_{t-\Delta t} \tag{3}$$

Houbolt 法:
$$\widetilde{M} = \left[2M + \frac{11 \Delta t}{6} C \right]$$
$$R_{t+\Delta t} = \Delta t^2 F_{t+\Delta t} + [5M + 3\Delta t C] X_t - \left[4M + \frac{3}{2} \Delta t C \right] X_{t-\Delta t} - \left[M + \frac{\Delta t}{3} C \right] X_{t-2\Delta t} \tag{4}$$

预估计-矫正的计算步骤如下:

(1) 由式(3)求得 $X'_{t+\Delta t}$,$\dot{X}'_{t+\Delta t}$

(2) 求得 $R'_{t+\Delta t}$

(3) 由式(4)求得 $\widetilde{X}_{t+\Delta t}$,$\dot{\widetilde{X}}_{t+\Delta t}$

(4) 迭代收敛判据 $\delta = | X'_{t+\Delta t} - \widetilde{X}_{t+\Delta t} | \leqslant \delta^*$

(5) 如果 $\delta \geqslant \delta^*$,松弛法修正 $X'_{t+\Delta t}$,$\dot{X}'_{t+\Delta t}$,重复步骤(2)~(4),当 $\delta \leqslant \delta^*$,令 $X_{t+\Delta t} = \widetilde{X}_{t+\Delta t}$,进入下一个 Δt 的计算步.

采用了由中心差分法和 Houbolt 法组合成的预估计-矫正时间积分法,使式(1)在时域上每个 Δt 上积分的迭代次数大为减少. 此外,式(2)中 $\widetilde{M}$ 是 $(4n \times 4n)$ 的矩阵,当转子离散节点数 n 较大时,如直接求 $\widetilde{M}$ 的逆阵会受到计算机容量的限制. 即便计算是可以进行的,计算速度和计算精度也不尽如人意,难以使大规模和高精度的计算得以实现. 本文作者首次成功地将块三对角矩阵追赶法用于式(2)的求解上,并设法将式(2)在 X 和 Y 两个方向上解耦,使原来对$(4n \times 4n)$ 的矩阵的求逆问题简化为对 n 个(2×2)的块矩阵的求逆,极大地加快了求解过程,同时使计算精度有一个质的提高. 对计算中每一 Δt 上需反复用到的每个轴承的非线性油膜力值的获取上,本文采取调用非线性油膜力数据库[4]的数据来替代费时的数值求解雷诺方程. 上述这些方法的采用,使大自由度的多跨转子-轴承系统的非线性仿真计算能在较大规模上进行和获得较高的计算精度.

2 不平衡质量的大小和分布对系统稳定性的影响

不平衡质量的存在对转子轴承系统有增稳作用,这种现象在许多实验已被多次观察到. 许多学者对这种现象进行了较为深入的研究,文献[1]关于刚性转子不平衡质量的大小对转子轴承系统的稳定性的影响的论证是近期在这方面较为卓有成效的研究.

本文作者以非线性仿真的方法,以文献[1]中算例的参数为参考,用均质光轴为计算模型(图2),考察了不平衡质量的大小以及分布对系统稳定性的影响. 作为对比,本文给出了相应的用作者编制的 ROTOR[5] 软件计算所得线性计算结果 n_1 和 n_{st}.

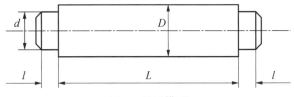

图 2 转子模型

转子和轴承的参数为:转子质量 $2m=870$ kg;$D=200$ mm;$L=2\,858.62$ mm. 轴承内径 $d=100$ mm;轴承长度 $l=100$ mm;轴承半径间隙 $C=0.1$ mm;轴承为 360°圆轴承. 润滑剂黏度 $\mu=0.018\,2$ Pa·s;且设:轴材密度 $\rho=7\,550$ kg/m³;弹性模量 $E=2.1\times10^{11}$. 线性计算所得,$n_1=1\,830$ r/min;$n_{st}=3\,500$ r/min. 非线性仿真计算时 L 段的离散数为20. 无量纲不平衡质量偏心距定义为:$\varepsilon=e/C$.

图 3 是设在转子的每个节点上突加均匀的不平衡质量偏心距 e' 的振动曲线. 图中横坐标 r 代表转次. 因其转子强迫振动的挠曲线与转子的一阶振型一致,增加不平衡质量会使系统的失稳转速下降. 当取 $\varepsilon=0.5$,系统失稳转速为 3 200 r/min. 在这个转速下突加不平衡质量激振出的半速涡动成分随转次的增加逐渐减少,使振动趋稳定. 但当取 $\varepsilon=0.5$,而转速高于此转速时,半速涡动成分随转次的增加而增加,使振动发散. 当取 $\varepsilon=0.8$,系统失稳转速也仅为 3 300 r/min(图 3b),低于线性失稳转速. 进一步加大取 $\varepsilon=3.0$ 和 $\varepsilon=5.0$ 计算也没能提高系统的失稳转速.

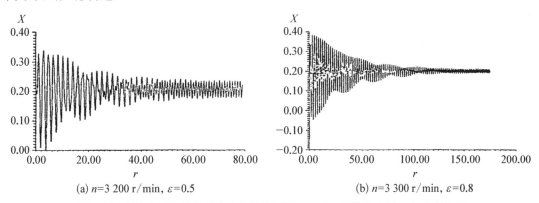

图 3 恒定转速下转子各节点突加均布不平衡质量,轴颈在水平方向响应曲线

图 4 是设在转子的中央节点突加不平衡质量的响应曲线. 此算例中转子强迫振动的挠曲线与转子的一阶振型接近但不完全一致,不平衡响应对系统的失稳转速的影响要比转子

强迫振动的挠曲线与转子的一阶振型完全一致的算例要略好.但不平衡量较大时,增加不平衡量的增稳作用还不明显,失稳转速还未能高于线性理论计算的失稳转速 n_{st}. 在转速 $n=3\,500$ r/min 时系统振动是发散的,不过发散的速度随不平衡量的增加而趋缓. 图 5 是 $n=3\,500$ r/min,转子中间节点突加不平衡质量的偏心距 $e=4.6$ mm 的响应曲线和峰-峰值曲线. 峰-峰值曲线中有"拍振"现象,"拍振"的幅值在逐渐衰减,表明冲击所激发出的半速涡动衰减的过程中,系统的振动状态波动于稳定和发散之间.

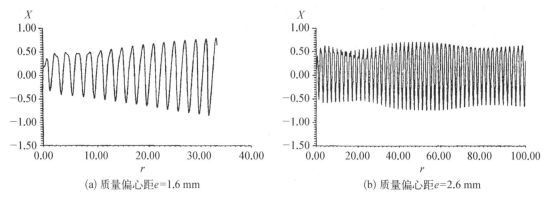

图 4 转速 $n=3\,500$ r/min 时中心节点突加质量偏心距,轴颈在水平方向响应曲线

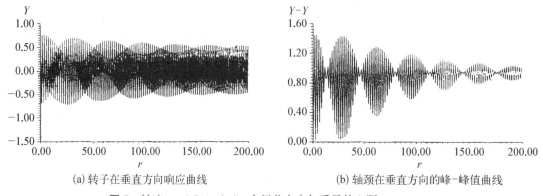

图 5 转速 $n=3\,500$ r/min,中间节点突加质量偏心距 $e=4.6$ mm

图 6 是设在转子的不同位置有不平衡质量的模拟升速时轴颈振动在 X 方向的峰-峰值曲线图,图中横坐标是角速度 Ω. 图 6(a)中强迫振动的挠曲线与转子的一阶振型接近,图 6(b)中强迫振动的挠曲线与转子的三阶振型接近,图 6(c)中强迫振动的挠曲线是一条空间曲线,在两个坐标方向上分别与转子的二阶和三阶振型接近. 从图中可以得知:不平衡质量引起强迫振动的挠曲线与转子的二阶以上的振型接近时,不平衡质量的存在和增加能有效地提高系统的稳定性. 而当强迫振动的挠曲线与转子的一阶振型接近时,不平衡质量的存在和增加对提高系统的稳定性无益. 尽管在图 5 的算例上看到了较为微弱的增稳效应,但在用相同的参数和计算条件进行模拟升速时,过一阶临界转速时会造成振幅太大的困难.

图 7 是在不同转速下突加不平衡量质量的响应曲线在 X 方向的峰-峰值曲线图,转子上不平衡质量的大小和分布同图 6(c)的算例中的相同. 其中 $n=3\,900$ r/min,峰-峰值曲线图中"拍振"的幅值逐渐衰减,系统逐渐趋稳;$n=4\,000$ r/min,峰-峰值曲线图中"拍振"的幅值保持不变,系统的振动状态波动于稳定和发散之间(类似现象在有的实验中曾有所发现);

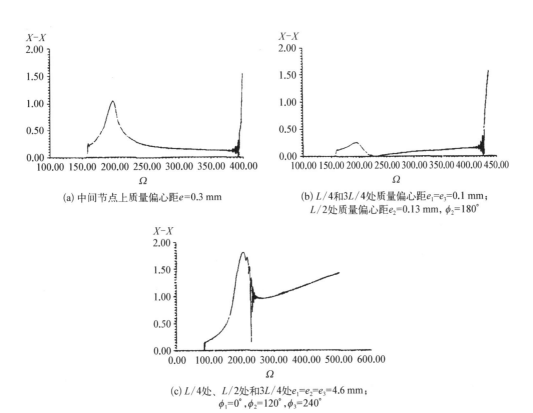

图 6 具有不同不平衡质量分布的转子在模拟升速过程中轴颈在水平方向振动的峰-峰值曲线

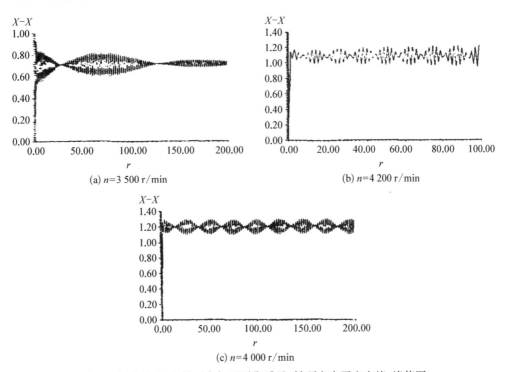

图 7 恒定转速下,转子突加不平衡质量,轴颈在水平方向峰-峰值图

$L/4$: $e_1 = 4.0$ mm, $\varphi_1 = 0°$
$L/2$: $e_1 = 4.6$ mm, $\varphi_2 = 120°$
$3L/4$: $e_1 = 4.0$ mm, $\varphi_3 = 240°$

$n=4\,200$ r/min,峰-峰值曲线图中"拍振"的幅值逐渐增大,系统逐渐趋于发散.

3 结论

对柔性转子而言:强迫振动的挠曲线与转子的一阶振型接近时,不平衡质量的存在和增加对提高系统的稳定性无益;不平衡质量引起强迫振动的挠曲线与转子的二阶以上的振型接近时,不平衡质量的存在和增加能有效地提高系统的稳定性.造成这种现象的原因是,转子-轴承系统的失稳转速是和系统的一阶固有振动有关的(线性失稳转速是系统特性值中实部大于零的根,往往是对应于系统的一阶固有复频率根),当强迫振动的挠曲线与转子的一阶振型接近时,容易激起系统的一阶振动,此时强迫振动力的增加是有限的,不足以抑制激发的油膜涡动.而当强迫振动的挠曲线与转子的二阶以上振型接近时,容易激起系统的二阶以上振动,一阶振动的成分在系统振动中不占主导地位,与一阶固有振动有关失稳现象不易发生或即便发生也容易被抑制.从实际转子来看,转子的一阶动平衡是做得较好的,残余的不平衡质量分布往往对二阶或二阶以上的振动作用才较明显.这种分布的残余的不平衡质量对转子轴承系统的稳定性提高起有益的作用.这是线性分析方法所无法考虑的.这也可能是某些实际转子能安全地工作于线性失稳转速之上的原因之一.

参 考 文 献

[1] 袁小阳.轴系的稳定裕度、非线性振动和动力性能优化.西安交通大学博士论文,1994.5.
[2] 李志刚,张直明.多跨转子轴承系统非线性动力学分析,PART1.待发表
[3] Bathe & Wilson. Numeerical Methods in Finite Element Analysis. John Wlley Publishers.
[4] 王文.非线性油膜力数据库研究.上海工业大学硕士论文,1993.1.
[5] 张直明.复套式转子-轴承系统固有复频率计算.上海工业大学学报,1990.

The Effect of Magnitude and Distribution of Unbalance on Stability of Flexible Rotor Journal Bearing System

Abstract: This paper studies in detail the effect of magnitude and distribution of rotor unbalance on the stability of rotor-journal bearing system. Systematically calculated results not only support the conclusion in certain literature about the effect of magnitude of unbalance of rigid rotor on system stability, but also show that for flexible rotor the distrbution of unbalance is also an important factor, possibly more important than the factor of unbalance magnitude. It is concluded that it will be unfavorable for system stability if the dynamic deflection shape of forced vibration caused by unbalance mass conforms with the first vibration mode of the rotor, whereas the system stability will be improved by properly increasing unbalance when the dynamic deflection shape of forced vibration conforms with the second or higher vibration mode of the rotor.

Approximate Tangent Plane Method for Calculating Surface Deformation in Elastic Contact Problems[*]

Abstract: A new numerical method for constructing a pressure distribution to calculate surface elastic deformation caused by normal contact pressure is developed in this paper. The pressure distribution over one of nonequidistant rectangles is fitted by an approximate tangent plane(ATP), which is formed by five pressure samples. Because the pressure distribution could be expressed as an one order linear polynomial, the iterative expression of elastic deformation deduced by this method is simple, and the numerical accuracy is higher.

Nomenclature

a	Hertzian contact radius, m	E_{rr}	$(d_{cal} - d_{th})/d_{th}$
a_i	half spacing of rectangular in x direction, m	F_{ij}^{kl}	influence coefficient
		$n \times m$	number of actual grid elements
A	contact area, m^2	$N \times M$	number of total grid elements
$A_{i,j}$	area of rectangular "i, j", m^2	$p_{i,j}$	pressure sample at point (x_i, y_j), Pa
b_j	half spacing of rectangular in y direction, m	R	radius of the ball, m
		x, y	coordinates in Cartesian coordinate system
$c_1 \sim c_3$	coefficients of ATP fitting polynomial		
d_{th}	theoretical values of the deformation, m	$s(x, y)$	original distant of the surface to the flat, m
$d(x_i, y_j)$	surface deformation at point (x_i, y_j), m	Δh_{max}	surface flatness in the contact region, m
E'	equivalent Young's modulus, Pa	ν_1, ν_2	Poisson's ratio
E_1, E_2	Young's modulus, Pa		

$$E' = \frac{2}{(1-\nu_1^2)/E_1 + (1-\nu_2^2)/E_2}$$

1 Introduction

The numerical solution of surface deformation in three-dimensional elastic contact problems is the basis of analysis of non-Hertzian contact problems and two-dimensional EHL problems. Most numerical methods of surface deformation calculation subjected to a

---

[*] In collaboration with Chen Xiaoyang and Shen Xuejin. Reprinted from *Journal of Shanghai University*, 1997, 1(2): 139-144, 154.

normal pressure distribution are based on the Boussinesq's solution[1]

$$d(x, y) = \frac{2}{\pi E'} \iint_\Omega \frac{p(\xi, \eta) \mathrm{d}\xi \mathrm{d}\eta}{\sqrt{(\xi-x)^2 + (\eta-y)^2}}. \tag{1}$$

There are three difficulties in numerical integration of the solution Eq. (1). The first arises from the weak singularity at $(x, y) = (\xi, \eta)$. The second is the required large storage capacity of the computer. If deformation at all centric points of the grids for an $n \times m$ nonequi-distant rectangular grid configuration are calculated, the storage space needed will be $n^2 \times m^2$, which is large even for up-to-date powerful personal computers. The third difficulty is the considerable amount of repeat calculations in three-dimensional elastic contact problems solved by an iteration procedure, for the deformations must be evaluated in each iteration.

Presently, the following techniques are used to overcome the above mentioned difficulties. Firstly, on the grid of the weak singularity (or even on all grids) a polynomial function is used to approximate the actual pressure distribution so that a closed-form analytical solution of the integration can be obtained. Secondly, the technique for dealing with the large storage capacity required is to control the total number of divided grids and to use the symmetry of the contact problem. Thirdly, in order to reduce the amount of the calculation, Eq. (1) is expressed as a linear combination of the pressure samples

$$d(x_k, y_l) = \frac{2}{\pi E'} \sum_l \sum_k F_{ij}^{kl} p_{i,j} \tag{2}$$

The coefficients F_{ij}^{kl} in Eq. (2) are defined as the deformations occurring at the point (x_k, y_l) due to a unit pressure $P_{i,j}(x, y)$ over the grid $A_{i,j}$, and are depend only on the geometric factors of the grids. Therefore they are called the influence coefficient. All the values of F_{ij}^{kl} need to be computed only once and the deformations can be calculated repeatedly by Eq. (2).

For many researchers it has been the main objective during the past two decades to look for a simple polynomial so that the real pressure distribution on a grid can be fitted accurately and the surface deformation can be calculated more quickly and involve less computer storage. Dowson and Hamrock[2] divided the contact area into equidistant rectangular grids and assumed the pressure on each grid to be a constant value which is a pressure sample at the center of the grid. Chang[3] divided the contact area into nonequidistant rectangular grids and also assumed that the pressure on each grid can be replaced by a constant value. However, the value is the mean of four pressure samples at the corners of the grid. In order to correct deformation error caused by the mean pressure on the surrounding grids, the distance between the deformation point and the grids was modified. In these two methods, the pressure distribution was replaced by blocks of uniform pressure, an analytical expression was used as the result of integration and the deformation of every point was expressed as a liner combination of the pressure samples. Biswas and Snidle[4] used an equidistant rectangular grid configuration and dealt with singularity by replacing the pressure distribution in that region by plane triangular surfaces. In Ref. [5] solution, the contact area was divided into nonequidistant rectangular grids and the pressure distribution was replaced by overlapping pressure "pyramids" in which the pressure was assumed to be linearly distributed in both directions. The method is therefore based upon the assumption that the pressure distribution corresponding to the overlapping volumes is balanced by the volumes from the "pyramids" in the pressure distribution. Biswas and Snidle[4] were the first to interpolate the pressure distribution in the singularity region by nine-node elements and biquadratic polynomials. Only the exact expression obtained for the rectangular region surrounding the singularity point was available, and an analytical form for the rest of the loaded regions was mentioned but not given. Using a piecewise biquadratic polynomial to approximate pressure distribution and adopting an influence coefficient on the whole domain, and intruducing the second equation of the two equivalent polynomials employed by Biswas and Snidle, Hou, Zhu and Wen[6], Jeng and Hamrock[7] obtained Eq. (2). Lin and Chu[8] also based on the first equation of Biswas and Snidle, obtained an analytical expression similar to Eq. (2).

Apart from the use of different pressure interpolation polynomials, all the methods mentioned above are conceptually the same. That is, the pressure interpolation polynomials are based only on the samples upon the approached area itself and not on those of the neighboring ones. The polynomials, and the values derived, are quite different from the real solution on the edges of the grids. Undoubtedly, for a given configuration, the use of a higher-order function for pressure interpolation yields a better numerical accuracy. However, this will not reduce the computational cost and storage requirement. That is the main reason why the simple uniform pressure interpolation is widely used, even in the MLMI method[9] although the fitting accuracy is not satisfactory. A new principle was proposed by Chen, Ma and Quan[10], in which the samples not only upon the approached area itself but also upon the neighboring ones are used. It is termed as "fitting larger and using smaller" (FLUS) principle in numerical method.

In this paper, according to the principle of FLUS, an ATP pressure interpolation polynomial on a nonequidistant rectangle is constructed and a new approximate tangent plane method is obtained. Using the new method, the calculating accuracy of surface deformation of elastic solids caused by a given normal pressure distribution is higher.

2 Representation of Pressure Distribution by ATP Method

If the contact area is divided into a number of nonequidistant rectangular grids as shown in Fig. 1, and the pressure value at the center of every grid has been known, the pressure sample $p_{i,j}$ at the center of grid $A_{i,j}$ can only give the information that the pressure on the grid varies in a certain range of the sample. We could not know how the pressure varies and what is its range. But, they can be determined by the difference and the relative position between the sample $P_{i,j}$ and the pressure samples around it, for example, with the forward difference, backward difference, or the centered difference. If the grids are divided fine enough, the slope of the fitting pressure on the grid is undoubtedly near the correct value, and the variational range between the fitting pressure and real pressure will be small enough.

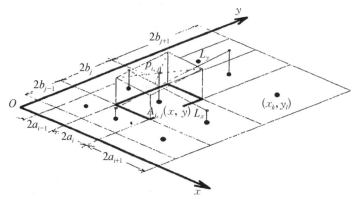

Fig. 1 The principle of ATP fitting method

In this solution the sample $p_{i,j}$ is used to be the reference value of pressure on grid $A_{i,j}$, the centered difference composed by $p_{i-1,j}$ and $p_{i+1,j}$ at the neighboring grids $A_{i-1,j}$ and $A_{i+1,j}$ is used to describe the slope in the x direction, and the centered difference composed by $p_{i,j-1}$ and $p_{i,j+1}$ at the neighboring grids $A_{i,j-1}$ and $A_{i,j+1}$ is the slope in the y direction. So the pressure distribution $p_{i,j}(x, y)$ over the fitted grid $A_{i,j}$ is represented by a linear polynomial:

$$p_{i,j}(x, y) = \frac{p_{i+1,j} - p_{i-1,j}}{a_{i+1} + 2a_i + a_{i-1}}(x - x_i) + \frac{p_{i,j+1} - p_{i,j-1}}{b_{j+1} + 2b_j + b_{j-1}}(y - y_j) + p_{i,j}$$
$$= c_1 x + c_2 y + c_3,$$
$$(x_i - a_i \leqslant x \leqslant x_i + a_i; \ y_j - b_j \leqslant y \leqslant y_j + b_j;$$
$$i = 1, 2, \cdots, n; \ j = 1, 2, \cdots, m) \quad (3)$$

and the boundary conditions are assumed to be

$$\begin{cases} p_{0,j} = p_{n+1,j} = p_{i,0} = p_{i,m+1} = 0, \\ a_0 = a_{n+1} = b_0 = b_{m+1} = 0. \end{cases} \quad (4)$$

The geometrical description of Eq. (3) is given in Fig. 1, in which the plane passed the point $(x_i, y_j, p_{i,j})$ and with normal line being the conormal line of both lines L_x and L_y is used to be an approximate tangent plane of the real pressure distribution at the point. Then the ATP function is used to be the pressure fitting polynomial of the grid $A_{i,j}$ $(x_i - a_i \leqslant x \leqslant x_i + a_i, \ y_j - b_j \leqslant y \leqslant y_j + b_j)$ containing the point (x_i, y_j). The new method can describe not only the pressure value at the point (x_i, y_j), as well as the slope of the real pressure distribution in both direction on the grid $A_{i,j}$.

By integrating Eq. (3) on the grid $A_{i,j}$ and summing over the entire domain, the force equilibrium can be written as

$$P = \sum_{i=1}^{n} \sum_{j=1}^{m} \iint_{A_{ij}} p_{i,j}(x, y) \mathrm{d}x\mathrm{d}y = \sum_{i=1}^{n} \sum_{j=1}^{m} p_{i,j} A_{i,j}. \tag{5}$$

The right-hand side of equation (5) is the same as the corresponding one in D-H method.

It is shown in Fig. 2 that the ATP method is better than the D-H method in the accuracy of fitting the pressure distribution on the fitted grid. And the computational error of elastic surface deformation mainly depends on the fitting accuracy.

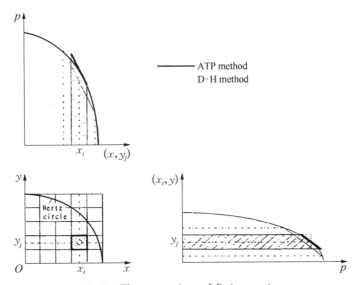

Fig. 2 The comparison of fitting section

3 Calculation of Surface Deformation

By substituting Eq. (3) into Eq. (1), the elastic surface deformation at point (x_k, y_l) caused by a piecewise ATP polynomial on the whole domain analyzed can be written as

$$\begin{aligned}
d(x_k, y_l) &= \frac{2}{\pi E'} \sum_{i=1}^{n} \sum_{j=1}^{m} \iint_{A_{ij}} \frac{p_{i,j}(x, y)\mathrm{d}x\mathrm{d}y}{\sqrt{(x-x_k)^2 + (y-y_l)^2}} \\
&= \frac{2}{\pi E'} \sum_{i=1}^{n} \sum_{j=1}^{m} \left[\frac{Q_{i,j} - U_i S_{i,j}}{a_{i+1} + 2a_i + a_{i-1}} (p_{i+1,j} - p_{i-1,j}) \right. \\
&\quad \left. + \frac{R_{i,j} - V_j S_{i,j}}{b_{j+1} + 2b_j + b_{j-1}} (p_{i,j+1} - p_{i,j-1}) + S_{i,j} p_{i,j} \right], \\
&\quad (k = 1, 2, \cdots, n; l = 1, 2, \cdots, m)
\end{aligned} \tag{6}$$

The analytical expressions of $Q_{i,j}$, $R_{i,j}$, $S_{i,j}$, U_i and V_j are given in Appendix.

Considering the boundary conditions of Eq. (4), the boundary lines are regarded as the grids without width, then the total number of grids is increased from $n \times m$ to $(n+2) \times (m+2) = N \times M$. Eq. (6) can be further changed into

$$d(x_k, y_l) = \frac{2}{\pi E'}\Bigg[\sum_{i=3}^{N}\sum_{j=2}^{M-1}\frac{Q_{i-1,j}-U_{i-1}S_{i-1,j}}{a_i+2a_{i-1}+a_{i-2}}p_{i,j} - \sum_{i=1}^{N-2}\sum_{j=2}^{M-1}\frac{Q_{i+1,j}-U_{i+1}S_{i+1,j}}{a_{i+2}+2a_{i+1}+a_i}p_{i,j}$$
$$+\sum_{i=2}^{N-1}\sum_{j=3}^{M}\frac{R_{i,j-1}-V_{j-1}S_{i,j-1}}{b_j+2b_{j-1}+b_{j-2}}p_{i,j} - \sum_{i=2}^{N-1}\sum_{j=1}^{M-2}\frac{R_{i,j+1}-V_{j+1}S_{i,j+1}}{b_{j+2}+2b_{j+1}+b_j}p_{i,j}$$
$$+\sum_{i=2}^{N-1}\sum_{j=2}^{M-1}S_{i,j}p_{i,j}\Bigg]$$
$$=\frac{2}{\pi E'}\sum_{i=1}^{N}\sum_{j=1}^{M}F_{i,j}^{k,l}p_{i,j},$$
$$(k=1, 2, \cdots, n; l=1, 2, \cdots, m). \quad (7)$$

It has been shown in Eq. (7) that the elastic surface deformations are expressed by linear combination of pressure samples. And the influence coefficients F_{ij}^{kl} are dependent on the grid geometry only.

4 Numerical Results

4.1 Accuracy

The accuracy of the ATP method is first evaluated by examining the elastic surface deformations at the center $(0, 0)$ and the corner (a, a) of a square area when a semi-infinite solid is subjected to a uniform pressure

$$p(x, y) = \begin{cases} 0.2 \text{ GPa}, & (-a \leqslant x \leqslant a, -a \leqslant y \leqslant a), \\ 0, & \text{other}. \end{cases} \quad (8)$$

Without loss of generality, the deformations are calculated when the contact region is divided into a number of uniform grids. The results are compared with the theoretical values as well as with those calculated by the D-H method.

Data of the deformations are given in Table 1. The theoretical values of the deformation are calculated according to Timoshenko and Goodier[11]. Both the ATP and the D-H method yields identical results with the theoretical values for the uniform pressure, no matter how coarse grid configuration used.

Table 1 Elastic surface deformations at the center and the corner of a square when a semi-infinite solid is subjected to a uniform pressure distribution defined in equation (8)

$n \times m$	$d(a, a)$ μm	E_{rr} ‰	$d(0, 0)$ μm	E_{rr} ‰	$n \times m$	$d(a, a)$ μm	E_{rr} ‰	$d(0, 0)$ μm	E_{rr} ‰
	D-H method					ATP method			
10×10	0.336 209	0.0	0.672 418	0.0	10×10	0.336 209	0.0	0.672 418	0.0
50×50	0.336 208	0.0	0.672 418	0.0	50×50	0.336 209	0.0	0.672 417	0.0
100×100	0.336 208	0.0	0.672 417	0.0	100×100	0.336 209	0.0	0.672 417	0.0

$E = 227.472\ 5$ GPa, $a = 0.170\ 375\ 8$ mm, $p(x, y) = 0.2$ GPa, $d_{th}(a, a) = 0.336\ 209$ μm, $d_{th}(0, 0) = 0.672\ 418$ μm

In the second evaluation, a hemispherical pressure distribution acting over a circle area is assumed

$$p(x,y) = \begin{cases} 0.822\,423\,6\sqrt{1-\left(\dfrac{x}{a}\right)^2-\left(\dfrac{y}{a}\right)^2}\ \text{GPa}, \\ (x^2+y^2 \leqslant a^2), \\ 0,\ \text{other}. \end{cases} \quad (9)$$

The results obtained from the ATP mathod and the D-H method are compared with the theoretical results obtained from Hertzian theory in the contact region, and compared with the results obtained from the D-H method with 260×260 grid number out of the contact region. Fig. 3 shows the percentage difference in elastic surface deformations when $n \times m = 10 \times 10$. Due to the symmetry of the contact problem, the elastic surface deformations are calculated along six radial directions in the 1/8 contact area, there are 41 calculated points on every line.

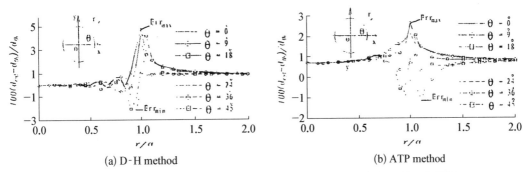

Fig. 3 Variations of E_{rr} along six radial directions by method D-H and ATP

From Fig. 3 it can be found that the large percentage difference in elastic surface deformation take place near the edge of contact area in both methods. It is due to the pressure being either zero if the center of the rectangular area shown in Fig. 2 is outside the contact, or of order 10^5 if the center of the rectangular area is within the contact. It is called "edge effect". In fact, this phenomenon is very serious at the large pressure gradients area, for example, in the outlet region of EHL and for some non-Hertzian contact problems. It can be seen from comparing Fig. 3a and Fig. 3b, the error of method ATP is less than method D-H.

In fact, the surface flatness in the Hertzian contact region provides another good comparison criterion. This aspect is extremely important at EHL of high load and finite line contacts, where the elastic surface deformation is two or three orders of magnitude larger than the film thickness. From Fig. 4, the flatness Δh_{max} is

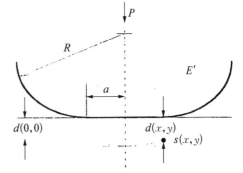

Fig. 4 The Hertzian contact between a teelball and a rigid flat

defined as

$$\Delta h_{max} = [d(x, y) + s(x, y)]_{max} - [d(x, y) + s(x, y)]_{min}$$
$$= [d(x, y) - d_{th}(x, y)]_{max} - [d(x, y) - d_{th}(x, y)]_{min},$$
$$(x^2 + y^2 \leqslant a^2). \tag{10}$$

Table 2 shows the calculated results of Δh_{max}. Considering the "edge effect", the calculating area is expanded from $(x^2+y^2 \leqslant a^2)$ to $(x^2+y^2 \leqslant 1.44\ a^2)$. The values of Δh_{max} obtained by the ATP method is about 1/2 of that by the D-H method. So the numerical accuracy of the ATP method is higher.

Table 2 Comparison of the surface flatness in a deformed contact obtained by two methods

$n \times m$	D-H method Δh_{max} nm	ATP method Δh_{max} nm	$n \times m$	D-H method Δh_{max} nm	ATP method Δh_{max} nm
6×6	123.07	38.24	50×50	4.46	3.21
10×10	69.63	35.07	80×80	2.47	1.74
20×20	21.21	10.94	160×160	0.80	0.41
30×30	11.95	6.72			

$E = 227.472\ 5$ GPa, $a = 0.170\ 375\ 8$ mm, $p_o = 0.822\ 423\ 6$ GPa, $R = 15$ mm.

4.2 Efficiency of calculation

Since the pressure fitting polynomials formed by the bilinear interpolating functions[5] or by the biquadratic interpolating functions[6-8] are composed of many terms, they result in more lengthy and complex mathematical expressions than the uniform pressure interpolation of the D-H method. The latter method is usually more efficient than the former ones in constructing the influence coefficient matrix of the same size. From theoretical analysis and calculation on an IBM-486 personal computer, it has been found that the CPU time consumed in constructing the influence coefficient matrix by the ATP method for various grid configurations is about two times as long as that by the D-H method.

The storage size required for the influence coefficient matrix of the ATP method is the same as that of the D-H method for the same grid configuration, being the least storage size among all of the methods mentioned above. This is very important for reducing computing time in the numerical solution of two-dimensional EHL problems by means of iteration.

5 Conclusions

According to the principle of FLUS, an ATP method for calculating elastic surface deformation has been introduced with a new contact pressure fitting approach. In the method, not only is the pressure sample on the grid considered, but also the relation to four samples around the grid. The form of the pressure fitting function is a first order linear polynomial, it can describe not only the pressure value over the fitted grid, but also the slopes of pressure in both direction over the grid approximately. The calculating accuracy using the ATP method is higher, especially, the surface flatness is less. It can improve the convergence in solving EHL problems, such as point contacts under heavy load and finite line contacts. The storage space required for the influence coefficient matrix by the method is the same as that of the D-H method. And the computational time required for the iteration in two-dimensional EHL problems is also the same as the D-H method.

References

[1] Johnson K. L. Contact Mechanics. Cambridge University Press, London, 1985.
[2] Dowson D. and Hamrock B. J. Numerical evaluation of the Surface Deformation of Elastic Solids Subjected to a Hertzian Contact Stress. ASLE Trans., 1976, 19: 279-286.
[3] Changc L., An efficient and accurate formulation of the surface-deflection matrix in elastohydrodynamic point contacts. Trans. of ASME, J. of Trib., 1989,111: 642-647.
[4] Biswas S. and Snidle R. W. Calculation of surface deformation in point contact EHD. Trans. of ASME, J. of Lub. Tech., 1997,99: 313-317.
[5] Ranger A. P., Ettles C. M. M. and Cameron A. The solution of the point contact elastohydrodynamic problem. Proc. Roy. Soc., London, 1975, A 346: 227-244.
[6] Hou K. P., Zhu D. and Wen S. Z. A new numerical technique for computing surface elastic deformation caused by a given normal pressure distribution. Trans. of ASME, J. of Trib., 1985, 107: 128-131.
[7] Jeng Y. R. and Hamerock B. J., The effect of surface roughness on elastohydrodynamically lubricated point contact. ASLE Trans., 1987, 30: 531-538.
[8] Lin J. F. and Chu H. Y. A numerical solution for calculating elastic deformation in elliptical-contact EHL of rough surface. Trans. of ASME, J. of Trib., 1991,113: 12-21.
[9] Lubrecht A. A. and Ioannides E. A fast of the dry contact problem and the associated subsurface stress field using multilevel techniques. Trans. of ASME, J. of Tribo. Tech., 1991, 113: 128-133.
[10] Chen X. Y., Ma J. J. and Quan Y. X. The approximate tangent plane fitting method of three-dimension contact pressure distribution. J. of Zhejiang University, 1994, 28(2): 239-243. (in Chinese)
[11] Timoshenko S. and Goodier J. N. Theory of Elasticity. 3rd ed. McGraw-Hill Book Co., New York, 1970.

Appendix

$$Q_{i,j} = \int_{U_i-a_i}^{U_i+a_i}\int_{V_j-b_j}^{V_j+b_j} \frac{x\,dx\,dy}{\sqrt{x^2+y^2}} = \frac{1}{2}[I_1(U_i+a_i, V_j+b_j) - I_1(U_i+a_i, V_j-b_j)$$
$$+ I_1(U_i-a_i, V_j-b_j) - I_1(U_i-a_i, V_j+b_j)$$
$$+ I_2(U_i+a_i, V_j+b_j) - I_2(U_i+a_i, V_j-b_j)$$
$$+ I_2(U_i-a_i, V_j-b_j) - I_2(U_i-a_i, V_j+b_j)], \tag{A1}$$

$$R_{i,j} = \int_{U_i-a_i}^{U_i+a_i}\int_{V_j-b_j}^{V_j+b_j} \frac{x\,dx\,dy}{\sqrt{x^2+y^2}} = \frac{1}{2}[I_1(V_j+b_j, U_i+a_i) - I_1(V_j+b_j, U_i-a_i)$$
$$- I_1(V_j-b_j, U_i-a_i) - I_1(V_j-b_j, U_i+a_i)$$
$$+ I_2(V_j+b_j, U_i+a_i) - I_2(V_j+b_j, U_i-a_i)$$

$$+ I_2(V_j - b_j, U_i - a_i) - I_2(V_j - b_j, U_i + a_i)], \tag{A2}$$

$$S_{i,j} = \int_{U_i - a_i}^{U_i + a_i} \int_{V_j - b_j}^{V_j + b_j} \frac{\mathrm{d}x\,\mathrm{d}y}{\sqrt{x^2 + y^2}} = I(U_i + a_i, V_j + b_j) - I(U_i + a_i, V_j - b_j)$$
$$+ I(U_i - a_i, V_j - b_j) - I(U_i - a_i, V_j + b_j), \tag{A3}$$

$$\begin{cases} U_i = x_i - x_k, \\ V_j = y_j - y_l. \end{cases} \tag{A4}$$

$$I(x, y) = x\ln(y + \sqrt{x^2 + y^2}) + y\ln(x + \sqrt{x^2 + y^2}), \tag{A5}$$

$$I_1(x, y) = y\sqrt{x^2 + y^2}, \tag{A6}$$

$$I_2(x, y) = x^2 \ln(y + \sqrt{x^2 + y^2}). \tag{A7}$$

弹性金属塑料瓦推力轴承的滑移问题研究*

摘　要：本文对流体力学的经典雷诺方程中固液界面无滑移假设对于弹性金属塑料瓦的适用性提出疑问，从理论上论证在一些表面能低的聚合物表面存在滑移的可能性，并进行对比实验．实验证明在聚四氟乙烯与润滑油的界面上存在滑移现象，发现滑移是在剪切速率和油膜厚度达到一定值才出现的，滑移速度随着转速的提高和膜厚的减小而增大．在分析以聚四氟乙烯为表面的弹性金属塑料轴瓦的润滑机理时，应当对雷诺方程进行修正，计入滑移因素的影响．

1. 引言

经典雷诺方程的一个重要假设是固液界面上没有滑移，认为在界面上油粘附在固体表面上，没有运动速度，这个假设在通常的流体润滑研究中都加以采用．由于通常轴瓦采用金属材料，而金属表面能高，液体极易粘附在其上，采用无滑移假设的雷诺方程的正确性在工程实践中得到证明．但是对于表面能很低、不黏性十分明显的聚四氟乙烯材料，它与润滑油的界面上滑移假设是否依然适用，就值得研究了．现在弹性金属塑料瓦推力轴承被广泛应用在大、中型水电机组上，其表面是一薄层聚四氟乙烯材料，因此研究聚四氟乙烯与润滑油固液界面上滑移是否存在，对于弹性金属塑料瓦推力轴承的设计研究就有非常重要的意义，而推力轴承是水电机组的关键部件．

另外，弹性金属塑料瓦推力轴承的一些试验现象用无滑移假设很难解释．这种瓦与原来采用的钨金瓦相比有一个特殊现象即瓦的温升很低，每块瓦的平均温升小于5℃，尺寸差别不大的钨金瓦与弹性金属塑料瓦的工作表面上测得的温度大致是一样的，表明油膜温度相差不大，并且真机实验证明两者因润滑油黏滞作用引起的发热量是大致相同的，大化电厂测得的氟塑料瓦的油膜温度比钨金瓦的低[1,2]．由于聚四氟乙烯的导热性很差，其导热系数为 0.24 W/(m·K)，而一般金属的导热系数比其大 300 多倍，因此油膜中的黏滞功耗产生的热量传到氟塑料瓦上的很少，使瓦体的温升低，绝大多数热量需靠润滑油的流动及镜板传热带走．由于二者发热量相同，使塑料瓦油膜内的油温比钨金瓦的油温要高，这与试验结果出现了矛盾．但是如果假设聚四氟乙烯与润滑油的界面上存在滑移，能使油膜内润滑油的流量增大，从而抵销了因瓦体热传导小产生的油膜温升，能很好的解释氟塑料瓦与钨金瓦油膜温度相差不大或者甚至更低的试验现象．这就需要用理论分析和实验来弄清和论证．本文工作即以此为目的．

本文从理论上论证滑移存在的可能性，并设计了对比实验，实验证明聚四氟乙烯与润滑油的界面上在一定的工况条件下出现滑移现象，并分析影响滑移大小的因素．

---

* 本文合作者：王小静、张国贤．原发表于《润滑与密封》，1997(4)：19-22,35．

2. 理论分析

图1所示的液珠,有三个介界面:固液(SL)、液汽(LV)和固汽(SV),因此有三个界面张力 γ_{SL}、γ_{SV}、γ_{LV},根据 Young 方程,它们之间的关系是:

$$\gamma_{SV} = \gamma_{SL} + \gamma_{LV}\cos\lambda \tag{1}$$

其中 λ 为平衡接触角. 当 $\lambda=0°$ 时,液体完全润湿固体表面,接触角越大,润湿性越差. 当 $\lambda>90°$ 时,液体由部分润湿变为不润湿;当 $\lambda=180°$ 时,液体变为完全不润湿,液珠在固体表面上形成一个小球. 水与金属的接触角为零度,水完全铺展在金属的表面;水与聚四氟乙烯的接触角是 98°~112°,润湿性很小.

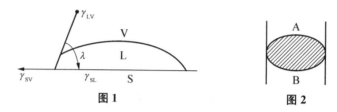

图1　　　　图2

考虑由 A、B 两种物质组成的一个圆柱体,如图2所示,若在 A—B 界面处划线将圆柱分开,则作用于单位界面面积上的功等于:

$$W_{AB} = \gamma_B + \gamma_A - \gamma_{AB} \tag{2}$$

此为粘附力,是两种物质粘结强度的度量.

若圆柱完全由一种物质组成,则分隔开圆柱对单位面积做的功等于:

$$W_{AA} = 2\gamma_A \tag{3}$$

此为内聚功,表示将一种物质分开的条件.

由于聚四氟乙烯是非极性材料,Fowkeb 方程为:

$$\gamma_{AB} = \gamma_A + \gamma_B - 2\sqrt{\gamma_A\gamma_B} \tag{4}$$

将式(4)代入式(2),得:

$$W_{AB} = 2\sqrt{\gamma_A\gamma_B} \tag{5}$$

要内聚功大于粘附功,即 $W_{AA} > W_{AB}$,

$$\gamma_A > \gamma_B \tag{6}$$

如果 A 相为油,B 相为聚四氟乙烯,一般油的界面张力是 30×10^{-3} N/m,聚四氟乙烯的界面张力为 18×10^{-3} N/m,式(6)可完全满足,即粘附断裂出现在内聚断裂之前,因此滑移现象有可能存在. 而金属的界面张力在 500×10^{-3} N/m 以上,内聚功远远小于粘附功,滑移假设不成立.

3. 试验设计

为了研究聚四氟乙烯界面上的滑移现象,设计了一组对比试验,用钢和聚四氟乙烯材料进行对比,观察滑移是否存在.

(1) 试验装置:整个试验装置安装在一台高精度的铣床上(见图 3). 钢制旋转盘 1 装在铣床的主轴上,可随主轴一起旋转,并且伸出端很短,保证了回转精度. 静止盘 2 固定在托盘 3 上,可以拆卸,分别用钢和聚四氟乙烯做成两个几何尺寸完全一致的静止盘进行对比试验,静止盘与旋转盘相对应表面的面积相等,托盘 3 通过三个等距的弹片 4 固定在支架 5 上,可以轻微转动,托盘只受工作表面润滑油摩擦力矩的影响以及弹片的反力矩作用,托盘的弹性悬挂方式兼有定位及显示力矩的功能,它避免了其他定位方式可能产生的摩擦力对其影响,保证所测得的摩擦力矩是润滑油作用在静止盘表面的. 三个支架 5 固定在底板上,底板连同固定在其上的各部件安置在铣床导轨上.

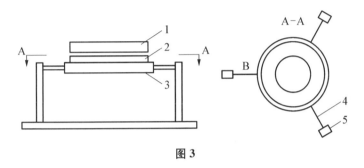

图 3

静止盘可以调节,使其与旋转盘的对应表面保持水平. 静止盘的中心开孔,使润滑油能顺利流入旋转盘与静止盘的间隙内,保证间隙内始终充满润滑油,并且使油膜内的温度可以降低,尽量减少由于发热引起的固体变形的影响.

在弹片上 B 的位置处,安装一个位移测量计,能测量 B 处的位移变化. 进行对比试验时,只需把不同材料的静止盘进行拆装,位移计的位置不需要移动,保证不同材料静止盘试验时是同一点的位移值进行对比,确保数据的可比性. 由润滑引起的摩擦力矩导致的位移很小,当静止盘轻微转动时在测量点引起的曲率变化可以忽略,位移计的读数精度达到 $1~\mu m$. 进行试验时,先记录 B 点的位移变化情况,然后利用砝码进行标定,得到位移与摩擦力矩的对应关系,由此得出静止盘所受的摩擦力矩大小.

在进行对比试验及标定时,环境温度保持不变,维持在 29℃,采用同一品种的润滑油,动力黏度为 $0.091~Pa \cdot s$,试验时采用四种转速,分别为 475 r/min、600 r/min、750 r/min 和 950 r/min. 调整静止盘与旋转盘的间隙,从 0.4 mm 到 0.06 mm,观察速度变化和油膜间隙变化对试验结果的影响.

(2) 试验原理:旋转盘相对于静止盘进行转动,油膜内的润滑油受到剪切作用,在静止盘上产生摩擦力矩,跟随着产生转动. 在静止盘与润滑油界面上有无滑移速度,对摩擦力矩的大小是有影响的. 现在分别用钢和聚四氟乙烯的静止盘作对比试验. 由于钢的润湿性好,易铺展,认为钢与油的界面不存在滑移现象,油分子粘附在钢表面上. 而聚四氟乙烯是低能面,吸附性差,以 V_s 表示聚四氟乙烯与油的界面上可能存在的滑移速度,现推导 V_s 对摩擦力矩大小的影响.

由连续方程:

$$\frac{1}{r}\frac{\partial}{\partial r}(rV) + \frac{1}{r}\frac{\partial u}{\partial \theta} + \frac{\partial w}{\partial z} = 0$$

动量方程：$\dfrac{\partial P}{r\partial\theta}=\mu\dfrac{\partial}{\partial z}\left(\dfrac{\partial u}{\partial z}\right)$，$\dfrac{\partial P}{\partial r}=\mu\dfrac{\partial}{\partial z}\left(\dfrac{\partial v}{\partial z}\right)$

由于没有楔形间隙和挤压作用，油膜中只有剪切流无压力流，

$$\dfrac{\partial P}{\partial\theta}=0,\ \dfrac{\partial P}{\partial r}=0$$

代入速度边界条件：

$$z=0,\ u=\omega r,\ v=w=0$$
$$z=h,\ u=V_s,\ v=w=0$$

速度分布：$u=\omega r\left(1-\dfrac{z}{h}\right)+V_s\dfrac{z}{h},\ v=0$

剪切应力：$\tau_\theta=\mu\dfrac{\partial u}{\partial z}=-\dfrac{\mu\omega r}{h}+\dfrac{\mu V_s}{h}$，

$\tau_r=0$

摩擦力矩：

$$M=-\iint\tau_\theta r^2\mathrm{d}\theta\mathrm{d}r=\dfrac{2\pi\mu}{h}\left[\dfrac{\omega}{4}(R_2^4-R_1^4)\right]-\dfrac{\mu}{h}\iint V_s\times r^2\mathrm{d}\theta\mathrm{d}r \tag{7}$$

其中，μ——动力黏度，ω——主轴转速，h——油膜厚度.

从摩擦力矩的表达式可知，如果滑移假设成立，采用聚四氟乙烯静止盘时的摩擦力矩应该小于采用钢静止盘时的摩擦力矩，二者的差值就是由滑移引起的.

4. 试验结果分析

环境温度保持在29℃，采用动力黏度为0.091 Pa·s的润滑油，在不同转速和不同间隙的工况下对钢静止盘与聚四氟乙烯静止盘的摩擦力矩进行测量，得出一组在不同转速下油膜膜厚与摩擦力矩的关系曲线，见图4.

在图4中，不同转速下，摩擦力矩与油膜膜厚的关系曲线形状大体一致，摩擦力矩随着膜厚的减小而增大，与油膜厚度成反比. 旋转速度的增加，使摩擦力矩增大.

图4(a)是主轴转速为475 r/min时静止盘所受的摩擦力矩与油膜厚度的关系曲线. 在不同间隙下，钢盘与聚四氟乙烯盘所受的摩擦力矩进行比较，发现在整个膜厚范围内，两条关系曲线几乎完全重合，两种材料的静止盘受到的摩擦力矩基本一样，说明在转速为475 r/min时无滑移现象.

图4(b)、图4(c)和图4(d)分别为600 r/min、750 r/min和950 r/min转速下摩擦力矩和油膜厚度的关系曲线. 图中可以明显地看到，油膜厚度从0.4 mm到0.25 mm时，聚四氟乙烯曲线与钢曲线重合在一起，膜厚从0.25 mm开始，随着油膜厚度的减小，两条曲线渐渐分开，膜厚达到0.2 mm时，两曲线分开明显，当膜厚逐渐减小，两个盘受到的摩擦力矩值相差越来越大，并且随着转速的提高，差值也增大. 转速600 r/min、膜厚0.06 mm的工况下，摩擦力矩的差值占总摩擦力矩的14.58%，当转速750 r/min、膜厚0.06 mm，差值占到

25.37%,当转速 950 r/min、膜厚 0.08 mm 时,差值占总摩擦力矩的 20.0%.

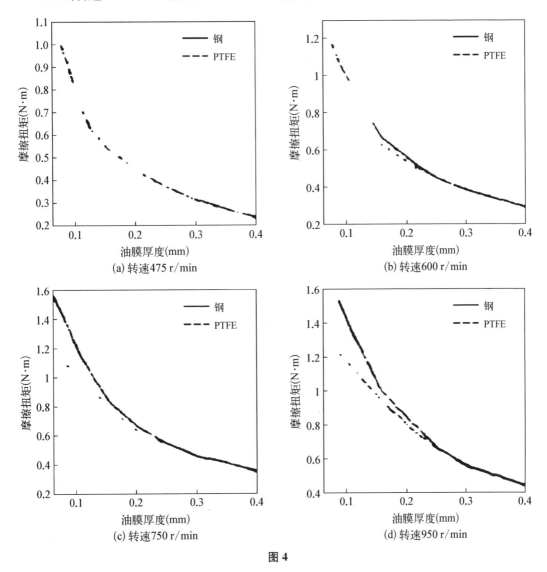

图 4

图 5 是三种转速下由滑移引起的摩擦力矩差值与油膜厚度的关系. 图中也呈现了摩擦力矩差值与膜厚成反比的关系,并随速度增大而增大. 图 6 是不同膜厚下摩擦力矩差值与转速的关系. 从这些数据可以看出,由滑移引起的摩擦力矩差值与转速和油膜厚度有关. 现在分析一下滑移速度与转速和膜厚的关系.

由滑移引起的摩擦力矩差值为:

$$\Delta M = \frac{\mu}{h} \iint V_s \times r^2 \mathrm{d}\theta \mathrm{d}r \tag{8}$$

因为滑移速度可能是运动表面线速度和膜厚的函数,本实验是采用两个圆盘作相对转动,圆盘上半径不同的各点线速度不同,所以各点的滑移速度应是径向坐标的函数. 用 A 表示滑移在盘上的综合影响:

$$A = \iint V_s \times r^2 \mathrm{d}\theta \mathrm{d}r \tag{9}$$

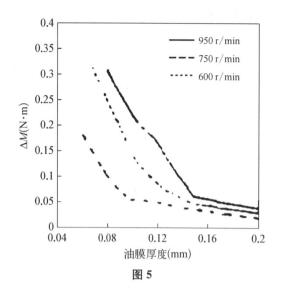
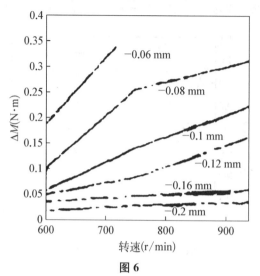

图 5   图 6

把式(9)代入式(8),得到:

$$A = \frac{h \Delta M}{\mu} \tag{10}$$

根据式(10),得到滑移综合量 A 与转速和膜厚的关系,如图 7、图 8.

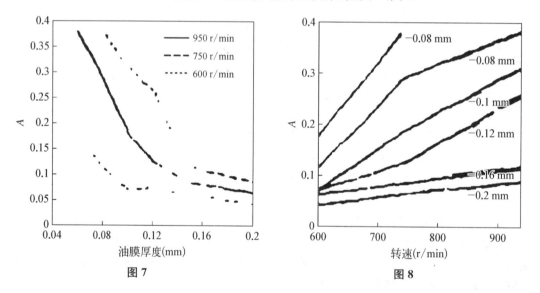

图 7   图 8

其中 A 的单位是 10^{-3} m^4/s. 从图 7 中可以看出,相同转速下滑移速度开始增长缓慢,当油膜厚度小到某一数值时,滑移速度曲线的斜率变大,随着膜厚的减小,滑移速度迅速增大,滑移速度与油膜厚度成反比.转速不同,滑移速度曲线率突然变大对应的油膜厚度也不同.其中 600 r/min 转速时,膜厚达到 0.1 mm 以后滑移速度增长迅速,750 r/min 转速时膜厚为 0.12 mm、950 r/min 转速时膜厚减小到 0.15 mm,滑移速度就开始明显增大了.说明转速越高,滑移速度开始增长迅速对应的膜厚越小.图 8 表明,转速越高,滑移速度越大,并随着膜厚的减小,滑移速度与转速的比值增长越快.

实验表明,滑移现象是在转速和油膜厚度达到某个临界值以后产生的.本次实验中,转速为 475 r/min 时,无滑移现象;但当转速在 600 r/min 以上,滑移现象发生,并随着转速的提高而滑移速度增大. 同样,油膜厚度大时,滑移不发生,当膜厚从 0.25 mm 开始,出现滑移,膜厚越小,滑移速度越大。这表明滑移速度的大小,与剪切速率或剪切应力有关. 剪切速率或剪切应力达到一定程度时,出现滑移;随着剪切速率越大,滑移速度越大.

5. 结论

本文根据弹性金属塑料瓦推力轴承的实验现象,并考虑到聚四氟乙烯材料的润湿性特别差,对雷诺方程无滑移假设在 PTFE 润滑问题上的适用性提出疑问,并对聚四氟乙烯与润滑油界面上的滑移行为进行研究. 首先从界面理论上论证滑移存在的可能性,又通过对比实验考察了在聚四氟乙烯与油的界面上的滑移规律,实验测得,因滑移引起的摩擦力矩最大占到总摩擦力矩的四分之一,说明滑移速度对润滑过程影响相当大,在润滑理论计算中必须计入滑移的影响. 并发现滑移是在剪切速率或剪切应力达到某个临界值后发生,随着剪切速率的增加而增加.

参 考 文 献

[1] 吴炳良,王建忠. 苏联弹性金属塑料推力轴瓦运行总结. 大电机技术,1992(1).
[2] 大化电厂单机 100 MW 发电机组 30 MN 级弹性金属塑料瓦推力轴承的研究.

Slip Study of the Plastic Thrust Bearing

Abstract: No-slip assumption in Reynolds equation of hydrodynamic lubrication is doubted. The possibility of slip in the interface between plastic and oil is verified. Slip phenomenon is found by comparative experiment. The factors affecting slip velocity is discussed. Slip velocity must be included in PTFE lubrication study.

大扰动情况下滑动轴承内瞬态油膜分布的研究*

由于各种原因造成大型转子系统的剧烈振动和失稳的问题逐渐引起注意,其中因油膜振荡造成的失稳现象最为典型.因此近几年广泛开展用质量守恒并计入空穴历史的观点来研究动载滑动轴承内油膜气穴分布的瞬态过程.

知道轴承内的气穴什么时候在什么地方出现是非常重要的.因为滑动轴承内流体润滑膜的气穴,影响轴承的性能及转子系统的稳定性.对润滑油膜分布的研究已有相当长的历史.1956年Cole和Hughes[1]首先使用玻璃做成的与实际间隙比相仿的轴承套,直接观察了静载滑动轴承内油膜分布的状态.后来White[2]、Marsh[3]等又相继做过各种不同工况下的油膜分布试验.1974年Dowson和Taylor[4]对稳态轴承载荷结果,即对稳态空穴的现象和性质有合理恰当的论述,但对动载轴承内油膜的分布和气穴的基本图像尚不清楚.1983年Jacobson和Hamrock[5]首先使用高速照相机对动载滑动轴承中的气穴进行了研究,认为同时存在气体气穴和蒸汽气穴,气体气穴从开始为单个气泡,然后越来越大,直到空穴的数量恒定,甚至能迁移到高压区不破裂.1991年、1993年D. C. Sun和D. E. Brewe[6,7]相继用高速摄像机对动态气穴进行研究.他们观察的结果是:(1) Jacobson和Hamrock的基本发现被证实,气穴气泡是在润滑流体内产生、长大,然后变尖破裂;(2) 小的轴承间隙、高的旋转速度和大的动偏心促使气穴出现,这些条件也引起气穴气泡变大,它的持续时间变长.

但对动载滑动轴承内瞬时油膜分布的概貌至今还不完全清楚,特别是理论和试验相结合的工作至今尚很不充分.

为此,本文对兼有静偏心和动偏心的圆柱轴承进行了计入质量守恒的理论分析及试验观察,并将理论和试验结果作了对比,得出了一些有意义的结论.这将为今后进一步深入研究以最终确定更合理的动载油膜边界条件,提供一定依据.

1 大扰动情况下轴承油膜分布的理论研究

轴颈中心 O_j 绕着偏离轴承中心 O_b 某一固定位置作圆形涡动,涡动频率为 ω_d,当动偏心相对较小时($\varepsilon_s > \varepsilon_d$)轴颈中心在轴承一侧涡动,而在大扰动情况下,即动偏心相对较大时($\varepsilon_s < \varepsilon_d$)涡动轨迹包容轴承中心,其几何关系如图1所示.

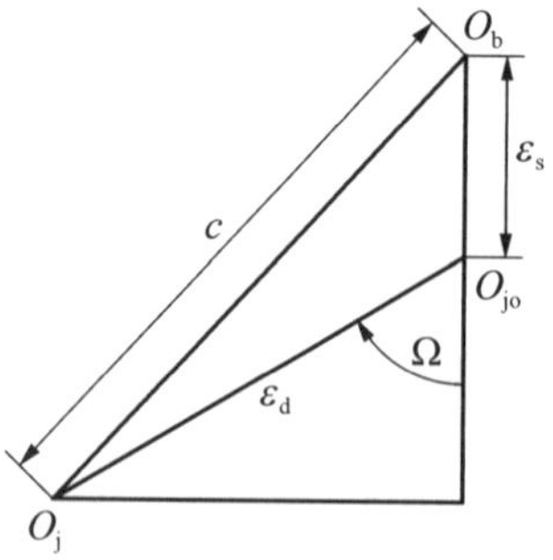

油膜厚度 H 为:

$$H = 1 - \varepsilon_x \sin(\lambda_\omega T) - \varepsilon_y \cos(\lambda_\omega T)$$
$$\varepsilon_x = -\varepsilon_d \sin(\lambda_\omega T) \quad \text{其中} \ \lambda_\omega = \omega_d / \Omega$$

图1

---

* 本文合作者:孙美丽、韩兆兵.原发表于《润滑与密封》,1997(5):7-9.

$$\varepsilon_y = -\varepsilon_s - \varepsilon_d \cos(\lambda_\omega T)$$

理论分析以下列轴承参数为对象：轴颈直径 $d=59.10$ mm,轴承直径 $D=59.40$ mm,长径比 $l/d=1.0$,静偏心率 $\varepsilon_s=0.2$,动偏心率 $\varepsilon_d=0.6$.

假设润滑油是牛顿流体,并使用质量守恒条件处理油膜边界.按照 Elord[8] 的方式,在油膜完整区,二维非稳态不可压缩的雷诺方程可表达为:

$$\frac{\partial}{\partial x}\left(\frac{h^3}{12\mu}\frac{\partial p}{\partial x}\right) + \frac{\partial}{\partial z}\left(\frac{h^3}{12\mu}\frac{\partial p}{\partial z}\right) = \frac{\partial(\theta h)}{\partial t} + \frac{\partial}{\partial x}\left(\frac{\mu\theta h}{2}\right)$$

在空穴区则有: $\frac{\partial(\theta h)}{\partial t} + \frac{\partial}{\partial x}\left(\frac{\mu\theta h}{2}\right) = 0$

式中 θ 定义为: $\theta = \begin{cases} 1 & \text{完整区} \\ \dfrac{V_f}{V_t} & \text{空穴区} \end{cases}$

其中 V_f 为油膜的体积, V_t 为单元体的体积.

无量纲化则为:

$$\frac{\partial}{\partial \Phi}\left(H^3 \frac{\partial P}{\partial \Phi}\right) + \left(\frac{l}{d}\right)^2 \frac{\partial}{\partial \lambda}\left(H^3 \frac{\partial P}{\partial \lambda}\right) = 6\frac{\partial(\theta H)}{\partial T} + 3\frac{\partial(\theta H)}{\partial \Phi}$$

$$2\frac{\partial(\theta H)}{\partial T} + \frac{\partial(\theta H)}{\partial \Phi} = 0$$

利用差分方法求解[9].

为计算瞬时油膜分布,以轴颈位于最大合成偏心时(即动偏心与静偏心同一方向)作为 0 时刻,见示意图 2.将轴颈回转周期分为 50 个时刻,同时将轴承内表面的一周 360°分成 50 等份,即 50 个弧段.每一个弧段为 7.5°.0～25 个弧段处于油膜发散区,即下游区.26～50 个弧段处于油膜收敛区,即上游区.计算每时刻油膜压力分布及油膜分布的情况.理论计算规定当压力 $p>0$ 时,有油膜存在.当 $p=0$ 时,则无承载油膜存在,即为油膜破裂.由理论计算结果知,最大压力峰值出现在第 46 个时刻,其压力分布图见图 3.当轴颈的动偏心方向运行在发散区域内时,虽然压力峰值降低,但有时会出现双压力峰.如压力峰值最小的第 16 个时刻,其压力分布图见图 4.理论计算的油膜分布与试验结果比较见图 5 所示.

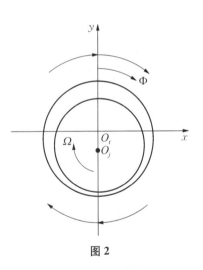

图 2

2 大扰动情况下油膜分布的试验研究

利用本文作者设计制造的滑动轴承静、动载荷试验台和试验规程进行试验[10].试验台如图 6 所示.其设计机理为:针对轴承在静载,动载及混合载荷情况下的工作状况,实验台设计考虑了只有静载、只有动载和两者兼有的混合载荷的情况,由于静载荷的大小体现在轴颈在轴承中的静平衡位置,因此调节轴颈在轴承中的不同静态偏心率,能观察到不同静载荷

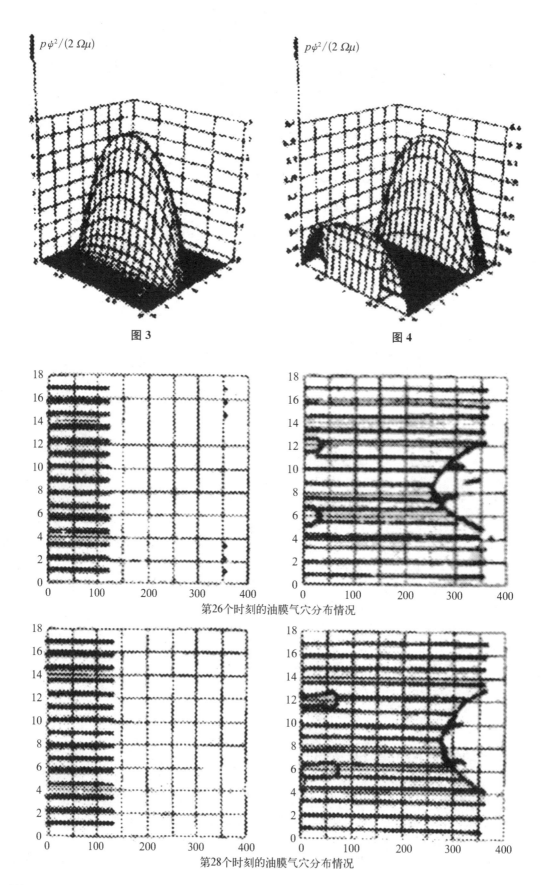

图 3　　　　　　　　　　图 4

第26个时刻的油膜气穴分布情况

第28个时刻的油膜气穴分布情况

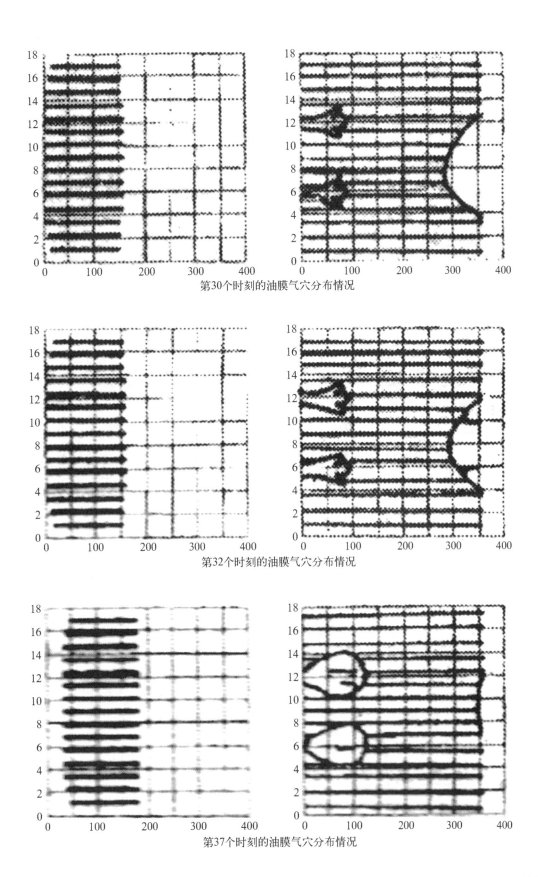

第30个时刻的油膜气穴分布情况

第32个时刻的油膜气穴分布情况

第37个时刻的油膜气穴分布情况

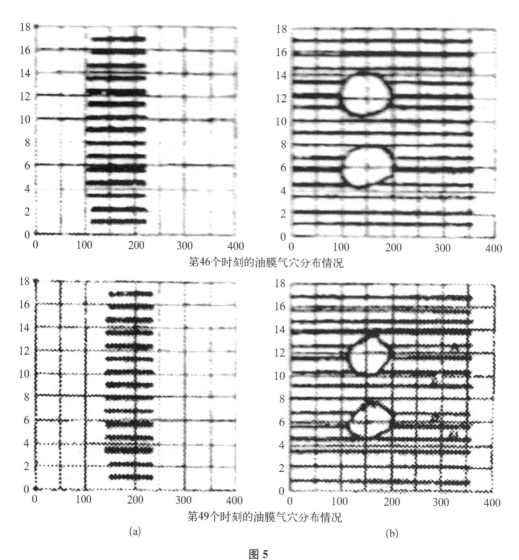

第46个时刻的油膜气穴分布情况

第49个时刻的油膜气穴分布情况

(a)   (b)

图 5

注：其中 a 列图为理论计算结果，b 列图为试验结果．黑点区域为油膜区域，空白区域为气穴区域．

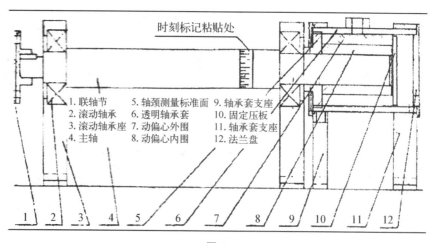

1. 联轴节　　5. 轴颈测量标准面　9. 轴承套支座
2. 滚动轴承　6. 透明轴承套　　　10. 固定压板
3. 滚动轴承座　7. 动偏心外围　　　11. 轴承套支座
4. 主轴　　　　8. 动偏心内围　　　12. 法兰盘

图 6

情况下静态油膜气穴的分布情况.对于动载,轴颈在轴承内的位置将随时间而变化,例如曲柄连杆式机械内轴承的工作情况.因此设计一机构使轴颈在轴承内的位置随时间的变化而变化,且利用调节动偏心率的大小来获得不同的动载荷,以观察不同动载荷情况下轴承内油膜气穴的状态.这一类装置在国内尚无人使用过.同时调节静偏心量 e_s 和动偏心量 e_d,将得到既有静载荷又有动载荷的混合载荷,以观察不同静、动载荷搭配情况下轴承内油膜破裂的情况.可分别调整静、动偏心量并可以模拟很大范围的各种轴承的工况.

造成静、动复合载荷的实验手段,通常有两类:一类是设定并加置静载荷和动载荷,在实验进行中测定轴颈在轴承中的动轨迹,同时观察记录油膜破裂情况;另一类是设定并形成静偏心和动偏心,实验中自动形成静、动复合载荷,观察并记录油膜破裂情况.后一种方案的优点在于静偏心量和动偏心量均可在实验运行前依靠实验台的结构设计和精确测量而得到精确的保证,从而保证了运行中油膜的动态几何形状可以精确地掌握.因此由于油膜动态几何形状方面的误差而影响到油膜动态边界的误差可以减少到最小.另一方面,在作对比性的理论计算时,有了动态几何形状的给定数据,就可以直接解算油膜动态边界和压力分布等,也不会有轴承载荷(等于油膜合力)方面的偏差对油膜动态边界的影响.因此,从实验验证理论假设的意义来说,后一种实验方案无疑远优于前一种.因此采取了后一种方案.

为使这一方案能发挥预想的效果,即保证轴颈在实验轴承内的动轨迹能精确实现,主轴的刚度、主轴支承轴承的精度和刚度以及实验轴承的外支承结构的刚度,都是至关重要的.在实验台结构设计上,对此给予了着重的考虑.

为与理论计算结果相对照,安装与之相对应的轴颈.建立所需的静偏心率 $\varepsilon_s=0.2$,动偏心率 $\varepsilon_d=0.6$.并在转动轴上对应于轴颈的动偏心方向处,作 360° 或 0° 的标记.以此点作起点,依次在全周上标出 24 个分格角度,标值的次序是使当轴颈转动时,由固定点观察所设标记呈角度增加的趋势,以便记录、推算出所拍照片是哪一个时刻的瞬态油膜分布的情况.按照试验规程建立所需动载滑动轴承试验工况,建立相应的转速 31.4 rad/s.打好强光源,待工况稳定后,使用 Praktica 型号照相机以千分之一秒拍照.

所得照片按事先所作标记进行时刻确定.由于油膜薄,彩色照片清晰.但复印成黑白的,由于对比度不够,不够清楚.所以把所得照片按事先设定的标记确定时刻后,非常粗略地定量绘图,与理论计算的结果比较,如图 5 所示.由于摄影是随机性的,未能连续排出所有时刻的油膜分布.现仅捕捉到的一系列分布图进行对比.

3 结论

(1) 在各个时刻,完整油膜区的上游边界位置,试验所得较理论计算所得向后推移了一定的角度.

(2) 试验所得连续油膜的分布区域较理论计算的为多.

(3) 试验结果重复性好.不同循环内观察到的同一时刻上的气穴分布几乎完全重合.这说明:动态气穴确有内在的物理规律.

(4) 在油膜的下游区域,由理论计算结果知,虽然压力峰较小,但有时有双压力峰出现,所得油膜分布较复杂.与试验所得油膜分布比较尚未发现有规律的对应性.

由于负压、油→气泡→油物理过程所需时间、表面张力等因素在计算中未加考虑,可能是造成计算与试验不符的原因,这有待以后进一步的研究.

参 考 文 献

[1] Cole, J. A. and Hughes, C. J., Oil Flow and Film Extent in Journal Bearing, Proc. Inst. Mech. Eng. 1956 Vol 170, 499-510.

[2] White, D. C., Squeeze Film Journal Bearing, Ph. D dissertation, Cambridge University, 1970.

[3] Marsh, H. 1974, Proc. of the 1st Leeds Lyon Sympo. on Tribology Held in the Institute of Tribology, 15-29.

[4] Dowson, D. and Taylor, C. M., Cavitation in Bearings. Ann. Rev. Fluid. Mech. 1979, 11, 35-66.

[5] B. O. Jacobson and B. J. Hamrock., High-Speed Motion Picture Camera Experiments of Cavitation in Dynamically Loaded Journal Bearings, Transactions of the ASME, Vol. 105, 446-452.

[6] D. C Sun and D. E. Brewe. A high Speed Photography Study of Cavitation in a Dynamically Loaded Journal Bearing, Journal of Tribology, April 1991, Vol. 113, 287-294.

[7] D. C. Sun and D. E. Brewe and P. B. Abel, Simultanlous Preasure Measurement and High-Speed Photographg Study of Canitation in a Dynamical Loaded Journal Bearing, Transactions of the ASME, Vol. 115, 88-95.

[8] H. G. Elrod, 1981, A Cavitation Algorithm, ASME Journal of Tribology, Vol. 103, No. 1, 398-354.

[9] 上海大学1996年韩兆兵硕士论文：油膜历史对轴心轨迹的效应研究.

[10] 上海大学1995年孙美丽硕士论文：360°滑动轴承静载和动载油膜气穴分布的研究.

[11] 张直明. 滑动轴承的流体动力润滑理论. 高教局出版社, 1986.

360°动、静载荷滑动轴承油膜分布实验台的设计及实验研究*

提　要：本文针对全轴承的定常和非定常工作情况，设计制造了一台对中旋转、不对中旋转及混合型旋转的轴承实验台．并在此实验台上进行了对中旋转、不对中旋转及混合型旋转轴承内油膜分布的实验研究，验证了 Jacobson and Hamrock 和 Sun and Brewe 的部分实验结果，并取得了一些有益的结论．

0　前言

由于滑动轴承具有制振、结构简单、安装方便、寿命长等许多独特的优点，因而被广泛地应用于汽轮机、发电机、透平压缩机等工业行业中．国内外大型旋转机械由于各种原因（地震、激励、不平衡、动载……）造成转子的剧烈振动和失稳的问题更是受到广泛重视，其动力学和稳定性问题至今仍是机械学中研究的一个热点．

油膜内由于气或蒸汽的出现，使连续流体状态产生分裂或破裂．知道轴承内气穴什么时候在什么地方产生具有重要意义．因为滑动轴承内流体润滑膜的气穴影响轴承的一些重要特性，因此对润滑油膜的研究已有很长的历史．1925 年，B. L. Newkirk 和 H. D. Taylor 首先在一柔性转子实验台观察到由轴承油膜引起的振动现象．1932 年 Swift 和 Stiebel 也开始了这个问题的研究，他们认为空穴区内压力为常数且与大气压力相差不大．1956 年 Cole 和 Hughes[1] 首先使用玻璃做成的与实际间隙比相仿的轴承套，直接观察了静载滑动轴承内油膜的状态，并获得一定数量的照片，他们的观察结果是：

（1）在间隙增大的区域里油膜破裂成细条状；轴承宽度内各处的开始破裂位置基本上是一致的．改变供油压力对于油膜破裂的位置没有什么影响．

（2）从油孔流出来的油与循环回来的破裂油膜汇合而逐渐铺开，直到覆盖住轴承的全部宽度，形成喇叭状过渡区．

其后 White(1970)、Marsh(1974) 等又相继做过各种不同工况下的油膜空穴实验．实验结果表明，空穴区油膜的形状并不总是条状，有时会出现蕨类植物状式的气穴状，它不仅存在于发散区，也可能出现在收敛区内．1974 年 Dowson 和 Taylor 对稳态轴承载荷的结果，即对稳态空穴的现象和性质有较合理而恰当的论述，但对不稳定载荷轴承内的油膜观测中，气穴的基本图像尚不清楚．1983 年 Jacobson 和 Hamrock[2] 首先用高速照相机对动载滑动轴承中的气穴进行了研究，认为同时存在气体气穴和蒸汽空穴，气体空穴开始为单个气泡，然后越来越大，直到空穴内气体的数量恒定，甚至能迁移到高压区不破裂，因此空穴内的压力有大气压力的许多倍．1991 年、1993 年，D. C. Sun 和 D. E. Brewe[3] 相继用高速摄像机对动

---

* 本文合作者：孙美丽．原发表于《上海大学学报（自然科学版）》，1997,3(5)：500 - 507．

态气穴进行研究,他们观察的结论是:

(1) Jacobson 和 Hamrock 的基本发现被证实,气穴气泡是在流体内产生,尺寸由小到大,然后变尖破裂.

(2) 小的轴承间隙、高的旋转速度和大的偏心率促使气穴出现.这些条件也引起气穴气泡变大,它的持续时间变长.

国内只有西安交通大学卢修连在作硕士期间采用激励的方法做了油膜动态破裂的实验研究.本文的工作针对旋转机械,设计制造一个动、静态油膜分布实验台,试验研究静、动载荷及混合载荷下轴承内油膜分布的规律,得出了一些有益的结论.这将为今后进一步研究滑动轴承油膜动态破裂的规律及与相应的油膜边界条件的关系提供一定的依据.

1 动、静载荷滑动轴承实验台的设计

1.1 设计机理

滑动轴承的工作状况无非是在静载、动载及混合载荷情况下工作,实验台设计将考虑只有静载、只有动载和两者兼有的混合载荷的情况.由于静载荷的大小体现在轴颈在轴承中的静平衡位置,因此调节轴颈在轴承中的不同静态偏心率,能观察到不同静载荷情况下静态油膜气穴的分布情况.对于动载,即轴颈在轴承内的位置将随时间而变化,例如曲柄连杆式机械内轴承的工作情况.因此设计一机构,使轴颈在轴承内的位置随时间的变化而变化,且利用调节动偏心率的大小来获得不同的动载荷以供观察.同时调节静偏心量 e_s 和动偏心量 e_d,将得到既有静载荷又有动载荷的混合载荷,可观察不同静、动载荷搭配情况下轴承内油膜破裂的情况.这一类装置在国内尚无人使用过.

观测静、动复合载荷的实验手段通常有两类.一类是设定并加置静载荷(例如转子重量)和动载荷(例如转子不平衡量),在实验进行中测定轴颈在轴承中的动轨迹,同时观察记录油膜破裂情况.另一类是设定并形成静偏心和动偏心,实验中自动形成静、动载荷以及复合载荷,观察并记录油膜破裂情况.后一种方案的优点在于静偏心量和动偏心量均可在实验运行前依靠实验台的结构设计和精确测量而得到,从而保证了运行中油膜的动态几何形状可以精确地被掌握.因此由于油膜动态几何形状方面的误差而影响到油膜动态边界的误差可以减少到最小.另一方面,在作对比性的理论计算时,有了动态几何形状的给定数据,就可以直接解算油膜动态边界和压力分布等,也不会有轴承载荷(等于油膜合力)方面的偏差对油膜动态边界的影响.因此,从实验验证理论假设的意义来说,后一种实验方案无疑远优于前一种.因此本文采取后一种方案.

为使这一方案能发挥预期的效果,即保证轴颈在实验轴承内的动轨迹能精确实现,主轴的刚度、主轴支承轴承的精度和刚度以及实验轴承的外支承结构刚度都至关重要.因此,在实验台结构设计上,对此给予了着重的考虑.

1.2 实验台设计和仪器设备

为研究滑动轴承静态和动态油膜气穴分布,设计的实验装置如图 1 所示,主要是由一个轴颈和一个有机玻璃轴承套组成.轴承套支撑在轴承座内,其轴承套的水平轴线在水平面和垂直面方向用机械的方法可调.主轴支撑在具有 D 级精度的 46305 型滚动轴承内,轴承座固定在钢基础上. 主轴的悬臂端安装有标准面和动偏心装置,动偏心的外圆为试验轴颈,可以检测动偏心 e_d 的精确量. 主轴的另一端用柔性连轴节与直流调速电机相联. 电机的最大转速为 2 000 r/min. 试验所用的转速用清华大学精仪系生产的 DVF-R 型数字矢量滤波器和 QH-85 涡流传感器进行标定. 两个角接触滚动轴承都被预紧,以减少轴的跳动,提高轴的旋转精度. 为保证测量精度,专门设计安装一标准测量板. 轴承的进油口设在最大静偏心间隙处,位于轴承的顶部,并在油内加入油溶性红色颜料. 为确定油膜在什么地方破裂,则使用固定的刻度作标记. 为便于说明,定义轴承内油膜上游区域为:从进油口的中心位置作为 0°,顺轴颈旋转方向到 180°;下游区域为:从 180° 起到 360°,即又回到进油口的中心位置. 内、外偏心圆端面的两个刻度盘同时使用可以确定动偏心量的粗值. 为保证静、动偏心率的精确度,不是单靠机械结构来保证,而是用仪表测量而定,可达到所用仪表精度水平.

轴颈:试验研究中使用 2 种轴颈,直径为 59.00 mm 和 59.10 mm. 长径比可以通过机械结构调整的方法得到所需的 1.0,0.8,0.5,0.3 等比值.

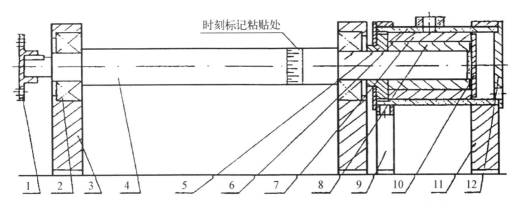

1—联轴节 2—滚动轴承 3—滚动轴承座 4—主轴 5—测量标准面 6—透明轴承套
7—动偏心外圆 8—动偏心内圆 9—轴承套支座 10—固定压板 11—轴承套支座 12—法兰盘

图 1 实验台的结构示意图

轴承套:采用水晶有机玻璃管料,套内的直径为 59.4 mm.

这样,轴颈与轴承之间的相对间隙数量级为百分之一. 相对间隙虽比一般轴承大,但仍满足雷诺方程成立条件,即 $\Psi \ll 1$. 对于实验轴承而言,较大的间隙可容纳较大的静偏心量和动偏心量,有利于保证其相对精确度,也利于压低其他可能误差(例如不圆度、粗糙度和运行时不可避免的微小跳动和振动)的相对影响. 相对间隙较大,对于油膜破裂后产生的气泡尺寸估计会有一定影响,即一般轴承中油膜破裂区中比较细而密的条状油流和空穴,在本实验中可能会集合成较宽大而稀疏的条状油流和空穴. 但根据质量守恒原理可知,其相对容量(即条状油流和空穴的总体百分比)应无影响,油膜破裂的边界位置也应无影响.

润滑油:研究中使用的润滑油是 20♯ 透平油,并在油中加入了红色油溶性颜料,以便拍照. 润滑油采用重力润滑,即利用自重流入实验腔.

测量仪器:轴颈的旋转速度用 GH-85 涡流传感器和 DVF-R 型数字矢量滤波器标定. 动偏心量使用百分表粗测,使用千分表精测.

照相设备:摄影需要强光,采用两只 1 kW 的汞灯同时照射. 照相机为 PRAKTICA 型.

综上所述,本文作者设计的滑动轴承静态、动态油膜气穴研究实验台具有下列特征:

(1) 转速利用控制箱无级可调,最高转速可达 2 000 r/min.

(2) 静偏心率在水平方向 0~1 可调,在垂直方向按需可调.
(3) 动偏心率在 0~1 可调.
(4) 侧泄量可测,且可以观测侧泄油管内流体流动的情况.
(5) 可在 0~360° 内进行油膜分布的可视观察研究.

2 实验程序

实验研究中可变的参数是相对间隙 C/R、长径比 L/D、静偏心率 ε_s、动偏心率 ε_d、旋转速度 n_j 和流动条件,一旦决定研究哪一种情况,必须仔细地按下面步骤进行:

(1) 安装所需轴颈,用百分表检测其旋转外圆,并沿测量标准面移动,仔细校核轴颈轴线的平行度.

(2) 松开支承实验轴承的两轴承座,利用标准垫块建立所需静偏心量,仔细校核轴承套轴线与轴颈轴线的平行度.并用塞尺检测轴承周围的间隙,核准后拧紧紧固螺栓.

(3) 转动偏心装置,从监视刻度盘上找到所需的动偏心量的大概位置,用百分表检测轴颈的外圆和标准面,以得到精确的动偏心量.

(4) 放所需润滑油到规定的高度,至少在前天晚上将油注入贮油桶,以便排除油内空气.试验时让油充满实验腔.

(5) 连接电动机,利用控制箱设定到给定的转速 n_j (r/min).

(6) 建立强光照明,然后转动装置,检查和调整 n_j 的数值.安装 PRAKTICA 型照相机,采用 1/1 000 s 的速度进行拍照.

(7) 清理回油桶,以便得到较精确的侧泄量.

3 实验结果和讨论

3.1 小扰动情况下油膜分布的研究

当长径比 $L/D=1.0$,间隙 $c=0.15$ mm,静偏心率 $\varepsilon_s=0.2$,动偏心率 $\varepsilon_d=0.1$,$n_j=900$ r/min,轴承的工作工况处于较小偏心量工况时,且静偏心量占主要成分.从所得照片可以看出:在此工况下轴承内油膜分布的情况,既呈现定常工况情况,即在上游区域存在油膜再生成的状态,有类似喇叭口的油膜分布情况;又呈现非定常工况的情况,有的照片在轴承内的 90° 附近有一椭圆形气泡存在,即在承载区域有气泡存在.因为若按定常工况情况来讲,在 0°~180° 左右应为承载区域,不应该有大气泡存在.这与 Cole 所获得的静载工况下的油膜分布明显不同.虽然彩色照片清晰,但由于油膜薄,对比度差,复印效果不佳,只能非常粗略地定量绘图表示,见图 2(a).以下同样处理.

当长径比 $L/D=1.0$,$c=0.15$ mm,$\varepsilon_s=0.4$,$\varepsilon_d=0.1$,定常工况占主导地位时,从照片可以看出,转速不同时轴承内下游区油膜气穴分布有明显的差异.与 Cole 所获得的静载工况下的油膜分布明显不同.当 $n_j=300$ r/min 时,油膜气穴出现在轴承中间约占 0.5 L 的地方,油流和气穴带都呈现较宽的条状,在油膜破裂区气穴带的宽度与油流的宽度约为 2:1,在差不多同一位置(约 186°)的地方整齐破裂,而在靠近轴承两端的地方油膜是连续的,见图 3(a).当 $n_j=1 450$ r/min 时,轴承内下游区的油膜气穴分布出现在轴承的中间部位约占

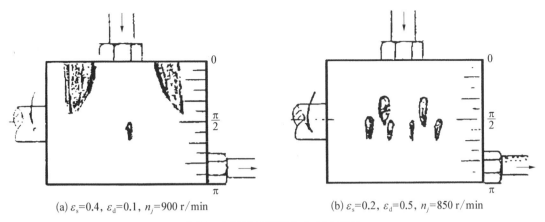

(a) $\varepsilon_s=0.4, \varepsilon_d=0.1, n_j=900$ r/min (b) $\varepsilon_s=0.2, \varepsilon_d=0.5, n_j=850$ r/min

图 2　上游区油膜气穴分布示意图

0.8L 的地方. 油流和气穴带呈现比 $n_j=300$ r/min 时窄很多的非常不规则的条状, 有点类似蕨类植物. 油流和气穴带的宽度之比约为 2∶1. 油膜差不多在同一角度位置的地方整齐破裂(约 184°). 在靠近轴承两端的地方油膜仍然是连续的, 见图 3(b).

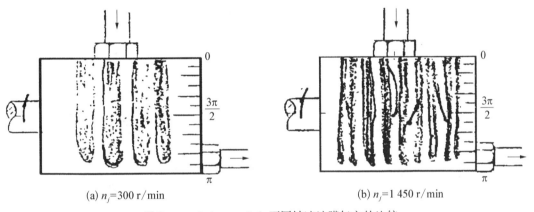

(a) $n_j=300$ r/min (b) $n_j=1\,450$ r/min

图 3　$\varepsilon_s=0.4, \varepsilon_d=0.1$, 不同转速油膜气穴的比较

当长径比 $L/D=1.0$, $c=0.20$ mm, $\varepsilon_s=0.4$, $\varepsilon_d=0.2$, $n_j=600$ r/min 时, 在 $-30°\sim 90°$ 的区域内, 轴承的中间部位出现比 $n_j=300$ r/min 时大些的气泡, 有时呈现长型. 在上游区有油膜破裂的情况, 细小气泡由下游区流入上游区, 同时向轴承两侧流动, 然后流入侧泄油管, 侧泄油管的上方有比 300 r/min 时大一些的气泡移动, 移动速度加快.

当 $n_j=850$ r/min 时, 在大约 $-70°\sim 150°$ 有气泡存在, 看见有气泡通过承载区而不破裂, 而又跟随轴颈到大约 $-70°\sim 150°$ 的区域, 即从静态的角度考虑, 再形成边扩大, 侧泄油管的气泡增大, 移动速度加快.

当 $n_j=1\,000$ r/min 时, 在 $-90°\sim 170°$ 区域有气穴存在, 其他类似 850 r/min 的情况.

当 $n_j=1\,450$ r/min 时, 在大约 $-90°\sim 175°$ 有气穴存在, 侧泄油管内的气泡明显增大, 移动加快.

当 $n_j=2\,000$ r/min 时, 在大约 $-97.5°\sim 180°$ 的区域有气穴存在, 侧泄油管内上的气泡明显增大, 移动加快.

显然在相同的轴承参数下, 气穴分布的区域随转速的升高而增大, 示意图见图 4.

当长径比为 $l/d=1.0$, $c=0.20$ mm, $e_s=0.08$ mm, $\varepsilon_s=0.4$, $\varepsilon_d=0.4$ 时, $n_j=300$ r/min,有时一开始就有大气泡出现,通过承载区不破裂,而又回到再生成边.但在同样的操作条件下,有时有气泡,有时没有.小气泡随轴颈旋转的同时流向轴承的两侧,流向及侧泄油管内的情况同上实验.

当 $n_j=600$ r/min 时,气泡大,明显通过承载区而不破裂.

当 $n_j=850、900、1000、1450、1500、1600、2000$ r/min 时,气穴的大小从圆到长,明显随转速变化,气穴的区域变动,有时直到180°时也存在气穴.

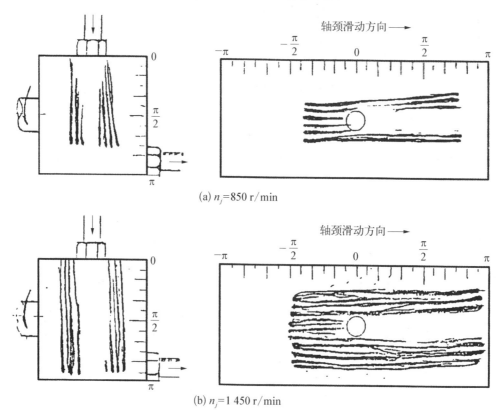

(a) $n_j=850$ r/min

(b) $n_j=1450$ r/min

图 4 $\varepsilon_s=0.4$, $\varepsilon_d=0.2$,不同转速油膜气穴分布的比较

3.2 大扰动情况下油膜分布的研究

当长径比为 $l/d=1.0$, $c=0.15$ mm, $\varepsilon_s=0.2$, $\varepsilon_d=0.5$, $n_j=850$ r/min,即非定常工况占主导地位时,在静态的收敛区域,从所得照片可以看出,在小于90°地方,轴承的中间位置出现两个明显可见的单个椭圆形气泡,这单个气泡基本处于稳定期,见图2(b).仔细看可以发现在大于90°的地方还存在肉眼可见的小气泡.在现场可以看见小气泡跟随轴颈的转动流向下游,并向轴承两端流去,侧泄油管内的上面气泡较多,并向侧泄油桶方向流去.从整体上看油膜在靠近轴承的两端是连续的.从所得照片可以看出,在90°的区域,轴承的中间位置出现四个明显可见的单个长椭圆形气泡,这些单个气泡基本处于破裂前期,见图5(a).

当 $n_j=900$ r/min 时,在静态的收敛区域,从照片可以明显看出大而长的气泡出现在轴承内约 60°~120°的区域上,这些单个气泡基本处于破裂前期,见图5(b),比 $n_j=850$ r/min 时大得多,从整体上看油膜虽然晃动较大但在靠近轴承的两端仍旧是连续的.在现场可以看见轴承内小气泡随轴颈的转动流向下游区和轴承的两端,侧泄油管内壁上面聚集的气泡较 $n_j=850$ r/min 时稍大、多,并向侧泄油桶方向流去.

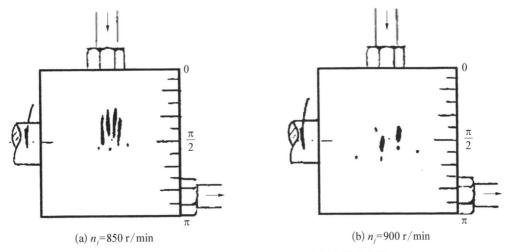

(a) $n_j=850$ r/min (b) $n_j=900$ r/min

图 5 $\varepsilon_s=0.2$, $\varepsilon_d=0.5$,上游区油膜气穴分布比较

当长径比为 $l/d=1.0$, $c=0.15$ mm, $\varepsilon_s=0.2$, $\varepsilon_d=0.6$, $n_j=300$ r/min 时,从所得一系列照片可以观察并分析出,由于动偏心量的挤压效应及转速的变化,承载油膜区域在轴承内周向漂移,这与静载工况完全不同.

从试验的整个过程来看,油膜的气穴分布是在大约 120°及以后的一些区域. 而从实验观察来看,最容易发生单个气泡的地方是在 0°~120°的地方. 特别是在转速较高时,实验中观察的气穴分布的区域更大,这必定使轴承内油膜承载力下降. 实验中观察的气穴分布的区域与理论计算结果有很大的差异. 这有待进一步的研究.

4 结论和存在的问题

由以上研究的有关的资料可以看出:

(1) 所设计的试验台在进行滑动轴承静、动载荷及混合载荷试验是切实可行的.
(2) 试验台设计机理正确,试验研究中出现的情况与设计前的设想基本相符.
(3) 当静偏心率占主导地位时,在低转速下,有类似静载工况的油膜分布,存在喇叭口状的过渡区,所以利用质量守恒理论处理的边界条件更贴近试验.
(4) 小扰动情况下,油膜气穴分布区域明显随转速变化. 转速变高,气穴区域变大.
(5) 大扰动情况下,由于挤压作用,油膜区域随动偏心的转动而在周向漂移,因而油膜分布的区域波动较大,并且气穴分布区域随转速的增大而增大,随偏心量的增大而增大.
(6) 只要有动偏心量存在,即使在相同的工况下,单个大气泡有时存在,有时不存在.
(7) 验证了 Jacobson and Hamrock 和 Sun and Brewe 的部分实验结果.

存在的问题:

(1) 由于有机玻璃轴承套刚度不足,影响测量精度,且其热膨胀系数较大,在长时间的实验中影响轴承间隙的精度.
(2) 由于各方面条件的限制,光学记录仪器还不够先进,因此在动态定量的记录上尚可进一步提高.

参 考 文 献

[1] Cole J A, Hughes C J. Oil Flow and Film Extent in Complete Journal Bearing. Proc Inst Mech Eng,1956,170: 499-510.
[2] Sun D C, Brewe D E. A High Speed Photography Study of Cavitation in a Dynamically Loaded Journal Bearing. Journal of Tribology, Trans ASME,1991,113: 287-294.
[3] 张直明. 滑动轴承的流体动力润滑理论. 高教局出版社,1986.
[4] 孙美丽. 360°滑动轴承静载和动载油膜气穴分布的研究.[硕士学位论文],上海大学,1995.

Experimental Study of the Film Distribution of Statically and Dynamically Loaded Cylinder Journal Bearing

Abstract: This paper aims at bearing in the working situations of full journal bearing under steady loads and dynamically loads and a test rig has been designed and built which can be run in centered whirl, off-centered whirl and mixed whirl. On this test rig, an experimental study has been carried out to check with part of experimental results of Jacobson and Hamrock and Sun and Brewe. The experimental results are also compared with theoretical calculations and some useful conclusions have been obtained.

有限长线接触弹流润滑研究的现状与展望[*]

摘　要：在已有的研究工作积累和文献调研的基础上,对有限长线接触弹流润滑研究的形成背景、发展过程、研究现状及其应用等进行了简要的综合介绍与分析评述,并指出了现有研究工作的特点、存在的问题和进一步研究的主攻方向及其必要性,展望了这方面研究工作的发展前景.

线接触副在工程实际中的应用相当广泛,大多存在于滚子类轴承中的滚动体与滚道之间,存在于啮合的渐开线齿轮的轮齿之间,存在于盘状凸轮与从动滚子之间,存在于摩擦式无级变速器的摩擦滚子之间.由于这些零部件的长度均为有限值,而且必须在润滑条件下工作,故此线接触副工作时接触压力的大小及其分布,以及润滑状况和成膜机理等,对接触副的承载能力、速度性能、工作温度和使用寿命都有很大的影响,这些都是进行零部件设计的重要依据.根据静弹性接触理论[1,2]分析认为,直母线轮廓有限长线接触副的接触压力分布不同于经典的 Hertz 无限长线接触理论的接触压力分布,在其端部会出现"边缘效应".理论分析和实践经验都表明[2,3],要减小或消除"边缘效应",就应当对线接触副的轮廓母线作适当的修形,从而产生了以改善接触副的摩擦学状态和延长其使用寿命为目标的新的摩擦学设计内容——凸度设计.凸度设计归属于精细几何设计,以微米计量,它包含凸型和凸度量这2个方面的内容,目前是以三维有限长静弹性接触理论[4]为设计的理论基础.

随着现代工业技术的高速发展,对线接触副提出的高速、重载和高可靠性的要求越来越高,通常需要它在高于 0.7 GPa 的 Hertz 接触压力下工作.在这样的工况下,润滑剂的特性将发生不同于常压状态下的变化,接触副的表面将产生与润滑油膜厚度相同数量级甚至比之更大的弹性变形.由于这两方面作用的偶合,产生了与普通流体润滑机理不同的弹性流体动力润滑(简称弹流,英文缩写为 EHL)作用.运用弹流理论[5-8],可以建立弹性接触表面几何形状、尺寸、材料性能、润滑流体黏度、润滑流体密度、表面速度、载荷、油膜厚度、压力分布、摩擦力和摩擦温升等参数的定量关系.在弹流润滑状态下,有限长线接触副的成膜机理,油膜压力的大小及其分布形式与静弹性接触条件下的是否有差别?差别如何?怎样计算以及如何设计才能使之达到最优工作状态等,都是应当进行深入研究解决的问题.

目前,对线接触副润滑状态的分析,主要还是依据无限长线接触和长椭圆接触条件下的弹流理论.但是,无限长线接触弹流理论不考虑端泄,而椭圆接触弹流考虑的则是二次曲面轮廓,这显然与有限长线接触副的轮廓形状不同,端泄状态也不相同.因此,这2种简化的弹流理论都只适用于分析线接触副中间部分的润滑状态.

但是,尽管考虑凸度等实际因素的有限长线接触弹流研究目前还处于艰难的探索过程中,然而毕竟已经取得了相当大的进展.为了推动对有限长线接触副润滑状态分析研究工作的发展,本文作者以自己多年研究工作的积累为基础,总结了国内外在有限长线接触弹流润滑方面已有的研究成果,并且提出了这个领域中今后有待进一步探索的几个主攻方向.

---

[*] 本文合作者：陈晓阳、马家驹.原发表于《摩擦学报》,1997,17(3)：281-288.

1 有限长线接触弹流研究的进展

1.1 实验技术

Gohar 等[9]最早对直母线滚子修形前后的弹流油膜厚度和膜形进行了滑动状态下的光干涉实验对比研究,给出了滚子在纯滑动过程中的光干涉图像,指出二者之间有明显的区别. Wymer 等[10]以镀膜技术[11]为基础,采用柔性局部加载技术,首次采用光干涉法研究了纯滚动条件下一圆锥滚子与玻璃盘接触所产生的线接触弹流. 实验结果给出了油膜形状和油膜厚度极为翔实的信息,显示了滚子端部修形与否对弹流油膜特性的影响和贫油润滑,以及滚子表面划痕等现象,并且清楚地表明了有限长线接触弹流油膜形状与长椭圆接触[12]之间的差别甚大,尤其滚子端部的油膜厚度远比滚子中部出口处的小.

值得强调指出的是,Wymer 等的工作不仅揭示了有限长线接触弹流有许多个性问题需要进行探索,奠定了开展这项研究工作的基础,而且也是弹流光干涉实验技术的一次重大突破,因为用一直径为 25.4 mm 的钢球与玻璃板接触时,要得到 0.7 GPa 的 Hertz 接触压力只需要施加 76 N 的载荷,而在用相同直径的滚子与玻璃板接触情况下,若要得到同样大小的 Hertz 接触压力,就需要施加 319 N/mm 的载荷,这意味着整个实验装置的结构更加复杂、技术难度更大、精度要求更高. 但遗憾的是他们只进行了一轮研究,没有就不同修形轮廓及不同凸度量对润滑油膜的影响进行考察,这是因为那时修形轮廓实际使用还比较少,加工还相当困难.

由于玻璃等透明材料不能承受高拉应力,而且局部加载技术的难度又比较大,后来其他人发表了几篇有关线接触光干涉方面的实验研究结果,但基本上都只限于模拟低速轻载或高速轻载工况[13-20],也都没有研究修形轮廓对润滑油膜的影响.

由于修形轮廓线接触副不仅是发展潮流,尤其近年来在我国的使用也日益增多,作者采用柔性局部加载技术,成功地运用光干涉实验方法对比研究了中等载荷条件下,2 种对数凸型滚子的凸度量对弹流油膜厚度和膜形的影响[21]. 实验结果表明,在给定的实验条件下,存在一个使滚子轴向油膜厚度和接触压力分布最为均匀的最佳凸度量. 这个最佳凸度量比运用静弹性接触力学方法在相同条件下求得的最佳凸度量大.

一般地说,人们对有限长线接触弹流所进行的实验研究采用的都是光干涉法. 可以将其分为 2 种类型:一类是模拟实际线接触副的工作情况[13-16,20],由于玻璃材料难以承受弯曲应力,只能在高速轻载条件下进行研究;另一类是研究平面玻璃与圆锥滚子之间的有限长线接触弹流问题[10,17-19,21],这类试验虽然与实际线接触副的工况不尽相同,但如采用柔性局部加载技术,就可以实现 Hertz 接触压力高达 0.7 GPa 重载条件下的弹流现象[10,21],故其是有限长线接触弹流机理研究的有效手段.

1.2 数值计算方法

尽管有限长线接触弹流与点接触弹流同属于二维弹流问题,其中有关后者的理论分析方法早在 70 年代中期就取得了突破性的进展[22-25],而对于前者的理论分析方法的研究进展却很缓慢. 有人曾经运用对无限长线接触解法的修正方法来研究有限长线接触问题,但只讨论了有限长线接触弹流的端部润滑油膜厚度,都没有涉及接触副的轮廓修形问题[26,27]. 此外,也有人将椭圆率为 8 的点接触弹流解近似用于有限长线接触的情况[23],但这只能用于椭球面接触轮廓,而不能用来分析有限长直母线接触和其他修形母线轮廓接触的弹流问题,因为它们端部的形状完全不同.

文献[10]报道的实验结果表明,滚子端部润滑油膜的厚度取决于局部的几何轮廓,如果端部未作修形,则润滑油膜非常薄.根据三维有限长静弹性接触理论[2],在直母线有限长滚子与半无限空间接触的情况下,其端部的接触压力呈奇异性分布,因而此处的弹流油膜厚度在理论上应当为零.

70 年代末和 80 年代初,有人在完成了一系列三维有限长静弹性接触压力数值计算的基础上,结合文献[23,25]报道的分析方法之优点,改用轴向非均匀网格划分法来适应滚子端部急剧变化的压力,研究了端部运用相切和相交圆弧修形的 2 种凸型滚子与半无限空间接触的弹流数值计算方法,并且分析了后一种凸型滚子修形圆弧半径对端部最小润滑油膜厚度的影响[28-30]. Mostofi 等[28]得出的结论是:"有限长滚子在富油润滑条件下的弹流最小油膜厚度和最大油膜压力都出现在滚子的端部.油膜厚度会受到经过选择的修形滚子的几何轮廓母线的影响,而最大油膜压力有可能会超过仅用静弹性接触理论分析得到的压力值."虽然他们只给出了一个完整的有限长线接触弹流数值解结果,而且这种结果在油膜形状和油膜厚度上均与文献[10]报道的实验结果差别较大,然而这无疑是人们在这个方向上迈出的可喜的一步.尤其值得一提的是,运用这种方法计算得到的滚子中部油膜厚度与通过实验测得的膜厚[31],以及与无限长线接触理论[32]或椭圆率为 8 的点接触理论[23]求得的膜厚吻合都比较好. Kuroda 等[33]运用有限元和 Newton-Rephson 混合法分析了两等长直母线滚子的弹流润滑问题,虽然所得结果有限,但其结论与文献[28]报道的相同.

尽管在 80 年代后期人们对有限长线接触弹流分析方法的研究几乎处于停顿状态,然而在进入 90 年代之后,有关这个问题的数值分析方法的研究报道又逐渐地多了起来,重新受到越来越多的摩擦学工作者的关注.我国有些学者运用逆解法分析了端部以相切和相交圆弧修形的直母线滚子弹流[34];运用顺解法研究了正反圆锥滚子弹流润滑问题[35];对端部用相切和相交圆弧修形的直母线和对数母线轮廓滚子弹流进行了数值分析,给出了两组滚动速度和凸度量对油膜压力分布、膜形和膜厚影响的分析结果[36-38].

国外有些学者运用差分法和 Newton-Rephson 混合法建立了端部以相切和相交圆弧修形的 2 种凸型滚子弹流数值计算的另一种方法,并且分析了前一种凸型滚子修形圆弧半径对端部最小油膜厚度的影响,虽然结论与文献[28]报道的相同,但其数值解结果比后者的更为合理. Evans 等[40-43]运用一种简化方法探讨了滚子侧泄对弹流膜厚的影响,随后又针对圆锥齿轮的弹流数值解法进行了详细研究,给出了一组与文献[20]报道的实验数据较为吻合的计算结果.

近年来,有人已经对圆锥滚子的弹流数值解法进行了初步研究[44],对直母线和对数母线轮廓滚子弹流的进一步深入研究也获得了突破性的进展[45,46],不仅能够数值分析中等载荷、中等速度、中等材料参数和凸度量对油膜压力分布、膜形和膜厚的影响,而且能够进行定量光干涉试验[21],所得到的结果与文献[10,21]报道的吻合较好,并且清楚地说明了凸度量对线接触副的端部弹流膜厚和油膜压力分布有很大的影响,指出进行凸度量设计应当考虑滚动速度和润滑剂性质等因素.

综上所述,对于有限长线接触弹流数值分析方法的研究在过去 20 多年中经历了 3 个不同的发展时期,已经为其进一步发展奠定了基础.目前,计算机性能虽然比 20 年前点接触弹流分析成功时提高了几个数量级,然而现有的有限长线接触弹流数值计算方法还只能求得理想条件(等温、Newton 流体和光滑表面)下若干工况时的解,这足以说明其难度很大,还有待人们进行更深入系统的大力研究.

2 现状分析

2.1 现有研究特点

虽然国内外不少学者对有限长线接触弹流进行了长期的研究,但不论是在试验研究方面,还是在数值分析方面,多数都只进行了一轮研究.尽管人们通过研究发现了许多线接触副润滑自身所特有的现象,然而运用已成熟的无限长线接触或点接触理想条件弹流理论却不能对这些现象作出令人信服的解释.其中,最主要的是不能够说明端部修形与否对其油膜压力分布和膜形所产生的极大影响,以及中部出现的贫油润滑等现象.由于对所发现的问题尚未进行系统深入的探索,人们目前还不能够透彻阐明这些问题.由此可见,关于理想条件下的有限长线接触弹流理论尚未建立.

有限长线接触弹流润滑的接触区域极小而狭长,其宽度一般都小于 0.2 mm,长宽比大于 30,而且润滑油膜相当薄,常处于贫油润滑状态.因此,除光干涉法外,采用其他现有的弹流膜厚或油膜压力测试方法很难进行试验研究.由于光干涉法可以直观、准确地观察到接触区域各处的油膜形状及其变化,能够精确标定进行定量研究,这也是其在现有有限长线接触弹流试验研究中被普遍采用的一个重要原因.

目前,对于有限长线接触弹流润滑数值分析采用的几乎都是顺解法,人们已经对多种凸型轮廓进行过分析计算,由于收敛性很差,每种凸型基本上都只能得到 2~3 种工况下的结果,这显然太少,难以说清楚问题,更无法得出回归公式.在当前的情况下,还没有一种计算方法的结果得到与之对应的相关实验的系统验证,然而这些方法的共同结论是:有限长线接触副中部的润滑油膜厚度及压力分布都与无限长线接触弹流理论一致;端部情况非但与后者不同,而且亦与长椭圆接触弹流理论相差较大;端部轮廓修形与否,以及修形量大小都对此处的润滑油膜厚度和压力分布的影响很大;现有算法的收敛性与点接触弹流分析计算方法的相比都很差.

值得一提的是,现在至少有 3 种计算方法所得结果的油膜形状与已有的光干涉实验照片吻合较好,而且变化趋势相同,这为今后的深入研究奠定了基础.

2.2 存在的问题

从技术角度看,有限长线接触光弹流试验的关键在于获得稳定和高质量的干涉图像,这取决于试验滚子的质量、干涉光线的强度、好的干涉测量系统,以及试验装置中驱动系统和加载系统的平稳性.

如前所述,由于接触区域极为狭长,还没有能够清晰地拍得超过滚子半长度的干涉测量系统,目前的试验结果仅限于局部而非全貌.试验滚子的表面粗糙度及滚子端部的倒角、球端面的轴线与滚子轴线的共线度都对图像质量影响很大,而滚子凸型与凸度量的加工质量又与前述几何参数负相关,因而很难获得合格的试验滚子.柔性局部加载装置中的柔性件制约了载荷的提高,目前最大接触压力仅能达到 0.75 GPa,而有限长线接触副主要是用于重载场合,所以目标值应当超过 1 GPa. 各凸型与凸度量对润滑膜形和膜厚影响的对比研究几乎还是空白,而这正是此项研究的核心内容之一,可以为优化最佳成膜条件下的线接触副提供实验依据并检验理论分析结果.

现有的表面弹性变形计算方法难以兼顾高精度与占用内存小之矛盾,而润滑膜厚精度是影响收敛性的关键因素之一. 由于有限长线接触弹流润滑计算模型本身的复杂性和计算机容量及运算速度的限制,虽然已经采取了利用对称性、控制划分单元数和用多项式函数来拟合局部区域上的压力分布等技巧,但是目前还没有一种能够简单而高精度地拟合接触压力分布函数,且在迭代求解中以较少的内存容量、高效率、高精度地计算表面弹性变形的好方法,这是限制出口区网格划分单元数的主要原因.

因为有限长线接触是非 Hertz 接触,其静弹性接触区域必须通过数值方法求出,所以接触区域参数 a 和 b 的精度均受制于网格划分的精度,而且彼此线性无关. 此外,由于凸度量与凸度量测量点的位置是除半径 R 外的另外 2 个决定表面轮廓的参数,故有限长线接触弹流润滑的数学模型比点接触弹流润滑的多 3 个自变量,这显然就给它的数值分析和无量纲化带来了很大的困难.

由于有限长线接触弹流润滑相对于椭圆接触弹流来说,其求解区域更为狭长,域内压力分布的光顺性更差,而且出口区和端部的压力梯度很大,特别是当凸度量较小时,因接触轮廓在端部的曲率变化大而产生膜厚分布的奇异性. 这些都是造成 Reynolds 方程强非线性的重要因素,虽然已经通过 Φ 变换或取几种不同数值计算方法的优点组成混合算法等加以改善,但效果并不大好,这是数值计算收敛性差的又一关键因素.

3 近期的主攻方向

根据上面的分析与讨论,并且结合弹流润滑研究的发展过程,作者认为在过去 30 多年的研究与探索中,人们已经为有限长线接触弹流研究的快速发展奠定了基础,预料在不远的将来,通过研究人员的共同努力,理想条件下的分析方法会有新的突破. 为此目的,应将以下几方面的工作作为近期研究的主攻方向:

在光弹流试验已揭示出有限长线接触弹流润滑基本个性特征的基础上,着重开展不同修形轮廓和不同凸度量对润滑油膜影响的对比研究,以期获得成膜状况最好条件下的线接触副最佳修形轮廓的实验依据,并在大量实验数据的基础上建立便于工程界使用的经验公式. 研究中还应注意考察贫油现象的成因并寻找相应的克服方法,因为供油方式对线接触副的功能发挥和使用性能影响很大.

分析现有各种数值计算方法的优缺点,并在此基础上针对一种或几种凸型建立能够适应某段工况的分析方法,进行不同凸度量对润滑油膜影响的计算,经过实验验证其正确性后,对此段工况内的成膜机理进行系统的分析研究,找出凸度量对润滑油膜影响的规律,分析其接触压力分布、凸度与静弹性接触状态下的差别,供凸度设计参考,并在大量数值计算数据的基础上得出分段回归公式,供工程界线接触副润滑设计参考.

加强简单高精度拟合接触压力分布函数的研究,并力求能在迭代求解中以较少的内存容量、高效率、高精度地计算接触表面的弹性变形,积极开展强非线性 Reynolds 方程收敛性和有限长线接触弹流无量纲参数群合理性的研究,寻找能够适应较宽工况条件的有效数值分析方法,为进一步建立理想条件下有限长线接触弹流理论奠定基础.

4 前景展望

从几何角度看,无限长线接触、点接触和有限长线接触仅仅是接触副轮廓参数的变化而已,而且前两种接触形式是后者向两端变化的极限特例,因而它们不涉及弹流本质.从逻辑上说,弹流理论不应被接触状态所割裂,即现有的各弹流理论可以而且应该统一.之所以被分开进行研究,是因为前两种情况属于Hertz接触范畴,数学模型简单,但其理论却显然难以解释中间那么宽广的变化问题,反之则相反.

由于工程问题自身发展的需要,有限长线接触弹流研究不仅突破了传统的Hertz状态润滑问题的讨论范围,而且将有助于全面弄清楚几何参数变化对高副成膜机理的影响,并为统一弹流理论的最终建立扫清障碍.

有限长线接触弹流理论不仅可从最佳接触压力分布角度,而且也能从最佳膜厚分布角度进行凸度量设计,因而对重载滚子轴承、齿轮、箔材轧辊和机械无级变速器等线接触副的廓型分析与润滑设计,改善机械的使用性能、提高其使用寿命都具有重要的实际意义,这无疑是考虑因素更为全面的凸度优化设计新理论.

在理想条件下的有限长线接触弹流理论的基础上,可以进一步研究在动态弹流润滑油膜压力场作用下接触副表面层应力的分布和变化规律,以及它们对滚动接触疲劳寿命的影响;也可以进一步考虑粗糙表面、非Newton流体、动载荷、贫油润滑和非等温等多种因素的综合影响,从而使其更加符合实际工况,为线接触副零件的动态设计提供理论依据和分析方法,这是进行此项研究的最终目标之一.

参 考 文 献

[1] Jonson K L. Contact Mechanics. London: Cambridge University Press, 1985: 129 - 152.
[2] 马家驹.线接触副凸度设计的研究.第五届全国摩擦学学术会议论文集(上册),武汉,1992:167 - 179.
[3] Reusner H. The logarithmic roller profile—the key to superior performance of cylindrical and taper roller bearings. Ball Bearing Journal, 1987, 230: 2 - 10.
[4] 马家驹.滚子凸度设计.轴承,1992(1):11 - 15.
[5] Dowson D, Higginson G R. Elastohydrodynamic Lubrication, the Fundamentals of Roller and Gear Lubrication. Oxford: Pergamon Press, 1977: 35 - 97.
[6] Hamerock B J, Dowson D. Ball bearing lubrication. New York: John Wiley & Sons Inc, 1981: 162 - 237.
[7] Gohar R. Elastohydrodynamics. Chichester: Ells Horwood Limited, 1988: 58 - 183.
[8] 温诗铸,杨沛然.弹性流体动力润滑.北京:清华大学出版社,1992:71 - 130.
[9] Gohar R, Cameron A. The mapping of elastohydrodynamic contacts. ASLE Trans, 1967, 10: 215 - 225.
[10] Wymer D G, Cameron A. Elastohydrodynamic lubrication of a line contact. Proc Instn Mech Engrs, 1974, 188: 221 - 238.
[11] Foord C A, Wedeven L D, Westlake F J, et al. Optical elastohydrohynamics. Proc Instn Mech Engrs, 1969, 184(1): 487 - 503.
[12] Bahadoran H, Gohar R. The oil film in elastohydrodynamic elliptical contacts. Wear, 1974, 29: 264 - 270.
[13] Bahadoran H, Gohar R. Oil film thickness in lightly-loaded roller bearings. J Mech Eng Sci, 1974, 16: 386 - 390.
[14] Gohar R. The lubrication of roller bearings subjected to couples. Tribol Int, 1975, 8: 21 - 26.
[15] Pemberton J C, Cameron A. An optical study of the lubrication of a 65 mm cylindrical roller bearing. J Lubr Technol(ASME Trans), 1979, 101: 327 - 337.
[16] Ford R A J, Foord C A. Studies on the separating oil film (EHD oil film thick ness) between the inner race and

rollers of a roller bearing. Mech Engng Trans, the Institution of Engineers, Australia, 1981: 140－144.
[17] Ren N, Zhu D, Wen S Z. Experimental method for quantitative analysis of transient EHL. Tribol Int, 1991, 24: 225－230.
[18] Hirata M, Cameron A. The use of optical elasto hydrodynamics to investigate viscosity loss in polymer-thickened oils. ASLE Trans, 1984, 27: 114－121.
[19] Dmytrychenko N, Aksyonov A, Gohar R, et al. Elastohydrodynamic lubrication of line contacts. Wear, 1991, 151: 303－313.
[20] Parkins D W, Rudd L. Thrust cone lubrication, Part Ⅲ: A test facility and preliminary measured data. Proc Instn Mech Engrs (Part J), 1996, 210(J2): 107－112.
[21] Chen X Y, Zhou S Q, Ma J J. Oil film thickness and shape in Lundberg's profile roller contacts. Proceedings of the 23rd Leeds-Lyon Sym on Tribol, Leeds, 1996: 415－422.
[22] Hamerock B J, Dowson D. Isothermal elastohydrohynamic lubrication of point contacts, Part Ⅰ: Theoretical formulation. J Lubr Technol (ASME Trans), 1976, 98: 223－229.
[23] Hamerock B J, Dowson D. Isothermal elastohydrodynamic lubrication of point contacts, Part Ⅲ: Fully flooded results. J Lubr Technol (ASME Trans), 1977, 99: 264－276.
[24] Hamerock B J, Dowson D. Isothermal elastohydrodynamic lubrication of point contacts, Part Ⅳ: Starvation results. J Lubr Technol (ASME Trans), 1977, 99: 15－23.
[25] Ranger A P, Ettles C M M, Cameron A. The solution of the point contact elastohydrodynamic problem. Proc Roy Soc (London), 1975, A346: 227－244.
[26] Bahadoran H, Gohar R. End closure in elastohydrodynamic line contact. J Mech Engng Sci, 1974, 16: 276－278.
[27] Hooke C J. The elastohydrodymamic lubrication of heavily loaded contacts. J Mech Engng Sci, 1977, 19: 149－156.
[28] Mostofi A, Gohar R. Elastohydrodynamic lubrication of finite line contacts. J Lubr Technol (ASME Trnas), 1983, 105: 598－604.
[29] Heydari M, Gohar R. The influence of axial profile on pressure distribution in radially loaded rollers. J Mech Engng Sci, 1979, 21: 381－388.
[30] Johns P M, Gohar R. Roller bearings under radial and eccentric loads. Tribol Int, 1981, 14: 131－136.
[31] Kannel J W. Comparison between predicted and measured axial pressure distribution between cylinders. J Lubr Technol (ASME Trans), 1974, 96: 508－514.
[32] Dowson D, Higginson G R. A numerical solution to the elastohydrodynamic problem. J Mech Engng Sci, 1959, 1: 7－15.
[33] Kuroda S, Arai K. Elastohydrodynamic lubrication between two rollers. J JSME, 1985, 28: 1367－1372.
[34] 崔子伟, 张和豪, 华东耘. 重载有限宽滚子弹流润滑的逆解法. 上海工业大学学报, 1991, 12: 303－311.
[35] 莫云辉. 正反圆锥滚子弹流润滑: [学位论文]. 上海: 上海工业大学机械系, 1994.
[36] 陈晓阳. 等温有限长线接触弹性流体动力润滑: [学位论文]. 杭州: 浙江大学机械系, 1993.
[37] Ma J J, Chen X Y, Liu X B. Elastohydrodynamic lubrication of Lundberg's profile roller contacts. Proc Int Sym on Tribol, Beijing, 1993: 132－138.
[38] 陈晓阳. 修形滚子凸型对润滑油膜的影响. 机械设计与研究, 1996, 2: 2－3.
[39] Park T J, Kim K W. The elastohydrodynamic lubrication of profiled cylindrical roller. Proceedings of Int Tribol Confe-rence (Yokohama), 1995, 2: 1043－1048.
[40] Evans H P, Snidle R W. Film thinning factor for rollers of finite width under elastohydrodynamics conditions. Wear, 1994, 175: 17－23.
[41] Barragan de Ling FdM, Evans H P, Snidle R W. Thrust cone lubrication, Part Ⅰ: Elastohydrodynamic analysis of conical rims. Proc Instn Mech Engrs, 1996, 210(J2): 85－96.
[42] Barragan de Ling FdM, Evans H P, Snidle R W. Thrust cone lubrication, Part Ⅱ: Elastohydrodynamic analysis of crowned rims. Proc Instn Mech Engrs, 1996, 210(J2): 97－105.
[43] Snidle R W, Evans H P. Elastohydrodynamics of gears. Proceedings of the 23rd Leeds-Lyon Sym on Tribol, Leeds,

1996: 271 - 280.

[44] Yamashita R, Dowson D, Taylor C M. An analysis of elastohydrodynamic film thickness in tapered roller bearings. Proceedings of the 23rd Leeds-Lyon Sym on Tribol, Leeds, 1996: 617 - 638.

[45] 陈晓阳. 有限长线接触弹性流体动力润滑机理的研究. 上海市青年科技启明星计划项目课题总结报告, 上海大学, 1996.

[46] 徐文. 滚子摩擦副弹流润滑特性研究: [学位论文]. 杭州: 浙江大学机械系, 1996.

Prospects for Elastohydrodynamic Lubrication of Finite Line Contacts

Abstract: The background, progress, and recent research state of elastohydrodynamic lubrication of finite line contacts are summarized and analyzed based on literature research and studies in author's laboratory. Then the features and problems of the research about this subject are pointed out. Finally, some suggestions about the future study and the prospects on this subject are proposed.

弹簧支承式推力轴承的热弹流研究[*]

提　要：本文对弹簧支承式推力轴承进行热弹流研究,联立求解广义雷诺方程、完整的三维能量方程、热传导方程、固体热弹变形方程和润滑油的温粘关系,计入了轴瓦热弹变形、弹簧压缩变形以及热效应的影响,提出了弹簧刚度等效的方法,研究各种参数变化对轴承性能的影响.计算结果表明,弹簧支承方式能降低最大油膜压力,使瓦面受载比较均匀,速度、载荷和进油温度变化对轴承性能有很大影响,对于弹簧支承方式,轴瓦的弹性变形大于热变形,热弹变形量达到甚至超过油膜厚度的数量级,变形的大小直接影响到轴承性能.

0　引言

推力轴承是水轮发电机组的关键部件,其设计优劣直接关系到机组能否正常运行.可倾瓦单支承推力轴承是一种常用轴承,在推力负荷不大及比压较低时是比较安全可靠的.随着水电机组的单机容量不断增大,推力轴承的负荷、瓦面积和比压不断增大,单支承可倾瓦的缺点日益明显,轴瓦严重变形,受力分布不好,造成油膜温度升高,严重时甚至造成油膜破坏、轴瓦烧损.我国的一些大型水电站如葛洲坝、白山、大化等都发生过烧瓦的恶性事故,造成巨大经济损失.为改善推力轴承性能,国际上用多点支承取代单支承方式已成趋势.弹簧支承式推力轴承是在托瓦上分布着上千个性能完全一致的小弹簧来承受轴向力,形成多点支承,见图1.弹簧支承式推力轴承避免了可倾瓦推力轴承受力集中的缺点,受力均匀,结构紧凑,自调整能力强,散热情况良好,对机组稳定性,减少振动、摆动有益处.

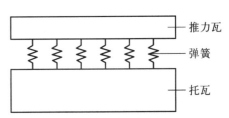

图1　弹簧支承式推力轴示意图

但由于这种轴承对载荷变化和初始油膜形状十分敏感,要可靠地分析和估计轴承性能,计算十分复杂,因此许多人都避免设计这种结构[1].迄今为止,据作者所查,国际上有关弹簧支承式推力轴承性能计算的报道非常少,仅有文献[2—5],而对其进行热弹流研究的只有文献[5].文献[5]讨论了轴瓦尺寸对轴承性能的影响,并指出对于薄瓦,弹簧支承范围特别重要,如果支承布满整个轴瓦,会产生大量的弹性变形导致出现空穴.但是文献[5]计算轴瓦变形时,采用克希霍夫薄板假设,使计算精度受到影响,并且文献[5]中有关弹簧刚度等效的方法有误,造成了距离弹簧近的节点等效刚度反而小的结果.

本文对弹簧支承式推力轴承进行了三维热弹流研究,计入轴瓦热弹变形和弹簧压缩变形因素,建立弹簧刚度等效的方法,分析各种工况参数变化对轴承性能影响.

---

[*] 本文合作者：王小静、张国贤.原发表于《上海大学学报(自然科学版)》,1997,3(增刊)：134 - 141.

1 理论分析

推力轴承工作运行时,润滑油膜会由于黏滞损耗产生热量,导致润滑油黏度降低,轴瓦产生热变形,同时,受到巨大的轴向推力,轴瓦产生弹性变形,弹簧产生压缩变形,以上因素使轴承承载力和油膜厚度发生变化.因此要能较接近实际地计算弹簧支承式推力轴承的性能,必须进行热弹流分析.

热弹流研究需联立求解的控制方程有:广义雷诺方程、完整的三维能量方程、固体热传导方程、固体热弹变形方程、油膜厚度方程及温粘关系等.

广义雷诺方程用于求解推力轴承的油膜压力场,计入了润滑油黏度的三维变化.广义雷诺方程如下:

$$\frac{\partial}{\partial r}\left(F_2 r \frac{\partial p}{\partial r}\right) + \frac{\partial}{\partial \theta}\left(F_2 \frac{\partial p}{r \partial \theta}\right) = \omega r \frac{\partial}{\partial \theta}\left(\frac{F_1}{F_0}\right), \tag{1}$$

式中: $F_0 = \int_0^h \frac{1}{\mu} dz$; $F_1 = \int_0^h \frac{z}{\mu} dz$; $F_2 = \int_0^h \frac{z(z-z_m)}{\mu} dz$; $z_m = \frac{F_1}{F_0}$.

润滑油膜的四周边界上压力为零.

完整的三维能量方程为

$$\rho c_p \left(u \frac{\partial T}{r \partial \theta} + v \frac{\partial T}{\partial r} + w \frac{\partial T}{\partial z}\right) = K \left[\frac{\partial^2 T}{r^2 \partial \theta^2} + \frac{\partial}{r \partial r}\left(r \frac{\partial T}{\partial r}\right) + \frac{\partial^2 T}{\partial z^2}\right] + \mu \left[\left(\frac{\partial u}{\partial z}\right)^2 + \left(\frac{\partial v}{\partial z}\right)^2\right]. \tag{2}$$

在油膜与轴瓦的界面上,保持热流量相等,即油膜中传出热流量等于轴瓦传入热流量:

$$K \frac{\partial T}{\partial z}\bigg|_{z=h} = K_B \frac{\partial T_B}{\partial z_B}\bigg|_{z_B=0}. \tag{3}$$

在油膜的入口区,存在不同温度的润滑油的混合现象.进入润滑油膜的油有两部分组成:一部分是从上瓦出来的热油跟随镜板一起运动,进入下瓦;另一部分是油槽中的冷油进入油膜.新补充的冷油的流量 Q_{sup} 与热油流入的流量 Q_{rec} 之和等于润滑油膜入口区流入的流量 Q_{in},即

$$Q_{\text{in}} = Q_{\text{sup}} + Q_{\text{rec}},$$

根据热量平衡,得到:

$$Q_{\text{in}} \times [T|_{\theta=0}]_{r,z} = Q_{\text{sup}} \times T_0 + Q_{\text{rec}} \times [T|_{\theta=\theta_0}]_{r,z}, \tag{4}$$

其中,T_0 为油槽温度,$[T|_{\theta=0}]_{r,z}$ 与 $[T|_{\theta=\theta_0}]_{r,z}$ 分别为油膜进油边与出油边的二维温度分布.由于热油进入下瓦的流量 Q_{rec} 未知,故假设一个混油比例系数 β,热油携带的流量等于混油比例系数与上瓦出油边流量的乘积,即 $Q_{\text{rec}} = \beta \times Q_{\text{out}}$.

轴瓦的热传导方程为拉普拉斯方程:

$$\frac{\partial^2 T}{r^2 \partial \theta^2} + \frac{\partial}{r \partial r}\left(r \frac{\partial T}{\partial r}\right) + \frac{\partial^2 T}{\partial z^2} = 0. \tag{5}$$

轴瓦与油槽中冷油接触,进行对流换热.

轴瓦受到压力和温度梯度的作用产生热弹变形,同时弹簧受到压力产生压缩变形.弹簧的压缩变形量与轴瓦热弹变形量耦合在一起,形成总变形量,使油膜厚度发生变化,各点的变形量不同,其膜厚值也不同.油膜厚度可写成如下表达式:$h(r, \theta) = h_0(r, \theta) + d(r, \theta)$,其中,$h_0$ 为基本膜厚分布,d 为总变形量.

每块轴瓦下面分布着几十个小弹簧,弹簧的分布位置各不相同且没有规律.进行变形计算时只能考虑节点载荷,每个弹簧的支反力需要等效到相邻节点上.论文[5]提出的弹簧刚度等效方法是:如果一个弹簧与相邻四个节点 A, B, C, D 的距离分别为 a, b, c, d,单个弹簧刚度为 K,则等效到 A 点的刚度值为 $\dfrac{a}{a+b+c+d} \times K$,其他节点依次类推.这样造成了距弹簧近的节点等效刚度反而小的错误结果.

本文利用形函数方法,对弹簧刚度进行等效,如图 2 所示.

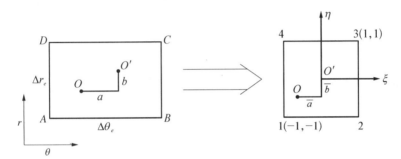

图 2 弹簧刚度等效方法示意图

设有一弹簧,刚度为 K_0,弹簧中心位置在任意点 O,相邻的四个节点 A, B, C, D,其网格中心在 O' 点,径向网格间距为 Δr_e,周向网格间距为 $\Delta \theta_e$,O 与 O' 沿径向和周向的距离分别为 a 和 b.将整体坐标转换成局部坐标,整体坐标中的 A 点对应于局部坐标中的 1 点,相应地 B 点对应于 2 点,依次类推.坐标转换的关系表达式为

$$\xi = \frac{\theta - \theta_{O'}}{\dfrac{\Delta \theta_e}{2}}, \quad \eta = \frac{r - r_{O'}}{\dfrac{\Delta r_e}{2}};$$

则各点的等效刚度可表达为

$$[K]_{A, B, C, D} = K_0 \times [N], \tag{6}$$

其中,$[N]$ 是形函数,$N_i = \dfrac{(1+\xi_i \xi)(1+\eta_i \eta)}{4}$,$\xi_i = -1, 1, 1, -1$;$\eta_i = -1, -1, 1, 1 (i = 1, 2, 3, 4)$.

将各节点对应的形函数代入式(6),可得到相邻节点上的弹簧等效刚度:

$$K_A = K_0 \times \frac{(1+\bar{a})(1+\bar{b})}{4}, \quad K_B = K_0 \times \frac{(1-\bar{a})(1+\bar{b})}{4},$$

$$K_C = K_0 \times \frac{(1-\bar{a})(1-\bar{b})}{4}, \quad K_D = K_0 \times \frac{(1+\bar{a})(1-\bar{b})}{4},$$

$$K_A + K_B + K_C + K_D = K_0,$$

其中

$$\bar{a} = \frac{a}{\frac{\Delta\theta_0}{2}}, \quad \bar{b} = \frac{b}{\frac{\Delta r_0}{2}}.$$

联立求解上述控制方程时,先假设初始油膜厚度分布,算出相应的油膜压力分布和油膜温度分布,求出轴瓦的热弹变形,计算新的油膜厚度.重复计算,直到压力分布和温度分布以及变形分布收敛.检查此时的油膜承载力是否与外负荷相等,若不相等,则修改初始膜厚.重复上述计算,直到油膜承载力与外负荷平衡为止.

2 性能分析

本文针对湖北清江隔河岩水利枢纽从加拿大 GE 公司引进的一套弹簧支承式扇形推力轴承进行性能分析,具体参数如下[6]:

瓦块数	20	进油温度	40℃
瓦张角	18°	油槽温度	40℃
轴瓦外径	3 605 mm	单瓦比压	3.326 MPa
轴瓦内径	2 608 mm	单瓦弹簧数	73
瓦厚	88 mm	弹簧刚度	$1.055\,5\times10^7$ N/m
镜板角速度	14.293 s^{-1}	弹簧直径	50.67 mm

2.1 油膜压力、温度、膜厚分布

根据上述工况、几何参数,进行热弹流分析,计算得轴承静特性为如下分布(见图3).

从上述等值线分布图可知,油膜压力峰值出现在径向中部偏向出油边的位置,油膜温度从进油边向出油边缓慢升高,最高值出现在出油边偏向外径侧距离轴瓦表面较近的位置,油膜厚度沿周向逐渐减小,形成一个收敛的油楔.其最大油膜压力为 7.5 MPa,品质系数 P_{max}/P_m 达到 2.22,最大油膜压力与平均油膜压力相差不大,最高油膜温度较低,为 61℃,最小膜厚大于 80 μm.上述计算结果表明该轴承的性能相当好.

2.2 速度变化对轴承性能的影响

图 4 表明,随着转速的提高,油膜温度、摩擦力和最小膜厚增加迅速,油膜压力也有增加,但增幅较小.这主要是由于转速增加,使油膜剪切率增大,黏滞损耗和发热量增大,导致油膜温度升高、摩擦力增大.

2.3 载荷变化对轴承性能的影响

从图 5 可以看出载荷对轴承性能影响非常显著.随着载荷增大,油膜压力升高,油膜厚度迅速减小,使流量相应减少,油膜温度上升.当载荷进一步增加时,除油膜压力继续与载荷成正比上升外,其他性能曲线变化趋缓,最高油膜温度位置逐渐移向出油边.载荷变化使膜厚分布变化明显,见图6,当载荷小时,热变形比较显著,最小膜厚出现在靠出油边的轴瓦中部;载荷大时,高的油膜压力使轴瓦弯曲变形加大,形成轴瓦中部下凹、内外径边上翘的形

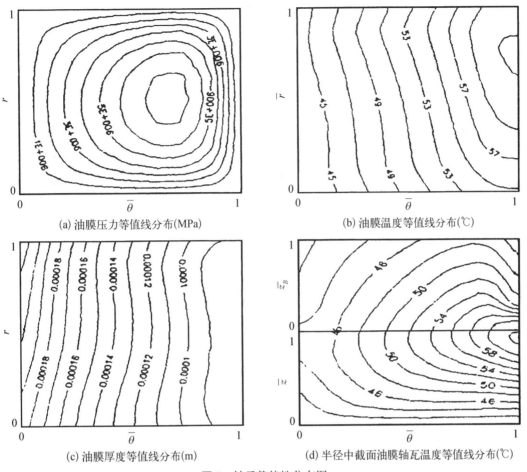

(a) 油膜压力等值线分布(MPa)

(b) 油膜温度等值线分布(℃)

(c) 油膜厚度等值线分布(m)

(d) 半径中截面油膜轴瓦温度等值线分布(℃)

图 3 轴承静特性分布图

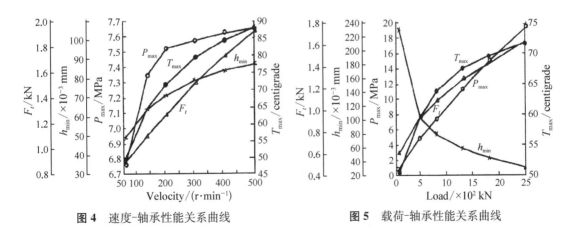

图 4 速度-轴承性能关系曲线　　**图 5** 载荷-轴承性能关系曲线

状,因此最小膜厚出现在靠出油边的内外径边上.

2.4 进油温度变化对轴承性能的影响

进油温度变化对油膜压力影响不大,但对轴承其他性能影响显著. 随着进油温度的升高,油膜温度迅速升高,黏度下降,膜厚减小,摩擦力减小(见图7).

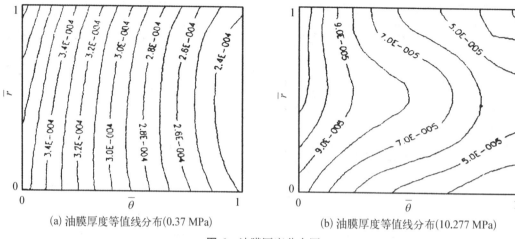

(a) 油膜厚度等值线分布(0.37 MPa)　　　(b) 油膜厚度等值线分布(10.277 MPa)

图 6　油膜厚度分布图

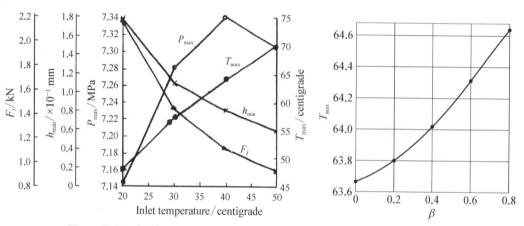

图 7　进油温度-轴承性能关系曲线　　　图 8　混油比例系数-最高油膜温度关系曲线

2.5　混油比例系数变化对轴承性能的影响

计算结果表明,混油比的大小与油膜压力、油膜厚度关系不大.混油比的变化对油膜温度场产生影响,从图 8 可知,混油比增大使油膜温度升高,混油比从 0 到 0.8,最高油膜温度上升 1 ℃.入口区的温度分布变化比较明显,混油比不同,油膜温度要相差几度.

2.6　弹性变形量与热变形量的比较

图 9 表明,弹簧支承时的弹性变形量十分大,最大变形出现在进油边沿径向中部,呈现出轴瓦径向中部变形大,而内外径边变形小的特点,最小变形出现在出油边的内外侧两端.弹簧支承时的热变形除了有向下的变形量外,还有向上的凸变形,凸变形出现在轴瓦径向中部偏向出油边的位置.把轴瓦的弹性变形与热变形进行比较,弹性变形量明显大于热变形量,弹性变形量最大达到 1 mm.由于热变形量和弹性变形量的大小都达到甚至超过油膜厚度的数量级,因此计算弹簧支承式推力轴承时,必须计算热弹变形,尤其要精确计算弹性变形.

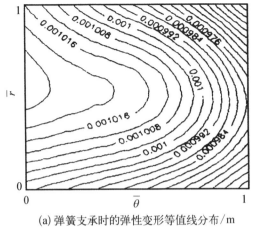
(a) 弹簧支承时的弹性变形等值线分布/m

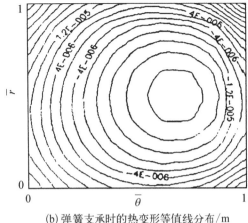
(b) 弹簧支承时的热变形等值线分布/m

图 9　变形分布图

3　结论

本文对弹簧支承式推力轴承进行三维热弹流求解,计入了热弹变形及弹簧压缩变形的影响,提出了弹簧刚度等效新方法,分析各种参数变化对轴承性能影响.现得出如下结论:

(1) 弹簧支承式推力轴承与传统的可倾瓦单支承推力轴承相比,具有受力均匀等优点.在本文算例中,弹簧支承式推力轴承的 $P_{\max}/P_m$ 仅为 2.22,而可倾瓦推力轴承的 $P_{\max}/P_m$ 要达到 3～3.5,弹簧支承方式降低了最大油膜压力值.

(2) 速度、载荷和进油温度对轴承性能影响显著.速度上升使油膜温度、摩擦力和膜厚增大,而油膜压力的变化幅度较小.载荷增大,使轴瓦的弹性变形增大,油膜压力增加迅速,膜厚开始迅速减小,随后由于油膜阻尼作用,减小缓慢.油膜压力对进油温度的变化不敏感,混油比例系数的变化对油膜温度场尤其是进油区域的温度有影响.

(3) 由于是弹簧支承方式,计算轴瓦的热弹变形更显重要.计算表明,弹性变形量要大于热变形量,热弹变形量达到甚至超过油膜厚度的数量级,变形的大小直接影响到轴承性能.

参 考 文 献

[1] Abramovitz S, Discussion in Ettles CM. Some Factors Affecting the Design of Spring Supported Thrust Bearings in Hydrodynamic Generators. ASME Jour of Trib, 1991, 113: 626 – 632.

[2] Ashour M E, et al. Elastic Distortion of a Large Thrust Pad on an Elastic Support. Tribology International, 1991, 24(5): 299 – 309.

[3] Sinha A N, Athre K, Biswas S. Spring-supported Hydrodynamic Thrust Bearing with Special Reference to Elastic Distortion Analysis. Tribology International, 1993, 26(4): 251 – 263.

[4] Sinha A N, Athre K, Biswas S. A Nonlinear Approach to Solution of Reynolds Equation for Elastic Distortion Analysis to Spring-Supported Thrust Bearing. STLE, Tribology Trans, 1994, 37(4): 802 – 810.

[5] Ettles C M. Some Factors Affecting the Design of Spring Supported Thrust Bearings in Hydrodynamic Generators. ASME, Jour of Trib, 1991, 113: 626 – 632.

[6] 廖贵成. 小弹簧式推力轴承的安装与运行. 湖北电力技术, 1994(1 – 2): 74 – 76.

TEHD Study of Spring Supported Thrust Bearings

Abstract: In the paper, TEHD study of spring supported thrust bearing is carried out. Generalized Reynolds equation, 3 - D energy equation, heat conduction equation and solid thermal-elastic distortion equation are solved. Thermal-elastic distortion of pad, displacement of springs and thermal effect are included. The method of spring equivalent stiffness is put forward. The effects of parameters variations on the performance of the pad are researched. The analysis shows that thermo-elastic distortion plays an important role in the operation of the pad.

多跨转子—滑动轴承系统非线性动力学仿真

摘　要：本文提出了一套新的关于大自由度的转子—轴承系统的非线性动力学仿真计算方法，其中包括块三对角矩阵追赶法、中心差分法＋Houbolt 法的预估计—校正时间积分法，系统方程在物理坐标 x 和 y 上的"解耦"，系统初始位移值的计算方法、计入基础参振的计算方法和非线性油膜力数据库的应用等．用上述方法详细地考察了柔性转子轴承系统不平衡质量大小和分布对转子轴承系统稳定性的影响问题．仿真计算的结果与对比性的实验结果较为吻合．

符 号 说 明

M_p, C_p, U_p, V_p, W_p——转子第 j 个离散节点上的质量、阻尼和三个刚度分矩阵；

X_p, F_p——转子第 j 个离散节点上的位移矢量和力矢量；

M, C, K——转子离散系统方程的总质量矩阵、总阻尼矩阵和总刚度矩阵；

X_t, F_t——转子离散系统方程在 t 时刻的位移矢量和力矢量；

$\tilde{M}$——转子离散系统方程时间积分的等效质量矩阵；

R_t——系统方程在 t 时刻的广义力矢量．

1　前言

高速旋转机械如大型汽轮发电机的转子轴承系统对动力性态有很高的要求，以确保安全工作．由于滑动轴承的油膜力具有强烈的非线性特性，目前常用的线性分析方法在某种程度上存在着很大的不足，对许多实际现象无法作出合理的解释．国内外越来越多的研究者开始把注意力转向对该系统的非线性动力学分析．但目前已有的各种算法在使用上各有不便．如：转换到模态坐标分析计算，需精确地计算系统方程的特征值和特征向量，对大型矩阵而言，有一定的难度，如直接采用有限元法，当系统的自由度较大时，即使在压缩自由度后仍会对计算机容量有较高的要求，计算速度也较难提高．由于必须同时迭代满足边界的约束条件和转子的位移和轴承的非线性油膜力的耦合关系，DT－TMM 在计算速度上和计算精度上较难完善．此外，计算中大量用到的非线性油膜力获取方法也需改进．由于上述原因，目前对转子—轴承系统的非线性动力学计算较多的局限于 Jeffcot 转子模型，即便是多跨转子模型，也是用较少的节点来描述．

本文作者提出了一套新的关于大自由度的转子—轴承—基础系统的非线性动力学计算方法．并用该方法考察了转子轴承系统各种非线性稳定性问题．仿真计算的结果与对比性的实验结果较为吻合．

---

* 本文合作者：李志刚．原发表于《自然杂志》，1997，19(增刊)：76－82．

2 系统广义动力学方程

转子—轴承—基础系统的广义动力学方程的建立方法可参照文献[1]. 也可用有限元法. 这两种方法所建立的广义动力学方程对我们要求解的问题而言,在矩阵结构上是相同的. 本文采用前一种方法.

第 j 节点的动力学方程可为:

$$M_j \ddot{X}_{jt} + C_j \dot{X}_{jt} + W_j X_{(j-1)t} + U_j X_{jt} + V_j X_{(j+1)t} = F_{jt} \quad (1)$$

系统的广义动力学方程可组装成:

$$M\ddot{X}_t + C\dot{X}_t + KX_t = F_t \quad (2)$$

其中系统的质量矩阵 M 和阻尼矩阵 C 分别是由节点的块质量矩阵 M_j 和块阻尼矩阵 C_j 构成的块对角矩阵,而系统的刚度矩阵 K 是由节点的刚度分矩阵 U_j, V_j, W_j 构成的块三对角矩阵. 即:

$$K = \begin{bmatrix} U_1 & V_1 & & & & & \\ W_2 & U_2 & V_2 & & & & \\ & \ddots & \ddots & \ddots & & & \\ & & W_j & U_j & V_j & & \\ & & & \ddots & \ddots & \ddots & \\ & & & & W_{n-1} & U_{n-1} & V_{n-1} \\ & & & & & W_n & U_n \end{bmatrix} \quad (3)$$

3 中心差分法+Houbolt 法的预估计-校正时间积分法

二阶问题的时间积分法有显式和隐式两大类[2],其形式分别为:

$$\text{显式}: MX_{t+\Delta t} = R_t \quad (4a)$$

$$\text{隐式}: MX_{t+\Delta t} = R_{t+\Delta t} \quad (4b)$$

显式时间积分法中主要有中心差分法. 隐式的有 Houbolt 法、Wilson θ 法和 Newmark β 法等. 文献[2]对这几种时间积分法的计算稳定性和计算精度问题作了较为详尽的分析. 从中可得知:中心差分法属条件稳定,其条件为: $\Delta t \leqslant \Delta t_{cr} = T_n / \pi$. T_n 是系统的最小振动周期. 由于转子是由连续介质构成,从理论上讲 $T_n \to 0$. 即使是离散化的转子模型,当轴段划分得小时, T_n 是相当小的. 从这个意义上讲,这种计算方法不宜单独用在对式(4b)问题的求解上. 但把中心差分法作为二阶隐式问题时间积分预估计方法是十分合适的. 其在每个 Δt 上预估计的初值和迭代矫正后的"精确值"相比有很高的精确度,可以有效地减少迭代次数,提高计算速度.

尽管 Newmark β 法和 Wilson θ 法在计算精度上优于 Houbolt 法,但受线性化的加速度假设的限制,应用范围和收敛性比不上 Houbolt 法. 作者认为,二阶隐式问题的时间积分法由中心差分法和 Houbolt 法组合应当优于现有其他方法,这已被作者系统的计算考察所证实[3].

4 块三对角矩阵的追赶法

当用式(4b)求解 $X_{t+\Delta}$ 时,$\tilde{M}$ 为块三对角阵,对 $\tilde{M}$ 的求逆无法直接转化为对块矩阵的求逆问题,如直接求 $\tilde{M}^{-1}$,当 n 较大时,即使计算机的内存允许,计算速度和计算精度也将受到较大的影响.本文采用块三对角矩阵追赶法,将上述结构的矩阵的求逆转化为 n 个块矩阵的求逆.工作量比原来的求($nm \times nm$)的矩阵 K 的逆阵要少得多,而且基本上不受 n 的大小的限制.

块三对角矩阵的追赶法求解具有式(2)结构的大型稀疏矩阵的极为有效的方法.文献[4-6]曾介绍此种方法.V. Castelli 用在解润滑油膜压力分布的"Column Method"的计算公式和块三对角矩阵的追赶法公式完全相同.但其定义也仅基于润滑油膜压力分布的列矢量概念.将块三对角矩阵的追赶法用在转子动力学问题的求解上,还未见文献记载.

5 系统方程的"解耦"简化

由于在式(4b)所涉及的问题中,广义力矢量 $R_{t+\Delta}$ 是由 $X_{t+\Delta}$ 决定的未知矢量,计算是一个预估计—矫正的过程.我们可以把使 x 和 y 耦合的项移到方程的右方归入广义力矢量项,使系统方程在等号左边在 x 和 y 两个坐标上解耦为如下形式:

$$M\ddot{X}_t + KX_t = F_{X_t} \tag{5a}$$

$$M\ddot{Y}_t + KY_t = F_{Y_t} \tag{5b}$$

其中 M 和 K 是(2×2)块矩阵构成的对角阵和三对角阵,式(4b)的求解已被简化为对由(2×2)的块矩阵构成的三对角阵施行"追赶法"的问题.(2×2)的矩阵的求逆可直接用公式计算,不仅计算工作量降为未经简化前的 1/4,计算精度还有一个质的飞跃.求解过程中的数值误差仅剩下计算机字长的截断误差,对整个计算过程中的变量采用双精度型时,截断误差非常之小,以至与离散误差或预估计—矫正过程中的迭代误差相比成为微不足道.

6 系统各离散节点位移初始值的确定方法

在对多跨转子—轴承系统进行非线性动力学分析的仿真计算时,系统各离散节点的位移初始值的确定是一个十分重要的问题.各轴承的载荷分配在这里是一个静不定问题,在转速和外载荷给定的情况下,各轴承的安装标高的确定须引入($m-2$)个定解条件.同时必须计入在这些定解条件下轴颈在各轴承中浮起量和分配到各轴承载荷之间的关系.由于轴颈在轴承中位移和轴承所承受的载荷呈非线性关系,这个静不定问题没有封闭解,只能进行迭代计算.且这个静不定问题是二维的,在 X 方向也存在位移和载荷分配的静不定问题,且与 Y 方向的静不定问题互相耦合,构成了复杂的固流耦合的二维静不定问题.非线性分析的仿真计算时,如果位移初始值的误差过大,就相当于使被计算的系统在计算开始的瞬间受到较大的阶跃载荷,这种载荷响应要经过较多的计算周期才能得以衰减,严重的甚至会使计算出错中断.为了防止这种现象在计算中出现和使整个计算过程有较合理的和较为精确的计算初值,即由静平衡状态开始,作者研究设计了一套求解上述固流耦合的二维静不定问题的精度高而速度快的计算方法.大大提高了整体仿真计算的效率,详见[3].

7 非线性油膜力数据库的应用

非线性动力学分析的仿真计算是在时间域上的逐点积分计算,因为系统的运动方程(2)式是隐式方程,在时域上的每一离散节点上必须对轴颈在轴承处的位移和轴承的油膜力进行反复迭代计算,轴承非线性油膜力值的精度和求取的速度将直接影响系统非线性动力学分析计算的精度和效率.目前国内外的研究者在涉及这个问题时采用的方法主要有两种:(1)直接用数值方法解 Reynolds 方程以获得非线性油膜力值;(2)采用近似计算方法获得非线性油膜力值,如采用短轴承理论或无限宽轴承理论、图线法等.

用数值计算方法获得非线性油膜力可以有较高的计算精度,但须耗费大量计算机时.尽管现在计算机技术发展日新月异,直接数值计算的方法因其庞大的计算工作量目前还只能用在一些简单的非线性分析计算上.近似计算的方法虽有较高的计算速度,计算精度上却不尽人意.作者在研究建立线性化的油叶型轴承单瓦性能数据库[7]的基础上指导研究建立了非线性油膜力数据库[8],并已由作者成功地应用于多跨转子—轴承系统非线性动力学分析的仿真计算中.

非线性油膜力数据库的数据是事先通过数值计算所得,其应用仅是检索调用插值计算的过程,无须再进行繁重的数值计算.故由数据库获得非线性油膜力数据的方法同时具有了直接数值计算时计算精度高和检索调用时快速高效的特点.数据库技术在多跨转子—轴承系统非线性动力学分析的仿真计算中的应用使大规模而又高精度的计算成为可能.

8 计算结果与实验对比

本文作者用上述方法详细地考察了转子轴承系统非线性的稳定性问题.其中主要包括:不平衡质量大小和分布对转子轴承系统稳定性的影响;某些转子轴承系统,当其转速高于线性失稳转速时,转子轴心可涡动于一个小的轨迹而不发散现象之原因的探索;多跨转子轴承系统在不同类别大激励下的动力性态研究等.由于文章篇幅限制,本文仅介绍有关考察不平衡质量大小和分布对转子轴承系统稳定性的影响的计算结果与实验的对比.

转子轴承系统线性失稳转速是通过求解系统动力学方程的齐次线性方程组得到,因此无法计入实际存在的转子不平衡质量造成的同步强迫振动的影响.许多实验和研究表明,不平衡质量的大小对系统稳定性有明显的作用.作者所作的大量计算结果除支持了文献中关于刚性转子不平衡质量大小对系统稳定性影响的结论外,更重要的还发现:对柔性转子而言,不平衡质量的分布是一个重要的影响因素,其程度可能要超过不平衡质量大小的因素.计算表明:强迫振动的挠曲线与转子的一阶振型接近时,不平衡质量的存在和增加能对提高系统的稳定性无益,与转子的二阶以上的振型接近时,不平衡质量的存在和适量增加能有效地提高系统的稳定性.造成这种现象的原因是,转子—轴承系统的失稳转速通常是和系统的一阶固有振动有关的,当强迫振动的挠曲线与转子的一阶固有振型接近时,容易激起系统的大幅度一阶振动,而强迫振动所引起的油膜特性的增稳效应则是有限的,不足以抑制激发的油膜涡动.当强迫振动的挠曲线与转子二阶以上振型接近时,一阶振动的成分在系统振动中不占主导地位,与一阶固有振动有关失稳现象不易发生或即便发生也容易被抑制.一般来讲,实际转子的一阶动平衡是做得较好的,残余的不平衡质量往往对二阶或二阶以上的振动作用才较明显.这种分布的残余不平衡质量对转子轴承系统的稳定性提高会起有益的作用.

图 1 是单跨三圆盘转子试验台的 3 个圆盘分别加不同偏心质量的模拟升速峰—峰值曲线.转子几何尺寸见图 2.仿真计算时采用的轴承类型和参数为:$2\times170°$椭圆轴承,椭圆度 $m=0.1$,$\psi_{\min}=0.0027$,$\psi=0.003$,$l/d\approx0.6$,润滑剂黏度 $\mu=0.03(\text{Pa}\cdot\text{s})$,和试验条件基本一致.所不同的是:试验轴承是下瓦 $180°$、上瓦 $150°$ 圆轴承,$\psi=3‰$,上瓦双侧进油,进油压力为 0.6 kgf/cm^2.考虑到进油压力会使轻载轴承的进油口附近轴承间隙呈缓慢开扩楔形区域建立局部的完整油膜区,当轴颈在轴承中涡动时,轴承的非线性特性会使这局部的完整油膜区的作用加大,起到增大轴承椭圆度的作用.作者采用以较小的椭圆度的椭圆轴承来等效计入进油压力效应的圆轴承.

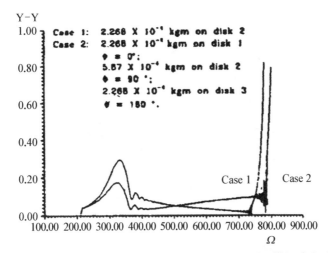

图 1 单跨三园盘转子试验台上各圆盘分别加不同的模拟升速图

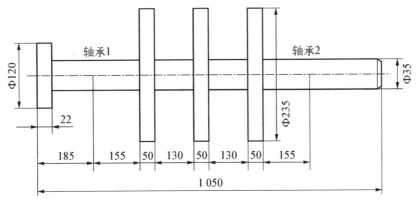

图 2 单跨多圆盘转子结构图

从仿真计算得知:激起二阶模态的偏心质量分布明显地提高了系统的失稳转速,而且使系统的失稳相对于激起一阶模态的偏心质量分布的失稳过程而言有一个渐进过程.仿真计算所得的两种不同偏心质量的分布下的升速时的峰—峰值曲线形状以及系统的临界转速,失稳转速与作者的实验曲线和数据较为吻合(图3).

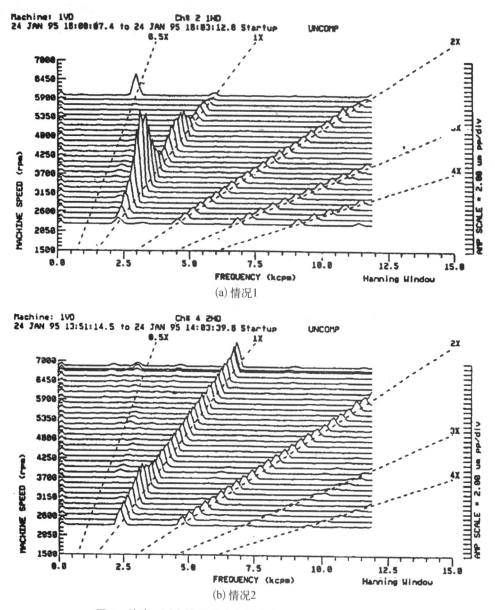

图 3　单跨三园盘转子试验台实验在垂直方向振动的瀑布图

9 结论

本文介绍了一套新的大自由度的转子—轴承—基础系统的非线性动力学计算方法. 提出中心差分法＋Houbolt 法的预估计—校正时间积分法,以解决二阶隐式非线性方程的时间积分问题;引入块三对角矩阵追赶法法成功地用于上述问题构成的大型矩阵的求解中;系统方程在物理坐标 x 和 y 上的"解耦"简化,使计算速度和计算精度得以进一步的提高. 大自由度的转子—轴承—基础系统的非线性动力学计算时系统各离散节点位移初始值的计算方法;计入基础参振的计算方法和非线性油膜力数据库在多跨转子—轴承—基础系统的非线性动力学计算中的应用.该方法的建立为大自由度的多跨转子—轴承—基础系统的非线性动力学性态的研究提供了一种高效而又精确的计算工具.

参 考 文 献

[1] 虞烈.轴承—转子系统的稳定性与振动控制研究.西安交通大学博士论文,1987.
[2] Bathe & Wilson. Numerical Methods in Finite Element Analysis, John Wiley Publishers.
[3] 李志刚.多跨转子—轴承系统非线性动力学研究.上海大学博士学位论文,1995.
[4] Golub G. H. & Van Loan, C. F. Matrix Computations The Johns Hopkins University Press,1983.
[5] George J. A. On Block Elimination for Sparse Linear Systems. SIAM J. Num. Anal., 1974,11:585-603.
[6] Varah J. M. On the Solution of Block-Tridiagonal Systems Arising from Certain Finite Difference Equations. Math. Comp., 1972, 26:859-868.
[7] Zhang, Z. Li Z. and Wang, W. Calculation of Lobe Type Hydrodymanic Bearings Aided by Database of Properties of Bush Segment. Proceedings of International Conference on Tribology 93, Beijing, China,1993:10.
[8] 王文.非线性油膜力数据库.硕士学位论文.上海工业大学,1991.

Non-Linear Dynamic Simulation of Multi-Span Rotor-Journal Bearing System

Abstract: A new calculation method of non-linear dynamics of rotor-bearing-fundation system of high degree of freedom is proposed in this paper, with the aim of studying the dynamic behavior of multi-span rotor supported by journal bearings subjected to large excitations. Triangular block matrix algorithm is successfully applied to solve the large matrix involved, central difference method is supplemented to Houbolt method to form a time integration method with estimation and correction to solve the time integration of second order implicit non-linear equation; simplification of system equation by decoupling between the physical coordinates x and y further improves the calculation accuracy and speed. Also introduced in the paper are: the method of calculation of the initial displacement of the system, and the application of datadase of non-linear oil film force in the non-linear dynamic calculation of rotor-bearing systems. The program realized with these techniques provides a calculation tool of high efficiency and accuracy for the study of non-linear dynamic behavior of multi-span rotor-bearing system with high degree of freedom.

用梁单元表达转子刚度时阶梯处过渡段的最佳等效参数*

摘 要：用有限元程序 ANSYS 5.4，采用 Solid 72 类型的单元，密布约 12 000 个节点，准确计算阶梯形轴段的弯曲挠度，据以反推确定梁单元模型中圆锥形过渡段的最佳半锥角，以及由此简化成的平均直径圆柱形过渡段的最佳长度。用最小二乘方法拟合得上述最佳值的线性公式，供转子动力学工程计算之用，以改进原有的过渡段参数的取值法。

1 引言

对实际转子进行转子动力学计算时，常常对各个离散的轴段采用梁单元作为模型，以求可以用效率比较高的方法来完成计算。实际转子往往由粗细不等的许多轴段构成，所以有很多阶梯。由于材料力学的平截面假设不适用于阶梯的附近部分，所以简单地按原结构划分梁单元，将过高地表达较粗轴段在阶梯附近部分的弯曲刚度。

工程计算时，对这个效应的一种计入方法，是在阶梯面偏粗轴段的一方，用一个等效圆锥形过渡轴段来代替实际结构(图1)，并再将它按平均直径 $0.5(D+d)$ 简化为一个短圆柱形单元来表达此处的弯曲刚度。

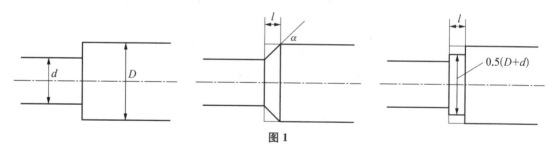

图 1

某些方法中，取这个等效圆锥段的半锥角为 $45°$。但未见有验证其合理性的报道。由于某些现代大型转子-滑动轴承系统已工作在稳定界限的边沿，其稳定性计算需要更仔细地计入各种附加因素，有必要对这个等效圆锥的最佳锥角进行考察，以求在用梁单元模化转子时能更准确地表达轴段的实际弯曲刚度。

本文所采用的考察方法，是以大型有限元程序 ANSYS 5.4 的三维有限元计算为依据，反推出梁单元模型中的等效圆锥轴段应当采用的半锥角，和进而转化成的过渡圆柱轴段应有的长度。这样的等效轴段梁单元模型，可以以 ANSYS 5.4 的精度来表达阶梯段的弯曲刚度。ANSYS 程序已被广泛公认为工程计算方面的一个具有相当权威性的有限元程序，在传统的固体力学计算范畴里，它常常可以代替实验。这也是本文以它的计算为依据的原因。

---

* 本文合作者：于军。原发表于《哈尔滨工业大学学报》1998,30(增刊)：253-256。

2 计算模型

图 2 表示本文所采用的计算模型. 其右端为插入端,左端受一弯矩作用.

用 ANSYS 5.4 计算其左端弯曲挠度 Y_{max}. 计算时采用 Solid 72 单元作为单元类型. 密布约 12 000 个节点. 可保证其计算误差可以忽略. 文末的附图简示单元和节点分布. Solid 72 为等参数四面体单元, 每个单元有四个节点, 每个节点有六个自由度.

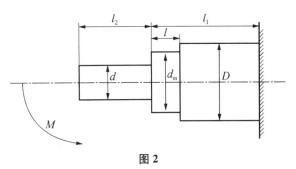

图 2

同时,将同一模型的较粗轴段在阶梯处的一小段替代为半锥角为 α 的过渡圆锥形轴段,其长度 l 等于 $0.5(D-d)/\tan\alpha$,再简化成长度同样为 l 而直径等于平均直径 $0.5(D+d)$ 的圆柱形过渡轴段,用材料力学中的梁弯曲公式表达其左端弯曲挠度. 即:

$$Y_{max}=\frac{M}{E}\left\{l\left(l_1+\frac{l}{2}\right)\left(\frac{1}{I}-\frac{1}{I_2}\right)+\left[\frac{l_1^2}{2I_1}+\frac{l_2(2l_1+l_2)}{2I_2}\right]\right\}$$

式中, $l=\dfrac{D-d}{2\tan\alpha}$; $I_1=\dfrac{\pi d^4}{64}$; $I=\dfrac{\pi(D+d)^4}{16\times 64}$; $I_2=\dfrac{\pi D^4}{64}$.

令它等于用有限元程序算出的 Y_{max},即可反推出圆锥形过渡轴段应有的半锥角 α 和圆柱形过渡轴段应有的长度 l.

如此推导出的半锥角 α 的确定公式为:

$$\tan^2\alpha\left[\frac{1}{2}\frac{l_1^2}{d^2}+\frac{l_2}{d}\frac{d^4}{D^4}\left(\frac{l_1}{d}+\frac{1}{2}\frac{l_2}{d}\right)-\frac{\pi E d^2}{64M}Y_{max}\right]+\frac{1}{2}\tan\alpha\cdot\frac{l_1}{d}\left(\frac{D}{d}-1\right)\left[\frac{16d^4}{(D+d)^4}-\frac{d^4}{D^4}\right]$$
$$+\frac{1}{8}\left(\frac{D}{d}-1\right)^2\left[\frac{16d^4}{(D+d)^4}-\frac{d^4}{D^4}\right]=0$$

上述计算在 $d/D=0.3\sim0.9$ 的范围内进行. 在轴段长度等于 $0.18\sim0.7$ 的范围内,轴段长度的取值不同对最佳半锥角的影响很小,差别不超过 $3°$.

相应地,可以算出等效圆柱形过渡轴段的最佳长度:

$$l=\frac{D-d}{2\tan\alpha}$$

或最佳长度比:

$$\frac{l}{D-d}=\frac{1}{2\tan\alpha}$$

3 计算结果

计算结果如表 1 所示.

表1 等效圆锥形轴段的最佳半锥角和圆柱形轴段的最佳长度比计算结果

d/D	0.3	0.4	0.5	0.6	0.7	0.8	0.9
l_1/d	3.000	3.000	7.000	5.833	5.000	4.375	3.899
l_2/d	11.667	8.750	7.000	5.833	5.000	4.375	3.899
$\bar{y}$	5.628	6.500	29.820	24.139	21.811	21.476	22.480
α	30.69	32.14	34.38	36.48	38.21	39.37	43.35
$l/(D-d)$	0.84	0.80	0.73	0.68	0.64	0.61	0.53

说明：$\bar{y}=\pi E d^2 Y_{max}/(64M)$，$Y_{max}$ 为 ANSYS 计算的最大挠度.

由计算结果可见，某些方法中对等效圆锥形过渡轴段采用45°的半锥角，有所偏大. 这会过高地表达阶梯处的弯曲刚度.

图3(a)示出等效圆锥轴段最佳半锥角 α 与直径比 d/D 的关系；图3(b)示出等效圆锥轴段最佳长度比 $l/(D-d)$ 与直径比 d/D 的关系.

由图可见，这些关系均十分近于直线关系. 经用最小二乘方法拟合得：等效圆锥形过渡轴段的最佳半锥角为

$$\alpha = 20.10 \frac{d}{D} + 24.32 \text{(deg)}$$

平均直径等效圆柱形过渡轴段的最佳长度比为

$$\frac{l}{D-d} = -0.50 \frac{d}{D} + 0.99$$

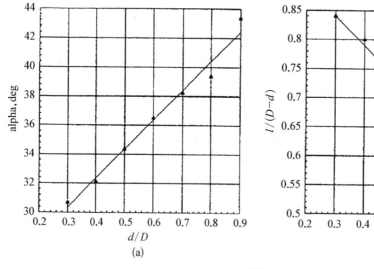
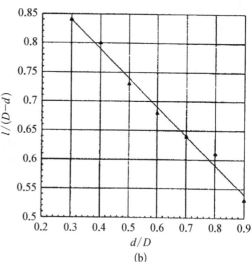

图3

4 结论

用梁单元模型作转子弯曲振动计算时，某些方法对等效圆锥形过渡轴段采用45°半锥角将过高地表达较粗轴段在阶梯处的弯曲刚度.

本文以 ANSYS 4.2 的准确计算为依据所推出的等效圆锥形过渡轴段半锥角值和平均直径等效圆柱形过渡轴段的最佳长度比，拟合得线性公式．用于转子弯曲振动等计算，可更精确地表达阶梯处的弯曲刚度．

参 考 文 献

[1] ANSYS. Modeling and Meshing Guide. 1998.
[2] ANSYS. Element Handbook. 1998.
[3] 宁俊. 材料力学. 上海工业大学，1989.
[4] 张直明. 转子动力学讲义. 上海工业大学，1990.

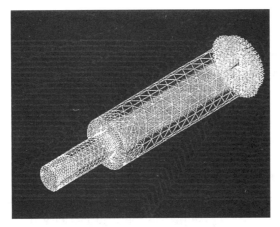

附图　单元及节点分布

The Optimal Equivalent Parameters for Expressing the Bending Stiffness of Stepped Rotor Segments by Beam Elements

Abstract：The bending deflection of stepped shaft segments are accurately calculated by FEM program ANSYS 5.4，with elements of the type SOLID 72，and 12,000 densely distributed nodes. The results are relied upon to derive the optimal half cone angle and the corresponding optimal length of the equivalent transitional cylindrical segment substituting for the stepped part during rotor dynamic calculations. The method of the least square is emplyed to obtain linear formulae for these optimal values，the adoption of which into rotor dynamics engineering calculations is capable of improving the parameter values assigned to the transitional segments.

Experimental Study of Active Magnetic Bearing on a 150 M³ Turbo Oxygen Gas Expander*

Abstract: This paper is concerned with the investigation, experiment and design analyses on the application of active magnetic bearings for a 150 m³ turbo oxygen gas expander having 1.16 kg weight and 30 mm diameter rotor, which was supported by two aerostatic bearings formerly. Now, the machine can work steady at a rotation speed of 92,000 r/min, and can be run up to a maximum rotation speed of 104,000 r/min.

1 Introduction

It is one of the important reasons for using active magnetic bearing to replace the aerostatic bearing in turbo oxygen gas expanders with rotation speed of 110,000 r/min and instantaneous speed of 130,000 r/min that the load capacity of the aerostatic bearing is inferior to that of the active magnetic bearing, and there also exists the cost of the air source, the higher precision needed for the machined parts, as well as the interest of the manufacturer for the other advantages of the active magnetic bearing, such as long working life, more stability, and freedom from oil or air seal technique[1-3]. This paper describes the design procedure for the bearing structure, parameter and control system, and also presents the experimental results.

2 Structure and Parameter Design

As limiting of the initial space of the expander and requiring of levitating technique, two radial magnetic bearings and a pair of thrust magnetic bearings are used to take the places of two aerostatic bearings. Each radial bearing has 8 magnetic poles, and their geometric and electrical parameters are listed in Table 1. The rotor including two wheels has about 1.16 kg weight and 210 mm length. The outer diameter of thrust plate is about 62 mm. The designed rotating speed is 108,000 r/min, and the instantaneous speed is required to reach 130,000 r/min. The sensors used for testing the displacement of the rotor center are 5 eddy current ones. Fig. 1 is the photograph of the radial and thrust bearings, and Fig. 2 shows the rotor with the magnetic bearings.

---

* In collaboration with Wang Xiping, Yu Liang, Wan Jingui and Cui Weidong. Reprinted from *Journal of Shanghai University*, 1998, 2(4): 334–336.

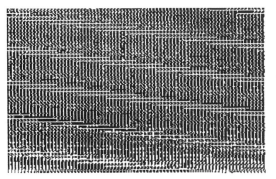
Fig. 1 Photograph of active magnetic bearings

Fig. 2 Photograph of a rotor with magnetic bearings

Table 1 Parameters of radial bearings

Parameters	Value	Parameters	Value
outer diameter(mm)	72	coil (tune/pole)	42
inner diameter(mm)	30	air clearance (mm)	0.16
thickness(mm)	15	DC resistance(Ω)	0.2
bias current(A)	0.7	inductance(mH)	1.8

3 Control System Investigation and Design

The design of control system is still based on the analogue PID controller[4]. The method and procedure of the design can be found in reference [5]. Independent control is adopted for each degree of freedom. The control diagram for each degree of freedom is shown in Fig. 3.

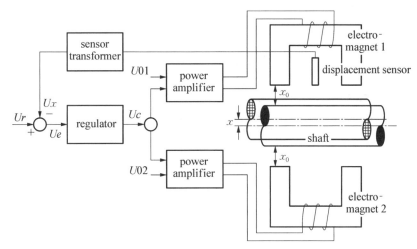
Fig. 3 Control diagram of one degree of freedom

The transfer function of the controller is

$$G(s) = \frac{\lambda A_S A_P (1+sT_P)(b_{2S}^2 + b_{1S} + b_0)}{(1+sT_S)(a_2 s^2 + a_{1S} + a_0)(P_2 s^2 + P_{1S} + P_0)}, \quad (1)$$

where λ is the gain of the power amplifier; A_S, the gain of the displacement sensor; A_P, the gain of the PID regulator; T_S, the time constant of the sensor transformer, and T_P, the time constant of the power amplifier. Otherwise, $a_0 \sim a_2$, $b_0 \sim b_2$ and $P_0 \sim P_2$ in Eq. (1) are related to the parameters of the PID regulator and power amplifier. The parameters of the PID regulator in 4 radial degrees of freedom are shown in Table 2. For example, substituting the parameters of the front bearing in horizontal direction (see Table 2) into Eq. (1), the transfer function can be written as follows:

$$G(s) = [2.08 \times 10^4 (1 + 1.59 \times 10^{-5} s)(5.05 \times 10^{-3} s^2 + 4.73s + 1)]/[s(1 + 3.18 \times 10^{-5} s)(3.91 \times (-4)s + 2.365)(5.4 \times 10^{-3} s^2 + 341.55s + 1)]. \quad (2)$$

Table 2 Parameters of analogue PID regulator in radial degrees of freedom

	A_S(V/V)	A_P(A/V)	T_i(s)	T_d(s)
front bearing, horizontal	6 790	3.064	2.365	1.97×10^{-3}
front bearing, vertical	9 440	3.330	2.595	8.65×10^{-4}
rear bearing, horizontal	3 800	3.330	2.190	1.94×10^{-3}
rear bearing, vertical	6 560	2.419	1.570	3.06×10^{-3}

Note: T_i and T_d are the time constants of the integrator and differentiator, respectively.

4 Experiment and Result Analysis

Fig. 4 shows the photograph of the expander rig in experiment. A few tests are performed on the expander rig, and the parameters of the controller are verified during the experiment. They are listed below.

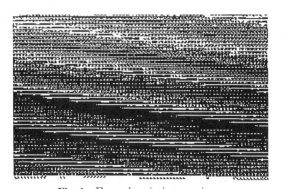

Fig. 4 Expander rig in experiment

(1) Dynamical balance test

The dynamical balance test of the rotor is carried out at the condition of rotation 3000 r/min, and the balancing precision reaches 2.5 mg · μm. So that, the rotary radius of the rotor center can be kept to less than 5 μm when rotation speed exceeds 15000 r/min.

(2) Vibration test

Rotor vibration is one vital problem for high rotating machines. In our experiment, however, no significant vibration of the rotor has been observed during experiment of overall rotation, except that the vibration in radius is about less than 20 μm at the rotation 12,000 r/min. The rig works safely.

(3) Bearing performance test

At first, the radial clearance of the bearing was made to be 80 μm. When the rotating speed exceeded 60,000 r/min, significant bearing temperature rise to about 70℃ was observed. Moreover, if there is no polymer layer on the inner surfaces of the bearing pole, the air resistance produced by the slits of the bearing core for rotating is observable, i. e., the existence of the slits of the radial bearing core had rather high influence on the bearing temperature and the rotational performance. Later, the gaps were changed up to 150 μm, and a type of polymer layer about 50 μm thickness is also used to coat the inner surface of the bearing pole.

(4) Power requirement test

Total power input to the controller is less than 600 W, in which the power supplied to the control circuit is about 100 W, and the rest is for the power amplifiers. This means that the power requirement for each degree of freedom is less than 100 W.

5 Conclusions

Analogue control system can be suitable for the active magnetic bearings, especially for the small size ones. The existence of the gaps of the radial bearing core can affect the rotating performance, and the polymer layer coating on inner surface of the radial bearings can contribute not only to decreasing the air friction, but also to improving the self-lubrication and further increasing the system stability.

References

[1] Canders W. R., Ueffing N., Schrader-Hausmann U., et al., MTG400: A magnetically levitated 400 kW turbo generator system for natural gas expansion. Proceedings of the 4th International Symposium on Magnetic Bearings, ETH Zurich, Switzerland, 1994: 435–440.

[2] Agahi R. R., Schroder, U. Industrial high speed turbo generator system for energy recovery. Proceedings of the 5th International Symposium on Magnetic Bearings, Kanazawa, Japan, 1996: 381–387.

[3] Antila M., Lantto E., Saari J., et al. Design of water treatment compressors equipped with active magnetic bearings, Proceedings of the 5th International Symposium on Magnetic Bearings, Kanazawa, Japan, 1996: 389–394.

[4] Wang Xiping, Parameter Design and Application Investigation of Electro-Magnetic Bearing System, Doctoral Dissertation of Xian Jiaotong University, 1994(in Chinese).

[5] Wang Xiping, Yuan Chongjun, Xie Youbai. Analysis on relation between the control parameters and stability of the electromagnetic bearing system. Chinese Journal of Mechanical Engineering, 1996,32(3): 65–69(in Chinese).

Effect of Geometry Change of Rough Point Contact due to Lubricated Sliding Wear on Lubrication*

Abstract: The geometry change of a single asperity due to lubricated wear was studied by an experimental simulation with a ball-on-disc set up. The wear leads to the formation of a tilted section at the tip of the ball, which is proved to be due to the presence of oil during the process. The effect of the geometry change of rough surface contacts due to wear was examined by a micro-EHL analysis. A non-Newtonian visco-plastic fluid model which includes the effect of a limiting shear strength was used.

1 Introduction

When two contact surfaces are loaded and run at different speeds under either dry or lubricated conditions, sliding wear inevitably occurs. A comparatively large wear rate is usually observed during the running-in stage of newly machined contact surfaces or at the start-up of a motion in which the contact is under boundary lubrication. The wear rate diminishes gradually during the running-in but with a much faster rate for the start-up case and this is often termed "dynamic wear" for referring to its dynamic nature. Once the running-in is completed and when a comparatively large oil film has been built up, the operation thus enters a stable mild wear stage. For contact surfaces running under partial-elastohydrodynamic lubrication (PEHL) conditions where the film thickness is of the same order of magnitude of surface roughness, dynamic wear occurs not only during running-in but also in the subsequent relatively stable mild wear stage due to the existence of asperity contacts. The wear is limited within the surface topography of the component and is termed "zero-wear". The topography of surfaces which is dependent on pressure distribution of the oil film and its thickness is changed during the wear process. In turn, the variation in film thickness alters the wear rate. Hence, the wear mechanism for contacts under PEHL regime is dynamic in nature even when under the relatively stable stage.

The phenomenon of dynamic wear such as running-in is sometimes explained as a result of the formation of a protective oxidation film during the process[1] but generally described as the effect of the variation of surface topography. The effect of alteration of surface topography on lubrication and wear under dry or lubricated conditions has attracted

---

* In collaboration with P. L. Wong, P. Huang and W. Wang. Reprinted from *Tribology Letters*, 1998,5: 265 – 274.

lots of attention in the last few decades[2]. Recently, Wang et al.[3] examined the changes in surface roughness and the relative mating of the two surfaces in lubricated sliding contact. It was found that the relative surface conformity increases during running-in. A simple concept for quantifying the degree of conformity of the contact surfaces based on surface roughness parameters was developed. Hu et al.[4] proposed a theoretical dynamic wear model that takes into consideration the rate of change of surface roughness under lubricated sliding wear and running-in. The model describes wear rate as a dependence of the change of surface roughness. The RMS roughness parameter is adopted in their model. Pawlus[5] considered the effects of cylinder surface topography by monitoring the changes of around twenty different statistical roughness parameters during the running-in stage of an engine. The initial and final surface topographies of a large number of cylinders after engine operation were examined. Patir and Cheng[6] studied numerically rough contacts under both EHL and PEHL conditions with their proposed average flow model which is based on a statistical approach. It yields only results containing the average effects. Recently, the concept of micro-EHL has been developed in which the detailed mechanism of the lubrication of individual or multiple asperities was studied. The micro-EHL model has been used to simulate deterministically the motion and interaction of lubricated contacts between rough surfaces, especially under heavily loaded conditions[7-9].

When considering the changes in topography, it is very common to use statistical roughness parameters in the analysis. For engineering practices, the use of statistical parameters is rational. However, in order to fundamentally understand the dynamic wear mechanism, it is important to study how an asperity changes and what its effect is during the process. In this work the change in geometry of a single asperity under sliding wear and its effect on lubrication were investigated. Sliding wear of a single asperity was simulated with a ball-on-disc setup. The comprehensive effect of the change in geometry of individual asperities and the nominal shape of the surface for lubricated rough contacts were analysed by the micro-EHL model.

2 Experiment

The wear of a single asperity under lubricated conditions was simulated with a stationary steel ball of 10 mm in diameter on a moving flat steel surface. The test was performed on a pin-on-disc machine and its schematic set up is shown in Figure 1. The ball was firmly attached to a stationary pin and loaded onto the rotating disc with an applied load. The contact potential was detected for monitoring the lubrication regime during the test. If two surfaces are fully in touch, there is no potential difference across, i.e., equal to zero. On the other hand, if the potential difference reaches the maximum value (40 mV), they are completely separated. Variations of contact potential between zero and the maximum value indicate the change in the amount of asperity contact.

The rotational disc setup can be replaced by a reciprocating pin-on-plate adapter which was adopted for checking the perpendicularity of the pin or pin holder on the flat plane. Before the pin-on-disc tests were carried out, a large pin of diameter 22 mm with flat end was used to perform a reciprocating test on the pin-on-plate adapter with slow speeds. The flat end surface of the pin and the plate were produced by fine grinding. The perpendicularity of the pin and the parallelism of the plate were checked with a coordinate measuring machine (CMM). The error of the former and the latter were found less than 0.0357° and 0.0040°, respectively. An uneven wear track as a result of a reciprocating test was observed from the pin surface indicating that the pin assembly was tilted. Hence, the perpendicularity of the pin holder assembly was rigorously adjusted and was confirmed by checking the flatness of the pin end with a dial gauge which was mounted on the sliding plate as the datum. Having ensured the squareness between the axis of the pin and the horizontal plane, lubricated ball on rotational disc tests were carried out. The main parameters of the test are listed in Table 1. The inlet and outlet directions were marked on the pin. After the wear test, the pin was unloaded from the pin-on-disc machine and mounted vertically on a high precision six jaws chuck. The wear track on the tip of the ball was measured by a Talysurf tester from the inlet to the outlet direction. Figure 2(a) shows a typical shape of the worn ball's tip after lubricated tests. The tip was flattened due to the material removal during the wear test. The section is not exactly horizontal (relative to the vertical axis of the pin). A wedge shape is formed. In order to find out if the wedge formation is attributable to bending of the pin due to friction, dry tests under similar running conditions were performed. Figure 2(b) depicts a trace of the worn ball's tip as a result of a dry wear test. No wedge formation is observed. Since the friction force generated under dry conditions is an order of magnitude greater than that of a lubricated test, a larger wedge shape would have occurred if bending of the pin was significant. However, the result of dry tests indicates that the wedge formation due to friction is inconceivable. The phenomenon has been confirmed by repeating the tests several times under both dry and lubricated conditions and it is certain that the formation of a wedge shape is due to the presence of lubricating oil in the test.

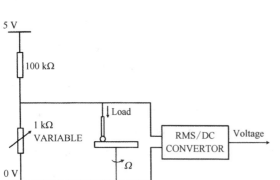

Fig. 1 Schematic diagram of test rig

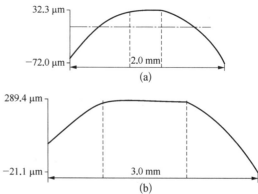

Fig. 2 (a) Worn profile of ball surface due to lubricated wear; (b) Worn profile of ball surface due to dry wear

Table 1 Experimental conditions

Ball radius	5 mm	Sliding speed	1.57 m/s
Loading	9.81 N	Elastic modulus	2.1×10^{11} N/m²

Figure 3 depicts the variation of contact potential difference and friction for a lubricated test running under the conditions listed in Table 1. The sharp change of the two curves indicates the pin was loaded at 143 s after the machine was started. The instantaneous response to the application of external load was that the potential dropped noticeably due to the sharp increase in the amount of asperity contacts. The curve increases with time with the rate of increase falling gradually. The frictional force has increased drastically to a comparatively high value with the application of external load. Then, the curve drops gradually by about 25% within 2 min. Beyond that, the frictional force maintains a rather constant value, which in fact indicates the process has reached a relatively steady state. Observations extracted from the experiment indicate that the lubrication of the contact has been improved by the change of the contact surfaces during running-in. The addition of

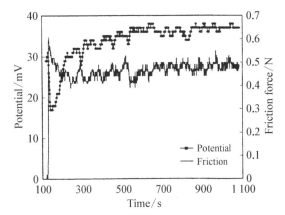

Fig. 3 Variations of contact potential and friction force during the test

external load leads to a significant increase in friction and reduction in contact potential. The contact, at that moment, is actually under boundary lubrication regime if any oil films exist between the two surfaces and there are a large amount of asperity contacts. Then, the amount of asperity contacts is reduced as illustrated with the decline of the friction curve. The running-in lasted for about 120 s. After that, the contact potential gradually settled at about 35 mV. That indicates the lubrication regime in the contact is under PEHL since the settling potential is less than 40 mV, which represents a complete separation of the two surfaces. After the test, the wear tracks both on the ball and disc surface were examined and no oxidation films were found, so that the effect of such an oxidation layer on the change of contact potential can thus be ruled out. Hence, the variation of potential difference can be accounted with the change in the condition of asperity contacts.

3 Mathematical Modelling

The experimental results showed that the improvement in lubrication between two contact surfaces under high loading can be attributed to the increase in contact area due to wear and the corresponding drop in contact pressures. The effect of the geometry change of a spherical surface contacted with a flat plane on lubrication is studied with EHL analysis and the comprehensive effect of roughness change due to the wear of individual asperities and geometry change of contact surfaces is analysed with a micro-EHL model.

The pressure in the contact is obtained from a solution of the Reynolds equation:

$$\frac{\partial}{\partial x}\left(\frac{\rho h^3}{\eta_e}\frac{\partial p}{\partial x}\right)+\frac{\partial}{\partial y}\left(\frac{\rho h^3}{\eta_e}\frac{\partial p}{\partial y}\right)=6u_s\frac{\partial \rho h}{\partial x} \tag{1}$$

with boundary conditions for inlet and outlet, respectively:

$$p\Big|_{x=x_1}=0;\ p\Big|_{y=y_1}=0;\ p\Big|_{y=y_2}=0;$$

$$p\Big|_{x=x_e}=0;\ \frac{\partial p}{\partial x}\Big|_{x=x_e}=0.$$

The non-dimensional inlet position X_1 for the computing zone is -2.5 and the outlet X_e which was determined in the solution process is 1.5. Y_1 and Y_2 are respectively -2 and 2 in the transverse direction. The film thickness between the rough surfaces is:

$$h(x,y)=h_0+\frac{x^2+y^2}{2R}+\frac{2}{\pi E'}\iint\frac{p(\xi,\lambda)\mathrm{d}\xi\mathrm{d}\lambda}{\sqrt{(x-\xi)^2+(y-\lambda)^2}}-\delta(x,y), \tag{2}$$

which includes three components: the geometrical gap size, elastic deformation and roughness. The load that the pressure distribution can carry is calculated from:

$$W=\iint_A p(x,y)\mathrm{d}x\mathrm{d}y. \tag{3}$$

The Roelands pressure-temperature-viscosity relationship[10] and the pressure-temperature-density relationship derived from the Dowson-Higginson formula[2] have been used to describe the influence of pressure and temperature on viscosity and density:

$$\eta=\eta_0\exp\left\{(\ln\eta_0+9.67)\times\left[-1+\left(1+\frac{p}{p_0}\right)^z\left(\frac{T-138}{T_0-138}\right)^{s_0}\right]\right\}, \tag{4}$$

$$\rho=\rho_0\left(\frac{p_1+1.34p}{p_1+p}\right)+d_0(T-T_0). \tag{5}$$

The viscosity and density are functions of temperature such that lubrication effect is affected by variations of temperature. The temperature distribution across the lubricating film can be obtained by the energy equation:

$$\rho c\left(u\frac{\partial T}{\partial x}+v\frac{\partial T}{\partial y}+w\frac{\partial T}{\partial z}\right)=k\frac{\partial^2 T}{\partial z^2}-\frac{T}{\rho}\left(u\frac{\partial p}{\partial x}+v\frac{\partial p}{\partial y}\right)\frac{\partial \rho}{\partial T}+\eta_e\left[\left(\frac{\partial u}{\partial z}\right)^2+\left(\frac{\partial v}{\partial z}\right)^2\right]. \tag{6}$$

The temperature boundary conditions are:

$$T(x_0,y,z)=T_0;\ T(x,y=\pm y_0,z)=T_0,\ 0\leqslant z\leqslant h.$$

The temperature of the upper and lower surface can be respectively computed[11] from:

$$T(x,y)\Big|_{z=0} = \frac{k_l}{k_s}\sqrt{\frac{k_s}{\pi c_s \rho_s u_1}} \int_0^x \frac{\partial T}{\partial z}\Big|_{z=0} \frac{d\xi}{\sqrt{x-\xi}} + T_0, \qquad (7)$$

$$T(x,y)\Big|_{z=h} = \frac{k_l}{k_s}\sqrt{\frac{k_s}{\pi c_s \rho_s u_2}} \int_0^x \frac{\partial T}{\partial z}\Big|_{z=h} \frac{d\xi}{\sqrt{x-\xi}} + T_0. \qquad (8)$$

It has been found that by using a Newtonian viscous model with exponentially increasing viscosity with pressure and with isothermal conditions are largely inaccurate in the determination of frictional forces in EHL analyses. It is mainly due to the thermal effects and the high pressure rheology of lubricants. Thus, the viscosity shown in the energy and Reynolds equations is an effective viscosity which depends on the fluid model. The non-Newtonian visco-plastic model is used in this analysis[12]. The constitutive equation for the shear stress and strain rate relation is:

$$\tau = \eta_e \dot{\gamma}, \qquad (9)$$

where η_e is the effective viscosity and can be expressed as:

$$\begin{aligned}\eta_e &= \eta, \quad \eta\dot{\gamma} \leqslant \tau_L, \\ \eta_e &= \frac{\tau_L}{\dot{\gamma}}, \quad \tau_L \leqslant \eta\dot{\gamma},\end{aligned} \qquad (10)$$

where τ_L is the limiting shear strength of lubricant and it is assumed to be a function of pressure only as:

$$\tau_L = \tau_0 + \beta p. \qquad (11)$$

For this work being a two-dimensional case, the resultant shear stress and strain rate used in equation (9) are calculated by:

$$\tau = \sqrt{\tau_{xz}^2 + \tau_{yz}^2}, \quad \dot{\gamma} = \sqrt{\dot{\gamma}_{xz}^2 + \dot{\gamma}_{yz}^2}. \qquad (12)$$

4 Numerical Technique

The Reynolds equation (1) is solved by the Gauss-Seidal iterative method for moderate pressures and by the Jacobi double polar iterative method for the high-pressure condition, such that convergence in calculating pressures can be obtained. The convergence criterion for pressure is:

$$\frac{\sum_i \sum_j |p_{ij}^{n+1} - p_{ij}^n|}{\sum_i \sum_j p_{ij}^{n+1}} \leqslant 1 \times 10^{-3}. \qquad (13)$$

The temperature distribution is calculated from the energy equation (6) using the Gauss-Seidal iterative method. The convergence criterion is:

$$\max_{i,j,k}\left(\frac{|T_{i,j,k}^{n+1} - T_{i,j,k}^{n}|}{T_{i,j,k}^{n+1}}\right) \leqslant 1 \times 10^{-4}. \tag{14}$$

The computation domain is $-2.5 \leqslant X \leqslant 1.5$ and $-2 \leqslant Y \leqslant 2$ in the sliding and transverse directions, respectively, with equally spaced 65×65 grids. The surface roughness data used in equation (2) can be either calculated with a mathematical function or input directly. Further, the elastic deformation shown in the film thickness equation (2) is computed by a multigrid numerical technique. Hence, the convergence rate is faster.

5 Modelling for Roughness

In this work both random and sinusoidal surface roughness profiles have been studied. The spherical rough surface was set stationary in imitation of the ball in the experiment.

The values of random roughness were obtained with a ground surface of $\sigma = 0.392$ and can be expressed as:

$$\delta = \delta_0 \alpha(x, y), \tag{15}$$

where α is the normalised roughness amplitude with values in the range of ± 1. The random roughness was confirmed to be isotropic and follow Gaussian distribution.

An isotropic surface roughness profile which takes the form of sinusoidal curves as:

$$\delta = \delta_0 \cos 2\pi x \cos 2\pi y \tag{16}$$

was also simulated. The RMS roughness σ is 0.508, which is statistically rougher than the random roughness profile.

The amount of wear, which is hereby termed "wear height" D, is expressed by the removal quantity of the asperity's tip as shown in Figure 4. The centre line represents the corresponding surface with zero roughness. The height of a worn asperity can be expressed as:

$$\begin{cases} \delta = \delta(x, y), & \delta \leqslant (\delta_0 - d), \\ \delta = (\delta_0 - d), & \delta \geqslant (\delta_0 - d). \end{cases} \tag{17}$$

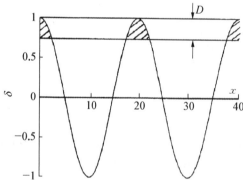

Fig. 4 Schematic diagram of wear of rough surface.

For surface having sinusoidal roughness, wear occurs in all asperities on the moving plane and the wear height of all asperities can be assumed the same while the wear height of different asperities varies across the stationary spherical surface. Further, the wear section can also be inclined with an angle of elevation.

6 Results and discussion

The analyses were carried out with the parameters shown in Table 2 in addition to the original experimental specification listed in Table 1. The meaning of the parameters can be found in the nomenclature.

6.1 Smooth contact

The isothermal and thermal EHL analyses of smooth surface (zero roughness) contacts were performed. The friction coefficient, the minimum film thickness and maximum pressure were obtained and listed in Table 3. It can be seen that the calculated friction coefficient drops with three orders of magnitude when the thermal effect has been taken into consideration. The minimum film thickness is reduced by about half and the maximum pressure increases by 14%. However, there is still a great discrepancy between the thermal result and the experimental friction coefficient. This can be attributed to two causes:

(1) The temperature boundary condition for the stationary surface cannot be determined in solving equations (7) and (8). Hence, the real experimental condition (pure sliding), for $s=2$, cannot be used in the calculation.

(2) The non-Newtonian effect has not been taken into consideration.

Table 2 Material parameters assumed in the theoretical solutions

c_l:	2 000 J/(kg·K)	s_0:	−1.1
c_s:	470 J/(kg·K)	T_0:	303 K
d_0:	−0.000 65	β:	0.036
E:	210 GPa	η_0:	0.028 3 Pa·s
k_l:	0.14 W/(m·K)	ρ_s:	7 850 kg/m^3
k_s:	46 W/(m·K)	ρ_0:	890 kg/m^3
p_0:	196 MPa	τ_0:	20 MPa
p_1:	59 MPa	ν:	0.3

Table 3 Thermal and isothermal EHL results for $W=9.81$ N

	Isothermal	Thermal
Friction coefficient f	101	0.35
Minimum film thickness $h_{\min}$ (μm)	0.068	0.038
Maximum pressure $p_{\max}$ (GPa)	1.00	1.14

The variation of friction coefficient in the range of the slide to roll ratio s from 0 to 1.8 with Newtonian and non-Newtonian fluids when under 9.81 N normal loading is shown in Figure 5. The figure depicts that for a small increase in s, the thermal Newtonian curve increases drastically to its maximum value and then drops steadily with the further increase in slide to roll ratio due to the thermal effect. Following the trend of the thermal

Newtonian curve, the friction coefficient at $s=2$ is about 0.1 which is still greater than the experimental results. Taking into consideration the effective viscosity and limiting shear stress, the calculated friction coefficient is much reduced. Figure 5 shows that the thermal non-Newtonian curve slightly increases in the beginning and then maintains steadily with $\mu=0.06$ which is much closer to the experimental results. For large slide to roll ratio, the non-Newtonian results are almost constant. Numerical results obtained with velocity and loading conditions similar to the current work by Hsiao and Hamrock[13] show the same trend of variation of thermal Newtonian and thermal non-Newtonian friction coefficient with slide to roll ratio. They adopted an energy model which considers two-dimensional heat flow in the bounding solid[14] such that the solution at slide to roll ratio equal to 2 can also be obtained. Their results show that the thermal non-Newtonian friction coefficients calculated with s from 1.9 to 2.0 are still following the trend of the curve and are only about 10% different from that at $s=1$.

6.2 Worn spherical contact

Figure 6 illustrates a worn spherical surface which is assumed as a section formed with a plane parallel to the y-axis. The wear section is characterised with the radius L and the angle of elevation θ. The function of the surface can be expressed as:

$$\begin{cases} f(x, y) = \dfrac{x^2 + y^2}{2R}, & x^2 + y^2 \geq l^2 + 2Rx\tan\theta, \\ f(x, y) = \dfrac{l^2}{2R} + x\tan\theta, & x^2 + y^2 \leq l^2 + 2Rx\tan\theta. \end{cases} \quad (18)$$

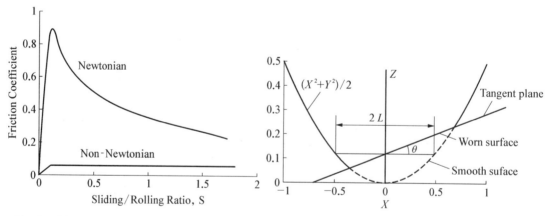

Fig. 5 Newtonian and non-Newtonian friction results against different slide to roll ratio

Fig. 6 Characterisation of worn profile of ball

Figure 7 shows the pressure distribution for surfaces with and without wear, calculated with the non-Newtonian EHL analysis. The radius of wear section is equal to half of the radius of the relevant Hertzian contact $L=0.5$ and its elevation is $-5°$. For the no wear case, the curve acquires the classical shape of a heavily loaded, pointcontact EHL problem such that a pressure spike exists near the outlet and its magnitude is smaller than the central maximum pressure. The shape of the pressure distribution of the worn surface is different. There are three pressure spikes found in the results. The last one corresponds to the pressure spike of the classical feature and the first two are results of the discontinuity of the curvature of the spherical surface due to the edge of the wear section. In the central part of the contact, the pressure distribution is convex downward. The second pressure spike, which corresponds to the minimum film thickness, acquires the maximum value and it is greater than that of the spherical surface without wear case. As a reference, pressure distributions for surfaces having different lengths of wear section under a small loading of 0.5 N are shown in Figure 8. The classical pressure spike for the undamaged surface is near the centre. Different from the heavily loaded results shown in Figure 7, the maximum pressure drops with the increase in the size of wear section.

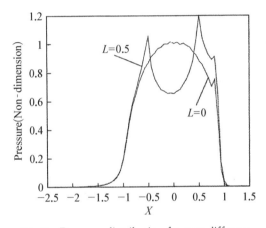

Fig. 7 Pressure distribution for two different wear section lengths L and $W=9.81$ N

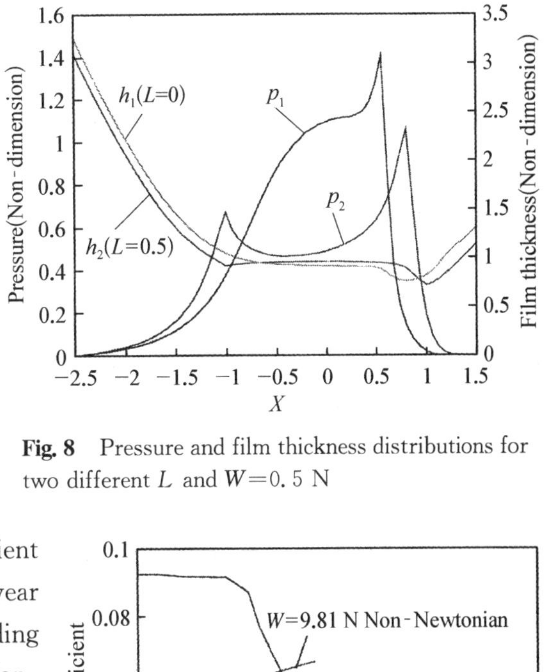

Fig. 8 Pressure and film thickness distributions for two different L and $W=0.5$ N

Figure 9 shows the friction coefficient calculated with different lengths of wear section under low and high loading conditions. For the highly loaded condition, the friction coefficient only slightly increases with the wear section length while that decreases prominently, especially in the range of $L = 0.5$ to 0.8, for the lightly loaded case. The magnitude of friction coefficient of the heavily loaded case is close to the experimental results. However, the gradual drop of the friction coefficient in the beginning

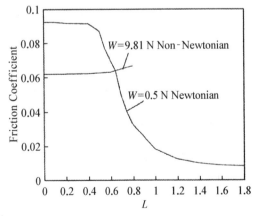

Fig. 9 Relation of friction coefficient and wear section length

of the experiment, in which L has small values, is not found in the numerical results for either low and heavy loading conditions. It is because in the beginning of the test, the operation is actually under boundary lubrication condition and the effect of the oil film is minimal. The results of friction coefficient, minimum film thickness and maximum pressure with different length of wear section for the heavily loaded condition are also tabulated in Table 4. It can be seen that the friction coefficient and maximum pressure increase, while the minimum film thickness diminishes with the increase in wear-section size. When the radius of wear section L is equal to 0.8, the friction coefficient is increased by 7.5%, the maximum pressure increased by 25% and the minimum film thickness dropped by 8.9%. Since the increase in friction coefficient and pressure and decrease in film thickness are not in favour of lubrication, it can be concluded that the increase in the size of a wear section of a smooth surface is not beneficial to lubrication. The effect of elevation from $-10°$ to $10°$ and with $L=0.5$ is studied and results are listed in Table 5. The maximum pressure only decreases by about 5% with the increase in positive angle of elevation. The effect of elevated angle is only marginal. Overall, the effect of the size of the wear section is greater than its angle of elevation.

Table 4 Variation of friction coefficient, minimum film thickness and maximum pressure with the length of wear section for $W=9.81$ N

L	f	$h_{min}/\mu m$	p_{max}/GPa
0.0	0.060 0	0.069 7	0.99
0.2	0.059 9	0.071 5	1.04
0.4	0.060 1	0.070 2	1.12
0.6	0.060 9	0.068 9	1.14
0.8	0.064 5	0.063 5	1.24

Table 5 Variation of friction coefficient, minimum film thickness and maximum pressure with different angle of elevation of wear section for $W=9.81$ N

θ/deg	f	$h_{min}/\mu m$	p_{max}/GPa
-10	0.060 2	0.069 8	1.28
-5	0.060 0	0.069 4	1.25
0	0.060 1	0.068 0	1.21
5	0.060 3	0.068 3	1.13
10	0.060 6	0.070 1	1.15

6.3 Rough surface contact

The effects of surface roughness on the friction coefficient, minimum film thickness and maximum pressure were considered. The results for different surfaces are listed in Table 6. Rough surface contacts obviously provide greater maximum pressure and smaller minimum film thickness. The friction coefficient is only slightly smaller for rough surface contacts. Since the minimum film thickness of rough surfaces is apparently smaller than that of smooth surfaces, it is proved that the existence of roughness reduces the lubrication effect. Considering that the sinusoidal roughness surface has the smallest minimum film thickness and the largest maximum pressure among the three cases and the RMS values of the simulated sinusoidal roughness is greater than that of random roughness, it can thus be concluded that the surface roughness has a negative effect on lubrication.

Table 6 Effect of different surfaces on friction coefficient, minimum film thickness and maximum pressure

	Smooth	Random	Sinusoidal
f	0.060	0.056	0.053
$h_{min}/\mu m$	0.070	0.043	0.032
p_{max}/GPa	0.99	1.81	2.56

Figure 10 shows the effect of roughness change of contact surfaces due to the wear of individual asperities on lubrication. Significant changes in the minimum film thickness and the maximum pressure can be observed. The former increases by 69% and 24% when $D = 1$ for the contact surfaces of sinusoidal and random roughness, respectively. It can be seen that the minimum film thickness increases and the maximum pressure decreases with the increase in the amount of asperity wear. Even though the calculated friction coefficient is slightly increased by around 5% for the two cases, the overall phenomenon indicates that the lubrication effect is improved with the increase in the wear of asperities.

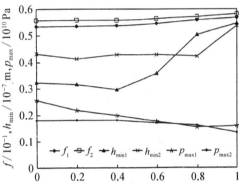

Fig. 10 Effect of different amount of "zero-wear" on lubrication

(1: sinusoidal; 2: random roughness.)

In the present micro-EHL analysis, the component of the frictional force due to the direct contact of the two surfaces has not been considered. It is obvious that for mixed lubrication, the frictional force, i.e., friction coefficient must be reduced following the increase in the amount of asperity wear and the film thickness.

6.4 From "zero-wear" to nominal wear

A rough surface can be considered as a combination of a rough profile and a nominal shape of the surface. The material removal process during the operation of two rough surface contacts generally starts with the "zero-wear". Subsequently, wear proceeds to change the nominal shape of the surface. The effect of geometry change of a smooth spherical surface incorporated with a Gaussian random roughness profile on lubrication was studied. The surface profile without any damage can be expressed as:

$$f(x, y) = \frac{x^2 + y^2}{2R} - \delta(x, y). \tag{19}$$

The worn surface profile which is characterised with an elevated section can be written as:

$$\begin{cases} f(x, y) = \dfrac{x^2+y^2}{2R} - \delta(x, y), \\ \dfrac{x^2+y^2}{2R} - \delta(x, y) \geqslant d + x\tan\theta, \\ f(x, y) = d + x\tan\theta, \\ \dfrac{x^2+y^2}{2R} - \delta(x, y) \leqslant d + x\tan\theta. \end{cases} \quad (20)$$

The solution of micro-EHL analysis of a contact of surfaces having Gaussian random profile with average nondimensional roughness amplitude of $\Delta=0.125$ are listed in Table 7. Figure 11 depicts the effect of the change in wear height on the friction coefficient, minimum film thickness, maximum pressure and average temperature increased. The calculated results are compared to those of undamaged surface shown in Table 7. Figure 11 shows that following the increase in the amount of wear on surface, the effect can be divided into four stages:

Table 7 Solution of micro-EHL analysis of contact surfaces of $\Delta=0.125$

f	$h_{min}/\mu m$	p_{max}/GPa	$\Delta T/K$
0.054	0.023 9	2.10	78.0

(1) For $0 \leqslant D < 0.6$, the effect of wear is very small. Since there are only few large asperities in a Gaussian distribution roughness and if no large asperities in or near the Hertzian contact area, wear is thus very minimal.

(2) For D in the range of 0.6 to 1, the lubrication parameters start changing. During this stage, it is limited to the "zero wear". Hence, the amount of wear is still very small. However, the wear is positive to the lubrication, since the minimum film thickness is increased and the maximum pressure is reduced.

(3) For D in the range of 1 to 1.8, the change in minimum film thickness is pronounced. During this stage, the relatively large asperities are removed and the wear of the nominal surface is started. However, the wear of the roughness profile is large while that of the nominal surface is still minimal. The minimum film thickness is increased most vigorously. The others are just slightly increased and the amount of increase is far less than that of the minimum film thickness. Hence, the lubrication benefits from the wear in this stage.

(4) For $D > 1.8$, the amount of wear for the nominal surface speeds up. When $D=2$, the wear height is equal to the maximum peak to valley value of asperity and also equivalent to the non-dimensional wear radius $L=1$. The effect of lubrication is dominated by the wear of the nominal surface. The minimum film thickness drops rapidly and is even less than that of undamaged surface. Further, the friction coefficient, maximum pressure and average temperature rise are kept increasing. Hence, the wear is negative to the lubrication.

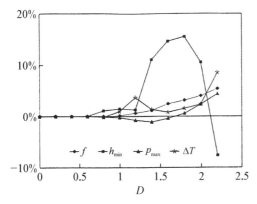
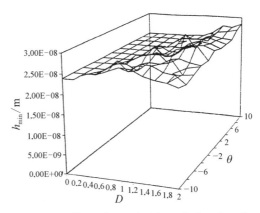

Fig. 11 Effect of different amount of wear on lubrication (Spherical surface with Gaussian random roughness.)

Fig. 12 Effect of wear height and elevation of wear section on the minimum film thickness

Figure 12 shows the variation of the minimum film thickness with the wear height and the section elevation. When the wear height is limited within the maximum roughness amplitude, the minimum film thickness has an increasing trend. Hence it is positive to the lubrication. For different angles of elevation, the amount of increase in the minimum film thickness with D is about the same. The increase is around 16% greater than that of the undamaged surface. However, the location of the start of the increase in the minimum film thickness curve and of its maximum point are different with different angles of elevation. When the angle is large and negative, the increase in the minimum film thickness starts with a smaller D value. The maximum point also happens earlier and the fall of the curve can be observed within the calculated range of D. When the angle of elevation is large and positive, the minimum film thickness keeps almost the same value of the undamaged surface for a large range of D. The maximum point is also located at a larger D value. Since large negative elevation is a result of the wear at the inlet, this indicates that the shape of the inlet has a significant effect on the change of the minimum film thickness. However, the changes of the other lubrication parameters — the maximum pressure, friction coefficient and the temperature increase — are only within 5%.

7 Conclusion

The geometry change of a single asperity due to lubricated wear was studied by an experimental simulation with a ball-on-disc set up. The wear leads to the formation of a tilted section at the tip of the ball, which is proved to be due to the presence of oil during the process. The effect of the geometry change of rough surface contacts due to wear was analysed and discussed. The results lead to the following conclusions:

(1) The calculated friction coefficient is close to the experimental results after having considered the thermal non-Newtonian effect in the analysis.

(2) For smooth surface contacts, the wear degrades the lubrication effect when under

the specified working conditions which can be considered as heavily loaded condition for Hertzian pressure equal to 1.0 GPa.

(3) The maximum pressure increased and the minimum film thickness decreased are noticeable due to the roughness of surfaces, i.e., surface roughness is negative to the lubrication when compared to smooth surface lubrication.

(4) The lubrication effect can be enhanced with the geometry change of asperities under wear within the limit of the surface topography. This is supported by the increase in the minimum film thickness and the decrease in the maximum pressure due to the increase in the wear of roughness.

(5) For considering a wear process starting from the "zero wear" to the significant change in the nominal shape of rough surface contacts, the minimum film thickness is most prominently varied with the amount of wear when compared to other lubrication parameters: the maximum pressure, friction coefficient and the temperature increased. For wear sections having different angles of elevation, the maximum pressure point and the start of the increase in the minimum film thickness occur at different amount of wear. Following the increase in the wear quantity, the wear process can be divided into four stages:

(i) The effect of wear on lubrication is only marginal.

(ii) The wear only occurs within the limit of the surface topography. The amount of wear is very small and the wear is beneficial to the lubrication.

(iii) The relatively large asperities are worn out and wear proceeds to the change of the nominal shape of the surface. The minimum film thickness starts to increase significantly due to the removal of large asperities but the geometry change of the surface is still minimal. The changes of other lubrication parameters are very small. During this stage, lubrication is enhanced by wear.

(iv) The lubrication effect is dominantly affected by the geometry change of the surface due to wear. The minimum film thickness drops rapidly. The friction coefficient, the maximum pressure and the average temperature rise are continuously increased. Hence, the wear is negative to the lubrication.

Acknowledgement

The authors wish to thank W. B. Lau, C. W. Mok and K. Y. Chung for their invaluable assistance in checking the alignment of the test rig and taking experimental measurements. They are also indebted to the City University of Hong Kong for the financial support of this work.

Nomenclature

b	half Hertzian length, m	d	roughness wear height, m
c	specific heat of lubricant, J/(kg·K)	d_0	constant in equation (5)

D	non-dimensional roughness wear height, $D = d/\delta_0$	v	speed in transverse direction, m/s
		w	speed in film thickness direction, m/s
E	Young's modulus, Pa	W	loading, N
E'	equivalent Young's modulus, $\frac{1}{E'} = \frac{1}{2}\left(\frac{1-\nu_1^2}{E_1}+\frac{1-\nu_2^2}{E_2}\right)$ Pa	x	coordinate in sliding direction, m
		X	non-dimensional coordinate, $X = x/b$
		y	coordinate in transverse direction, m
		Y	non-dimensional coordinate, $Y = y/b$
f	friction coefficient	z	coordinate in film thickness direction, m
H	non-dimensional film thickness, $H = hR/b^2$	Z	non-dimensional coordinate, $Z = z/b$
		β	constant in equation (11)
h	film thickness, m	δ	roughness amplitude, m
h_0	central film thickness, m	δ_0	maximum roughness amplitude, m
i, j, k	sub-coordinate of point x, y, z	Δ	non-dimensional amplitude, $\Delta = \delta_0 R/b^2$
k	thermal conductivity, W/(m·K)	γ	shear rate, s^{-1}
l	radius of wear section, m	η	viscosity of lubricant, Pa·s
L	non-dimensional radius of wear section, $L = l/b$	η_0	viscosity at ambient pressure and temperature, Pa·s
n	iterative number	ρ	density of lubricant, kg/m^3
N	number of node points	ρ_0	density at ambient pressure and temperature, kg/m^3
p	pressure, Pa		
p_0	constant in equation (4)		
p_1	constant in equation (5)	θ	angle of normal to wear section and z axis, degree or radian
p_H	maximum Hertzian pressure, Pa		
P	non-dimensional pressure, $P = p/p_H$	σ	RMS roughness, $\sigma = \sqrt{\frac{1}{N}\sum_{i=1}^{N}z_i^2}$
R	equivalent radius of curvature, $\frac{1}{R} = \frac{1}{R_1}+\frac{1}{R_2}$ m	τ	shear stress, Pa
		τ_L	limiting shear stress, Pa
		ν	Poisson ratio
s	slide to roll ratio, $s = \frac{2(u_1-u_2)}{(u_1+u_2)}$	Subscript:	
		0	nominal value
s_0	constant in equation (4)	1	lower surface
u_s	surface velocity, u_1+u_2 m/s	2	upper surface
T	average temperature in film thickness direction, K	l	fluid or liquid lubricant
		e	equivalent value
u	speed in sliding direction, m/s	s	solid surface
U	non-dimensional speed, $U = \frac{\eta_0(u_1+u_2)}{2ER}$		

References

[1] M. Suzuki and K. C. Ludema, Trans. ASME, J. Tribol., 109 (1987) 587.

[2] D. Dowson and G. R. Higgison, Elasto-Hydrodynamic Lubrication (Pergamon, Oxford, 1977).

[3] F. Wang, P. Lacey, R. S. Gates and S. M. Hsu, Trans. ASME, J. Tribol., 113 (1991) 755.

[4] Y. Z. Hu, N. Li and K. Tonder, Trans. ASME, J. Tribol., 113 (1991) 499.

[5] P. Pawlus, Wear 209 (1997) 69.

[6] N. Patir and H. S. Cheng, Trans. ASME, J. Lub. Tech., 100 (1978) 12.

[7] C. C. Kweh, H. P. Evans and R. W. Snidle, Trans. ASME, J. Tribol. , 110 (1998) 421.

[8] L. Chang, Trans. ASME, J. Tribol. 114 (1992) 186.

[9] M. J. Patching, H. P. Evans and R. W. Snidle, STLE, Tribol. Trans. , 39 (1996) 595.

[10] C. J. A. Roelands, Correlational aspects of the viscosity-temperature-pressure relationships of lubricating oils, Ph. D. Thesis, Technical University Delft, Delft, The Netherlands (1966).

[11] H. S. Carslaw and J. C. Jaeger, Conduction of Heat in Solids (Oxford, 2nd Ed. , 1959).

[12] P. Huang and S. Z. Wen, Acta Tribologica, 2 (1994) 23.

[13] H. S. Hsiao and B. J. Hamrock, Trans. ASME, J. Tribol. , 116 (1994) 559.

[14] H. S. Hsiao and B. J. Hamrock, Trans. ASME, J. Tribol. , 114 (1992) 540.

Jeffcot 转子-滑动轴承系统不平衡响应的非线性仿真*

摘 要：本文用动力仿真法考察了 Jeffcot 转子-椭圆轴承系统的不平衡响应. 计入了轴承油膜力的非线性. 仿真计算前,先以非定常雷诺方程和雷诺破膜条件为依据,生成了轴瓦非定常油膜力数据库. 用龙格-库塔法对运动方程作步进积分,同时反复对轴瓦力数据库进行插值以获得轴承力的瞬时值. 考察了支撑于一对椭圆轴承上的 Jeffcot 转子的不平衡响应. 所得的动力学行为以及转子和轴颈的涡动轨迹,均与线性动力学(以轴承的线性化动特性系数为依据)所得的结果相比较. 两者虽在很小的不平衡量下吻合良好,但凡当不平衡量不是很小时就有显著差别. 可见有必要计入油膜力的非线性,特别是当需要计算大不平衡下的不平衡响应时.

符号说明

c_{min}	轴承最小半径间隙(m)	x_j, y_j	以 c_{min} 为参考的轴颈中心坐标无量纲值
d	轴承直径(m)	x_r, y_r	以 c_{min} 为参考的转子中心坐标无量纲值
e_u	转子质量中心的偏心距(m)	μ	润滑油的动力黏度(Pa·s)
E_u	质量中心的相对偏心(e_u/c_{min})	F	轴承的静载荷(N)
f	轴在自重下的静挠度(m)	ω	转子角速度
η	轴的相对挠度(f/c_{min})	ω_k	转子固有频率
l	轴承长度(m)	Ω	相对速度(ω/ω_k)
So_k	以转子固有频率为参考的轴承 Sommerfeld 数 $So_k = \dfrac{F\psi_{min}^3}{dl\mu\omega}$	ψ_{min}	轴承的最小间隙化 $\psi_{min} = \dfrac{c_{min}}{r}$

0 前言

在工程实践中,常常用线性动力理论来计算转子-滑动轴承系统的不平衡响应,即:计算时以线性化的轴承动力特性(轴承的八个刚度和阻尼)来表达轴承油膜的动态力[1]. 但油膜力实际上是非线性的动力元素,因此这样的线性化不可避免地要导致不平衡响应计算中的误差. 本文目的在于用非线性和线性动力学两种计算来考察不平衡响应,并作比较,以明确其异同.

1 线性分析

本文以 Jeffcot 转子-轴承系统(图1)为考察对象.

---

* 本文合作者：王德强. 原发表于《振动与冲击》,1999,18(1)：57-62,93.

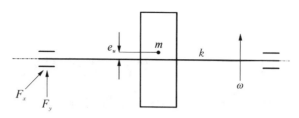

图 1 Jeffcot 转子-滑动轴承系统

运动方程为

$$m\ddot{x}_r + k(x_r - x_j) = \omega^2 m e_u \sin\omega t \tag{1}$$

$$m\ddot{y}_r + k(y_r - y_j) = \omega^2 m e_u \cos\omega t - mg \tag{2}$$

$$k(x_r - x_j) = 2F_x \tag{3}$$

$$k(y_r - y_j) = 2F_y \tag{4}$$

将轴承力线性化,并以轴承的八个线性化动力特性系数来表达:

$$F_x = k_{xx}\Delta x_j + k_{xy}\Delta y_j + b_{xx}\dot{x}_j + b_{xy}\dot{y}_j \tag{5}$$

$$F_y = k_{yx}\Delta x_j + k_{yy}\Delta y_j + b_{yx}\dot{x}_j + b_{yy}\dot{y}_j \tag{6}$$

将上述方程组无量纲化,用文[1]中所述之方法解算.

2 非线性仿真

轴承油膜力对于轴颈运动参数实际上是非线性的,特别当轴颈运动到轴承间隙边界附近时更有强烈的非线性.今以下述两式表达其非线性关系:

$$F_x = F_x(\Delta x_j, \Delta y_j, \dot{x}_j, \dot{y}_j) \tag{7}$$

$$F_y = F_y(\Delta x_j, \Delta y_j, \dot{x}_j, \dot{y}_j) \tag{8}$$

用此两式替代方程组中的式(5)和(6),即得非线性运动方程组.

在给定的转速下,由轴颈和转子中心的任意位置开始,用变步长四阶龙格-库塔法对运动方程组积分,获得轴颈和转子中心的坐标序列,即其离散表达的动轨迹[2].当动轨迹稳化为同步于轴转速的周期涡圈时,记录下轴颈和转子中心的涡动轨迹,并由之推算出振幅.

非线性关系(7)和(8)可以用不同方式来表达,例如对每块轴瓦在轴颈相对于该瓦的瞬时运动参数下解算雷诺方程,对压力分布积分以求得轴瓦力分量,然后将其矢量合成而求得整个轴承的力分量.也可以用简化方式来表达式(7)和(8),例如取其泰勒展开的头几项来表达,或其至采用基于短轴承理论和长轴承理论的表达式.为使动力仿真更切近实际,同时又提高计算效率,本文采用了[2]中的技术,即:对给定的轴瓦几何参数生成一个轴瓦油膜力的专用表格式数据库,来体现轴瓦无量纲力分量与三个决定性无量纲参数(轴颈相对于该轴瓦的偏心率、偏位角、挤压/旋转比)之间的非线性函数关系.这个数据库是依靠解算非定常雷诺方程和雷诺边界条件而生成的,所以数据库中的油膜力值准确地体现了这种框架中的理论而不含简化(除了极微小的插值误差外).在对运动方程组作步进积分时,每一步都只需从这个数据库快速而精确地插读油膜力瞬时值.由此保证了仿真的高效和准确,方有可能获得丰富而满意的结果.

3 仿真算例

本文的考察对象中,轴承为椭圆轴承,其长径比为 0.8,预负荷系数为 2/3,轴瓦的张角为 150°,轴承以转子固有频率为参考的 Sommerfeld 数为 0.4,轴的相对柔度 $f/c_{\min}$ 为 0.2.

不平衡量为:转子质量中心偏离轴的几何轴线一个距离 e_u,相对不平衡量 $e_u/c_{\min}$ 在 0.01～0.5 范围内取值,相对速度 ω/ω_k 取为 0.2 以上.

4 计算结果

图 2 表示不平衡量很小时($e_u/c_{\min}=0.01$),用非线性仿真和用线性理论得到的转子振幅与轴转速的关系.非线性和线性响应的振幅曲线相当接近.非线性仿真所得的临界转速和共振放大倍数与线性动力学计算结果相差不大.其间少量差别,主要不是由不同理论所致,更大程度上是由仿真计算中的数值误差所引起,这种误差在振幅很小的场合更容易表现出来.两种理论所得的轴颈和转子中心的涡动轨迹也相互颇为近似,如图 3 和 4 所示.非线性动力学所得的涡动圈大体上是椭圆形的,与线性动力学的结果相仿.

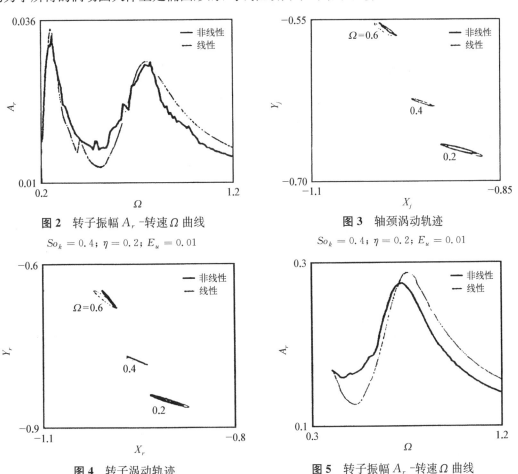

图 2 转子振幅 A_r-转速 Ω 曲线
$So_k=0.4;\eta=0.2;E_u=0.01$

图 3 轴颈涡动轨迹
$So_k=0.4;\eta=0.2;E_u=0.01$

图 4 转子涡动轨迹
$So_k=0.4;\eta=0.2;E_u=0.01$

图 5 转子振幅 A_r-转速 Ω 曲线
$So_k=0.4;\eta=0.2;E_u=0.1$

图 5 示出 $e_u/c_{min}=0.1$ 的转子振幅曲线. 非线性和线性动力学结果之间的差别略为明显了一些, 不过总体上还是相似的. 非线性仿真所得的临界转速略为降低了一些, 共振振幅有所降低. 图 6 示出这种工况下的轴颈涡动轨迹, 图 7 示出转子的涡动轨迹. 由轴颈涡动轨迹可见, 当相对转速为 0.4 时, 非线性涡动轨迹的下端有一尖利的转角, 相应的转子涡动轨迹则已经由椭圆圈变成了 8 字形圈. 相对速度为 0.1 时的轴颈涡动轨迹明显地偏到了线性涡动轨迹的右方, 相应的非线性转子涡动轨迹与线性涡动轨迹之间, 在长轴方向和幅度上差别更为明显了.

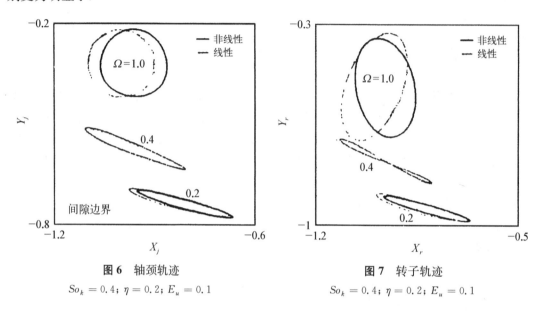

图 6 轴颈轨迹
$So_k=0.4; \eta=0.2; E_u=0.1$

图 7 转子轨迹
$So_k=0.4; \eta=0.2; E_u=0.1$

图 8 和图 9 示出 $e_u/c_{min}=0.3$ 时的轴颈和转子涡动轨迹. 非线性结果与线性结果之间的差别与 $e_u/c_{min}=0.1$ 时的相似, 但更显著一些.

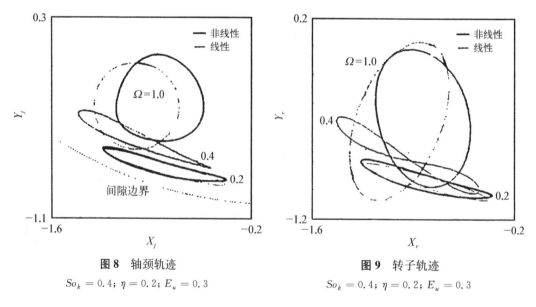

图 8 轴颈轨迹
$So_k=0.4; \eta=0.2; E_u=0.3$

图 9 转子轨迹
$So_k=0.4; \eta=0.2; E_u=0.3$

图 10 示出 $e_u/c_{min}=0.5$ 时的转子振幅曲线. 由图可见, 即使在转速远超过线性临界转速 0.76 后, 转子振幅仍持续上升. 图 11 示出这种工况下的轴颈涡动轨迹. 相对转速为

0.7时的非线性轴颈轨迹的形状已与线性轨迹相去甚远.特别值得注意的是,轴颈轨迹的下端受到轴承间隙边界的严格限制,而相应的线性轴颈轨迹已超越了间隙边界,这显然是不合理的,从而也清楚地表明了线性动力学在大不平衡量情况下的不适用性.图12示出这种工况下的转子轨迹,相对转速为0.7时的非线性轨迹比线性轨迹长得多,因此振幅大得多,似乎可以这样来理解:轴颈轨迹的两端已伸展到了贴近间隙边界的地方,轴颈在此处的运动受到了间隙边界的法线方向上极大的油膜刚度,这使得转子的行为接近于一根在这个方向受到刚性支撑的柔性转子,因此当轴的速度接近转子固有频率(相对转速接近1)时,转子在这个方向上的振幅急剧上升.

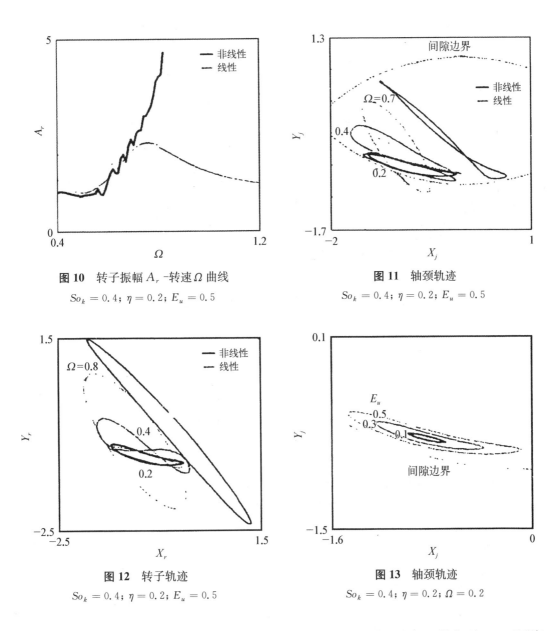

图 10 转子振幅 A_r-转速 Ω 曲线
$So_k = 0.4;\ \eta = 0.2;\ E_u = 0.5$

图 11 轴颈轨迹
$So_k = 0.4;\ \eta = 0.2;\ E_u = 0.5$

图 12 转子轨迹
$So_k = 0.4;\ \eta = 0.2;\ E_u = 0.5$

图 13 轴颈轨迹
$So_k = 0.4;\ \eta = 0.2;\ \Omega = 0.2$

图13~图15示出某一个转速下,轴颈轨迹如何随着不平衡量由0.1增大到0.5而逐渐变得离椭圆形状越来越远.图16~图18示出相应的转子轨迹变化.

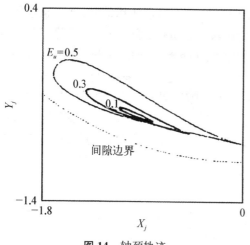

图 14 轴颈轨迹
$So_k = 0.4; \eta = 0.2; \Omega = 0.4$

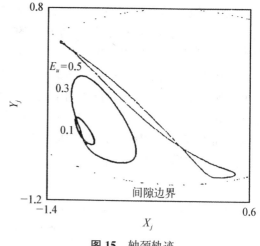

图 15 轴颈轨迹
$So_k = 0.4; \eta = 0.2; \Omega = 0.7$

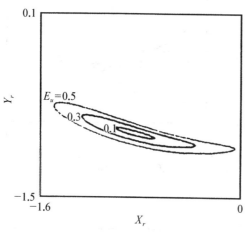

图 16 转子轨迹
$So_k = 0.4; \eta = 0.2; \Omega = 0.2$

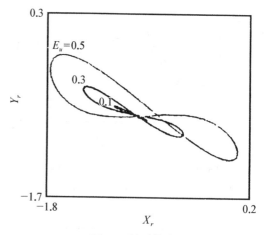

图 17 转子轨迹
$So_k = 0.4; \eta = 0.2; \Omega = 0.4$

5 结论

考察了支撑于椭圆轴承上的Jeffcot转子的不平衡响应. 在一个速度范围内,不同的不平衡量下,用非线性仿真和线性动力学理论计算了该系统的不平衡振动,对比两种计算的结果,可以得到下列结论:

(1) 当不平衡量极小时,两种理论所得的动力性态互相接近.

(2) 随着不平衡量的增大,轴颈和转子的非线性轨迹形状越来越偏离线性轨迹,临界转速和振幅值也发生相应的差别.

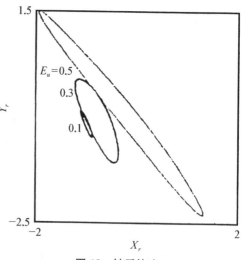

图 18 转子轨迹
$So_k = 0.4; \eta = 0.2; \Omega = 0.7$

（3）在很大不平衡量的激励下,当转速趋近转子固有频率时,轴颈非线性轨迹延展到了轴承间隙边界的近旁,此处的法向油膜刚度极高,使转子的行为相似于在这个方向上受到刚性支撑的柔性转子.转子振幅急剧升高,临界转速远离其线性计算值.

（4）线性动力学可以用来计算转子滑动轴承系统在很小不平衡量下的不平衡响应,当不平衡量不是很小时,应当用非线性动力学,特别是当不平衡量很大并且轴的转速接近转子固有频率时.

参 考 文 献

[1] Glienicke J, Leonhard M. Stabilitaelsprobleme bei der Lagerung schnellanfender Wellen, Bericht der Universitaet Karlsruhe, Juni 1981.
[2] Wang W, Zhang Z. Calculation of Journal Dynamic Locus Aided by Database of Non-stationary Oil Film Force of Single Bush Segment. Proc. of Asia-Pacific Vibration Conference 93,Vol. 2, Nov,1993, Kitakiushu, Japan.

Nonlinear Simulation of Unbalance Response of Jeffcot Rotor-bearing System

Abstract: Unbalance response of Jeffcot rotor-ellipse bearing system, considering the non-linear oil film force of the bearing, is studied by means of dynamic simulation in the paper. For numerical similation, a database of nonlinear oil film force for a pad has been set up previously, based upon dynamic Reynolds equations and Reynolds boundary conditions. The motion equations are integrated step by step by Rung Kuta method and the oil film force in the database is interpolated to get instantaneous value of bearing force. The dynamic behavior obtained, together with the whirl orbits of the rotor and journal, are compared with those by linear theory. It is noticed that the difference is significant under large unbalance response and therefore, the importance of considering the nonlinear oil film force becomes evident.

Improving the Performance of Spring-supported Thrust Bearing by Controlling Its Deformations[*]

Abstract: The performance of the spring-supported thrust bearing is studied with three-dimensional thermo-elastic hydrodynamic lubrication theory. The generalized Reynolds equation, the energy equation, the heat conduction equation, and the thermo-elastic deformation equation are solved simultaneously using the combination of the finite difference method and finite element method. Thermo-elastic deformation plays an important role in the performance of the spring-supported thrust bearing. Several factors such as spring pattern, pad thickness and initial pad geometry are analyzed. The results show that the above factors influence the performance of the bearing significantly. Suggestions based on the results are put forward to assist design considerations. © 2000 Elsevier Science Ltd. All rights reserved.

Nomenclature

A, B	coefficient, Eq. (4)		m/s
c_p	lubricant specific heat, J/(kg·K)	z	coordinate in the direction of film thickness, m
F_0	integral of Reynolds equation, m/(Pa·s)		
F_1	integral of Reynolds equation, m²/(Pa·s)	z_B	coordinate in the direction of pad thickness, m
F_2	integral of Reynolds equation, m³/(Pa·s)		
h	film thickness, m	θ	coordinate in circumferential direction, deg
h_a	heat transfer coefficient, W/(m²·K)		
k, k_B	thermal conductivity of oil and pad, W/(m·K)	ω	angular velocity of the slider, 1/s
		ρ	lubricant density, kg/m³
p	oil film pressure, Pa	ν	lubricant kinematic viscosity, m²/s
r	coordinate in the radial direction, m	μ	lubricant viscosity, Pa·s
T	film temperature, ℃	$[K]$	matrix of whole stiffness
T_B	pad temperature, ℃	$\{Q\}$	matrix of external load
T_a	oil sump temperature, ℃	$\{U\}$	matrix of deformation
u, v, w	film velocity in the directions of θ, r, z,		

1 Introduction

With the ever increasing dimension and loading of thrust bearings in modern hydro-generators, the pad with multiple support appears to offer greater advance in its

---

[*] In collaboration with Xiaojing Wang, Guoxian Zhang. Reprinted from *Tribology International*, 1999, 33: 713–720.

characteristics. The spring supporting form is one kind of multiple support in which thousands of springs are arranged under the thrust pads to bear the axial weight, as shown in Fig. 1. The spring-supported thrust bearing has good self-adjustment and heat dissipation. It is also of benefit with respect to vibration in running[1].

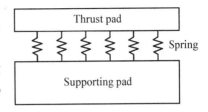

Fig. 1 Structure of spring-supported thrust bearing

To the best knowledge of the authors, the information on the performance of spring-supported thrust bearings is limited. Ashour[2] and Sinha[3,4] studied spring-supported thrust bearings including the effect of elastic distortion. Ettles[5] discussed effects of pad size and spring arrangement on the pad's behavior by thermo-elasto-hydrodynamic analysis. A rather simplified way of solving the pad distortion was adopted in this paper, where Kirchhoff plate assumption and a method of spring stiffness equivalence were used.

Refinement of analysis is evidently possible, and corresponding results will undoubted lead to more plausible conclusions, including those concerning the effects of such parameters as spring distribution, pad thickness and initial surface geometry.

2 Governing equations

In a 3-D TEHD analysis, the governing relations are: the generalized Reynolds equation, the 3-D energy equation, the heat conduction equation, and the pad thermo-elastic distortion calculation.

The generalized Reynolds equation is:

$$\frac{\partial}{\partial r}\left(F_2 r \frac{\partial p}{\partial r}\right) + \frac{\partial}{\partial \theta}\left(F_2 \frac{\partial p}{r\partial \theta}\right) = \omega r \frac{\partial}{\partial \theta}\left(\frac{F_1}{F_0}\right) \tag{1}$$

where,

$$F_0 = \int_0^h \frac{1}{\mu} dz, \quad F_1 = \int_0^h \frac{z}{\mu} dz, \quad F_2 = \int_0^h \frac{z(z - z_m)}{\mu} dz, \quad z_m = \frac{F_1}{F_0}.$$

The film pressure becomes ambient on the pad boundaries.

The energy equation with the lubricating film:

$$\rho c_p \left(u \frac{\partial T}{r\partial \theta} + v \frac{\partial T}{\partial r} + w \frac{\partial T}{\partial z}\right) = k \left[\frac{\partial^2 T}{r^2 \partial \theta^2} + \frac{\partial}{r\partial r}\left(r \frac{\partial T}{\partial r}\right) + \frac{\partial^2 T}{\partial z^2}\right] + \mu \left[\left(\frac{\partial u}{\partial z}\right)^2 + \left(\frac{\partial v}{\partial z}\right)^2\right] \tag{2}$$

The heat conduction equation within the pad:

$$\frac{\partial^2 T_B}{r^2 \partial \theta^2} + \frac{\partial}{r\partial r}\left(r \frac{\partial T_B}{\partial r}\right) + \frac{\partial^2 T_B}{\partial z_B^2} = 0 \tag{3}$$

At the film-pad interface, heat flux and temperature are matched:

$$k\left.\frac{\partial T}{\partial z}\right|_{z=h}=k_B\left.\frac{\partial T_B}{\partial z_B}\right|_{z_B=0}, \text{ and } T\Big|_{z=h}=T_B\Big|_{z_B=0}$$

Along the pad exterior surfaces, the convective boundary condition

$$-k_B\frac{\partial T_B}{\partial n}=h_a(T_a-T_B)$$

is used, where h_a is heat transfer coefficient.

The temperature effect on the oil viscosity is great pronounced. In this work the ASTM Walter formula is used to describe viscosity temperature relation:

$$\lg\lg(\nu+0.6)=A-B\lg T \tag{4}$$

With the kinematic viscosity values of a certain oil at two different temperatures, the two coefficients A and B can be solved.

The thrust pad is treated as three dimensional solid. On the basis of the principle of virtual work, strain-displacement equations and Hooke's law, the thermal-elastic deformation equation of the pad can be expressed:

$$[K]\{U\}=\{Q\} \tag{5}$$

where $[K]$ is the matrix of the whole stiffness and $\{Q\}$ is the external loading matrix superposed by the film pressure loading and the pad thermal loading. The displacement of the back surface of the pad is the value of the compressive deformation of springs. The displacement boundary conditions on other pad surfaces are free.

Thousands of springs with the same stiffness are arranged irregularly under thrust pads, therefore it is necessary to allocate the reaction force of each spring to its neighbour nodes in deformation calculation. The authors put forth an reasonable equivalent method of spring stiffness based on shape function method in reference [6]. This equivalent method is used in this research.

Compressive deformation of springs and thermo-elastic distortion of pads are superposed to form the vertical displacement of the pad surface, and contribute directly to the local film thickness:

$$h(r,\theta)=h_0(r,\theta)+d(r,\theta) \tag{6}$$

Where, $h_0(r,\theta)$ denotes the initial value of local film thickness and $d(r,\theta)$ the local superposed deformation.

3 Numerical solutions

The governing equations were solved numerically using the combination of the finite difference and finite element methods. The finite element method (FEM) was used to compute the thermo-elastic distortion of pad and the compressive deformation of springs. Other governing equations were solved using finite difference method (FDM). The Reynolds equation was solved using the SOR method, and temperature in the fluid and solid were computed simultaneously as a whole.

The solutions were obtained using iterative procedure. Briefly, for an assumed temperature and film distribution, the Reynolds equation, energy and heat conduction equations were evaluated. Having updated values for the lubricant temperature, a new viscosity field was computed which was subsequently used in the computation of the Reynolds equation. The iterations were repeated until oil pressure and fluid-solid temperature converged. Next, deformation equation was solved to obtain new film distribution. The above steps were executed repeatedly until the entire system converged and the loading capacity of the thrust bearing was equal to the axial weight.

4 Result analysis

In the present study, the hydro-generator spring-supported thrust bearing reported by Liao[1] is analyzed using the 3-D program. The dimensions and operating conditions are: number of pads 20, pad angle 18°, outer diameter 3.605 m, inner diameter 2.608 m, pad thickness 0.088 m, speed 136 r/min, oil reserve temperature 40℃, pad average pressure 3.326 MPa.

4.1 The effect of spring distribution

In order to analyze the effect of spring distribution on the performance of bearing, eight different patterns are considered, as shown in Fig. 2. In pattern 1 - 4, the number of springs per pad is 73, which is the same number as reported by Liao[1]. In pattern 5 - 8, the springs' number is unlimited. The corresponding bearing performance are calculated and compared in Fig. 3.

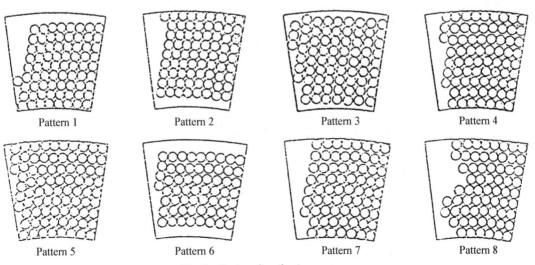

Fig. 2 Spring distribution patterns

It can be seen from Fig. 3 that different spring distri-butions result in quite different bearing performance. The difference of maximum film temperature reaches about 5℃ and the difference of minimum film thickness attains nearly 20 μm in these spring distributions. The values of pressure, friction and flow are also significantly different.

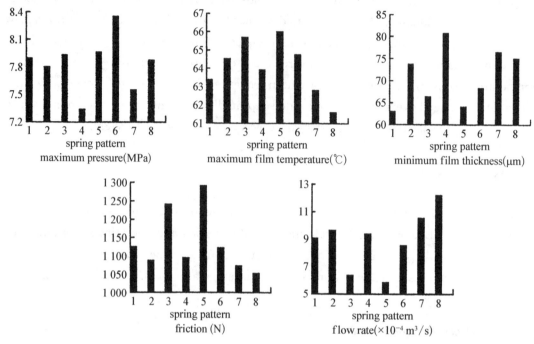

Fig. 3 Pad's performance with various spring patterns

In patterns 3 and 5, the springs are arranged to the full extent of the pad. In the inlet zone the deformation of the pad is small, at the meanwhile, large deformation appears in the pad center. The film wedge shape is comparatively poor. This phenomenon causes small inlet flow, high film temperature, thin film thickness and high value of friction. This spring distribution which springs are arranged to the full extent of the pad appears to be the worst choice.

In pattern 6, the springs are arranged somewhat symmetrically in the radial direction. The high supporting stiffness at the mean radius makes the deformation here small, and the low supporting stiffness at both the extreme radii makes the local deformation great. Compared with the performance of other patterns, pattern 6 has the highest film pressure, rather high temperature and thin film thickness.

Decreasing the spring amount in the inlet zone, as in patterns 4 and 7, creates low stiffness and big deformation here. Oil wedge is thus easily formed. Cool inlet oil flows into the film more easily and makes the film temperature lower.

As a further improvement of pattern 7, pattern 8 is formed by removing some springs at the mean radius in the inlet zone, to further ease the oil wedge formation. The calculated behavior of this pattern appears to be very satisfactory.

4.2 The effect of pad thickness

The pad deformation depends directly on pad thickness. Increasing the pad thickness results in high pad stiffness and small elastic deformation. At the meanwhile, the thicker pad causes the higher temperature difference between the upper and back pad surfaces and large thermal deformation. On the contrary, decreasing the pad thickness can cause low pad stiffness and improve heat dissipation. Elastic deformation increases and thermal deformation decreases.

Figs. 4 and 5 are the characteristics of the bearings with different pad thickness. In Fig. 4, the pad thickness is given a small value of 0.05 m. The isobar of pressure is long and narrow. The film temperature and friction are high. The film thickness distribution is unsatisfactory. The pad center sinks and the inner and outer radial sides turn upwards. The influence of elastic deformation is greater than that of thermal deformation.

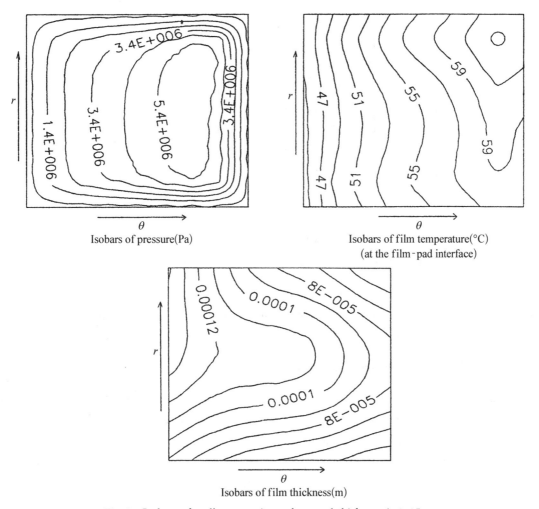

Fig. 4 Isobars of pad's properties, where pad thickness is 0.05 m

In Fig. 5, the pad thickness has a value of 0.3 m. The minimum film thickness appears at the mean radius near the oil outlet. Because high thermal gradient causes big thermal deformation, the pad is convex.

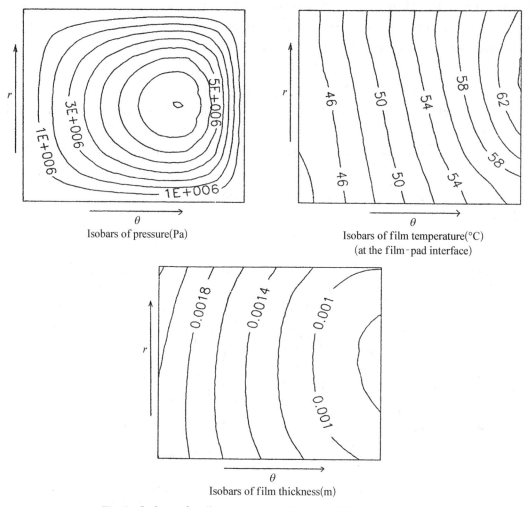

Fig. 5 Isobars of pad's properties, where pad thickness is 0.3 m

Fig. 6 shows that the relationship between pad thickness and pad performance is not linear. Between a thin pad and a thick one, there exists an optimum pad thickness to obtain the excellent performance.

4.3 The effect of pad surface initial geometry

In spring-supported thrust bearings, small deformation appears at the inlet zone where pressure is low, and large deformation occurs at the pressure center. This will change the film shape in the divergent sense, and decrease the load carrying capacity. Therefore, an initial geometry of pad form aiming at an initial convergence of the oil film is favorable. Nine cases are considered as shown in Fig. 7, and their calculated results are compared in Fig. 8.

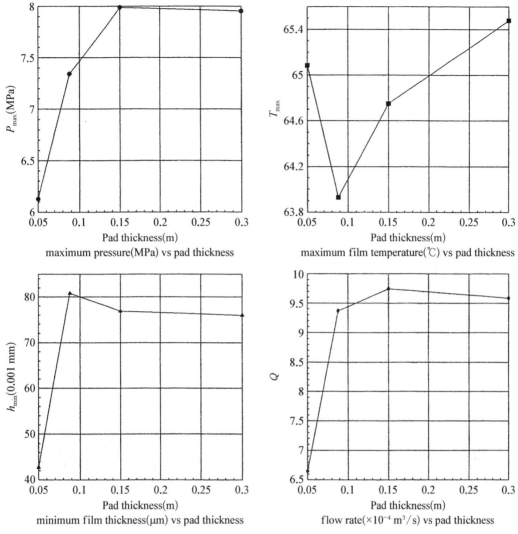

Fig. 6 Pad's performance under different pad thickness

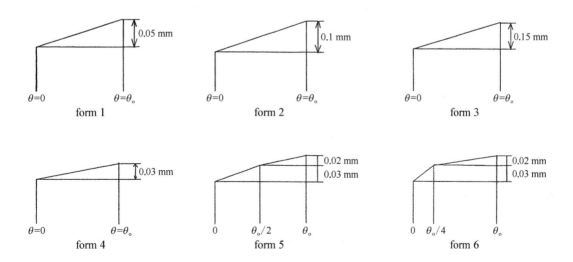

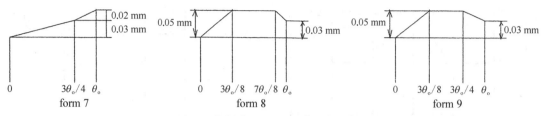

Fig. 7 Initial geometry of pad surface

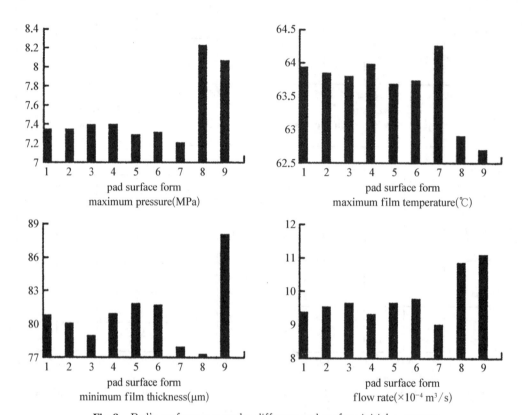

Fig. 8 Pad's performance under different pad surface initial geometry

Forms 1 - 4 are flat pad surfaces with different initial inclinations. Fig. 8 shows that there are no significant differences in their properties. Initial inclination of flat pad surface has little effect on bearing properties.

From a general comparison of the properties of forms 5 - 9, it can be seen that, a pad surface initial geometry which gives high inlet convergence can increase the oil flux and film thickness and reduce the film temperature. An initial geometry which provides exit divergence can decrease the film temperature significantly and increase the film thickness, at the expense of increasing the film pressure for the loading area decreasing.

5 Conclusions

3-D thermo-elasto-hydrodynamic performance of spring-supported thrust bearing is analyzed. The results show that spring distribution pattern, pad thickness and pad surface influence the pad performance significantly. Suggestions are given to aid the design of spring-supported thrust bearings:

1. Uniform distribution of springs in the full extent of the pad results in bad performance. Removing the springs in the inlet zone, especially those at the mean radius and in its vicinity, can be beneficial to the oil wedge formation and thus to the pad behavior.

2. In spring-supported thrust bearings, the pad surface initial geometry must be designed to form a convergent oil wedge. A large inlet convergence and an exit divergence of the pad surface can effectively improve the pad performance.

3. Pad thickness must be calculated carefully. Either a too great thickness or a too thin one is disadvantageous to the bearing behavior.

Acknowledgements

The authors gratefully acknowledge the financial support by Chinese National Natural Science Foundation (59775037) and Science Foundation of Shanghai Education Committee (98QN48) during this research.

References

[1] Liao G. Mounting and running of spring-supported thrust bearings. Hubei Elect Technol, 1994,1-2: 74-6 (In Chinese).

[2] Ashour ME, Athre K, Biswas S, Nath Y. Elastic distortion of a large thrust pad on an elastic support. Tribol Int, 1991,24(5): 299-309.

[3] Sinha AN, Athre K, Biswas S. Spring-supported hydrodynamic thrust bearing with special reference to elastic distortion analysis. Tribol Int, 1993,26(4): 251-263.

[4] Sinha AN, Athre K, Biswas S. A nonlinear approach to solution of Reynolds equation for elastic distortion anlysis to spring-supported thrust bearing. STLE, Tribol Trans, 1994,37(4): 802-810.

[5] Ettles CM. Some factors affecting the design of spring supported thrust bearings in hydrodynamic generators. ASME, J Tribol, 1991,113: 626-632.

[6] Wang X, Zhang Z, Zhang G. TEHD study of spring-supported thrust bearings. In: Proceedings of ASIATRIB'98, 1998: 30-34.

计入非牛顿效应的曲轴轴承的混合润滑分析*

摘　要：分析了剪切变薄的非牛顿流变学特性和两表面都具有的纵向、横向和各向同性粗糙度对动载有限宽径向滑动轴承性能的综合影响. Christensen 的粗糙表面流体动力润滑的随机模型和 Greenwood-Tripp 接触压力的计算模型用于处理粗糙问题，并考虑了磨合对粗糙高度分布的影响. 幂律流体模型用来表征剪切变薄的流变学特征，质量守恒的油膜破裂算法用于雷诺方程的求解. 计算结果表明，粗糙度总是减小最小名义油膜厚度，并使油膜压力在接触区剧烈振荡，其幅值大于光滑表面时周期内的最大名义油膜压力. 名义最小油膜厚度在纵向粗糙时最大，横向粗糙时最小. 粗糙纹理相同时，相同粗糙结构下的名义最小油膜厚度在牛顿流体时大于不同粗糙结构时的相应值，在非牛顿流体情况下，结论相反. 混合润滑的轴承性能受粗糙纹理和结构、幂律指数、轴承几何结构、轴颈质量及运行工况的综合影响.

符号说明

A——轴承表面积；

A_c——接触总面积；

B, D, R——轴承的宽度、直径、半径；

C——径向间隙；

E'——等效弹性模量，$\dfrac{1}{E'}=\dfrac{1}{2}\left(\dfrac{1-\nu_1^2}{E_1}+\dfrac{1-\nu_2^2}{E_2}\right)$；

$\ddot{e}_x, \ddot{e}_y$——轴颈在 x, y 方向的加速度；

$F_{\text{oil}x}, F_{\text{oil}y}$——油膜承载力在 x, y 方向的分量；

F_x, F_y——轴承负荷在 x, y 方向的分量；

h——名义油膜厚度；

h_T——实际油膜厚度，$h_T=h-\delta_1-\delta_2$；

M——轴颈等效质量；

n——幂律指数；

p, p_c——油膜压力、接触压力；

t——时间；

U_1, U_2——轴瓦和轴颈表面的切向速度，$U=U_1+U_2$，$U_{21}=U_2-U_1$；

χ, z——周向和轴向坐标；

λ——宽径比，B/D；

δ_1, δ_2——轴瓦和轴颈自中心线的粗糙高度；

ε——偏心率；

η, β——粗糙峰的表面密度、曲率半径；

ν_1, ν_2——轴瓦和轴颈的泊松比；

Φ^*——两表面峰值分布和的高斯概率密度；

σ——粗糙高度的均方根值；

σ^*——Φ^* 的均方根值；

μ——流体黏度（牛顿流体）和黏稠系数（幂律流体）；

$\overline{(\ )}$——表明对该变量取数学期望值.

引言

曲轴轴承的最小油膜厚度与其表面粗糙度具有相同的数量级，发动机正常运行时，轴承不可避免地处于混合润滑状态. 添加剂使润滑剂具有剪切变薄的非牛顿特性，典型的幂律指数是 $0.9^{[1]}$. 因此，研究粗糙度和非牛顿的流变学特性对曲轴轴承混合润滑性能的影响有助

---

* 本文合作者：张朝. 原发表于《内燃机学报》，1999，17(3): 303–307.

于精确预测轴承性能,这是本文研究的内容.

本文采用 Christensen 的粗糙表面流体动力润滑的随机模型[2]处理两表面均具有纵向(表面粗糙纹理与表面运动方向平行)、横向(表面粗糙纹理与表面运动方向垂直)和各向同性(表面粗糙纹理无方向性)粗糙的问题并考虑磨合对粗糙高度分布的影响.用幂律模型表征剪切变薄的非牛顿特性,分析表面粗糙纹理和结构,幂律指数对轴承性能的综合影响.

1 数学分析

3 种粗糙纹理的幂律流体的雷诺方程为[3]:

$$\frac{\partial \rho \bar{h}_T}{\partial t} + \frac{\partial}{\partial x}\left(\rho \varphi_A - \frac{\rho \varphi_B}{12\mu n}\frac{\partial \bar{p}}{\partial x}\right) + \frac{\partial}{\partial z}\left(-\frac{\rho \varphi_C}{12\mu}\frac{\partial \bar{p}}{\partial z}\right) = 0 \tag{1}$$

对应于纵向、各向同性和横向的粗糙纹理 φ_i 依次为:

$$\varphi_A: \frac{U_{21}^{n-1} U \bar{h}_T}{2}, \frac{U_{21}^{n-1} U \bar{h}_T}{2}, U_{21}^{n-1}\left[\frac{U}{2}\frac{\overline{h_T^{-(1+n)}}}{\overline{h_T^{-(2+n)}}} - \frac{U_1\overline{\delta_1 h_T^{-(2+n)}} + U_2\overline{\delta_2 h_T^{-(2+n)}}}{\overline{h_T^{-(2+n)}}}\right]$$

$$\varphi_B: \overline{h_T^{2+n}}, \overline{h_T^{2+n}}, 1/\overline{h_T^{-(2+n)}}$$

$$\varphi_C: 1/\overline{h_T^{-(2+n)}}, \overline{h_T^{2+n}}, \overline{h_T^{2+n}}$$

式中: $h_T = h - \delta_1 - \delta_2$; $\overline{(\ )} = \int_{-\infty}^{\delta_2'}\int_{-\infty}^{\delta_1'}(\ )f_1(\delta_1)f_2(\delta_2)\mathrm{d}\delta_1\mathrm{d}\delta_2$, $f_1(\delta_1)$ 和 $f_2(\delta_2)$ 为 δ_1 和 δ_2 的概率密度函数.对完全流体润滑, $\delta_1' = \delta_{1\max}/C$, $\delta_2' = \delta_{2\max}/C$; 对混合润滑, δ_1' 和 δ_2' 依轴瓦和轴颈的弹性模量比求得.

视流体为不可压缩,采用类似 Elrod 空穴法[4],定义空穴度 λ,开关函数 g 如下:

在完全油膜区 ($\lambda \geqslant 0$): $\lambda = \bar{p}$; $g = 1$

在油膜破裂区 ($\lambda \leqslant 0$): $\lambda = \theta - 1$; $g = 0$

式中: θ 为油膜比例.将 λ 和 g 代入方程(1),有

$$\frac{\partial}{\partial t}\{[1+(1-g)\lambda]\bar{h}_T\} + \frac{\partial}{\partial x}\left\{\varphi_A[1+(1-g)\lambda] - \frac{\varphi_B g}{12\mu n}\frac{\partial \lambda}{\partial x}\right\} + \frac{\partial}{12\mu \partial z}\left\{-\varphi_C g\frac{\partial \lambda}{\partial z}\right\} = 0 \tag{2}$$

方程(2)各项的离散可见文献[4].

将 Greewood 和 Tripp 的计算接触压力的方法[5]用于本文的研究:

$$p_c(\bar{h}/\sigma^*) = K'E'F_{5/2}(\bar{h}/\sigma^*) \tag{3}$$

接触负荷和接触面积可由下式求解:

$$F_c = 2BRK'E'\int_\varphi^{180} F_{5/2}(\bar{h}/\sigma^*)\cos(180-\varphi)\mathrm{d}\varphi \tag{4}$$

$$A_c = 2BR\pi^2(\eta\beta\sigma^*)^2\int_\varphi^{180} F_2(\bar{h}/\sigma^*)\mathrm{d}\varphi \tag{5}$$

$$\begin{cases} K' = \dfrac{8\sqrt{2}}{15}\pi(\eta\beta\sigma^*)^2\sqrt{\dfrac{\sigma^*}{\beta}} \\ F_n(u) = \displaystyle\int_u^{180}(s-u)^n \Phi^*(s)\mathrm{d}s \\ \varphi' = \arccos\left(\dfrac{3\sigma^* + \Delta_s - C}{C\varepsilon}\right) \end{cases} \quad (6)$$

式中:Δ_s 为粗糙高度中心线与粗糙峰分布中心线之间的距离,可从下式获得:

$$3\sigma^* + \Delta_s = \delta_{1\max} + \delta_{2\max} \quad (7)$$

参考图1,轴颈的运动方程为:

$$M\ddot{e}_x = F_{\text{oil}x} + F_{ex} + F_{cx} \quad (8)$$

$$M\ddot{e}_y = F_{\text{oil}y} + F_{ey} + F_{cy} \quad (9)$$

$$F_{\text{oil}x} = -\int_A \bar{p}\cos\theta \mathrm{d}A \quad (10)$$

$$f_{\text{oil}y} = -\int_A \bar{p}\sin\theta \mathrm{d}A \quad (11)$$

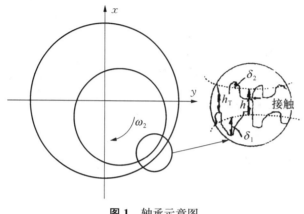

图 1 轴承示意图

方程(2)由差分法求解,周向和轴向的网格数分别为 80 和 24,由于对称性仅分析半宽轴承. 采用 SOR 法且收敛误差为 0.000 05. 方程(8)和(9)由 4 阶精度的 Runge-Kutta 法同步积分求解,时间步长为 1°CA. 用辛卜生法积分求解 $F_{\text{oil}x}$, $F_{\text{oil}y}$, F_{cx} 和 F_{cy}. 对任给初始条件,用周期性条件完成动载问题的求解.

2 结果与讨论

工况条件见表1;图2是从发动机主轴承表面磨合后测得的粗糙高度分布的概率密度函数,在本研究中用于轴瓦和轴颈表面的粗糙分布. 对工程表面来说,K' 的范围是 0.000 03~0.003 及 $\Delta_s \geqslant 0$,选择 K' 和 Δ_s 时考虑以更好地表现接触负荷的影响. 主轴承负荷见图3. 图4~图6是粗糙纹理与结构对油膜的影响情况. 表3和图5中的"S""L""I""T"依次表示光滑、纵向、各向同性和横向粗糙表面.

表 1 工况条件

B	2.1×10^{-2} m	D	7.2×10^{-2} m
C	5.0×10^{-4} m	p_0	3.5×10^5 N·m^{-2}
M	400 kg	E'	46.5 GPa
K'	0.003	Δ_s	0
U_1	0	U_2	13.56 m·s^{-1}
n	1(牛顿流体)		0.924 6(幂律流体)
μ	0.003 Pa·s (牛顿流体)		0.012 06 Pa·s^n (幂律流体)
供油	半油槽(0°~180°)		

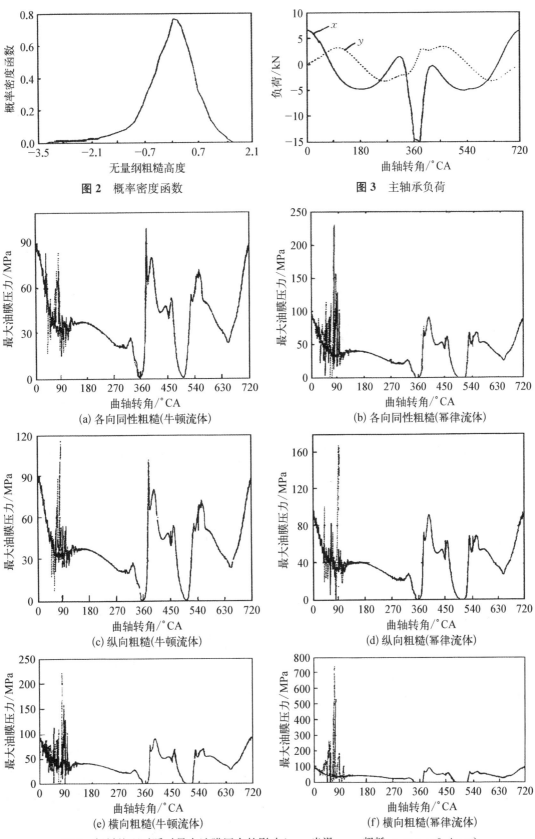

图2 概率密度函数

图3 主轴承负荷

图4 粗糙纹理对瞬时最大油膜压力的影响(——光滑,……粗糙,$\sigma_1=\sigma_2=0.4\ \mu m$)

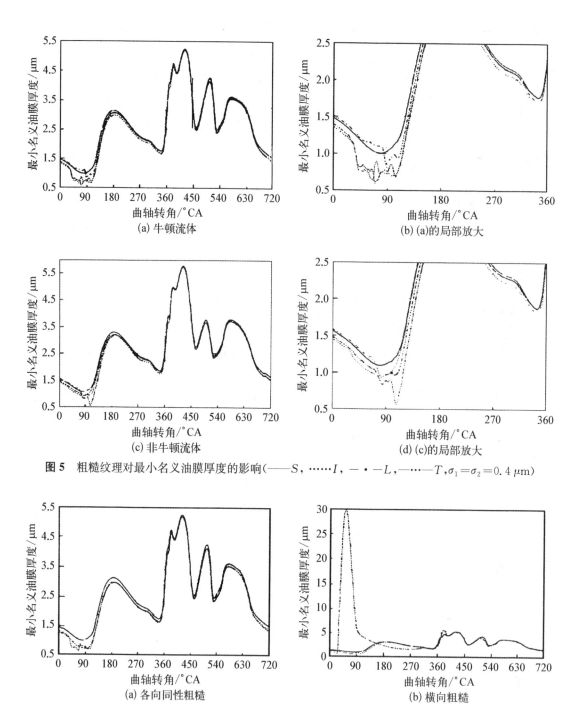

图 5 粗糙纹理对最小名义油膜厚度的影响(——S,……I,— · —L,— — —T,$\sigma_1=\sigma_2=0.4\ \mu m$)

图 6 粗糙结构对最小名义油膜厚度影响(——光滑,……粗糙,$\sigma_1=\sigma_2=0.4\ \mu m$;— · —$\sigma_1=0.7\ \mu m$, $\sigma_2=0.1\ \mu m$;牛顿流体)

与光滑表面轴承相比,粗糙度总是减小最小名义油膜厚度 h_{min}. 对牛顿和幂律流体来说,纵向粗糙度的 h_{min} 最大,横向粗糙度的 h_{min} 最小. 当粗糙纹理相同时,相同粗糙结构 ($\sigma_1=\sigma_2=0.4\ \mu m$) 的 h_{min} 在牛顿流体时大于不同粗糙结构($\sigma_1=0.7\ \mu m$,$\sigma_2=0.1\ \mu m$) 的 h_{min},而在幂律流体的情况下,其结论相反.

与光滑表面轴承相比,粗糙度总是使对应于曲轴转角的最大油膜压力的数学期望值在接触区剧烈波动,其幅值通常要大于光滑表面的周期中最大的油膜压力的数学期望值 $\bar{p}_{max}$,显然,它将严重影响轴承油膜压力的分布状况. 如表 2 所示,磨合使表面粗糙高度的分布偏离高斯分布,并使其最大粗糙峰值小于后者. 其磨合表面的 δ_{imax} 是根据图 2 算得的,而高斯分布的 δ_{imax} 是由式 $\delta_{imax}=3\sigma_i$ 算得. 显然,在混合润滑状态下,这种偏离将严重影响两表面粗糙接触的程度.

表 2 σ_i 与 δ_{imax} 相互关系(A——磨合表面,B——高斯表面)

σ_1	σ_2	δ_{1max}		δ_{2max}		$\delta_{1max}+\delta_{2max}$	
		A	B	A	B	A	B
0.7	0.1	1.17	2.10	0.16	0.30	1.28	2.40
0.4	0.4	0.64	1.20	0.64	1.20	1.28	2.40

表 3 粗糙度和剪切变薄对主轴承性能的影响

	$h_{min}/\mu m$							
	S	I	L	T	S_c	I_c	L_c	T_c
$\sigma_1=\sigma_2=0.4\ \mu m$	1.01	0.65	0.86	0.61	1.10	0.93	0.94	0.57
$\sigma_1=0.7\ \mu m, \sigma_2=0.1\ \mu m$	1.01	0.73	0.92	0.71	1.10	0.52	0.93	0.49
	$\bar{p}_{max}/MPa$							
$\sigma_1=\sigma_2=0.4\ \mu m$	98	231	167	740	100	93	116	110
$\sigma_1=0.7\ \mu m, \sigma_2=0.1\ \mu m$	98	194	141	163	100	109	123	294

S, I, L, T——光滑、纵向、各向同性、横向粗糙表面(牛顿流体)
S_c, I_c, L_c, T_c——光滑、纵向、各向同性、横向粗糙表面(幂律流体)

上面分析对应于给定的轴承几何特征、轴颈质量、幂律指数、运行工况,当这些条件改变时,上述结论可能与式(3),(4),(6)有所不同. 因此,预测轴承性能时需考虑上述因素的综合影响.

3 结论

粗糙度总是减小最小名义油膜厚度,并使油膜压力在接触区剧烈震荡,其幅值大于光滑表面时的周期中最大的油膜压力. 名义最小油膜厚度在纵向粗糙时最大,横向粗糙时最小.

粗糙度及剪切变薄的非牛顿效应对轴承性能的影响依赖于粗糙纹理和结构、幂律指数、轴承几何特征、轴颈质量及运行工况.

磨合改变粗糙高度的分布,这将影响混合润滑时粗糙峰的接触程度,从而影响接触负荷.

参考文献

[1] Rastogi A, Gupta R K. A ccounting for Lubricant Shear Thinning in the Design of Short Journal Bearings[J]. Journal of Rheology, 1991, 35(4).
[2] Christensen H. Stochastic Models for Hydrodynamic Lubrication of Rough Surfaces [C]. Proc Instn Mech Engrs, 1970, 184(55): 1013-1026.
[3] Zhang Chao, Qiu Zugan. Analysis of Two-Sided Roughness and Non-New tonian Effects in Dynamically Loaded

Finite Journal Bearings [C]. Proc of the International Tribology Conference, Yokohama, 1995: 1005-1010.
[4] Zhang Chao, Zhang Zhiming. Analysis of Crankshaft Bearings in Mixed Lubrication Including Mass Conserving Cavitation [C]. Proceedings of Second International Conference on Hydrodynamic Bearing-Rotor System Dynamics, Xi'an, 1997: 18-24.
[5] Greewood J A, Tripp J H. The Contact of Two Nominally Flat Rough Surfaces [C]. Proc Inst Mech Eng London, 1971, 185: 625-633.

Analysis of Crankshaft Bearings in Mixed Lubrication Including Non-Newtonian Effects

Abstract: The comprehensive effects of non-Newtonian shear thinning characteristics and two-sided purely longitudinal, transverse and isotropic roughness on dynamically loaded finite journal bearings in mixed lubrication are numerically analyzed. Christensen's stochastic model of hydrodynamic lubrication of rough surfaces and the asperity contact model developed by Greenwood and Tripp are used to treat the roughness problem. Running-in effect on roughness height distribution is considered. Shear thinning is characterized by the Power-law fluid model. A mass conserving cavitation algorithm is used to solve the modified Reynolds equation. It is shown that roughness always decreases the minimum nominal film thickness and increases oscillations of the maximum nominal film pressure in the mixed lubrication regions, resulting in its maximum values bigger than those in the smooth cases. The minimum nominal film thickness is biggest in the longitudinal roughness and the smallest in the transverse case. For the same roughness texture, in the Newtonian fluid, the minimum nominal film thickness of the same roughness structure is bigger than that of the different roughness structure, and it is opposite in the shear thinning case. The bearing performances of crankshaft bearings in mixed lubrication are affected seriously by the roughness texture and structure, power law index, features of the nominal geometry, journal mass, and operation conditions.

计入入口冲击压力的弹簧支承式推力轴承热弹流研究*

摘　要：提出了计算一维入口冲击压力的方法，对弹簧支承式推力轴承进行三维热弹流研究，讨论弹簧支承式推力轴承在不同工况参数下性能的变化. 研究表明，入口冲击压力对弹簧支承式推力轴承的性能影响显著，并揭示了在不同转速和载荷下轴承的各项性能曲线的变化情况.

0　前言

随着水电机组的单机容量不断增大，推力轴承的负荷、瓦面积和比压不断增大，为改善推力轴承性能，用多点支承取代单支承方式已成趋势. 弹簧支承式推力轴承是在托瓦上分布着上千个性能完全一致的小弹簧来承受轴向力，形成多点支承，具有受力均匀、自调整能力强等特点. 据作者所查，有关弹簧支承式推力轴承性能计算的报道较少[1-6]，而对其进行热弹流研究的有论文[4-6]. 考虑到入口冲击压力对固定瓦和可倾瓦的性能有显著影响，因此在研究弹簧支承式推力轴承的性能时，不仅要考虑热弹变形和温度场的变化，还有必要计入入口冲击压力的影响. 本文采用二维动量守恒方程计算进油压力分布，用三维热弹流理论研究计入入口冲击压力效应的弹簧支承式推力轴承的性能.

1　理论分析

推力轴承工作运行时，润滑油膜会由于黏滞损耗产生热量，导致润滑油黏度降低，轴瓦产生热变形，同时，受到巨大的轴向推力，轴瓦发生弹性变形，弹簧产生压缩变形. 因此要能较接近实际地计算弹簧支承式推力轴承的性能，必须进行热弹流分析.

广义雷诺方程计入了润滑油黏度的三维变化，用于求解推力轴承的油膜压力场.

$$\frac{\partial}{\partial r}\left(F_2 r \frac{\partial p}{\partial r}\right) + \frac{\partial}{\partial \theta}\left(F_2 \frac{\partial p}{r \partial \theta}\right) = \omega r \frac{\partial}{\partial \theta}\left(\frac{F_1}{F_0}\right) \tag{1}$$

式中　$F_0 = \int_0^\delta \frac{1}{\mu} dz$　　$F_1 = \int_0^\delta \frac{z}{\mu} dz$

$F_2 = \int_0^\delta \frac{z(z-z')}{\mu} dz$　　$z' = \frac{F_1}{F_0}$

p——油膜压力
ω——镜板角速度
r, θ, z——径向、周向和膜厚方向坐标

---

* 本文合作者：王小静、张国贤. 原发表于《机械工程学报》，2000，36(2)：51-55.

δ——油膜厚度

μ——润滑油动力黏度

F_0, F_1, F_2, z'——广义雷诺方程系数

完整的三维能量方程为

$$\rho c \left(u \frac{\partial T}{r \partial \theta} + v \frac{\partial T}{\partial r} + w \frac{\partial T}{\partial z} \right) = k \left[\frac{\partial^2 T}{r^2 \partial \theta^2} + \frac{\partial}{r \partial r} \left(r \frac{\partial T}{\partial r} \right) + \frac{\partial^2 T}{\partial z^2} \right] + \mu \left[\left(\frac{\partial u}{\partial z} \right)^2 + \left(\frac{\partial v}{\partial z} \right)^2 \right] \tag{2}$$

式中　T——油膜温度

　　　ρ——润滑油密度

　　　c——比热容

　　　k——传热系数

u, v, w——周向、径向和膜厚方向的速度

油膜与轴瓦的界面上,保持热流量相等,即油膜中传出的热流量等于轴瓦传入热流量

$$k \frac{\partial T}{\partial z} \bigg|_{z=\delta} = k_b \frac{\partial T_b}{\partial z_b} \bigg|_{z_b=0} \tag{3}$$

式中　T_b——轴瓦温度

　　　k_b——轴瓦传热系数

　　　z_b——瓦厚方向坐标

在油膜的入口区,存在不同温度润滑油的混合现象. 进入润滑油膜的油有两部分: 一部分是从上瓦出来的热油跟随镜板一起运动,进入下瓦; 另一部分是油槽中的冷油进入油膜. 新补充的冷油流量 $q_{V,s}$ 与热油流入的流量 $q_{V,r}$ 之和等于润滑油膜入口区流入的流量 $q_{V,i}$. 根据热量平衡,得到

$$q_{V,i} \cdot T_1(r, z) = q_{V,s} \cdot T_0 + q_{V,r} \cdot T_2(r, z) \tag{4}$$

式中　T_0——油槽温度

$T_1(r, z)$——油膜进油边温度分布

$T_2(r, z)$——油膜出油边温度分布

轴瓦的热传导方程为拉普拉斯方程

$$\frac{\partial^2 T_b}{r^2 \partial \theta^2} + \frac{\partial}{r \partial r} \left(r \frac{\partial T_b}{\partial r} \right) + \frac{\partial^2 T_b}{\partial z^2} = 0 \tag{5}$$

轴瓦受到压力和温度梯度的作用产生热弹变形,同时弹簧受到压力产生压缩变形. 弹簧的压缩变形量与轴瓦热弹变形量耦合在一起,形成总变形量,使油膜厚度发生变化.

不考虑入口冲击压力时,油膜压力场的边界条件是四周边界上压力为零. 现考虑冲击压力,边界条件应改为进油边的油膜压力为一维分布,其余边界的油膜压力为零.

入口冲击压力用动量守恒原理计算. 对于一维油膜压力场其计算方法如下: 对图1中的 $APOCDE$ 围成的区域进行动量和冲量计算,假设 AP 和 AE 上作用环境压力,DC 上压力呈线性变化,从环境压力变化到进油压力,在 PO 上的黏性力很小,忽略不计. 得到的动量方程为

$$\int_0^{\delta_2}\rho u^2 \mathrm{d}z - \int_0^{\delta_1}\rho u_i^2 \mathrm{d}z + \int_0^{\delta_3} p_{\mathrm{amb}}\mathrm{d}z - \int_0^{\delta_1} p_i \mathrm{d}z - \int_{\delta_1}^{\delta_3}\left[\frac{p_{\mathrm{amb}}-p_i}{\delta_3-\delta_1}(z-\delta_1)+p_i\right]\mathrm{d}z = 0 \quad (6)$$

式中 p_{amb}——环境压力

p_i——进油压力

u——运动表面速度

u_i——进油边油膜速度

δ_1——入口处膜厚

δ_2——运动表面所携带的油层厚度

δ_3——流线开始垂直于运动表面时的位置

根据文献[8]的研究结果,δ_3 可以足够准确地用 $3\delta_1$ 代替.

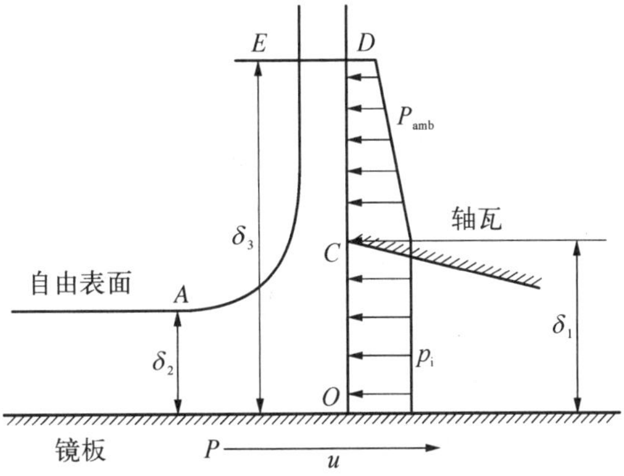

图 1 轴瓦入口前端示意图

将以上的一维油膜压力场分析扩展到沿油膜周向和径向的二维油膜压力场分析,进油压力 p_i 是沿径向的一维分布. 现假设其为抛物线分布,径向中部进油压力为最大,在内外径处油膜压力为零.

$$p_i(r) = p_{i,c}\cdot\left[1-\left(\frac{2r-R_1-R_2}{R_2-R_1}\right)^2\right] \quad (7)$$

式中 $p_{i,c}$——进油边径向中部压力值

R_1, R_2——轴瓦内径和外径

考察图 1 中的 APOCDE 区域和轴瓦宽度,求解二维动量方程,可得到 $p_{i,c}$ 的表达式

$$p_{i,c} = \frac{2\rho\left(\omega^2\int_{R_1}^{R_2} r^2\delta_2 \mathrm{d}r - \int_{R_1}^{R_2}\int_0^{\delta_1} u_i^2 \mathrm{d}z\mathrm{d}r\right) + p_{\mathrm{amb}}\int_{R_1}^{R_2}(\delta_1+\delta_3)\mathrm{d}r}{\int_{R_1}^{R_2}(\delta_1+\delta_3)\left[1-\left(\frac{2r-R_1-R_2}{R_2-R_1}\right)^2\right]\mathrm{d}r} \quad (8)$$

假设运动表面所携带的油层厚度为上瓦出油边的油膜厚度.

2 结果与讨论

本文采用的算例是清江隔河岩水利枢纽引进的一套弹簧支承式扇形推力轴承,具体工况和几何参数如下[9]:瓦块数 20,瓦张角 18°,轴瓦外径 3.605 m,轴瓦内径 2.608 m,瓦厚 0.088 m,镜板转速 136 r/min,进油温度 40℃,油槽温度 40℃,单瓦比压 3.326 MPa,单瓦弹簧数 73,弹簧刚度 10.56 MN/m.

2.1 油膜压力、温度分布

根据上述工况几何参数,得到油膜压力和温度分布,见图 2 和图 3. 图 2 表明,油膜压力峰值偏向出油边,压力分布沿径向基本对称. 在图 2a 中,不计入进油压力时进油边油膜压力为零,图 2b 中,考虑了流体动量的作用,进油边油膜压力的最大值为 0.53 MPa,占最大油膜压力的 7.34%. 图 2b 与图 2a 比较,计入进油压力的最大油膜压力比不计进油压力的小

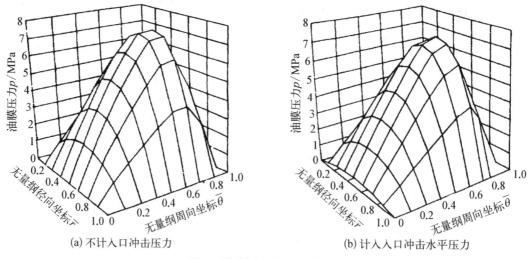

(a) 不计入口冲击压力　　　　　　　(b) 计入入口冲击水平压力

图 2　油膜压力分布（p/MPa）

0.1 MPa，并且油膜压力峰略偏向出油边.

从图 3 的名义半径处油膜和轴瓦的温度分布看，分布形状大致相同，最高油膜温度出现在出油边近轴瓦表面处，有进油压力可明显降低油膜和瓦体温度，在入口区和瓦体背面下降更为显著，最高油膜温度可下降 2℃ 左右.

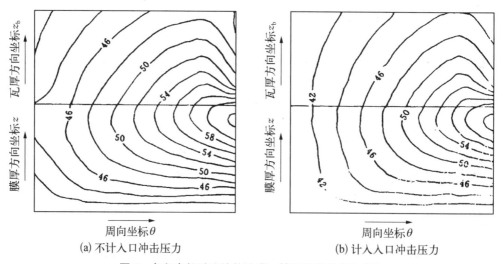

(a) 不计入口冲击压力　　　　　　　(b) 计入入口冲击压力

图 3　名义半径平面处的油膜—轴瓦温度分布（t/℃）

2.2　速度变化对轴承性能的影响

图 4、图 5 表明，随着转速的提高，最高油膜温度、最小膜厚、摩擦功耗和流量增加迅速. 转速小于 130 r/min 时最大油膜压力增加较快，但之后随着转速的提高最大油膜压力增加缓慢. 由于转速增加，使油膜剪切率增大，黏滞损耗和发热量增大，导致油膜温度升高、摩擦功耗和流量增大.

进油压力与最大油膜压力的比值先随转速的增加而增加，之后在 16%～18% 之间变化. 计入进油压力效应时，最大油膜压力有所降低. 与不计进油压力的计算结果相比，计入进

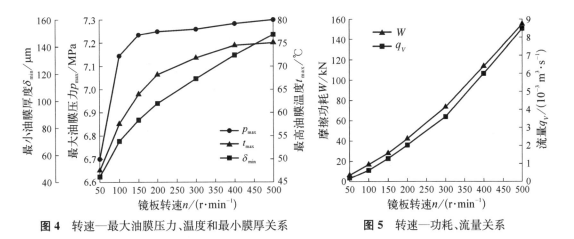

图4 转速—最大油膜压力、温度和最小膜厚关系

图5 转速—功耗、流量关系

油压力能降低油膜温度.速度越高,油膜温度越高,计入进油压力与不计进油压力计算的油膜温度差值越大.同时,计入进油压力使摩擦功耗下降,并随速度的增大,摩擦功耗下降量不断增大.入口区油膜压力增大使该区域的弹簧压缩变形加剧,从而增厚油膜厚度,增大流量,降低油膜和轴瓦温度和摩擦功耗,提高轴承承载力.

2.3 载荷变化对轴承性能的影响

从图6、图7可以看出载荷对轴承性能影响非常显著.随着载荷增大,油膜压力升高,油膜厚度迅速减小,使流量相应减少,油膜温度上升.当载荷进一步增加时,最高油膜温度位置逐渐移向出油边.

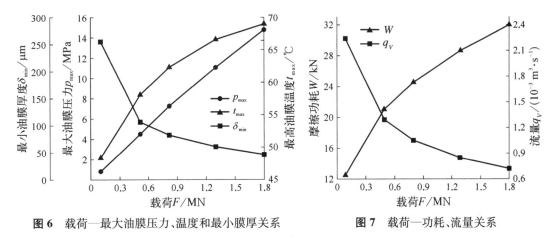

图6 载荷—最大油膜压力、温度和最小膜厚关系

图7 载荷—功耗、流量关系

图8是载荷为1.8 MN时的油膜厚度分布,油膜沿周向呈收敛型分布,计入入口冲击压力时的油膜厚度值明显大于不计时的膜厚值,图8b膜厚的收敛形状比图8a更为平滑.

3 结论

(1)进油压力对弹簧支承式推力轴承的性能影响显著,入口区油膜压力增大致使该区域的弹簧向下压缩变形增大,增加油膜厚度尤其是入口区膜厚,最大油膜压力降低,油膜温度和轴瓦温度降低,流量增大和摩擦功耗减少.

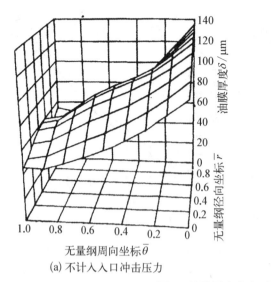 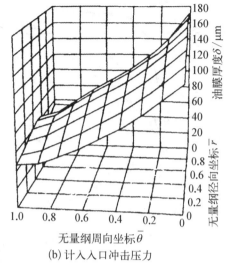

(a) 不计入入口冲击压力　　　　　　(b) 计入入口冲击压力

图 8　油膜厚度分布(载荷为 1.8 MN)

(2) 随着转速的提高,最高油膜温度、最小膜厚、摩擦功耗和流量增加迅速.速度变化时,进油压力对轴承性能的影响程度不同,速度上升使进油压力增大.与不计进油压力的计算相比,转速提高使最高油膜温度较大下降,最小膜厚的增量和摩擦功耗的减少量增加.

(3) 随着载荷增大,油膜压力升高,油膜厚度迅速减小,流量相应减少,油膜温度上升.

参考文献

[1] Ashour M E, Athre K, Biswas S, el at. Elastic distortion of a large thrust pad on an elastic support. Tribology International, 1991, 24(5): 299-309.

[2] Sinha A N, Athre K, Biswas S. Spring-supported hydrodynamic thrust bearing with special reference to elastic distortion analysis. Tribology International, 1993, 26(4): 251-263.

[3] Sinha A N, Athre K, Biswis S. A nonlinear approach to solution of Reynolds equation for elastic distortion analysis to spring-supported thrust bearing. STLE Tribo. Trans., 1994, 37(4): 802-810.

[4] Ettles C M. Some factors affecting the design of spring supported thrust bearings in hydrodynamic generators. ASME J. Tribo., 1991, 113: 626-632.

[5] 王小静,张直明,张国贤,等.弹簧支承式推力轴承的热弹流研究.上海大学学报,1997,3: 134-141.

[6] 王小静,张直明,张国贤.影响弹簧支承式推力轴承性能的几个因素:润滑与密封,1998,3: 24-27.

[7] Rodkiewicz M, Kim K W, Kennedy J S. On the significance of the inlet pressure build-up in the design of tilting-pad bearings. ASME J. Tribo., 1990, 112(1): 17-22.

[8] Tipei N. Flow characteristics and pressure head build-up at the inlet of narrow passages. ASME J. Lubr. Tech., 1978, 100(1): 47-55.

[9] 廖贵成.小弹簧式推力轴承的安装与运行.湖北电力技术,1994,(1,2合刊): 74-76.

Study of Spring-supported Thrust Bearings Inclusive of Fore-region Pressure Build-up

Abstract: A solution of one dimensional fore-region pressure build-up is put forward. The performance of spring-supported thrust bearing is carried out with 3-D TEHD theory inclusive of fore-region pressure. The proposed solution system uses an efficient hybrid numerical approach which FEM and FDM were used for governing equations. The effects of fore-region pressure build-up and the variation of some operating conditions on the performance of the pad are researched.

复合型紊流润滑理论模式的研究*

摘　要：对复合型紊流润滑理论模式和国际上通用的几种紊流润滑理论模式进行比较研究，针对纯 Couette 流动和兼有压力梯度与剪切运动的复杂流动 2 种流场，用各种紊流润滑模式进行计算分析，并与不同雷诺数下时均速度的现有试验数据对比. 研究表明：与其他紊流模式比较，复合型紊流润滑模式能准确分析不同工况的流场，与试验数据最为吻合；在低雷诺数下，复合型紊流模式由于理论基础的坚实性，仍能很好地适用，当用于既有高雷诺数又有低雷诺数的润滑膜时优点尤其明显.

随着机器向高速大功率方向发展，许多大型高速轴承和密封装置在紊流工况下运转. 现有的许多紊流润滑模式或仅适用于高速轻载工况，或以有限范围内的经验公式为依据，而以紊流能量为依据的单一模式难以同时计入润滑油膜的各个特点. 张运清等[1]提出了新的紊流润滑理论——模式复合型紊流润滑理论模式，即在近壁区采用低紊流雷诺数的 k-ε 理论、在紊流核心区采用代数雷诺应力理论. 该模式能从理论上更准确地描述润滑油膜的特点. 本文对复合型紊流润滑理论模式和国际上通用的几种紊流模式进行比较，用上述各种紊流模式对平行平板流动和闭口槽复杂流场流动的速度分布进行预测，并与试验数据分析比较，讨论各种紊流润滑模式的优劣.

1　常用紊流润滑模式理论

由于紊流方程组的不封闭性，在紊流研究发展过程中，形成了各种各样的紊流润滑模式理论. 在工程实际应用中，最为广泛的理论有零方程模式的 Constantinescu 理论、Ng-Pan 理论、Hirs 整体流动理论以及两方程模式的 k-ε 理论等[2-4].

Constantinescu[2]以 Prandtl 混合长度理论为依据建立了紊流润滑理论. 该模式理论使用简便，但含有混合长度理论的缺陷. 在压力脉动作用下，Prandtl 的动量输运为常值的假设不再成立，周向流动与轴向流动被人为解耦，而且忽视了过渡区的存在. 所以该模式只能近似适用于以单向 Couette 流为主的高速轻载轴承.

Ng 等[3]建立了以 Reichardt 壁面律为依据的紊流润滑模式理论，用局部切应力代替壁面切应力，并采用了小扰动简单 Couette 流线性假设. 但由于壁面律的常切应力层假设并不适用于实际润滑流场，因此，Ng-Pan 的方法基本上也只适用于高速轻载的场合，不适用于高压供油、大偏心率和大的轴向流动的复杂流场的润滑问题.

Hirs[4]紊流润滑模式理论以壁面剪应力与体积流平均速度之间的经验关系为依据，并采用将剪切流当量转化成压力流的处理方法；作者仅从纯剪切流和纯压力流的经验数据出发，大胆地推广到所有的复杂流态中去. 该模式理论作为经验性理论，当应用场合不同于经验公式的条件时，就必然存在不可避免的误差.

---

* 本文合作者：王小静、孙美丽. 原发表于《摩擦学学报》，2000, 20(2): 127-130.

两方程 $k\text{-}\varepsilon$ 模式[5]通过求解紊动能方程和能量耗散方程来确定涡粘系数,避免了上述模式中的人为假设,在理论上更趋合理,成功解决了很多紊流问题. 但该理论求解复杂,在紊动能方程和能量耗散方程模式化的过程中还存在着一些不确定因素,并且该模式理论没有直接考虑固体壁面对紊流脉动的阻滞效应. 这些因素使直接用该模式处理润滑问题时难免带来误差.

Lee 等[6]提出了一种复合模式,即在紊流核心区采用代数雷诺应力模式,在近壁区采用混合长度理论. 该方法虽然利用了代数雷诺应力模式适用于紊流核心区的优点,但近壁区采用的混合长度理论含有许多缺陷.

2 复合型紊流模式理论

在复杂紊流流场中,润滑油膜通常含有 1 个主导方向的剪切流和 2 个方向的压力流,并且三者可属同一数量级. 考虑到紊流流场的复杂性,难以采用单一模式来准确地描述包括壁面区和核心区在内的整个润滑流场的特征.

张运清等[1]在分析比较前人工作的基础上,提出了一种新的紊流模式理论即复合型紊流模式理论. 按照该理论,在近壁区采用低紊流雷诺数的 $k\text{-}\varepsilon$ 方程,而在紊流核心区采用代数雷诺应力模式. 根据试验数据[7,8],提出近壁区和核心区的划分以紊流切应力占总切应力的 95% 来确定,同时保证在分界面上剪应力和速度场连续. 这个模式既直接考虑固壁效应的影响,又考虑了近壁区黏性效应的影响,从而使这种紊流模式理论具有较上述通用紊流润滑模式更先进合理、考虑因素更全面的特点,能更全面地计入润滑膜的各个特点. 当然,如果直接用这个模式来解算润滑问题,将比采用以往的所有模式都更加复杂;为此,选择 Kaneko 等[9]的数据库支持技术,使其成为既具有更完善的理论基础,又便于用来解算润滑问题的理论方法.

3 计算结果分析

用 Constantinescu、Ng-Pan、$k\text{-}\varepsilon$ 和 Lee-Kim 等常用紊流模式和文献[1]的复合型紊流润滑模式理论对平行平板纯 Couette 流动和闭口槽复杂流场流动的两个流场进行理论计算,分析流场的速度分布情况,并与试验数据进行对比.

3.1 平行平板纯 Couette 流动

Leutheusser 等[7]对平行平板流动进行了试验研究. 其试验结果以及用通用紊流润滑模式和复合型紊流润滑模式对该流场的时均速度计算结果示于图 1. 可以看出,随着雷诺数的增大,流场完全进入紊流流态,速度分布不再沿膜厚呈线性分布,而是呈"S"形分布,即在静止面和运动面附近速度梯度变大,而在膜厚中部速度变化平缓[图 1(a)]. 在此条件下,除 Constantinescu 模式以外,其余几种模式的计算结果与试验数据均较为吻合. 图 1(b)中所示的紊流流场进一步发展,复合型紊流模式与试验结果最为吻合;$k\text{-}\varepsilon$ 模式过分估计了紊流切应力的影响,Constantinescu 模式的计算结果对应的紊流发展不充分;Lee-Kim 模式由于在近壁区采用了混合长度理论,其计算结果也因此受到影响,计算的紊流切应力过小;相比之下,Ng-Pan 模式的计算结果与试验结果较为接近.

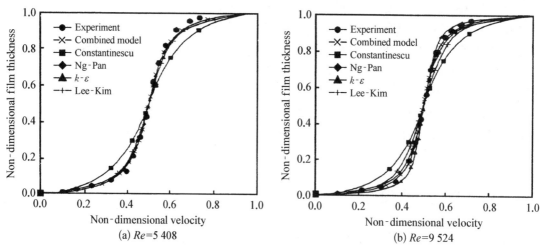

Fig. 1 Velocity distributions across the film under plane Couette flow
图 1 纯 Couette 流动下沿膜厚方向无量纲速度分布

从平行平板流动的计算结果可以看出,在不同雷诺数下,与其他模式相比,Constantinescu 模式的计算结果偏离试验值较大,这是由于其较弱的理论基础混合长度理论造成的,导致了该紊流模式的计算精度不高.

3.2 闭口槽复杂流场流动

Tsanis 等[8]对截面流量为 0 的复杂流场进行了试验研究,其所研究的流场中同时存在剪切流和压力流,将其试验结果以及用通用紊流润滑模式和复合型紊流润滑模式对该流场的时均速度计算结果示于图 2. 可见,随着雷诺数的提高,紊流深度逐步发展. 无论是在低雷诺数还是在高雷诺数下,复合型紊流润滑模式的计算结果与试验数据最为吻合. 在中、低雷诺数时,k-ε 模式的计算结果与试验结果较为接近;当雷诺数很高时,采用 k-ε 模式则紊流切应力的作用被高估,与试验数据有一定偏差,这是由于其理论模式未能很好地计入固壁面对紊流脉动的阻滞效应造成的. 在雷诺数较小时,采用 Constantinescu 模式所得到的计算结果与试验数据较为接近;当雷诺数进一步提高、反向压力梯度进一步增大时,计算值与试验数

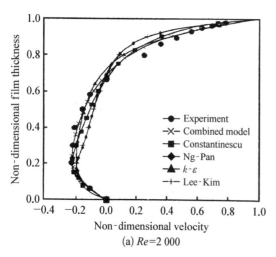
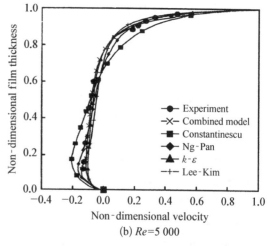

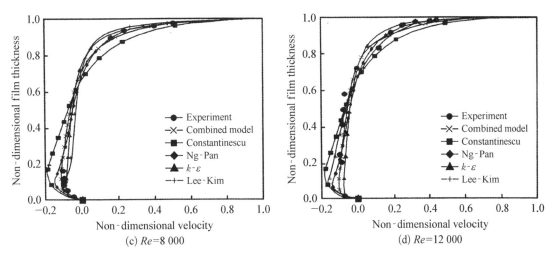

Fig. 2 Velocity distributions across the film under countercurrent flow
图2 闭口槽复杂流场流动下沿膜厚方向无量纲速度分布

据相差较大,紊流流态的发展被低估. 按照 Ng-Pan 模式和 Lee-Kim 模式计算的速度分布在近静止面附近的形状与试验结果相差较大,这是由于流场中存在压力梯度造成的. 并且在低雷诺数时,这两种紊流模式计算的速度分布与试验结果之间存在较大差距.

4 结论

a. 无论是在纯剪切流还是有压力梯度的复杂流动下,复合型紊流润滑模式都能用以较为准确地分析流场,与其他各种紊流模式相比,其相应的计算结果与试验数据吻合较好.

b. 由于 k-ε 模式未能很好地计入固壁面对紊流脉动的阻滞效应,因此虽然其在中低雷诺数下相应的计算结果与试验结果较为一致,但当雷诺数很高时,因忽略固壁阻滞效应致使紊流切应力的作用被高估,导致相应的结果与试验数据之间出现一定偏差.

c. 在低雷诺数工况下,复合型紊流模式具有较好的适用性,特别是对既有高雷诺数又有低雷诺数的润滑膜尤为如此.

d. 复合型紊流润滑模式在理论基础和计算精度上都优于其他紊流模式,但直接应用其原始模式时,计算最为繁复且耗时长. 因此,采用一定的计算技术如数据库支持技术将有助于其应用推广.

参 考 文 献

[1] 张运清,张直明. 应用高级紊流模式对雷诺润滑方程的解算方法[J]. 上海工业大学学报,1994,15:337-342.
[2] Constantinescu V N. Analysis of Bearing Operating in Turbulent Regime[J]. Trans ASME Ser D, 1962, 82: 139-151.
[3] Ng C W, Pan C H T. A Linearized Turbulent Lubrication Theory[J]. Trans ASME Ser D, 1965, 87: 675-688.
[4] Hirs G G. A Bulk-Flow Theory for Turbulence in Lubricant Films[J]. Trans ASME Ser F, 1973, 95: 137-146.
[5] Jones W P, Launder B E. The Calculation of Low Reynolds Number Phenomena with a Two-Equation Model of Turbulence [J]. Int J of Heat and Mass Transfer, 1973, 16: 1119-1126.
[6] Lee D W, Kim K W. Turbulent Lubrication Theory Using Algebraic Reynolds Stress Model in Finite Journal Bearings with Cavitation Boundary Conditions [J]. JSME International Journal Series II, 1990, 33: 200-207.

[7] Leutheusser H J, Aydin E M. Plane Couette Flow between Smooth and Rough Walls [J]. Experiments in Fluids, 1991, 11: 302-312.

[8] Tsanis I K, Leutheusser H J. The Structure of Turbulent Shear-induced Countercurrent Flow [J]. J Fluid Mech, 1988, 189: 531-552.

[9] Kaneko S, Hori Y, Tanaka M, et al. Static and Dynamic Characteristics of Annular Plain Seals [J]. I Mech E, 1984, C278/84: 205-214.

A Combined Theory for Turbulence in Lubricating Films

Abstract: Several commonly used turbulent lubrication models are compared with the combined turbulent lubrication model. These models are applied to a pure Couette flow and a combined shear-pressure flow with zero net flow rate to predict the velocity distribution under different Reynolds number. The calculating results based on these models are compared with existing experimental data. As the results, the calculated results based on the combined turbulent model agree well with the experimental ones. The combined turbulent model is also applicable to situations of low Reynolds number. Therefore, for lubricating films involving Reynolds numbers ranging from low to high values, this model could find promising application.

电磁轴承在透平膨胀机中的应用研究进展*

摘 要：电磁轴承是目前发展迅速的新一代支承部件,在国外已有许多应用实例,透平机械是其主要应用领域之一.介绍此项技术的主要特点及在国内透平机械应用领域中的研究进展情况,结合我们的获得的试验结果,分析讨论试验过程中出现的部分现象及产生的主要原因.

现代旋转机械已日趋向高速化发展,而支承部件的影响则是不可忽视的.小型制氧透平膨胀机的转速一般很高,可达 100 000 r/min 以上,此时通常采用静压空气轴承.而在制氧透平膨胀机中使用电磁轴承取代静压空气轴承的主要原因之一,是后者的承载能力比前者的要小.另外,使用静压空气轴承还存在着空气的损失和对部件加工精度的要求较高.同时,制氧透平膨胀机的制造商也注意到了电磁轴承的其他优点,例如,免维护、工作寿命长、高稳定性、不需要空气密封措施等,这方面的应用实例[1-3]是企业欲采用此项技术,推出新型制氧透平膨胀机的一个因素.国外对此项技术的基础性研究已趋于完成[4],国内的基础性研究也获得了初步的结果[5].以此为基础,近期我们在这方面的研究又有新的突破,在一台 150 m³ 的制氧透平膨胀机中,应用电磁轴承支承转子获得初步成功,试验转速已达 104 000 r/min,稳定工作转速为 92 000 r/min,工作状态稳定,效果良好.

1 结构与参数设计

受原有的制氧透平膨胀机安装尺寸和悬浮技术所需的条件限制,本试验样机采用 2 个径向电磁轴承和 1 个轴向电磁轴承取代原来的 2 个静压空气轴承.径向电磁轴承为 8 磁极周向分布,轴向电磁轴承为同心圆槽结构,技术参数见表 1. 150 m³ 制氧透平膨胀机的转子重 1.16 kg,长约 210 mm,推力盘外径为 62 mm,采用双推力盘结构.转子位移的在线监测采用 5 个电涡流式位移传感器分别对 5 个自由度独立进行,传感器的电压灵敏度为 20 mV/μm.按有关电磁轴承承载力计算方法可知,径向电磁轴承的承载力约为 150 N/个,轴向电磁轴承的承载力是 600 N.整机设计工作转速为 108 000 r/min,要求达到瞬时转速 130 000 r/min,并运行 3 min 系统不失稳.图 1 为试验采用的电磁轴承的照片.图 2 为转子轴承的安装结构照片.

表 1 电磁轴承参数

参 数 名 称	径向轴承	轴向轴承	参 数 名 称	径向轴承	轴向轴承
轴承外径(mm)	72	72	每磁极线圈(匝)	42	80
轴承内径(mm)	30	30	平衡气隙(mm)	0.12	0.08
铁芯厚度(mm)	15	15	直流电阻(Ω/磁极)	0.2	0.3
偏磁电流(A)	1.0	12	线圈电感(mH/磁极)	1.8	0.8

---

* 本文合作者：汪希平、于良、万金贵.原发表于《中国机械工程》,2000,11(4)：379-381.

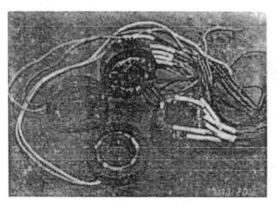
图 1 径向和轴向电磁轴承

图 2 转子轴承的安装结构

2 控制系统的设计

对于系统的控制器设计,仍然采用模拟 PID 控制器.其设计的方法和过程见文献[5]中的有关内容.这里采用的是五自由度独立控制方式(图 3 显示了一个自由度控制系统的原理).采用独立控制方式的原因是,对于本试验样机,转速在 130 000 r/min 以下,系统的陀螺效应可忽略不计.

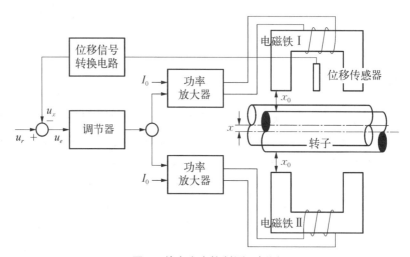

图 3 单自由度控制原理框图

控制器的传递函数

$$G(s)=\frac{\lambda A_s A_p (1+sT_p)(b_2 s^2 + b_1 s^2 + b_0)}{(1+sT_s)(a_2 s^2 + a_1 s + a_0)(P_2 s^2 + P_1 s + P_0)} \quad (1)$$

式中,λ 为功率放大器的增益系数;A_s 为位移传感器的增益系数;A_p 为 PID 调节器的增益系数;T_s 为位移传感器的时间常数;T_p 为功率放大器的时间常数;a_0、a_1、a_2、b_0、b_1、b_2、P_0、P_1、P_2 分别为与 PID 调节器和功率放大器的参数有关的系数.

表 2 给出了在 4 个径向自由度上的 PID 调节器的参数,用表中前轴承水平自由度的参数代入式(1),该自由度的系统传递函数

表 2 径向电磁轴承的模拟 PID 调节器参数

自由度	A_s(V/V)	A_p(A/V)	T_i(s)	T_d(s)
工作端水平	6 790	3.064	2.365	1.97×10^{-3}
工作端垂直	9 440	3.330	2.595	8.65×10^{-4}
风机端水平	3 800	3.330	2.190	1.94×10^{-3}
风机端垂直	6 560	2.419	1.570	3.06×10^{-3}

注：T_i 和 T_d 分别为积分和微分调节器的时间常数.

$$G(s)=[2.08\times10^4(1+1.59\times10^{-5}s)(5.05\times10^{-3}s^2\\+4.73s+1)]/[s(1+3.18\times10^{-5}s)(3.91\times10^{-4}s\\+2.365)(5.4\times10^{-3}s^2+341.55s+1)] \quad (2)$$

图 4 为系统控制器的一个自由度单元的电路板照片. 图 5 为五自由度控制器的外观照片.

图 4 控制器中的控制单元电路

图 5 五自由度电磁轴承系统控制器

3 试验与结果分析

图 6 为试验中的透平膨胀机样机照片，稳定运行转速 92 000 r/min，稳定运行时间为 1.5 h. 我们对此样机进行了部分性能测试和控制器参数的校正，并在试验中观测到一些现象，表明电磁轴承要达到工程应用还有一些问题需要解决.

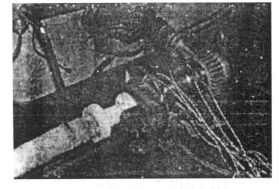

图 6 试验中的制氧透平膨胀机样机

3.1 转子动平衡试验

为保证样机能在高转速下平稳运行，本试验样机的转子在特定条件下进行了动平衡处理，其最终的动平衡精度达到 2.5 mg·cm. 此条件下，转子中心的旋转半径在转速 15 000 r/min 以上时均小于 5 μm.

3.2 振动试验

对于高速旋转机械，转子的振动是影响系统的稳定性的一个重要因素. 在我们的试验过程中,转子在低速($<$20 000 r/min)时有一个较大的振动区;在 12 000 r/min 附近,转子中心的振动幅度约 20 μm. 除此以外,转子均未见有明显的振动现象,试验样机工作平稳.

3.3 轴承特性试验

第 1 次试验用的径向电磁轴承半径间隙为 80 μm,当转速达到 60 000 r/min 时,样机外壳的温度上升至 50℃左右. 据分析,发生这一现象的主要原因是电磁轴承的内表面结构不平度引起空气阻力的增大. 为解决这一问题,一是将电磁轴承的气隙增加到 150 μm;二是在轴承的内表面涂以特殊的耐高温有机材料,提高表面的平整度;三是采用水冷却装置. 采取上述措施后,轴承的温升问题便完全解决.

本试验中,径向轴承的单位面积承载能力为 0.35 N/mm^2,轴向轴承的单位面积承载能力可达 0.7 N/mm^2.

3.4 系统功耗试验

本试验样机的系统控制器设计输入功率为 600 W,其中,系统控制元件消耗的功率 100 W,而每自由度功率输出单元的所需功率不会超过 100 W.

3.5 转子动力学特性分析

由于本试验样机的要求转速较高,为保证系统运行的稳定性,利用传统转子动力学分析的方法,结合电磁轴承的刚度阻尼特性分析方法,对样机进行转子动力学特性分析,考察系统运行的稳定性. 一是理论计算样机轴承转子系统的固有频率和临界转速分布;二是试验测定样机轴承转子系统的固有频率和临界转速. 结果表明,该系统的 4 个低频临界转速均在 20 000 r/min 以下,其对应的对数衰减率均大于零,试验也证明了样机转子可以安全越过这些临界转速. 计算结果还表明,系统存在一个 105 000 r/min 的失稳转速,因为,此时的对数衰减率小于零,所以系统无法越过这个转速,这也是试验未达到预定设计转速的原因. 通过分析计算结果可知,该失稳转速是转子主导型固有频率所致,改变控制器参数对此固有频率的位置几乎无影响,只能通过修改转子的几何尺寸才能达到目的.

4 结论

模拟控制器能够完全适合电磁轴承系统的控制要求,特别是对那些尺寸较小的电磁轴承. 电磁轴承内径的表面形状及平整度对其特性有明显的影响,其中又以对轴承体的发热和空气阻力等方面影响更为显著,采用增大电磁轴承的有效工作间隙、用耐高温的材料对轴承体的内表面进行涂覆处理,以改善表面的平整度,以及采用水冷等强迫制冷手段,都可减小这些影响.

电磁轴承转子系统的稳定性一直是这种新型支承部件真正进入工程应用领域需要解决的关键问题之一. 我们的前期工作仅对系统的静态稳定性做了初步的研究[5],这次试验中,尽管观察到了一些振动现象,但其产生的原因和机理,以及对系统稳定性的影响程度等问题,将有待于今后更深入的工作去解决.

参 考 文 献

[1] Canders W R, Ueffing N, Schrader-Hauamann U et al. MTG400: A Magnetically Levitated 400 kW Turbo

Generator System for Natural Gas Expansion. In: Gerhard Schweitzer. Proceedings of the Fourth International Symposium on Magnetic Bearings. ETH Zurich, Switzerland, 1994: 435-440.

[2] Agahi R R, Schroder U. Industrial High Speed Turbogenerator System for Energy Recovery. In: Fumio Matsumura. Proceedings of the Fifth International Symposium on Magnetic Bearings, Kanazawa, Japan, 1996: 381-387.

[3] Antila M, Lantto E, Saari J. et al. Design of Water Treatment Compressors Equipped with Active Magnetic Bearings. In: Fumio Matsumura Proceedings of the Fifth International Symposium on Magnetic Bearings, Kanazawa, Japan, 1996: 389-394.

[4] 汪希平. 电磁轴承系统的参数设计与应用研究: [博士学位论文]. 西安: 西安交通大学, 1994.

[5] 汪希平, 袁崇军, 谢友柏. 电磁轴承系统控制参数与稳定性的关系分析. 机械工程学报, 1996, 32(3): 65-69.

Development of Applied Research of Active Magnetic Bearing on a Gas Expander

Abstract: Active magnetic bearings developed quickly are a new type of support components, and there are a lot of application examples overseas. Turbomachinery is as one application domain. This paper describes some main characteristics of this technology and the evolvement of domestic application of AMB in the domain. Some test results are presented here, some phenomena occured during the experiments and the main causes are also discussed.

A Comparison of Flow Fields Predicted by Various Turbulent Lubrication Models With Existing Measurements[*]

Abstract: Flow field predictions of various turbulent lubrication models are compared with the existing experimental data of turbulent Couette flow and shear-induced countercurrent flow.

1 Introduction

In modern machinery, more and more fluid lubrication films get into complicated turbulent regime with large pressure gradients in the circumferential and/or axial directions, resulting in flow fields significantly different from a Couette one. Numerous models have been developed for turbulent lubrication problems, the best known ones being Constantinescu's, Ng-Pan's, Hirs', and the k-ε model, mostly for flow condition not far from Couette.

Constantinescu[1] used Prandtl's mixing length theory to describe the turbulence stress. Ng-Pan[2] used Reichardt's Wall Law to express turbulent eddy viscosity. Both models have turbulent lubrication equations derived with linearization, resulting in turbulence coefficients depending solely on the local Reynolds number R_h, and are theoretically justifiable for Couette flows superimposed by mild pressure-induced components. Hirs' model[3] was developed with a rather far extension of the empirical data of pure shear flow and pure pressure flow to cover all complex flows. The k-ε model[4] is theoretically more advanced, being based on equations of turbulence kinetic energy k and turbulence energy dissipation rate ε, yet it does not take account of the damping effect of the mating walls. The algebraic Reynolds stress model proposed by Ljuboja and Rodi[5] takes in this effect, but is unsuitable for the viscous sublayers. Lee and Kim[6] brought out a physically more perfect model which combines the algebraic Reynolds stress model for the turbulent kernel and the Prandtl's mixing length theory for the near-wall regions, yet containing the evidently imperfect match in physical level of the two constituent models.

An improved combined Reynolds stress model was proposed by Zhang and Zhang[7], where the algebraic Reynolds stress model is used for the turbulent kernel and low turbulence Reynolds number k-ε model for the near-wall regions, thus bringing the two involved models to similar physical levels.

---

[*] In collaboration with Xiaojing Wang and Meili Sun. Reprinted from *Journal of Tribology, Transactions of the ASME*, 2000,122(2): 475–477.

2 A Brief Description of Combined Reynolds Stress Model for Lubrication

The governing equations of algebraic Reynolds stress model are

$$\frac{\partial}{\partial y}\left(\frac{\nu_t}{\sigma_k}\frac{\partial k}{\partial y}\right)+P_d-\varepsilon=0 \qquad (1)$$

$$\frac{\partial}{\partial y}\left(\frac{\nu_t}{\sigma_\varepsilon}\frac{\partial \varepsilon}{\partial y}\right)+C_{\varepsilon 1}P_d\frac{\varepsilon}{k}-C_{\varepsilon 2}\frac{\varepsilon^2}{k}=0 \qquad (2)$$

where

$$\nu_t = 0.09G_\mu \frac{k^2}{\varepsilon^2}, \quad P_d = \nu_t\left[\left(\frac{\partial u}{\partial y}\right)^2+\left(\frac{\partial w}{\partial y}\right)^2\right],$$

and

$$G_\mu = \frac{1+\frac{3}{2}\frac{C_2 C_2'}{1-C_2}f}{1+\frac{3}{2}\frac{C_1'}{C_1}f}\cdot\frac{1-2\frac{C_2 C_2' P_d/\varepsilon}{C_1-1+C_2 P_d/\varepsilon}f}{1+2\frac{C_1'}{C_1+P_d/\varepsilon-1}f}, \quad C_2'=0.3,$$

$$C_{\varepsilon 1}=1.44, \quad C_{\varepsilon 2}=1.92, \quad \sigma_k=1.0, \quad \sigma_\varepsilon=1.3,$$

$$f=\frac{k^{3/2}}{C_w y \varepsilon}, \quad C_w=3.72, \quad C_1=1.8, \quad C_2=0.6, \quad C_1'=0.6.$$

The governing equations of low Reynolds number k-ε model are

$$\frac{\partial}{\partial y}\left[\left(\nu+\frac{\nu_t}{\sigma_k}\right)\frac{\partial k}{\partial y}\right]+\nu_t\left[\left(\frac{\partial u}{\partial y}\right)^2+\left(\frac{\partial w}{\partial y}\right)^2\right]-\varepsilon-\frac{2\nu k}{y^2}=0 \qquad (3)$$

$$\frac{\partial}{\partial y}\left[\left(\nu+\frac{\nu_t}{\sigma_\varepsilon}\right)\frac{\partial \varepsilon}{\partial y}\right]+C_{\varepsilon 1}\nu_t\left[\left(\frac{\partial u}{\partial y}\right)^2+\left(\frac{\partial w}{\partial y}\right)^2\right]\frac{\varepsilon}{k}$$
$$-C_{\varepsilon 2}[1-0.3\exp(R_k^2)]\frac{\varepsilon^2}{k}-2\nu\left(\frac{\partial \varepsilon^{1/2}}{\partial y}\right)^2=0 \qquad (4)$$

where

$$\nu_t = C_m[1-\exp(-A_m R_k)]\frac{k^2}{\varepsilon}, \quad R_k=\frac{k^2}{\nu\varepsilon}, \quad C_m=0.09,$$

$$C_\mu=0.09, \quad \sigma_k=1.0, \quad \sigma_\varepsilon=1.3, \quad C_{\varepsilon 1}=1.45, \quad C_{\varepsilon 2}=2.0.$$

For Newtonian fluids, $A_m=0.0015$.

Bused on an optimal fitting to the experimental data from Leutheusser et al.[8] and Tsanis et al.[9], the transition between different regions is determined by the turbulent stress reaching a value of 95 percent of the total stress, and continuity of eddy viscosity and velocity is demanded at the junctions.

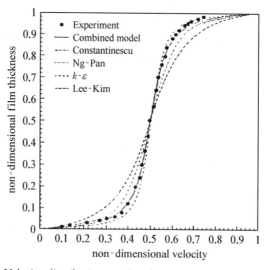

Fig. 1 Velocity distributions under plane Couette flow ($Re=9\,524$)

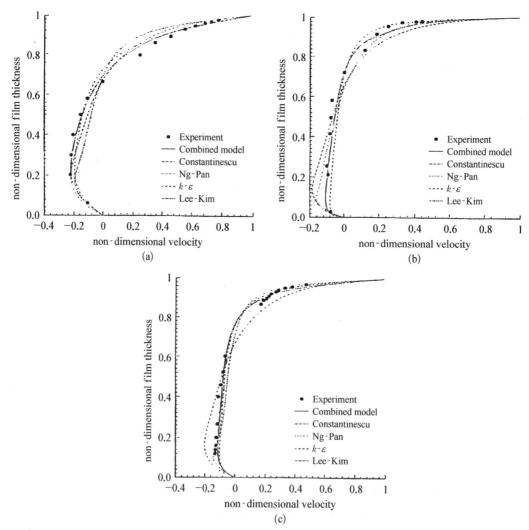

Fig. 2 Velocity distributions across the film under countercurrent flow (a) $Re=2\,000$ (b) $Re=5\,000$ (c) $Re=12\,000$

3 Analysis

In order to make fundamental evaluations of the various turbulent lubrication models, they are used to predict the flow fields of a pure Couette flow and a shear-induced countercurrent flow with zero cross-sectional net flow where results of carefully conducted experiments are available from Leutheusser and Aydin[8] and Tsanis and Leutheusser[9].

For Constantinescu's model and Ng-Pan's model, the eddy viscosity can be readily obtained, and the velocity are solved from velocity formulas. For the other models, the system of governing equations is solved iteratively.

Figure 1 is the velocity distribution in pure Couette flow with a Reynolds number 9524. The combined Reynolds stress model agrees well with the experimental result. Ng-Pan's model is also close to the experimental curve, k-ε model overestimates the effect of turbulent stress, while Constantinescu's model and Lee-Kim's model underestimate the depth of turbulence in the film.

Figures 2(a) - 2(c) show the comparisons of velocity distributions across the film under different Reynolds numbers in countercurrent flow. With the increase of Reynolds number, the depth of turbulence in the film develops continuously.

From these figures, it can be seen that the combined Reynolds stress model corresponds with experimental results well, whether under low Reynolds number or under a high one. The results of k-ε model are close to experimental curve under low and middle Reynolds numbers. But when the Reynolds number is high, it overestimates the turbulent development. An apparent cause is its negligence of the effect of the wall's damping. Constantinescu's model deviates from the experimental results rather obviously, and underestimates the turbulence development, especially under high Reynolds number and large counter pressure gradient. The forms of velocity distributions near the stationary wall calculated by Ng-Pan's model and Lee-Kim's model deviate from the experimental curve perceptibly. In the case of Ng-Pan's model, it is possibly attributable to the existence of significant pressure gradient in the film; and in the case of Lee-Kim's model, the imperfect match of constituent models and inadequate solidness of transition criterion are possibly responsible.

The combined Reynolds stress model is not only physically based but also without superfluous simplification such as linear perturbation from Couette flows, and therefore can be expected to be applicable to lubricating films with existence of strong pressure flows in circumferential and/or axial directions. However, the application of this model has the inconvenience of lengthy calculation if the original differential equations are directly solved. A more practical way is to adopt the technique of Kaneko et al.[10] which realizes the nonlinear model into a general dependence of the turbulence coefficients on the three determinant parameters, namely, the local Reynolds number R_h, and the nondimensional pressure gradients B_x and B_z in both circumferential and axial directions, and then aiding with it to solve the lubrication equation, as proposed by Zhang and Zhang[7].

4 Conclusion

The combined Reynolds stress model and Ng-Pan's model rank the foremost in the comparisons. The first named is justified not only by its satisfactory correspondence with existing experimental results of flow field measurement, hut also held to he prospective to have wider applicability in view of its more generalized physical considerations.

Acknowledgments

The authors gratefully acknowledge the financial support by the Chinese National Natural Science foundation (59775035) and the Science Foundation of Shanghai Education Committee (98QN48) during this research.

Nomenclature

B_x, B_z = nondimensional pressure gradient in circumferential and axial directions
k = turbulence kinetic energy [m^2/s^2]
R_h = local Reynolds number
R_k = turbulent Reynolds number

u, w = circumferential and axial velocity [m/s]
y = coordinate in film thickness direction [m]
ε = turbulence energy dissipation rate [m^2/s^3]
ν = kinematic viscosity [m^2/s]
ν_t = turbulent eddy viscosity [m^2/s]

References

[1] Constantinescu, V. N., 1962. "Analysis of Hearing Operating in Turbulent Regime," ASME J. Basic Eng., 84, pp. 139–151.

[2] Ny, C. W., and Pan, C. H. T., 1965, "A Linearized Turbulent Lubrication Theory," ASME J. Basic Eng., 87, pp. 675–688.

[3] Hirs, G. G., 1973, "A Bulk-Flow Theory for Turbulence in Lubricant Films," ASME J. Lubr. Technol., 95, pp. 137–146.

[4] Jones, W. P., and Launder, B. E., 1973, "The Calculation of Low Reynolds Number Phenomena with a Two-Equation Model of Turbulence," Int. J. Heat Mass Transf., 16, p. 1139.

[5] Ljuboja, M., and Rodi, W., 1980, "Calculation of Turbulent Wall Jets with an Algebraic Reynolds Stress Model," ASME J. Turbomach., 102, pp. 350–356.

[6] Lee, D. W., and Kim, K. W., 1990, "Turbulent Lubrication Theory Using Algebraic Reynolds Stress Model in Finite Journal Bearings with Cavitation Boundary Conditions," JSME, International Journal, Series 11, 33, No. 2, pp. 200–207.

[7] Zhang, Y. Q., and Zhang, Z. M., 1995, "A New Model of Theoretical Analysis of Turbulent Lubrication Using a Combined Model of Turbulence," Tribology, 15, No. 3, pp. 271–275 (in Chinese).

[8] Leutheusser, H. J., and Aydin, E. M., 1991, "Plane Couette Flow Between Smooth and Rough Walls," Exp. Fluids, 11, pp. 302–312.

[9] Tsanis, I. K., and Leutheusser, H. J., 1988, "The Structure of Turbulent Shear-induced Countercurrent Flow," J. Fluid Mech., 189, pp. 531–552.

[10] Kaneko, S., Hori, Y., and Tanaka, M., 1984, "Static and Dynamic Characteristics of Annular Plain Seals," C27884, I Mech E, pp. 205–214.

THD Analysis of High Speed Heavily Loaded Journal Bearings Including Thermal Deformation, Mass Conserving Cavitation, and Turbulent Effects*

Abstract: Theoretical and experimental THD analyses of high speed heavily loaded journal bearings are presented. Numerical solutions include thermal deformation, mass conserving cavitation and turbulent effects. The pressure and temperature distributions, the eccentricity ratio, and the flow rate are measured. Agreement between theoretical results and experimental data is satisfactory.

1 Introduction

Under high speed heavily loading operation conditions, the journal bearings frequently experience turbulent flow, high film temperature gradients due to viscous dissipation, and significant thermal deformation. Hence, in order to make accurate prediction of bearing performance parameters, a THD (thermal hydrodynamic lubrication) analysis including thermal deformation, mass conserving cavitation and turbulent effects is imperative.

Thermal effects in hydrodynamic journal bearings have been investigated theoretically and experimentally by many researchers, on which Khonsari[1] and Pinkus[2] have presented comprehensive reviews, respectively. Recent theoretical studies dealing with thermal effects concentrated on mass conserving cavitation[3], turbulent effects[3,4], thermal deformation effects[5,6], transient effects[7,8], and dynamic considerations[9]. Bouchoule et al.[10], Flack, Kostrzewsky, and Taylor[11], Ma and Taylor[12], and Fitzgerald and Neal[13] presented experimental studies on thermal effects. For the THD analysis including thermal deformation and turbulent effects, no result has yet been available in literature.

This paper presents a THD model of the modified pocket journal bearing including thermal deformation, mass conserving cavitation and turbulent effects, and presents new experimental data which compare to theory.

---

* In collaboration with Chao Zhang and Zixia Yi. Reprinted from *Journal of Tribology*, Transactions of the ASME, 2000,122(3): 597 – 602.

2 Title of Heading

2.1 Pocket Bearing

The modified pocket bearing is presented in Fig. 1. The preloading unsymmetrical profile has two curvature radius R_1 and R_2 for the lower and upper half bushings. The nominal film thickness equations for these half bushings are, respectively:

$$h = C + e\cos\phi + \overline{OO_2}\cos\varphi + \sqrt{R_2^2 - \overline{OO_2}\sin^2\varphi} - R_1,$$
$$h = C + e\cos\phi + \overline{OO_1}\cos\varphi + \sqrt{R_2^2 - \overline{OO_1}\sin^2\varphi} - R_1,$$

where $\varphi = 90 + \alpha$ and $0 \leqslant \alpha \leqslant 180$ deg. α is circumferential angle beginning from a horizontal level.

2.2 Basic Equations

The nondimensional generalized Reynolds equation for a turbulent, incompressible Newtonian fluid under steady-loading condition can be written as follows:

$$\frac{\partial}{\partial\theta}\left[F_\theta \frac{\partial(G\gamma)}{\partial\theta}\right] + \frac{1}{\lambda^2}\left[F_z \frac{\partial(G\gamma)}{\partial\bar{z}}\right] = \frac{\partial}{\partial\theta}\{F_u[1 + (1-G)\gamma]\} \tag{1}$$

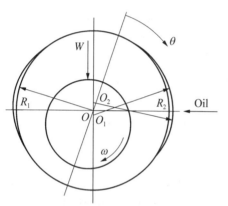

Fig. 1 Geometry of the modified pocket journal bearing

where

$$F_0 = \int_0^1 \frac{1}{\bar{\mu}} d\bar{y}, \quad F_1 = \int_0^1 \frac{\bar{y}}{\bar{\mu}} d\bar{y}, \quad F_\theta = \frac{\bar{h}^3}{K_\theta}\int_0^1 \frac{\bar{y}}{\bar{\mu}}\left(\bar{y} - \frac{F_1}{F_0}\right) d\bar{y},$$

$$F_u = \bar{h}\left(1 - \frac{F_1}{F_0}\right), \quad F_z = \frac{\bar{h}^3}{K_{\bar{z}}}\int_0^1 \frac{\bar{y}}{\bar{\mu}}\left(\bar{y} - \frac{F_1}{F_0}\right) d\bar{y},$$

$$K_x = 1 + 0.00133 Re^{0.9}, \quad K_z = 1 + 0.00036 Re^{0.96}, \quad Re = \frac{\rho U h}{\mu}.$$

K_x and K_z are the turbulent correction factors given by Ng et al.[14], Re a Reynolds number. G and γ denote a cavitation index and a void fraction, respectively, for the mass conserving cavitation[15,16]. $\bar{\mu}$ is assumed to vary according to the Vogel's formula as follows:

$$\bar{\mu} = C_1 \exp[C_2/(\bar{T} + C_3)],$$

where C_1, C_2, and C_3 are the temperature-viscosity coefficients.

The pressure boundary conditions are that it is the atmospheric pressure at bearing external boundaries, the supply pressure at the inlet zone, and nil in the cavitation regime, respectively.

The energy, heat conduction, and thermal deformation equations are solved in the mid-plane of the bearing.

The nondimensional energy equation is[4]:

$$\bar{u}\frac{\partial \bar{T}}{\partial \theta}+\bar{v}\frac{\partial \bar{T}}{\partial \bar{y}}=\frac{1}{P_e}\frac{1}{h^2}\frac{\partial}{\partial \bar{y}}\left[\left(1+\bar{\mu}\frac{\varepsilon_m}{\nu}\frac{P_r}{P_r^t}\right)\frac{\partial \bar{T}}{\partial \bar{y}}\right]+\frac{N_e}{Re}\frac{\bar{\mu}}{h^2}\left(1+\frac{\varepsilon_m}{\nu}\right)\left(\frac{\partial \bar{u}}{\partial \bar{y}}\right)^2, \quad (2)$$

where

$$P_e=\frac{\rho C_p \omega C^2}{k_0}, \quad P_r=\frac{P_e}{Re}, \quad N_e=\frac{R^2 \omega^2}{C_p T_0},$$

$$P_r^t=\frac{\varepsilon_m}{\varepsilon_H}, \quad Re=\frac{\rho \omega C^2}{\mu_0}$$

P_e, P_r, N_e, P_r^t, and Re are Peclet, Prandtl, Eckert, turbulent Prandtl, and modified Reynolds numbers, respectively.

The nondimensional conduction equation for the bushing is

$$\frac{1}{\bar{r}}\frac{\partial}{\partial \bar{r}}\left(\bar{r}\frac{\partial \bar{T}_B}{\partial \bar{r}}\right)+\frac{1}{\bar{r}^2}\frac{\partial^2 \bar{T}_B}{\partial \theta^2}=0. \quad (3)$$

The journal is treated as an isothermal component[5] due to its high rotational speed.

The boundary conditions on temperature are as follows. At the oil-shaft interface, the constant temperature is given by the nil net heat flux to the shaft. At the oil-bushing interface, the continuity conditions of the temperature and the heat flux are employed. On the outer surface of the bushing, free convection is used. At the inlet zone, the uniform temperature is assumed and determined based on the conservation of the mass and energy in the groove[5].

Since the shaft is treated as an axisymmetrical isothermal component, its thermal displacement is obtained by the classical thermoelasticity relations[5]. Because most of the bearing outer surface is exposed to the ambient, the outer surface is treated as free to expand due to thermal swelling, while the thermal deformation of its inner surface is determined using the thermal flexibility matrix[5]. The thermal deformations are used for the film thickness.

2.3 Numerical Procedure

The number of nodes of the film and bushing are 81 and 25 nodes in the circumferential and axial directions, 11 and 8 across the film and the bushing thickness, respectively. The generalized Reynolds equation is solved by finite difference and Gauss-Seidel methods with over relaxation (cf. [16]). The energy equation in the film is solved in the rectangular film field[17] by an implicit finite difference method and the Richtmyer technique[18] when a reverse flow occurs. The Laplace heat conduction equation for the bushing is solved by finite element method. A finite element package ANSYS is employed to obtain the thermal flexibility matrix, which needs to be calculated only once for a given bearing and is used in the global iteration to calculate the bushing thermal deformation.

The governing equations are integrated by iterative technique and the scheme is similar to that [19]. The convergence criterion is that relative discrepancy between two successive iterations is less than 10^{-3}.

3 Description of Test Rig

Figure 2 shows a schematic of the test rig which is similar to that used by Flack et al.[11] The test bearing is a modified pocket bearing, which material is steel, lined with alloy $ZS_n S_{b12} C_{u6} C_{d1}$, and rests on the journal of a rigid rotor between two high precision ball bearings supporting the rotor. The rotation accuracy is within 2.3 μm. Two separate filtered, recirculating, temperature and pressure controlled lubricant feed systems provide oil to the test bearing and support bearings respectively. External load is applied by a nitrogen-filled bellows directly onto the test bearing housing. Nonparalled lateral movement of the bearing resulting from misalignment of the bearing loading system is minimized by the use of axial tensioned wires attached to the ends of bearing housing. A DC motor, via a multiplying gearbox and a toothed belt, provides a continuously variable journal speeding in the range 1 500 - 10 000 r/min.

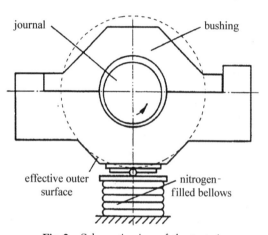

Fig. 2 Schematic view of the test rig

Four eddy current displacement probes are mounted vertically and horizontally in the front and rear of the test bearing respectively, giving the eccentricity ratio and attitude angle. The oil pressure distribution is measured to within 2 percent using 18 holes drilled through the bearing housing and each hole is connected via a small copper tube to a high precision manometer. The temperature distribution is measured to within 0.5℃ by 25 copper-constantan thermocouples. The location of the thermocouple and pressure holes in the bushing is shown in Fig. 3. The hot junction of the thermocouples is very close to the bushing inner surface. The inlet pressure is measured by a strain gauge pressure transducer. The load is measured by a high precision manometer connected to the air bellows. The lubricant flow rate into the test bearing is measured to within 1 percent by a turbinetype flow meter. Speed is recorded by an impulse tachometer and controlled electronically within 0.2 percent.

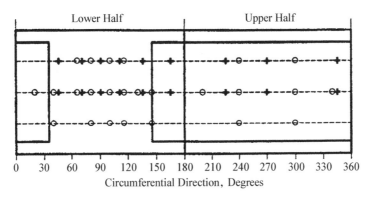

Fig. 3 Diagram showing thermocouple and pressure hole locations in the bushing and grooves (○ thermocouple; +pressure hole)

4 Results and Discussion

The bearing geometry and operating conditions are given in Table 1. Because the averaged bushing thickness is bigger than the bearing radius, and because the thermal conductivity and inertia of the bushing are much higher than those of the oil, the bushing outer surface is treated as a circle, based on the same surface area for heat transfer to simplify simulation. All measurements are performed under steady-state conditions after thermal equilibrium is reached.

Table 1 Bearing geometry and operating conditions

Journal radius, R	80 mm	Journal thermal expansion coefficient	$1.2 \times 10^{-5}\,°C^{-1}$
curvature radius R_1	79.965 mm	Journal thermal conductivity	45 W/(m·K)
curvature radius R_2	80.22 mm	Bushing thermal expansion coefficient	$1.8 \times 10^{-5}\,°C^{-1}$
$\overline{OO_1}$	0.25 mm	Bushing thermal conductivity	190 W/(m·K)
$\overline{OO_2}$	0.43 mm	Bushing elasticity modulus	70 GPa
Radial clearance, C	130 μm	Lubricant	No. 22 turbine oil
Effective external radius	170 mm	Lubricant viscosity at 40°C	0.045 4 Pa·s
Bearing length, B	142 mm	Temperature-viscosity coefficients in	0.001 773 76,
supply groove axial length	114 mm	Vogel's formula, C_1, C_2 and C_3	19.787 7, 2.123 71
Oil supply pressure	0.15 MPa	Lubricant density, ρ	860 kg/m³
Ambient temperature, T_a	19°C	Lubricant specific heat, C_p	2 000 J/(kg·°C)
Applied load	70 kN	Lubricant thermal conductivity, k_0	0.13 W/(m·K)

The distribution of the film pressure, temperature, and thermal deformation in the bearing mid-plane for two different operating rotational speeds are shown in Fig. 4 along with the experimental results.

Figures 4(a)–4(d) show that theoretical and experimental results concerning pressure and temperature in the oil-bushing interface are in good agreement and that the predicted temperature distributions are closer to the experimental measurements when considering the thermal deformation. Nevertheless, some discrepancies on the film temperature around the inlet and cavitation exist as shown in Figs. 4(c) and 4(d). It is due to uncertainties in the mixing of the recirculating oil and supply oil and the value of lubricant and bearing properties chosen for the simulation, the penetration depth of thermocouples, and the effect of the temperature fade in the negative pressure region[20], which is not considered in this study.

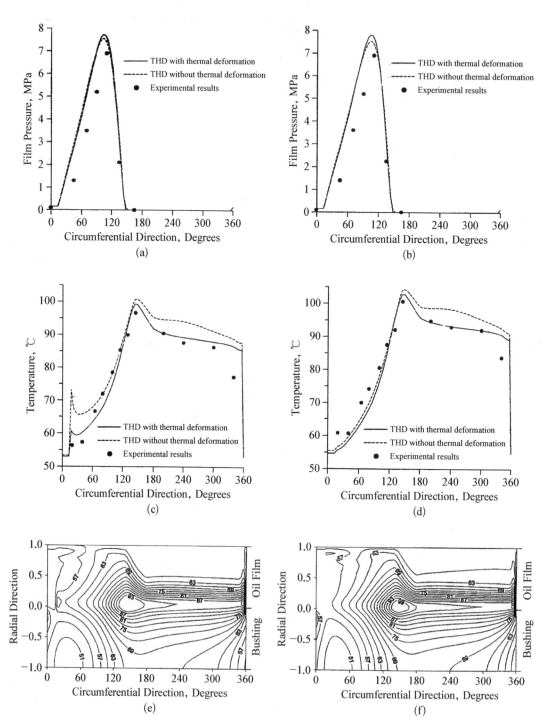

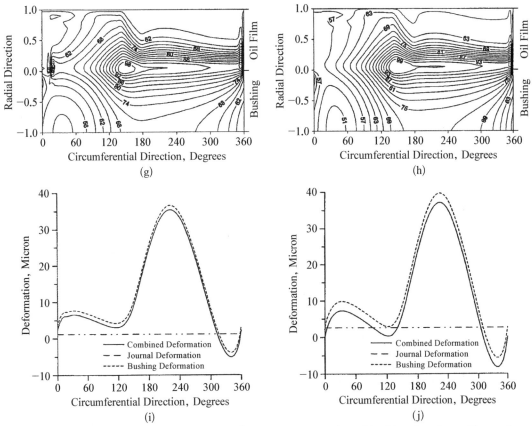

Fig. 4 Distribution of the film pressure, temperature in the oil-bushing interface, film and bushing temperature, and thermal deformation in the bearing mid-plane (load=70 kN for speed 6 000 r/min, inlet oil temperature 40.2℃; for speed 7 000 r/min, inlet oil temperature 40.5℃): (a) speed 6 000 r/min; (b) speed 7 000 r/min; (c) speed 6 000 r/min; (d) speed 7 000 r/min; (e) speed 6 000 r/min, with thermal deformation; (f) speed 7 000 r/min, with thermal deformation; (g) speed 6 000 r/min, without thermal deformation; (h) speed 7 000 r/min, without thermal deformation; (i) speed 6 000 r/min; and (j) speed 7 000 r/min.

The temperature contours in the oil and the bushing are shown in Figs. 4(e)–4(h). The maximum temperatures occur in the oil-bushing interface and in the neighborhood of the minimum film thickness along the circumferential direction. In general, the radial temperature gradient and the temperature in the oil-bushing interface are bigger in the cavitation region along the circumferential direction than in the full film region.

Figures 4(i) and 4(j) show that the thermal deformation is much bigger in the bushing inner surface than in the journal outer surface. The thermal deformation of the bushing inner surface varies significantly in the circumferential direction, especially in the cavitation region, where the inner surface compresses and dilates significantly and sharp. Since the differences between the bushing inner surface temperature and the ambient temperature are much bigger in the cavitation region than in the full film region the thermal deformation of the bushing inner surface is much bigger in the cavitation region than in the full film region. The thermal deformation reduces the film temperature slightly in the full film region and considerably in the cavitation region, and it makes the predicted results more close to the experiment data.

The leaking flow rates predicted theoretically are more than those of the experimental data and are slightly increased by the thermal deformation due to an increase in the clearance as shown in Table 2.

Table 2 Comparison between the theoretical and experimental results

	A	B	C	A	B	C
Speed (RPM)		6 000			7 000	
T_{in} (℃)		40.2			40.5	
ε	0.63	0.60	0.51	0.61	0.60	0.48
ϕ (Degrees)	56.4	56.3	66.2	58.1	57.5	72.1
h_{min} (micron)	47.8	50.0	64.4	50.7	50.0	67.9
Oil Flow (l/min)	17.4	17.2	14.5	19.1	18.0	16.1

A and B: Theoretical results with and without thermal deformation;
C: Experimental results

A large discrepancy on eccentricity ratio and the attitude angle between the theoretical and experimental data is noted in Table 2 and it is due to the applied external load direction which should be vertical for the simulation and may be not exactly in the test. Another reason for the difference is the differential dilation of the bushing, its housing and the journal as shown by Ferron et al.[21].

Table 3 and Figs. 5(a)-5(d) present a comprehensive study of the parameters. Bearing data are same as those listed in Table 1. With the increase of speed, the temperature increases significantly in the cavitation area, while slightly in the full film area because the cool oil is supplied into the full film area and the equivalent thermal conductivity of the fluid in the cavitation area is much smaller than that in the full film area. Dilation of the bushing inner surface in the cavitation region increases considerably with the increase of speed and slightly with the increase of the applied load. Effects of the speed are small on the eccentricity ratio and the film pressure and considerable on the leaking flow.

Table 3 Effects of load, speed, and thermal deformation on bearing performances (inlet oil temperature 40.5℃; * and ** denote results with and without thermal deformation, respectively)

Load(kN)	Speed(r/min)	Mode	ε	p_{max} (MPa)	T_{max} (℃)	Oil Flow(l/min)
70	8 000	*	0.59	7.67	104.8	23.2
		**	0.57	7.37	107.2	20.9
	9 000	*	0.59	7.61	109.1	25.8
		**	0.55	7.34	108.4	23.4
	10 000	*	0.58	7.50	111.8	28.8
		**	0.54	7.29	113.5	25.6
	11 000	*	0.57	7.40	114.7	31.4
		**	0.52	7.25	117.1	27.7

(续表)

Load(kN)	Speed(r/min)	Mode	ε	p_{max}(MPa)	T_{max}(℃)	Oil Flow(l/min)
90	8 000	*	0.64	10.2	109.6	23.5
		* *	0.62	9.72	112.4	21.8
	9 000	*	0.63	10.1	114.5	25.3
		* *	0.60	9.65	115.2	23.9
	10 000	*	0.62	10.0	116.8	28.4
		* *	0.59	9.54	118.4	26.3
	11 000	*	0.61	9.93	119.7	31.6
		* *	0.57	9.50	121.3	(28.6)

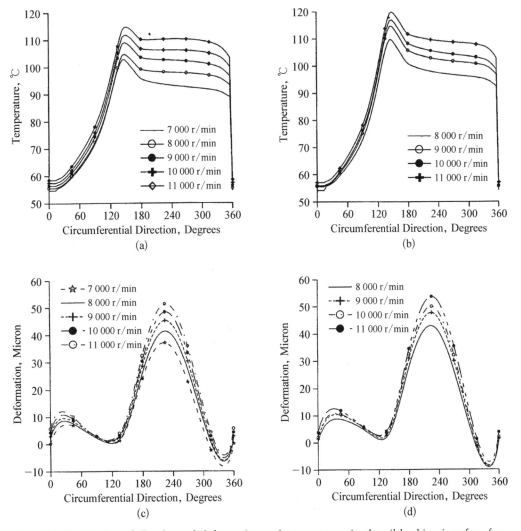

Fig. 5 Distribution of the thermal deformation and temperature in the oil-bushing interface for different load and speed (inlet oil temperature 40.5℃): (a) load 70 kN; (b) load 90 kN; (c) load 70 kN; and (d) load 90 kN.

5 Conclusions

From this study, the following conclusions can be drawn:

1. The maximum temperatures occur in the oil-bushing interface and in the neighborhood of the minimum film thickness. The radial temperature gradient is bigger in the cavitation region than in the full film region.

2. The thermal deformation is much bigger in the bushing inner surface than in the journal external surface. The thermal deformation of the bushing inner surface is much bigger in the cavitation region than in the full film region, and it reduces the oil temperature slightly in the full film region but significantly in the cavitation region.

3. With increase of the rotational speed, the temperature and thermal deformation in the bushing inner surface increase significantly in the cavitation area, while slightly in the full film area.

4. With an increase in the external load, the temperature increases considerably, while the thermal deformation and the film pressure increase slightly[22].

Acknowledgments

The authors would like to thank financial support for this research by the National Natural Science Foundation of the People's Republic of China under grant No. 59375198. Many thanks are due to the institute of Shanghai Turbine Works for their help during experiments.

Nomenclature

B = bearing length
C = nominal radial clearance
C_p = lubricant specific heat
D = bearing diameter
e = eccentricity
G = cavitation index
h = film thickness, $\bar{h} = h_T/C$
k_0 = lubricant thermal conductivity
p = film pressure, $\bar{p} = p\psi^2/(\mu_0 \omega)$
R = bearing radius
r = radial coordinate, $\bar{r} = r/R$
T, T_0 = oil temperature and apparent oil temperature, $\bar{T} = T/T_0$
T_{in}, T_B = inlet oil temperature and bush temperature, $\bar{T}_B = T_B/T_0$
U = tangential surface velocity of the journal, $U = R\omega$

u, v = fluid velocities in the x and y directions, $\bar{u} = u/U$, $\bar{V} = V/(\psi U)$
x, y, z = coordinates of circumferential, radius, and axial directions, $\theta = x/R$, $\bar{y} = y/h$, $\bar{z} = 2z/B$
ω = angular velocity of the journal
ρ = lubricant density
μ, ν = dynamic and kinematic viscosity, $\bar{\mu} = \mu/\mu_0$
ε_m, ε_H = eddy viscosity for momentum and heat transfer
λ = width/diameter ratio
ψ = relative clearance, C/R
γ = void fraction in cavitation zone and $\bar{p}$ in full film zone
ϕ = attitude angle

References

[1] Khonsari, M. M., 1987, "A Review of Thermal Effects in Hydrodynamic Bearings: Part II—Journal Bearings," ASLE Trans., 30, p. 26.

[2] Pinkus, O., 1990, Thermal Aspect of Fluid-Film Tribology, ASME Press, New York.

[3] Mittwollen, N., and Glienicke, J., 1990, "Operating Conditions of Multi-Lobe Journal Bearings Under High Thermal Loads," ASME J. Tribol., 112, pp. 330–340.

[4] Bouard, L., Fillon, M., and Frene, J., 1996, "Thermohydrodynamic Analysis of Tilting-Pad Journal Bearings Operating in Turbulent Flow Regime," ASME J. Tribol., 118, pp. 225–231.

[5] Khonsari, M. M., and Wang, S. H., 1991, "On The Fluid-Solid Interaction in Reference to Thermoelastohydrodynamic Analysis of Journal Bearings," ASME J. Tribol., 113, pp. 398–404.

[6] Fillon, M., Desbordes, H., Frene, J., and Chan Hew Wai, C., 1996, "A Global Approach of Thermal Effects Including Pad Deformations In Tilting-Pad Journal Bearings Submitted to Unbalance Load," ASME J. Tribol., 118, pp. 169–174.

[7] Khonsari, M. M., and Wang, S. H., 1992, "Notes on Transient THD Effects in a Lubricating Film," STLE Tribology Transaction, 35, pp. 177–183.

[8] Paranjpe, R. S., 1996, "A Study of Dynamically Loaded Engine Bearings Using a Transient Thermohydrodynamic Analysis," STLE Tribology Transaction, 39, pp. 636–644.

[9] Tucker, P. G., and Keogh, P. S., 1996, "On the Dynamic Thermal State in a Hydrodynamic Bearing With a Whirling Journal Using CFD Techniques," ASME J. Tribol., 118, pp. 356–363.

[10] Bouchoule, C., Fillon, M., Nicolas, D., and Barresi, F., 1996, "Experimental Study of Thermal Effects in Tilting-Pad Journal Bearings at high Operating Speeds," ASME J. Tribol., 118, pp. 532–538.

[11] Flack, R. D., Kostrzewsky, G. J., and Taylor, D. V., 1993, "A Hydrodynamic Journal Bearing Test Rig with Dynamic Measurement Capabilities," STLE Tribology Transaction, 36, pp. 497–512.

[12] Ma, M. T., and Taylor, C. M., 1996, "An Experimental Investigation of Thermal Effects in Circular and Elliptical Plain Journal Bearings," Tribol. Int., 29, No. 1, pp. 19–26.

[13] Fitzgerald, M. K., and Neal, P. B., 1992, "Temperature Distribution and Heat Transfer in Journal Bearings," ASME J. Tribol., 114, pp. 122–130.

[14] Ng et al.

[15] Elrod, H. G., 1981, "A Cavitation Algorithm," J. Lubr. Technol. 103, No. 3, pp. 350–354.

[16] Zhang, C., and Zhang, C., 1997, "Analysis of Crankshaft Bearings in Mixed Lubrication Including Mass Conserving Cavitation," Proc. of the 2nd International Conference on Hydrodynamic Bearing-Rotor System Dynamics, Xi'an, China, pp. 18–24.

[17] Ezzat, H., and Rhode, S., 1973, "A Study of the Thermohydrodynamic Performance of Finite Slider Bearings," J. Lubr. Technol. 95, pp. 298–307.

[18] Richtmyer, R. D., 1957, Difference Methods for Initial Value Problems, Interscience Publishers, Inc., New York, pp. 101.

[19] Boncompain, R., Fillon, M., and Frene, J., 1986, "Analysis of Thermal Effects in Hydrodynamic Bearings," ASME J. Tribol., 108, pp. 219–224.

[20] Booser, E. R., and Wilcock, D. F., 1988, "Temperature Fade in Journal Bearing Exit Region," ASLE Trans., 31, pp. 405–410.

[21] Ferron, J., Frene, J., and Boncompain, R., 1983, "A Study of the Thermohydrodynamic Performance of a Plain Journal Bearing Comparison Between Theory and Experiments," ASME J. Tribol., 105, pp. 422–428.

[22] Taylor, C. M., and Dowson, D., 1973, "Turbulent Lubrication Theory: Application to Design," ASME Paper 73-LubS-10.

Application of the Non-Stationary Oil Film Force Database*

Abstract: The technique of non-stationary oil film force database for hydrodynamic bearing is introduced and its potential applications in nonlinear rotor-dynamics are demonstrated. Through simulations of the locus of the shaft center aided by the database technique, nonlinear stability analysis can be performed and the natural frequency can be obtained as well. The easiness of "assembling" the individual bush forces from the database to form the bearing force, makes it very convenient to evaluate the stability of various types of journal bearings. Examples are demonstrated to show how the database technique makes it possible to get technically abundant simulation results at the expense of very short calculation time.

1 Introduction

For modern rotor systems supported by fluid film bearings, the traditional linear theory, which use linear stiffness and damping coefficients to represent the dynamic characteristics of the bearing, is not capable of explaining why the system can run at much higher speed than the linear stability threshold. The linear theory also can not explain why certain systems, originally stable, can go unstable when sufficiently large external excitation is suddenly applied, *etc*. Thus, the nonlinear stability study of the rotor-bearing system have attracted more and more attentions, and a convenient way is to calculate the locus of the shaft center and then check the stability of the system. The locus is obtained by the numerical time-integration of the shaft center motion, in which the transient oil film force is needed at each step of the integration. There are two ways to obtain the nonlinear oil film force, the first is direct solving of the transient Reynolds equation[1] and the second is using approximate methods such as the short bearing theory[2,3] or some nonlinear dynamic model using higher-order stiffness and damping coefficients[4]. In order to obtain the locus, the time-integration should be performed over a sufficiently long time span. Thousands of integration steps are needed at least and the transient Reynolds equation is repeatedly solved for the same number of times. Obviously, the direct solving method is highly time-consuming, and very inconvenient and even inapplicable for most rotor systems except the simplest ones. Though the approximate method is fast, its accuracy is very questionable, and its application is of limited value.

---

* In collaboration with Wang Wen, Chen Xiaoyang. Reprinted from *Journal of Shanghai University* (*English Edition*), 2001, 5(3): 230–233.

In 1993, the authors put forward a database technique to solve the problem of fast determination of accurate non-stationary oil film force[5]. It was proved that the database technique not only had practically the same accuracy as the direct solving method, but also shortened the computation time dramatically. In this paper, the simple Jeffcot rotor model was used to demonstrate the potential application of this technique on nonlinear rotor-dynamics. The nonlinear dynamic behavior of the rotor is investigated on the simple model and the results were discussed.

2 Dynamic Equation for Jeffcot Rotor Model

The single mass elastic rotor (Jeffcot rotor) is schematically shown in Fig. 1. The dynamic equations of the rotor-bearing system are:

$$m\ddot{x}_m = -k(x_m - x_j) - m\delta\omega^2 \sin(\omega t + \varphi),$$
$$m\ddot{y}_m = -mg - k(y_m - y_j) - m\delta\omega^2 \cos(\omega t + \varphi),$$
$$f_x(x_j - x_b, y_j - y_b, \dot{x}_j - \dot{x}_b, \dot{y}_j - \dot{y}_b) = \frac{1}{2}k(x_m - x_j),$$
$$f_y(x_j - x_b, y_j - y_b, \dot{x}_j - \dot{x}_b, \dot{y}_j - \dot{y}_b) = \frac{1}{2}k(y_m - y_j).$$

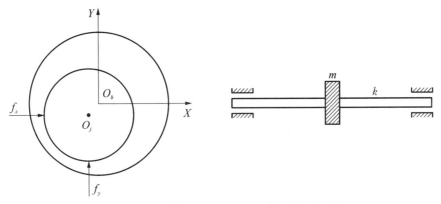

Fig. 1 Jeffcot rotor supported on fluid film bearings

In the above equations, $m\delta$ represents the level of the unbalance, and f_x and f_y are components of the non-stationary oil film force. Runge-Kutta method of 4th order with variable steps is adopted for the time-integration of the dynamic equations. Table 1 lists the main parameters used in the simulation.

Table 1 Main parameters of the rotor and bearings

Bearing parameters	$L=0.05$ m, $D=0.1$ m, $\eta=5.6\times10^{-3}$ Pa·s
	$c=5\times10^{-4}$ m (clearance of the bearing)
Rotor	$m=15$ kg, $k=2.94\times10^5$ N/m, $\omega_k=140$ s^{-1}
Dimensionless definition	Unbalance: $AA = m\delta/mc$
	$X = x/c, Y = y/c, A_x = a_x/c$

3 Application of Non-Stationary Oil Film Force Database

3.1 Nonlinear stability analysis and natural frequency

One of the advantage of the non-stationary oil film force database is fast and accurate determination of the non-stationary oil force. This makes it possible to simulate the unbalance response of the rotor-bearing system in a short time. Through analysis of the shaft center loci at different speeds, the stability of the rotor-bearing system can be known and the natural frequency can be obtained.

A series of the typical simulated loci of the unbalance response at different speeds are shown in Fig. 2. The journal bearings are cylindrical bearings with a 10 degree oil inlet groove at the top. The unbalance AA is equal to 0.1. From Fig. 2, it can be seen that the amplitudes at speed $\omega=70$ and at $\omega=140$ are smaller than the amplitude at $\omega=120$. This implies that the critical speed is in the range of $\omega=70 - 140$. In Fig. 2, the locus of unbalance response at $\omega=215$ is not a simple ellipse as the others. It has an ω-shaped crossed loop, and this implies the unbalance response consists not only of a component with the same frequency as the rotation speed, but also has components with other frequencies, and among them the most prominent one has a frequency roughly equal to half of the running speed. However, the amplitude of the crossed loop is still rather small at this speed. Fig. 3 shows the oil whip locus when the speed increases to $\omega=231$. Here the journal locus does not converge and the rotor-bearing system goes unstable.

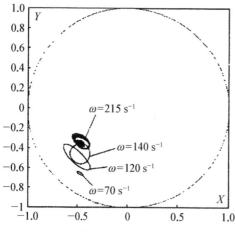
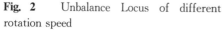

Fig. 2 Unbalance Locus of different rotation speed

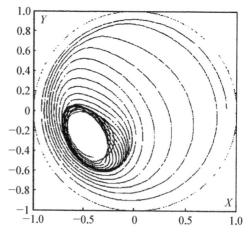

Fig. 3 Oil Whip Locus at the rotation speed $\omega=231$ s^{-1}

In order to analyze the locus of the unbalance response, its frequency spectrum is obtained by FFT analysis of the locus. 512 uniformly time-stepped points (totaling 0.512 second) on the locus are picked out for this analysis. The obtained amplitude-frequency spectra are shown in Figs. 4 and 5. Fig. 4 shows the spectrum at lower speeds of

$\omega=70$, 120 and 140, there is only one dominant component with the same frequency as the rotation speed. But at higher speeds of $\omega=215$ and 231, the vibration has two prominent components as shown in Fig. 5, one of which has a frequency equal to the rotation speed, and the other has a frequency about half of the rotation speed and therefore is termed as "half-frequency" component. When the rotation speed increases, the frequency of the latter component does not increase, but the intensity increases sharply and it is this component that induces the system to go unstable and cause the oil whip shown in Fig. 3. The explanation lies in the coincidence of the half-frequency over the natural frequency of the rotor-bearing system, which is about 110 s^{-1} as shown in Fig. 5. The induced resonance makes the amplitude rise drastically and the system to become unstable.

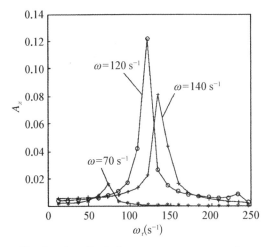 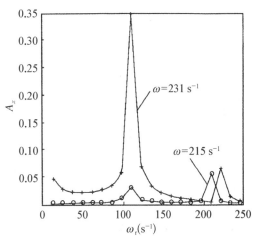

Fig. 4 Amplitude-frequency spectrum at low speed of $\omega=70$, 120, 140 s^{-1}

Fig. 5 Amplitude-frequency spectrum at high speed of $\omega=215$, 231 s^{-1}

The above demonstrations show how the unbalance response and stability of the system can be known from the locus simulation, which is very fast and convenient by the aid of the database technique. By FFT analysis of the unstable locus, the natural frequency of the system is easily obtained.

3.2 Stability analysis of different types of bearing

In the database technique, a special program is used for the fast generation of a database of oil film force for an individual bush. The oil film force of a whole bearing, whatever lobe-type, can be obtained by simply assembling the oil film forces of all the involved individual bushes. The easiness and fastness of this method makes the database technique very advantageous for calculating all the lobe-typed fluid film bearings.

Fig. 6 shows the unbalance response of the rotor supported by elliptical bearings. The two bushes of the bearing have 150 degree angular spans, with oil inlets at both sides. The ellipticity coefficient is 0.5. All the other parameters are same as those of the cylindrical bearing listed in Table 1. In Fig. 6, at speed $\omega=120$, the locus is a typical, nearly elliptical unbalance response, and there is only one predominant vibration frequency component. When the speed increases to $\omega=220$, it can be seen from the locus that half-frequency vibration have emerged. When the speed comes to $\omega=280$, oil whip happens and the system goes unstable. Fig. 7 shows the unbalance loci of the rotor supported by three-lobe bearings. The span angle of each bush is 100 degrees and the pre-load coefficient is 0.5. All the other geometric parameters are same as those of the cylindrical bearing. In Fig. 7, at speed $\omega=220$, the half-frequency vibration component still does not emerge, and the unstable speed is $\omega=281$. By comparison of Fig. 6 with Fig. 7, it can be easily concluded that the stability of the three-lobe bearing is better than the elliptical bearing. Moreover, both of them are much stabler than the cylindrical bearing.

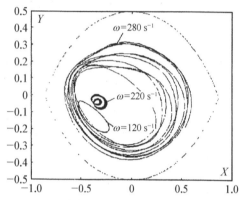

Fig. 6 Unbalance loci with elliptical bearing

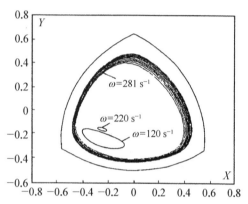

Fig. 7 Unbalance loci with three-lobe bearings

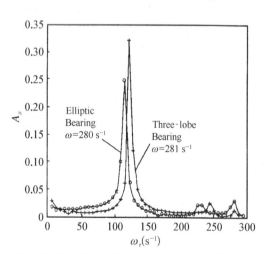

Fig. 8 Amplitude-frequency spectrum of the oil whip loci

Fig. 8 shows the amplitude-frequency spectra obtained from the oil whip loci of Fig. 6 and Fig. 7. More details of the oil whip vibration can be known from Fig. 8. Take the amplitude-frequency curve of the elliptical bearing as example, it can be seen that the oil whip consists of three components. The predominant one is the vibration of the natural frequency, the second is the vibration of the multi-natural frequency and the third is the vibration of the rotation speed. At variance with Fig. 5, multi-natural frequency component emerges in oil whip in the cases of elliptical bearing and three-lobe bearing. From Fig. 8, the natural frequency (ω_n) of the rotor-bearing system supported on elliptical and three-lobe bearings are approximately 115 s^{-1} and 122 s^{-1} respectively.

Table 2 is the stability comparison of the cylindrical bearing, elliptical bearing and three-lobe bearing. All the results are obtained with an unbalance of $AA=0.1$. Table 2 shows that the stability of elliptical and three-lobe bearings are much better than the cylindrical bearing, and the three-lobe one is the best.

Table 2 Stability comparison of different types of bearing

Bearing type	Cylindrical	Elliptical	Three-lobe
Natural frequency (ω_n)	110	115	122
Unstable speed (ω_{st} oil whip)	231	280	281

4 Conclusions

Applications of the technique of non-stationary oil film force database in nonlinear rotor-dynamics are demonstrated. By applying the database technique, the oil film force of the hydrodynamic bearing can be accurately calculated in very short time. Thus the locus of the rotor center can be easily simulated. By analyzing the locus, the knowledge concerning the intensity of unbalance response, the composition of the vibration, when the half-frequency vibration emerges and when the oil whip happens, can all be easily obtained. The natural frequency of the rotor-bearing system can also be easily obtained through the amplitude-frequency spectrum of the oil whip locus. The stability of different types of hydrodynamic bearing are also analyzed, taking advantage of the easiness of assembling the bush forces to get the bearing forces. All these simple demonstrations show that the database technique has cleared a way for nonlinear rotor-dynamics by solving the problem of fast and accurate determination of the non-stationary oil film force. This technique has also been adopted by various researchers in applications to nonlinear stability analysis of real complex rotor-bearing system and has shown great potential, and more and more such applications are expected in the time to come.

References

[1] Hori Y, Kato T. Earthquake-induced instability of a rotor supported by oil film bearings. J. Vib. Acoust., Trans. ASME, 1990, 112: 160 – 165.
[2] Barrett L E et al. The dynamic analysis of journal bearing using a finite length correction for short bearing theory. Topics in Fuild Film Bearing and Rotor Bearing System Design and Optimization, ASME, 1978.
[3] Hattori H. Dynamic analysis of a rotor-journal bearing system with large dynamic loads (stiffness and damping coefficient variations in bearing oil film). JSME International, 1993, 36(2): 251 – 257.
[4] Chu C S, Wood K L, Busch-Vishniac I J. A nonlinear dynamic model with confidence bounds for hydrodynamic bearings. J. of Tribology, 1998, 120(July): 595 – 604.
[5] Wang W, Zhang Z M. Nonlinear oil film force database. Journal of Shanghai University of Technology, 1993, 14(4): 299 – 305(in Chinese).

Analysis on Dynamic Performance for Active Magnetic Bearing-Rotor System*

Abstract: In the application of active magnetic bearings (AMB), one of the key problems to be solved is the safety and stability in the sense of rotor dynamics. The project related to the present paper deals with the method for analyzing bearing rotor systems with high rotation speed and specially supported by active magnetic bearings, and studies its rotor dynamics performance, including calculation of the natural frequencies with their distribution characteristics, and the critical speeds of the system. One of the targets of this project is to formulate a theory and method valid for the analysis of the dynamic performance of the active magnetic bearing-rotor system by combining the traditional theory and method of rotor dynamics with the analytical theory and design method based on modern control theory of the AMB system.

When active magnetic bearing (AMB) is going to its applications, its performance of rotor dynamics and stability is one of the problems that attract most experts' attention. So far, many learners all over the world have been persistently studying this problem[1-5]. The results show that the performances of AMB system have a crucial effect on its stability. Bearing Institution of Shanghai University has been studying in this field since 1994. Through theoretical analysis and test observation, the performances of AMB rotor system has been learned, and this will set a basis for the complete resolution of this problem[6].

This paper deals with the rotor system with high rotation speed and specially supported by AMBs, and studies its performance of rotor dynamics, including the calculation of natural frequencies, critical rotating speeds and its performances in the distribution of the system. Meantime, it is also concerned with setting up the parameters of the controller and the influence of the AMB structure on the system stability. And here, a kind of analysis method is established by combining the traditional theory and method of rotor dynamics with the analysis theory and design method based on modern control theory of the AMB system. As an application of the analysis results, a showpiece of 150 m^3 turbo oxygen gas expander was manufactured and experimented, and the test results are equal precisely to the theory analysis. This would be used for the further analysis and design of performances of the active magnetic bearing. This paper briefly introduces our works in this field, including theoretical analysis and system model on dynamic performance for AMB-rotor system, a kind of analysis method with its results and a showpiece with the

---

* In collaboration with YAN Hui-yan, WANG Xi-ping, ZHU Li-jin and WAN Jin-gui. Reprinted from *Journal of Shanghai University* (English Edition), 2001, 5(3): 234-237.

results.

In order to realize complete levitation in space, a rotor needs some forces offered from five degrees of freedom. This means that two radial bearings and one axial bearing are needed. These bearings, rotors and a set of electronic controller constitute a complete AMB system. It is usually called the magnetic levitation system with five degrees of freedom. It is a complicated system combined with machinery, electricity and electronics, magnetism. Therefore it is difficult to set up a precise mathematical model of the system. The common way is to simplify it to be a single degree of freedom system so that it can be analyzed easily. This way is suitable for a rigid rotor, and has been proved to be valid[7]. But for the flexible rotor, the error is much more than a rigid one, and the key problem is that the natural frequencies and critical rotating speeds of the system can not be calculated and analyzed. And the result of the analysis tells nothing about the rotation stability of the system. The critical rotating speeds of the flexible rotor system are usually low and the working speed is higher than the first critical speed of the system. So the analysis for it is necessary.

1 Foundation of Dynamics of Active Magnetic Bearing System

In the system, an eddy current sensor and a linear analog controller with current strategy are used. The structure of the analyzed system and its analyzing coordinates are shown in Fig. 1. It is supposed that there is no coupling between the freedoms.

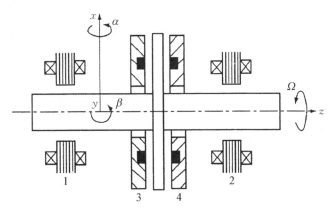

Fig. 1 Analyzing coordinate of a bearing rotor system

(1, 2—radial bearing; 3, 4—axial bearing)

In traditional rotor dynamics theory a bearing rotor system is always described with a kinematics equation with n degrees of freedom, which includes quality matrix $[M]$ of rotor, stiffness and damping matrix $[K]$ and $[C]$ of bearings. To analyze the stability, calculate the critical speeds and unbalance response of the system, all the eigenvalue and eigenvector of the equation must be counted. There are many ways to calculate eigenvalue and eigenvector of the equation[10]. The key to be described here is how to establish the kinematics equation of AMB system, that is, how to obtain the matrix $[K]$ and $[C]$. And they are explained with 4 stiffness coefficients, K_{xx}, K_{xy}, K_{yx}, K_{yy} and 4 damping coefficients, C_{xx}, C_{xy}, C_{yx}, C_{yy} in the traditional bearing rotor dynamics theory.

In 1980, H. Haberman and Liard presented that AMB has a complex stiffness curve like a "bath" shape[7]. Now it is known that the "bath" curve is actually a approximation of a frequency response curve of system. But it is unuseful in calculating the performance of mechanics. In 1986, R. Humphris and other experts deduced equivalent stiffness and damping coefficients from transfer function of AMB system. It is also a basis of analyzing the performance of stiffness and damping of AMB system in this paper. Four cross stiffness and damping coefficients can be got through the installing structure of the displacement sensors, and the other coefficients by the frequency response curve of the controller[8, 9].

2 Analysis of the Performance of Rotor Dynamics of System

It is known from traditional rotor dynamics theory[10] that the stiffness and damping coefficients of sliding bearing are the functions of angle rotation speed of the rotor, and the coefficients are the functions of whirl frequency but not the rotation speed in the AMB system. This is a main difference between the two systems. Meanwhile, there is another fact that the whirl frequency can not be predetermined. And it is a way to set some frequency values in advance for analysis and calculation. It must be replied whether these frequency values exist or not, and what type of the vibration should be determined by the calculating results.

The analysis object of rotor dynamics is usually the simplified mechanical model of rotor system. In this paper, the rotor is simplified in the same way for the AMB system.

The simplification of AMB is refered to determine correctly the stiffness and damping coefficients of active magnetic bearing system. This is the key to calculate the critical speeds of traditional bearing-rotor system or the AMB-rotor system. Like sliding bearings, magnetic forces in AMB can be also expressed by 8 equivalent stiffness and damping coefficients as follows:

$$\begin{Bmatrix} F_{Ix} \\ F_{Iy} \end{Bmatrix} = \begin{bmatrix} c_{xx} & c_{xy} \\ c_{yx} & c_{yy} \end{bmatrix} \begin{Bmatrix} \dot{x} \\ \dot{y} \end{Bmatrix} + \begin{bmatrix} k_{xx} & k_{xy} \\ k_{yx} & k_{yy} \end{bmatrix} \begin{Bmatrix} x \\ y \end{Bmatrix}.$$

The force analysis of shafts, determination of boundary conditions, nondimension of parameters, determination of vibration equation of system and searching of complex eigenvalue are the same as the traditional analysis method of rotor dynamics. Here the solving process is omitted.

The curves in Fig. 2 and Fig. 3 are the typical calculated results in theory. In Fig. 2 the points, at which the curves cross with the 45° diagonal, are corresponding to the whirl frequency (speed), which are the critical speeds of isochronous positive whirl of the system. The logarithmic decrement values are shown in Fig. 3.

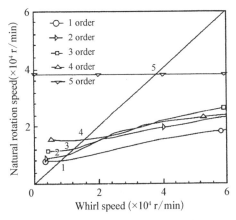
Fig. 2 Curves of critical speeds

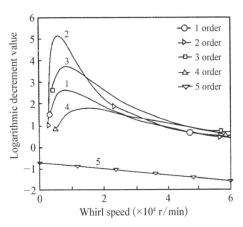
Fig. 3 Curves of logarithmic decrement rate

3 Application and Practice

By using the theory and method above, a five-degree of freedom magnetic levitation system controller was developed. The controller is used in theoretic analysis, calculation and the test for a 150 m³ turbo oxygen gas expander. Fig. 4 shows the rotor and bearing parts. The results of both theoretic analysis and experiment are equal precisely each other. The main parameters of the rig are listed in Table 1. And Table 2 shows a group of the critical speeds and logarithmic decrement values. Fig. 5, Fig. 6 and Fig. 7 show the rotor axle center locus near critical speeds, which are monitored by two oscillographs.

Fig. 4 Photo of the rotor and the AMB of 150 m³ turbo oxygen gas expander

Table 1 Parameters of the bearings

	Radial bearing	Thrust bearing		Radial bearing	Thrust bearing
Outer diameter	72 mm	72 mm	Turns of coil	42	80
Inner diameter	30 mm	30 mm	Air gap	0.17 mm	0.2 mm
Thickness	15 mm	15 mm	DC resistance	0.2 Ω	0.3 Ω
Bias current	1.0 A	1.2 A	Inductance	1.8 mH	0.8 mH

Table 2 A group of critical speeds (n_{cr}) and logarithmic decremen values (ζ)

order	1	2	3	4	5	6
$n_{cr}(\times 10^3 \text{ r/min})$	8.050	9.430	12.240	15.380	105.52	105.54
ζ	2.647	4.634	3.540	1.800	−0.004 6	0.001

(Considering bearing's anisotropism and crossing dynamic performances)

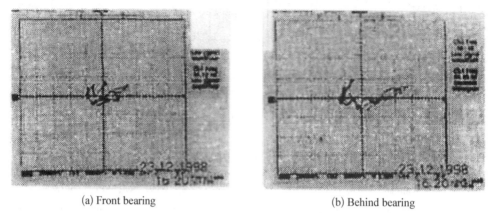

(a) Front bearing (b) Behind bearing

Fig. 5 Moving track of axis center at the rotation speed of 7 200 r/min

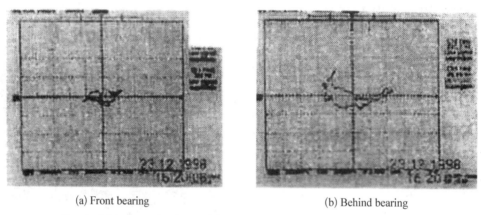

(a) Front bearing (b) Behind bearing

Fig. 6 Moving track of axis center at the rotation speed of 14 600 r/min

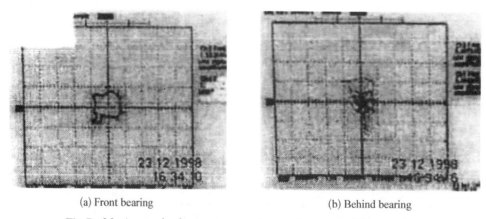

(a) Front bearing (b) Behind bearing

Fig. 7 Moving track of axis center at the rotation speed of 94 000 r/min

The highest axis rotation speed of this expander was reached up to 104 000 r/min, the stable rotation speed 96 000 r/min. Fig. 8 is the photo of the turbo oxygen gas expander in the experiment.

Fig. 8 Photo of 150 m³ turbo oxygen gas expander in testing

4 Conclusions

(1) To get the stiffness and damping coefficients, the transfer function of the system controller must be known. When a series of supposed whirl frequency signals are input into the controller, the coefficients depend mainly on the function and those cross coefficients related to the instlling feature of the displacement sensors.

(2) To obtain the critical speeds of the system, it is important that there must be the points on the critical speed curves crossing with the 45° diagonal. Though the points on the curves mean the critical speeds of the system in an oil-film bearing rotor system, only the cross points are truth for AMB rotor system. This is the main difference between two types of the systems.

(3) In a traditional bearing-rotor system, the curves of logarithmic decrement rate and damping value can qualitatively show the capability for restraining the vibrations. But in an AMB-rotor system, only those values in the cross point can be used to analysis of the system, as the abscissa of the curves is whirl frequencies. When the rotation speed of the rotor increases from low frequency area to high, it must sweep all of the critical speeds of the system. If a natural frequency can not be met with any whirl frequency calculated by the theory, it is known as "not truth" for the system. This means that the angle speed of rotor does not resonate with the natural frequency. On the contrary, if met, It is said "truth" and resonant.

(4) It is known that the "true points" represent the vibration type of every critical speed, and can be used to determine the natural frequencies, the vibration decrement rate and the stability of the system. But, it must be pointed out that the stability of the system does not depend completely on angle rotation speed. Under this theoretical frame, the unstable speed of the system does not exist. It is either stable forever or unstable from low speed. The conclusions above are deduced based on linear system. Those are the questions worth further studing how whirling tracks diverge and develop, whether its limit tracks existence and what size it is after losing linear stability.

References

[1] Chen H M, Wilson D. Stability analysis for rotors supported by active magnetic bearing. Proceeding of the 2nd International Symposium on Magnetic Bearings, 1990: 325-328.

[2] Dhar D, Barrett L E. Design of magnetic bearings for rotor systems with harmonic excitations. Transactions of the ASME, Journal of Vibration and Acoustics, 1993, 115: 359-365.

[3] Wang Hongli, Zhang Xinsheng, Liu Yong, Zhang Anxin. Chinese Journal of Mechanical Engineering, 1994, 30(6): 41-46.

[4] Wang Haiying. Stability analysis for rotors supported by active magnetic bearing. Master Dissertation of Xian Jiaotong University, Xian, China, 1995 (in Chinese).

[5] Tang Zhonglin, Feng Zhihua, Huang Xiaowei. A method for analyzing flexible rotors supported by active magnetic bearings. Chinese Journal of Mechanical Engineering, 1999, 35(2): 97-100.

[6] Wan Jingui. Dynamic behavior analysis for active magnetic bearing-rotor system. Master Dissertation of Shanghai University, Shanghai, China. 1999 (in Chinese).

[7] Wang Xiping. Parametric design and application investigation of electro-magnetic bearing system, Doctorate Dissertation of Xian Jiaotong University, Xian, China, 1994 (in Chinese).

[8] Wang Xiping. Study on stiffness and damping performances for electro-magnetic bearings. Chinese Journal of Applied Mechanics, 1997, 14(3): 95-100 (in Chinese).

[9] Wang Xiping, Cui Weidong. Investigation on contactless displacement sensors for active magnetic bearings. Journal of Shanghai University (Natural Science), 1998, 4(1): 54-60 (in Chinese).

[10] Xu Longxiang. Dynamic Design of High Rotation Speed. National Defence Industry Press, Beijing, 1997.

主动磁轴承转子系统动力学特性的研究[*]

摘　要：以电磁轴承支承下的高速转子系统为分析对象,以传统的转子动力学分析理论和方法为基础,结合电磁轴承系统分析理论和方法,研究电磁轴承支承条件下转子系统的动力学特性.应用基于系统传递函数的刚度、阻尼特性分析理论和方法,建立用于分析电磁轴承转子系统动力学特性的理论和方法.并对一个实用型 150 m³ 制氧透平膨胀机样机的研制和试验进行指导和结果验证.

0　前言

电磁轴承(Active magnetic bearing, AMB)在我国的研究历史可追溯到 20 世纪 70 年代中叶.清华大学、西安交通大学、南京航空航天大学、哈尔滨工业大学、天津大学以及上海大学等高校仍在进行这方面的研究.尽管基本原理方面的研究均已完成,但普遍存在的轴承转子系统在高速运行时的稳定性问题成为这一技术推广应用的瓶颈问题.虽然也做了一些研究工作[1-4],但对此问题系统深入地研究在国内至今未见有详细的报道.

对电磁轴承早期的研究对象和试验模型多为刚性转子系统,故振动问题并不突出.随着研究的深入,模型扩展到柔性转子系统中,转子转速可能在一个或几个临界转速之上,这意味着转子从静止到达工作转速的过程中,必须越过这些临界转速,这是 AMB 技术的进步对转子动力学理论研究提出的新课题.分析研究电磁轴承转子系统动力学特性的困难在于电磁轴承刚度和阻尼特性的确定,因为电磁轴承是一个复杂的机、电、磁耦合系统,无法使用传统的滑动轴承或滚动轴承的刚度、阻尼系数计算方法;另外,电磁轴承的刚度、阻尼特性是涡动频率而不是转速的函数;转子位移信号是靠位移传感器传送给控制系统,而产生控制力的电磁轴承与拾取位移信号的传感器在轴上的安装位置常不一致,因此传感器并不能准确反映轴承处转子的位移,这些都是电磁轴承转子系统的动力学特性与传统概念间的不同之处.

对于电磁轴承系统稳定性的研究,早期有 Salm 和 Schweitzer,后来的 H. M. Chen 是研究较多的学者之一[5].国内也有不少的学者开始此项内容的研究[1-3].上海大学轴承研究室在此方面的研究已获得一些有意义的成果,并应用于一台 150 m³ 制氧透平膨胀机的研制、试验和结果验证[2].

1　电磁轴承及其系统动力学分析基础

1.1　系统模型与工作原理

工业应用的电磁轴承转子系统包括 2 个径向轴承、1 个轴向轴承、1 个控制器及

---

[*] 本文合作者：汪希平、朱礼进、于良、王文、王小静、万金贵.原发表于《机械工程学报》,2001,37(11)：7-12.

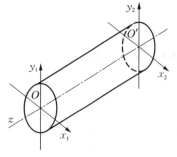

图1 转子的5自由度控制示意图

转子,称为5自由度磁悬浮系统,5个自由度如图1所示[6].

通常各个自由度上的控制电路在结构上没有差别,分析时常以单自由度磁悬浮闭环控制回路作为对象.图2即为1个单自由度磁悬浮轴承系统框图.当转子中心发生偏移时,转子位移量由位置信号传感器拾取后送到调节电路,经适当的增益和相位调节后送入功率放大电路,最终输出一个控制电流调节电磁铁的磁场力,使转子回到正常悬浮位置.

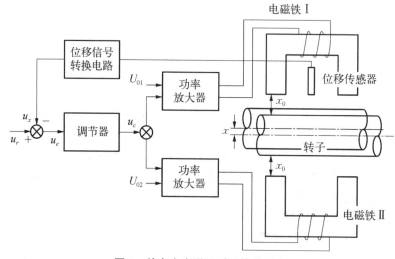

图2 单自由度磁悬浮系统的组成

u_x, u_r, u_e——位移 x 产生的信号电压,位置参考电压和误差信号电压
u_c, U_{01}, U_{02}——控制器输出电压及偏磁电流预置电压

1.2 基本方程

对单自由度磁悬浮系统,仅考虑转子平移运动和外干扰力 F 的影响时,转子质心的线性运动方程可表示为

$$m\ddot{x} = k_1 x - k_2 i_c + F \tag{1}$$

式中 k_1——电磁轴承系统的位移刚度

$k_1 = \mu_0 S_0 N_0^2 I_0^2 / x_0^3$

k_2——电磁轴承系统的电流刚度系数

$k_2 = \mu_0 S_0 N_0^2 I_0^2 / x_0^2$

μ_0——真空磁导率

S_0——气隙截面积

N_0——电磁线圈的匝数

x_0——转子处于平衡位置时的轴承气隙长度

x——转子质心的实际偏移距离

I_0——偏磁电流分量

i_c——位移量 x 引起的控制电流分量

因为系统控制器总可以自动平衡恒定负载,故将零位移参考位置取在转子重力作用下的静平衡位置,对转子的动态分析不会产生任何影响.

在选取了合适的位移传感器及功率放大电路之后,调节电路成为控制器的重要功能部件,它的设计不仅关系着电磁轴承的成败,而且对系统的转子动力学特性起着决定性的作用.本文讨论的是当系统采用电涡流位移传感器,控制器采用线性模拟电路时的动力学特性分析方法.

1.3 系统动力学理论基础

本文采用传统转子动力学分析方法作为理论基础[7-9],对电磁轴承转子系统采用相应的"离散节点的集总质量模型";分析研究转子轴承系统的动力学行为的方法采用传递矩阵法.

传统的转子动力学理论在分析轴承转子系统时,不论是计算系统的稳定性、临界转速还是不平衡响应,都归纳为计算系统方程的特征值与特征矢量,具体计算有很多种方法,在此不予赘述.轴承转子系统的方程中包含有转子的质量矩阵 m、轴承的刚度矩阵 k 和阻尼系数矩阵 c.通常,系统质量矩阵 m 的获得并不困难,应用传统的转子动力学方法分析电磁轴承转子系统的动力学特性,关键的问题是如何获得准确的轴承刚度阻尼特性,即电磁轴承系统的刚度矩阵 k 和阻尼系数矩阵 c.

应当指出的是,多数情况下,高速转子系统的稳定性可按线性问题处理.但当系统中的非线性因素不能忽视时,以线性理论为基础计算的结果将与实际情况有较大差别,许多实际现象用线性理论也无法做出合理的解释,这时就应该对转子轴承系统作非线性动力学分析.本文仅讨论线性稳定性的分析方法,且不考虑机器支座(基础)对振动的影响.

2 电磁轴承转子系统动力学特性分析

2.1 5自由度电磁轴承转子系统及其特点

5自由度电磁轴承转子系统的结构见图3,当取图中所示的坐标系时,系统控制器中的

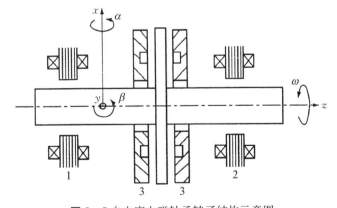

图 3 5自由度电磁轴承转子结构示意图

1,2—径向电磁轴承;3—轴向电磁轴承

5个控制电路单元分别控制着转子的5自由度的运动：在x,y,z方向上的3个平动以及绕径向方向x,y的2个转动α及β.

当轴向轴承的力矢量作用在过转子质心的轴线上时，z方向上的运动和x与y方向上的运动彼此独立，轴向运动的分析可以与4个径向自由度分离，且相应的计算可按单自由度系统来处理.本文设讨论的系统满足上述条件，分析时将不包括轴向自由度.这一点与滑动轴承转子系统相似[2].

当采用单个径向传感器或传感器的安装位置有误差，径向自由度间会存在耦合[2,10]，结合控制实施情况，可将xz平面内的运动与yz平面内的运动分开，即把系统分为两个解耦的子系统，一个代表了平面内的运动，变量为x和β；另一个则代表了yz平面内的运动，变量为y和α.这2个平面内的运动情况完全相似，可取其中一个平面，如xz平面来分析.这一点也类似于滑动轴承转子系统.

2.2 电磁轴承转子系统刚度、阻尼系数的确定方法

1986年，R. Humphris等从电磁轴承系统的传递函数导出了等效刚度和等效阻尼系数.这是本文分析电磁轴承刚度阻尼特性的主要理论依据[11].

设转子的位移信号x为输出变量，电磁力F_x为输入变量，系统控制器的传递函数为$G(s)$，系统的闭环传递函数框图如图4所示.

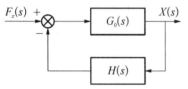

图4 不同输入输出量条件下系统的闭环传递函数简化框图

图中$G_0(s)=\dfrac{1}{ms^2-k_1}$；$H(s)=k_2G(s)$.则系统的闭环传递函数为

$$X(s)=\frac{F_x(s)}{ms^2-k_1+k_2G(s)} \tag{2}$$

式中　$X(s)=L[x(t)]$
　　　　$F_x(s)=L[F_x(t)]$
　　　　L——拉普拉斯变换的符号
　　　　s——拉普拉斯算子的符号

由上式可获得系统的等效刚度、阻尼系数，以k_{eq}表示系统的等效刚度，以C_{eq}表示系统的等效阻尼系数，则

$$\left.\begin{aligned}k_{ep}&=k_2\text{Re}[G(j\omega)]-k_1\\C_{ep}&=\frac{k_2\text{Im}[G(j\omega)]}{\omega}\end{aligned}\right\} \tag{3}$$

式中　ω——输入控制器的信号角频率(rad/s)
　　　　$\text{Re}[G(j\omega)],\text{Im}[G(j\omega)]$——$G(j\omega)$的实部和虚部

由式(3)可求得系统中一个径向电磁轴承的刚度k_{xx},k_{yy}和阻尼系数C_{xx},C_{yy}.而交叉刚度系数k_{xy},k_{yx}和交叉阻尼系数C_{xy},C_{yx}则与选用的传感器类型和安装等具体情况有关[11].当系统传感器采用单方向安装方式，系统的交叉刚度、阻尼系数可写为

$$\left.\begin{array}{l} C_{yx} = \pm x_R \dfrac{k_2 \operatorname{Im}\{G_c(j\omega)\}}{\omega} = \pm x_R C_{yy} \\ k_{yx} = \pm x_R k_2 \operatorname{Re}\{G_c(j\omega)\} = \pm x_R k_{yy} \end{array}\right\} \quad (4)$$

式中 $x_R = \dfrac{x_0}{2R}$

R——转子的半径,正负号则对应于 $x < 0$ 和 $x \geqslant 0$

同理可求出系统另一个径向电磁轴承的 4 个系数: C_{xx}, C_{xy}, k_{xx}, k_{xy}, 这样求出的 8 个系数就构成一组径向电磁轴承完整的动特性系数.

2.3 电磁轴承系统动特性系数计算的特点

上述分析说明,电磁轴承的刚度系数阻尼系数除与系统的固有参数 k_1、k_2 有关,还与控制器的传递函数和信号角频率 ω 有关. 应当说明的是,ω 是转子的涡动角频率而不是转子的转动角频率,因此电磁轴承动特系数是转子涡动角频率而不是通常转速的函数,这是电磁轴承与传统滑动轴承在动力学特性上的重大区别.

3 理论分析与计算结果

3.1 系统的简化与模化

实际的转子结构往往比较复杂,为分析方便,转子动力学分析的对象通常是转子系统的简化力学模型. 限于篇幅,本文仅扼要介绍采用转子的集总质量模型分析电磁轴承转子系统动力学特性的方法,详细的分析过程可参看文献[2]. 类似于滑动轴承,电磁轴承力也用上述 8 个等效刚度和等效阻尼系数来表示,即

$$\begin{Bmatrix} F_x \\ F_y \end{Bmatrix} = \begin{bmatrix} C_{xx} & C_{xy} \\ C_{yx} & C_{yy} \end{bmatrix} \begin{Bmatrix} \dot{x} \\ \dot{y} \end{Bmatrix} + \begin{bmatrix} k_{xx} & k_{xy} \\ k_{yx} & k_{yy} \end{bmatrix} \begin{Bmatrix} x \\ y \end{Bmatrix} \quad (5)$$

电磁轴承支承的转子上第 j 段受力方程可采用类似于滑动轴承转子系统的分析方法获得. 为计算方便,在计算过程中也将各物理量作量纲一处理[2],这些在此均不予赘述. 当转子的两端既不承受力,也不承受力矩时,边界条件可由下式确定

$$\begin{Bmatrix} M \\ F_{sh} \end{Bmatrix}_0^R = \begin{Bmatrix} M \\ F_{sh} \end{Bmatrix}_n^L = 0 \quad (6)$$

式中 M——边界轴段受到的弯矩
   F_{sh}——边界轴段受到的剪切力

上标 R,L 分别表示轴段的右侧与左侧,下标 0 和 n 表示转子两端轴段的编号.

最终得到的系统动力学基本方程也类似于滑动轴承转子系统

$$\boldsymbol{m}_t \{\ddot{x}\} + \boldsymbol{c}_t \{\dot{x}\} + \boldsymbol{k}_t \{x\} = 0 \quad (7)$$

式中 $\boldsymbol{m}_t$, $\boldsymbol{c}_t$, $\boldsymbol{k}_t$——系统的总质量矩阵、总阻尼系数矩阵及总刚度系数矩阵

3.2 特征行列式的建立

与传统转子动力学分析方法相似,电磁轴承转子系统的特征行列式的分析基础也是第 i 个轴段模型(详见文献[2,3,8,9,10]). 在该节点上,可求出系统的状态矢量传递矩阵 $\boldsymbol{A}_i$ 为

$$\boldsymbol{A}_i = \begin{bmatrix} 1 & l_i & b_i & -c_i & 0 & 0 & 0 & 0 \\ 0 & 1 & a_i & -b_i & 0 & 0 & 0 & 0 \\ 0 & J_i\gamma^2 & 1+a_iJ_i\gamma^2 & -l_i-b_iJ_i\gamma^2 & 0 & 0 & 0 & 0 \\ F_i & l_iF_i & b_iF_i & 1-c_iF_i & G_i & l_iG_i & b_iG_i & -c_iG_i \\ 0 & 0 & 0 & 0 & 1 & l_i & b_i & -c_i \\ 0 & 0 & 0 & 0 & 0 & 1 & a_i & -b_i \\ 0 & 0 & 0 & 0 & 0 & J_i\gamma^2 & 1+a_iJ_i\gamma^2 & -l_i-b_iJ_i\gamma^2 \\ H_i & l_iH_i & b_iH_i & -c_iH_i & k_i & l_ik_i & b_ik_i & 1-c_ik_i \end{bmatrix}$$

(8)

式中　γ——复特征值(复频率)

　　　l_i——轴段分段长度

　　　J_i——第 i 节轴段质量惯性矩

　　　$a_i = \dfrac{l_i}{EI_i}$　$b_i = \dfrac{l_i^2}{2EI_i}$　$c_i = \dfrac{l_i^3}{6EI_i}$

　　　E——轴段材料的弹性模量

　　　I_i——第 i 节轴段截面二次矩

　　　m_i——集中质量(包括轴段的及附加的)

　　　$F_i = m_i\gamma^2 + c_{xx}\gamma + k_{xx}$

　　　$G_i = c_{xy}\gamma + k_{xy}$

　　　$H_i = c_{yx}\gamma + k_{yx}$

　　　$k_i = m_i\gamma^2 + C_{yy}\gamma + k_{yy}$

　　　$i = 1 \sim n$ 是转子的分段数

由此特征行列式组构成的系统动力学分析方程组,可确定 γ 的各个复根,即电磁轴承转子系统的各个复固有频率.

3.3 实例设计与计算机仿真

透平机械是电磁轴承重要而广阔的应用领域之一. 本文用以验证上述理论分析研究结果的试验样机是为苏州某低温设备有限公司研制的一台 150 m³ 制氧透平膨胀机,其转子由 5 自由度电磁轴承系统完全悬浮支承. 在上述样机的系统转子动力学分析过程中,该转子等效处理为 49 个节点(即 $n=49$),转子各单元段结构参数和结构尺寸略. 电磁轴承系统控制器由实用 PID 电路、一阶惯性环节电路和 PID 电压电流功放电路组成. 按文献[6]的方法可求出系统各向同性系统的稳定参数.

由于各向同性系统稳定性较差,故通常在此基础上进行各向异性系统的调试. 表 1 是一组调试后获得的各向异性系统的稳定参数. 表 2 是应用表 1 所列稳定参数计算获得的一组

数值结果,图5是计算结果的曲线图.其中,图5(a)为临界转速变化曲线,图5(b)为系统刚度特性曲线,图5(c)为对数衰减率曲线,图5(d)则为系统阻尼特性曲线.

表1 调试后各通道不同(各向异性)的控制器参数表之一

参数名称 测量位置	积分时间常数 T_i(s)	微分时间常数 T_d(s)	比例环节增益 A_p(V/V)	微分环节常数 ε	信号转换增益 A_s(V/V)
前端水平	2.365	1.970×10^{-3}	3.064	0.082	6 790
前端垂直	2.595	8.650×10^{-4}	3.330	0.082	9 440
后端水平	2.190	1.940×10^{-3}	3.330	0.082	3 800
后端垂直	1.570	2.395×10^{-3}	2.419	0.082	6 560

表2 一组临界转速和对数衰减率计算结果

	1阶	2阶	3阶	4阶	5阶	6阶
n_{cr}	8 050	9 480	12 240	15 400	105 520	105 540
ζ	2.647	4.622	3.539	1.800	−0.005	0.001

注:n_{cr}——临界转速(r/min);ζ——对数衰减率.

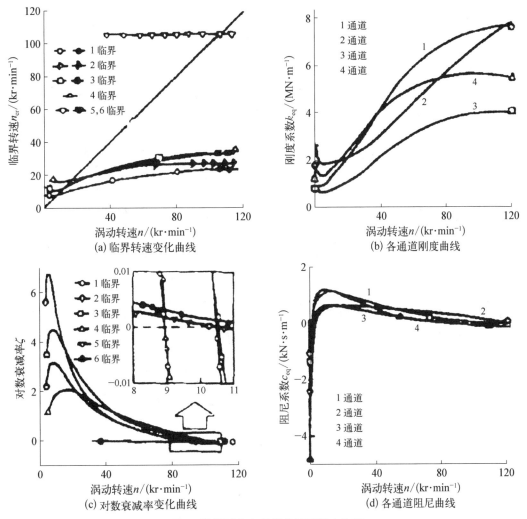

(a) 临界转速变化曲线
(b) 各通道刚度曲线
(c) 对数衰减率变化曲线
(d) 各通道阻尼曲线

图5 样机试验中的部分转子轴心轨迹

4 试验过程及现象分析

试验用制氧透平膨胀机样机的转子与轴承的有关参数及系统特性参数列于表 3. 该机转子的最高转速达 104 kr/min, 稳定试验转速 98 kr/min, 回转精度小于 $10\ \mu\text{m}/r$(r 为半径), 动刚度约为 $16\ \text{N}/\mu\text{m}$.

表 3 与试验有关的参数

转子质量 m	1.16 kg	线圈匝数 N_0	42/极
气隙宽度 x_0	0.17 mm	偏磁电流 I_0	1 A/极
磁极数	8	位移刚度 k_1	0.22 MN/m
气隙面积 S_0	132 μm^2	电流刚度 k_2	37.42 N/A

对于试验结果和在样机试验中遇到的一些问题和观察到的一些特殊现象,限于篇幅,本文仅对其中部分与转子动力学有关的结果进行分析讨论.

计算获得的系统固有频率和临界转速与试验结果基本吻合. 图 6 是试验中获得的转子在越过临界转速时的部分轨迹图,对照表 2 即可证实这一点. 由于在系统的临界转速上转子无法稳定运行,所以图中的结果实际上都是在略偏离临界转速时观测得到的.

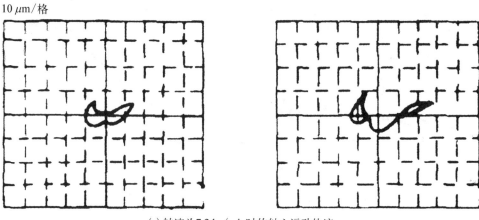

(a) 转速为7.2 kr/min时的轴心运动轨迹

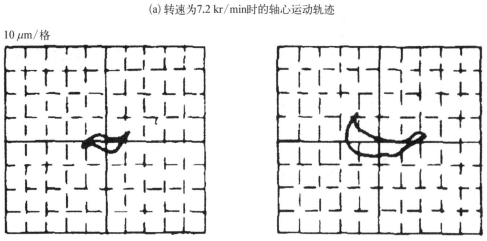

(b) 转速为14.6 kr/min时的轴心运动轨迹

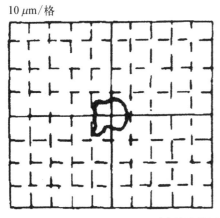
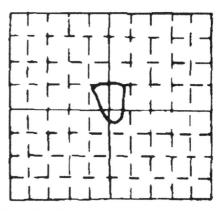

(c) 转速为 94 kr/min 时的轴心运行轨迹
(左图为前轴承,右图为后轴承)

图 6 样机试验中的部分转子轴心轨迹

分析上述计算结果可知,各向同性和各向异性系统的临界转速和稳定性之间存在很大的差别;而考虑交叉动特性系数与不考虑交叉动特性系数的各向异性轴承计算结果几乎无差别.这与理论分析结果是一致的.

图 5(a) 的 6 条固有频率变化曲线与 45°线都有一个交点,且均在计算转速范围内,因此系统有 6 个临界转速(见表 2),其中前 4 个集中在 10 kr/min 左右,为轴承主导型临界转速,后两个约 105 kr/min,为转子主导型临界转速.

在图 5(a) 中各对数衰减率曲线后段系统阻尼已变负,但曲线上只有与 45°线的交点为现实临界转速点,故开始变为不稳定的涡动频率与系统的稳定性无关,只有临界转速点处的对数衰减率才能判断对应模态的稳定性.可见前 4 个临界转速都是稳定的;而后 2 个临界转速已处于不稳定状态.

图 5(c) 中前 4 个对数衰减率曲线形状与图 5(d) 中对应的通道阻尼系数变化曲线大致相似,并于相近的频率范围段(对应转速 8~11 kr/min)过零参考线变负值.而 5、6 阶对数衰减率曲线变化很小,说明它们受轴承阻尼系数的影响很小;但过零的位置和前四阶的基本一样,说明电磁轴承转子系统各阶振动的稳定性与轴承阻尼直接相关.

综合分析计算结果,得出以下结论:不考虑径向和轴向振动的耦合,计算该电磁轴承转子系统的弯曲振动时,获得的轴承主导型临界转速的位置主要由系统控制器的特性决定,且均可安全越过;而转子主导型临界转速的位置无法通过改变轴承结构参数和控制器参数来明显改变,只有通过改变转子的几何尺寸才能达到此目的[9].有关试验的详细情况可参看有关文献.

5 结论

电磁轴承转子系统临界转速和稳定性的计算,可在控制系统稳定性理论分析方法获得系统稳定的控制参数后,根据各向同性系统预算电磁轴承的动特性系数和系统的临界转速;再根据经验值,估计各向异性的动特性系数,用以指导实际调试,然后由调试结果修正动特性系数,为确定实际控制器各向异性时的刚度、阻尼系数提供理论依据,为最终获得系统的临界转速分布和稳定性分析奠定基础. 在考虑系统交叉动特性系数时, 则应视系统的具体情况而定, 对转速较低的系统, 通常可以不计交叉刚度、阻尼系数的影响. 本试验观察到的临界转速与计算所得值相差在 10% 以内, 说明本文介绍的磁悬浮支承转子系统动力学分析方法是正确可行的, 它将为电磁轴承的工业应用提供有力的理论分析、计算和设计工具.

参 考 文 献

[1] 王海英. 电磁轴承—转子系统的稳定性分析: [硕士学位论文]. 西安: 西安交通大学, 1995.
[2] 万金贵. 电磁轴承转子系统的动力学特性分析: [硕士学位论文]. 上海: 上海大学, 1999.
[3] 唐钟麟, 冯志华, 黄晓蔚. 电磁轴承柔性转子系统分析方法. 机械工程学报, 1999, 35(2): 97-100.
[4] 王洪礼, 张新生, 刘勇, 等. 磁浮轴承的控制和动态过程研究. 机械工程学报, 1994, 30(6): 41-46.
[5] Chen H M, Wilson D. Stability analysis for rotors supported by active magnetic bearings. In: Proceedings of the 2nd International Symposium on Magnetic Bearings. 1990: 325-328.
[6] 汪希平. 电磁轴承系统的参数设计与应用研究: [博士学位论文]. 西安: 西安交通大学, 1994.
[7] 钟一谔, 何衍宗, 王正, 等. 转子动力学. 北京: 清华大学出版社, 1987.
[8] 张文. 转子动力学理论基础. 北京: 科学出版社, 1990.
[9] 徐龙祥. 高速旋转机械轴系动力学设计. 北京: 国防工业出版社, 1994.
[10] 汪希平, 崔卫东. 电磁轴承用非接触式位移传感器的研究. 上海大学学报(自然科学版), 1998, 4(1): 54-60.
[11] 汪希平. 电磁轴承系统的刚度阻尼特性分析. 应用力学学报, 1997, 14(3): 95-100.

Investigation on Dynamic Performance of Active Magnetic Bearing Rotor System

Abstract: When AMB is going to the investigation on its application, here have a lot of problems that must be solved. At first, one of the problems is the safety and stability or called as rotor dynamics. Rotor system with high rotation speed and specially supported is dealt with by active magnetic bearings, and studies its performance of rotor dynamics on the way of the combination traditional theory and method of rotor dynamics with the analyzing theory and design method of AMB system. Herein establishes a kind of method for investigating the AMB rotor dynamics on the basic theory of stiffness and damping of the system. As an application of the result on the analysis of above theory and method, a showpiece of 150 m^3 turbo oxygen gas expander is manufactured and experimented.

电磁轴承及其系统设计方法[*]

摘　要：电磁轴承是磁悬浮原理在机械领域中的一个应用实例，其工作机理是依靠电磁力使转子非接触地"支承"于轴承体内，有运动阻力小、无接触摩擦和磨损、功耗低以及寿命长等优点．国外在这方面已有成熟的产品应用于生产实际．介绍了电磁轴承及其系统设计的特点、系统控制器的设计方法、刚性转子系统的设计方法和柔性转子系统的动力学设计方法，并提供了几个实际设计的数据和部分试验结果．

0　前言

电磁轴承是新一代非接触支承部件，在国外已经开始进入工业应用阶段．在国内，有关电磁轴承的研究在不断升温后距工业应用仍有较大的差距．究其原因，主要是对电磁轴承的设计缺乏成熟和系统的理论基础．尽管国外在这方面的理论研究已获得了很大的成果，但详细描述设计理论和方法的文章不多见．

电磁轴承及其支承的转子本质上是一个多自由度运动体的位置控制系统．由于采用了电子控制器和位移传感器而使其设计过程涉及诸多的领域，应用于高速旋转机械时，还须考虑系统转子动力学问题；而作为一个机械支承部件，进行相关的机械原理及设计方面的工作也是必需的．

本文简要介绍有关电磁轴承转子系统的特点和设计方法，说明在电磁轴承设计过程中的一些常用步骤，期望为分析设计电磁轴承转子系统提供较为系统的理论和方法．限于篇幅，本文各节仅扼要其思想，详细的介绍可参考文后的相关文献．

1　电磁轴承及其转子系统方程

稳定地悬浮运转着的转子需对其5自由度的运动（绕转轴的转动除外）实施有效的控制，通常称为5自由度磁悬浮系统．由于其轴向控制与径向控制间的耦合影响可以通过工艺和技术措施得到明显的改善，故在建立此类系统的方程时，常常不考虑轴向的影响，这样做在一般工况下是可行的[1]．

设电磁轴承支承的转子在空间的坐标由图1所示．转子是刚性的，不发生几何变形．图中：O为转子的质心点位置；ω_z是绕转子轴线转动的角速度；

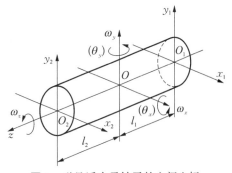

图1　磁悬浮支承转子的空间坐标

---

[*] 本文合作者：汪希平、章东义、张纲、朱礼进．原发表于《机械工程学报》，2002，38(5)：1-6．

ω_x 和 ω_y 分别对应过质心的两个坐标轴方向上的角速度;θ_x 和 θ_y 则为其对应的角度值. O_1 和 O_2 分别是两个径向轴承的支承点,电磁轴承产生的恢复力分别沿 O_1 和 O_2 为中心的 x_1、y_1 和 x_2、y_2 方向.

$K_x = J_z\omega_z \times \omega_x$ 和 $K_y = J_z\omega_z \times \omega_y$ 为陀螺力矩. 而作用在轴承座上的陀螺压力为

$$\begin{cases} p_{kx} = \dfrac{J_z\omega_z\omega_x}{l_1+l_2} \\ p_{ky} = \dfrac{J_z\omega_z\omega_y}{l_1+l_2} \end{cases} \tag{1}$$

考虑陀螺效应时,电磁轴承转子系统的特性可由下述方程描述

$$\begin{cases} mx_1'' = F_1 - F_3 + f_{x1} \\ mx_2'' = F_5 - F_7 + f_{x2} \\ my_1'' = F_2 - F_4 + \dfrac{mg}{2} + f_{y1} \\ my_2'' = F_6 - F_8 + \dfrac{mg}{2} + f_{y2} \\ J_x\theta_x'' = (F_2 - F_4)l_1 - (F_6 - F_8)l_2 - J_z\omega_z\theta_y' + f_{y1}l_1 - f_{y2}l_2 \\ J_y\theta_y'' = -(F_1 - F_3)l_1 + (F_5 - F_7)l_2 + J_z\omega_z\theta_x' - f_{x1}l_1 + f_{x2}l_2 \end{cases} \tag{2}$$

$$\begin{cases} F_1 - F_3 = \dfrac{\mu_0 S_0 N^2}{4}\left[\left(\dfrac{I_0 - i_{x1}}{\delta_0 - x_1}\right)^2 - \left(\dfrac{I_0 + i_{x1}}{\delta_0 + x_1}\right)^2\right] \\ F_5 - F_7 = \dfrac{\mu_0 S_0 N^2}{4}\left[\left(\dfrac{I_0 - i_{x2}}{\delta_0 - x_2}\right)^2 - \left(\dfrac{I_0 + i_{x2}}{\delta_0 + x_2}\right)^2\right] \\ F_2 - F_4 = \dfrac{\mu_0 S_0 N^2}{4}\left[\left(\dfrac{I_0 - i_{y1}}{\delta_0 - y_1}\right)^2 - \left(\dfrac{I_0 + i_{y1}}{\delta_0 + y_1}\right)^2\right] \\ F_6 - F_8 = \dfrac{\mu_0 S_0 N^2}{4}\left[\left(\dfrac{I_0 - i_{y2}}{\delta_0 - y_2}\right)^2 - \left(\dfrac{I_0 + i_{y2}}{\delta_0 + y_2}\right)^2\right] \end{cases} \tag{3}$$

式中 $F_1 \sim F_8$——两个径向电磁轴承中 4 自由度方向上的 8 个电磁力,其中 1、3 为 x_1 坐标轴上的电磁力;2、4 为 y_1 坐标轴上的电磁力;5、7 是 x_2 坐标轴上的电磁力;6、8 是 y_2 坐标轴中的电磁力. 标号小的电磁力对应坐标轴的正方向

x_1,x_2,y_1,y_2——对应于各坐标轴上转子的位移量

f_{x1},f_{x2},f_{y1},f_{y2}——两个径向电磁轴承中对应坐标轴方向上的外干扰力,并规定以坐标轴方向为力的正方向

m——每个轴承支承的转子质量. 这里设两个径向轴承各支承转子重量的一半

g——重力加速度

J_x,J_y,J_z——转子绕 3 个坐标轴转动的转动惯量,由于其对称性,故有 $J_x = J_y$

l_1,l_2——两个径向电磁轴承支承点到转子质心的距离

ω_z——转子转动的角速度

N——每磁极上线圈的匝数

I_0——每磁极偏磁电流

δ_0——转子平衡时半径气隙宽度

式(2)中,前面 4 个是转子的质点运动方程,后面两个是转子的刚体转动方程. 将方程线性化,并令所有外干扰力为零,则方程为

$$
\begin{cases}
x_1'' = \dfrac{1}{m}(C_1 x_1 - C_2 i_{x1}) \\
x_2'' = \dfrac{1}{m}(C_1 x_2 - C_2 i_{x2}) \\
y_1'' = \dfrac{1}{m}(C_1 y_1 - C_2 i_{y1}) \\
y_2'' = \dfrac{1}{m}(C_1 y_2 - C_2 i_{y2}) \\
\theta_x'' = \dfrac{1}{J_x}[(C_1' y_1 - C_2' i_{y1})l_1 - (C_1' y_2 - C_2' i_{y2})l_2] - \dfrac{J_z \omega_z}{J_x}\theta_y' \\
\theta_y'' = \dfrac{1}{J_y}[-(C_1' x_1 - C_2' i_{x1})l_1 + (C_1' x_2 - C_2' i_{x2})l_2] + \dfrac{J_z \omega_z}{J_y}\theta_x'
\end{cases}
\tag{4}
$$

式(4)是研究、分析和设计电磁轴承及其转子系统的基础. 式中,$C_1' = C_1/m$,$C_2' = C_2/m$. C_1 和 C_2 分别为系统的位移刚度系数和电流刚度系数,是电磁轴承系统的固有系数[2].

2 系统控制器的设计方法

由于转动方程中包含了 4 个通道间的耦合信号,设计过程将较为复杂. 当系统控制器结构完全对称时,设计过程可分为两个部分:即单自由度磁悬浮系统的设计与稳定性分析方法,包括代数法[3]、根轨迹法[4]等,以及系统的陀螺效应耦合设计方法. 前者是设计一个电磁轴承系统必不可少的,而后者一般在特殊情况下进行.

将系统的质点运动方程进行拉普拉斯变换,得系统的线性传递函数方程组

$$
\begin{cases}
s^2 X_1(s) = \dfrac{1}{m}[C_1 X_1(s) - C_2 I_{x1}(s)] \\
s^2 X_2(s) = \dfrac{1}{m}[C_1 X_2(s) - C_2 I_{x2}(s)] \\
s^2 Y_1(s) = \dfrac{1}{m}[C_1 Y_1(s) - C_2 I_{y1}(s)] \\
s^2 Y_2(s) = \dfrac{1}{m}[C_1 Y_2(s) - C_2 I_{y2}(s)]
\end{cases}
\tag{5}
$$

这是 4 个独立的方程,由此设计得到的是独立控制方式的磁悬浮系统. 取其中任一个方程作为分析对象,构成单自由度磁悬浮系统,其闭环控制框图如图 2 所示. 设系统控制环节的传递函数为 $G(s)$,则

$$
\begin{cases}
G_0(s) = \dfrac{1}{ms^2 - C_1} \\
H(s) = C_2 G(s)
\end{cases}
\tag{6}
$$

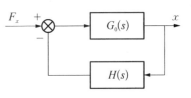

图 2 单自由度磁悬浮系统的闭环控制框图

按照控制系统的分析方法,系统的闭环传递函数为

$$G_c(s) = \frac{G_0(s)}{1+G_0(s)H(s)} = \frac{G_0(s)}{1+C_2G_0(s)G(s)} \tag{7}$$

由此方程可确定系统达到基本稳定所必需的参数设置范围,当系统采用 PID 控制器时,需设置的参数有控制器的比例环节增益系数 K_p、微分环节时间常数 T_d、一阶惯性环节的时间常数 t_d 以及积分环节的时间常数 T_i[2,3]. 其中时间常数 t_d 决定了系统控制频率的带宽. 而对系统的根轨迹进行系统参数的预设计也是可行的方法之一[4].

3 系统的重要技术指标与分析

电磁轴承设计的大部分工作是在非机械领域中进行的,作为一种机械支承部件,应用时必须考虑其机械特性与相关的技术指标,其中较为重要的有系统的刚度阻尼指标、承载能力、回转精度指标等. 作为一个控制系统,还有其他一些指标,如控制器的响应速度、传感器的灵敏度以及功率放大器的特性以及系统的稳定性等. 这里仅介绍部分机械指标的分析与设计方法. 其余内容可参看文献[2].

3.1 电磁轴承的刚度与阻尼特性

仔细分析式(7)可以看出,它与刚度有同样的量纲. 按照电子学中的方法绘出其幅频特性曲线,即俗称的"浴盆"曲线[5]. 实际上,这个曲线对系统的动力学特性分析是毫无意义的. 但由此方法可以获得系统的等效刚度系数和等效阻尼系数分别为[5]

$$K_e = C_2 \text{Re}\{G(j\omega)\} - C_1 \tag{8}$$

而电磁轴承的等效阻尼系数为

$$C_e = C_2 \frac{\text{Im}\{G(j\omega)\}}{\omega} \tag{9}$$

式中 $\text{Re}\{G(j\omega)\}$,$\text{Im}\{G(j\omega)\}$——控制器传递函数 $G(j\omega)$ 的实部和虚部

式(8)、(9)是分析电磁轴承支承条件下的转子系统动力学特性的有效方法之一[5].

对于电磁轴承的刚度阻尼系数,通常都会像考虑滚动轴承和滑动轴承那样是越大越好,但分析研究以及试验的结果说明并非如此. 对于一个具体的电磁轴承转子系统,有一个最优的刚度和阻尼特性系数,这时系统获得的动力学特性可以达到最优. 这一点正在得到人们的认可.

3.2 电磁轴承的承载能力

承载能力 p_g 是电磁轴承非常重要的一个技术指标. 通常在轴承具体设计前,可以按照其有效几何投影面积 S' 估算,即

$$p_g = S' p_{w,p}$$

式中 p_g——比承载力,约为 0.3~0.5 MPa

当一个电磁轴承系统设计完成后,则可按下式估算

$$p_g = C_2 I_{cmax}$$

式中 I_{cmax}——系统可以提供的最大控制电流

3.3 回转精度

对于机床主轴等机械,其回转精度也是主要的技术指标之一. 电磁轴承支承下的转子回转精度与诸多的因素有关,对此问题作精确的定量分析尚有一定的难度. 据实践经验,回转精度与电磁轴承的设计工作气隙有较密切的联系. 通常转子中心回转轨迹的半径不会大于轴承平衡时气隙的十分之一. 例如,当轴承的标称气隙半径为 0.2 mm 时,其转子中心的回转轨迹半径一般不会大于 0.02 mm.

4 柔性转子设计方法及其特点

上述分析过程均未考虑转子的特征,即均将转子视为刚性体. 工程上实际应用的多为柔性转子,当转速达到一定数值时,其固有频率、系统稳定性、系统何时失稳乃至陀螺效应等问题都会显露出来. 因此,如何对此类问题进行分析和必要的设计也是电磁轴承能否成功应用于工程实际的重要环节.

4.1 磁悬浮转子系统动力学的分析与设计

这里采用的主要分析设计思想是将传统的转子动力学理论分析方法和电磁轴承系统的刚度阻尼特性分析方法结合[6],作为磁悬浮支承转子系统动力学分析的理论基础.

一般的步骤如下:

将轴承转子系统描述为具有 n 自由度的运动方程,方程中包含有转子的质量矩阵 m、轴承的刚度阻尼矩阵 k 和 c.

计算系统的特征值与特征矢量,求得系统的固有频率和对应的对数衰减率,由此分析系统的稳定性.

上述计算方法和过程有许多的文献可以查阅,在此不予赘述. 这里仅简要介绍如何获得上述计算必需的电磁轴承系统刚度阻尼特性系数,包括 4 个刚度系数 k_{xx}、k_{xy}、k_{yx}、k_{yy} 和 4 个阻尼系数是 c_{xx}、c_{xy}、c_{yx}、c_{yy}. 这是解决这一问题的关键所在.

式(8)、(9)是计算系数 k_{xx}、k_{yy}、c_{xx} 和 c_{yy} 的依据,而 k_{xy}、k_{yx}、c_{xy} 和 c_{yx} 是否存在取决于位移传感器安装形式[7,8],其数值可由下式决定

$$k_{xy} = \pm \frac{R}{2\delta_0} C_2 \text{Re}\{G(j\omega)\} \quad \begin{cases} \text{当 } x < 0, \text{取 "+"} \\ \text{当 } x \geq 0, \text{取 "−"} \end{cases}$$

$$k_{yx} = \pm \frac{R}{2\delta_0} C_2 \text{Re}\{G(j\omega)\} \quad \begin{cases} \text{当 } y < 0, \text{取 "−"} \\ \text{当 } y \geq 0, \text{取 "+"} \end{cases}$$

$$c_{xy} = \pm \frac{R}{2\delta_0} \frac{C_2 \text{Im}\{G(j\omega)\}}{\omega} \quad \begin{cases} \text{当 } x < 0, \text{取 "+"} \\ \text{当 } x \geq 0, \text{取 "−"} \end{cases}$$

$$c_{yx} = \pm \frac{R}{2\delta_0} \frac{C_2 \text{Im}\{G(j\omega)\}}{\omega} \quad \begin{cases} \text{当 } y < 0, \text{取 "−"} \\ \text{当 } y \geq 0, \text{取 "+"} \end{cases}$$

式中 R——转子在传感器测量面处的半径

ω——转子的涡动频率

滑动轴承的刚度阻尼特性系数是转子角速度的函数,而对于电磁轴承转子系统,其刚度阻尼特性系数是转子涡动频率的函数,这是两者的重大区别.另外,转子涡动频率的无法预知使分析计算变得复杂些.最终计算得到的是如图3所示的系统固有频率曲线及其对应的对数衰减率曲线(图4),它们都是转子涡动频率的函数.特别需要说明的是,与滑动轴承不同,电磁轴承转子系统的固有频率曲线上,仅有那些与45°对角线相交的点,才是系统真实的固有频率,对应的转速即为系统的临界转速,而此时系统是否稳定则取决于其对应的对数衰减率数值的大小与符号.

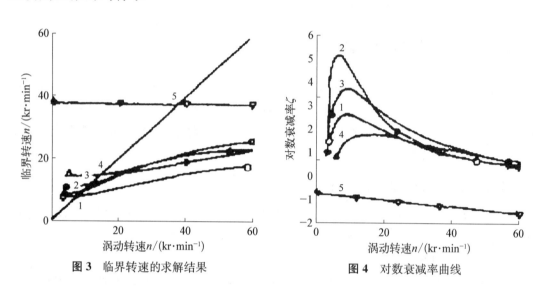

图3 临界转速的求解结果　　　　图4 对数衰减率曲线

4.2 陀螺效应对系统分析设计的影响

由式(4)即可确定陀螺效应对电磁轴承系统的影响[9].经过对式(4)的分析简化,考虑陀螺效应时的电磁轴承转子系统可由图5中的传递函数框图表示.图中的系数 K 即系统的陀螺效应系数,标志着系统陀螺效应的大小,同时也给控制器解耦陀螺效应设计提供了理论依据.陀螺效应系数可由下式确定

$$K=\frac{J_z\omega_z}{J_r}$$

式中 J_z——轴转动惯量(kg·m²)

J_r——极转动惯量(kg·m²)

ω_z——转动角速度(rad/s)

图5中的 θ_r 为控制器设定的转子转动的角度值,且系数 $a=C_1/m$.

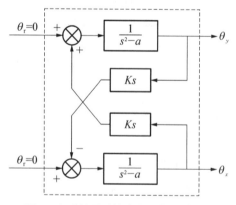

图5 电磁轴承系统中的陀螺效应框图

5 设计实例简介

应用上述分析设计理论和方法,先后对数台工业用样机及实验室试验台进行了设计,其中包括一个磁悬浮支承转子系统试验台[2]、一个航空发动机模型试验台、一个150 m³透平制氧膨胀机样机[10]以及目前正在组装的高速磁悬浮支承磨床主轴. 下面简要介绍设计的过程.

根据前述方法,对电磁轴承转子系统首先进行控制器参数的稳定域计算,得到的是各向同性的系统参数,最终系统需通过调试达到各向异性. 表1是150 m³透平制氧膨胀机样机调试后获得的各向异性控制器参数之一. 使用这组参数时,样机的转速可达98 kr/min.

表1 调试后各通道各向异性的控制器参数之一

测量位置\参数	T_i	T_d	A_p	ε	A_s	T_s
前端水平	2.365	1.970×10⁻³	3.064	0.082	6 790	
前端垂直	2.595	8.650×10⁻⁴	3.330	0.082	9 440	3.18×10⁻⁵
后端水平	2.190	1.940×10⁻³	3.330	0.082	3 800	
后端垂直	1.570	2.395×10⁻³	2.419	0.082	6 560	

注:T_i——积分器时间常数(s);T_d——微分器时间常数(s);A_p——控制器增益系数(V/V);ε——微分器惯性时间常数(s);A_s——位移传感器增益系数(V/V);T_s——位移传感器惯性时间常数(s).

由于该样机的工作转速很高,因此再按上述方法进行轴承转子系统的动力学特性分析,以了解系统在高速运行时的稳定性. 表2是其计算结果. 从表中可以看出,系统存在105 kr/min左右的失稳转速. 这一点在试验中已得到了验证. 分析计算表明,这一失稳转速只能通过改变转子的几何结构改变之,因此属于转子主导型临界转速.

表2 一组临界转速和对数衰减率计算结果

	一阶	二阶	三阶	四阶	五阶	六阶
临界转速 n/(kr·min⁻¹)	8.05	9.48	12.24	15.40	105.52	105.54
对数衰减率 ζ	2.647	4.622	3.539	1.800	−0.004 8	0.001

第三步是对系统的陀螺效应进行分析计算,以确定是否需要进行陀螺效应控制补偿. 表3是150 m³透平制氧膨胀机样机和航空发动机模型试验台的轴承转子系统陀螺效应分析的结果. 表中,A、B分别为150 m³透平制氧膨胀机样机和航空发动机试验台的转子. 由上述结果可以看到,尽管透平制氧膨胀机转子的转速高达108 kr/min,但其陀螺效应是可以忽略不计的.

表3 陀螺效应的计算实例

	J_z	J_r	ω_z	K
A	5.244×10⁻⁵	10.170 520	11 309.733	0.058 317 6
B	9.818×10¹¹	3.997×10¹⁰	2 094.395 1	51 447.654

6 结论

电磁轴承支承转子系统的设计是一个复杂的过程,其中主要的可分为:以轴承几何与电气参数为主的系统控制器的参数设计、以转子结构和轴承动特性系数为主的动力学性能设计以及特定条件下的陀螺效应解耦设计.除此之外,一个完整的电磁轴承转子系统的设计还包括必要的机械设计、磁路与电磁场设计以及电子电路的设计等.一个稳定运行的电磁轴承转子系统除了精确地设计外,系统的调试环节也是至关紧要的,有时甚至对成功与否起到决定性的作用.

上述设计思想和方法经实践证明是可行的.同时,证实了电磁轴承与滑动轴承类似,其动力学特性也可用8个等效刚度系数和等效阻尼系数来表示,上述工作为全面开展电磁轴承支承条件下的转子动力学问题研究和在工程中推广应用电磁轴承奠定了理论基础.本文未涉及数字控制系统的设计方法.

参 考 文 献

[1] 汪希平,张直明,于良,等.轴向磁悬浮轴承的力学特性分析.应用力学学报,2000,17(3):29-34.
[2] 汪希平.电磁轴承系统的参数设计与应用研究:[博士学位论文].西安:西安交通大学,1994.
[3] 汪希平,袁崇军,谢友柏.电磁轴承系统控制参数与稳定性的关系分析.机械工程学报,1996,32(3):65-69.
[4] 汪希平,崔卫东,刘令.经典控制理论在电磁轴承系统控制中的应用.控制理论与应用,1997,14(6):837-841.
[5] 汪希平.电磁轴承系统的刚度阻尼特性分析.应用力学学报,1997,14(3):95-100.
[6] 万金贵.电磁轴承转子系统的动力学特性分析:[硕士学位论文].上海:上海大学,1999.
[7] 汪希平,崔卫东.电磁轴承用非接触式位移传感器的研究.上海大学学报(自然科学版),1998,4(1):54-60.
[8] 汪希平,万金贵.轴向磁悬浮轴承用非接触式差动电感位移传感器的实验研究.仪器仪表学报,1998,19(6):615-619.
[9] 汪希平,陈学军.陀螺效应对电磁轴承系统设计的影响.机械工程学报,2001,37(4):48-52.
[10] 汪希平,张直明,于良,等.电磁轴承在透平制氧膨胀机中的应用研究进展.中国机械工程,2000,11(4):379-381.

Active Magnetic Bearing and Its Systematic Design Methods

Abstract:Active magnetic bearing (AMB) is as an application example of using magnetic levitates theory in mechanical domain. Its working principle involves contactless support of the shaft by the bearings through electrical magnetic force. So, AMB has some good features, such as no touch friction and wear, small motion resistance, lower power supply and long utility time. Overseas, AMB technology has reached mature level and increasingly products, which set up with AMB, can be found. The ways of design and process AMB rotor system are described. Here a number of design methods of AMB are dealt with, which mainly include the features in the systematic design, methods of design for the system controller, rigid shaft system and flexible shaft system. Some numeric results are also presented.

周隙密封紊流润滑研究*

摘　要：用先进的复合型紊流润滑理论研究流场复杂的周隙密封紊流润滑性能,提出了紊流润滑参数数据库的方法,解决了复合型紊流润滑理论求解复杂、烦琐的难点.研究表明,复合型紊流润滑理论能很好地描述周隙密封的润滑性能,数值计算结果与实验数据吻合.

周隙密封被广泛应用于各类旋转机械中,具有较大的相对间隙,随着机器向高速、大功率方向发展,周隙密封间隙内的流动多为紊流,其动态特性对高速转子的稳定性有重要影响.其流态十分复杂,有周向旋转运动产生的剪切流,径向偏心产生的周向压力流,还有沿密封轴向的压差产生的轴向压力流,这三种不同性质的流动构成了周隙密封的复合流场.显而易见,再用通常的层流状态下的流体润滑理论来描述周隙密封的特性已不能满足实际工程需要.对紊流工况下的周隙密封必须以紊流润滑理论进行分析,紊流润滑理论模式的先进性对密封性能的计算精度有着直接影响.

已有的紊流润滑理论或以剪切流主导为推导依据(如 Constantinescu[4],Ng-Pan[2]等)仅适用于高速轻载工况;或以有限范围内的经验公式为依据(如 Hirs[3])缺乏推广应用的坚实依据并已被证明为具有理论缺陷;或以紊流能量为依据(如 k-ε 理论[4]等),但单一模式难以同时计入润滑油膜的各个特点.经研究发现,代数雷诺应力模式能很好地反映润滑膜中紊流核心部分的流动特性,并能计入壁面对其紊流强度的抑制作用,而低紊流雷诺数 k-ε 理论能反映近壁区的流动特性及黏性效应.考虑到单一模式很难同时计入润滑油膜必有的各个特点,作者提出了复合型紊流润滑理论[5],来更准确地描述润滑流场.在本文中,即采用复合型紊流润滑理论研究流场复杂的周隙密封的紊流性能.

1　理论分析

1.1　复合型紊流润滑理论

复合型紊流润滑理论的主要思想是在近壁区采用低紊流雷诺数 k-ε 方程,在紊流核心区采用代数雷诺应力模式,根据基础性实验数据[6,7],提出近壁区和核心区的划分以紊流切应力占总切应力的 95％ 来确定,同时保证在分界面上剪应力和速度场连续.

1.1.1　适用于近壁区的 k-ε 模式方程(不计 Dk/Dt,$D\varepsilon/Dt$ 项)

$$\frac{\partial}{\partial y}\left[\left(\nu+\frac{\nu_t}{\sigma_k}\right)\frac{\partial k}{\partial y}\right]+\left[\left(\frac{\partial u}{\partial y}\right)^2+\left(\frac{\partial w}{\partial y}\right)^2\right]-\varepsilon-\frac{2\nu k}{y^2}=0$$
$$\frac{\partial}{\partial x}\left[\left(\nu+\frac{\nu_t}{\sigma_\varepsilon}\right)\frac{\partial \varepsilon}{\partial y}\right]+C_{\varepsilon 1}\nu_t\left[\left(\frac{\partial u}{\partial y}\right)^2+\left(\frac{\partial w}{\partial y}\right)^2\right]\frac{\varepsilon}{k} \tag{1}$$

---

* 本文合作者：王小静、孙美丽. 原发表于《太原重型机械学院学报》,2002,23(增刊):7-10.

$$-C_{\varepsilon 2}[1-0.3\exp(R_k^2)]\frac{\varepsilon^2}{k}-2\nu\left[\frac{\partial\varepsilon^{\frac{1}{2}}}{\partial y}\right]=0 \tag{2}$$

其中：$\nu_t=C_\mu\dfrac{k^2}{\varepsilon}=C_m[1-\exp(-A_mR_k)]\dfrac{k^2}{\varepsilon}$，$R_k=\dfrac{k^2}{\nu\varepsilon}$，$C_m=0.09$，$C_\mu=0.09$，$\sigma_k=1.0$，$\sigma_\varepsilon=1.3$，$C_{\varepsilon 1}=1.45$，$C_{\varepsilon 2}=2.0$，$A_m=0.0015$

1.1.2 适用于紊流核心区的代数雷诺应力方程（不计 Dk/Dt，$D\varepsilon/Dt$ 项）

$$\frac{\partial}{\partial y}\left(\frac{\nu_t}{\sigma_k}\frac{\partial k}{\partial y}\right)+p_d-\varepsilon=0 \tag{3}$$

$$\frac{\partial}{\partial y}\left(\frac{\nu_t}{\sigma_\varepsilon}\frac{\partial\varepsilon}{\partial y}\right)+C_{\varepsilon 1}p_d\frac{\varepsilon}{k}-C_{\varepsilon 2}\frac{\varepsilon^2}{k}=0 \tag{4}$$

其中：

$$\nu_t=0.09G_\mu\frac{k^2}{\varepsilon},\quad p_d=\nu_t\left[\left(\frac{\partial u}{\partial y}\right)^2+\left(\frac{\partial w}{\partial y}\right)^2\right]$$

$$G_\mu=\frac{1+\dfrac{3}{2}\dfrac{C_2C_2}{1-C_2}f}{1+\dfrac{3}{2}\dfrac{C_1}{C_1}f}\cdot\frac{1-2\dfrac{C_2C_2}{C_1-1+C_2}\dfrac{p_d}{\varepsilon}f}{1+2\dfrac{C_1}{C_1+\dfrac{p_d}{\varepsilon}-1}f}$$

$C_2=0.3$，$C_{\varepsilon 1}=1.44$，$C_{\varepsilon 2}=1.92$，$\sigma_k=1.0$，

$\sigma_k=1.3$，$f=\dfrac{k^{\frac{3}{2}}}{C_w y\varepsilon}$，$C_w=3.72$，

$C_1=1.8$，$C_2=0.6$，$C_1=0.6$

作者将复合型紊流润滑理论与国际上通用紊流润滑理论从理论基础完备性以及与基础性实验的吻合性进行比较[5]，研究表明，复合型紊流润滑理论既考虑了固壁效应的影响，又考虑近壁区黏性效应的作用，能更全面地计入润滑膜的各个特点，适用于流动复杂的紊流流场. 无论在纯剪切流和有压力梯度的复杂流场下，在不同雷诺数、不同工况参数下，复合型紊流润滑理论模式都能准确分析流场，与其他紊流模式相比，数值结果与实验数据最为吻合. 在低雷诺数甚至层流工况下，复合型紊流润滑模式由于理论基础的坚实性，仍能很好适用.

1.2 周隙密封紊流雷诺方程

由 Navier-Stokes 方程和连续方程出发，并采用 Boussinesq 假设，推导得到用以描述流场压力分布的无量纲紊流雷诺方程由 Navier-Stokes 方程和连续方程出发，并采用 Boussinesq 假设，推导得到用以描述流场压力分布的无量纲紊流雷诺方程：

$$\frac{\partial}{\partial \varphi}\left[\bar{G}H^3\frac{\partial \bar{p}}{\partial \varphi}\right]+\left(\frac{D}{L}\right)^2\frac{\partial}{\partial \bar{z}}\left[\bar{G}H^3\frac{\partial \bar{p}}{\partial \bar{z}}\right]=\frac{\partial}{\partial \varphi}(\bar{F}H) \tag{5}$$

其中：$\bar{G}=\frac{\bar{A}_2}{\bar{A}_1}\bar{B}_1-\bar{B}_2$，$\bar{F}=1-\frac{\bar{B}_1}{\bar{A}_1}$，$\bar{A}_1=\int_0^1\frac{1}{1+\bar{\nu}_t}\mathrm{d}\bar{y}$

$\bar{A}_2=\int_0^1\frac{\bar{y}}{1+\bar{\nu}_t}\mathrm{d}\bar{y}$，$\bar{B}_1=\int_0^1\int_0^{\bar{y}}\frac{1}{1+\bar{\nu}_t}\mathrm{d}\bar{y}\mathrm{d}\bar{y}$

$\bar{B}_2=\int_0^1\int_0^{\bar{y}}\frac{\bar{y}}{1+\bar{\nu}_t}\mathrm{d}\bar{y}\mathrm{d}\bar{y}$

在密封的轴向进油边,流体在进入密封间隙时假设受到瞬间轴向加速,进入密封间隙的流体的压力因此小于进入密封之前的压力. 两者之间的压降值 Δp 依决于 Bernoulli 方程：

$$\Delta p=\frac{1}{2}(1+\zeta)\rho[(\alpha_w w_m)^2+(\alpha_u u_m)^2]_{z=0} \tag{6}$$

其中,入口损失系数 ζ 设为 0.45；系数 w_m 和 u_m 为油膜厚度方向轴向和周向的平均速度. 轴向和周向的预旋因子 α_w 和 α_u 分别假设为 0.9 和 0.1.

1.3 数值计算

在求解周隙密封紊流润滑性能时,需求解紊流雷诺方程. 紊流雷诺方程中包含紊流运动黏性系数 ν_t,而 ν_t 值是由紊流润滑理论模式所决定的. 复合型紊流润滑理论虽然很先进,但求解十分复杂、费时,需将两种紊流模式根据当地剪切应力的大小分别计算紊动能和涡扩散系数,并且互相耦合,具有强烈的非线性效应. 求解紊流雷诺方程时,每一个迭代步骤的每个节点上都需计算 ν_t 的沿膜厚方向的分布,整个计算过程十分繁复,且收敛性较差. 如何在保持优良理论精度的前提下使该紊流理论应用于工程实际,是急需解决的一个关键问题. 作者提出了建立紊流润滑参数数据库来解决这一难题. 以相似理论为指导,在雷诺数、周向压力梯度和轴向压力梯度这三个决定性参数的实用范围内,建立复合型紊流润滑理论模式中各个积分参数的实用数据库,从而使计算过程大大简化而计算精度得以保证.

2 结果与讨论

用复合型紊流润滑理论对周隙密封的紊流润滑性能进行研究,主要针对两个周隙密封紊流实验的实验数据进行对比.

2.1 与 Marquette, Childs and Andres[8] 实验数据考核

Marquette, Childs and Andres 对周隙密封紊流润滑性能进行了实验研究,测得了一系列在不同转速、不同偏心和轴向压差下的实验数据. 实验的具体参数如下：

直径 $D=76.29$ mm　　长度 $L=34.93$ mm

半径间隙 $C=0.11$ mm

图1～图3为不同工况参数下密封侧泄量随偏心率的变化情况. 图1的工况为转速

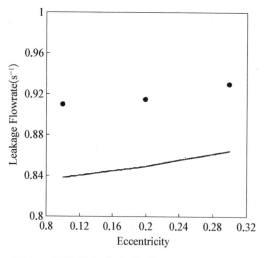

图1 侧泄量与偏心率关系(17 400 r/min, 5.52 MPa)

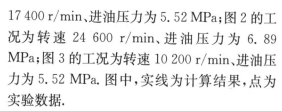

17 400 r/min、进油压力为 5.52 MPa;图2的工况为转速 24 600 r/min、进油压力为 6.89 MPa;图3的工况为转速 10 200 r/min、进油压力为 5.52 MPa.图中,实线为计算结果,点为实验数据.

图1～图3分别为不同转速和进油压力的侧泄量与偏心率的关系曲线.主轴转速较高,轴向压差较大,周隙密封油膜间隙内的流场情况十分复杂,有大的轴向压力流动和大的周向剪切流动,雷诺数分别为 2 350～5 670,流场处于完全紊流状态.从图1～图3的比较情况可以看出,用复合型紊流润滑理论的计算结果与实验数据能很好吻合.

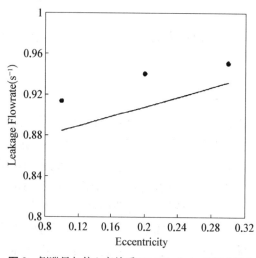

图2 侧泄量与偏心率关系(24 600 r/min, 6.89 MPa)

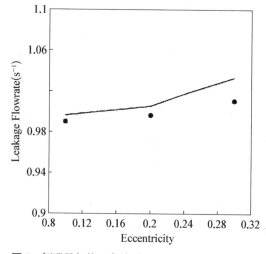

图3 侧泄量与偏心率关系(10 200 r/min, 5.52 MPa)

2.2 与上海大学周隙密封实验台实验数据对比

上海大学周隙密封实验台的基本参数如下:
直径 $D=99.15$ mm　　长度 $L=70$ mm
半径间隙 $C=0.435$ mm
运动黏度 $\gamma=8.31\times10^{-6}$ m^2/s(16℃)
　　　　 $\gamma=4.32\times10^{-6}$ m^2/s(40℃)

图4～图7分别为不同转速、不同环境温度、不同进油压力下密封内油膜压力沿密封轴向的变化情况(从密封上方顺时针 90°位置),实线为计算结果、点为实验结果.

上述计算结果与上海大学周隙密封实验台测得的数据吻合得非常好.由于此时工况的雷诺数较小,复合型紊流润滑理论仍能很好地描述周隙密封的性能.且在密封轴向进油边的压力与实验结果很好吻合,表明理论模型能很好预计流体进入周隙密封油膜间隙时产生的压降和预旋效应.

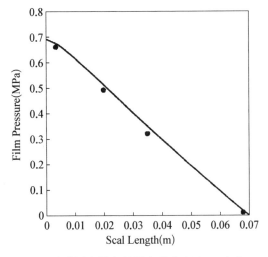

图 4 油膜压力沿密封轴向分布(1 200 r/min, 12℃,0.763 MPa)

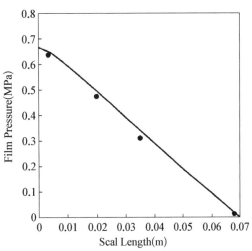

图 5 油膜压力沿密封轴向分布(1 800 r/min, 12℃,0.733 MPa)

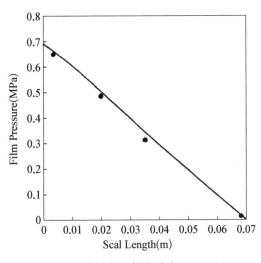

图 6 油膜压力沿密封轴向分布(2 400 r/min, 10.5℃,0.747 MPa)

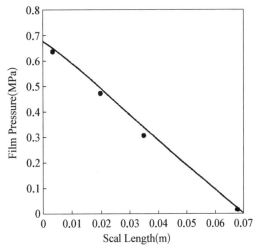

图 7 油膜压力沿密封轴向分布(3 000 r/min, 11℃,0.723 MPa)

3 结论

(1) 将先进的复合型紊流润滑理论应用于周隙密封的性能研究,从实验数据的吻合性来看,复合型紊流润滑理论能很好地描述周隙密封的复杂流场,为周隙密封的设计提供了有力的手段.

(2) 研究结果再一次验证了复合型紊流润滑理论的先进性,适合于不同雷诺数的复杂流场.

(3) 紊流润滑参数数据库的建立使先进的复合型紊流润滑理论应用于工程实际成为可能.

参 考 文 献

[1] Constantinescu V N. Analysis of Bearing Operating in Turbulent Regime[J]. Trans. ASME, Ser. D, 1962, 82: 139-151.

[2] Ng C W, Pan C H T. A Linearized Turbulent Lubrication Theory[J]. Trans. ASME, Ser. D, 1965, 87: 675-688.

[3] Hirs G G. A Bulk-Flow Theory for Turbulence in Lubricant Films[J]. Trans. ASME, Ser. F, 1973, 95: 137-146.

[4] Jones W P, Launder B E. The Calculation of Low Reynolds Number Phenomena with a Two-Equation Model of Turbulence[J]. Int. J. Heat and Mass Transfer, 1973, 16: 1119.

[5] Wang X, Zhang Z, Sun M. A Comparison of Flow Fields Predicted by Various Turbulent Lubrication Models with Existing Measurements[J]. Transactions of the ASME, Journal of Tribology, 2000, 122(2): 475-477.

[6] Leutheusser H J, Aydin E M. Plane Couette Flow Between Smooth and Rough Walls[J]. Experiments in Fluids 1991, 11: 302-312.

[7] Tsanis I K, Leutheusser H J. The Structure of Tuitulent Shear-induced Countercurrent flow[J], J. Fluid Mech, 1988, 189: 531-552.

[8] Marquette O R, Childs D W, Andres L S, Eccentricity effects on the rotordynamic coefficients of plain annular seals: theory versus experiment[J]. Trans. ASME, Journal of Tribology, 1997, 119: 443-448.

Turbulence Models of Hydrodynamic Lubrication*

Abstract: The main theoretical turbulence models for application to hydrodynamic lubrication problems were briefly reviewed, and the course of their development and their fundamentals were explained. Predictions by these models on flow fields in turbulent Couette flows and shear-induced countercurrent flows were compared to existing measurements, and Zhang & Zhang's combined k-ε model was shown to have surpassingly satisfactory results. The method of application of this combined k-ε model to high speed journal bearings and annular seals was summarized, and the predicted results were shown to be satisfactory by comparisons with existing experiments of journal bearings and annular seals.

1 Introduction

It is well known that fluid flows in seals and bearings turn from laminar regime into turbulent one when their Reynolds number becomes higher than a critical value. The earliest theories of hydrodynamic lubrication have been based on the presumption of existence of laminar flow regime. In turbulent regime, however, the characteristics of seals and bearings differ from the predictions by theories so based. Special lubrication theories based on contemporary turbulence models since the later half of 1900's were therefore developed to take the effects of turbulence into consideration.

In practical applications to lubrication problems, Constantinescu[1] was the first to apply Prandtl's mixing-length theory to develop a lubrication theory in the turbulent regime. Several years later, Ng-Pan[2] applied the law of wall to develop another theory which became the most widely adopted one in many countries till now, and Hirs[3] presented yet another theory based on a generalization of basic empirical relationships. Various other theories were also developed in succeeding years, though most of them are not much different in physical essence from the above three theories.

All these theories have more-or-less satisfactory results when applied to lubrication problems with the near-Couette turbulent flows, such as those existing in lightly loaded high-speed bearings. Yet with the development of modern machinery, the flow in the oil films of heavily loaded hydrodynamic bearings have significantly deviated from the Couette flow, since the large pressure gradients in both the longitudinal and lateral directions can

---

* In collaboration with WANG Xiao-Jing, SUN Mei-Li. Reprinted from *Journal of Shanghai University* (English Edition), 2003, 7(4): 305 – 314.

cause the pressure-induced flow constituents reach levels comparable to the shear-induced one. High pressure annular seal is another example where the co-existence of strong pressure gradient in the axial direction and the intensive shear action in the circumferential direction makes the resultant flow drastically different from the Couette flow. Lubrication theories based on more sophisticated turbulence models are therefore desired, in order that they can deal with theoretical firmness not only the near-Couette flows, but also complex flows with pressure-induced flow constituents commensurable to the shear-induced one.

Accordingly, the more advanced models such as the k-ε model[4] and algebraic Reynolds stress (ARS) model[5], which do not rely on the existence of a near-Couette condition, were later introduced into lubrication problems.

Lee-Kim[6] combined the ARS model with the mixing-length theory, to comply with the co-existence of the turbulence kernel in the mid-part of the lubrication film and the usually equally significant sub-layer flows in the near-wall regions.

Zhang & Zhang[7] proposed a further improvement by combining the ARS model with the low turbulence Reynolds stress k-ε model, and setting the transition criterion to fit Leutheusser's experimental measurements of complex turbulent lubrication flow fields.

2 Brief Descriptions of Various Existing Models

2.1 Constantinescu's theory

Prandtl's mixing length theory of turbulence dissipation was adopted by Constantinesu, and the turbulence shear stress was assumed to be

$$-\overline{\rho u'v'} = \rho l_m^2 \left|\frac{\partial \bar{u}}{\partial y}\right| \frac{\partial \bar{u}}{\partial y}, \quad -\overline{\rho v'w'} = \rho l_m^2 \frac{\partial \bar{w}}{\partial y} \left|\frac{\partial \bar{u}}{\partial y}\right|$$

or in other words, the turbulence viscosity is assumed to be:

$$\nu_t = l_m^2 \left|\frac{\partial \bar{u}}{\partial y}\right|$$

The mixing length l_m was assumed to be a linear function of the local distance to the nearer wall:

$$l_m = \chi y, \quad (0 \leqslant y \leqslant h/2)$$
$$l_m = \chi(h-y), \quad (h/2 \leqslant y \leqslant h)$$

where the constant χ can be taken as 0.2 - 0.4, or calculated by $\chi = 0.125 Re^{0.07}$.

Using these assumptions and making simplifications consisting mainly of linearization based on the assumption of the flow condition being not far deviated from a Couette one, Constantinesu derived a Reynolds' formulation of his theory:

$$\frac{\partial}{\partial x}\left(\frac{h^3}{\mu K_x}\frac{\partial p}{\partial x}\right) + \frac{\partial}{\partial z}\left(\frac{h^3}{\mu K_z}\frac{\partial p}{\partial z}\right) = \frac{U}{2}\frac{\partial h}{\partial x} + V_2 - V_1$$

where the turbulence factors are calculated by the following formulae:

$$K_x = 12 + 0.53(\chi^2 Rh)^{0.725}$$

$$K_z = 12 + 0.296(\chi^2 Rh)^{0.65}$$

This theory, although universally acknowledged as the pioneering one in the theories of turbulent lubrication, has been almost completely replaced by the later developed, more firmly based theories. Although improvements have been made to this type of theory by various authors, it is hard to attain a very satisfactory description of the complex behavior of turbulent flow by relying on a model so simplified.

2.2 Ng-Pan's theory

This theory was based on Reichardt's law of wall:

$$\frac{\nu_t}{\nu} = k\left[y^+ - \delta_L^+ \tanh\left(\frac{y^+}{\delta_L^+}\right)\right]$$

where ν—kinetic viscosity; y^+—non-dimensional distance from wall, $y^+ = yV^*/\nu$; V^*—velocity of wall friction: $V^* = (\tau_w/\rho)^{1/2}$.

Based on experimental measurements, the values of the empirical constants were taken to be

$$\delta_L^+ \approx 10.7, \ k = 0.4$$

Ng-Pan used local shear stress instead of wall stress to express the local velocity of friction, and assumed that the flow field could be expressed by a first-order perturbation of the turbulent Couette flow. The following equations were derived when inertia terms are neglected:

$$\frac{\partial}{\partial y}\left[\mu\left(1+\frac{\nu_t}{\nu}\right)\frac{\partial \bar{u}}{\partial y}\right] - \frac{\partial \bar{p}}{\partial x} = 0,$$

$$\frac{\partial}{\partial y}\left[\mu\left(1+\frac{\nu_t}{\nu}\right)\frac{\partial \bar{w}}{\partial y}\right] - \frac{\partial \bar{p}}{\partial z} = 0$$

Substituting the above equations into the continuity equation and integrating across the film thickness, a turbulent lubrication equation of Reynolds' formulation was derived as

$$\frac{\partial}{\partial x}\left(\frac{h^3}{K_x \mu}\frac{\partial p}{\partial x}\right) + \frac{\partial}{\partial z}\left(\frac{h^3}{K_z \mu}\frac{\partial p}{\partial z}\right) = \frac{U}{2}\frac{\partial h}{\partial x}$$

where the turbulence coefficients were obtained as $K_x = 12 + 0.0136\, Re^{0.9}$ and $K_z = 12 + 0.0043\, Re^{0.96}$.

2.3 Low turbulence Reynolds number k-ε model

Based on a dimensional analysis, Jones and Launders derived an expression of the turbulence kinetic viscosity as

$$\nu_t = C_\mu \frac{k^2}{\varepsilon}$$

where k—turbulence kinetic energy, ε—rate of energy dissipation and $\varepsilon \propto k^{3/2}/L$, L—characteristic length and C_μ—a non-dimensional value.

For low turbulence Reynolds number, the k-ε model was expressed by

$$\frac{Dk}{Dt} = \frac{1}{\rho}\frac{\partial}{\partial y}\left[\left(\nu+\frac{\nu_t}{\sigma_k}\right)\frac{\partial k}{\partial y}\right] + \nu_t\left[\left(\frac{\partial u}{\partial y}\right)^2 + \left(\frac{\partial w}{\partial y}\right)^2\right] - \varepsilon - \frac{2\nu k}{y^2}$$

$$\frac{D\varepsilon}{Dt} = \frac{\partial}{\partial y}\left[\left(\nu+\frac{\nu_t}{\sigma_\varepsilon}\right)\frac{\partial \varepsilon}{\partial y}\right] + C_{\varepsilon 1}\nu_t\left[\left(\frac{\partial u}{\partial y}\right)^2 + \left(\frac{\partial w}{\partial y}\right)^2\right]\frac{\varepsilon}{k}$$

$$- C_{\varepsilon 2}[1-0.3\exp(R_k^2)]\frac{\varepsilon^2}{k} - 2\nu\left(\frac{\partial \varepsilon^{1/2}}{\partial y}\right)^2$$

where $\varepsilon_m = C_\mu k^2/\varepsilon = C_m[1-\exp(-A_m R_k)]k^2/\varepsilon$ and the turbulence Reynolds number $R_k = k^2/\nu\varepsilon$. The constants are $C_m = 0.09$, $C_\mu = 0.09$, $\sigma_k = 1.0$, $\sigma_\varepsilon = 1.3$, $C_{\varepsilon 1} = 1.45$ and $C_{\varepsilon 2} = 2.0$. For Newtonian fluids, $A_m = 0.0015$.

2.4 Lee-Kim's combined method

Lee and Kim combined the algebraic Reynolds stress model for the turbulence kernel with Prandtl's mixing length theory for the near-wall regions.

The governing equations of algebraic Reynolds stress model were taken as

$$\frac{\partial}{\partial y}\left(\frac{\nu_t}{\sigma_k}\frac{\partial k}{\partial y}\right) + P_d - \varepsilon = 0$$

$$\frac{\partial}{\partial y}\left(\frac{\nu_t}{\sigma_\varepsilon}\frac{\partial \varepsilon}{\partial y}\right) + C_{\varepsilon 1}P_d\frac{\varepsilon}{k} - C_{\varepsilon 2}\frac{\varepsilon^2}{k} = 0$$

where $\nu_t = 0.09 G_\mu \frac{k^2}{\varepsilon}$, $P_d = \nu_t\left[\left(\frac{\partial u}{\partial y}\right)^2 + \left(\frac{\partial w}{\partial y}\right)^2\right]$, and

$$G_\mu = \frac{1+\frac{3}{2}\frac{C_2 C_2'}{1-C_2}f}{1+\frac{3}{2}\frac{C_1'}{C_1}f}\frac{1-2\frac{C_2 C_2' P_d/\varepsilon}{C_1-1+C_2 P_d/\varepsilon}f}{1+2\frac{C_1'}{C_1+P_d/\varepsilon-1}f}, \quad C_2'=0.3,$$

$$C_{\varepsilon 1}=1.44, \ C_{\varepsilon 2}=1.92, \ \sigma_k=1.0, \ \sigma_\varepsilon=1.3, \ f=\frac{k^{3/2}}{C_w y \varepsilon},$$

$$C_w=3.72, \ C_1=1.8, \ C_2=0.6, \ C_1'=0.6.$$

2.5 Zhang & Zhang's combined turbulence k-ε Model

Zhang & Zhang developed another combined turbulent lubrication k-ε model, which was different from the previous ones in three aspects: 1. Low turbulence Reynolds number

k-ε model was used for the near-wall regions, in combination with algebraic Reynolds stress model for the turbulence kernel. It is believed that the combination of two models of similar physical levels would bring forth better results. 2. Continuities of both eddy viscosity and velocity between the two models were demanded. 3. The transit from the first model to the second one was triggered by the criterion that the turbulence stress reaches 95% of the total stress. The last named criterion was established by the optimal fitting of the resulting flow field to Tsanis-Leutheusser's measured data in countercurrent flow experiments[8] and also checked well with Leutheusser-Aydin's Couette flow experiments[9]. It is believed that a model capable of accurately predicting flow fields in both the Couette and countercurrent flow conditions ought to be reliably applicable to general cases of turbulent hydrodynamic lubrication.

3 Comparison of Theoretical Models with Existing Flow Field Experiments[10]

The above models have been applied to predict the flow fields under the turbulent Couette flows and shear-induced countercurrent flows with zero cross-sectional net flow at different Reynolds numbers. They were compared to Tsanis-Leutheusser's and Leutheusser-Aydin's corresponding experimental data to check their validity, and the results are shown in Figs. 1–4.

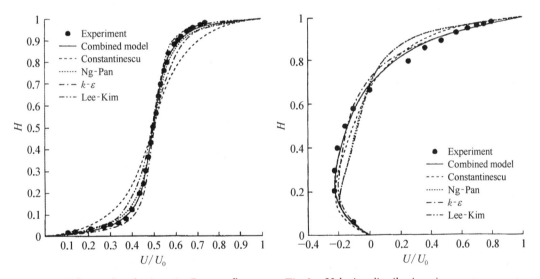

Fig. 1 Velocity distributions in Couette flow, $Re = 9\,524$

Fig. 2 Velocity distributions in countercurrent flow, $Re = 2\,000$

All the figures show the measured velocity distributions and the velocity distributions predicted by various models.

Fig. 1 shows the comparison in the pure Couette flow with a Reynolds number 9524. Zhang & Zhang's combined model agrees well with the experimental results. Prediction by Ng-Pan's model is also close to the experimental results. The k-ε model overestimates the effect of turbulence stress, while Lee-Kim's model underestimate the depth of turbulence in the film.

Figs. 2 - 4 show the comparisons in countercurrent flows with different depths of turbulence as indicated by the corresponding Reynolds numbers. It can be seen that Zhang & Zhang's model accords well with the experimental results in all three cases. The predictions by k-ε model are close to experimental results at low and middle Reynolds numbers, but overestimates the turbulence effect at high Reynolds number apparently due to its negligence of the wall damping effect. Constantinescu's model deviates from the experimental results rather obviously, underestimating the turbulence development, especially at high Reynolds number and large counter pressure gradient. Not very significant but still perceptible deviations of predictions by Ng-Pan's and Lee-Kim's models from the experimental results can be noticed in the forms of the velocity distribution curves. Inadequate consideration of the existence of significant pressure gradient in the case of Ng-Pan's model, and the imperfect match of constituent models and inadequate solidness of transition criterion in the case of Lee-Kim's model, have possibly to take the responsibility.

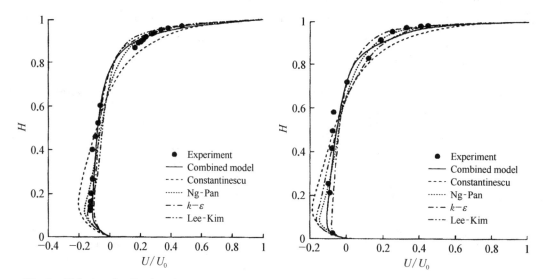

Fig. 3 Velocity distributions in countercurrent flow, $Re = 5\,000$

Fig. 4 Velocity distributions in countercurrent flow, $Re = 12\,000$

From the above comparison, Zhang & Zhang's model and Ng-Pan's model give the best satisfaction. The former is justified not only by its satisfactory agreement with existing experimental data, but also held to be prospective to have wider applicability in view of its more generalized physical considerations.

4 Reynolds Equation in Turbulent Regime

Based on the Navier-Stokes and continuity equations, together with the Boussinesq assumption, the following Reynolds equation for turbulent flow regime can be obtained:

$$\frac{\partial}{\partial x}\left[G\left(\frac{\partial p}{\partial x}+\rho\bar{f}\right)\right]+\frac{\partial}{\partial z}\left[G\left(\frac{\partial p}{\partial z}+\rho\bar{g}\right)\right]=\frac{\partial}{\partial x}(UFh)$$

where

$$G=\frac{A_2}{A_1}B_1-B_2, \quad F=1-\frac{B_1}{A_1 h}$$

$$A_1=\int_0^h \frac{1}{\nu+\nu_t}dy, \quad A_2=\int_0^h \frac{y}{\rho(\nu+\nu_t)}dy$$

$$B_1=\int_0^h\int_0^y \frac{1}{\nu+\nu_t}dydy, \quad B_2=\int_0^h\int_0^y \frac{y}{\rho(\nu+\nu_t)}dydy,$$

$$f(x,y)=\frac{1}{h}\left(\frac{\partial I_{xx}}{\partial x}+\frac{\partial I_{xz}}{\partial z}\right),$$

$$\bar{g}(x,z)=\frac{1}{h}\left(\frac{\partial I_{xz}}{\partial x}+\frac{\partial I_{zz}}{\partial z}\right),$$

$$I_{xx}=\int_0^h u^2 dy, \quad I_{xz}=\int_0^h uw dy, \quad I_{zz}=\int_0^h w^2 dy,$$

$$u=\left(\frac{\partial p}{\partial x}+\rho f\right)\left[\int_0^{y_1}\frac{ydy}{\rho(\nu-\nu_t)}-\frac{A_2}{A_1}\int_0^{y_1}\frac{dy}{\nu+\nu_t}\right]+U\left(1-\frac{1}{A_1}\int_0^{y_1}\frac{dy}{\nu+\nu_t}\right),$$

$$w=\left(\frac{\partial p}{\partial z}+\rho\bar{g}\right)\left[\int_0^{y_1}\frac{ydy}{\rho(\nu-\nu_t)}-\frac{A_2}{A_1}\int_0^{y_1}\frac{dy}{(\nu-\nu_t)}\right]$$

For journal bearings and annular seals, when the inertia terms are neglected, the Reynolds equation is non-dimensionalized as

$$\frac{\partial}{\partial\varphi}\left(\bar{G}H^3\frac{\partial P}{\partial\varphi}\right)+\left(\frac{D}{L}\right)^2\left(\bar{G}H^3\frac{\partial P}{\partial Z}\right)=\frac{\partial}{\partial\varphi}(\bar{F}H)$$

where

$$\bar{G}=\frac{\bar{A}_2}{\bar{A}_1}\bar{B}_1-\bar{B}_2, \quad \bar{F}=1-\frac{\bar{B}_1}{\bar{A}_1},$$

$$\bar{A}_1=\int_0^1 \frac{1}{1+\bar{\nu}_t}dY, \quad \bar{A}_2=\int_0^1 \frac{Y}{1+\bar{\nu}_t}dY,$$

$$\bar{B}_1=\int_0^1\int_0^Y \frac{1}{1+\bar{\nu}_t}dYdY, \quad \bar{B}_2=\int_0^1\int_0^Y \frac{y}{1+\bar{\nu}_t}dYdY,$$

$$\bar{u}=B_x\left(-\int_0^{Y_1}\frac{YdY}{1+\bar{\nu}_t}+\frac{\bar{A}_2}{\bar{A}_1}\int_0^{Y_1}\frac{dY}{1+\bar{\nu}_t}\right)+1-\frac{1}{\bar{A}_1}\int_0^{Y_1}\frac{dY}{1+\bar{\nu}_t},$$

$$\bar{w}=B_z\left(-\int_0^{Y_1}\frac{YdY}{1+\bar{\nu}_t}+\frac{\bar{A}_2}{\bar{A}_1}\int_0^{Y_1}\frac{dY}{1+\bar{\nu}_t}\right)+1-\frac{1}{\bar{A}_1}\int_0^{Y_1}\frac{dY}{1+\bar{\nu}_t},$$

$$\bar{w} = B_z \left(-\int_0^{Y_1} \frac{Y \mathrm{d}Y}{1+\nu_t} + \frac{\bar{A}_2}{\bar{A}_1} \int_0^{Y_1} \frac{\mathrm{d}Y}{1+\bar{\nu}_t} \right),$$

$$B_x = -\frac{h^2}{\mu U} \frac{\partial p}{r \mathrm{d}\varphi} = -H^2 \frac{\partial P}{\partial \varphi},$$

$$B_z = -\frac{h^2}{\mu U} \frac{\partial p}{\partial z} = -H^2 \frac{D}{L} \frac{\partial P}{\partial Z},$$

$$Y = \frac{y}{h}, \quad \bar{\nu}_t = \frac{\nu_t}{\nu}, \quad H = \frac{h}{c}, \quad P = \frac{p}{p_0}, \quad p_0 = \frac{\mu \omega R^2}{c^2}, \quad Z = \frac{z}{L/2}$$

In the above equations, D—diameter of bearing or seal, R—radius of the same, L—length, c—radial clearance, μ—dynamic viscosity, ω—angular speed of shaft.

With the help of the above equations, the turbulence coefficients $\bar{G}$ and $\bar{F}$ can be calculated at each point (x, z) in the oil film, the pressure, shear stress and velocity distributions in the oil film can be solved, and the static and dynamic properties of the bearing/seal can be integrated.

5 Database and Formulae Fitting of Turbulence Coefficients[11,12]

For engineering application purposes, the above calculation would be too complicated if performed directly with the equation system. In order to make practical applications more feasible, a technique similar to that of Kaneko *et al*'s is adopted.

From a similarity analysis, it can be seen that the flow pattern across the film thickness at a point (x, z) is solely determined by three non-dimensional parameters, namely, the local Reynolds number R_h, the non-dimensional pressure gradient in the circumferential direction B_x and that in the axial direction B_z. Consequently, the non-dimensional coefficients $\bar{G}$ and $\bar{F}$ are also solely determined by R_h, B_x and B_z. A tabular database of these turbulence coefficients can therefore be generated by systematic calculations within sufficiently wide ranges of R_h, B_x and B_z, which are discretized into the following nodal values:

For $R_h = 800 - 5\,800$, step 500; $5\,800 - 20\,800$, step $1\,000$; $20\,800 - 30\,800$, step $2\,000$.

For $B_x = -75 - -35$, step 20; $-35 - -15$, step 10; $-15 - 15$, step 5; $15 - 35$, step 10; $35 - 75$, step 20.

For $|B_z| = 0 - 90$, step 15; $90 - 160$, step 35; $160 - 370$, step 70.

For each point (x, z) in the oil film, the values of the turbulent coefficients can be interpolated from the database by

$$\alpha(R_h, B_x, B_z) = \sum_{k=1}^{3} \sum_{j=1}^{3} \sum_{i=1}^{3} \left(\prod_{\substack{s=1 \\ s \neq k}}^{3} \frac{R_h - R_{hs}}{R_{hk} - R_{hs}} \prod_{\substack{m=1 \\ m \neq j}}^{3} \frac{B_x - B_{xm}}{B_{xj} - B_{xm}} \prod_{\substack{n=1 \\ n \neq k}}^{3} \frac{B_z - B_{zn}}{B_{zi} - B_{zn}} \right) \alpha_{ijk}$$

Alternatively, a technique of formula fitting to the database can be used to obtain empirical formulae for the turbulence coefficients, which can be yet easier to apply than directly interpolation from the database.

The following formulae are thus obtained:

$$\frac{1}{\bar{G}} = 12 + 0.0136(R_h - 800)^{0.9} + 0.269 |B_x|^{0.93} + 0.844 |B_z|^{0.65}$$
$$+ 0.05(R_h - 800)^{0.35} |B_z|^{0.74} + 0.00032(R_h - 800)^{0.85} |B_x|^{0.95}$$
$$- 0.0705 |B_x|^{0.0911} \cdot |B_x|^{0.0355} - 0.35(R_h - 800)^{0.35} |B_x|^{0.05} |B_z|^{0.2}$$

$$F = 0.5 + \mathrm{sign}(B_x)(-5.51 \times 10^{-7} |B_x|^3 + 8.6812 \times 10^{-5} |B_x|^2 - 4.078$$
$$\times 10^{-3} |B_x|) + \mathrm{sign}(B_x)(-0.018944 |B_x|^{0.5} + 0.00425 |B_x| - 2.502$$
$$\times 10^{-5} |B_x|^2) \times [-0.01717(R_h - 800)^{0.5} - 0.0013(R_h - 800) + 1.0$$
$$\times 10^{-5}(R_h - 800)^{1.5}] + \mathrm{sign}(B_x)(0.0062 |B_x|^{0.15} |B_z|^{0.25}) + \mathrm{sign}(B_x)$$
$$(0.0007 |B_x|^{0.5} - 0.00013 |B_x| + 2.5 \times 10^{-7} |B_x|^2) \times (-0.1 |B_z|^{0.5}$$
$$- 0.01 |B_z| + 3.0 \times 10^{-4} |B_z|^2) \times [0.1(R_h - 800)^{0.5}$$
$$- 5.2 \times 10^{-6}(R_h - 800)^{1.5}]$$

Figs. 5 – 8 show the values of the turbulence coefficients as calculated by the fitted formulae against those interpolated from the database. The former are shown by continuous curves, while the latter by curves covered with heavy dots. In Fig. 5, prediction by Ng-Pan's theory is also shown by a dotted curve for comparison.

Fig. 5 corresponds to the pure Couette flow. For Reynolds number below 800, the flow will be laminar, and $\bar{G}$ should be 12, or its reciprocal should be 0.08333. It can be seen that the values predicted by Zhang & Zhang's model are also satisfactory in the range of low Reynolds number, while that predicted by Ng-Pan somewhat overestimate the turbulence development in this range.

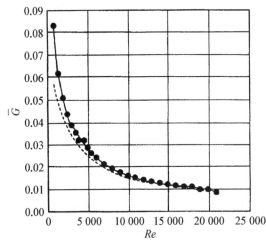
Fig. 5 Values of $\bar{G}$ in dependence on Re ($B_x = B_z = 0$)

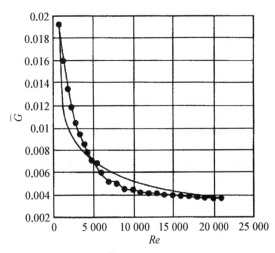
Fig. 6 Values of $\bar{G}$ in dependence on Re ($B_x = 0$, $B_z = 370$)

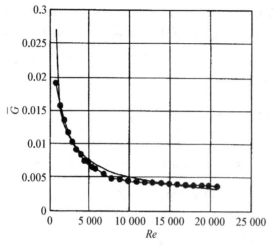

Fig. 7 Values of $\bar{G}$ in dependence on Re ($B_x = 75$, $B_z = 370$)

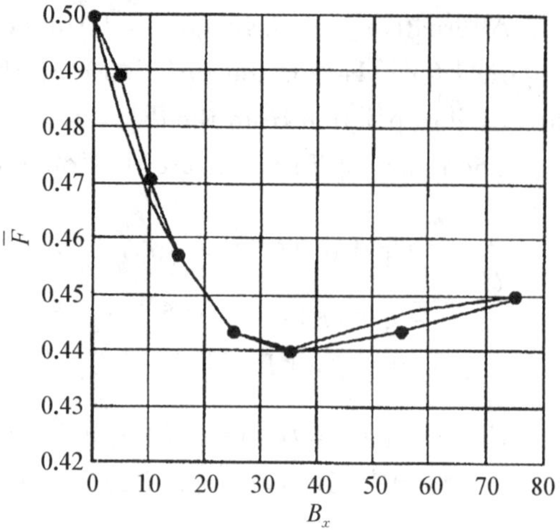

Fig. 8 Values of F in dependence on B_x ($R_h = 800$, $B_z = 0$)

6 Comparisons with Existing Experiments[13]

6.1 Comparison with Smith and Fuller's experiment of journal bearing[14]

The parameters of the test bearing are: $D = 3$ in, $L = 3$ in, $\psi = 0.00293$, $\mu = 1.26 \times 10^{-7}$ Lb·s/in.

Figs. 9 – 12 show the experimental data and predictions by Zhang & Zhang's model. Predictions by Ng-Pan's model are also shown for comparison.

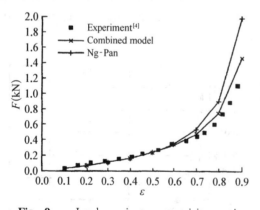

Fig. 9 Load against eccentricity ratio (speed=3 000 r/min)

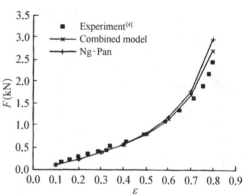

Fig. 10 Load against eccentricity ratio (speed=7 450 r/min)

Satisfactory results have been obtained by Zhang & Zhang's model as shown by all these comparisons.

6.2 Comparison with Kato and Hori's experiment of journal bearings[15]

The parameters of the test bearings are: $D = 210$ mm, $L/D = 0.95$, $\psi = 0.0019$ and

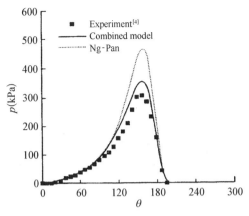

Fig. 11 Pressure distribution (speed = 3 000 r/min)

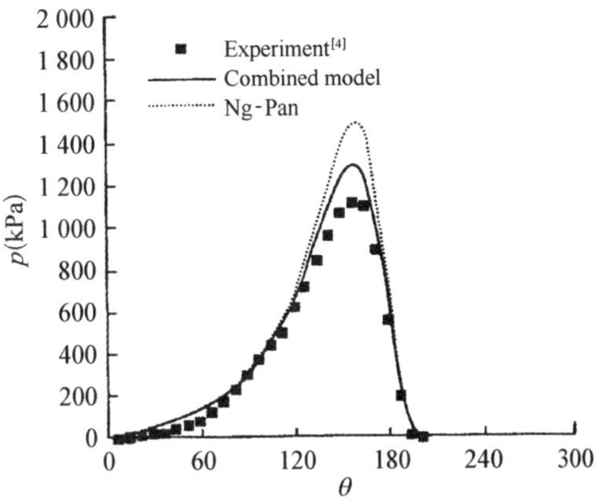

Fig. 12 Pressure distribution (speed = 7 450 r/min)

0.003 6.

Figs. 13 and 14 show the experimental data and predictions by Zhang & Zhang's model. The comparisons are satisfactory.

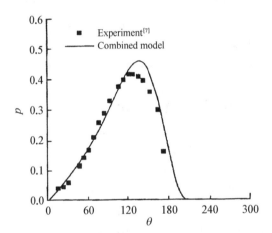

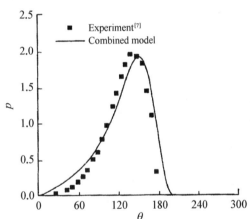

Fig. 13 Non-dimensional pressure distribution ($Re = 2\,000$)

Fig. 14 Non-dimensional pressure distribution ($Re = 6\,000$)

6.3 Comparison with Kaneko-Hori-Tanaka's experiment of annular seal[16]

The parameters of the test seal are: $D = 70$ mm, $L = 35$ mm, $c = 0.175$ mm, $n = 4\,080$ r/min, inlet pressure $= 0.7$ MPa.

Figs. 15 and 16 show the comparisons of static properties.

The comparisons of static properties can be deemed as more or less satisfactory.

Comparisons on dynamic properties are not so satisfactory. While this might point to the necessity of further investigation of turbulence model for non-stationary condition, it should also be reminded that significant discrepancies between theory and experiments on bearing dynamic properties are not uncommon even in the case of laminar flow regime.

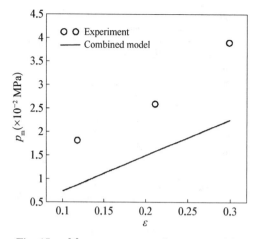
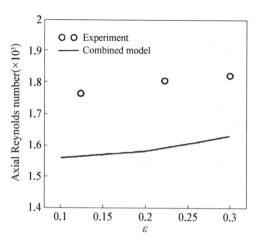

Fig. 15 Mean pressure against eccentricity ratio

Fig. 16 Axial Reynolds number against eccentricity ratio

6.4 Comparison with Marquette-Childs-Andres's experiment of annular seal[17]

The parameters of the test seal are: $D=76.29$ mm, $L=34.93$ mm, $c=0.11$ mm.

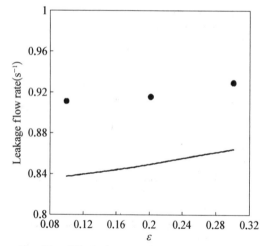

Fig. 17 Side leakage against eccentricity ratio (speed=17 400 r/min, $p_{in}=5.52$ MPa)

Figs. 17 – 19 show the comparisons of leakage flow rates.

The comparisons are fairly satisfactory.

7 Concluding Remarks

With the deepening of knowledge over flow characteristics in turbulent lubricating oil films, increasingly more turbulence models have been applied to take in the complicated physical essence of these films and to strengthen the physical basis of the governing equations, in order to make the predictions more satisfactory and to have wider range of applicability. Better correlation with experimental results has been continuously sought for in the development of models. Tsanis-Leutheusser's fundamental measurements have been found to be especially valuable in helping the formation of Zhang & Zhang's combined model. Extensive comparisons with existing experiments show the Zhang & Zhang's model surpassingly satisfactory in rather wide application ranges. Computing technique has been used to generate a database of the turbulence coefficients for the turbulent Reynolds equation, and empirical formulae have been fitted to the database. The practical application of such a rather sophisticated model is thus made feasible.

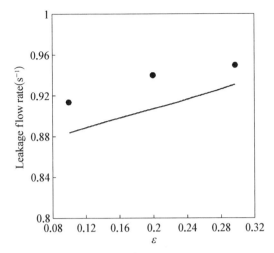 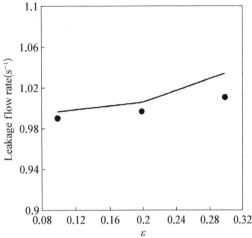

Fig. 18 Side leakage against eccentricity ratio (speed=24 600 r/min, p_{in}=6.89 MPa)

Fig. 19 Side leakage against eccentricity ratio (speed=10 200 r/min, p_{in}=5.42 MPa)

References

[1] Constantinescu VN. On turbulent lubrication [J]. Proc. Inst. Mech. Eng., 1959, 173: 881-889.

[2] Ng C W, Pan C H T. A linearized turbulent lubrication theory[J]. Trans. ASME Ser. D, 1965, 87: 675-688.

[3] Hirs G G. A bulk flow theory for turbulence in lubricant films[J]. Trans. ASME Ser. F, 1973, 95: 137-146.

[4] Jones W P, Launder B E. The calculation of low Reynolds number phenomena with a two-equation model of turbulence[J]. Int. J. Heat and Mass Transfer, 1973, 16: 1119.

[5] Ljuboja M, Rodi W. Calculation of turbulent wall jets with an algebraic Reynolds stress model [J]. Trans. ASME J. Fluids Eng., 1980, 102: 350-356.

[6] Lee D W, Kim K W. Turbulent lubrication theory using algebraic Reynolds stress model in finite journal bearings with cavitation boundary conditions[J]. JSME International Journal Ser. Ⅱ, 1990, 33: 200-207.

[7] Zhang Y Q, Zhang Z M. A new method of analysis of turbulent lubrication theory — combined turbulence model of lubrication[J]. Journal of Tribology, 1995, 15: 271-275(in Chinese).

[8] Tsanis I K, Leutheusser H J. The structure of turbulent shear-induced countercurrent flow [J]. J. Fluid Mech., 1988, 189: 531-552.

[9] Leutheusser H J, Aydin E M. Plane Couette flow between smooth and rough walls[J]. Experiments in Fluids, 1991, 11: 302-312.

[10] Wang X J, Zhang Z M. A comparison of flow fields predicted by various turbulent lubrication models with existing measurements [J]. Trans. ASME J. Tribology, 2000, 122: 475-477.

[11] Wang X J, Zhang Z M, Sun M L. Formation of database of turbulent lubrication parameters [A]. Research Report 3 of National Natural Science Foundation Project "Research of Combined Turbulent Lubrication Theory"[R]. Shanghai University, Shanghai, China, 2000, 1-6(in Chinese).

[12] Wang X J, Zhang Z M, Sun M L. Investigation on formula fitting of turbulent coefficients[A]. Research Report 5 of National Natural Science Foundation Project 'Research of Combined Turbulent Lubrication Theory' [R]. Shanghai University, Shanghai, China, 2000, 1-7(in Chinese).

[13] Wang X J, Zhang Z M, Sun M L. Comparison of turbulent lubrication theoretical predictions with existing experiments[J]. Research Report 4 of National Natural Science Foundation Project "Research of Combined Turbulent Lubrication Theory"[R], Shanghai University, Shanghai, China, 2000, 1-11 (in Chinese).

[14] Smith M I, Fuller D D. Journal bearing operation at superlaminar speeds[J]. Trans. ASME, 1956, 78: 469-474.

[15] Kato T, Hori Y. Pressure distribution in a journal bearing lubricated by drag reducing liquids under turbulent conditions [J]. Proc. JSLE International Tribology Conference, 1985, Tokyo, Japan, 571-576.

[16] Kaneko S, Hori Y, Tanaka M. Static and dynamic characteristics of annular plain seals[J]. I Mech E, 1984, C279/84: 205-214.

[17] Marquette O R, Childs D W, Andres L S. Eccentricity effects on the rotordynamic coefficients of plain annular seals: theory versus experiment[J]. Trans. ASME J. of Tribology, 1997, 19: 443-448.

Theory of Cavitation in an Oscillatory Oil Squeeze Film*

Abstract: The present knowledge on cavitation in oil film bearings is reviewed, and the inadequacy of the nowadays widely used method of designing dynamically loaded oil film bearings is pointed out. A new model of dynamic cavitation in submerged oil film bearings is proposed. The model preserves mass conservation in the cavitation region and allows the occurrence of tensile stresses in the oil film. The role of surface tension and contact angle is considered and incorporated in the model and their influences brought forth quantitatively. The model is applied to a parallel-plate oscillatory squeeze film bearing. Results derived from the model compare favorably with available measurements reported in the literature.

Nomenclature

Symbols topped with bars are dimensionless.
a = radius of thrust plate
$\bar{C}_a$ = cavitation number, $= \eta \omega a / \sigma$
$h(t)$ = gap between thrust plates, $\bar{h} = h/h_o$
h_o = mean gap during oscillations
h_{past} = explained in the text following Eq. (20), $\bar{h}_{past} = h_{past}/h_o$
$\dot{h} = dh/dt = (h_o \omega) d\bar{h}/d\bar{t}$
p = oil film pressure, $\bar{p} = p/p_a$
p_a = ambient pressure in the oil pool
p_c = threshold pressure at the onset of cavitation, $\bar{p}_c = p_c/p_a$
p_v = cavitation pressure, taken to be the vapor pressure of oil, $\bar{p}_v = p_v/p_a$
Q = volume flow rate, defined in Eq. (12)
$R(t)$ = boundary of cavitation region, $\bar{R} = R/a$
R_o = assumed size of cavitation nucleus, $\bar{R}_o = R_o/a$

R' = another principal radius of the meniscus, given in Eq. (2) and shown in Fig. 1
$\dot{R} = dR/dt = (a\omega) d\bar{R}/d\bar{t}$
t = time, $\bar{t} = \omega t$
(r, z) = radial and cross-film coordinates, Fig. 1, $\bar{r} = r/a$
(v_r, v_z) = velocity components of oil film flow
$\bar{\alpha}$ = dynamic contact angle
$\bar{\alpha}_{0+}$, $\bar{\alpha}_{0-}$ = static contact angle values
$\bar{\beta}$ = factor explained in the text following Eq. (18)
$\bar{\delta} = h_o/a$, the small parameter on which lubrication theory is based
ε = amplitude of oscillations, $\bar{\varepsilon} = \varepsilon/h_o$
η = dynamic viscosity coefficient of oil
$\bar{\lambda}$ = squeeze number, $= (\eta \omega / p_a)(a/h_o)^2$
σ = surface tension coefficient of oil
ω = frequency of oscillations

1 Introduction

When an oil-lubricated journal bearing is in steady-state operation, the oil film ceases

---

* In collaboration with D. C. SUN, WANG WEN, CHEN XIAOYANG and SUN MEILI. Reprinted from *Tribology Transactions*, 2008, 51: 332 – 340.

to be continuous in the divergent part of the bearing clearance. The phenomenon is known as (stationary) cavitation. Features of stationary cavitation are well known (Dowson and Taylor[1]), viz. the cavitation region commences slightly downstream of the minimum bearing clearance and closes before the maximum bearing clearance is reached; inside the cavitation region, ambient pressure prevails and the oil film breaks down into thin streamers sheared by the moving bearing surface. However, when the journal bearing works under a variable load, the (dynamic) cavitation phenomenon is less well understood. Apparently, dynamic cavitation presents itself in different forms according to the operating environment of the bearing. If the bearing is vented to the ambient atmosphere air may be entrained into the oil film, which then becomes a nonhomogeneous mixture of air bubbles and oil. San Andres and Diaz studied the problem extensively in the case of open-end squeeze film dampers (San Andres and Diaz[2]), and many other publications cited therein, plus a series of video clips entitled, "Dynamic Forced Performance of Fluid Film Bearings Operating with Air Entrainment". If the bearing is submerged in oil, a different form of cavitation occurs. Sun and Brewe[3] and Sun, et al.[4] studied the case of a submerged bearing with a transparent sleeve performing non-centered whirl about a fixed journal. In their experiment, pressure traces were taken through transducers mounted on the surface of the fixed journal, while a high-speed movie was taken through the transparent sleeve with the camera's field of view covering the transducers as well as the whole life span of the cavitation region. They reported the following findings: (a) For a given operating condition, if cavitation occurred, it was confined to one region, which evolved from birth to collapse during a cycle of the sleeve's whirl. The cavitation region contained residual oil filaments of the fractured film. (b) The pressure in the cavitation region was close to (absolute) zero, while tensile stresses were measured in the oil film outside the cavitation region. (c) Cavitation sometimes did not occur for the same operating condition, indicating that the occurrence of cavitation depended also on factors other than the bearing's operating parameters, such as the whirling speed and amplitude. If cavitation did not occur, the magnitudes of tensile stresses in the oil film were measured to be larger than otherwise.

From the above description it may be said that the mechanisms underlying dynamic cavitation are still unclear. Consequently, the accuracy in predicting the bearing performance under dynamic loads may be limited. Yet dynamically loaded bearings are widely used, including important machine components such as the main and connecting-rod bearings in reciprocating internal combustion engines. The current prevailing method of computing the performance of dynamically loaded bearings is through the use of various cavitation algorithms, first proposed by Elrod and Adams[5] and subsequently developed by many others; e.g., Kumar and Booker[6]. These cavitation algorithms are computational implementations of the JFO theory (Jakobsson and Floberg[7]; Olsson[8]) that essentially contains three elements: (a) A uniform pressure prevails in the cavitation region (to be called the cavitation pressure), which may be set as zero or the ambient pressure, whichever is appropriate. (b) At the upstream (or rupture) boundary of the cavitation region a zero pressure-gradient condition (or Reynolds boundary condition) is prescribed. (c) At the downstream (or reformation) boundary of the cavitation region a flow continuity condition is applied. By its construction the JFO theory preserves mass conservation in the cavitation region and results in the pressure level everywhere in the oil film not lower than the cavitation pressure. The application of zero pressure-gradient condition at the rupture boundary may be justified in the case of stationary cavitation, because matching the flow rate of the streamers in the cavitation region with the flow rate in the full-film region, which consists of both shear flow and pressure-gradient flow, requires the pressure gradient to vanish at the boundary. However, in the case of dynamic cavitation there is no experimental evidence of shear flow of streamers in the cavitation region. In fact, the residual oil filaments observed in the cavitation region do not constitute a continuous film. Then, it is not evident why the pressure gradient at the rupture boundary should vanish. Besides, the JFO theory does not reconcile with the experimental evidence that tensile stresses could be present in the oil film. Therefore, the applicability of the JFO theory, or the various cavitation algorithms derived from it to analyze the performance of dynamically loaded bearings, may be limited.

In this paper a new model is proposed for the phenomenon of dynamic cavitation in submerged oil film bearings. The model preserves mass conservation in the cavitation region and allows the occurrence of tensile stresses in the oil film. The simple configuration of an oscillatory oil squeeze film bounded between two parallel circular plates is chosen for analysis, so as to avoid the whirling motion and complicated geometry of a journal bearing. This problem has been studied extensively: by Hays and Feiten[9] experimentally and analytically; by Rodrigues[10] analytically and experimentally; by Kuroda and Hori[11] * experimentally; by Parkins and May-Miller ([12], experimentally, and many publications from the Cranfield Institute of Technology cited therein); by Boedo and Booker[13] and by Optasanu and Bonneau[14], both analytically. The experience gained from these earlier studies guided us in the design of a perhaps more thoughtful test apparatus, to be reported separately (Sun, et al.[15]), as well as to include a few more features in the analytical model. In particular, the role played by surface tension and contact angle in the phenomenon is addressed and their influences brought forth quantitatively.

2 Cavitation Model

Consider a squeeze film bearing submerged in an oil pool, Fig. 1. The bearing consists of two horizontal plates, the lower plate being fixed while the upper one (circular, with

---

* We became aware of this work from Professor Hori's recent publication, *Hydrodynamic Lubrication*, Springer-Verlag, pp. 145 – 149 (2006).

radius a) made to oscillate with a given frequency ω and an amplitude ε. The gap between the plates may be described as:

$$h = h_o - \varepsilon \sin\omega t \quad \text{for } t \geqslant 0 \tag{1a}$$

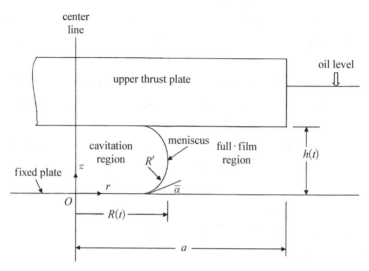

Fig. 1 Schematic of a cavitated axisymmetric oil squeeze film

The fluctuation of oil film pressure at the center ($r=0$) is the largest if the oil film does not cavitate. Let it be assumed that the cavitation starts here from a nucleus of size R_o, when the center pressure decreases to a threshold value p_c. A circular cavitation region, at a uniform cavitation pressure p_v and containing broken oil filaments inside, grows from the assumed nucleus size outward to a peak value, then the refilling oil from the surrounding pool forces the cavitation region to shrink and eventually vanish at $r=0$. Outside the cavitation region the oil film is continuous. As the cavitation region grows, the radial velocity distribution in the film thickness direction at and near the cavitation boundary is unlikely to be uniform. The largest of this velocity is recognized as that of the cavitation boundary, because only in this way no cavity can be entrained into the full-film region. Thus, the slower moving part of the oil film is "overtaken" by the fast-moving boundary and left in the cavitation region, which forms the observed residual oil filaments. These filaments are left stationary on the bearing surfaces since there is no pressure gradient inside the cavitation region. When the cavitation region shrinks, these oil filaments join the refilling oil to become a part of the continuous film again. The equilibrium of forces at the interface determines the oil pressure on the full-film side. The location of the cavitation boundary, denoted by $R(t)$, is determined by appropriate flow continuity conditions that take into account the different oil film responses during the expansion and shrinking stages of the cavitation region.

Sun and Brewe[16] assessed the time of filling a void with the vapor of the surrounding liquid and that of filling the void by diffusion of dissolved gas in the liquid. It was found that the time of liquid evaporation was much shorter than the characteristic time of the dynamic operation of oil film bearings, whereas the time needed for the release of dissolved gas in a liquid into the void was much longer than the characteristic time of the dynamic operation of oil film bearings. Hence, the release of dissolved gas in oil into the cavitation region may be neglected, and the cavitation region may be considered full of oil vapor. The cavitation pressure p_v is therefore taken to be the vapor pressure of the oil.

3 Role of Surface Tension and Contact Angle

The interface between the cavitation and the full-film regions is a moving meniscus, whose principal radii of curvature may be denoted as R and R', as shown in Fig. 1. If the meniscus makes equal contact angle $\bar{\alpha}$ with both the top and the bottom plates, then R' is related to $\bar{\alpha}$ and the oil film thickness h by:

$$R'\cos\bar{\alpha} = \frac{h}{2} \tag{2}$$

The mechanical equilibrium at each point of the meniscus may be expressed as (Defay and Prigogine[17]):

$$p_v - \left(p - 2\eta \frac{\partial v_r}{\partial r}\right) = \sigma\left(\frac{1}{R} + \frac{1}{R'}\right) \tag{3}$$

where σ denotes the surface tension coefficient of oil, η the dynamic viscosity coefficient of the oil, and the parentheses on the LHS of the equation represent the (compressive) normal stress on the full-film side. The viscous normal stress term, which often appears in the formulation of bubble dynamics (e.g., Plesset and Prosperetti[18]; Prosperetti[19]), is of the order $(h_o/a)^2$ relative to the oil film pressure p in the situation of fluid film lubrication and may be neglected. Other terms representing the dynamics of the cavity are not included in Eq. (3) because our main interest is in the oil film behavior rather than bubble dynamics. By incorporating the expression for R' from Eq. (2), the above equation may be rearranged to read:

$$p_{r=R} = p_v - \sigma\left(\frac{1}{R} + \frac{2\cos\bar{\alpha}}{h}\right) \tag{4}$$

One sees that, if p_v is close to zero, the oil film pressure can be negative; i.e., it can become a tensile stress. Of the two surface tension terms, the one associated with the cross-film curvature of the meniscus dominates, because h is usually much smaller than R, which characterizes the size of the cavitation region.

The contact angle of a moving meniscus is different from that of a stationary one and depends on the moving velocity (de Gennes[20]). If $\dot{R}>0$, the angle is smaller than a static value $\bar{\alpha}_{0+}$ and approaches zero for large $\dot{R}$; but if $\dot{R}<0$, it is greater than a static value $\bar{\alpha}_{0-}$ and approaches 180° for large $|\dot{R}|$. There is a slight hysteresis between the two static contact angle values. A large amount of measured contact angle data were correlated by

Jiang, et al. [21] into the following formulas:

$$\tanh\left[4.96\left(\frac{\eta \dot{R}}{\sigma}\right)^{0.702}\right] = \tanh\left[4.96\left(\bar{C}_a \frac{d\bar{R}}{d\bar{t}}\right)^{0.702}\right] = \frac{\cos\bar{\alpha}_{0+} - \cos\bar{\alpha}}{\cos\bar{\alpha}_{0+} - 1} \quad \text{if } \dot{R} > 0 \quad (5a)$$

$$\tanh\left[4.96\left(\frac{\eta |\dot{R}|}{\sigma}\right)^{0.702}\right] = \tanh\left[4.96\left(\bar{C}_a \left|\frac{d\bar{R}}{d\bar{t}}\right|\right)^{0.702}\right] = \frac{\cos\bar{\alpha}_{0-} - \cos\bar{\alpha}}{\cos\bar{\alpha}_{0-} + 1} \quad \text{if } \dot{R} < 0 \quad (5b)$$

Figure 2 shows the typical behavior of the contact angle as a function of $\dot{R}$. (In this plot the static contact angle values are around 45°) The cavitation number $\bar{C}_a$ is usually large enough to render the value of the tanh function near one. Accordingly, $\cos\bar{\alpha} \approx 1$ (or $\bar{\alpha} \approx 0$) when the cavitation region expands and $\cos\bar{\alpha} \approx -1$ (or $\bar{\alpha} \approx \pi$) when it shrinks. Then, Eq. (4) shows that the oil film pressure adjacent to the meniscus is less than p_v when the cavitation region expands and greater than p_v when it shrinks. Therefore, the measured pressure trace at a given point should display a dip in pressure just before the rupture boundary arrives, whereas it should display no such pressure dip just after the reformation boundary passes over.

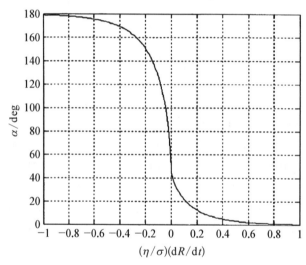

Fig. 2 Contact angle dependence on the meniscus velocity

4 Analysis

With the approximations based on small film thickness ($\bar{\delta} \ll 1$), the flow in the full-film region ($R < r < a$) is governed by the following:

continuity equation
$$\frac{1}{r}\frac{\partial}{\partial r}(rv_r) + \frac{\partial v_z}{\partial z} = 0 \quad (6)$$

r-momentum equation
$$\frac{\partial p}{\partial r} = \eta \frac{\partial^2 v_r}{\partial z^2} \quad (7)$$

z-momentum equation
$$\frac{\partial p}{\partial z} = 0 \quad (8)$$

velocity boundary conditions

$$v_r = 0 \text{ and } v_z = 0 \text{ at } z = 0 \tag{9a}$$

$$v_r = 0 \text{ and } v_z = \dot{h} \text{ at } z = h \tag{9b}$$

pressure boundary condition at ambient

$$p = p_a \text{ at } r = a \tag{10}$$

Equation (8) shows that the pressure is a function of r and t only. Hence, Eq. (7) can be integrated (twice) w. r. t. z, and by using the velocity boundary conditions in (9), to give:

$$v_r = -\frac{z(h-z)}{2\eta}\frac{\partial p}{\partial r} \tag{11}$$

For later discussions it is convenient to define a volume flow rate Q as:

$$Q \equiv 2\pi r \int_0^h v_r \, dz = -\frac{\pi r h^3}{6\eta}\frac{\partial p}{\partial r} \tag{12}$$

Equation (6) may be integrated w. r. t. z across the oil film thickness, and by using the velocity boundary conditions in (9), to give:

$$\frac{\partial Q}{\partial r} + 2\pi r \dot{h} = 0 \tag{13}$$

which, with the substitution of the expression for Q as given in Eq. (12), transforms to the celebrated Reynolds equation:

$$\frac{\partial}{\partial r}\left(r\frac{\partial p}{\partial r}\right) - 12\eta r \frac{\dot{h}}{h^3} = 0 \tag{14}$$

This second-order ODE in r (where t is just a parameter) can be readily solved to give:

$$p - 3\eta r^2 \frac{\dot{h}}{h^3} = A \ln r + B \tag{15}$$

where A and B are two integration functions of t.

During the time period when cavitation is absent, the two integration functions A and B can be determined by the pressure boundary condition (10) and the condition of symmetry at $r = 0$. The pressure is then given by:

$$p = p_a - 3\eta(a^2 - r^2)\frac{\dot{h}}{h^3} \text{ for } 0 \leqslant r \leqslant a \tag{16a}$$

When cavitation is present, A and B are determined by the pressure boundary conditions (10) and (4). The resulting expression for the pressure is:

$$p = 3\eta r^2 \frac{\dot{h}}{h^3} - 3\eta \frac{\dot{h}}{h^3}\frac{1}{\ln\frac{a}{R}}\left[a^2 \ln\frac{r}{R} + R^2 \ln\frac{a}{r}\right] + \frac{1}{\ln\frac{a}{R}}$$

$$\times \left\{ p_a \ln \frac{r}{R} + \left[p_v - \sigma\left(\frac{1}{R} + \frac{2\cos\bar{\alpha}}{h}\right) \right] \ln \frac{a}{r} \right\} \text{ for } R \leqslant r \leqslant a \qquad (17a)$$

During the expansion of the cavitation region, the postulated flow continuity condition is:

$$Q_{r=R} = 2\pi R(1-\bar{\beta})h\dot{R} \text{ for } \dot{R} > 0 \qquad (18)$$

where $\bar{\beta}$ is a numerical factor between 0 and 1. Since $\dot{R}$ is taken to be the largest v_r, the volume expansion rate of the cavitation region, $2\pi Rh\dot{R}$, is more than the volume flow rate as given by Eq. (12). Thus, the former should be reduced by a factor $(1-\bar{\beta})$ as shown in the flow continuity condition. If the radial flow velocity given by Eq. (11) is still accurate at and near the cavitation boundary, the value of $\bar{\beta}$ is readily calculated to be 1/3. Equation (18) also reveals that the oil left behind the cavitation boundary has an equivalent film thickness $\bar{\beta}h$. By substituting into Eq. (18) the expressions of Q and p from Eqs. (12) and (17a), respectively, and rearranging, one obtains:

$$\frac{d}{dt}(R^2) = \left[\frac{a^2 - R^2}{\ln \frac{a^2}{R^2}} - R^2 \right] \frac{\dot{h}}{(1-\bar{\beta})h} - \frac{h^2}{(1-\bar{\beta})3\eta \ln \frac{a^2}{R^2}}$$

$$\times \left\{ p_a - \left[p_v - \sigma\left(\frac{1}{R} + \frac{2\cos\bar{\alpha}}{h}\right) \right] \right\} \text{ for } \dot{R} > 0 \qquad (19a)$$

During shrinking of the cavitation region, because the residual oil filaments in the cavitation region join the inward-flowing continuous film, the inward volume flow rate need only match a reduced volume reduction rate of the cavitation region. The corresponding flow continuity condition, therefore, takes the form:

$$Q_{r=R} = 2\pi R(h - \bar{\beta}h_{past})\dot{R} \text{ for } \dot{R} < 0 \qquad (20)$$

where h_{past} denotes the previous gap thickness at $r=R$ when $\dot{R} > 0$. By substituting the expressions of Q and p from Eqs. (12) and (17a), respectively, into Eq. (20), one obtains:

$$\frac{d}{dt}(R^2) = \left[\frac{a^2 - R^2}{\ln \frac{a^2}{R^2}} - R^2 \right] \frac{\dot{h}}{h - \bar{\beta}h_{past}} - \frac{h^3}{3\eta(h - \bar{\beta}h_{past})\ln \frac{a^2}{R^2}}$$

$$\times \left\{ p_a - \left[p_v - \sigma\left(\frac{1}{R} + \frac{2\cos\bar{\alpha}}{h}\right) \right] \right\} \text{ for } \dot{R} < 0 \qquad (21a)$$

Using the dimensionless variables and parameters defined in the Nomenclature, Eqs. (16a), (17a), (19a), and (21a) take the following dimensionless form:

$$\bar{p} = 1 - 3\bar{\lambda} \frac{1}{\bar{h}^3} \frac{d\bar{h}}{d\bar{t}}(1-\bar{r}^2) \text{ for } 0 \leqslant \bar{r} \leqslant 1 \qquad (16b)$$

$$\bar{p} = 1 - 3\bar{\lambda} \frac{1}{\bar{h}^3} \frac{d\bar{h}}{d\bar{t}}(1-\bar{r}^2) + 3\bar{\lambda} \frac{1}{\bar{h}^3} \frac{d\bar{h}}{d\bar{t}}(1-\bar{R}^2) \frac{\ln \bar{r}}{\ln \bar{R}}$$

$$-\left\{1-\left[\bar{p}_v-\frac{\bar{\lambda}\bar{\delta}^2}{\bar{C}_a}\left(\frac{1}{\bar{R}}+\frac{2\cos\bar{\alpha}}{\bar{\delta}\bar{h}}\right)\right]\right\}\frac{\ln\bar{r}}{\ln\bar{R}} \quad \text{for } \bar{R}\leqslant\bar{r}\leqslant 1 \qquad (17b)$$

$$\frac{d}{d\bar{t}}(\bar{R}^2)=-\left[\frac{1-\bar{R}^2}{\ln(\bar{R}^2)}+\bar{R}^2\right]\frac{1}{(1-\beta)\bar{h}}\frac{d\bar{h}}{d\bar{t}}+\frac{\bar{h}^2}{3\bar{\lambda}(1-\beta)\ln(\bar{R}^2)}$$

$$\times\left\{1-\left[\bar{p}_v-\frac{\bar{\lambda}\bar{\delta}^2}{\bar{C}_a}\left(\frac{1}{\bar{R}}+\frac{2\cos\bar{\alpha}}{\bar{\delta}\bar{h}}\right)\right]\right\} \quad \text{for } \frac{d\bar{R}}{d\bar{t}}>0 \qquad (19b)$$

$$\frac{d}{d\bar{t}}(\bar{R}^2)=-\left[\frac{1-\bar{R}^2}{\ln(\bar{R}^2)}+\bar{R}^2\right]\frac{1}{\bar{h}-\beta\bar{h}_{past}}\frac{d\bar{h}}{d\bar{t}}+\frac{\bar{h}^3}{3\bar{\lambda}(\bar{h}-\beta\bar{h}_{past})\ln(\bar{R}^2)}$$

$$\times\left\{1-\left[\bar{p}_v-\frac{\bar{\lambda}\bar{\delta}^2}{\bar{C}_a}\left(\frac{1}{\bar{R}}+\frac{2\cos\bar{\alpha}}{\bar{\delta}\bar{h}}\right)\right]\right\} \quad \text{for } \frac{d\bar{R}}{d\bar{t}}<0 \qquad (21b)$$

where $\cos\bar{\alpha}$ is given by Eqs. (5). The dimensionless form of the gap function that drives the cavitation phenomenon is

$$\bar{h}=1-\bar{\varepsilon}\sin\bar{t} \quad \text{for } \bar{t}\geqslant 0 \qquad (1b)$$

To solve for $\bar{R}(\bar{t})$ one needs to integrate ODE (19b) from an initial value R_o, which is taken to be the size of the cavitation nucleus. Therefore, the problem involves seven dimensionless parameters, viz. $\bar{R}_o$, $\bar{p}_v$, $\bar{\delta}$, $\bar{C}_a$, $\bar{\alpha}_o$ (including its hysteresis), $\bar{\lambda}$, and $\bar{\varepsilon}$. The dimensionless cavitation threshold pressure $\bar{p}_c$ is related to $\bar{R}_o$ and $\bar{p}_v$.

A comment on the conditions of determining the rupture boundary may be in order. If, in addition to (4) and (10), one also applies the zero pressure-gradient condition:

$$\frac{\partial p}{\partial r}=0 \text{ at } r=R \qquad (22)$$

to the solution given in (15), the rupture boundary can be determined along with the integration functions A and B. The resulting equation for R is

$$\left[\frac{a^2-R^2}{\ln\frac{a^2}{R^2}}-R^2\right]\frac{\dot{h}}{h^3}-\frac{1}{3\eta\ln\frac{a^2}{R^2}}\left\{p_a-\left[p_v-\sigma\left(\frac{1}{R}+\frac{2\cos\bar{\alpha}}{h}\right)\right]\right\}=0 \qquad (23)$$

Then R can be determined algebraically, without invoking the continuity condition (18), which results in an ODE for R. In other words R at any instant is solely determined by h and $\dot{h}$ at the instant. In fact, by comparing Eq. (23) with ODE (19a), one can see that the R obtained from Eq. (23) is the maximum R given by the ODE. Therefore, by using any of the existing cavitation algorithms to treat the present problem, one would obtain an instant (at $t=0$) finite-size cavitation region, instead of one growing gradually from a cavitation nucleus, and just compute the shrinking of the region. This result was demonstrated in an earlier work (Boedo and Booker [13]; Fig. 10).

5 Method of Computation

The oil film response was computed numerically starting from $\bar{t}=0$ for two cycles of gap variation as prescribed by Eq. (1b). The center pressure was calculated according to Eq. (16b) until it decreased to $\bar{p}_c$, when the cavitation was considered to commence. Then, Eq. (19b) was integrated by using a MATLAB function "ode23" (23) from an assumed nucleus size $\bar{R}_o$. When $d\bar{R}/d\bar{t}$ became negative, Eq. (21b) was integrated until $\bar{R}$ decreased to zero, when the cavitation was considered to have ended. After $\bar{R}$ was obtained for a time step, $d\bar{R}/d\bar{t}$ was evaluated by a backward difference, which was then used to calculate $\bar{\alpha}$ according to Eqs. (5). The slight difference between the two static contact angle values was ignored. The pressure distribution in the oil film in the absence or presence of cavitation was calculated according to Eq. (16b) or (17b), respectively.

6 Results and Discussion

The following dimensions, operating conditions, and assumed oil properties were selected for computation:

$a = 25$ mm, $h_o = 0.25$ mm, $p_a = 101.3$ kPa, $\omega = 25$ Hz,
$\varepsilon = 0.125$ mm, $\eta = 0.096$ Pa·s, $\sigma = 0.035$ N/m, $p_v = 30$ Pa,
$R_o = 0.1$ mm, $\bar{\alpha}_o = 14$ deg

which result in the following values of the dimensionless parameters:

$\bar{R}_o = 0.004$, $\bar{p}_v = 0.0003$, $\bar{\delta} = 0.01$,
$\bar{C}_a = 10.77$, $\bar{\lambda} = 1.489$, $\bar{\varepsilon} = 0.5$

The computation was carried out by using $\bar{\beta} = 1/3$ and $\bar{p}_c = -0.006$, the latter obtained from Eq. (4) with the above R_o, h_o, p_v, and $\bar{\alpha}_o$ values. The results are shown in Figs. 3–6.

Figure 3 shows the controlled gap variation and the corresponding variations of center pressure and cavitation size. It is seen that the center pressure peaks before the gap decreases to its minimum, and the refilling process begins during the increase of the gap. The increase of R is milder than what is reported in the literature (Optasanu and Bonneau[14]; Figs. 4–6; Ku and Tichy[22]; Figs. 5a-d). It is believed that the faster growth of the rupture boundary in these works was caused by the application of the zero pressure-gradient boundary condition. On the contrary, Hays and Feiten ([9]; Figs. 10 and 11) used a continuity condition (same as Eq. (19a) but without the surface tension term) instead, and found a mild growth of the rupture boundary. Figure 3(b) shows that at the end of cavitation there is a sudden pressure rise. This is because once cavitation ends the pressure is determined from Eq. (16b) with the $\bar{h}$ and $d\bar{h}/d\bar{t}$ values at that

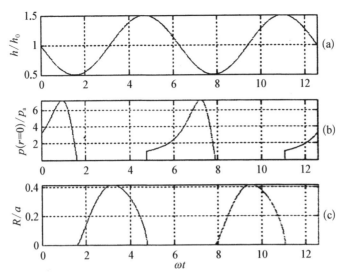

Fig. 3 Variation of (a) film thickness, (b) pressure at center, and (c) cavitation boundary with time ($\bar{\lambda}=1.489$, $\bar{C}_a=10.77$, $\bar{\varepsilon}=0.5$)

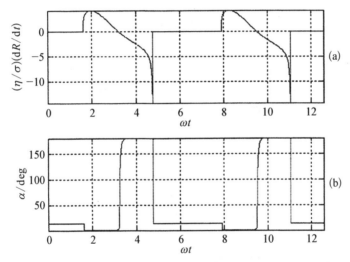

Fig. 4 Variation of (a) meniscus velocity and (b) contact angle with time ($\bar{\lambda}=1.489$, $\bar{C}_a=10.77$, $\bar{\varepsilon}=0.5$)

moment. Hence, the pressure slope is discontinuous at the transition.

The corresponding variations of the meniscus velocity and contact angle are shown in Fig. 4. It is seen that the contact angle essentially alternates between zero and 180° during cavitation. In the absence of cavitation, no meniscus is present, then the contact angle loses its meaning. During such period the assumed static contact angle value is plotted in Fig. 4(b) to distinguish the period from when cavitation is present.

A few snapshots of the pressure distribution in the absence and presence of cavitation are shown in Fig. 5(a) and Fig. 5(b), respectively. The curve "e" in Fig. 5(a) shows a situation where cavitation is imminent. At this instant the film thickness has passed its minimum, and squeeze velocity has become pulling velocity. Curves "a" and "b" in

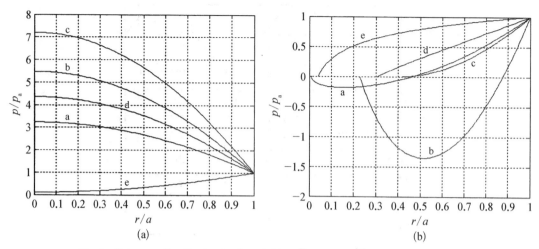

Fig. 5 Pressure distribution in the oil film ($\bar{\lambda}=1.489$, $\bar{C}_a=10.77$, $\bar{\varepsilon}=0.5$)

(a) In the absence of cavitation a: $\omega t=0$; b: $\omega t=0.503$; c: $\omega t=1.005$; d: $\omega t=1.370$; e: $\omega t=1.621$. (b) In the presence of cavitation a: $\omega t=1.646$; b: $\omega t=2.249$; c: $\omega t=3.255$; d: $\omega t=4.084$; e: $\omega t=4.738$.

Fig. 5(b) are taken during the growth of the cavitation region. Notice that the pressure gradient at the rupture boundary is not zero, and the tensile stress occurs in a significant portion of the oil film. Ignoring this tensile stress would introduce appreciable error in assessing the damping capacity of the squeeze film. Curves "d" and "e" in Fig. 5(b) are taken during the shrinking of the cavitation region. During this period the pressure gradient is positive everywhere in the full-film region, consistent with the refilling process. Notice that curve "e" shows a situation where the cavitation region is about to collapse. The steep pressure gradient at the reformation boundary at this instant explains the large meniscus speed occurring at the verge of cavitation collapse, as shown in Fig. 4(a). The curve "c" in Fig. 5(b) is taken when the cavitation region is at its maximum. Notice that the pressure gradient at the cavitation boundary at this instant is zero.

The pressure traces at selected locations are shown in Fig. 6, where the corresponding pressure traces without the occurrence of cavitation are also plotted for comparison. Figure 6(a) shows the pressure trace at a location very close to the center, $r/a=0.05$. It is seen that before the rupture boundary arrives at this location the pressure here dips below p_v. After the refilling process forces the reformation boundary to recede past this location, but before the collapse of the cavitation region, the pressure trace here is given by Eq. (17b). Eventually, with the disappearance of cavitation everywhere in the oil film, the pressure trace joins the one given by Eq. (16b). Figure 6(b) shows the pressure trace at $r/a=0.2$, which is still located within the maximum radius that cavitation can reach. It is seen that the magnitude of tensile stress is more pronounced before the rupture boundary arrives, and the transition of pressure from the cavitated segment to the uncavitated one is less abrupt. Figure 6(c) shows the pressure trace at $r/a=0.5$, which lies outside the maximum radius that cavitation can reach. The segment of the

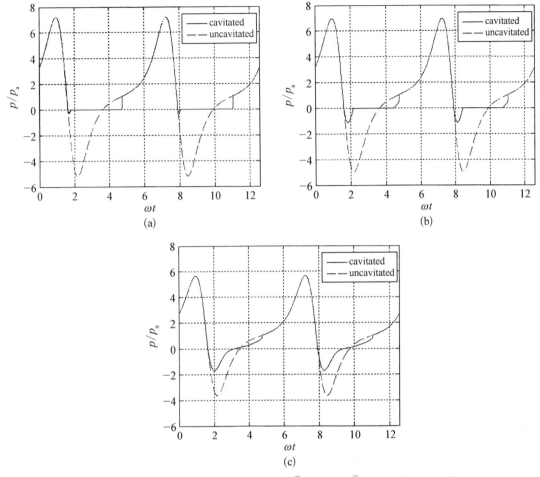

Fig. 6 Variation of pressure with time ($\bar{\lambda}=1.489$, $\bar{C}_a=10.77$, $\bar{\varepsilon}=0.5$)
(a) at $r/a = 0.05$, (b) at $r/a = 0.2$, (c) at $r/a = 0.5$.

pressure trace that deviates from the dashed curve is clearly given by Eq. (17b). Note that, while the range of pressure variation is smaller at this location, the magnitude of the tensile stress is larger.

The similarity between these computed pressure traces and the measured ones reported in the literature, e. g., by Parkins and May-Miller ([12]; Fig. 3); by Ku and Tichy ([22]; Fig. 4); and by Sun, et al. ([4]; Figs. 8 – 13), is evident. The computed magnitudes of the tensile stresses are on the same order of those measured in Sun, et al.[4] These magnitudes are explicable because tensile stresses are produced by pulling one bearing surface away from another, and the fluid film lubrication mechanism can only generate such magnitudes for relatively large bearing clearances. The qualitative agreement with literature results provides preliminary validity of the proposed model. More detailed comparisons between the model predictions and direct experimental results are contained in Sun, et al.[15]

6.1 Effect of $\bar{\varepsilon}$ and $\bar{\lambda}$

The results of a smaller amplitude of oscillation ($\bar{\varepsilon}=0.3$) are shown in Fig. 7. In comparison with the case of $\bar{\varepsilon}=0.5$, the maximum cavitation size is reduced and the cavitation duration is shorter. Pressure traces at the three chosen locations in this case preserve the same features as given in Fig. 6. They are grouped together in Fig. 8. The results of a smaller squeeze number ($\bar{\lambda}=0.893$) are shown in Figs. 9 and 10. This case was obtained by choosing $\omega=15$ Hz; hence, the cavitation number was also changed. The results of reducing the squeeze number are qualitatively the same as those of reducing the oscillatory amplitude.

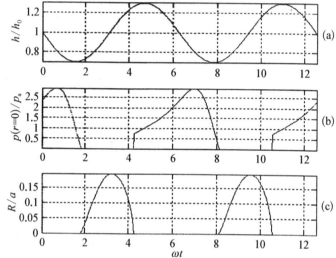

Fig. 7 Variation of (a) film thickness, (b) pressure at center, and (c) cavitation boundary with time ($\bar{\lambda} = 1.489$, $\bar{C}_a = 10.77$, $\bar{\varepsilon} = 0.3$)

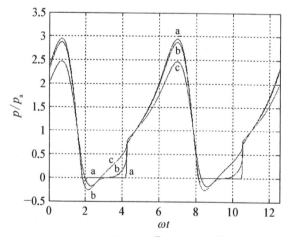

Fig. 8 Variation of pressure with time ($\bar{\lambda} = 1.489$, $\bar{C}_a = 10.77$, $\bar{\varepsilon} = 0.3$)

a: at $r/a=0.05$; b: at $r/a=0.2$; c: at $r/a=0.5$.

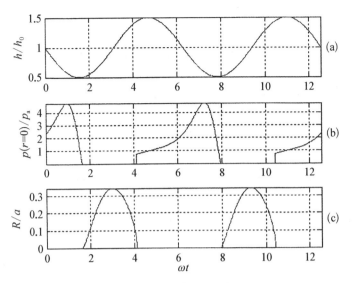

Fig. 9 Variation of (a) film thickness, (b) pressure at center, and (c) cavitation boundary with time ($\bar{\lambda} = 0.893$, $\bar{C}_a = 6.463$, $\bar{\varepsilon} = 0.5$)

6.2 Effect of $\bar{C}_a$ and $\bar{\alpha}$

Both these parameters are associated with surface tension. As mentioned in the section on the role of surface tension and contact angle, $\bar{C}_a$ is usually large enough to render $\bar{\alpha}$ alternating between 0 and π. The magnitude of the dominant surface tension term relative to pressure can be determined from Eq. (4) to be of the order $\sigma/(p_a h_o)$. When σ is unusually large and h_o exceedingly small, the influences of the surface tension coefficient and contact angle can be significant. But in the present example their effects are negligible.

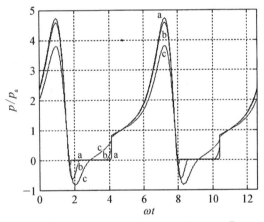

Fig. 10 Variation of pressure with time ($\bar{\lambda} = 0.893$, $\bar{C}_a = 6.463$, $\bar{\varepsilon} = 0.5$)

a: at $r/a = 0.05$; b: at $r/a = 0.2$; c: at $r/a = 0.5$.

6.3 Effect of $\bar{p}_v$ and $\bar{R}_o$

In the computed examples the cavitation pressure is taken to be 30 Pa because this value represents the vapor pressure level of oil at ordinary temperatures. The size of the cavitation nucleus is taken to be 0.1 mm because this is a size naked eyes can resolve. While such a choice of the nucleus size may appear naïve, detailed calculations for a range of nuclear sizes reveal no pronounced effects on the cavitation development regarding its size, shape, and duration. It is recognized that the nucleus size is crucially related to the threshold pressure, p_c, which controls the onset of cavitation. But once the liquid is ruptured, vapor pressure prevails in the cavitation region and the $\bar{R}_o$ value used for computing the subsequent cavitation development becomes uncritical. More discussions on the nucleus size aspect are contained in Sun, et al.[15], in conjunction with the comparison between the model predictions and the measured results.

6.4 Effect of $\bar{\beta}$

This parameter is introduced because of our ignorance of the exact distribution of the radial velocity in the film thickness direction at the rupture boundary. If the radial velocity profile given in Eq. (11) is still applicable there, then $\bar{\beta}$ is readily calculated to be 1/3, which is used in the computed examples. With this parameter the cavitation model also brings forth a plausible explanation for the observed presence of residual oil filaments in the cavitation region. Quantitatively, $\bar{\beta}=0$ would mean all the oil, whereas $\bar{\beta}=1$ would mean no oil, is driven out of the cavitation region. The latter case, however, is equivalent to prescribing the zero pressure-gradient condition (22), as is evident from Eqs. (12) and (18), which then suppresses the occurrence of tensile stresses in the oil film.

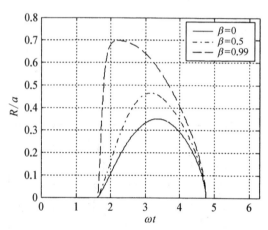

Fig. 11 Variation of cavitation boundary with time for several $\bar{\beta}$ values ($\bar{\lambda} = 1.489$, $\bar{C}_a = 10.77$, $\bar{\varepsilon} = 0.5$)

The influence of $\bar{\beta}$ on cavitation development is quite pronounced. This is shown in Fig. 11. As $\bar{\beta}$ is increased toward 1, the maximum cavitation size becomes larger, and the rupture boundary expands more rapidly. When $\bar{\beta} = 1$, Eq. (19b) becomes singular and may not be used. In this case, Eq. (23) can be used to solve for the rupture boundary.

7 Concluding Remarks

A new model that describes the phenomenon of dynamic cavitation in submerged oil film bearings is proposed. The model allows the occurrence of tensile stresses in the oil film by employing a flow continuity condition at the rupture boundary of the cavitation region. The work is aimed at obtaining an improved method for analyzing the performance of dynamically loaded bearings. Admittedly the model is a primitive one, because it only addresses the life cycle of a single cavitation region, which begins at a place where the fluctuation of pressure is the largest in an unruptured oil film. Besides, the model makes no attempt to address the dynamical aspects of bubble development, and neither the thermodynamics nor the chemical aspects of nucleation process. Our main interest is in the oil film behavior in the full-film region.

The model is analyzed for the simple geometry of a parallelplate oscillatory squeeze film. The results obtained from this model, on the variation of pressure in the oil film, compare favorably with available measurements reported in the literature, as mentioned previously. However, a direct validation of the model by experiment is deemed necessary. Hence, a carefully designed parallel-plate squeeze film bearing rig was built and experimental work was carried out. This companion experimental study is reported in Sun, et al.[15]

Acknowledgments

The research was jointly supported by China's State Key Basic Research Development Program, under Grant G1998020317 - 4. 2; and Shanghai's Baiyulan Foundation for Talents in Science and Technology, under Grant 200328. The authors also gratefully acknowledge Professor Fang Minglun, Senior Provost of Shanghai University, for his encouragement as well as space and equipment support.

References

[1] Dowson, D. and Taylor, C. M. (1979), "Cavitation in Bearings," Ann. Rev. Fluid Mech., 11, pp. 35 - 66.
[2] San Andres, L. and Diaz, S. E. (2003), "Flow Visualization and Forces from a Squeeze Film Damper Operating with Natural Air," ASME Journal of Tribology, 125, pp. 325 - 333.
[3] Sun, D. C. and Brewe, D. E. (1991), "A High Speed Photography Study of Cavitation in a Dynamically Loaded Journal Bearing," ASME Journal of Tribology, 113, pp. 287 - 294.
[4] Sun, D. C., Brewe, D. E. and Abel, P. B. (1993), "Simultaneous Pressure Measurement and High-Speed Photography Study of Cavitation in a Dynamically Loaded Journal Bearing," ASME Journal of Tribology, 115, pp. 88 - 95.
[5] Elrod, H. G. and Adams, M. L. (1974), "A Computer Program for Cavitation and Starvation Problems," Proceedings of Leeds-Lyon Conference on Cavitation, Paper II(ii).
[6] Kumar, A. and Booker, J. F. (1991), "A Finite Element Cavitation Algorithm," ASME Journal of Tribology, 113, pp. 276 - 286.
[7] Jakobsson, B. and Floberg, L. (1957), "The Finite Journal Bearing, Considering Vaporization," Trans. Chalmers University of Technology, No. 190.
[8] Olsson, K. (1965), "Cavitation in Dynamically Loaded Bearings," Trans. Chalmers University of Technology, No. 308.
[9] Hays, D. F. and Feiten, J. B. (1964), "Cavities Between Moving Parallel Plates," Cavitation in Real Liquids, R. Davies, Ed., Elsevier, pp. 122 - 137.
[10] Rodrigues, A. N. (1970), "An Analysis of Cavitation in a Circular Squeeze Film and Correlation with Experimental Results," PhD Thesis, Cornell University, Ithaca, New York.
[11] Kuroda, S. and Hori, Y. (1978), "An Experimental Study on Cavitation and Tensile Stress in a Squeeze Film," Journal of Japan Society of Lubrication Engineers, 23(6), pp. 436 - 442. (in Japanese)
[12] Parkins, D. W. and May-Miller, R. (1984), "Cavitation in an Oscillatory Oil Squeeze Film," ASME Journal of Tribology, 106, pp. 360 - 367.
[13] Boedo, S. and Booker, J. F. (1995), "Cavitation in Normal Separation of Square and Circular Plates," ASME Journal of Tribology, 117, pp. 403 - 410.

[14] Optasanu, V. and Bonneau, D. (2000), "Finite Element Mass-Conserving Cavitation Algorithm in Pure Squeeze Motion, Validation/Application to a Connecting-Rod Small End Bearing," ASME Journal of Tribology, 122, pp. 162–169.

[15] Sun, Meili, Zhang, Zhiming, Chen, Xiaoyang, Wang, Wen, Meng, Kai and Sun, D. C. (2008), "Experimental Study of Cavitation in an Oscillatory Oil Squeeze Film," Tribology Transactions, 51, pp. 341–350.

[16] Sun, D. C. and Brewe, D. E. (1992), "Two Reference Time Scales for Studying the Dynamic Cavitation of Liquid Films," ASME Journal of Tribology, 114, pp. 612–615.

[17] Defay, R. and Prigogine, I. (1966), Surface Tension and Adsorption, Translated by D. H. Everett, Longmans.

[18] Plesset, M. S. and Prosperetti, A. (1977), "Bubble Dynamics and Cavitation," Annual Review of Fluid Mechanics, 9, pp. 145–185.

[19] Prosperetti, A. (1982), "A Generalization of the Rayleigh-Plesset Equation of Bubble Dynamics," Physics of Fluids, 25, No. 3, pp. 409–410.

[20] de Gennes, P. G. (1985), "Wetting: Statics and Dynamics," Reviews of Modern Physics, 57(3), pp. 827–863.

[21] Jiang, T. S., Oh, S. G. and Slattery, J. C. (1979), "Correlation for Dynamic Contact Angle," J. of Colloid and Interface Sci, 69, p. 74.

[22] Ku, C. P. and Tichy, J. A. (1990), "An Experimental and Theoretical Study of Cavitation in a Finite Submerged Squeeze Film Damper," ASME Journal of Tribology, 112, pp. 725–733.

[23] The Math Works, Inc. (2005), MATLAB, Version 7.1.

Experimental Study of Cavitation in an Oscillatory Oil Squeeze Film[*]

Abstract: A displacement-controlled parallel-plate squeeze film test apparatus was designed and built to study the cavitation of oil films. Simultaneous pressure measurement and high-speed photography were conducted in the experiment. The study was aimed at comparing the measured results with the prediction of a recently proposed cavitation model (Sun, et al. [1]). Among the findings are: (a) When cavitation did not occur, the measured pressure traces agreed well with the theory of non-cavitated oil squeeze film. In particular, tensile stresses of significant magnitudes were measured in the oil film. (b) When cavitation occurred, the measured pressure traces agreed well with the prediction of the new cavitation model. In particular, tensile stresses were measured in the oil film just before the arrival of the expanding cavitation front. (c) The measured cavitation duration correlated well with the new cavitation model. But the measured growth of the cavitation region was faster, and its maximum size was larger, than the model prediction. An analysis of the size effect of the cavitation nuclei, which were likely the entrained air bubbles, provided plausible explanations for the non-occurrence or occurrence of cavitation, as well as for some of the discrepancies between the measured results and model predictions.

Nomenclature

a = radius of parallel-plate thrust bearing
h_m = minimum gap between the thrust plates, = $h_o - \varepsilon$
h_o = mean gap during oscillations
p = oil film pressure (in absolute scale)
p_a = ambient pressure in the oil pool
p_c = threshold pressure at the onset of cavitation
p_{in} = pressure inside an air bubble
p_{out} = pressure outside an air bubble
p_v = oil vapor pressure
$R(t)$ = radius of cavitation region
R_o = assumed size of cavitation nucleus
$\hat{R}$ = radius of air bubble
t = time
r = radial coordinate
$\bar{\beta}$ = a dimensionless parameter used in the new cavitation model (Sun, et al. [1])
$\bar{\gamma}$ = adiabatic exponent, = 1.4 for air
ε = oscillatory amplitude
σ = surface tension coefficient of oil
ω = oscillatory frequency

1 Introduction

The paper describes an experimental study aimed at validating a new cavitation model (Sun et al. [1]). The new model was proposed for improving the existing methods of

---

[*] In collaboration with SUN MEILI, CHEN XIAOYANG, WANG WEN, MENG KAI and D. C. SUN. Reprinted from *Tribology Transactions*, 2008, 51: 341 – 350.

computing the performance of dynamically loaded bearings, which fail to produce the observed tensile stresses in the oil film.

The presence of tensile stresses in the oil films of journal bearings has been known for many years. Dyer and Reason[2] presented an extensive literature review on the subject and measured tensile stresses up to 740 kPa in a steadily loaded journal bearing. Their test bearing was steadily loaded, but their pressure sensor was mounted on the rotating journal. As the sensor went through a "termination region" and, presumably, experienced the separation of the opposing bearing surface, it registered tensile stresses. They noted the sporadic nature of the phenomenon, in the sense that two types of behaviors both existed, the normal one being the "termination" pressure equal to the oil vapor pressure, and the unusual one where tensile stresses manifested. Likewise, by mounting a pressure transducer on the rotating journal, Nakai and Okino[3] detected tensile stresses in the oil film of steadily loaded journal bearings. Natsumeda and Someya[4] measured significant tensile stresses with a pressure pickup mounted on the rotating journal while the bearing sleeve was loaded. In the case of static load, the maximum tensile stress detected was 0.6 MPa, and in the case of dynamic load the maximum one was 1.2 MPa. Sun, et al.[5] carried out a simultaneous pressure measurement and high-speed photography study of cavitation in a dynamically loaded journal bearing. The largest tensile stress measured in their study was 140 kPa. They further observed that (a) when cavitation occurred the pressure in the cavitation region was close to the oil vapor pressure, while outside the cavitation region tensile stresses existed; (b) cavitation sometimes did not occur for the same operating conditions; then the tensile stresses measured were larger than otherwise; and (c) the non-occurrence or occurrence of cavitation was not predictable but was quite persistent once a type of behavior was established. The above observations may be considered credible, yet they are still not explicable.

On the other hand, analyses to date of dynamically loaded oil film bearings do not predict tensile stresses. The prevailing method used in such analyses is through various cavitation algorithms (e.g., Elrod and Adams[6]; Kumar and Booker[7]). In these cavitation algorithms a zero pressure-gradient condition (or Reynolds boundary condition) is prescribed at the upstream (or rupture) boundary of the cavitation region, which results in the pressure level everywhere in the oil film being not lower than the cavitation pressure. Recently, a new model of dynamic cavitation was proposed (Sun, et al.[1]) that replaced the zero pressure-gradient boundary condition with a flow continuity condition, and in this way allowed the occurrence of tensile stresses in the oil film. The model was applied to the simple configuration of a parallel-plate oscillatory oil squeeze film bearing, and the results of the analysis compared favorably with the available measurements reported in the literature. However, a direct validation of the model by experiment was deemed necessary; hence, the present work.

Experimental studies of cavitation in parallel-plate oscillatory oil squeeze film bearings have been carried out by previous researchers. Hays and Feiten[8] photographed the growth and collapse of cavitation but did not measure the pressure in the oil film. Kuroda and Hori[9]* detected tensile stresses at many points in the oil film, with the largest magnitude being 2.5 bar. Parkins and May-Miller[10] (and many publications from the Cranfield Institute of Technology cited therein) photographed the cavitation pattern and measured the pressure in the oil film but did not detect tensile stresses in the oil film. Chen, et al. [11] built a well-thought-out parallel-plate squeeze film test apparatus and with it obtained clear photographs of cavitation as well as measured tensile stresses in the oil film. But, because the oscillatory motion was driven with an electromagnetic exciter, as was done in Parkins and May-Miller[10], the displacement versus time relationship was rather involved, and a direct correlation among the displacement, the measured pressure, and the photographed cavitation pattern was difficult to establish. Hence, the experiment was repeated with another modification of the test apparatus, viz. the oscillatory motion was driven with a meticulously designed crank-linkage mechanism. This paper describes the new design and the measured results obtained with the new test apparatus.

2 Test Apparatus

A schematic view of the apparatus is shown in Fig. 1(a). The squeeze film bearing consisted of a 65-mm-diameter lower steel flat plate[9] and a 50-mm-diameter upper glass plate[10]. The lower thrust plate, where the pressure transducers[6] were installed, was fixed to the base plate[8]. Three displacement transducers[7] were mounted on the base plate, which in turn was fixed to the stationary housing[4]. The upper glass plate, made from an optical flat approximately 25 mm thick, was fixed to the movable frame[5]. The parallelism between the lower and upper thrust plates was controlled by two levels of taut steel wires (three in each level), one end of which was attached to the movable frame and the other end anchored to the stationary housing by

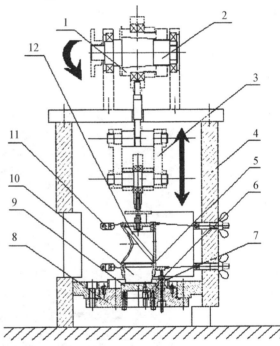

Fig. 1(a) Schematic of the parallel-plate squeeze film test apparatus

1. eccentric sleeve, 2. eccentric shaft, 3. linkage box, 4. stationary housing (of outside diameter 275 mm and height 330 mm), 5. movable frame, 6. pressure transducers, 7. displacement transducers, 8. base plate, 9. lower thrust plate, 10. upper glass plate fixed to the movable frame, 11. pull-rods and taut steel wires, 12. 45° mirror supported on the stationary housing.

---

* We became aware of this work after our experiment was concluded, from Professor Hori's recent publication, *Hydrodynamic Lubrication*, Springer-Verlag, pp. 145–149 (2006).

means of threaded pull rods[11]. By means of an adjustable coupling, the movable frame was hung on a linkage box[3], which was connected to the eccentric sleeve[1] by means of a crank. The driving system, mounted on the stationary housing, consisted of a variable-frequency motor and an eccentric shaft[2], whose relative position with the eccentric sleeve[1] could be adjusted (to provide the desired oscillatory amplitude) and then fixed by a jam nut. The true vertical motion of the movable frame was essentially assured by the taut steel wires. Tolerance and backlash in the crank-linkage mechanism were carefully controlled to limit the error in the displacement setting to within several micrometers.

The squeeze film bearing was submerged in an oil pool confined by a ring wall (not shown) extended about 8 mm above the squeeze film. The oil used was ISO-VG #68 whose kinematic viscosity and density values were, respectively, 57.42 mm^2/s and 865 kg/m^3 at 40℃ and 7.56 mm^2/s and 832.3 kg/m^3 at 100℃.

A 45° mirror[12] was placed above the upper thrust plate but attached to the stationary housing for viewing cavitation in the oil film. Illumination of the test chamber was provided with light led through optical fibers from an intense lamp. Hence, the oil viscosity was not affected by the heat released from the lamp. A photograph of the test apparatus is shown in Fig. 1(b).

Fig. 1(b) A photograph of the test apparatus

Usually two pressure transducers were installed on the lower thrust plate, one at its center and the other 25 mm away from the center. During the adjustment of parallelism before each test, the position of the top thrust plate was unavoidably shifted relative to the lower thrust plate; as a result, the center pressure transducer was not located at the center of the squeeze film. The exact locations of these transducers must be determined from the photographs of the cavitation, which covered the edge of the squeeze film.

3 Instrumentation

Differential pressure transducers were used. The transducers had a full-scale pressure of 100 psi with 45 psi overload capability and a sensitivity of 6.67 mV/psi. These transducers were calibrated in air using a precision pressure gauge. Each transducer was calibrated twice, once with the pressure source connected to the transducer's normal sensing end and the second time to the other end of the transducer. The two calibration curves were found to match into a line with good linearity. In this way the transducers could be used to measure both positive and negative pressures. The transducers were press fit into holes manufactured in the lower thrust plate, and the side clearances surrounding the transducers were sealed with epoxy. The diameter of pressure sensing holes on the plate surface was 0.5 mm. To prepare the transducers for measurement, the cavities above them were filled with oil. The lower thrust plate (with the transducers installed) was placed in a dessicator, and the side of the plate facing the oil film was submerged in oil. The pressure in the dessicator was first pumped down to the limit of the vacuum gage, which was 0.06 Pa, to evacuate the cavities. Then, the pressure was gradually returned to the atmosphere and the oil allowed to enter the cavities. The process was repeated several times to make sure that the cavities were completely filled with oil. Once filled, the oil remained in the cavities even if the lower thrust plate was flipped upside down, due to surface tension. However, if during a test the cavitation region contained a minute amount of air and spread over a pressure sensing hole, air could enter the cavity and annihilate the transducer's ability to measure tensile stresses. When this happened, the above-described preparation procedure must be repeated. Thus, the failure of a pressure transducer in registering tensile stresses did not necessarily mean that tensile stresses did not occur in the oil film.

The displacement transducers were of the eddy current type, with a natural frequency of 10 kHz. Because different transducers reacted differently to the material of the movable frame, it was necessary to find the linear range of the transducers before performing the calibration, which was done with the use of a micrometer. The sensitivity of the displacement transducers was determined to be 10.2 mV/μm. Once calibrated, any one of the three transducers could be used to measure the vertical displacement of the upper thrust plate. But all three were needed for adjusting the parallel between the lower and upper thrust plates. The values of the minimum bearing gap (to be denoted as h_m) were set with shims. The precise values of the oscillatory amplitude (to be denoted as ε) were determined with a dial gauge.

A CCD with a highest recording rate of 1 000 frames/s was used for photographing. A low-speed CCD was also used to monitor cavitation in real time. Displacement and pressure signals were picked up with a data acquisition system. The synchronization of CCD photographing and displacement and pressure signals was controlled by LABVIEW software.

4 TEST Procedure and Results

For a test, the oscillatory amplitude was first fixed, then the parallel between the thrust plates and the minimum gap were set simultaneously. The oil pool was then filled. Tests were conducted about 24 h after every time the oil pool was filled, to allow any entrained air to escape. A test was run for a series of low to high, then high to low, frequencies. At each frequency, the low-speed CCD was used to monitor the ongoing of cavitation. The command was given at a chosen moment to begin the data acquisition process. In every second 1250 displacement signals, 1250 pressure signals and 250 or 125 (for lower frequencies) photos were taken. The offset between a pair of displacement and pressure signals was about 0.4 ms. The range of oscillatory frequencies tested covered from 2 Hz to 25 Hz. Selected results are presented in the following.

4.1 Pressure Measurement*

At a given oscillatory amplitude, the oil film did not cavitate at low frequencies. Figure 2(a) shows a case in which cavitation did not occur up to a frequency of 5 Hz. The location of the pressure transducer, at $0.177a$ from the center of the squeeze film, was determined after the test run by measuring the cavitation photos. The pressure at this location dipped well into the tensile stress range (greater than 1.5 bar). Figure 2(b) shows the calculated results in this case. The measured displacement in Fig. 2(a) is evidently not the smooth sinusoidal curve shown in Fig. 2(b). The measured pressure trace exhibited the same shape as that given by the theory of an uncavitated oil squeeze film. The agreement between the measured and predicted maximum tensile stresses was excellent. However, the measured maximum pressure was considerably lower than the theoretical prediction (by almost 1 bar). This discrepancy is more likely due to the deviation from ideal of the displacement curve, especially near the minimum film thickness. The discrepancy could also be, to a lesser extent, due to the ratio of film thickness to the thrust plate size being not small enough for lubrication theory to be accurate. In this test run the oil film cavitated at oscillatory frequencies higher than 5 Hz. When the frequency was reduced to 5 Hz again, however, the oil film remained cavitated. The measured results are shown in Fig. 3(a). When the pressure transducer hole was inside the cavitation region, the measured pressure was about absolute zero. Before the cavitation region spread over the hole, a tensile stress of about 0.6 bar was measured. The calculated results based on the new cavitation model are shown in Fig. 3(b). The model predicted the correct shape of the pressure trace but a smaller tensile stress (about 0.3 bar).

---

* The pressure presented in all the figures is in the absolute scale (NOT as gauge pressure).

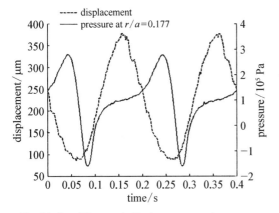

Fig. 2(a) Measured displacement and pressure variation in an uncavitated oil film ($\omega = 5$ Hz, $h_o = 234$ μm, $\varepsilon = 139$ μm)

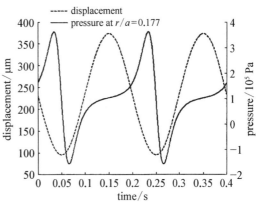

Fig. 2(b) Calculated pressure variation in an uncavitated oil film ($\omega = 5$ Hz, $h_o = 234$ μm, $\varepsilon = 139$ μm)

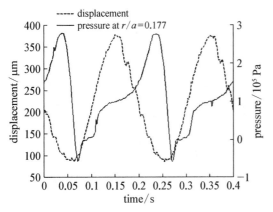

Fig. 3(a) Measured displacement and pressure variation in a cavitated oil film. ($\omega = 5$ Hz, $h_o = 234$ μm, $\varepsilon = 139$ μm)

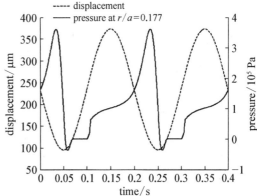

Fig. 3(b) Calculated pressure variation in a cavitated oil film. ($\omega = 5$ Hz, $h_o = 234$ μm, $\varepsilon = 139$ μm)

The above-described features also prevailed at higher frequencies. Figure 4(a) shows a case of an uncavitated oil film at 15 Hz, and Fig. 4(b) shows the corresponding calculated behavior. Shortly afterwards during this test run the oil film cavitated; Fig. 5(a) displays the measured results, and Fig. 5(b) shows the corresponding model prediction. The measured displacement deviated considerably from the ideal sinusoidal behavior. Probably the tolerances embedded in the various components of the displacement-control mechanism were amplified at higher frequencies. The measured pressure traces were also erratic. In the uncavitated oil film the measured maximum tensile stress exceeded 2.3 bar, whereas no tensile stress was measured in the cavitated oil film.

Figure 6(a) shows a case where two pressure traces were simultaneously taken, one near the center (0.018a) and another about halfway between the center and the edge of the squeeze film (0.518a). The center transducer was located within the cavitation region; it measured near absolute zero pressure. The side transducer was occasionally covered by the cavitation region as revealed by the low-speed CCD recording. During the first cycle of

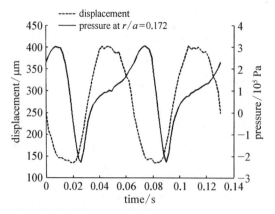

Fig. 4(a) Measured displacement and pressure variation in an uncavitated oil film. ($\omega = 15$ Hz, $h_o = 279$ μm, $\varepsilon = 139$ μm)

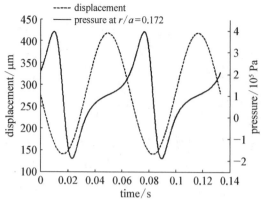

Fig. 4(b) Calculated pressure variation in an uncavitated oil film. ($\omega = 15$ Hz, $h_o = 279$ μm, $\varepsilon = 139$ μm)

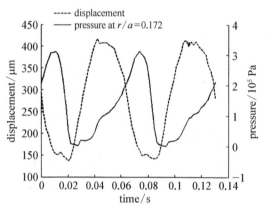

Fig. 5(a) Measured displacement and pressure variation in a cavitated oil film ($\omega = 15$ Hz, $h_o = 279$ μm, $\varepsilon = 139$ μm)

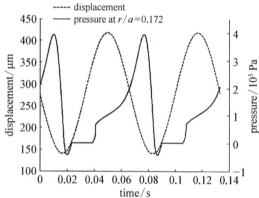

Fig. 5(b) Calculated pressure variation in a cavitated oil film ($\omega = 15$ Hz, $h_o = 279$ μm, $\varepsilon = 139$ μm)

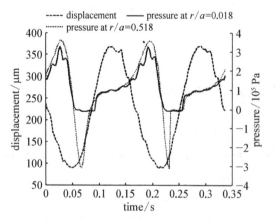

Fig. 6(a) Measured displacement and pressure variations in a cavitated oil film ($\omega = 6$ Hz, $h_o = 224$ μm, $\varepsilon = 139$ μm)

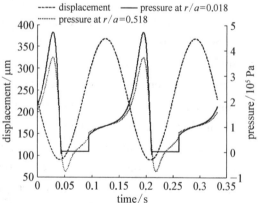

Fig. 6(b) Calculated pressure variations in a cavitated oil film ($\omega = 6$ Hz, $h_o = 224$ μm, $\varepsilon = 139$ μm)

oscillation in Fig. 6(a), this side transducer was clearly not in the cavitation region, and it measured a large tensile stress (exceeding 3 bar). During the second cycle of oscillation it was obviously covered by the cavitation region, and the measured pressure trace exhibited the characteristic dip into the tensile stress range before the arrival of the cavitation front. In subsequent cycles (not shown), this transducer did not measure any tensile stress even though the CCD recording clearly showed that it was outside the cavitation region. According to the new cavitation model, this side transducer was always outside the cavitation region, Fig. 6(b). Besides, the measured maximum pressure at the side transducer was greater than that at the center transducer, which was contrary to the model prediction.

4.2 Visual Study

Several cavitation photos consecutively taken during the test run described in Fig. 6(a) are given in Fig. 7. The oscillatory frequency was 6 Hz; 20 photos were taken in each cycle, and the 9 photos in this figure were #6 (0.042 s) to #14 (0.108 s) of the 20, counting from the beginning (t=0) of a cycle. The sequence (with its time instant indicated below each photo) shows the appearance of a cavitation region at #6 (0.042 s), its growth to the largest size at #8 (0.058 s), its decay during #9 (0.067 s) through #11 (0.083 s), and down to a small residual size (#12, at 0.092 s, through #14, at 0.108 s). Details of the cavitation development may appear obscure in static photos but were clear in the low-speed CCD recording. The white spot in the first quadrant of these photos was the side transducer hole, which was clearly outside the cavitation region.

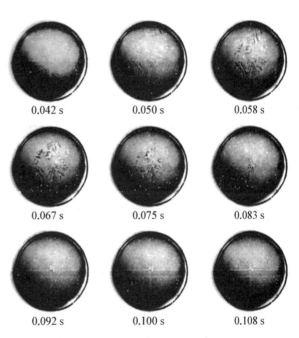

Fig. 7 Photos of cavitation during one cycle of oscillation in the case of Fig. 6(a)

The size of the cavitation region was determined by visually measuring the envelope encircling the residual oil filaments. This size, measured from the photos of Fig. 7, is plotted as open circles in Fig. 8, where theoretical curves for three $\bar{\beta}$ values are also shown. The numerical factor $\bar{\beta}$ (between 0 and 1) is introduced in the new cavitation model to characterize the amount of residual oil filaments left behind the expanding cavitation front. That is, $\bar{\beta}=0$ means no residual oil filament is left behind, whereas $\bar{\beta}=1$ means all the oil remains in the cavitation region. The model analysis further shows that the latter

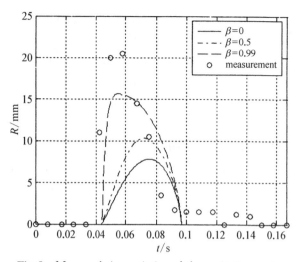

Fig. 8 Measured size variation of the cavitation region during one cycle of oscillation in the case of Fig. 6(a)

case amounts to prescribing the zero pressure-gradient boundary condition at the expanding cavitation front, with the consequence that no tensile stress can arise in the full film region. From Fig. 8 it appears that the measured results agreed better with the model prediction of $\bar{\beta}=1$. But this would also imply the absence of tensile stresses in the oil film, which would be in discord with experimental evidence. Besides, the measured cavitation size was considerably larger than the model prediction. This discrepancy will be discussed in a later section in connection with the possible presence of numerous cavitation sites in the oil film.

Figure 8 shows that cavitation disappeared at 0.125 s, then reappeared at 0.133 s. It was observed from the low-speed CCD recording that at this instant an air bubble emerged from the center transducer hole. The observation was significant, because it provided evidence that air bubbles could enter the transducer holes. Since an air bubble could not transmit tensile stresses, when this happened the transducer would not measure tensile stresses. Hence, the failure of a pressure transducer in registering tensile stresses did not necessarily mean that tensile stresses did not occur in the oil film. This could be why the side transducer, in the case described in Fig. 6(a), did not measure tensile stresses in subsequent cycles of oscillation even though it was located outside the cavitation region.

4.3 Longer Time Behavior

The above description of events hints that history played a role in the oil film behavior. Indeed there existed a trend that, under the same operating conditions, an uncavitated oil film would tend to become cavitated, and initially measured tensile stresses would tend to disappear after more oscillations. The drift from one type of behavior to another took place in a timescale of several minutes. Several additional longer time features were also observed: (a) Continuously monitoring the development of the cavitation region revealed that, when its maximum size in a cycle exceeded a certain magnitude, the maximum size would gradually grow until reaching the oil pool. After that the maximum size would return to its initial magnitude, and the process repeated itself. Thus, the variation of cavitation size sometimes appeared to possess both a small period (the oscillatory period) and a large one containing a large number of oscillations. (b) After the cavitation region became vented to the oil pool, tensile stresses were no longer measured and the range of pressure fluctuations was also reduced.

5 Discussion

Despite the general success of the new cavitation model, several discrepancies exist between the model predictions and measurements. These include the non-occurrence or occurrence of cavitation at apparently the same operating conditions and the drift from one type of behavior to the other. Explanations for these discrepancies might be drawn from a consideration of the size effect of cavitation nuclei. A preliminary analysis is attempted in the following.

Consider that initially there is a spherical cavitation nucleus of radius R_o at $r=0$, and the nucleus is filled with oil vapor. The mechanical equilibrium of the vapor bubble can be described as (Defay and Prigogine[12]):

$$p_v - p_c = \frac{2\sigma}{R_o} \tag{1}$$

where p_v is the vapor pressure of oil, σ is the surface tension coefficient of oil, and p_c is the threshold pressure; i.e., the low oil pressure outside the bubble ready for the nucleus to grow. In this preliminary analysis the various works involved in the nucleation process (Blander and Katz[13]) are ignored. If $R_o=0.1$ mm, $p_v=30$ Pa, $\sigma=0.035$ N/m, Eq. (1) yields $p_c=-670$ Pa, which is only slightly below the absolute zero. The examples given in the new cavitation model (Sun, et al.[1]) are based on these values. On the other hand, if $R_o=0.1$ μm, the same calculation would yield $p_c=-7$ bar. Then, unless the squeeze film produces at $r=0$ a low pressure of this magnitude (i.e., a tensile stress of 7 bar), no cavitation would occur and tensile stresses would appear in the oil film.

Cavitation nuclei in an oil squeeze film are more likely the entrained tiny air bubbles. In this case Eq. (1) may be written as

$$p_{in} - p_{out} = \frac{2\sigma}{\hat{R}} \tag{2}$$

where p_{in} is the pressure inside an air bubble, p_{out} is pressure outside the bubble, and $\hat{R}$ is its radius. If no oil vapor enters the bubble during the increase of $\hat{R}$, the variation of p_{in} may be assumed to obey the equilibrium thermodynamic relations. In the two limiting cases of isothermal and adiabatic processes the relations are $p_V = \text{const}$ and $pV^{\bar{\gamma}} = \text{const}$, respectively, where V denotes the bubble volume and $\bar{\gamma}$ the adiabatic exponent ($=1.4$ for air). Hence, for an isothermal process,

$$p_{in}\hat{R}^3 = p_{in,o}R_o^3 \tag{3a}$$

and for an adiabatic process,

$$p_{in}\hat{R}^{4.2} = p_{in,o}R_o^{4.2} \tag{3b}$$

where R_o is the radius of an air bubble (in the ambient condition) serving as the cavitation nucleus, and $p_{in,o}$ is its inside pressure. By eliminating p_{in} between Eqs. (2) and (3a) or between Eqs. (2) and (3b), one obtains for an isothermal process,

$$p_{out} = \frac{p_{in,o}}{(\hat{R}/R_o)^3} - \left(\frac{2\sigma}{R_o}\right)\frac{1}{(\hat{R}/R_o)} \qquad (4a)$$

or for an adiabatic process,

$$p_{out} = \frac{p_{in,o}}{(\hat{R}/R_o)^{4.2}} - \left(\frac{2\sigma}{R_o}\right)\frac{1}{(\hat{R}/R_o)} \qquad (4b)$$

The parameter $p_{in,o}$ in Eqs. (4) can be obtained by substituting the ambient pressure p_a for p_{out} and R_o for $\hat{R}$ into Eq. (2); i. e.,

$$p_{in,o} - p_a = \frac{2\sigma}{R_o} \qquad (5)$$

The relations between p_{out} and $\hat{R}/R_o$ as given by Eqs. (4) are plotted in Figs. 9(a) and 9(b) for $R_o = 0.1$ μm and $R_o = 0.1$ mm, respectively.

Consider the case $R_o = 0.1$ μm. Figure 9(a) shows that the isothermal curve has a minimum at $\hat{R} = 1.853 R_o$ with $p_{out} = -2.518$ bar. As p_{out} decreases from a positive value, the growth of the bubble is very small. When p_{out} reaches the minimum, the bubble becomes unstable and can grow without bound even at very small negative pressure values. Therefore, it may be appropriate to take the threshold pressure value to be -2.518 bar. Similarly, the adiabatic curve has a minimum at $\hat{R} = 1.633 R_o$ with $p_{out} = -3.265$ bar, and one may take $p_c = -3.265$ bar. In either case, cavitation would not occur unless the squeeze film produces at $r=0$ a low pressure of such large magnitudes. On the other hand, if the nucleus size is 0.1 mm, as used for the calculations in the cavitation model, Fig. 9(b) shows that the minimum of the isothermal curve is at $\hat{R} = 20.9 R_o$ with $p_{out} = -22.3$ Pa; and the minimum of the adiabatic curve is at $\hat{R} = 7.43 R_o$ with $p_{out} = -71.8$ Pa. In either case, cavitation would occur when the pressure at $r=0$ becomes just a little lower than the absolute zero.

The above analysis shows that the oil film's ability to sustain tensile stresses depends on the size of cavitation nuclei. At the beginning of a test run, either the oil is free of nuclei or any existing nuclei are of very small size; then cavitation does not occur. With continuing oscillations, some air is bound to be entrained into the oil. As an entrained air bubble grows in size, the oil film there can no longer sustain tensile stresses and cavitation occurs. For a full squeeze film (i. e., without visible cavitation), the fluctuation of pressure at the center is the largest; hence, it is reasonable to consider that cavitation starts from there. Once the nucleus is large enough for cavitation to grow, the near absolute-zero pressure outside the nucleus is associated with a wide range of bubble sizes (Figs. 9). Then it would not matter much what initial size one should use to compute the development of the cavitation region. At this stage the interaction between the cavitation region and the full-film flow is the main mechanism. Thus, the prescription of an appropriate continuity equation at the rupture boundary is a key element in the cavitation model.

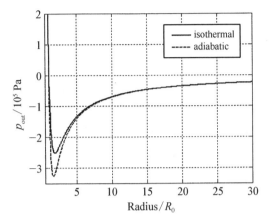

Fig. 9(a) Pressure outside an air bubble of radius $R_o = 0.1\ \mu m$ at the ambient condition

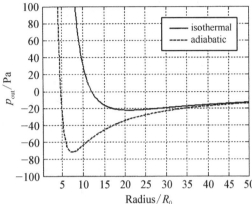

Fig. 9(b) Pressure outside an air bubble of radius $R_o = 0.1\ mm$ at the ambient condition

As oscillations continue, air entrainment could take place anywhere in the squeeze film. The entrained air bubbles could join the expanding cavitation region or become cavitation sites themselves if the combination of the bubble sizes and the local pressure fluctuations is right. When this happens one would see large and rapidly growing cavitation extent. This could explain why the observed growth of the cavitation region was faster, and its maximum size was larger, than the model prediction.* Recall that the new cavitation model (Sun, et al.[1]) deals with the idealized situation that a single nucleus exists at the center of the squeeze film and cavitation starts and spreads from there. While such a model appears reasonable in describing the early stage of cavitation development, it cannot simulate the subsequent events when numerous air bubbles are present in the oil film and their associated threshold pressures are all near the absolute zero, in which case no tensile stresses can arise anywhere in the oil film.

In an earlier study (Sun, et al.[5]) it was noted that the nonoccurrence or occurrence of cavitation was quite persistent once a type of behavior was established. In the present study, however, we observed instead a drift from one type of behavior to the other. We believe this discrepancy is due to the difference in the bearing geometry. The entrained air bubbles in a horizontal parallel-plate squeeze film can migrate from one place to another, causing the drift in oil film behavior. In a journal bearing with horizontal bearing axis, the entrained air bubbles tend to accumulate in the top part of the bearing clearance. In such a situation, persistent cavitation patterns and tensile stresses were observed in the bottom part of the bearing clearance (Sun, et al. [5]; Fig. 13).

---

* A quote from Professor Hori (Ibid., p. 3) may be relevant: "Cavitation is not the result of growth of a single bubble, but of many bubbles that originate at various points and quickly grow, and ... cover almost the whole area ... of the squeeze surface."

6 Summary

A displacement-controlled parallel-plate squeeze film test apparatus was designed and built to study the cavitation of oil films. The various features of the test apparatus are described. Simultaneous pressure measurement and high-speed photography were conducted during the oscillations of the squeeze film. The measured pressure traces and observed cavitation development were compared with the prediction of a recently proposed cavitation model (Sun, et al.[1]). The findings of the study are as follows:

(a) When cavitation did not occur, the measured pressure traces agreed well with the theory of non-cavitated oil squeeze film. In particular, tensile stresses of significant magnitudes were measured in the oil film.

(b) When cavitation occurred, the measured pressure traces agreed well with the prediction of the new cavitation model. Specifically, the shape of the measured pressure trace agreed with the model prediction; tensile stresses were measured in the oil film just before the arrival of the expanding cavitation front; and the pressure in the cavitation region was very close to the absolute zero. (The pressure transducers used could not resolve the low oil vapor pressure from the absolute zero pressure.)

(c) The observed cavitation pattern and shape could be better described as one region with residual oil filaments of the fractured film inside than as a nonhomogeneous mixture containing disconnected cavities.

(d) The measured cavitation duration correlated well with the new cavitation model. But the measured growth of the cavitation region was faster, and its maximum size was larger, than the model prediction.

(e) The variation of cavitation size sometimes possessed both a small period (the oscillatory period) and a large one containing a large number of oscillations.

(f) When the cavitation region was vented to the surrounding oil pool, tensile stresses were not measured, either because the air entrainment made the oil film ready to cavitate or because the air entered the pressure sensing hole(s) and annihilated the transducer's ability to measure tensile stresses.

(g) An analysis of the size effect of cavitation nuclei provided an estimate of the oil film's ability to sustain tensile stresses. If the sizes of cavitation nuclei remain in the submicrometer range, it is unlikely that an oil squeeze film would cavitate. When air entrainment causes the nuclei to grow to the submillimeter range, cavitation would occur near the absolute-zero pressure level.

(h) The growth of entrained air bubbles in multiple sites in the oil film, and their possible migration, might explain the discrepancy between the measured cavitation pattern (its expansion rate and maximum size) and the model prediction, as well as the drift from one type of oil film behavior to another.

Acknowledgments

The research was jointly supported by China's State Key Basic Research Development Program, under Grant G1998020317 - 4. 2; and Shanghai's Baiyulan Foundation for Talents in Science and Technology, under Grant 200328. The authors also gratefully acknowledge Prof. Fang Minglun, Senior Provost of Shanghai University, for his encouragement as well as space and equipment support.

References

[1] Sun, D. C., Wang, W., Zhang, Z., Chen, X. and Sun, M. (2008), "Theory of Cavitation in an Oscillatory Oil Squeeze Film," Tribology Transactions, 51, pp. 332 - 340.

[2] Dyer, D. and Reason, B. R. (1976), "A Study of Tensile Stresses in a Journal-Bearing Oil Film," Journal of Mechanical Engineering Science, 18, pp. 46 - 52.

[3] Nakai, M. and Okino, N. (1976), "Tensile Stress in Journal Bearings," Wear, 39, pp. 151 - 159.

[4] Natsumeda, S. and Someya, T. (1986), "Negative Pressures in Statically and Dynamically Loaded Journal Bearings," Proc. Leeds-Lyon Symposium on Tribology, Paper III(ii), pp. 65 - 72.

[5] Sun, D. C., Brewe, D. E. and Abel, P. B. (1993), "Simultaneous Pressure Measurement and High-Speed Photography Study of Cavitation in a Dynamically Loaded Journal Bearing," ASME Journal of Tribology, 115, pp. 88 - 95.

[6] Elrod, H. G. and Adams, M. L. (1974), "A Computer Program for Cavitation and Starvation Problems," Proceedings of Leeds-Lyon Conference on Cavitation, Paper II(ii).

[7] Kumar, A. and Booker, J. F. (1991), "A Finite Element Cavitation Algorithm," ASME Journal of Tribology, 113, pp. 276 - 286.

[8] Hays, D. F. and Feiten, J. B. (1964), "Cavities Between Moving Parallel Plates," in Cavitation in Real Liquids, Ed. R. Davies, Elsevier, pp. 122 - 137.

[9] Kuroda, S. and Hori, Y. (1978), "An Experimental Study on Cavitation and Tensile Stress in a Squeeze Film," Journal of Japan Society of Lubrication Engineers 23(6), pp. 436 - 442. (in Japanese)

[10] Parkins, D. W. and May-Miller, R. (1984) "Cavitation in an Oscillatory Oil Squeeze Film," ASME Journal of Tribology, 106, pp. 360 - 367.

[11] Chen, Xiaoyang, Sun, Meili, Wang, Wen, Sun, D. C., Zhang, Zhiming and Wang, Xiaojing (2004), "Experimental Investigation of Time-Dependent Cavitation in an Oscillatory Squeeze Film," Science in China G (Physics, Mechanics & Astronomy), 47, pp. 107 - 112.

[12] Defay, R. and Prigogine, I. (1966), Surface Tension and Adsorption, Translated by D. H. Everett, Longmans.

[13] Blander, M. and Katz, J. L. (1975), "Bubble Nucleation in Liquids," AIChE Journal, 21, pp. 833 - 848.

径向浮环动静压轴承稳定性研究*

摘　要：以径向浮环动静压轴承为研究对象，针对轴颈和浮环建立了统一的动力学方程，用 Routh-Hurwitz 准则推导了径向浮环轴承的稳定性判据. 用有限元法计算了某结构高速径向浮环动静压轴承的刚度系数和阻尼系数，在此基础上得到了不同偏心率下的失稳转速. 由计算结果可以看出，浮环轴承具有极佳的稳定性，且随着偏心率的增加，失稳转速迅速提高. 文章在高速浮环轴承稳定性整体建模和分析方面有较大的参考意义.

动静压浮环轴承具有摩擦功耗低、精度高、寿命长、稳定性好等突出优点，在航空航天、空分及精密加工机床等高速、超高速旋转机械上广泛应用[1-3]. 国内外很多学者对不同工况下的动静压轴承进行了大量研究，尤其是在非线性领域[4-6]和高速运转时稳定性方面[7-9]. 对于浮环轴承，康召辉等[10]研究了浮环涡动特性，得到了稳态时浮环质心运动轨迹为椭圆形的结论；郭红等[11]采用将浮环轴承内外两层油膜刚度系数和阻尼系数进行串并联的方法进行建模，对不同转速和偏心率下浮环动静压轴承的稳定性进行了仿真和试验. 本文考虑浮环轴承内外层油膜之间的相互作用，针对轴颈和浮环分别建立动力学方程，得到系统完整的动力学模型. 在有限元仿真的基础上，给出了径向浮环轴承的稳定性判据，并计算了特定结构参数轴承在不同偏心率下的失稳转速. 为高速浮环轴承的稳定性建模和计算分析提供了一定的理论基础.

1　径向浮环轴承控制方程

图 1 为径向浮环轴承结构示意图，轴颈和浮环之间形成内膜，浮环和轴瓦之间形成外膜. 内膜采用四腔结构，外膜采用五腔结构，内外膜每个腔都设置有深腔和浅腔. 深腔具有静压效应，浅腔和封油边具有动压效应.

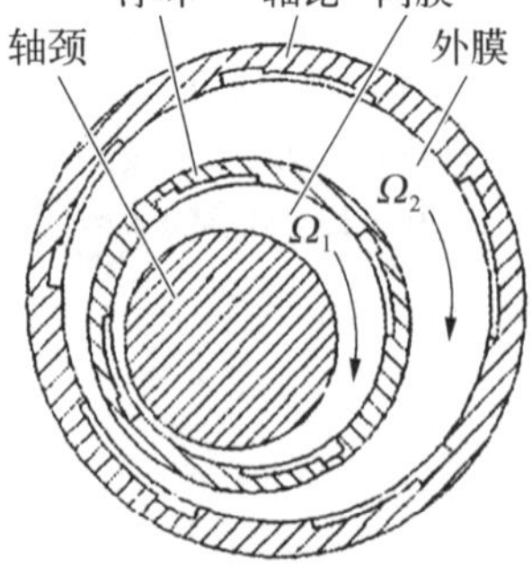

1.1　控制方程

取无量纲因子如下，并用 Φ 表示圆周方向坐标，λ 表示轴向方向坐标，可得到内外膜无量纲动态 Reynolds 方程：

$$\bar{p}_i = \frac{p_i}{p_s},\ \bar{h}_i = \frac{h_i}{c_i},\ \varepsilon_i = \frac{e_i}{c_i},\ \bar{h}_{si} = \frac{h_{si}}{c_i},$$

$$\bar{h}_{qi} = \frac{h_{qi}}{c_i},\ BM_1 = \frac{\mu(\Omega_1 + \Omega_2)D_1^2}{p_s c_1^2},$$

图 1　向心浮环轴承结构

---

* 本文合作者：郭红、岑少起、陈昌婷. 原发表于《振动与冲击》，2012，31(17)：17-21.

$$BM_2 = \frac{\mu \Omega_2 D_2^2}{p_s c_2^2}, \quad A_1 = \frac{1}{1+\frac{\Omega_2}{\Omega_1}}, \quad A_2 = 1$$

$$\frac{\partial}{\partial \Phi_i}\left(\bar{h}_i^3 \frac{\partial \bar{p}_i}{\partial \Phi_i}\right) + \left(\frac{D_i}{L_i}\right)^2 \frac{\partial}{\partial \lambda_i}\left(\bar{h}_i^3 \frac{\partial \bar{p}_i}{\partial \Phi_i}\right) = \frac{3}{2} BM_i \frac{\partial \bar{h}_i}{\partial \Phi_i} + 3 A_i BM_i (\dot{\varepsilon}_i \cos\Phi_i + \varepsilon_{0i}\dot{\theta}_i \sin\Phi_i)$$

(1)

1.2 边界条件

如图 2 所示，向心浮环内外膜满足如下压力边界条件和深腔节流器流量平衡条件.
压力边界条件为：

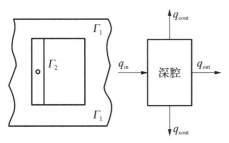

图 2 压力边界条件及流量平衡条件

$$\left.\begin{array}{ll} \bar{p}_j = 0 & j \in \Gamma_1 \\ \bar{p}_j = \bar{p}_{r \cdot m} & j \in \Gamma_2 \\ \bar{p}_j = 0, \frac{\partial \bar{p}_j}{\partial \Phi_j} = 0 & j \in \Gamma_3 \end{array}\right\}$$

(2)

式中：Γ_1 为轴承两端面；Γ_2 为第 m 个深腔边界；Γ_3 为油膜破裂边；$\bar{p}_{r \cdot m}$ 为第 m 个深腔无量纲压力.
无量纲深腔流量平衡方程为：

$$\frac{1-\bar{p}_{r \cdot m}}{\bar{R}_{j \cdot m}} = (\bar{q}_{\text{out}} - \bar{q}_{\text{in}} + \bar{q}_{\text{cout}} + \bar{q}_{\text{cout}})_m$$

(3)

式中 $\bar{R}_{j \cdot m}$ 为第 m 个深腔节流器无量纲液阻；$\bar{q}_{\text{out}}$，$\bar{q}_{\text{in}}$，$\bar{q}_{\text{cout}}$ 为第 m 个深腔无量纲流量.

$$\bar{q}_{\text{in}} = \int_{\lambda_2}^{\lambda_3}\left[6BM_i\left(\frac{L_i}{D_i}\right)^2 \bar{h}_i - \bar{h}_i^3 \frac{\partial \bar{p}_i}{\partial \Phi_i}\right]_{\varphi=\varphi_2} d\lambda$$

$$\bar{q}_{\text{out}} = \int_{\lambda_2}^{\lambda_3}\left[6BM_i\left(\frac{L_i}{D_i}\right)^2 \bar{h}_i - \bar{h}_i^3 \frac{\partial \bar{p}_i}{\partial \Phi_i}\right]_{\varphi=\varphi_3} d\lambda$$

$$\bar{q}_{\text{cout}} = -\int_{\Phi_2}^{\Phi_3}\left[\lambda_i \bar{h}_i^3 \frac{\partial \bar{p}_i}{\partial \lambda}\right]_{\lambda=\lambda_3} d\Phi$$

1.3 油膜厚度

内外膜无量纲油膜厚度可表示为：

$$\bar{h}_i = \begin{cases} 1+\varepsilon_i \cos(\Phi_i - \theta_i) & \text{封油边} \\ 1+\varepsilon_i \cos(\Phi_i - \theta_i) + \bar{h}_{si} & \text{深腔} \\ 1+\varepsilon_i \cos(\Phi_i - \theta_i) + \bar{h}_{qi} & \text{浅腔} \end{cases}$$

(4)

1.4 浮环平衡工作条件

满足内外膜作用到浮环的力和力矩平衡条件时，浮环即可以一定的转速平衡运转.

$$F_{r1}=F_{r2} \tag{5}$$

$$M_{r1}=M_{r2} \tag{6}$$

在径向浮环轴承控制方程、压力边界条件及节流器流量平衡方程基础上,采用小扰动法对控制方程进行有限元仿真,计入压力边界条件和流量平衡条件,满足浮环平衡工作条件,可得到不同转速、不同偏心率下内外膜的无量纲刚度系数 $\bar{k}_{mni}=k_{mni}/(p_s D_i^2/c_i)$,$(m,n=x,y)$ 和阻尼系数 $\bar{b}_{mni}=b_{mni}/(p_s D_i^2/c_i\Omega_i)$,$(m,n=x,y)$,这是进行稳定性分析的前提.

式中:D_i 为轴承直径(mm);L_i 为轴承长度(mm);h_i 为油膜厚度(mm);c_i 为油膜间隙(mm);Ω 为轴颈角速度(1/s);Ω_2 为浮环角速度(1/s);e_i 为偏心距(mm);h_{si} 为深腔厚度(mm);h_{qi} 为浅腔厚度(mm);ε_0 为静平衡位置偏心率;θ_i 为偏位角;μ 为润滑油黏度(Pa·s);p_i 为油膜压力(Pa);p_{si} 为供油压力(Pa);F_{ri} 为内外膜作用到浮环的力(N);M_{ri} 为内外膜作用到浮环的力矩(N·mm);k_{mni},$(m,n=x,y)$(N·mm^{-1}) 为刚度系数;b_{mni},$(m,n=x,y)$(N·s·mm^{-1}) 为阻尼系数.

下标 $i=1$ 表示内膜参数,下标 $i=2$ 表示外膜参数;带上画线者为无量纲参数,其余为有量纲参数. 其他未标注者同一般润滑理论规范.

2 向心浮环轴承稳定性分析

浮环轴承内外膜动力学模型如图 3 所示,设轴颈质量为 $2m_R$,不计浮环质量. 内层油膜的动力特性系数为 k_{xx1},k_{xy1},k_{yx1},k_{yy1},b_{xx1},b_{xy1},b_{yx1},b_{yy1};外层油膜的动力特性系数为 k_{xx2},k_{xy2},k_{yx2},k_{yy2},b_{xx2},b_{xy2},b_{yx2},b_{yy2}. 作用在轴颈上的简谐变动力为 F_x,F_y;轴颈的简谐变动位移 x_1,y_1;浮环的简谐变动位移 x_2,y_2.

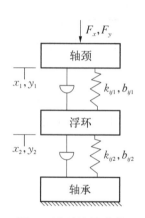

图 3 浮环轴承内外膜动力学模型

轴颈运动方程为:

$$\left.\begin{aligned}&m_R\ddot{x}_1+b_{xx1}(\dot{x}_1-\dot{x}_2)+k_{xx1}(x_1-x_2)+\\&b_{xy1}(\dot{y}_1-\dot{y}_2)+k_{xy1}(y_1-y_2)=F_x\\&m_R\ddot{y}_1+b_{yx1}(\dot{x}_1-\dot{x}_2)+k_{yx1}(x_1-x_2)+\\&b_{yy1}(\dot{y}_1-\dot{y}_2)+k_{yy1}(y_1-y_2)=F_y\end{aligned}\right\} \tag{7}$$

浮环运动方程为:

$$\left.\begin{aligned}&b_{xx2}\dot{x}_2+k_{xx2}x_2+b_{xy2}\dot{y}_2+k_{xy2}y_2-b_{xx1}(\dot{x}_1-\dot{x}_2)-\\&k_{xx1}(x_1-x_2)-b_{xy1}(\dot{y}_1-\dot{y}_2)-k_{xy1}(y_1-y_2)=0\\&b_{yx2}\dot{x}_2+k_{yx2}x_2+b_{yy2}\dot{y}_2+k_{yy2}y_2-b_{yx1}(\dot{x}_1-\dot{x}_2)-\\&k_{yx1}(x_1-x_2)-b_{yy1}(\dot{y}_1-\dot{y}_2)-k_{yy1}(y_1-y_2)=0\end{aligned}\right\} \tag{8}$$

式(7)式(8)解的一般形式可表示为:$x_1=x_{10}e^{vt}$,$y_1=y_{10}e^{vt}$,$x_2=x_{20}e^{vt}$,$y_2=y_{20}e^{vt}$,式中下标 0 表示振幅,v 为复数,代入式(7)、式(8)得到:

$$\left.\begin{array}{l}(m_R v^2 + b_{xx1}v + k_{xx1})x_{10} + (b_{xy1}v + k_{xy1})y_{10} - \\ (b_{xx1}v + k_{xx1})x_{20} - (b_{xy1}v + k_{xy1})y_{20} = 0 \\ (b_{yx1}v + k_{yx1})x_{10} + (m_R v^2 + b_{yy1}v + k_{yy1})y_{10} - \\ (b_{yx1}v + k_{yx1})x_{20} - (b_{yy1}v + k_{yy1})y_{20} = 0 \\ (b_{xx1}v + k_{xx1})x_{10} + (b_{xy1}v + k_{xy1})y_{10} - \\ [(b_{xx1} + b_{xx2})v + k_{xx1} + k_{xx2}]x_{20} - \\ [(b_{xy1} + b_{xy2})v + k_{xy1} + k_{xy2}]y_{20} = 0 \\ (b_{yx1}v + k_{yx1})x_{10} + (b_{yy1}v + k_{yy1})y_{10} - \\ [(b_{yx1} + b_{yx2})v + k_{yx1} + k_{yx2}]x_{20} - \\ [(b_{yy1} + b_{yy2})v + k_{yy1} + k_{yy2}]y_{20} = 0\end{array}\right\} \quad (9)$$

取无量纲转子质量 $M_R = m_R \left/ \dfrac{p_s D_1^2}{c_1 \Omega_1^2} \right.$；内膜无量纲刚度 $\bar{k}_{ij1} = k_{ij1}/K_{a1}(i,j=x,y)$，无量纲因子 $K_{a1} = \dfrac{p_s D_1^2}{c_1}$；内膜无量纲阻尼 $\bar{b}_{ij1} = b_{ij1} \left/ \dfrac{p_s D_1^2}{c_1(\Omega_1 + \Omega_2)} \right. (i,j=x,y)$；外膜无量纲刚度：$\bar{k}_{ij2} = k_{ij2} \left/ \dfrac{p_s D_2^2}{c_2} \right. (i,j=x,y)$；外膜无量纲阻尼 $\bar{b}_{ij2} = b_{ij2} \left/ \dfrac{p_s D_2^2}{c_1 \Omega_2} \right. (i,j=x,y)$；取 $S = \dfrac{v}{\Omega_1}$，式(9)两边同除 K_{a1}，得到无量纲表达式：

$$\left.\begin{array}{l}(M_R S^2 + B_{xx1}S + K_{xx1})x_{10} + (B_{xy1}S + K_{xy1})y_{10} - \\ (B_{xx1}S + K_{xx1})x_{20} - (B_{xy1}S + K_{xy1})y_{20} = 0 \\ (B_{yx1}S + K_{yx1})x_{10} + (M_R S^2 + B_{yy1}S + K_{yy1})y_{10} - \\ (B_{yx1}S + K_{yx1})x_{20} - (B_{yy1}S + K_{yy1})y_{20} = 0 \\ (B_{xx1}S + K_{xx1})x_{10} + (B_{xy1}S + K_{xy1})y_{10} - \\ [(B_{xx1} + B_{xx2})S + K_{xx1} + K_{xx2}]x_{20} - \\ [(B_{xy1} + B_{xy2})S + K_{xy1} + K_{xy2}]y_{20} = 0 \\ (B_{yx1}S + K_{yx1})x_{10} + (B_{yy1}S + K_{yy1})y_{10} - \\ [(B_{yx1} + B_{yx2})S + K_{yx1} + K_{yx2}]x_{20} - \\ [(B_{yy1} + B_{yy2})S + K_{yy1} + K_{yy2}]y_{20} = 0\end{array}\right\} \quad (10)$$

式中：

$$K_{ij1} = \bar{k}_{ij1}, \ B_{ij1} = \bar{b}_{ij1}\dfrac{\Omega_1}{(\Omega_1+\Omega_2)}, \ K_{ij2} = \bar{k}_{ij2}\left(\dfrac{c_1 D_2^2}{c_2 D_1^2}\right), \ B_{ij2} = \bar{b}_{ij2}\left(\dfrac{c_1 \Omega_1 D_2^2}{c_2 \Omega_2 D_1^2}\right)$$

式(10)非平凡解存在条件为系数矩阵的行列式等于零,将式(10)的行列式展开,得到特征方程为：

$$\alpha_0 S^6 + \alpha_1 S^5 + \alpha_2 S^4 + \alpha_3 S^3 + \alpha_4 S^2 + \alpha_5 S + \alpha_6 = 0 \quad (11)$$

其中：

$$\alpha_0 = M_R^2 D_{2,s}$$

$$\alpha_1 = M_R^2 G_s + M_R(D_{1,1}D_{2,2} + D_{2,1}D_{1,2})$$
$$\alpha_2 = M_R^2 R_{2,s} + D_{2,1}D_{2,2} + M_R(D_{2,1}R_{1,2} + D_{2,2}R_{1,1} + D_{1,2}G_1 + D_{1,1}G_2)$$
$$\alpha_3 = M_R(R_{1,2}G_1 + R_{1,1}G_2 + E_{1,1}R_{2,2} + E_{1,2}R_{2,1}) + D_{2,1}G_2 + D_{2,2}G_1$$
$$\alpha_4 = M_R(R_{2,1}R_{1,2} + R_{1,1}R_{2,2}) + D_{2,1}R_{2,2} + D_{2,2}R_{2,1} + G_1 G_2$$
$$\alpha_5 = R_{2,2}G_1 + R_{2,1}G_2$$
$$\alpha_6 = R_{2,1}R_{2,2}$$

其中：
$$R_1 = K_{xx} + K_{yy}$$
$$R_2 = K_{xx}K_{yy} - K_{xy}K_{yx}$$
$$R_{2,s} = (K_{xx1} + K_{xx2})(K_{yy1} + K_{yy2}) - (K_{xy1} + K_{xy2})(K_{yx1} + K_{yx2})$$
$$D_1 = B_{xx} + B_{yy}$$
$$D_2 = B_{xx}B_{yy} - B_{xy}B_{yx}$$
$$D_{2,s} = (B_{xx1} + B_{xx2})(B_{yy1} + B_{yy2}) - (B_{xy1} + B_{xy2})(B_{yx1} + B_{yx2})$$
$$G = B_{xx}K_{yy} + B_{yy}K_{xx} - B_{xy}K_{yx} - B_{yx}K_{xy}$$
$$G_s = (B_{xx1} + B_{xx2})(K_{yy1} + K_{yy2}) + (B_{yy1} + B_{yy2})(K_{xx1} + K_{xx2})$$
$$- (B_{xy1} + B_{xy2})(K_{yx1} + K_{yx2}) - (B_{yx1} + B_{yx2})(K_{xy1} + K_{xy2})$$

对于径向浮环动静压轴承，$n=6$，由 Routh-Hurwitz 准则，当 $\alpha_0 > 0$ 时，油膜涡动为稳定的充要条件是：

$$\alpha_i > 0 \quad (i = 1, 2, \cdots, 6)$$
$$\begin{vmatrix} \alpha_1 & \alpha_3 & \alpha_5 & 0 & 0 \\ \alpha_0 & \alpha_2 & \alpha_4 & \alpha_6 & 0 \\ 0 & \alpha_1 & \alpha_3 & \alpha_5 & 0 \\ 0 & \alpha_0 & \alpha_2 & \alpha_4 & \alpha_6 \\ 0 & 0 & \alpha_1 & \alpha_3 & \alpha_5 \end{vmatrix} > 0 \tag{12}$$

即可得到径向浮环动静压轴承稳定性判别条件：

$$\left.\begin{aligned}
& D_{2,s} > 0 \\
& M_R^2 G_s + M_R(D_{1,1}D_{2,2} + D_{2,1}D_{1,2}) > 0 \\
& M_R^2 R_{2,s} + D_{2,1}D_{2,2} + M_R(D_{2,1}R_{1,2} + D_{2,2}R_{1,1} + D_{1,2}G_1 + D_{1,1}G_2) > 0 \\
& M_R(R_{1,2}G_1 + R_{1,1}G_2 + D_{1,1}R_{2,2} + D_{1,2}R_{2,1}) + D_{2,1}G_2 + D_{2,2}G_1 > 0 \\
& M_R(R_{2,1}R_{1,2} + R_{1,1}R_{2,2}) + D_{2,1}R_{2,2} + D_{2,2}R_{2,1} + G_1 G_2 > 0 \\
& R_{2,2}G_1 + R_{2,1}G_2 > 0 \\
& R_{2,1}R_{2,2} > 0 \\
& \alpha_1\alpha_2\alpha_3\alpha_4\alpha_5 + 2\alpha_1^2\alpha_2\alpha_5\alpha_6 - \alpha_1\alpha_2\alpha_3^2\alpha_6 - \alpha_1\alpha_2^2\alpha_5^2 - \alpha_1^2\alpha_4^2\alpha_5 + \alpha_1^2\alpha_3\alpha_4\alpha_6 + \\
& 2\alpha_0\alpha_1\alpha_4\alpha_5^2 - \alpha_1^3\alpha_6^2 - 3\alpha_0\alpha_1\alpha_3\alpha_5\alpha_6 - \alpha_0\alpha_3^2\alpha_4\alpha_5 + \alpha_0\alpha_3^3\alpha_6 + \alpha_0\alpha_2\alpha_3\alpha_5^2 - \alpha_0^2\alpha_5^3 > 0
\end{aligned}\right\} \tag{13}$$

因此,只要求出了浮环轴承内外膜的无量纲刚度系数和阻尼系数,即可按照上述稳定性条件判断浮环轴承是否处于稳定状态.

除了按照判别条件式(13)进行稳定性判断之外,还可以计算浮环轴承的失稳转速.特征方程(11)有三对根,它们为共轭复根.三对根数实部负值越大,油膜越稳定.在临界状态下,根的实部为零.对于工况条件和结构参数一定的浮环动静压轴承,刚度系数和阻尼系数与转速及偏心率有关,某一特定的转速对应一自由振动的运动方程.计算时可采用迭代法,在选定偏心率下求出方程的三对复数解,如果解的实部全为负,则轴承—转子系统是稳定的,以一定的步长增加系统的转速,直到解的实部变为零,则此转速即为系统选定偏心率下的失稳转速.

3 向心浮环轴承稳定性分析算例

表1为图1向心浮环动静压轴承结构参数.

表1 向心浮环轴承结构参数

	内 膜	外 膜
腔数	4	5
直径/mm	80	92
轴向宽度/mm	80	70
轴向两边封油边宽度/mm	10/10	10/10
深腔所占角度/(°)	15	12
浅腔所占角度/(°)	50	40
浅腔深度/mm	0.03	0.03
深腔深度/mm	0.25	0.25
半径间隙/mm	0.025	0.023

取润滑油黏度 $\mu = 4.475 \times 10^{-3}$ Pa·s,供油压力 $p_s = 1$ MPa,经有限元计算后可得到该轴承压力分布、静特性参数和刚度阻尼等动特性参数.

图4为轴颈转速 10 000 r/min 时内外膜在偏心率 0.15 下的压力分布.由图中可以明显看出,内膜四腔出现四个压力峰值,外膜五腔呈现五个压力峰值,动压效应明显;深腔部分腔压稳定,为静压效应.

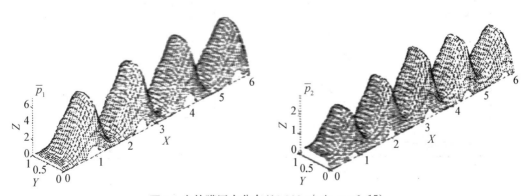

图4 内外膜压力分布(10 000 r/min, $\varepsilon = 0.15$)

图 5～图 8 为不同转速下内外膜部分无量纲刚度系数和阻尼系数随偏心率变化曲线,对于动静压轴承来说,无量纲动态特性参数在不同转速和不同偏心率下都是不同的,正确求解这些系数是浮环轴承稳定性分析的前提.

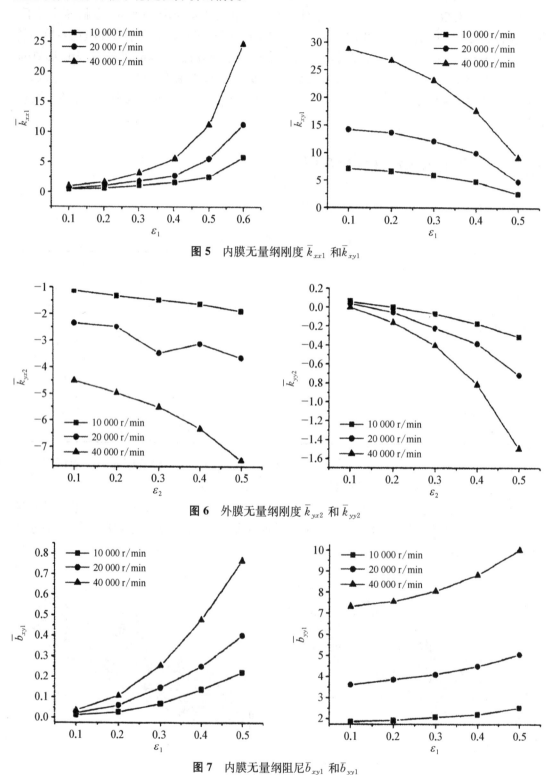

图 5　内膜无量纲刚度 $\bar{k}_{xx1}$ 和 $\bar{k}_{xy1}$

图 6　外膜无量纲刚度 $\bar{k}_{yx2}$ 和 $\bar{k}_{yy2}$

图 7　内膜无量纲阻尼 $\bar{b}_{xy1}$ 和 $\bar{b}_{yy1}$

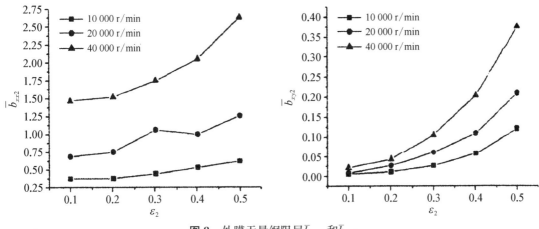

图8 外膜无量纲阻尼 $\bar{b}_{xx2}$ 和 $\bar{b}_{xy2}$

在刚度系数和阻尼系数计算的基础上,可以得到该浮环轴承不同偏心率下的失稳转速,如表2所示.可以看出,浮环轴承稳定性极佳,且随着偏心率的增加,失稳转速上升很快.因此高速时可以采取减小长径比、降低供油压力等措施来增大浮环轴承工作的偏心率,从而提高其失稳转速.

表2 浮环轴承失稳转速

偏心率	0.15	0.2	0.25	0.3	0.35	0.4	0.45
失稳转速×10^3/(r·min^{-1})	345	400	440	515	635	670	690

4 结论

(1)考虑浮环轴承内外两层油膜的相互作用,针对轴颈和浮环建立系统统一的动力学方程,在此基础上给出了径向浮环动静压轴承的稳定性判据.

(2)对特定结构参数的径向浮环动静压轴承进行了有限元计算,得到了不同转速下刚度系数和阻尼系数随偏心率变化的曲线.在此基础上,计算了不同偏心率下轴承的失稳转速,且浮环轴承失稳转速随着偏心率的增加而迅速提高.

(3)增加浮环轴承工作的偏心率可大幅提高其稳定性,浮环轴承具有极好的稳定性,在航空航天等领域具有广阔的应用前景.

参考文献

[1] 康召辉,任兴民,黄金平,等. 浮环轴承系统中浮动环作用机理研究[J]. 振动工程学报,2009,5(22):533-537.
KANG Zhao-hui, REN Xing-min, HUANG Jin-ping, et al. Research of mechanism of a floating ring in the floating ring bering system[J]. Journal of Vibration Engineering, 2009, 5(22):533-537.

[2] 万召,孟光,荆建平,等. 燃气轮机转子-轴承系统的油膜涡动分析[J]. 振动与冲击,2011,30(3):38-41,52.
WAN Zhao, MENG Guang, JING Jian-ping, et al. Analysis on oil whirl of gas turbine rotor-bering system [J]. Journal of Vibration and Shock, 2011, 30(3):38-41,52.

[3] 岑少起,郭红,薛东岭,等. 透平膨胀机组动静压浮环轴承稳定性研究[J]. 机械强度,2002,24(1):32-34.
CEN Shao-qi, GUO Hong, XUE Dong-ling, et al. Stability study of radial-thrust floating-ring hybrid bearing [J]. Journal of Mechanical Strength, 2002,24(1):32-34.

[4] 成玫,孟光,荆建平. 转子-轴承-密封系统的非线性振动特性[J]. 上海交通大学学报,2007,41(3):398-403.
CHENG Mei, MENG Guang, JING Jian-ping. The nonlinear dynamical behaviors of a rotor-bearing-seal system [J]. Journal of Shanghai Jiaotong University,2007,41(3):398-403.

[5] 吕延军,虞烈,刘恒. 流体动压滑动轴承-转子系统非线性动力特性及稳定性[J]. 摩擦学学报,2005,25(1):61-66.
LU Yan-jun, YU lie, LIU Heng. Stability and nonlinear dynamic behavior of a hydrodynamic journal bearing-rotor system[J]. Tribology, 2005,25(1):61-66.

[6] 郭建萍,邱鹏庆,崔升,等. 有限长轴承非稳态油膜力建模及非线性油膜失稳[J]. 振动工程学报,2001,14(1):7-12.
GUO Jian-ping, QIU Peng-qing, CUI Sheng, et al. Study on unsteady nonlinear oil-film force model for finite journal bearings and it's oil-film whil analysis [J]. Journal of Vibration Engineering,2001,14(1):7-12.

[7] Guo H, Lai X M, Cen S Q, Theoretical and experimental study of constant restrictor deep/shallow pockets hydrostatic/hydrodynamic conical bearing [J]. ASME, Journal of Tribology,2009,131(4):041701-041707.

[8] Ene N M, Dimofte F, Keith J T G. A stability analysis for a hydrodynamic three-wave journal bearing [J]. Tribology International,2008,41:434-442.

[9] 杨金福,杨昆,付忠广,等. 转子滑动轴承系统中油膜谐波振荡过程的试验研究[J]. 燃气涡轮试验与研究,2007,20(3):42-47.
YANG Jin-fu, YANG Kun, FU Zhong-guang, et al. Experimental rearch on the harmonic oil whip in sliding bearing-rotor system [J]. Gas Turbine Experiment and Research,2007,20(3):42-47.

[10] 康召辉,任兴民,王鸷,等. 浮环轴承系统中浮动环涡动运动研究[J]. 振动与冲击,2010,29(8):195-197.
KANG Zhao-hui, REN Xing-min, WANG Zhi, et al. Research of floating ring whirl in the floating ring bering system[J]. Journal of Vibration and Shock,2010,29(8):195-197.

[11] Guo H, Lai X M, Wu X L, et al. Performance of flat capillary compensated deep/Shallow pockets hydrostatic/hydrodynamic journal-thrust floating ring bearing [J]. Tribology Transcations,2009,52(2):204-212.

Stability of a Journal Floating Ring Hybrid Bearing

Abstract: The unitized dynamic model of journal and floating ring for a journal floating ring hydrostatic/hydrodynamic bearing was established. The stability criterion of the journal floating ring bearing was deduced using Routh-Hurwitz method. The stiffness and damping coefficients under different rotating speeds and eccentricities for a high speed floating ring bearing were calculated with finite element method. On this base, the variation of threshold speed versus eccentricity was acquired. The results showed that a floating ring bearing has excellent stability and the threshold speed grows quickly with increase in eccentricity; the study results provide a reference for model integrity and stability analysis of a journal floating ring hybrid bearing.

An Explicit Solution for the Elastic Quarter-space Problem in Matrix Formulation*

Abstract: This paper presents a fast and convenient algorithm for the solution of the elastic quarter-space contact problem, which uses discretization to form matrices to realize the overlapping solution process for the elastic quarter-space as developed by Hetenyi [Hetenyi, M., 1970. A general solution for the elastic quarter space, Trans. ASME Journal of Applied Mechanics 37E(1), 70 – 76]. This proposed method provides an explicit solution which is as yet absent in existing literatures. The generated matrices are only related to the mesh structure and Poisson's ratio while unrelated to the loading, such that they can be applied to different loading cases. Hence, the present method offers a possibility for substantially improving the efficiency of those numerical iterative analyses, such as the elastohydrodynamic lubrication of the contact in an elastic quarter-space. Verification of the present method was accomplished by comparison with the existing quarter-space results.

1 Introduction

The contact problem in an elastic quarter-space is quite common in practical mechanical systems such as the contact of rail-wheels, cam-followers, gears and roller bearings. The common characteristic of these contacts is that there are free end surfaces near the contact or loading region. Due to the difficulty of solving this problem in a purely analytical way, a number of ingenious ways were developed by various researchers. Of special significance is Hetenyi's ingenious concept of using overlapped half spaces to solve this problem (Hetenyi, 1970). A system of two coupled integral equations is formed to express the involved mathematical relations. The existing methods of solution of this system can be roughly classified into the following categories: 1. Numerical solutions performed in the physical space, which are numerical iteration (Hetenyi, 1970) and direct numerical solution (Hanson and Keer, 1990); 2. Transformation into a transformed space, followed by numerical solution, and inverse transformation of the result. Two kinds of transformation are applied, namely, Fourier transformation (Sneddon, 1971; Keer et al., 1983) and Mellin transformation (Hecker and Romanov, 1993); 3. Ritz's method based numerical solution (Guenfoud et al., 2010). The present paper strives to

---

* In collaboration with W. Wang and P. L. Wong. Reprinted from *International Journal of Solids and Structures*, 2013, 50: 976 – 980.

follow Hetenyi's and Keer's works in the aspect of solution methodology of the first category. Its main difference from the existing methods lies in the employment of matrix formulation to the discretized equations. This approach is benefited from the flexibility of matrix operations to gain, within Hetenyi's concept of reflection and overlapping, an explicit solution and characteristics unique of the elastic quarter space in the form of matrices.

As a first stage of applying the present method to attack this type of problems, the stress singularity caused by the edge loading is not considered. This does not affect its applicability to the problems without edge loads, for example the elastohydrodyanmic lubrication problem which takes the boundary condition of zero pressure and excludes any non-zero boundary pressure (Zhu et al., 2009).

2 Analysis

The general loading case of the quarter space can be reduced to an equivalent case of quarter space with two loads applied respectively normal onto the top and side surfaces by overlapping of half space or half spaces (Yu et al., 1996). The case of a quarter space with normal loadings on both surfaces can further be split into two overlapped quarter spaces with one surface normally loaded and the other free. The last case can therefore be looked upon as the hard kernel of the general loading case in this sense, and its solution can be seen as a key to the solution of the general loading case. It is taken as the basic form of elastic quarter-space problem in Hetenyi's and also the present paper, as shown in Fig. 1.

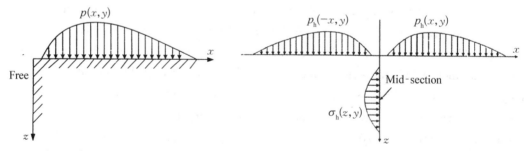

Fig. 1 Basic elastic quarter-space problem   **Fig. 2** Horizontal (H) half-space symmetrically loaded

Geometrically, the quarter-space can be taken as the half of a half-space model. Fig. 2 and 3 show schematically the horizontal (H) and the vertical (V) half-spaces respectively. In both of these half-spaces, the symmetrical loads on the two sides cause only normal stresses but not any shear stresses on the respective mid-sections.

The numerical treatment begins with discretization of both H- and V-surfaces. Fig. 4 (a) shows the effective part of the H-surface discretized into a set of k_h rectangles. One of the rectangles, the ith rectangle, is shown in Fig. 4 (b). The coordinates of its centre are denoted by (x_i, y_i), and its length and width are respectively $2\alpha_i$ and $2\beta_i$. The V-surface is discretized into a set of k_v rectangles, either geometrically similar to H- with z replacing x, or independently.

The loads on the surfaces are approached by piece-wise distributions. The values of the distributed loads on H- and V- surfaces are denoted by their values at the centers of the rectangles $(p_h)_i = p_h(x_i, y_i)$ and $(p_v)_j = p_v(z_j, y_j)$. The mid-sectional normal stresses caused by the loads with their mirror images are similarly denoted by their values at the rectangle centers $(\sigma_h)_j = \sigma_h(z_j, y_j)$ and $(\sigma_v)_i = \sigma_v(x_i, y_i)$. The arrays of these four sets of pressure and stress values form four vectors $\boldsymbol{P}_h$, $\boldsymbol{P}_v$, $\boldsymbol{S}_h$ and $\boldsymbol{S}_v$. The applied load distribution can also be approached by a piecewise distribution and expressed by a vector $\boldsymbol{P}$.

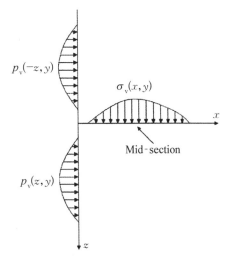

Fig. 3 Vertical (V) half-space symmetrically loaded

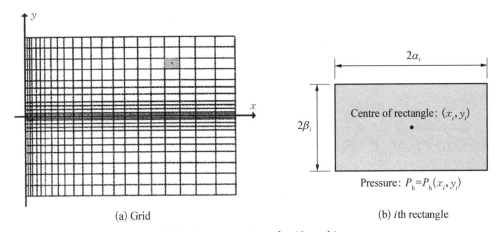

(a) Grid (b) ith rectangle

Fig. 4 Demonstration of grid meshing

$$\boldsymbol{P} = \begin{bmatrix} p_1 \\ p_2 \\ \vdots \\ p_k \end{bmatrix}; \quad \boldsymbol{P}_h = \begin{bmatrix} (p_h)_1 \\ (p_h)_2 \\ \vdots \\ (p_h)_{k_h} \end{bmatrix}; \quad \boldsymbol{P}_v = \begin{bmatrix} (p_v)_1 \\ (p_v)_2 \\ \vdots \\ (p_v)_{k_v} \end{bmatrix};$$

$$\boldsymbol{S}_h = \begin{bmatrix} (\sigma_h)_1 \\ (\sigma_h)_2 \\ \vdots \\ (\sigma_h)_{k_v} \end{bmatrix}; \quad \boldsymbol{S}_v = \begin{bmatrix} (\sigma_v)_1 \\ (\sigma_v)_2 \\ \vdots \\ (\sigma_v)_{k_h} \end{bmatrix}$$

The stress $\sigma_h(z_j, y_j)$ at rectangle j produced by the constant pressure $p_h(x_i, y_i)$ on the ith rectangle and its image $p_h(-x_i, y_i)$ can be expressed as:

$$\sigma_h(z_j, y_j) = m_{ij} p_h(x_i, y_i) \quad (1)$$

where the coefficient m_{ij} is determined by the Love's solution (Love, 1929). The 2-dimensional array of all the coefficients m_{ij} forms a 'reflecting' matrix M:

$$M = \begin{bmatrix} m_{11} & m_{21} & \cdots & m_{k_h 1} \\ m_{12} & m_{22} & \cdots & m_{k_h 2} \\ \vdots & \vdots & & \vdots \\ m_{1k_v} & m_{2k_v} & \cdots & m_{k_h k_v} \end{bmatrix}$$

The relation between the stress vector S_h and the load vector P_h can be expressed by the following equation:

$$S_h = M \cdot P_h \tag{2}$$

Similarly, the vector S_v can also be related to the vector P_v by another 'reflecting' matrix N:

$$S_v = N \cdot P_v \tag{3}$$

The right part of H-space is overlapped on the lower part of V-space. The stresses on the boundaries of the resulting quarter space must fulfill the given conditions, and two simultaneous equations are therefore formed:

$$-P_h + S_v = -P_h + N \cdot P_v = -P \tag{4}$$

$$-P_v + S_h = -P_v + M \cdot P_h = 0 \tag{5}$$

Eq. (5) gives $P_v = M \cdot P_h$. Substitution of this into Eq. (4) results in a decoupled equation for P_h:

$$P_h - N \cdot (M \cdot P_h) = P \tag{6}$$

Since $N \cdot (M \cdot P_h) = (N \cdot M) \cdot P_h$, Eq. (6) can be written as:

$$(I - N \cdot M) \cdot P_h = P \tag{7}$$

Thus, an explicit solution for the load P_h is readily obtained:

$$P_h = A \cdot P \tag{8}$$

where $A = (I - N \cdot M)^{-1}$. Similarly,

$$P_v = B \cdot P \tag{9}$$

where $B = M \cdot (I - N \cdot M)^{-1}$ or $B = M \cdot A$.

With P_h and P_v known, the stress and deformation distributions within the respective half-spaces can be obtained by applying the Love's solution again, and superposed to give the stress and deformation distributions within the quarter-space.

It is also interesting to note that the above solution is actually the limit of infinite Hetenyi's iteration. To show this, the matrix form is applied to Hetenyi's process. The nth iterative results are:

$$P_h^{(n)} = \left[I + \sum_{i=1}^{n}(N \cdot M)^i\right] \cdot P \text{ and}$$

$$P_v^{(n)} = M \cdot \left[I + \sum_{i=1}^{n}(N \cdot M)^i\right] \cdot P \tag{10}$$

The bounded character and convergence of such iterations were proven by Hetenyi (1970). It is evident that the limits of infinite iterations are identical to Eqs. (8) and (9).

The "reflecting" matrices M and N, and therefore also the matrices A and B, are solely related to the mesh structure and Poisson's ratio, while independent of the loading. They are special properties unique of the elastic quarter space, extracted from the reflection and overlapping within Hetenyi's concept. Once generated, these matrices can be stored and employed in different loading cases of the same Poisson's ratio and geometrically similar mesh structure of the surfaces. This could provide substantial convenience in treating different loading cases, and particularly convenient in treating problems where repetitive calculations of the same quarter-space are necessary.

3 Case study

A case is selected from Hanson and Keer (1990), and the results obtained by the present solution are compared with those of Hanson and Keer. The very detailed loading condition and abundance of results provided by Hanson and Keer (1990) are very suitable for verification of the present method.

A half hemispherical loading is applied to the top surface of an elastic quarter-space, with the spherical centre located at the edge. The load can be described by the following formulae:

$$p(x, y) = p_0 \frac{\sqrt{a^2 - x^2 - y^2}}{a}, \sqrt{x^2 + y^2} \leqslant a \tag{11}$$
$$p(x, y) = 0, \sqrt{x^2 + y^2} > a$$

where a denotes the radius of the semi-circular region of the pressure distribution, and p_o is the maximum pressure at the centre of the semi-circle. Poisson's ratio, ν, is taken to be 0.3. The H-surface is discretized into a set of 5,015 uneven-sized rectangles. In the x-direction starting from $x=0$, the width of the first rectangle 2α is taken to be $0.01a$, and the width of the succeeding rectangles is expressed in the form of a rising geometric series with a factor of 1.1 till $0.1a$, and then with a factor of 1.05. In the y-direction, the first rectangle centre is placed at $y=0$ and its width 2β taken as $0.05a$. The width of the succeeding rectangles in both positive and negative y-directions initially increases with a geometric series with a factor of 1.1 and then 1.05 in the same way as above. The discretization continues until it covers a sufficiently large region with $0 \leqslant x \leqslant 9.7a$ and $-9.7a \leqslant y \leqslant 9.7a$.

The very fine meshing in the x-direction from the edge is employed to minimize the error caused by imperfectly treating the stress singularity at the edge involved in the reflecting action due to the edge loading by a distributed normal loading. It can be seen from the results that this measure is fairly effective with respect to the precision of global distributions of various stress components. The discretization of V-surface is identical to that of the H-surface with z in the place of x. Therefore, the 'reflecting' matrices M and N are identical matrices of $5\,015 \times 5\,015$, and the matrix A becomes $A = (I - M^2)^{-1}$ in Eq. (8).

The required loads on H- and V-half surfaces are calculated from the load vector P by Eqs. (8) and (9), and the results are shown in Figs. 5 and 6.

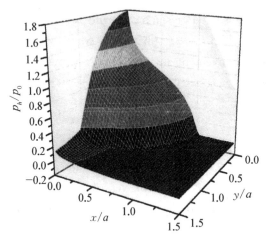

Fig. 5 Distribution of p_h on H-surface   **Fig. 6** Distribution of p_v on V-surface

Normal stresses within the quarter-space are calculated from the loads on H- and V-surfaces together with their images. The stress distributions on the x-z plane were obtained with the present matrix formulation and are plotted in contour together with the results of Hanson and Keer (1990) in Fig. 7. Comparing the two sets, it shows that the present results are in good correlation with Hanson and Keer's. The matrix solution is thus

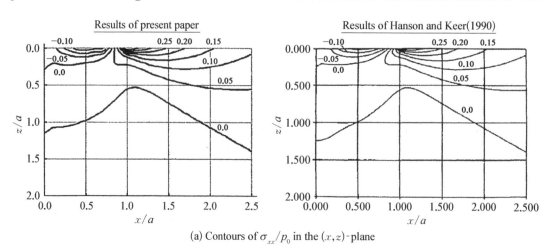

(a) Contours of σ_{xx}/p_0 in the (x,z)-plane

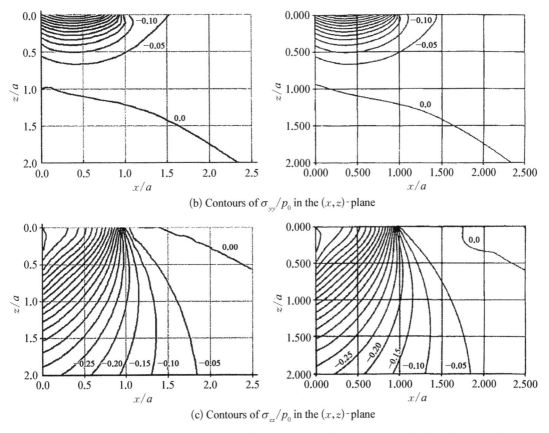

(b) Contours of σ_{yy}/p_0 in the (x,z)-plane

(c) Contours of σ_{zz}/p_0 in the (x,z)-plane

Fig. 7 Comparison of results of stress distributions obtained by present method and extracted from Hanson and Keer (1990)

validated. The present solution should be conveniently applicable in practical elastic quarter-space problems which have no or insignificant effect of edge loading.

4 Conclusion

(1) The matrix formulation is employed to express the numerical equations for the elastic quarter-space, and an explicit solution is achieved, which has not been found in existing literatures. This solution is shown to be identical to the limit of Hetenyi's iteration.

(2) A special solution process is formed, wherein the involvement of the load is pushed to the last moment, after the generation of the final matrices. These matrices can be seen as characteristics unique of the elastic quarter space within Hetenyi's concept of overlapping. They can be directly applied to different given loadings. This facilitates fast and efficient calculations of the elastic stress and deformation fields of a quarter-space problem, particularly when repetitive calculations are necessary.

Acknowledgement

The authors would like to express their gratitude to the City University of Hong Kong for the financial support under SRG7002693.

References

[1] Guenfoud, S., Bosakov, S. V., Laefer, D. F., 2010. A Ritz's method based solution for the contact problem of a deformable rectangular plate on an elastic quarter-space. International Journal of Solids and Structures 47 (14 – 15), 1822 – 1829.

[2] Hanson, M. T., Keer, L. M., 1990. A simplified analysis for an elastic quarter-space. Journal of Mechanics and Applied Mathematics 43 (4), 561 – 587.

[3] Hecker, M., Romanov, A. E., 1993. The stress fields of edge dislocations near wedgeshaped boundaries and bonded wedges. Material Science and Engineering A164, 411 – 414.

[4] Hetenyi, M., 1970. A general solution for the elastic quarter space. Transactions on ASME, Journal of Applied Mechanics. 37 E(1), 70 – 76.

[5] Keer, L. M., Lee, J. C., Mura, T., 1983. Hetenyi's elastic quarter space problem revisited. International Journal of Solids and Structures 19 (6), 497 – 508.

[6] Love, A. E. H., 1929. The stress produced in a semi-infinite solid by pressure on part of the boundary. Philosophical Transactions of the Royal Society of London Series A 228, 377 – 420.

[7] Sneddon, I. N., 1971. Fourier transformation solution of the quarter plane problem in elasticity. File PSR-99/6, Applied Mathematics Research Group, North Carolina, State University.

[8] Yu, C.-C., Keer, L. M., Moran, B., 1996. Elastic – plastic rolling-sliding contact on a quarter space. Wear 191, 219 – 225.

[9] Zhu, D., Ren, N., Wang, Q. J., 2009. Pitting life prediction based on a 3D line contact mixed EHL analysis and subsurface von Mises stress calculation. Transactions on ASME Journal of Tribology 131 (4), 1 – 8.

An Explicit Matrix Algorithm for Solving Three-dimensional Elastic Wedge Under Surface Loads*

Abstract: Analytical solutions for three-dimensional wedge problems are difficult to obtain. It thus leads inevitably to the use of alternative methods such as the finite element method in practice. We here present an explicit matrix algorithm for solving 3D wedge problems under general surface loads: arbitrarily distributed normal and shear loads. The methodology is based on the concept of overlapping two half-spaces formed by the surfaces of a wedge and all calculations are based on half-space equivalent loads. We show that the equivalent loads on the two half-spaces are directly related to the product of the original loads on the wedge and transformation matrices. The transformation matrices are functions of wedge angle, mesh structure, and Poisson's ratio, but not the applied load. Hence, a 3D wedge problem can be solved using those classical solutions for half-spaces once the equivalent loads are obtained. Stress analyses of elastic wedge with different wedge angles are conducted. Results of three special cases: wedge angle of 170°, 90° and 60° are compared with half-space results, the published data of quarter-space and FEM, respectively. The new algorithm is validated by the good correlation shown in the comparison. The effect of wedge angle on internal stresses is also discussed.

Nomenclature

I unit matrix

M reflecting matrix for stress calculation on plane X2 - Y2 induced by equivalent loads on plane X1 - Y1

N reflecting matrix for stress calculation on plane X1 - Y1 induced by equivalent loads on plane X2 - Y2

P_1 vector of applied normal load on plane X1 - Y1 in X1 - Y1 - Z1 coordinate

Q_{X1} vector of applied shear load in X1 direction on plane X1 - Y1 in X1 - Y1 - Z1 coordinate

Q_{Y1} vector of applied shear load in Y1 direction on plane X1 - Y1 in X1 - Y1 - Z1 coordinate

P_2 vector of applied normal load on plane X2 - Y2 in X2 - Y2 - Z2 coordinate

Q_{X2} vector of applied shear load in X2 direction on plane X2 - Y2 in X2 - Y2 - Z2 coordinate

Q_{Y2} vector of applied shear load in Y2 direction on plane X2 - Y2 in X2 - Y2 - Z2 coordinate

$\bar{P}_1$ vector of equivalent normal load on plane X1 - Y1 in X1 - Y1 - Z1 coordinate

$\bar{Q}_{X1}$ vector of equivalent shear load in X1 direction on plane X1 - Y1 in X1 - Y1 - Z1 coordinate

$\bar{Q}_{Y1}$ vector of equivalent shear load in Y1 direction on plane X1 - Y1 in X1 - Y1 - Z1 coordinate

$\bar{P}_2$ vector of equivalent normal load on plane X2 - Y2 in X2 - Y2 - Z2 coordinate

$\bar{Q}_{X2}$ vector of equivalent shear load in X2 direction on plane X2 - Y2 in X2 - Y2 - Z2 coordinate

$\bar{Q}_{Y2}$ vector of equivalent shear load in Y2 direction on plane X2 - Y2 in X2 - Y2 - Z2 coordinate

PP_v, $\overline{PP}_v$ mirrored pair of equivalent load on

---

* In collaboration with L. Guo, W. Wang and P. L. Wong. Reprinted from *International Journal of Solids and Structures*, 2017,000: 1 - 12.

vertical half-space in reference (Zhang et al., 2013)

PP_h, $\overline{PP}_h$ mirrored pair of equivalent load on horizontal half-space in reference (Zhang et al., 2013)

S_Z vector of normal stress on plane X2 - Y2 in X2 - Y2 - Z2 coordinate induced by equivalent loads on plane X1 - Y1

S_X vector of shear stress in X2 direction on plane X2 - Y2 in X2 - Y2 - Z2 coordinate induced by equivalent loads on plane X1 - Y1

S_Y vector of shear stress in Y2 direction on plane X2 - Y2 in X2 - Y2 - Z2 coordinate induced by equivalent loads on plane X1 - Y1

T_Z vector of normal stress on plane X1 - Y1 in X1 - Y1 - Z1 coordinate induced by equivalent loads on plane X2 - Y2

T_X vector of shear stress in X1 direction on plane X1 - Y1 in X1 - Y1 - Z1 coordinate induced by equivalent loads on plane X2 - Y2

T_Y vector of shear stress in Y1 direction on plane X1 - Y1 in X1 - Y1 - Z1 coordinate induced by equivalent loads on plane X2 - Y2

a radius of the semi-circular load distribution in the case study

k number of rectangular grids on plane X1 - Y1

l number of rectangular grids on plane X2 - Y2

m coefficient for stress calculation

p_0 the maximum of the load distribution in case study

$(p_1)_i$ applied normal load at the center of the ith rectangle on plane X1 - Y1

$(q_{x1})_i$ applied shear load in X1 direction at the center of the ith rectangle on plane X1 - Y1

$(q_{y1})_i$ applied shear load in Y1 direction at the center of the ith rectangle on plane X1 - Y1

$(\bar{p}_1)_i$ equivalent normal load at the center of the ith rectangle on plane X1 - Y1

$(\bar{q}_{x1})_i$ equivalent shear load in X1 direction at the center of the ith rectangle on plane X1 - Y1

$(\bar{q}_{y1})_i$ equivalent shear load in Y1 direction at the center of the ith rectangle on plane X1 - Y1

θ wedge angle

1 Introduction

Traditional contact mechanics are based on Hertz's (1882) theory, in which two basic assumptions are to be satisfied. First, the contact or the loaded area should be much smaller than the dimension of the contact bodies. Second, the contact area should be far from any edge surface. Cases satisfying these conditions can be approximated as contact problems of half-space or half-plane, depending on the types of load distribution. The internal stress of loaded half-spaces can be obtained directly from Boussinesq's (1885) solution for point load and Love's (1929) solution for distributed load. Although many practical cases satisfy Hertz criteria, some cases, such as rail-wheel systems, roller bearings, and cutting tools, do not fulfill the second assumption. For these cases, the contact or loaded areas are not far from the edge surface of the contact bodies. The edge surface may affect the contact stress and the internal stress field. Therefore, solving such problems is more complex compared with half-plane or half-space. Generally, the geometry of a side edge surface in connection with a loaded surface can be modeled as a wedge with a specific wedge angle (<180°). Such a contact problem can be considered as two-dimensional only if the applied load is evenly distributed and of infinite length in the direction parallel to the apex edge of the wedge. It becomes three-dimensional if the applied load is a concentrated or any other type of distributed load and it is generally termed as a three-dimensional wedge problem (Fig. 1).

If the wedge angle is equal to 90°, it becomes a quarter-space problem, which is a special case of a wedge. Several effective solutions have been developed for three-dimensional stress analysis of a loaded quarter-space. Hetenyi (1970) developed an iterative scheme of overlapping two half-spaces with mirrored loads to free the originally unloaded surface from any normal or shear stress. He considered a concentrated

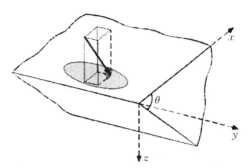

Fig. 1 Three-dimensional wedge problem (The wedge is of infinite length in y-direction, and with general load distributed in the shaded spot)

normal load on the top surface in the analysis. Keer et al. (1983) followed Hetenyi's overlapping half-space idea and formulated a coupled pair of integral equations which were numerically solved by applying a Fourier transform. They considered both normal and tangential loads. Obviously, their method is only limited to quarter-space problems with loads which can be Fourier-transformed. Hanson and Keer (1990) overcame this limitation by developing a direct method to solve the two coupled integral equations, such that any type of loads can be considered. Guilbault (2011) derived a correction factor to approximate the solution of quarter-space instead of going through the iterative process of Hetenyi. Recently, Zhang et al. (2013) approached the limit of Hetenyi's iteration using a smart matrix method and obtained a direct and efficient solution to the quarter-space problem. These aforementioned methods for solving the stress distribution of quarter-space are based on Hetenyi's overlapping half-space concept. Furthermore, Bower et al. (1987) applied finite element method for quarter-space to study the ratcheting limit of the plastic deformation of rails. Hecker and Romanov (1993) solved the stress distribution of quarter-space using Mellin transform. Guenfoud et al. (2010) used Ritz's method to obtain the displacement solution of quarter-space.

Regarding the stress solution to a general wedge problem, early efforts mainly targeted to two-dimensional cases, which considered only the evenly distributed load parallel to the apex edge of the wedge. Sneddon (1951) and Uflyand (1965) reviewed the application of Mellin transform to solve the two-dimensional wedge problem in detail. However, the three-dimensional wedge problem remains largely unexplored in the literature. The earliest report on the analysis of a three-dimensional wedge problem can be found in Russian literature, and the report was translated by Uflyand (1965). The formulation of the Paphovich-Neuber potential functions was developed for a general wedge as in Uflyand (1965). Knowing the potentials allows the elastic field to be derived. Efimov and Efimov (1986) evaluated the potential functions for a point normal load on a surface of an incompressible wedge with an arbitrary angle. Similar evaluations of the potential functions were also conducted independently by Hanson and Keer (1991). Hanson et al. (1994) extended the work to accommodate not only a point normal load but also a point tangential load on a surface of an incompressible wedge. However, these analyses were only limited to determining the potential functions and no solutions were provided. Hanson (1995) is the first to derive the closed-form expressions for an entire internal elastic field of a three-dimensional wedge under a concentrated normal force and a point moment load on the surface of the wedge.

For all the aforementioned studies, loads on the wedge surface were pre-defined, i. e., the problem was tackled with given stress boundary conditions on the surfaces of the wedge. On the other hand, the three-dimensional wedge problem was also approached by considering its contact with another rigid body. This approach is with a fixed contact shape boundary condition. Gerber (1968), who may be the first to study a smooth contact between a rigid body and an elastic quarter-space, solved the stress distribution in a quarter-space pressed by a flat rectangular punch. Keer et al. (1984) provided numerical results of a quarter-space loaded with a rigid cylindrical indenter using integral transform techniques. Hanson and Keer (1991) solved numerically the contact stresses between a spherical indenter and a quarter-space in a study of the edge effect on the rail-wheel contact. Hanson and Keer (1995) evaluated the effect of proximity of indentation to the wedge apex on contact stress distribution, contact area geometry, and contact compliance. Wang et al. (2012) used an equivalent inclusion method to study the contact of a rigid sphere and two joined quarter-spaces. They reported that the von Mises stress is higher in quarter-space than half-space. Recently, Zhang et al. (2015, 2017) studied the effect of two free edge surfaces by considering a contact between a rigid roller and an elastic body of finite width using Hetenyi's overlapping half-space concept and the problem was solved using the matrix method of Zhang et al. (2013).

Up to now, very few stress results in three-dimensional wedge are available in the literature probably because of the challenges involved in solving complex integral equations. Zhang et al. (2013) provided an alternative, which is an efficient matrix algorithm to directly solve a quarter-space problem. The present work adopts a similar approach of Zhang et al. (2013) to tackle general three-dimensional elastic wedge problems, as such, the stress analysis of three-dimensional wedge is conducted using matrix operations, and explicit solutions can be obtained. As mentioned, there are two different types of the wedge problem. This paper derives the solution to the stress problem of three-dimensional wedge with given loads, i. e., a fixed stress boundary condition applied.

2 Analysis

A recent study of Zhang et al. (2013) proved that a quarterspace problem can be readily and directly solved with matrix operations through the superposition of the solutions of two mutually orthogonal half-spaces loaded with mirrored-pairs of equivalent loads (Fig. 2). These two equivalent loads can be obtained directly by the product of the original load and transformation matrices. This study started with the same principle of replacing the originally applied load with the equivalent loads of half-spaces to perform the stress analysis of three-dimensional elastic wedge with an arbitrary angle. Hetenyi's overlapping half-space concept for solving a quarter-space problem considers the application of a coupled pair of loads which are the one on the quarter-space and its mirrored image (Fig. 2). Thus, the mid-plane of the half-space is always free from any shear stress. However, for a general wedge problem (Fig. 1), none of the two surfaces of the wedge can be the mid-plane of the half-spaces formed by the extension of the wedge surfaces. Therefore, both normal and shear stresses on the two planes of the wedge should be considered because of the absence of symmetry.

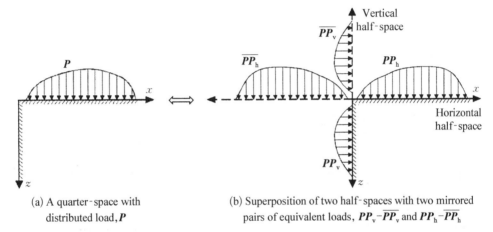

(a) A quarter-space with distributed load, P

(b) Superposition of two half-spaces with two mirrored pairs of equivalent loads, PP_v-$\overline{PP_v}$ and PP_h-$\overline{PP_h}$

Fig. 2 Reformulation of a quarter-space problem using two overlapping half-spaces with mirrored pairs of equivalent load, i.e. quarter-space problem can be solved using classical contact stress formulae of half-space with the mirrored pair of equivalent loads (Zhang et al., 2013)

Fig. 3(a) depicts the originally applied loads (normal p and shear q) on a wedge. Fig. 3(b) shows the definition of equivalent loads (normal $\bar{p}$ and shear $\bar{q}$) and the corresponding induced stresses of the equivalent loads on the plane surfaces of the wedge. The coordinate systems of X1 - Y1 - Z1 and X2 - Y2 - Z2 are defined in Fig. 3(b), which also illustrates the overlapping of the two half-spaces formed by the extension of the plane surfaces of the wedge. Numerical analysis started with the discretization of the two surfaces of the wedge. Fig. 4(a) shows the effective grid meshing of the top surface (I) of the wedge. The surface was divided into k rectangular grids. The ith rectangle is shown in Fig. 4(b). The dimensions of the ith rectangle are $2\alpha_i$ and $2\beta_i$ in the X1- and Y1-directions, respectively. The coordinate of its center is denoted as (x_{1i}, y_{1i}) in the coordinate system X1 - Y1 - Z1. Similarly, the other surface of the wedge was discretized into a set of l rectangles. The loads applied on the two surfaces were approached by piecewise distributions. The values of the distributed loads on the two surfaces are represented by the values at the centers of rectangles, such as $(p_1)_i = p_1(x_{1i}, y_{1i})$, $(q_{x1})_i = q_{x1}(x_{1i}, y_{1i})$, $(q_{y1})_i = q_{y1}(x_{1i}, y_{1i})$ for the top surface (I) and $(p_2)_j = p_2(x_{2j}, y_{2j})$, $(q_{x2})_j =$

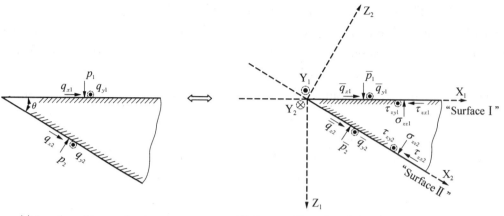

(a) A wedge with normal and shear loads (p and q)

(b) Superposition of two half-spaces with equivalent loads ($\bar{p}$ and $\bar{q}$) and corresponding induced stresses

Fig. 3 Solution to the loaded wedge using the superposition of solutions of half-space equivalent loads

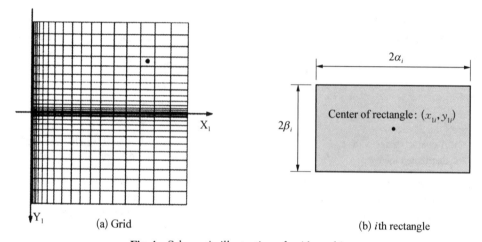

(a) Grid

(b) ith rectangle

Fig. 4 Schematic illustration of grid meshing

$q_{x2}(x_{2j}, y_{2j})$, $(q_{y2})_j = q_{y2}(x_{2j}, y_{2j})$ for the edge surface (II). Therefore, the normal and shear loads on the two surfaces of the wedge can be defined as $\boldsymbol{P}_1$, $\boldsymbol{Q}_{X1}$, $\boldsymbol{Q}_{Y1}$ and $\boldsymbol{P}_2$, $\boldsymbol{Q}_{X2}$, $\boldsymbol{Q}_{Y2}$ in vector format, as expressed respectively in Eqs. (1) and (2).

$$\boldsymbol{P}_1 = \begin{bmatrix} (p_1)_1 \\ (p_1)_2 \\ \vdots \\ (p_1)_k \end{bmatrix} ; \boldsymbol{Q}_{X1} = \begin{bmatrix} (q_{x1})_1 \\ (q_{x1})_2 \\ \vdots \\ (q_{x1})_k \end{bmatrix} ; \boldsymbol{Q}_{Y1} = \begin{bmatrix} (q_{y1})_1 \\ (q_{y1})_2 \\ \vdots \\ (q_{y1})_k \end{bmatrix} \qquad (1)$$

$$\boldsymbol{P}_2 = \begin{bmatrix} (p_2)_1 \\ (p_2)_2 \\ \vdots \\ (p_2)_l \end{bmatrix} ; \boldsymbol{Q}_{X2} = \begin{bmatrix} (q_{x2})_1 \\ (q_{x2})_2 \\ \vdots \\ (q_{x2})_l \end{bmatrix} ; \boldsymbol{Q}_{Y2} = \begin{bmatrix} (q_{y2})_1 \\ (q_{y2})_2 \\ \vdots \\ (q_{y2})_l \end{bmatrix} \qquad (2)$$

In the X1 - Y1 - Z1 coordinate system, the corresponding equivalent loads, $\bar{\boldsymbol{P}}_1$, $\bar{\boldsymbol{Q}}_{X1}$, and $\bar{Q}_{Y1}$, acting on the imaginary half-space formed by the extension of the top surface (I) of plane X1 - Y1 are also expressed with piece-wise distribution (Eq. (3)). The stresses at rectangle j on the plane X2 - Y2 in the imaginary half-space induced by these equivalent loads on the plane X1 - Y1 are denoted by their values at the centers of the rectangles, namely, $(\sigma_{zz2})_j = \sigma_{zz2}(x_{2j}, y_{2j})$, $(\tau_{zx2})_j = \tau_{zx2}(x_{2j}, y_{2j})$, and $(\tau_{zy2})_j = \tau_{zy2}(x_{2j}, y_{2j})$. These induced stresses of the plane X2 - Y2 form three vectors, namely, $\boldsymbol{S}_Z$, $\boldsymbol{S}_X$, and $\boldsymbol{S}_Y$ (Eq. (4)).

$$\bar{\boldsymbol{P}}_1 = \begin{bmatrix} (\bar{p}_1)_1 \\ (\bar{p}_1)_2 \\ \vdots \\ (\bar{p}_1)_k \end{bmatrix} ; \quad \bar{\boldsymbol{Q}}_{X1} = \begin{bmatrix} (\bar{q}_{x1})_1 \\ (\bar{q}_{x1})_2 \\ \vdots \\ (\bar{q}_{x1})_k \end{bmatrix} ; \quad \bar{\boldsymbol{Q}}_{Y1} = \begin{bmatrix} (\bar{q}_{y1})_1 \\ (\bar{q}_{y1})_2 \\ \vdots \\ (\bar{q}_{y1})_k \end{bmatrix} \quad (3)$$

$$\boldsymbol{S}_Z = \begin{bmatrix} (\sigma_{zz2})_1 \\ (\sigma_{zz2})_2 \\ \vdots \\ (\sigma_{zz2})_l \end{bmatrix} ; \quad \boldsymbol{S}_X = \begin{bmatrix} (\tau_{zx2})_1 \\ (\tau_{zx2})_2 \\ \vdots \\ (\tau_{zx2})_l \end{bmatrix} ; \quad \boldsymbol{S}_Y = \begin{bmatrix} (\tau_{zy2})_1 \\ (\tau_{zy2})_2 \\ \vdots \\ (\tau_{zy2})_l \end{bmatrix} \quad (4)$$

Based on the principle of superposition in the theory of elasticity (Timoshenko and Goodier, 1951), stresses σ_{zz2}, τ_{zx2}, and τ_{zy2}, at the center of rectangle j, (x_{2j}, y_{2j}), on the plane X2 - Y2 in the imaginary half-space induced by the equivalent distributed load on the i th rectangle on the plane X1 - Y1, can be expressed as:

$$\sigma_{zz2}(x_{2j}, y_{2j}) = m_{zzij}\bar{p}_1(x_{1i}, y_{1i}) + m_{zxij}\bar{q}_{x1}(x_{1i}, y_{1i}) + m_{zyij}\bar{q}_{y1}(x_{1i}, y_{1i}) \quad (5)$$

$$\tau_{zx2}(x_{2j}, y_{2j}) = m_{xzij}\bar{p}_1(x_{1i}, y_{1i}) + m_{xxij}\bar{q}_{x1}(x_{1i}, y_{1i}) + m_{xyij}\bar{q}_{y1}(x_{1i}, y_{1i}) \quad (6)$$

$$\tau_{zy2}(x_{2j}, y_{2j}) = m_{yzij}\bar{p}_1(x_{1i}, y_{1i}) + m_{yxij}\bar{q}_{x1}(x_{1i}, y_{1i}) + m_{yyij}\bar{q}_{y1}(x_{1i}, y_{1i}) \quad (7)$$

where coefficients from m_{zzij} to m_{yyij} can be determined by Love's (1929) solution for normal load and Ahmadi et al. (1987) for shear load (Solutions for normal and shear stresses are given in the Appendix). The first subscript of m signifies the induced stress at the j th rectangle on the surface II (z: normal stress; x and y: shear stresses in X2-and Y2-directions). The second subscript represents the equivalent load on the i th rectangle on the surface I (similarly, z: normal load; x and y: shear loads in X1-and Y1-directions). These coefficients are only related to the grid meshing, Poisson's ratio of the material and the wedge angle. The two-dimensional array of the coefficients forms the following nine reflecting matrices: $\boldsymbol{M}_{ZZ}$, $\boldsymbol{M}_{ZX}$, $\boldsymbol{M}_{ZY}$, $\boldsymbol{M}_{XZ}$, $\boldsymbol{M}_{XX}$, $\boldsymbol{M}_{XY}$, $\boldsymbol{M}_{YZ}$, $\boldsymbol{M}_{YX}$ and $\boldsymbol{M}_{YY}$.

$$\boldsymbol{M}_{ZZ} = \begin{bmatrix} m_{zz11} & m_{zz21} & \cdots & m_{zzk1} \\ m_{zz12} & m_{zz22} & \cdots & m_{zzk2} \\ \vdots & \vdots & & \vdots \\ m_{zz1l} & m_{zz2l} & \cdots & m_{zzkl} \end{bmatrix} \quad (8)$$

$$\boldsymbol{M}_{ZX} = \begin{bmatrix} m_{zx11} & m_{zx21} & \cdots & m_{zxk1} \\ m_{zx12} & m_{zx22} & \cdots & m_{zxk2} \\ \vdots & \vdots & & \vdots \\ m_{zx1l} & m_{zx2l} & \cdots & m_{zxkl} \end{bmatrix} \quad (9)$$

$$\vdots \qquad\qquad \vdots$$

$$\boldsymbol{M}_{YY} = \begin{bmatrix} m_{yy11} & m_{yy21} & \cdots & m_{yyk1} \\ m_{yy12} & m_{yy22} & \cdots & m_{yyk2} \\ \vdots & \vdots & & \vdots \\ m_{yy1l} & m_{yy2l} & \cdots & m_{yykl} \end{bmatrix} \quad (16)$$

The three induced stress vectors, $\boldsymbol{S}_Z$, $\boldsymbol{S}_X$, $\boldsymbol{S}_Y$, on the plane X2 – Y2 due to the equivalent load vectors $\bar{\boldsymbol{P}}_1$, $\bar{\boldsymbol{Q}}_{X1}$, $\bar{\boldsymbol{Q}}_{Y1}$, on the plane X1 – Y1 can be calculated based on the half-space model by:

$$\boldsymbol{S}_Z = \boldsymbol{M}_{ZZ} \cdot \bar{\boldsymbol{P}}_1 + \boldsymbol{M}_{ZX} \cdot \bar{\boldsymbol{Q}}_{X1} + \boldsymbol{M}_{ZY} \cdot \bar{\boldsymbol{Q}}_{Y1} \quad (17)$$

$$\boldsymbol{S}_X = \boldsymbol{M}_{XZ} \cdot \bar{\boldsymbol{P}}_1 + \boldsymbol{M}_{XX} \cdot \bar{\boldsymbol{Q}}_{X1} + \boldsymbol{M}_{XY} \cdot \bar{\boldsymbol{Q}}_{Y1} \quad (18)$$

$$\boldsymbol{S}_Y = \boldsymbol{M}_{YZ} \cdot \bar{\boldsymbol{P}}_1 + \boldsymbol{M}_{YX} \cdot \bar{\boldsymbol{Q}}_{X1} + \boldsymbol{M}_{YY} \cdot \bar{\boldsymbol{Q}}_{Y1} \quad (19)$$

Similarly, the equivalent normal and shear load vectors, $\bar{\boldsymbol{P}}_2$, $\bar{\boldsymbol{Q}}_{X2}$, and $\bar{\boldsymbol{Q}}_{Y2}$, on the surface X2 – Y2 of half-space induces stress vectors, $\boldsymbol{T}_Z$, $\boldsymbol{T}_X$, and $\boldsymbol{T}_Y$, on the plane X1 – Y1. The relationship between these equivalent loads and stresses of half-space model are expressed as:

$$\boldsymbol{T}_Z = \boldsymbol{N}_{ZZ} \cdot \bar{\boldsymbol{P}}_2 + \boldsymbol{N}_{ZX} \cdot \bar{\boldsymbol{Q}}_{X2} + \boldsymbol{N}_{ZY} \cdot \bar{\boldsymbol{Q}}_{Y2} \quad (20)$$

$$\boldsymbol{T}_X = \boldsymbol{N}_{XZ} \cdot \bar{\boldsymbol{P}}_2 + \boldsymbol{N}_{XX} \cdot \bar{\boldsymbol{Q}}_{X2} + \boldsymbol{N}_{XY} \cdot \bar{\boldsymbol{Q}}_{Y2} \quad (21)$$

$$\boldsymbol{T}_Y = \boldsymbol{N}_{YZ} \cdot \bar{\boldsymbol{P}}_2 + \boldsymbol{N}_{YX} \cdot \bar{\boldsymbol{Q}}_{X2} + \boldsymbol{N}_{YY} \cdot \bar{\boldsymbol{Q}}_{Y2} \quad (22)$$

Overlapping the two sets of half-space solutions and satisfying the stress boundary conditions of the wedge (original loads acting on the two wedge surfaces) yields:

$$-\boldsymbol{P}_1 = -\bar{\boldsymbol{P}}_1 + \boldsymbol{T}_Z = -\bar{\boldsymbol{P}}_1 + \boldsymbol{N}_{ZZ} \cdot \bar{\boldsymbol{P}}_2 + \boldsymbol{N}_{ZX} \cdot \bar{\boldsymbol{Q}}_{X2} + \boldsymbol{N}_{ZY} \cdot \bar{\boldsymbol{Q}}_{Y2} \quad (23)$$

$$-\boldsymbol{Q}_{X1} = -\bar{\boldsymbol{Q}}_{X1} + \boldsymbol{T}_X = -\bar{\boldsymbol{Q}}_{X1} + \boldsymbol{N}_{XZ} \cdot \bar{\boldsymbol{P}}_2 + \boldsymbol{N}_{XX} \cdot \bar{\boldsymbol{Q}}_{X2} + \boldsymbol{N}_{XY} \cdot \bar{\boldsymbol{Q}}_{Y2} \quad (24)$$

$$-\boldsymbol{Q}_{Y1} = -\bar{\boldsymbol{Q}}_{Y1} + \boldsymbol{T}_Y = -\bar{\boldsymbol{Q}}_{Y1} + \boldsymbol{N}_{YZ} \cdot \bar{\boldsymbol{P}}_2 + \boldsymbol{N}_{YX} \cdot \bar{\boldsymbol{Q}}_{X2} + \mathrm{N}_{YY} \cdot \bar{\boldsymbol{Q}}_{Y2} \quad (25)$$

$$-\boldsymbol{P}_2 = -\bar{\boldsymbol{P}}_2 + \boldsymbol{S}_Z = -\bar{\boldsymbol{P}}_2 + \boldsymbol{M}_{ZZ} \cdot \bar{\boldsymbol{P}}_1 + \boldsymbol{M}_{ZX} \cdot \bar{\boldsymbol{Q}}_{X1} + \boldsymbol{M}_{ZY} \cdot \bar{\boldsymbol{Q}}_{Y1} \quad (26)$$

$$-\boldsymbol{Q}_{X2} = -\bar{\boldsymbol{Q}}_{X2} + \boldsymbol{S}_X = -\bar{\boldsymbol{Q}}_{X2} + \boldsymbol{M}_{XZ} \cdot \bar{\boldsymbol{P}}_1 + \boldsymbol{M}_{XX} \cdot \bar{\boldsymbol{Q}}_{X1} + \boldsymbol{M}_{XY} \cdot \bar{\boldsymbol{Q}}_{Y1} \quad (27)$$

$$-\boldsymbol{Q}_{Y2} = -\bar{\boldsymbol{Q}}_{Y2} + \boldsymbol{S}_Y = -\bar{\boldsymbol{Q}}_{Y2} + \boldsymbol{M}_{YZ} \cdot \bar{\boldsymbol{P}}_1 + \boldsymbol{M}_{YX} \cdot \bar{\boldsymbol{Q}}_{X1} + \boldsymbol{M}_{YY} \cdot \bar{\boldsymbol{Q}}_{Y1} \quad (28)$$

Thus, the explicit solution to the equivalent loads on the two wedge surfaces in matrix form is:

$$\begin{bmatrix} \bar{P}_1 \\ \bar{Q}_{X1} \\ \bar{Q}_{Y1} \\ \bar{P}_2 \\ \bar{Q}_{X2} \\ \bar{Q}_{Y2} \end{bmatrix} = \begin{bmatrix} I & 0 & 0 & -N_{ZZ} & -N_{ZX} & -N_{ZY} \\ 0 & I & 0 & -N_{XZ} & -N_{XX} & -N_{XY} \\ 0 & 0 & I & -N_{YZ} & -N_{YX} & -N_{YY} \\ -M_{ZZ} & -M_{ZX} & -M_{ZY} & I & 0 & 0 \\ -M_{XZ} & -M_{XX} & -M_{XY} & 0 & I & 0 \\ -M_{YZ} & -M_{YX} & -M_{YY} & 0 & 0 & I \end{bmatrix}^{-1} \begin{bmatrix} P_1 \\ Q_{X1} \\ Q_{Y1} \\ P_2 \\ Q_{X2} \\ Q_{Y2} \end{bmatrix} \quad (29)$$

Eq. (29) indicates that the equivalent loads can be directly calculated by the product of the originally applied loads and a transformation matrix. Once the equivalent loads on the two wedge surfaces are acquired, the internal stress field of the wedge can be obtained with the superposition of the two sets of half-space solutions of the equivalent loads, as schematically illustrated in Fig. 3.

3 Verification of the Present Method

Three cases were examined to verify the proposed matrix method and prove its accuracy with the following conditions. Poisson's ratio is set to 0.3. The X1 - Y1 surface was discretized into 8500 uneven-sized rectangles with 85 units in the y-direction and 100 units in the x-direction. The same grid meshing was applied to surface X2 - Y2. Discretization covers a region of $0 \leqslant x \leqslant 10.1a$, and $-9.7a \leqslant y \leqslant 9.7a$. The calculated stress distributions are shown in X1 - Y1 - Z1 coordinate. Two surface loading types are used to verify the proposed method. A half-semispherical normal load distribution, as schematically shown in Fig. 5, which was used by Hanson and Keer (1990), is adopted firstly, such that some of our results can be compared with theirs. The load is applied to the X1 - Y1 surface ($z_1 = 0$) of this elastic wedge with the maximum, p_0, at the origin (0, 0, 0). The load is expressed as:

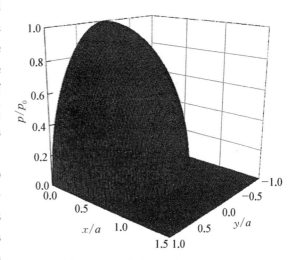

Fig. 5 Illustration of the applied load on the top surface of the wedge structure

$$p(x_1, y_1) = p_0 \frac{\sqrt{a^2 - x_1^2 - y_1^2}}{a}, \quad \sqrt{x_1^2 + y_1^2} \leqslant a$$

$$p(x_1, y_1) = 0, \quad \sqrt{x_1^2 + y_1^2} > a \quad (30)$$

where a is the radius of the half-semispherical load distribution.

3.1 Wedge angle: 170°

The first case is a wedge structure with 170° wedge angle. It is close to 180° such that the new algorithm can be verified by comparing its results with those of the classical half-space. Firstly, the algorithm is executed to obtain the six equivalent loads on the two planes of the wedge as expressed in Eq. (29). Fig. 6 depicts the six equivalent loads on the plane X1 - Y1 (Fig. 6(a - c)) and X2 - Y2 (Fig. 6(d - f)). In fact, the equivalent load $\bar{\boldsymbol{P}}_1$ which is in the direction normal to the top surface of the wedge (plane X1 - Y1) deviates very little from the original applied load as shown in Fig. 5. The other equivalent loads are of much smaller magnitude. The resultant stress is thus dominated by the equivalent load $\bar{\boldsymbol{P}}_1$. The internal stress field of the wedge is the summation of stresses calculated with the individual equivalent load using the classical half-space formulae as illustrated in the Appendix.

Since the wedge angle is close to 180°, the stress distribution should be very close to that of half-space under the same surface loads. Fig. 7 shows the calculated stress distribution in the $y = 0$ plane. The dotted line denotes the edge (or boundary) of the

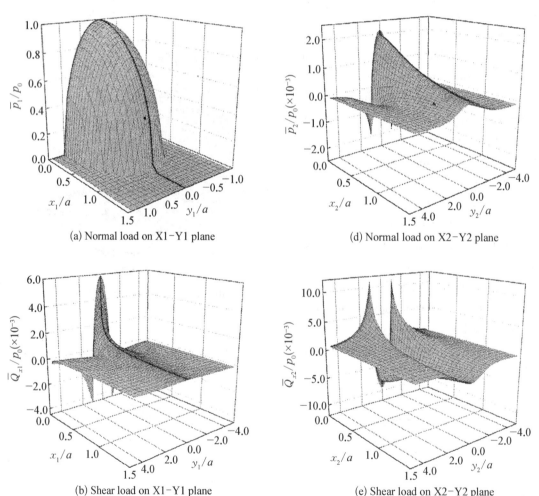

(a) Normal load on X1−Y1 plane   (d) Normal load on X2−Y2 plane

(b) Shear load on X1−Y1 plane   (e) Shear load on X2−Y2 plane

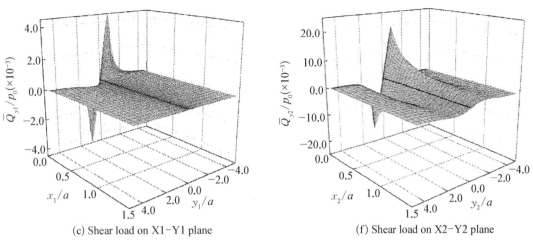

(c) Shear load on X1-Y1 plane

(f) Shear load on X2-Y2 plane

Fig. 6 Six equivalent loads on plane X1-Y1 (a), (b), (c) and X2-Y2 (d), (e) and (f)

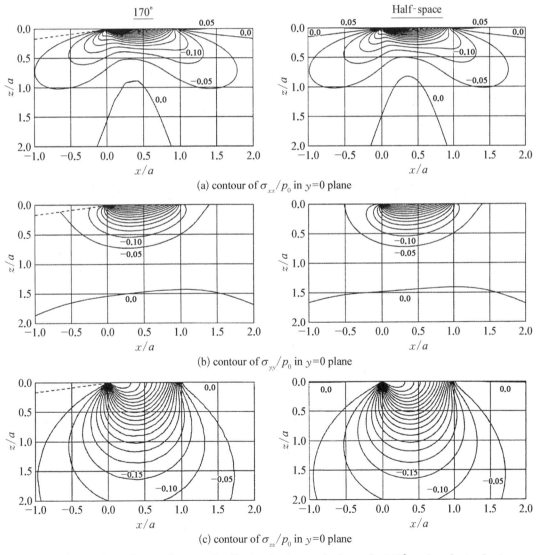

(a) contour of σ_{xx}/p_0 in $y=0$ plane

(b) contour of σ_{yy}/p_0 in $y=0$ plane

(c) contour of σ_{zz}/p_0 in $y=0$ plane

Fig. 7 Comparison of normal stress distribution in the $y=0$ plane of a 170° wedge obtained using the matrix method and of a half-space

wedge. The classical results of half-space are also presented in Fig. 7 for comparison. It is clear that the normal stress distributions of the 170° wedge are almost the same with those of a half-space, which confirm our prediction and the proposed algorithm.

Based on the work of Hamilton and Goodman (1966) and Bryant and Keer (1982), von Mises yield criterion can be used to predict the mechanical failure taking the form of plastic yielding and deformation. It can be obtained from the square root of the second invariant of the deviator stress tensor, i. e.

$$J_2^{1/2} = \sqrt{\tau_{xy}^2 + \tau_{yz}^2 + \tau_{xz}^2 + \frac{1}{6}[(\sigma_{xx} - \sigma_{yy})^2 + (\sigma_{yy} - \sigma_{zz})^2 + (\sigma_{zz} - \sigma_{xx})^2]} \qquad (31)$$

Yielding occurs when $J_2^{1/2}$ reaches the yield stress. The shake-down analysis (Johnson, 1987) and failure judgement of an elastic body (Bower et al., 1987) also follow this principle.

Fig. 8 illustrates the comparison of $J_2^{1/2}/p_0$ in the $y = 0$ plane of the 170° wedge obtained using the matrix algorithm and the half-space model. It shows that the shape of $J_2^{1/2}/p_0$ of the two structures coincides well except the small difference next to the boundary of wedge. The maximum value of $J_2^{1/2}/p_0$ for 170° wedge and half-space with the current meshing are 0.341 and 0.345, respectively.

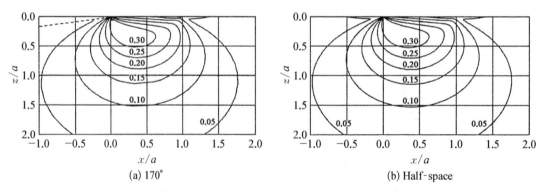

Fig. 8 Comparison of $J_2^{1/2}/p_0$ in the $y = 0$ plane of a 170° wedge obtained using the matrix algorithm and of a half-space

3.2 Wedge angle: 90°

The second case was extracted from Hanson and Keer (1990). They provided stress solutions of a loaded quarter-space, which is a three-dimensional problem of a wedge with 90° wedge angle. Normal stresses and $J_2^{1/2}/p_0$ in the quarter-space were calculated with the proposed matrix method. Stress distributions and $J_2^{1/2}/p_0$ on the plane of $y = 0$ were obtained and plotted in contours along with the results of Hanson and Keer (1990) in Figs. 9 and 10. Generally, the results of present study coincide satisfactorily with those of Hanson and Keer (1990). However, it should be noticed that there are some differences near the edges, especially $J_2^{1/2}/p_0$. The maximum magnitude of $J_2^{1/2}/p_0$ is 0.551 with the current method, which is slightly higher than 0.546 as obtained by Hanson and Keer (1990). The differences may be attributed to various meshing used.

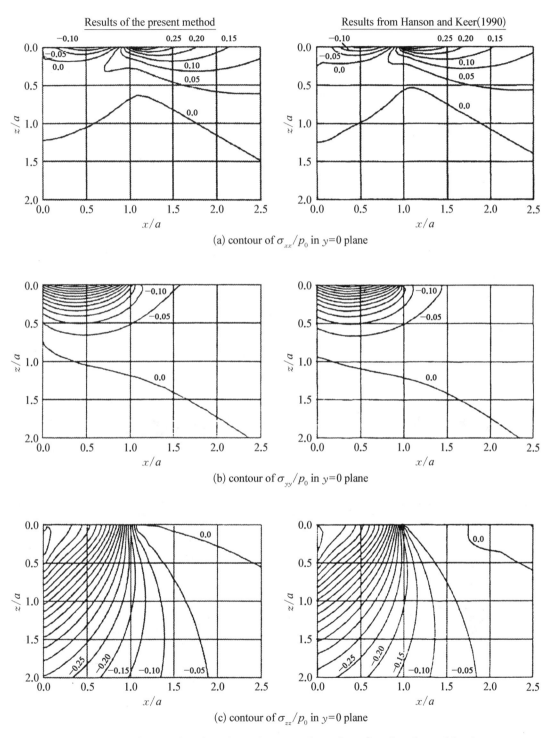

Fig. 9 Comparison of stress distributions in the $y=0$ plane of a 90° wedge obtained by the current algorithm and extracted from Hanson and Keer (1990)

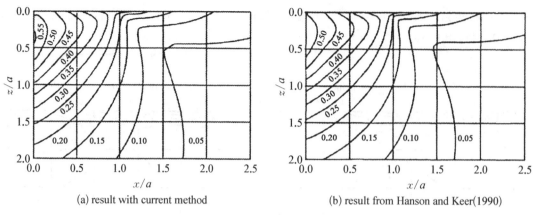

(a) result with current method (b) result from Hanson and Keer(1990)

Fig. 10 Comparison of $J_2^{1/2}/p_0$ in the $y=0$ plane of a 90° wedge obtained by the current algorithm and extracted from Hanson and Keer (1990)

3.3 Wedge angle: 60°

The third case is an elastic wedge with 60° wedge angle. The surface load described with Eq. (30) was applied on its top surface. Figs. 11 and 12 show, respectively, the calculated normal stress distributions and $J_2^{1/2}/p_0$ with the current matrix algorithm and FEM. It is clear that the results obtained with these two methods are almost the same. The magnitude and the gradient of the normal stresses and $J_2^{1/2}/p_0$ are quite high near the tip because of the small wedge angle, indicating that the failure appears easily there.

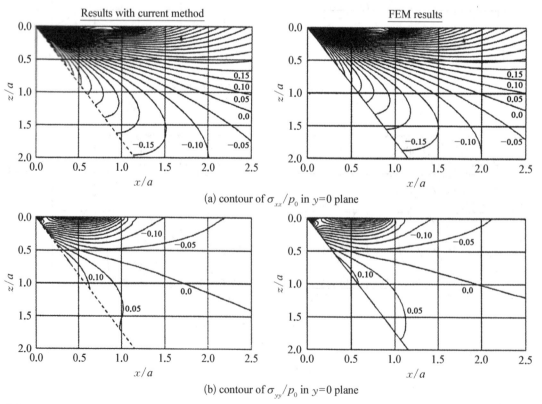

(a) contour of σ_{xx}/p_0 in $y=0$ plane

(b) contour of σ_{yy}/p_0 in $y=0$ plane

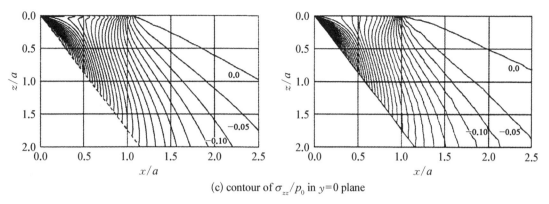

(c) contour of σ_{zz}/p_0 in $y=0$ plane

Fig. 11 Comparison of normal stress distributions in the $y=0$ plane of a 60° wedge obtained using the matrix algorithm and FEM

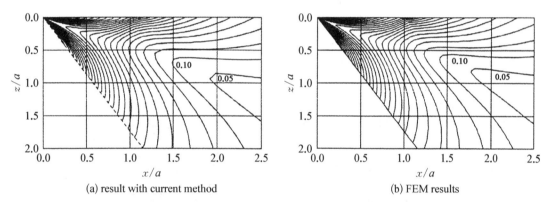

(a) result with current method (b) FEM results

Fig. 12 Comparison of $J_2^{1/2}/p_0$ in the $y=0$ plane of a 60° wedge obtained using the matrix algorithm and FEM

The whole contact body should be divided into small elements with FEM. However, only the two side surfaces of the wedge structure should be meshed with current method. Therefore, it is much simpler and more suitable to study the wedge problems with varying surface loads. Furthermore, the transformation matrix in Eq. (29) is only related with the wedge angle, mesh structure and the elastic properties of contact body but not the applied load. The transformation matrix can be determined and saved in advance for subsequent calculations with different loads. Therefore, the explicit matrix method is quite suitable to be embedded into those studies requiring iterative calculations of wedge deformation or stress, such as the elastohydrodynamic lubrication analysis of roller bearings.

3.4 A wedge under simultaneous loads on its two planes

To simulate a more general wedge problem, a quarter-space with the same normal load distribution on its two surfaces was studied. The normal load distribution was also extracted from Hanson and Keer (1990) and the one applied on the top surface is expressed as,

$$p(x_1, y_1) = p_0 \frac{\sqrt{a^2 - (x_1 - a)^2 - y_1^2}}{a}, \sqrt{(x_1-a)^2 + y_1^2} \leqslant a \quad (32)$$

$$p(x_1, y_1) = 0, \sqrt{(x_1-a)^2 + y_1^2} > a$$

The same load distribution, as described with Eq. (32) except x_1, y_1 being replaced with x_2, y_2, was simultaneously applied on the side surface (X2 - Y2 plane). The calculated results are shown in Figs. 13 and 14. Corresponding results obtained by Hanson and Keer (1990) were also plotted for comparison.

Clearly, the proposed method was verified by the satisfactory coincidence of results obtained by the two methods. In detail, the maximum modulus of σ_{xx}/p_0 is 1.0 and just locates at (0, 0, 1.0), which is the center of the side semispherical surface load. Because the current problem is symmetrical about the plane $x_1 = z_1$, the result of σ_{zz} can be extracted from Fig. 13(a) and is thus not plotted here. For the same reason, the distribution of σ_{yy} in $y=0$ plane is also symmetrical about $x_1 = z_1$ (as shown in Fig. 13(b)), which verifies the calculated results indirectly. The maximum value of $J_2^{1/2}/p_0$ under current conditions is 0.341 at (0.5, 0, 0.5).

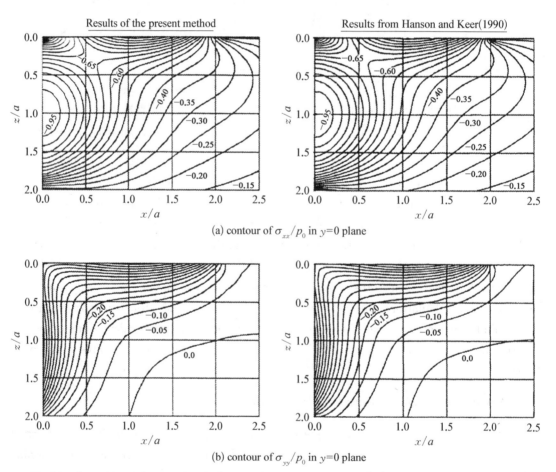

(a) contour of σ_{xx}/p_0 in $y=0$ plane

(b) contour of σ_{yy}/p_0 in $y=0$ plane

Fig. 13 Comparison of stress distributions in the $y=0$ plane of a 90° wedge obtained by the current method and extracted from Hanson and Keer (1990)

The good correlation of the results obtained by the matrix algorithm and those published data and FEM outputs, as illustrated in Figs. 7 – 14, verifies the proposed method.

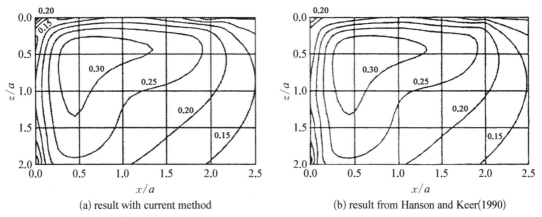

(a) result with current method (b) result from Hanson and Keer(1990)

Fig. 14 Comparison of $J_2^{1/2}/p_0$ in the $y = 0$ plane of a 90° wedge obtained by the current method and extracted from Hanson and Keer (1990)

4 Effect of wedge angle on internal normal stresses in $y=0$ plane

Fig. 15 shows the difference in the normal stresses in the $y = 0$ plane with wedge angle. The maximum p_o of the half-semispherical distributed-load, as expressed in Eq. (30), is at the origin (0, 0, 0). As shown in the example of σ_{xx}, the maximum tensile stress located on the top surface next to the right boundary of the applied load increases with decreasing wedge angle. The maximum σ_{xx} for the wedge with 170° is 0.089 p_0 on the top surface, but the maximum σ_{xx} reaches 0.41p_0 for quarter-space. The compressive stress on the top surface decreases with the wedge angle. Specially, the compressive stress on the top surface totally changes into tensile stress when the wedge angle is <90° This finding is reasonable given the lack of enough support under the top surface for such wedge structure with angle <90° For the normal stress σ_{xx} inside the contact body, the trend can be separated into two parts. For wedges with angle >90°, the area of compressive stress shrinks to the apex of the wedge with decreasing wedge angle. However, the area of compressive stress increases under the load area with the wedge angle decreasing from 90° Furthermore, the compressive stress along the free surface decreases slightly with depth for the wedge with 60°.

The stresses σ_{yy} are totally compressive when the wedge angle is >90° However, tensile stress appears next to the free surface when the wedge is small enough, such as the wedge with 60° Additionally, the effect of wedge angle on σ_{yy} in the corner is apparent. The gradient of σ_{yy} at the corner drops gradually with the decrease in wedge angle from 170° to 90° However, this trend is reversed once the wedge angle reaches below 90° A similar finding was observed for normal stress σ_{zz}.

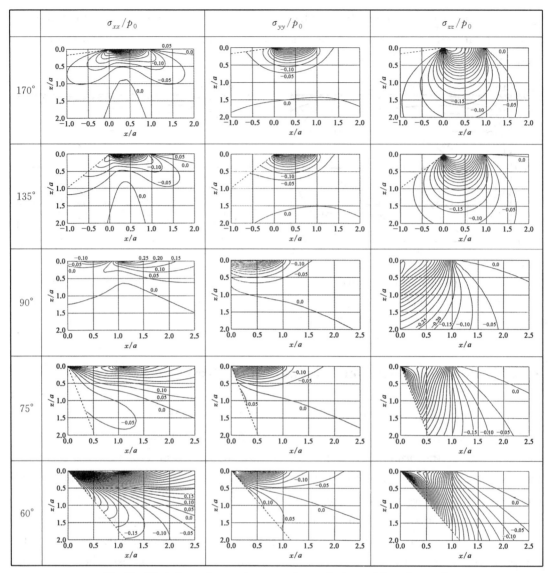

Fig. 15 Variations of normal stresses in $y=0$ plane with wedge angle

5 Conclusions

In this paper, a matrix algorithm was developed to solve the three-dimensional elastic wedge problem under surface loads. Such solution has not yet been reported. By multiplying the original applied loads with the derived transformation matrix, the equivalent loads in half-space can be obtained directly and the wedge problem is transformed into a half-space problem. The transformation matrix is related only to the wedge angle, mesh scheme, and contact material. Once the transformation matrix of a given wedge is obtained, the stresses induced by various distributed loads can readily be solved. The matrix method was verified using three cases (i. e., 170°, 90° and 60° wedges).

Acknowledgment

This research work is fully supported by the Research Grants Council of Hong Kong (Project no. CityU11213914).

Appendix

The following half-space stress solutions are extracted from the appendix of Yu and Bhushan (1996). It is not only for the convenience of the readers, but also for the provision of correct formulae since a typo was found in their paper Fig. A1 shows a semi-infinite space subjected to a rectangle patch load on the top surface with uniform normal and tangential loads.

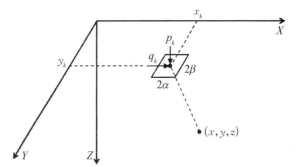

Fig. A1

A1 Normal load

The stress field at an internal point (x, y, z) caused by an evenly distributed normal load of a rectangular grid with 2α length in x-direction and 2β width in y-direction on the $z=0$ surface of a semi-infinite space (Love, 1929) is:

$$\sigma_{ij}^k = \frac{p_k}{2\pi}[A_{ij}(\bar{x}+\alpha, \bar{y}+\beta, z) + A_{ij}(\bar{x}-\alpha, \bar{y}-\beta, z)$$
$$- A_{ij}(\bar{x}+\alpha, \bar{y}-\beta, z) - A_{ij}(\bar{x}-\alpha, \bar{y}+\beta, z)]$$

where p_k is the normal uniform load applied over the kth rectangle, $\bar{x}=x-x_k$, $\bar{y}=y-y_k$, and $(x_k, y_k, 0)$ is the coordinate of the center of the kth rectangle. The functions A_{ij} are described as:

$$A_{xx}(x, y, z) = 2\mu\left[\tan^{-1}\left(\frac{xz}{yR}\right) - \tan^{-1}\left(\frac{x}{y}\right)\right] - \tan^{-1}\left(\frac{y}{x}\right) + \tan^{-1}\left(\frac{yz}{xR}\right) + \frac{xyz}{R(x^2+z^2)}$$

$$A_{yy}(x, y, z) = 2\mu\left[\tan^{-1}\left(\frac{yz}{xR}\right) - \tan^{-1}\left(\frac{y}{x}\right)\right] - \tan^{-1}\left(\frac{x}{y}\right) + \tan^{-1}\left(\frac{xz}{yR}\right) + \frac{xyz}{R(y^2+z^2)}$$

$$A_{zz}(x, y, z) = -\tan^{-1}\left(\frac{y}{x}\right) - \tan^{-1}\left(\frac{x}{y}\right) + \tan^{-1}\left(\frac{yz}{xR}\right) + \tan^{-1}\left(\frac{xz}{yR}\right)$$
$$- \frac{xyz}{R(x^2+z^2)} - \frac{xyz}{R(y^2+z^2)}$$

$$A_{xy}(x, y, z) = -(1-2\mu)\ln(R+z) - \frac{z}{R}$$

$$A_{xz}(x, y, z) = \frac{z^2 y}{R(x^2 + z^2)}$$

$$A_{yz}(x, y, z) = \frac{z^2 x}{R(y^2 + z^2)}$$

where $R = \sqrt{x^2 + y^2 + z^2}$ and μ is the Poisson's ratio.

A2 Shear load

The stress field at an internal point (x, y, z) caused by an evenly distributed shear load (in x-direction) of the rectangular grid on the top surface of a semi-infinite space (Ahmadi. et al. 1987) is:

$$\sigma_{ij}^k = \frac{q_k}{2\pi}[B_{ij}(\bar{x}+\alpha, \bar{y}+\beta, z) + B_{ij}(\bar{x}-\alpha, \bar{y}-\beta, z) \\ - B_{ij}(\bar{x}+\alpha, \bar{y}-\beta, z) - B_{ij}(\bar{x}-\alpha, \bar{y}+\beta, z)]$$

where q_k is the shear uniform load applied over the kth rectangle, $\bar{x} = x - x_k$, $\bar{y} = y - y_k$, and $(x_k, y_k, 0)$ is the coordinate of the center of the kth rectangle. The functions B_{ij} are described as

$$B_{xx}(x, y, z) = 2\ln(R+y) + \frac{x^2 y}{R(x^2+z^2)} + (1-2\mu)\frac{y}{R+z}$$

$$B_{yy}(x, y, z) = 2\mu\ln(R+y) - \frac{y}{R} + (1-2\mu)\frac{y}{R+z}$$

$$B_{zz}(x, y, z) = \frac{z^2 y}{R(x^2+z^2)}$$

$$B_{xy}(x, y, z) = \ln(R+x) - \frac{x}{R} + (1-2\mu)\frac{x}{R+z}$$

$$B_{xz}(x, y, z) = -\tan^{-1}\left(\frac{xy}{Rz}\right) + \frac{xyz}{R(x^2+z^2)}$$

$$B_{yz}(x, y, z) = -\frac{z}{R}$$

where $R = \sqrt{x^2 + y^2 + z^2}$ and μ is the Poisson's ratio.

References

[1] Ahmadi, N., Keer, L. M., Mura, T., Vithoontien, V., 1987. The interior stress-field caused by tangential loading of a rectangular patch on an elastic half-space. ASME J. Tribol. 109, 627–629.

[2] Boussinesq, J., 1885. Applications Des Potentiels a L'etude De L'equilibre Et Du Inouvement Des Solides Elastiques. Gauthier-Villars, Paris.

[3] Bower, A. F., Johnson, K. L., Kalousek, J., 1987. A ratchetting limit for plastic deformation of a quarter-space under rolling contact loads. In: Proceedings of International Symposium on Contact Mechanics and Wear of Rail/Wheel Systems II. University of Waterloo, pp. 117–132.

[4] Bryant, M. D., Keer, L. M., 1982. Rough contact between elastically and geometrically identical curved

bodies. ASME J. Appl. Mech. 49, 345–352.

[5] Efimov, A., Efimov, D., 1986. Concentrated actions applied to an elastic incompressible wedge. Mech. Solids 21, 83–86.

[6] Gerber, C. E., 1968. Contact Problems for the Elastic Quarter-Plane and for the Quarter-space. Stanford University Doctoral Dissertation.

[7] Guenfoud, S., Bosakov, S. V., Laefer, D. F., 2010. A Ritz's method based solution for the contact problem of a deformable rectangular plate on an elastic quarter-space. Int. J. Solids Struct. 47, 1822–1829.

[8] Guilbault, R., 2011. A fast correction for elastic quarter-space applied to 3D modeling of edge contact problems. ASME J. Tribol. 133, 031402.

[9] Hamilton, G. M, Goodman, L. E., 1966. The stress field created by a circular sliding contact. ASME J. Appl. Mech. 33, 371–376.

[10] Hanson, M. T., 1995. Elastic fields resulting from concentrated loading on a three-dimensional incompressible wedge. ASME J. Appl. Mech. 62, 557–565.

[11] Hanson, M. T., Keer, L. M., 1990. A simplified analysis for an elastic quarter-space. Q. J. Mech. Appl. Math. 43, 561–587.

[12] Hanson, M. T., Keer, L. M., 1991. Analysis of edge effects on rail-wheel contact. Wear 144, 39–55.

[13] Hanson, M. T., Xu, Y., Keer, L. M., 1994. Stress analysis for a three-dimensional incompressible wedge under body force or surface loading. Q. J. Mech. Appl. Math. 47, 141–158.

[14] Hanson, M. T., Keer, L. M., 1995. Mechanics of edge effects on frictionless contacts. Int. J. Solids Struct. 32, 391–405.

[15] Hecker, M., Romanov, A., 1993. The stress fields of edge dislocations near wedge-shaped boundaries and bonded wedges. Mater. Sci. Eng. A 164, 411–414.

[16] Hertz, H., 1882. On the contact of elastic solids. J. Reine Angew. Math. 92, 156–171.

[17] Hetenyi, M., 1970. A general solution for the elastic quarter space. ASME J. Appl. Mech. 37, 70–76.

[18] Johnson, K. L., 1987. Contact Mechanics and Wear of Rail/Wheel Systems II. University of Waterloo Press, pp. 83–98.

[19] Keer, L. M., Lee, J. C., Mura, T., 1983. Hetenyi's elastic quarter space problem revisited. Int. J. Solids Struct. 19, 497–508.

[20] Keer, L. M., Lee, J. C., Mura, T., 1984. A contact problem for the elastic quarter space. Int. J. Solids Struct. 20, 513–524.

[21] Love, A. E. H., 1929. The stress produced in a semi-infinite solid by pressure on part of the boundary. Philos. Trans. R. Soc. Lond. A 228, 377–420.

[22] Sneddon, I. N., 1951. Fourier Transforms. McGraw-Hill, New York.

[23] Timoshenko, S., Goodier, J. N., 1951. Theory of Elasticity. McGraw-Hill, New York.

[24] Uflyand, Y. S., 1965. Survey of Articles on the Applications of Integral Transforms in the Theory of Elasticity. North Carolina State University at Raleigh, Applied Mathematics Research Group, File No. PSR-24/6, pp. 342–383.

[25] Wang, Z. J., Jin, X. Q., Keer, L. M., Wang, Q., 2012. Numerical methods for contact between two joined quarter spaces and a rigid sphere. Int. J. Solids Struct. 49, 2515–2527.

[26] Yu, M. H., Bhushan, B., 1996. Contact analysis of three-dimensional rough surfaces under frictionless and frictional contact. Wear 200, 265–280.

[27] Zhang, H. B., Wang, W. Z., Zhang, S. G., Zhao, Z. Q, 2015. Modeling of finite-length line contact problem with consideration of two free-end surfaces. ASME J. Tribol. 138, 021402.

[28] Zhang, H. B., Wang, W. Z., Zhang, S. G., Zhao, Z. Q, 2017. Modeling of elastic finite-length space rolling-sliding contact problem. Tribol. Int. 113, 224–237.

[29] Zhang, Z. M., Wang, W., Wong, P. L., 2013. An explicit solution for the elastic quarter-space problem in matrix formulation. Int. J. Solids Struct. 50, 976–980.

Surface Normal Deformation in Elastic Quarter-space*

Abstract: An efficient and explicit solution for the surface deformation of quarter-space under normal load is developed using the concept of flexibility matrix, which serves like springs in response to loads. Quarter-space is characterized by the unbounded side surface, such as in roller bearings and gears. The solution method is verified using a typical case. The edge effect on surface deformation under three load types namely, Hertzian point, flat cylindrical punch and Hertzian line, are evaluated. The effect can be considerable if the applied load is close to edge. The flexibility matrix is constant for a given case. Hence, the solution method is highly efficient, and particularly suitable for quarter-space problems which require iterative calculations, such as elastohydrodynamic lubrication analyses.

1 Introduction

Acquiring the elastic deformation of contact surfaces is important in engineering. The solution process also needs to be fast and efficient for certain applications that require iterative calculations for the surface deformation, such as the analysis of tribo-pairs operating under the elasto-hydrodynamic lubrication (EHL) regime. Some common engineering components, such as roller bearings, gears and camfollowers, are characterized by the existence of free edge surfaces. Contact problems of these components are, in fact, more accurately modeled by elastic quarter-space (Fig. 1(a)). Nevertheless, the available solutions of elastic quarter-space are very complex, such that the elastic half-space model (semi-infinite body) is widely adopted for calculating contact stress and deformation in practical mechanical systems, such as those aforementioned applications, for their contact solutions are readily obtained with Bussinessq or Love formulae[1,2]. The assumption of semi-infinite body model is obviously not satisfied in these practical cases. For example, in the contact of gears and roller bearings, the length of the gear tooth and bearing roller are finite. Thus, the effect of free edge surfaces cannot be ignored and these components cannot be taken as semi-infinite bodies. The elastic quarter-space model is, indeed, more appropriate.

Hetenyi[3] tackled the quarter-space problem with the concept of iteratively overlapping mutually orthogonal half-spaces with mirrored load pairs till fulfilling the

---

* In collaboration with W. Wang, L. Guo and P. L. Wong. Reprinted from *Tribology International*, 2007, 114: 358 – 364.

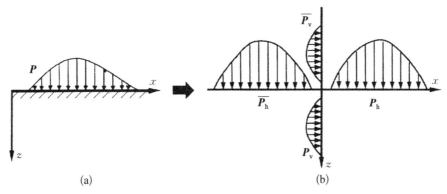

Fig. 1 Quarter-space solutions equivalent to overlapping half-spaces

boundary conditions of the quarter-space. Keer et al. [4] utilized Hetenyi's overlapping half-space idea and derived two integral equations to describe the quarter-space problem. They solved the equations with Fourier transform. Nevertheless, their method can only be applied to cases where the load can be Fourier-transformed. Later on, Hanson and Keer[5] overcame this limitation with a direct numerical solution of the quarter-space by solving two dimensional integral equations. Thus, any load type can be considered. The difference in the stress obtained with quarter-space and half-space models was studied. As pointed out in [5], the magnitude and position of the maximum stress calculated with quarter-space and half-space vary, especially when the load is located in the immediate neighborhood of the free end. Guilbault[6] made use of a correction factor which multiplies the Hetenyi's mirrored loads to simultaneously correct the influence of stresses on elastic deformation of a quarterspace, which provides a much faster solution than a complete Hetenyi process. This correction factor method was applied by Najjari and Guilbault[7] to investigate the edge effect in EHL analysis of roller bearings. Nevertheless, this method gives only approximate solutions. The present authors[8] have recently obtained the limit of Hetenyi's iteration with a matrix method and developed an explicit solution to the stress field of quarter-space problems. The aforementioned methods are all based on the overlapping half-space concept of Hetenyi. Apart from these, Bower et al. [9] adopted finite element method (FEM) to analyze the ratcheting limit of rail's plastic deformation. Hecker and Romanov[10] applied Mellin transform to solve the stress distribution of quarter-space. Ritz's method was also applied by Guenfoud et al. [11] to obtain the displacement solution of quarter-space.

There is another perspective of the contact problem of quarter-space by considering its contact with a rigid body. Gerber[12] was the first to study a contact between a rigid body and an elastic quarter-space. He obtained the stress distribution in a quarter-space which is pressed by a rectangular punch. Keer et al. [13] studied a quarter-space loaded with a rigid cylindrical indenter by integral transform techniques. Hanson and Keer[14] solved the contact stresses between a spherical indenter and a quarter-space. Wang et al. [15] studied the problem of a quarter-space in contact with a rigid sphere using equivalent inclusion method. Zhang et al.[16,17] analyzed the contact of a rigid roller and a finite-length elastic body. To include the effect of the free edge surfaces of the elastic body, they applied the method of Zhang et al.[8] in the study. However, the shear stresses on a free end surface as induced by the mirrored loads on the plane of the other end surface cannot be eliminated, i.e. it does not fulfill the zero stress boundary condition of the free surface.

The elastic deformation of quarter-spaces is needed in the solution of many engineering problems, such as rail/wheel contacts[14] and EHL analyses of roller contacts[18-20]. The analyses of these application examples require iterative calculations. Thus, it requires not only accurate but also efficient solution for surface deformation of elastic quarter-space. In this paper, the solution of surface deformation of a quarter-space is developed resembling a matrix of springs. The deformation of a spring is obtained by simply dividing the load over its stiffness, or multiplying the load with its flexibility. If such a simple process can be implemented into the calculation loops of the above examples, the solution process would be significantly simplified and shortened. The complete calculation times can thus be much lower, especially if a great many times of iteration is needed. In order to realize this, a characteristic property of elastic quarter-space concerning its stiffness or flexibility must be known prior to entering the loop of calculation. This characteristic property must be independent of the load, and derived without the knowledge of the current load. The present paper achieves this aim by extending our recently proposed technique[8] for quarter-space solutions, such that an explicit form of the flexibility matrix of the elastic quarter-space is derived. This flexibility matrix is independent of the load, so that it can be used in every loop of the iteration process. The elastic surface deformation can be immediately obtained by simply multiplying this matrix with the current load distribution. The theoretical derivation of the flexibility matrix for the elastic surface deformation with the quarter-space model is presented. The solution method is also validated through a special case study. The difference in the surface normal deformation calculated with quarter-space and half-space models are not yet investigated comprehensively. Therefore, the results of the surface normal deformation of quarter-and half-space models under different typical loads are also presented and discussed.

2 Solution of Surface Normal Deformation with Quarter-space Model

2.1 Derivation of flexibility matrix

A quarter-space problem with a distribution load P on the top surface as shown in Fig. 1(a) can be solved by making use of the solutions of two mutually orthogonal half-spaces as shown in Fig. 1(b), which is based on the overlapping half-space idea of Hetenyi[3]. To solve a quarter-space problem with matrix formulation[8], the horizontal and vertical surfaces of the quarter-space are discretized with rectangular meshes of different sizes. Fig. 2(a) shows schematically the mesh pattern on the top surface. The region near the free edge surface is discretized with finer meshes in order to enhance the accuracy of the deformation results close to the free edge. The solution of Fig. 1(a) is obtained by superimposing the half-space solutions of load-pairs: $P_h, \bar{P}_h$ and $P_v, \bar{P}_v$ ($\bar{P}_h$ and $\bar{P}_v$ are mirror loads of P_h and P_v, respectively). Making use the stress boundary conditions of the original quarter-space (Fig. 1(a)): P on the top surface and zero stress on the vertical side surface, explicit solutions of the equivalent load P_h and P_v of the half-spaces (Fig. 1(b)) can be readily obtained by [8],

$$P_h = A \cdot P \qquad (1)$$

$$P_v = B \cdot P \qquad (2)$$

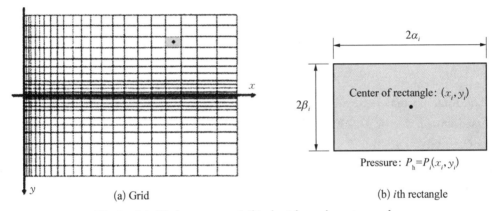

Fig. 2 (a) Mesh pattern and (b) the ith patch on top surface

The Appendix shows how the two coefficient matrices A and B can be calculated. All pressure distributions are in matrix format as,

$$P = \begin{bmatrix} p_1 \\ p_2 \\ \vdots \\ p_{k_h} \end{bmatrix}; \quad P_h = \begin{bmatrix} (p_h)_1 \\ (p_h)_2 \\ \vdots \\ (p_h)_{k_h} \end{bmatrix}; \quad P_v = \begin{bmatrix} (p_v)_1 \\ (p_v)_2 \\ \vdots \\ (p_v)_{k_v} \end{bmatrix}$$

where k_h and k_v are, respectively, the number of mesh on the top and vertical side surfaces.

For the calculation of the surface normal (vertically downward) deformation of the quarter-space induced by the distributed load P as shown in Fig. 1(a), there is no effect generated by the load-pair, P_v and its mirror image $\bar{P}_v$ due to the symmetry of P_v and $\bar{P}_v$ with respect to the x-axis (the originally loaded surface of the quarter-space) in Fig. 1(b), Therefore, only the surface normal deformation caused by P_h and $\bar{P}_h$ is considered.

The elastic deformation of a point on a plane due to a uniformly load patch on the same surface is given by the classical Love solution[2]. The normal deformation at point j on the top contact surface, which is caused by a uniform pressure, $P_h(x_i, y_i)$, on a patch i and its mirror image, $P_h(-x_i, y_i)$, on the top surface, can be expressed as[5],

$$\delta_{ji}(x_j, y_j) = \frac{P_h(x_i, y_i)}{4\pi G}\{[B_z(x_j - x_i + \alpha_i, y_j - y_i + \beta_i)$$
$$+ B_z(x_j - x_i - \alpha_i, y_j - y_i - \beta_i) - B_z(x_j - x_i + \alpha_i, y_j - y_i - \beta_i)$$
$$- B_z(x_j - x_i - \alpha_i, y_j - y_i + \beta_i)] + [B_z(x_j + x_i + \alpha_i, y_j - y_i + \beta_i)$$
$$+ B_z(x_j + x_i - \alpha_i, y_j - y_i - \beta_i) - B_z(x_j + x_i + \alpha_i, y_j - y_i - \beta_i)$$
$$- B_z(x_j + x_i - \alpha_i, y_j - y_i + \beta_i)]\} = (c_{ji} + \bar{c}_{ji})P_h(x_i, y_i) \quad (3)$$

and

$$B_z(x, y) = 2(1-\nu)[y\ln(R+x) + x\ln(R+y)] \quad (4)$$

where G is the shear modulus and ν is the Poisson's ratio; x_i and y_i are the coordinates of the center of ith pressure patch; x_j and y_j are the coordinates of deformation point; α and β are the half-length and half-width of the pressure patch (Fig. 2(b)) and $R = (x^2 + y^2)^{1/2}$. The total deformation at point j is thus given as,

$$\delta_j(x_j, y_j) = \sum(c_{ji} + \bar{c}_{ji})P_h(x_i, y_i) \quad (5)$$

The total normal deformation caused by the entire load vectors $\boldsymbol{P}_h$ and $\bar{\boldsymbol{P}}_h$ can be expressed with a vector $\boldsymbol{D}$. All coefficients $c_{ji} + \bar{c}_{ji}$ are also expressed in the form of a matrix $\boldsymbol{C}$. Thus,

$$\boldsymbol{D} = \boldsymbol{C} \cdot \boldsymbol{P}_h \quad (6)$$

Combining with Eq. (1) yields,

$$\boldsymbol{D} = \boldsymbol{F} \cdot \boldsymbol{P} \quad (7)$$

where $\boldsymbol{F} = \boldsymbol{C} \cdot \boldsymbol{A}$. Matrix $\boldsymbol{F}$ links the originally applied load $\boldsymbol{P}$ to the normal deformation vector $\boldsymbol{D}$. The normal deformation of the top surface of the elastic quarter-space can be readily obtained by multiplying the pressure $\boldsymbol{P}$ with $\boldsymbol{F}$. Matrix $\boldsymbol{F}$, which is related only to the contact material and surface meshing but remains constant for different loads, can be calculated using Eq. (3) with an unit pressure distribution for a given quarter-space problem. $\boldsymbol{F}$ is, thus, the flexibility matrix of the elastic quarter-space. This approach is very convenient for engineering calculation.

2.2 Verification of calculation for elastic deformation

A Fortran program that calculates the elastic surface deformation of a quarter-space and a half-space was written. The solution of half-space is directly obtained with the applied load using the classical theory of Love[2]. The program was validated with an example of a Hertzian contact, as illustrated in Fig. 3. The radius of the Hertzian contact is a. The dimensions of the block are $A_x = 20a$, $A_y = 40a$ and $A_z = 20a$. The vertical y-z plane is taken to be the targeted free side surface and its effect is studied. A_y is double in length of A_x. The number of mesh on the top surface (x-y plane) is 25,070 (230 × 109). The vertical side surface (y-z plane) has the same number of mesh (109 × 230). As illustrated in Fig. 2(a), finer mesh resolution is chosen in the regions close to the free side

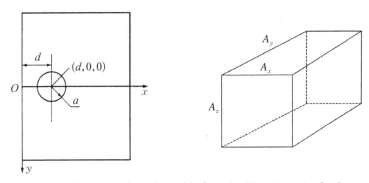

Fig. 3 Example of an elastic block under Hertzian point load

surface and along the x-axis where the applied load is located.

A semi-elliptical pressure distribution according to the Hertzian point contact theory is assumed and applied at a distance of d from the free edge along the x-axis on the top surface, as shown in Fig. 3. The Hertz equation, which provides analytical solutions for a point contact on an elastic half-space, has sufficient precision to give deformation and stress if a is much smaller than the surface size and the load center is far from the free side surface. For this example, a is taken as 0.5 mm. The Hertzian pressure distribution for a point contact is expressed as,

$$P(r) = P_{max}(1-(r/a)^2), \; r \leqslant a \tag{8}$$

The normal surface deformation is given as [1],

$$\begin{aligned} u_z &= \frac{1-\nu^2}{E} \frac{\pi \cdot P_{max}}{4a}(2a^2 - r^2), \; r \leqslant a \\ u_r &= \frac{1-\nu^2}{E} \frac{P_{max}}{2a}\{(2a^2 - r^2)\sin^{-1}(a/r) + ra\sqrt{1-(a/r)^2}\}, \; r > a \end{aligned} \tag{9}$$

Table 1 lists the parameters of the Hertz contact. This example has the center of the Hertzian pressure distribution at the mid-span of the block, i.e. $d = 10a$, for the validation of the computer program. Since the load is away from the two free edges and the contact area of radius a is far less than the top surface area of the block, the half-space criterion for the use of the Hertz theory is valid. Therefore, the surface normal deformation (as the output of the developed computer program) calculated with the quarter-space and the half-space models should be the same or very much similar to the analytical solution of Hertz as expressed in Eqn. (9). The normal deformation of the top surface along the x-axis is depicted in Fig. 4. The good correlation of the three curves as shown in Fig. 4(b) proved the assumption and the developed program. Fig. 4(b) is an enlarged plot of the maximum deformation in the Hertzian contact region (from $9a$ to $11a$), which also shows that the deformation calculated using the quarter-space model is slightly larger than that of the half-space model or Hertz theory. The differences are attributed to the effect of the two free end surfaces.

Table 1 Parameters of the Hertzian point contact

Max. Hertzian pressure: $P_{max} = 0.5$ GPa
Elastic modulus: $E = 201$ GPa,
Poisson's ratio: $\nu = 0.3$
Contact radius: $a = 0.5$ mm

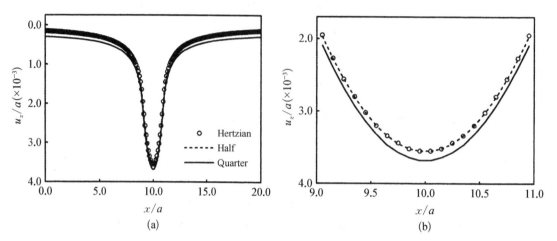

Fig. 4 Surface normal deformation: comparison of half-space model, quarter-space model and Hertz theoretical solution

3 Effect of Free Side Surface on Surface Normal Deformation

To evaluate the effect of a vertical free surface on the surface normal deformation, pressure distributions of three typical contacts, namely Hertzian point, flat cylindrical punch and Hertzian line, are chosen for case studies.

3.1 Hertzian point contact

The surface normal deformations under a Hertzian point contact with the same parameters listed in Table 1 but different d values were calculated. The load distribution is expressed with Eq. (8). The load centers were, respectively, located at $d = a$, $2a$, $3a$, $4a$ and $5a$ (compared to the full length of the block, $20a$). The effect of the free side surface at $x = 0$ was studied. It was assumed that the free surface on the other side at $x = 20a$ produces no significant effect due to it being far from the load center. Hence, finer mesh resolution was adopted in the region close to the free edge at $x = 0$ as depicted in Fig. 2(a). Fig. 5 shows deformations along the x-axis on the top surface ($z = 0$), which were directly calculated using the derived flexibility matrix. To highlight the effect of free edge surface, the same set of results calculated using the half-space model is also illustrated. Finite element modeling (using ANSYS) was also complied and results are presented in Fig. 5 to demonstrate the accuracy of the matrix solution of quarter-space model. The insets are blow-ups of the load center region where the deformation curves

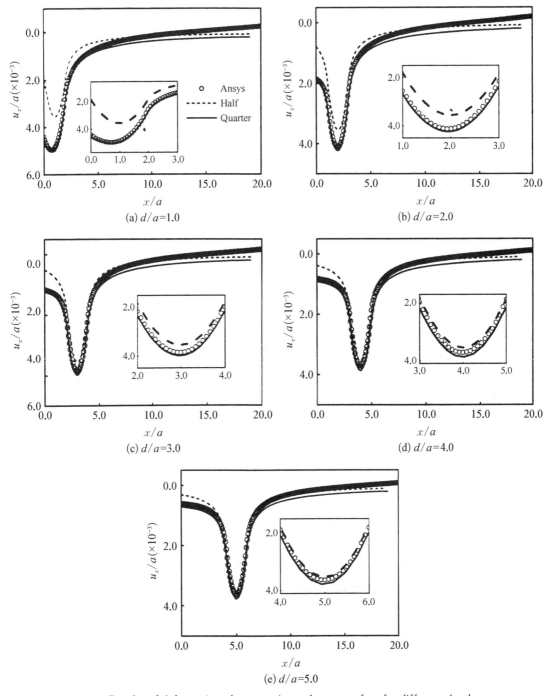

Fig. 5 Results of deformation along x-axis on the top surface for different d values

acquire the maximum values.

The quarter-space results coincide well with the ANSYS results, which substantiates the correctness and accurateness of the developed Fortran program. The minor difference between the two is due to the mesh difference. The meshing in ANSYS is 3-dimensional and it is difficult to divide the contact body into too many grids. The presented quarter-space treatment is semi-analytical and its results would converge to the exact solution if the resolution of the mesh is increased.

The differences between the results of quarter-space (solid curve) and half-space (large dotted curve) are significant. The actual quarter-space deformation is bigger than that of half-space model due to the effect of the free side surface. One can consider that there exists no free surface in the half-space model and its structure is thus stronger than a quarter-space. Such a difference becomes increasingly obvious as the load center approaches the side edge. Furthermore, when the pressure center is shifted towards the free surface, as shown in Fig. 5(a), the maximum deformation point moves away from the pressure center towards the free end. This result contradicts the prediction of the Hertz theory. Fig. 6 shows the differences between the magnitude and the position of the maximum deformation with different locations of pressure center in the x-axis. The differences rise sharply when the pressure center is located close to the free surface. When the center of pressure is located at $d=a$ from the edge, the error of the half-space deformation can be as large as 30% (Fig. 6(a)) and the position shift of the maximum deformation (Fig. 6(b)) reaches 28.4% of the load radius a. The effect of free edge surface fades gradually when the center of pressure shifts away from the free end. The location of maximum deformation almost coincides with the pressure center when d reaches $2a$. The effect of free edge surface on the magnitude of deformation remains significant even when the center of pressure is relatively far ($d=5a$) from the free end, as shown in Fig. 6(a).

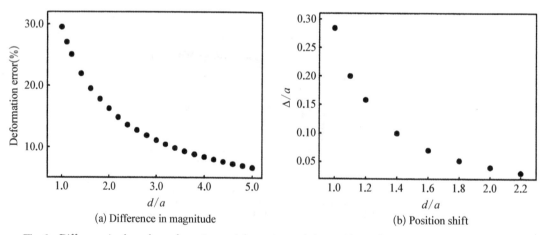

Fig. 6 Difference in the values of maximum deformation and the positions of quarter-and half-space models

3.2 Flat cylindrical punch contact

The quarter-space loaded with a uniform pressure distribution which simulates the contact of a flat cylindrical punch is also investigated. The pressure is given as,

$$P(r)=0.5 \text{ GPa}, r \leqslant a$$
$$P(r)=0, r > a \tag{10}$$

Results were obtained with the pressure center located at $d=a$, $3a$ and $5a$ along the x-axis on the top surface (full length of the block: $20a$). Fig. 7 illustrates the deformed top surfaces obtained with the quarter-space model as well as the half-space model. The results show that if the pressure center is located close to the free edge, the deformed profile deviates significantly from that predicted by the half-space model. The effect of free edge side can be considerable. For example, the maximum deformation that reaches to $0.00757a$ occurs at the free surface when cylindrical pressure is applied at $d=a$, which is much larger than the corresponding $0.00454a$ of half-space model. Fig. 7 depicts that the half-space model gives smaller deformation when compared with the quarter-space, and the maximum deformation locates right at the pressure center. These results indicate failure of the half-space model to provide accurate results when the load is applied near the free surface. Fig. 8 shows the deviations in maximum deformation for various distances between the center of pressure and free surface. More than 40% error in maximum deformation is found in the half-space model when the distance between the load center and the free surface is equal to the radius of the cylindrical load distribution.

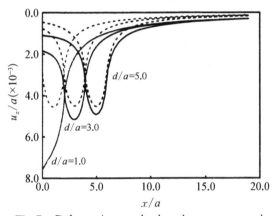

Fig. 7 Deformation results based on quarter-and half-space models under cylindrical load distribution (dashed line: half-space; solid line: quarter-space; full length: $20a$)

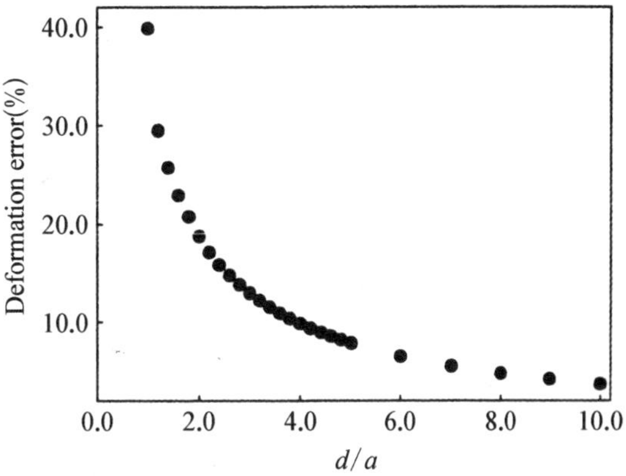

Fig. 8 Maximum deformation deviations for various locations of applied load

3.3 Hertzian line contact

The third case is a Hertzian line contact, which simulates an elastic roller contact. The Hertz contact radius a is 0.5 mm in the y-direction. The pressure distribution is given as,

$$P(y) = 0.5\sqrt{1-\left(\frac{y}{a}\right)^2} \text{ GPa}, \; |y| \leqslant a \quad (11)$$
$$P(y) = 0, \; |y| > a$$

Deformation results were calculated with three different loading lengths of $4a$, $8a$ and $20a$. The center of these loads is fixed at the center of the block (mid-span, $d=$

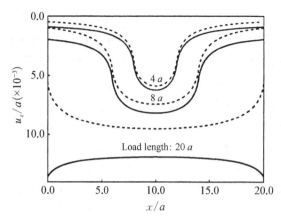

Fig. 9 Deformation of a block of width $20a$ under Hertz line contact in quarter-and half-space models (dashed line: half-space; solid line: quarter-space)

$10a$). These loading cases are axisymmetric about the mid-section at $x=10a$. Only the half-length of the distributed load was adopted in the calculation of the surface deformation of the block. The mirror image of that calculated surface deformation gives the deformation due to the other half of the load. Superimposing the two deformation results gives the surface deformation of the applied load. Fig. 9 shows the calculated deformation with the quarter-space and half-space models. When the line Hertzian load of $4a$ in length is applied at the center and its edges are far from the two free side surfaces at $x=0$ and $20a$, the difference in the half-space and quarter-space deformed profile is small. With increasing the loading length to $8a$, i. e. the free side surfaces are closer to the load, the difference is increased. When the length of the line load is extended to $20a$, as same as the length of the elastic block, the quarter-space deformation profile depicts the maximum deformation at the two ends whereas the half-space attains the maximum deformation at the mid-length (the block center as well as the load center). Results show that when the load is close to the free side surface, the influence of the side surface is considerable and the use of half-space model may lead to significant errors.

4 Conclusions

An accurate and fast method for the solution of surface normal deformation in quarter-space is introduced in this paper. The surface normal deformation can be obtained directly by multiplying the original normal load on the top surface of the quarter-space with a flexibility matrix. The flexibility matrix, which is constant for a given discretized quarter-space, can be retrieved and used for different loads. This facilitates fast and accurate calculation for the quarter-space normal surface deformation.

This paper has evaluated three different load types on a quarter-space, namely Hertzian point, flat cylindrical punch and Hertzian line. The maximum load is up to 0. 5 GPa and the material of the quarter-space under consideration is taken as steel. It is found that if the load is close to a free edge surface, the effect of the free surface can be considerable. For the specified cases, the error in the maximum deformation can be up to 40% (for a load located very close to the edge) if the inappropriate half-space model is adopted. Furthermore, the effect of the free edge surface leads to the position shift of the maximum deformation point and asymmetric deformation profile.

Acknowledgement

The study described in this paper was fully supported by a grant from the Research Grants Council of Hong Kong Special Administrative Region, China (Project No. CityU11213914).

Appendix

The stresses acting on the vertical (y-z plane) and the top planes (x-y plane) of the quarter-space as the results of the load-pairs: P_h, $\bar{P}_h$ and P_v, $\bar{P}_v$, respectively, as shown in Fig. 1 can be expressed in vector matrix format as,

$$\boldsymbol{\sigma}_{yz} = \boldsymbol{M} \cdot \boldsymbol{P}_h \tag{A1}$$

$$\boldsymbol{\sigma}_{xy} = \boldsymbol{N} \cdot \boldsymbol{P}_v \tag{A2}$$

where $\boldsymbol{M}$ and $\boldsymbol{N}$ are two-dimensional coefficient matrices, and they are directly determined by the Love theory[2]. Overlapping the solutions of the two half-spaces in Fig. 1(b) and considering the boundary conditions yields,

$$-\boldsymbol{P}_h + \boldsymbol{\sigma}_{xy} = -\boldsymbol{P}_h + \boldsymbol{N} \cdot \boldsymbol{P}_v = -\boldsymbol{P} \tag{A3}$$

$$-\boldsymbol{P}_v + \boldsymbol{\sigma}_{yz} = -\boldsymbol{P}_v + \boldsymbol{M} \cdot \boldsymbol{P}_h = 0 \tag{A4}$$

Combining (A3) and (A4) gives,

$$\boldsymbol{P}_h = (\boldsymbol{I} - \boldsymbol{N} \cdot \boldsymbol{M})^{-1} \cdot \boldsymbol{P} \tag{A5}$$

Comparing with Eq. (1),

$$\boldsymbol{A} = (\boldsymbol{I} - \boldsymbol{N} \cdot \boldsymbol{M})^{-1}$$

Similarly, from Eq. (2),

$$\boldsymbol{B} = \boldsymbol{M} \cdot (\boldsymbol{I} - \boldsymbol{N} \cdot \boldsymbol{M})^{-1}$$

References

[1] Johnson KL. Contact mechanics. Cambridge: Cambridge University Press, 1985.
[2] Love AEH. The stress produced in a semi-infinite solid by pressure on part of the boundary. Philos Trans R Soc Lond Ser A, Contain Pap A Math Or Phys Character, 1929, 228: 377–420.
[3] Hetenyi M. A general solution for the elastic quarter space. J Appl Mech, 1970, 37: 70–76.
[4] Keer L, Lee J, Mura T. Hetenyi's elastic quarter space problem revisited. Int J Solids Struct, 1983, 19: 497–508.
[5] Hanson M, Keer L. A simplified analysis for an elastic quarter-space. Q J Mech Appl Math, 1990, 43: 561–587.
[6] Guilbault R. A fast correction for elastic quarter-space applied to 3D modeling of edge contact problems. J Tribol, 2011, 133: 031402.
[7] Najjari M, Guilbault R. Edge contact effect on thermal elastohydrodynamic lubrication of finite contact lines. Tribol Int, 2014, 71: 50–61.

[8] Zhang Z, Wang W, Wong P. An explicit solution for the elastic quarter-space problem in matrix formulation. Int J Solids Struct, 2013,50: 976-980.

[9] Bower AF, Johnson KL, Kalousek J. A ratchetting limit for plastic deformation of a quarter-space under rolling contact loads. In: Proceedings international symp on contact mechanics and wear of rail/wheel systems II. The Netherlands: University of Waterloo Press, 1987: 117-132.

[10] Hecker M, Romanov A. The stress fields of edge dislocations near wedge-shaped boundaries and bonded wedges. Mater Sci Eng: A, 1993,164: 411-414.

[11] Guenfoud S, Bosakov S, Laefer DF. A Ritz's method based solution for the contact problem of a deformable rectangular plate on an elastic quarter-space. Int J Solids Struct, 2010,47: 1822-1829.

[12] Gerber CE. Contact problems for the elastic quarter-plane and for the quarter-space. UMI Diss. Services, 1968.

[13] Keer LM, Lee JC, Mura T. A contact problem for the elastic quarter space. Int J Solids Struct, 1984, 20: 513-524.

[14] Hanson M, Keer L. Analysis of edge effects on rail-wheel contact. Wear, 1991,144: 39-55.

[15] Wang Z, Jin X, Keer LM, Wang Q. Numerical methods for contact between two joined quarter spaces and a rigid sphere. Int J Solids Struct, 2012,49: 2515-2527.

[16] Zhang H, Wang W, Zhang S, Zhao Z. Modeling of elastic finite-length space rolling-sliding contact problem. Tribol Int, 2016.

[17] Zhang H, Wang W, Zhang S, Zhao Z. Modeling of finite-length line contact problem with consideration of two free-end surfaces. J Tribol, 2016,138: 021402.

[18] Liu X, Yang P. Analysis of the thermal elastohydrodynamic lubrication of a finite line contact. Tribol Int, 2002,35: 137-144.

[19] Liu XL, Yang PR. Numerical analysis of the oil-supply condition in isothermal elastohydrodynamic lubrication of finite line contacts. Tribol Lett, 2010,38: 115-124.

[20] Sun H, Chen X. Thermal EHL analysis of cylindrical roller under heavy load. In: Proceedings of IUTAM symposium on elastohydrodynamics and micro-elastohydrodynamics. Springer, 2006: 107-120.

Study on the Free Edge Effect on Finite Line Contact Elastohydrodynamic Lubrication*

Abstract: This paper incorporated a recently developed algorithm, which directly gives the elastic deformation of loaded quarter-space (QS) bodies, into a finite line contact elastohydrodynamic lubrication (EHL) analysis. Half-space (HS) and QS models were adopted in the analysis and the results were compared with the data obtained from conventional optical EHL tests of steel rollers on a glass disc. This study found that contact deformation was dominated by glass disc deformation. Thus, the use of a steel roller on a glass disc in conventional optical EHL technique is not appropriate for studying the free edge surface effect. Free edge surfaces reduce the magnitude of pressure spikes near the ends of contact. Rollers with different end profiles were also evaluated.

Nomenclature

A, B	Transformation matrixes for equivalent loads calculation in Ref. [25].		horizontal HS
b	Half-length of Hertzian line contact, m	R	Radius of roller, m
E_1, E_2	Elastic modulus of roller and plane, Pa	R_y	dub-off or crown radius (Fig. 12)
E'	Reduced elastic modulus, $[(1-\nu_1^2)/(2E_1)+(1-\nu_2^2)/(2E_2)]^{-1}$, Pa	$s(x, y)$	Profile function of roller, m
		U	Entrainment speed, m/s
		U^*	Non-dimensional entrainment speed, $U_{\eta 0} = E'R$
F	Flexibility matrix for deformation calculation in QS model	w	Applied load, N
h	Film thickness, m	x, y	Coordinates
h_o	Central film thickness in EHL, m	α	Barus viscosity-pressure coefficient, m^2/N
H	Non-dimensional film thickness, hR/b^2		
L	Length of roller, m	$\delta(x, y)$	Deformation at point (x, y), m
p	Fluid pressure, Pa	η	Dynamic viscosity of lubricant, Pa·s
p_o	Maximum Hertzian contact pressure, Pa	η_0	Ambient dynamic viscosity of lubricant, Pa·s
P_v, $\bar{P}_v$	Mirrored pair of equivalent load on vertical HS	ρ	Density of lubricant, kg/m^3
		ρ_0	Ambient density of lubricant, kg/m^3
P_h	Mirrored pair of equivalent load on	ν_1, ν_2	Poisson's ratio of roller and plane

---

* In collaboration with L. Guo, W. Wang and P. L. Wong. Reprinted from *Tribology International*, 2017, 116: 482−490.

1 Introduction

Finite line contact elastohydrodynamic lubrication (EHL) is present in typical engineering applications, such as roller bearings, gears, and camfollowers. The main feature of such EHL is the existence of free edge surfaces at both ends of the contact. Cameron's group[1-3] was the first to conduct an experimental study on finite line contact EHL. Their study used two types of rollers, namely, blended and unblended steel rollers. The lubrication conjunction of the rollers with a glass disc was captured and analysed using optical interferometry. Typical optical interferograms of lubricated roller contacts were presented. The experimental results showed that the minimum film thickness occurs near the ends of the roller. Soon after, Bahadoran and Gohar[4] solved a finite line contact EHL problem of an unblended roller by modifying an infinite line contact solution. Their solutions were satisfactorily consistent with Wymer and Cameron's experimental results[3]. Later, Mostofi and Gohar[5] obtained the solution of finite line contact EHL, wherein a cylindrical roller with axially profiled ends rolling on a plane was considered by solving a two-dimensional Reynolds equation. They reported that the maximum pressure and minimum film thickness occur near the beginning of the end profiling. By utilizing finite difference and the Newton-Raphson method, Park and Kim[6] solved the finite line contact EHL problem and drew the same conclusion as Mostofi and Gohar[5]. Kushwaha et al.[7] also confirmed Wymer and Cameron's experiment[3] with their theoretical solution. In their study, the edge stress discontinuities and lubricant flow near the edge were analysed in detail. Particularly, the lubrication behaviour under misaligned condition was also considered in their analysis. Later, the finite line contact EHL under transient conditions was also studied by Kushwaha and Rahnejat[8] and the phenomenon of squeeze caving was found in their study. Liu and Yang[9] conducted a thermal-EHL analysis of a finite line contact using multi-level method and found that the maximum pressure, maximum film temperature, and minimum film thickness appear at the end region of the roller. To reduce stress concentration and pressure peak near the roller end, several researchers examined the end profiling of the roller. For example, Johns and Gohar[10] studied the roller axial profile effect on the contact pressure distribution and footprint shape under dry contact condition. In their study, modified Lundberg, circular arc, dub-off and crown profiles were considered for the cases of aligned and misaligned rollers. Chen et al.[11] conducted experiments to analyse how the Lundberg (logarithmic) profile would affect the thickness and shape of the oil. Sun and Chen[12] theoretically studied the lubrication behaviours of rollers using the Lundberg profile. They proposed that the Lundberg profile should be changed by multiplying the Lundberg function with a coefficient (larger than 1.0) to obtain the optimal EHL effect. Zhu et al.[13] investigated the effects of contact length and profile on the stress concentration and the EHL film thickness. He et al.[14] analysed the effect of the radius of round corner on the finite line contact plasto-EHL of rollers. They reported an optimal range of the corner radius under the studied working conditions.

In all the aforementioned finite line contact EHL analyses, the elastic deformation of contact bodies were calculated using either the classical Boussinesq[15] or Love[16,17] formulae (for point and distributed load, respectively). The free edge surface effect on the deformation was ignored because the two classical formulae are based on the assumption that the loaded body is semi-infinite (commonly known as the half-space (HS) model in contact mechanics). The geometry near the end of a roller contact is accurately modelled

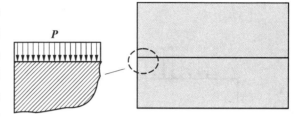

Fig. 1 Illustration of a typical QS problem-roller on roller

as a quarter-space (QS) as shown in Fig. 1. It is reasonable to believe that the free edge surface affects the elastic deformation of the loaded surface, i. e., the thickness of the lubricating film. Thus, the finite line contact EHL can be appropriately studied using the QS model. However, obtaining an analytical solution of QS problem is difficult despite the development of several original numerical methods. Hetenyi[18] addressed a QS problem under a point normal load on its top surface. He proposed the use of two mutually perpendicular HSs under mirrored load-pairs, as shown in Fig. 2. The application of mirrored load-pairs gives zero shear stress on the mid-plane of the HS, thereby fulfilling the shear stress-free boundary conditions of the original two surfaces of the QS (the mid-plane of the two perpendicular HSs corresponding to the two plane surfaces of the QS). The normal stresses on the mid-plane can be rectified by overlapping the two HSs iteratively until the boundary stress criteria of the QS surfaces are achieved. Keer et al.[19] simplified Hetenyi's iterative solution scheme by using a coupled pair of integral equations. Solving the coupled integral equations is not easy and Keer et al.[19] applied Fourier transform in the solution process. Later, Hanson and Keer[17] developed a direct method to get the solution of the two coupled integral equations. Guilbault[20] proposed an approximate method to obtain the solution of a QS using a corrected load. Najjari and Guilbault[21] conducted the first study of free edge effect on finite line contact EHL using the correction factors proposed by Guilbault[20]. The finite element method[22], Mellin transform[23] and Ritz's method[24] were also applied to solve QS problems. Recently, Zhang et al.[25] successfully obtained the limit of Hetenyi's iteration. Thus, QS problems can be directly solved based on the equivalent mirrored load-pairs on the two mutually orthogonal HSs (as depicted in Fig. 2). This efficient solution method[25] was soon adopted by another research group, Zhang et al.[26], to solve contact problems with two free end surfaces. Subsequently, Zhang et al.[27] carried out a finite line contact EHL analysis using the method proposed in their previous study[26]. They applied the discrete convolution-fast Fourier transform (DC-FFT) method of Liu et al.[28] to speed up the calculation in their study. The effect of free edge surfaces at the two ends of the contact was demonstrated. The work of Zhang[26,27], however, deviated from the original null shear stress boundary conditions since there are shear stresses induced on the planes corresponding to the two end surfaces when the equivalent loads were applied.

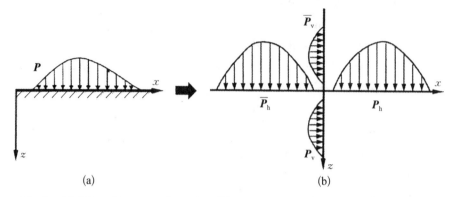

Fig. 2 (a) QS problem transformed to (b) superposition of two mutually orthogonal HSs loaded with mirrored load-pairs

More recently, the present authors[29] further developed the explicit QS solution method of Zhang et al.[25] and derived a flexibility matrix that enables a direct deformation calculation of the QS top surface by simply multiplying the applied load with it. The free edge effect on the deformation was studied in detail. It is proved that the free edge can lead to the increase in the deformation of the top surface compared with that of the HS model under the same load. Especially for the case of an infinite body with two free ends under a distributed load across its width, the deformed body calculated based on the QS model is in opposite shape of that obtained with the HS model (an example extracted from Ref. [29] is illustrated in Fig. 3). The flexibility matrix derived by Wang et al.[29] provides an explicit and efficient deformation solution for contact bodies containing free

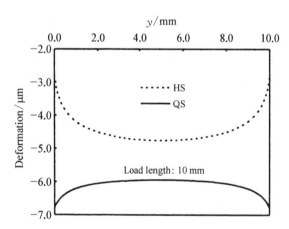

Fig. 3 Differences in deformed shape of a finite block of 10 mm width under Hertzian line contact calculated with HS and QS models (excerpted from Fig. 9 of [29])

edges. The present study incorporates this matrix in a finite line contact EHL analysis to study the free edge effect on lubrication.

2 Derivation of Flexibility Matrix

The derivation of the flexibility matrix is briefly described to facilitate understanding of the algorithm. Fig. 2 shows the principle of the explicit overlapping method proposed by Zhang et al. [25]. The original load (P) applied on the top surface of the QS can be transformed into two equivalent load-pairs (P_h, $\bar{P}_h$ and P_v, $\bar{P}_v$) on mutually orthogonal HSs. The superposition of two equivalent load-pairs fulfils the stress boundary conditions of the QS, where the stresses on the side and top surfaces are null and $-P$, respectively.

All loads are presented in vector forms after discretization. Equivalent load can be expressed as the product of the originally applied load and a transformation matrix as

$$P_h = A \cdot P \tag{1}$$

$$P_v = B \cdot P \tag{2}$$

P_v and $\bar{P}_v$ produce no effect on the normal deformation (in the z-direction) of the top surface because they are symmetrical about the top surface (x-axis in Fig. 2). Therefore, only P_h and $\bar{P}_h$ are considered in the calculation of the deformation of the top surface in the z-direction, and the deformation can be expressed as,

$$D = C_1 \cdot P_h + C_2 \cdot \bar{P}_h \tag{3}$$

The elements of matrices C_1 and C_2 can be obtained directly from the Love solution[16,17] for distributed load. P_h and $\bar{P}_h$ are axisymmetric about the free edge surface (z-axis in Fig. 2). Therefore, C_1 and C_2 can be combined together, i.e.

$$D = C \cdot P_h \tag{4}$$

Combining Eqs. (1) and (4) yields the relationship between the normal deformation of the top surface and the originally applied load as,

$$D = F \cdot P \tag{5}$$

where $F = C \cdot A$ and is defined as the flexibility matrix. Hence, the deformation can be calculated directly by multiplying the flexibility matrix with the originally applied load. Both P and D vectors are represented with a single column matrix and their dimension is equal to the total number of mesh points. The flexibility matrix F is thus a square matrix. (More details about the discretization can be referred to our earlier paper [29].)

The flexibility matrix can be determined in advance and used in the subsequent iterative calculation of film thickness in the EHL analysis because it is only related with the mesh structure and material properties of the contact bodies. Note that the equivalent loads do not need to be physically determined because the calculation of the deformation D is directly obtained from applied load P as expressed in Eq. (5).

3 Finite Line Contact EHL Study

3.1 Governing equations

Fig. 4 shows two main reduced geometrical types of finite line contact EHL. Fig. 4(a) schematically resembles the contact of roller bearings. The contact can be typically simplified as a roller (termed as QS body) on a semi-infinite solid body (HS body). Their deformation should be calculated separately using the QS and HS models. Typical examples of the contact type shown in Fig. 4(b) include cam-followers and gears. Contact occurs between a roller (QS body) and a semi-infinite plane of the same width (QS body). Both the deformation of the roller and the plane should be calculated using the QS model.

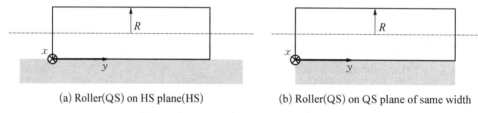

(a) Roller(QS) on HS plane(HS)　　　(b) Roller(QS) on QS plane of same width

Fig. 4 Two reduced contact types (QS/HS and QS/QS)

The current analysis is limited to isothermal conditions because the main focus of this study is free edge surface effect. The lubrication behaviour under transient conditions is not studied here. Hence, the isothermal Reynolds equation is used as,

$$\frac{\partial}{\partial x}\left(\frac{\rho h^3}{\eta}\frac{\partial p}{\partial x}\right)+\frac{\partial}{\partial y}\left(\frac{\rho h^3}{\eta}\frac{\partial p}{\partial y}\right)=12u\frac{\partial(\rho h)}{\partial x} \tag{6}$$

The boundary conditions for solving Eq. (6) are:

$$\begin{cases} p(x_{in}, y)=p(x_{out}, y)=p(x, y_{in})=p(x, y_{out})=0 \\ p(x, y)\geqslant 0, \ (x_{in}<x<x_{out}, \ y_{in}<y<y_{out}) \end{cases}$$

The isothermal viscosity-pressure relationship proposed by Roelands[30] is adopted here,

$$\eta=\eta_0\exp\{(\ln\eta_0+9.67)[(1.0+5.1\times 10^{-9}p)^{z_0}-1.0]\} \tag{7}$$

The constant z_0 can be obtained as,

$$z_0=\frac{\alpha}{5.1\times 10^{-9}(\ln\eta_0+9.67)}$$

where α is the Barus viscosity-pressure coefficient.

The density-pressure-temperature relationship of Dowson-Higginson[31] under isothermal condition is,

$$\frac{\rho}{\rho_0}=1.0+\frac{0.6\times 10^{-9}p}{1.0+1.7\times 10^{-9}p} \tag{8}$$

Load balance is expressed as,

$$\iint p\,dx\,dy = w \tag{9}$$

The film thickness is composed of three parts, namely, central film thickness, roller profile, and the deformation of contact bodies, which can be expressed as,

$$h(x, y) = h_0 + s(x, y) + \delta(x, y) \tag{10}$$

where $s(x, y)$ is the mathematical function of the roller profile and $\delta(x, y)$ is the sum of the deformation of the roller and the plane at point (x, y). In conventional finite line contact EHL analysis, $\delta(x, y)$ is calculated using classical formulae derived from the HS model. In this study, the free end surface (of the roller in Fig. 4(a) and of the roller and the plane in Fig. 4(b)) is taken into consideration using the QS model and $\delta(x, y)$ is solved using the flexibility matrix.

3.2 Numerical approach

To show the edge surface effect and improve calculation accuracy, the contact area was divided into uneven grids in the calculation. In the x-direction, the calculated domain is $[-5.0b, 5.0b]$ and the number of mesh is 120. In the y-direction, only half of the contact length is considered because of symmetry. The mesh size in the y-direction is 200 for the full contact length. Normalization was performed before numerical analysis. The non-dimensional expressions of the parameters are shown in nomenclature. To guarantee convergence under heavy loads, the semisystem method proposed by Ai[32] was adopted for the numerical calculation of the Reynolds equation.

3.3 Comparison with experiments

To verify the developed program, a case under the same working conditions of Refs. [2,3] was first studied. In the optical tests of Wymer and Cameron, the EHL contact was conformed between a steel roller and a glass disc. The related parameters are listed in Table 1.

Table 1 Relevant parameters of Cameron's experiments

Parameter	Value
Mean radius of roller R, (mm)	4.06
Length of roller L, (mm)	13.7
Young's modulus of roller E_1, (GPa)	206
Poisson's ratio of roller ν_1	0.3
Young's modulus of glass disc E_2, (GPa)	75.8
Poisson's ratio of glass disc ν_2	0.25
η_0, (Pa·s)	0.57
α, (m^2/N)	2.76×10^{-8}
Load w, (N)	1 552 (349 lbf, and p_o=0.73 GPa)

The selected case for the present comparison is an unblended conical roller (cone angle: 7.9°) on a glass disc from Refs. [2,3]. To simplify the calculation, the case was reduced into the model of a cylindrical roller of the mean radius in a lubricated rolling contact with a semi-infinite plane (the contact type of QS/HS, as shown in Fig. 4(a)). To show the free edge surface effect, the deformation of the roller was calculated with both QS and HS models. The calculated EHL results were compared with the experimental data. The results are shown in Figs. 5 and 6.

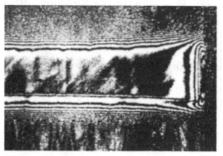

(a) Interferogram obtained from Refs. [2, 3]

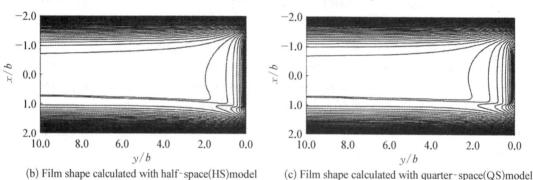

(b) Film shape calculated with half-space(HS)model   (c) Film shape calculated with quarter-space(QS)model

Fig. 5 Film shapes calculated with HS and QS models and interferogram excerpted from Refs. [2,3] ($U^* = 82 \times 10^{-11}$)

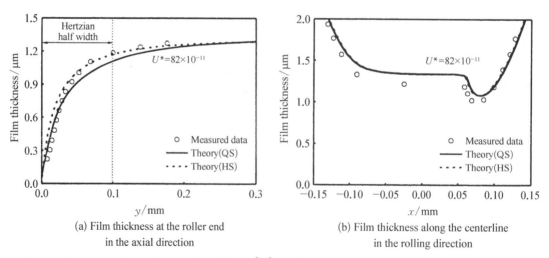

(a) Film thickness at the roller end in the axial direction   (b) Film thickness along the centerline in the rolling direction

Fig. 6 Comparison of experimental film thickness[2,3] and film thickness calculated with HS and QS models

Fig. 5(a) shows the interferogram with $U^* = 82 \times 10^{-11}$ extracted from Refs. [2,3]. Fig. 5(b) and (c) show the film shapes calculated using the HS and QS models. Both calculated film shapes are consistent with the experiment data (Fig. 5 (a)). Fig. 6 shows the comparison of the measured film thickness[2,3] and the calculated film thickness with the two different models along the end closure in the axial direction and the centreline. Differences in the calculated film thickness between the two models are considerably small and they are consistent with the measured data of Wymer and Cameron[2,3]. That the two models produce almost the same film thickness on the centreline is because it is far from the free end surface. However, the fact that the film shape (Fig. 5) and thickness (Fig. 6) near the end surface are also almost identical using the two models remains interesting. The free end surface does not seem to affect the behaviour of finite line contact EHL. However, it should be noticed that the contact was between a glass disc and a steel roller in Wymer and Cameron's experiment[2,3]. Moreover, the main deformation occurred on the glass disc, which has only approximately one-third of the Young's modulus of steel. The glass disc can be considered as a semi-infinite body and its deformation can be calculated using the HS model. However, in most engineering applications (such as bearings and gears), the contact is composed of two steel bodies. The individual deformations of these steel bodies are significant in determining total deformation. To illustrate the free end surface effect in these engineering applications, the analysis was repeated with the same working conditions listed in Table 1. However, the material was changed from glass to steel. Thus, some parameters are modified, such as $U^* = 43.2 \times 10^{-11}$ and the Hertz maximum pressure p_0 is 1.0 GPa.

The comparison of film shapes and pressure profiles calculated with HS and QS models, as shown in Figs. 7 and 8, shows that significant differences exist at the roller edge for the contact between a steel roller and a steel plane. Especially, the gradient of film thickness near the edge with QS model (Fig. 7(b)) is considerably smaller than that of HS model (Fig. 7(a)). Apart from the classical EHL pressure spike that exists at the entrainment outlet of the contact, another apparent pressure spike at the roller edge can be observed in the conventional HS results in Fig. 8(a). However, the results obtained using the QS model only show a small pressure spike at the roller edge and the pressure distribution along the roller axis is quite even (Fig. 8(b)). Film thickness along the centreline and the edge of the roller in the entrainment direction are illustrated in Fig. 9(a). The central film profiles calculated using the QS and HS models are approximately the same because the side edge is far away and its effect is negligible. However, difference in film thickness is apparent near the edge with these two models. The minimum film thicknesses calculated at the edge using the HS and the QS models are 0.086 μm and 0.343 μm, respectively. This finding shows that the predicted minimum film thickness using HS model is much smaller than its practical value, i.e. the result calculated with QS model. The reason is that the calculated deformation near the edge is smaller if the free edge effect is ignored (HS model) [29]. The variations of film thickness and pressure along the axial direction at the roller end region are shown in Fig. 9(b). The maximum pressure of HS is $1.69p_o$, which overestimates the realistic QS maximum pressure of $1.18p_o$ by 43.2%. Therefore, conventional EHL analysis based on the HS model would overestimate the pressure near the roller end. The free edge surface of the roller would reduce the magnitude of pressure spike and increase the minimum film thickness at the end of the roller.

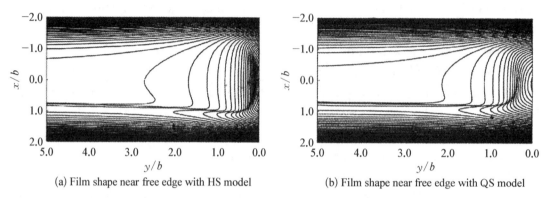

(a) Film shape near free edge with HS model (b) Film shape near free edge with QS model

Fig. 7 Comparison of film shape calculated with HS and QS models for a contact between a steel roller and a steel infinite plane (QS/HS contact type)

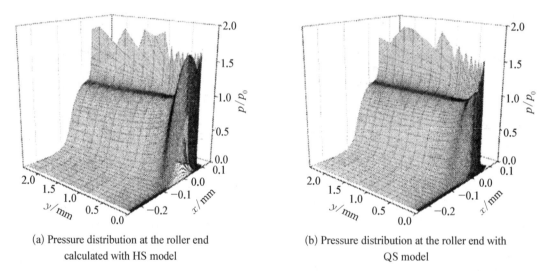

(a) Pressure distribution at the roller end calculated with HS model (b) Pressure distribution at the roller end with QS model

Fig. 8 Pressure distribution calculated with HS and QS models for a contact between a steel roller and a steel infinite plane (QS/HS contact type)

3.4 Identifying the optimal roller profile

In practice, it is common to use profiled rollers to alleviate the stress concentrate (i.e. the pressure spike) at the roller end. Previous studies[11-14] examined finite line contact EHL with different roller profiles using the HS model. These studies assumed that free edge surface effect does not exist. To evaluate roller profiles accurately, the present study conducted analyses using the QS model. The EHL of a steel roller with different end profiles and a steel plane (QS/HS contact type illustrated in Fig. 4(a)) under the working conditions listed in Table 1 was studied firstly.

3.4.1 Lundberg profile

Lundberg [33] proposed the optimal profile in the axial direction (described by Eq. (12)) from the perspective of steady pressure distribution under static and dry conditions. Sun and Chen[12] challenged it based on their thermal-EHL results of rollers with the Lundberg profile, which depicted apparent pressure spikes near the roller end. However, their thermal-EHL analysis was based on the HS model and the effect of the free edge surface was disregarded. Thus, the Lundberg profile is hereby evaluated. Eq. (11) describes the Lundberg profile function in the contact area.

$$s(x, y) = R - \sqrt{(R-\delta)^2 - x^2} \tag{11}$$

where

$$\delta = \frac{2.0w}{\pi E'L} \ln\left(1 - \left(\frac{2(y-0.5L)}{L}\right)^2\right) (0 < y \leqslant 0.5L) \tag{12}$$

$$\delta = \frac{2.0w}{\pi E'L}\left(1.1932 + \ln\left(\frac{L}{2b}\right)\right) (y=0)$$

Fig. 10 depicts the calculated pressure distribution with the Lundberg profile using the HS and QS models. The HS results shown in Fig. 10(a) depict a pressure spike at the roller end. The appearance of pressure spike in the HS results motivated Sun and Chen[12] to modify the Lundberg profile. However, the QS results in Fig. 10(b) did not show pressure spikes at the roller end because of free edge surface effect. Moreover, pressure distribution in the axial direction appears smooth. Fig. 11 shows the HS and QS results of pressure and film thicknesses along the axial direction at the roller end. The results obtained with an unblended roller are also depicted in Fig. 11 as reference. The QS result with the Lundberg profile shows improvements, in terms of lower pressure spikes and larger minimum film thicknesses than that of unblended roller. This difference is attributed to the implementation of the roller end profile. Considering the Lundberg profile results, Fig. 11(a) shows that the HS maximum pressure near the edge is $1.09p_o$, whereas the QS maximum pressure near the edge is $0.96p_o$, which is less than the maximum Hertzian contact pressure and the pressure is gradually distributed along the roller's central axis. Correspondingly, the magnitude of the QS minimum film thickness is higher than that of the HS model, as shown in Fig. 11(b). Furthermore, the minimum film thickness of QS with the Lundberg roller is 0.832 μm, which is 2.4 times higher than that of the unblended roller (0.343 μm, the red line in Fig. 11(b)). However, this level remains thinner than the minimum film thickness at the outlet in the rolling direction (Fig. 9(a)). Therefore, the Lundberg profile can greatly improve the finite line contact EHL behaviour and can be seen as the optimal choice for EHL of the contact between a roller and a HS plane.

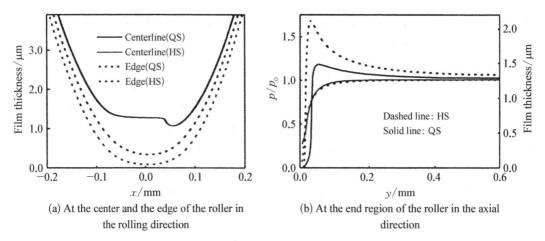

(a) At the center and the edge of the roller in the rolling direction

(b) At the end region of the roller in the axial direction

Fig. 9 Film thickness and pressure calculated with HS and QS models for a contact between an unblended steel roller and a steel plane

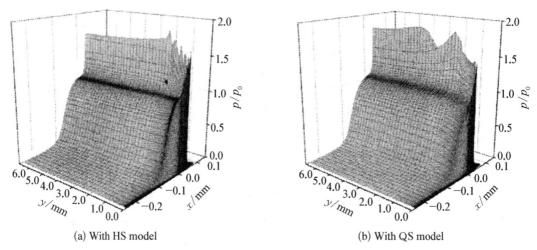

(a) With HS model

(b) With QS model

Fig. 10 Pressure distribution with Lundberg profile

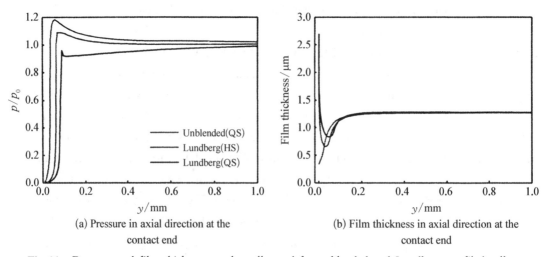

(a) Pressure in axial direction at the contact end

(b) Film thickness in axial direction at the contact end

Fig. 11 Pressure and film thickness at the roller end for unblended and Lundberg profiled rollers calculated with HS and QS models

3.4.2 Roller with dub-off profile

Fig. 9(b) shows that the pressure distribution along the axial direction of an unblended roller is even except near the edge. Therefore, improving the lubrication with a dub-off modification, as shown in Fig. 12(a), is reasonable. The dub-off profile is defined by two parameters, namely, R_y and d. The profile of such a roller can be described using Eq. (13). Similarly, the optimal values of R_y and d depend on the working conditions, especially the load of contact. Under the present working conditions, d is set to 0.3 mm, which is near the range where the pressure starts to increase for an unblended roller (Fig. 9(b)).

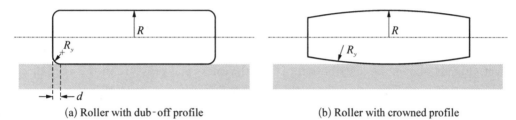

(a) Roller with dub-off profile (b) Roller with crowned profile

Fig. 12 Two different roller profile types

$$\begin{cases} s(x, y) = R - \sqrt{R^2 - x^2}, & (d < y \leqslant 0.5L) \\ s(x, y) = R - \sqrt{\delta^2 - x^2}, & (0 \leqslant y \leqslant d) \end{cases} \quad (13)$$

where $\delta = R - (R_y - \sqrt{(R_y^2 - (y-d)^2)})$.

Fig. 13 shows the film thickness and the pressure distribution along the axis (y) of the roller at $x = 0$ with different R_y. Some evident phenomena can be observed from this result. As shown in Fig. 13(a), the duboff profile can considerably increase the minimum film thickness near the edge compared with the unblended roller profile. Minimum film thickness increases with a decrease in R_y. For example, the minimum film thickness is 0.782 μm with $R_y = 45$ mm, which is considerably higher than that of the unblended

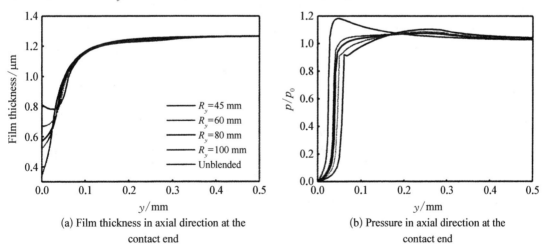

(a) Film thickness in axial direction at the contact end

(b) Pressure in axial direction at the contact end

Fig. 13 Film thickness and pressure at the contact end for rollers with dub-off profile of different radii ($d = 0.3$ mm)

roller. Simultaneously, pressure spike near the edge decreases with R_y, as shown in Fig. 13(b). However, reducing the R_y continuously is not a good idea because another pressure concentration will appear near the start of profiling at approximately 0.3 mm (Fig. 13(b)) if R_y is too small. This conclusion is consistent with that of Mostofi and Gohar[5], although the HS model was adopted in their study. Under the present working conditions, designing R_y in the range of 45 mm – 80 mm is acceptable.

3.4.3 Crowned roller

$$s(x, y) = R - \sqrt{\delta^2 - x^2} \tag{14}$$

where $\delta = R - (R_y - \sqrt{R_y^2 - (y - 0.5L)^2})$.

The crowned roller was shown in Fig. 12(b) and its profile can be described using Eq. (14). Similar to the Lundberg profile, the optimal value of R_y depends on the load of contact. Fig. 14 shows the distribution of film thickness and pressure in the axial direction at the end closure using a crowned roller with different crown radii R_y. A roller with a small crown radius (for example, $R_y = 15$ m) can increase the minimum film thickness and suppress the pressure spikes at the edge. However, the bearing load capacity is largely concentrated on the central part of the roller if the crown radius is too small, which is not favourable for a lubrication design. Pressure distribution becomes increasingly even in the axial direction with the increase of crown radius, but the maximum pressure is still located at the centre of contact. When the crown radius is big enough (for example, $R_y = 30$ m), film thickness becomes increasingly small near the edge (Fig. 14(a)), thereby leading to pressure spike in that area (Fig. 14 (b)). However, this pressure spike is smaller than that of an unblended roller (Fig. 14 (b)). Competition exists between the evenness of pressure and increase in the minimum film thickness or decrease in the pressure spikes near the edge with a crowned profile, i.e., obtaining a relatively even pressure distribution and a higher minimum film thickness at the roller edge is contradictory. Designing the crown radii in the range of 15 m – 25 m is acceptable under the present conditions.

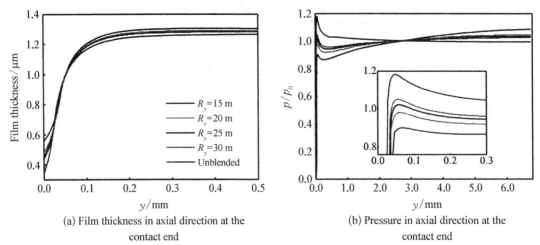

(a) Film thickness in axial direction at the contact end

(b) Pressure in axial direction at the contact end

Fig. 14 Film thickness and pressure at the roller end for rollers with different crown radii

3.4.4 Double QS case (gears)

In some cases (such as gears), the contact bodies are reduced into a finite line roller on a QS plane similar to the contact type of QS/QS shown in Fig. 4(b). The free surfaces of the roller and the plane affect the total deformation simultaneously. Thus, the QS model is adopted for the deformation calculation of these two bodies in the EHL analysis.

Fig. 15 depicts the calculated pressure distribution with an unblended roller on a QS plane. Surprisingly, almost no pressure peaks exist near the free edge, unlike in the case with the unblended roller on a HS plane (as shown in Fig. 8(b)). The changes of pressure along the axis (y) are also extracted and shown in Fig. 16(a). It is found that the maximum pressure with an unblended roller on a QS plane is almost equal to the Hertzian maximum pressure. This observation means that the two free surfaces (that of the roller and that of the plane) can completely eliminate pressure spikes by increasing deformation, which is beneficial to

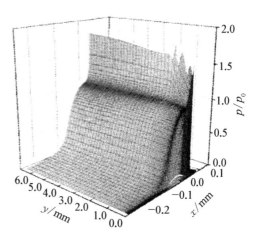

Fig. 15 Pressure distribution at the roller end for an unblended roller on a plane of same width (QS/QS contact type).

the lubrication of such contacts. Correspondingly, the minimum film thickness (0.543 μm) near the edge is higher than that of the unblended roller on the HS plane case (0.343 μm) (Fig. 16(b)). Therefore, under the present working conditions, the unblended roller can be seen as an optical profile design for the roller on a QS plane contact from the viewpoint of even pressure distribution in EHL. It should be noted that a large error (84.2% of the minimum film thickness) would be obtained if the HS model is adopted for the EHL analysis of an unblended roller on a QS plane contact.

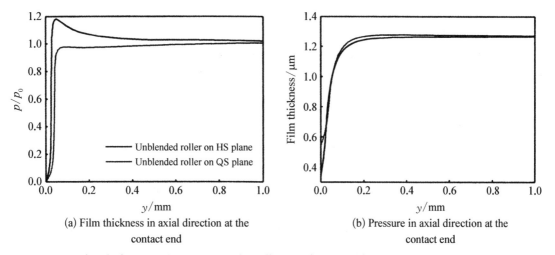

(a) Film thickness in axial direction at the contact end

(b) Pressure in axial direction at the contact end

Fig. 16 Film thickness and pressure at the roller end for an unblended roller on a plane of same width (QS/QS contact type).

4 Conclusion

A flexibility matrix was introduced into the deformation calculation in EHL analysis. The free edge effect on finite line contact EHL was examined. The lubrication behaviours with different roller profiles were analysed. Some important conclusions can be summarized as follows.

1. A steel roller-on-glass disc test rig is not appropriate to study the free edge effect on a finite line contact EHL because the main deformation would appear on the HS glass plane.

2. The free end surface does not play a significant role in determining the lubrication behaviour in the centre of the line contact under the current working conditions.

3. Using the HS model for the finite line contact EHL analysis with an unblended roller would overestimate the magnitude of pressure spikes near the edge. The free surface can reduce pressure concentration through larger deformation, which is beneficial to lubrication behaviour.

4. For roller on HS plane case, the Lundberg profile is quite a good choice for finite line contact EHL design. The pressure distribution with the Lundberg profile is even along the axial direction. Apparent pressure spikes are not observed near the edge, which contradicts the previous analyses that adopted the HS model[12].

5. For roller on QS plane contact, no pressure spikes exist near the free edge just using an unblended roller for the two free surfaces from two contact bodies promoting the deformation.

6. In all cases, minimum film thickness appears near the free edge surface even when the free surface effect is considered.

Acknowledgement

This research was fully supported by the Research Grants Council of Hong Kong (Project No. CityU11213914).

References

[1] Stejskal EO, Cameron A. Optical interferometry study of film formation in lubrication of sliding and/or rolling contacts. NASA Lewis Res Cent 1972. NASACR-120842.

[2] Wymer DG. Elastohydrodynamic lubrication of a rolling line contact. PhD thesis. University of London, 1972.

[3] Wymer DG, Cameron A. Elastohydrodynamic lubrication of a line contact. Proc Institution Mech Eng, 1974,188: 221-38.

[4] Bahadoran H, Gohar R. Research note: end closure in elastohydrodynamic line contact. J Mech Eng Sci, 1974,16: 276-8.

[5] Mostofi A, Gohar R. Elastohydrodynamic lubrication of finite line contacts. ASME J Lubr Technol, 1983, 105: 598-604.

[6] Park TJ, Kim KW. Elastohydrodynamic lubrication of a finite line contact. Wear, 1998,223: 102 - 9.

[7] Kushwaha M, Rahnejat H, Gohar R. Aligned and misaligned contacts of rollers to races in elastohydrodynamic finite line conjunctions. Proc Institution Mech Eng Part C J Mech Eng Sci, 2002,216: 1051 - 70.

[8] Kushwaha M, Rahnejat H. Transient concentrated finite line roller-to-race contact under combined entraining, tilting and squeeze film motions. J Phys D Appl Phys, 2004,37: 2018.

[9] Liu XL, Yang PR. Analysis of the thermal elastohydrodynamic lubrication of a finite line contact. Tribol Int, 2002, 35: 137 - 44.

[10] Johns PM, Gohar R. Roller bearings under radial and eccentric loads. Tribol Int, 1981,14: 131 - 6.

[11] Chen XY, Shen XJ, Xu W, Ma JJ. Elastohydrodynamic lubrication studies on effects of crowning value in roller bearings. Journal of Shanghai University, 2001, 5: 76 - 81. English Edition.

[12] Sun HY, Chen XY. Thermal EHL analysis of cylindrical roller under heavy load. IUTAM Symposium Elastohydrodyn Micro-elastohydrodynamics, 2006: 107 - 20. Springer Netherlands.

[13] Zhu D, Wang J, Ren N, Wang Q. Mixed elastohydrodynamic lubrication in finite roller contacts involving realistic geometry and surface roughness. ASME J Tribol, 2012,134: 011504 - 10.

[14] He T, Wang J, Wang Z, Zhu D. Simulation of plasto-elastohydrodynamic lubrication in line contacts of infinite and finite length. ASME J Tribol, 2015,137: 041505 - 12.

[15] Boussinesq J. Applications des potentiels a l'etude de l'equilibre et du mouvement des solids elastiques. 1885. Paris.

[16] Love AEH. The stress produced in a semi-infinite solid by pressure on part of the boundary. Philosophical Trans R Soc Lond A, 1929,228: 377 - 420.

[17] Hanson MT, Keer LM. A simplified analysis for an elastic quarter-space. Q J Mech Appl Math, 1990, 43: 561 - 87.

[18] Hetenyi MM. A general solution for the elastic quarter space. ASME J Appl Mech, 1970,37: 70 - 6.

[19] Keer LM, Lee JC, Mura T. Hetenyi's elastic quarter space problem revisited. Int J Solids Struct, 1983, 19: 497 - 508.

[20] Guilbault R. A fast correction for elastic quarter-space applied to 3D modeling of edge contact problems. ASME J Tribol, 2011,133: 031402 - 10.

[21] Najjari M, Guilbault R. Edge contact effect on thermal elastohydrodynamic lubrication of finite contact lines. Tribol Int, 2014,71: 50 - 61.

[22] Bower AF, Johnson KL, Kalousek J. A ratchetting limit for plastic deformation of a quarter-space under rolling contact loads. Contact Mechanics and Wear of Rail/ Wheel Systems II. University of Waterloo Press, 1986: 117 - 31.

[23] Hecker M, Romanov A. The stress fields of edge dislocations near wedge-shaped boundaries and bonded wedges. Mater Sci Eng A, 1993,164: 411 - 4.

[24] Guenfoud S, Bosakov SV, Laefer DF. A Ritz's method based solution for the contact problem of a deformable rectangular plate on an elastic quarter-space. Int J Solids Struct, 2010,47: 1822 - 9.

[25] Zhang ZM, Wang W, Wong PL. An explicit solution for the elastic quarter-space problem in matrix formulation. Int J Solids Struct, 2013,50: 976 - 80.

[26] Zhang HB, Wang WZ, Zhang SG, Zhao ZQ. Modeling of finite-length line contact problem with consideration of two free-end surfaces. ASME J Tribol, 2015,138: 021402 - 10.

[27] Zhang HB, Wang WZ, Zhang SG, Zhao ZQ. Elastohydrodynamic lubrication analysis of finite line contact problem with consideration of two free end surfaces. ASME J Tribol, 2016,139: 031501 - 11.

[28] Liu SB, Wang Q, Liu G. A versatile method of discrete convolution and FFT (DC-FFT) for contact analyses. Wear, 2000,243: 101 - 11.

[29] Wang W, Guo L, Wong PL, Zhang ZM. Surface normal deformation in elastic quarter-space. Tribol Int, 2017,114: 358 - 64.

[30] Roelands CJA. Correlation aspects of viscosity-temperature-pressure relationship of lubricating oils. PhD

thesis. Netherlands: Delft University of Technology, 1966.
[31] Dowson D, Higginson GR. Elasto-hydrodynamic lubrication. Pergamon Press, 1977.
[32] Ai XL. Numerical analyses of elastohydrodynamically lubricated line and point contacts with rough surfaces by using semi-system and multigrid methods. PhD thesis. Northwestern University, 1993.
[33] Lundberg G. Elastische berührung zweier Halbräume. Forsch dem Geb Ingenieurwes A, 1939,10: 201-11.

国际会议、全国会议论文集论文

Effects of Pad Elastic Deformation on Tilting Pad Bearing Properties and Rotor System Behavior[*]

Abstract: This paper investigates the effects of pad and pivot elastic deformation on tilting pad journal bearings and dynamic behavior of rotor supported on them. Small harmonic oscillations are assumed in deriving the dynamic EHL equations of individual pads and the formulae of pad dynamic properties. The finite difference equations for steady state and dynamic incremental pressure distributions of oil film are solved by SOR method. The deformations of pad are calculated with a flexibility matrix based on plane strain curved beam. Global forward iterations with under-relaxation are performed to get the simultaneous solutions of pressure and deformation. The dynamic properties of deformable tilting pads are obtained in dependence on the whirl ratio. The effective pad stiffness and especially the pad damping, are seen to be significantly reduced by pad and pivot deformations. A database of deformable pad properties covering a wide range of practical pad parameters is established. The bearing dynamic properties are assembled therefrom. Effects of pad elastic deformation on rotor system dynamics are shown to be unneglible under circumstances.

Nomenclature

A Ratio of amplitude to residual eccentricity of mass centre

b_{rr} and B_{rr} Pad radial damping and its nondimensional value: $B_{rr} = b_{rr}\psi^3/(2\eta l)$

B_{ij}^* ($i, j = t, r$)   Nondimensional dampings of "fixed pad"

b_{st} and B_{st} Threshold negative damping and its nondimensional value: $B_{st} = b_{st}\psi_{min}^3/(2\eta l)$

c Pad clearance, difference between radii of pad and journal

c_{min} Minimum radial clearance of bearing

d Journal diameter

E Elasticity modulus

e_b and e_{rO} Radial eccentricities of journal relative to bearing and pad

F and $F_{t,b}$ Bearing load and frictional force

$\bar{F}_{t,b}$ Nondimensional frictional force of bearing:

$\bar{F}_{t,b} = F_{t,b}\psi_{min}/(\eta\omega dl)$

F_r and F_t Pad radial and frictional forces

$\bar{F}_r$ and $\bar{F}_t$ Nondimensional values of above:
$\bar{F}_r = F_r\psi^2/(\eta\omega dl)$; $\bar{F}_t = F_t\psi/(\eta\omega dl)$

f Static deflection of rotor due to its own weight:
$f = m_r g/k_r$

$f(\phi, \phi_1)$ Radial deflection of pad at ϕ under unit radial force at ϕ_1

h and H Film thickness and its non-dimensional value: $H = h/c$

h_{min} and $h_{min,b}$ Minimum film thicknesses of pad and bearing

H_O and H_d Static component and deformation amplitude of H

δH and (δH) Increment of H due to deformation and its static component

K_δ Deformation factor of pad: $K_\delta = 0.75(1-$

---

[*] In collaboration with X. WU and Z. ZHEN. Reprinted from *Proceedings of the JSLE International Tribology Conference*, 1985.

$\nu^2)\eta\omega/(T_1 E\psi^3 \alpha_J)$

k_r and k_p Stiffnesses of rotor and pivot

k_{rr} and K_{rr} Pad radial stiffness and its nondimensional value: $K_{rr} = k_{rr}\psi^3/(2\eta\omega l)$

K_{ij}^* $(i, j = t, r)$ Nondimensional stiffnesses of "fixed pad"

K_{st} Non-dimensional threshold cross coupling stiffness: $K_{st} = k_{st}\psi_{\min}^3/(2\eta\omega_k l)$

l Length of journal or width of bearing

M Harmonic complex amplitude of H

m Preload factor of bearing: $m = \delta/c$

m_r Mass of rotor

p and P Pressure and its non-dimensional value: $P = p\psi^2/(\eta\omega)$

P_O and Q Static component and complex amplitude of P

Q_s and $Q_{s,b}$ Double-sided leakages of pad and bearing

$\bar{Q}_s$ and $\bar{Q}_{s,b}$ Non-dimensional values of above: $\bar{Q}_s = 2Q_s/(\omega rcl); \bar{Q}_{s,b} = 2Q_{s,b}/(\omega rc_{\min}l)$

r Radius of journal

So_b and So_k Bearing Sommaerfeld numbers: $So_b = F\psi_{\min}^2/(\eta\omega dl)$; $So_k = F\psi_{\min}^2/(\eta\omega_k dl)$

t and T Time and its non-dimensional value: $T = t\omega$

T_1 Ratio of pad thickness to pad neutral diameter

z and Z Axial coordinate and its non-dimensional value: $Z = 2z/l$

α_J Factor of curvature: $\alpha_J = 1 + 3T_1^2/5 + 3T_1^4/7 + \cdots$

α_p Pivot deformation factor: $\alpha_p = 2\eta\omega l/(k_p\psi^3)$

β Angular span of pad

γ Offset coefficient of pivot position

δ Built-in eccentricity of pad centre relative to bearing centre

δ_p Deformation of pivot

$\varepsilon_{tO}, \varepsilon_{rO}$ Steady state tangential and radial eccentricity ratios of journal relative to pad

$\Delta\varepsilon_t, \Delta\varepsilon_r$ Complex amplitudes of the above

η Dynamic viscosity

ν Poisson's ratio

θ Angular position of pivot

Ω Relative frequency of perturbation or whirl ratio: $\Omega = \omega_i/\omega$

ω and $\bar{\omega}$ Angular speed of shaft and its ratio to ω_k: $\bar{\omega} = \omega/\omega_k$

ω_1 and $\bar{\omega}_1$ Angular frequency of perturbation and its ratio to ω_k: $\bar{\omega}_1 = \omega_1/\omega_k$

ω_{cr} Critical angular speed

ω_k Natural angular frequency of rigidly supported rotor: $\omega_k = \sqrt{k_r/m_r}$

ψ and $\psi_{\min}$ Relative clearance: $\psi = 2c/d$; $\psi_{\min} = 2c_{\min}/d$

Lower index b denotes quantity belonging to bearing

Lower index s denotes derivative with respect to s ($s = \Delta\varepsilon_t$ or $\Delta\varepsilon_r$)

1 Introduction

Bearing researchers have shown increased interest over the effect of tilting pad and pivot deformation on rotor system dynamic behavior in recent years. Based on an assessment of the present state of investigation on this problem as represented by[1-6], [7] selected a corrected plane strain curved beam model as the basis of calculation of pad deformation, since it combines adequate reasonability with efficiency of calculation. As a continuation, this paper will formulate basic equations of small harmonic dynamic EHL of radial tilting pad in supplement to their steady state counter parts, the method of solution, and the formulae of dynamic properties. Comprehensive calculations will be done to set up a database of properties of individual deformable pad. This will form a basis for assembling the complete bearing properties. These will be coupled with equations of motion of Jeffcott rotors, and the effect on rotor dynamic behavior will be predicted and compared with practical experience.

2 Basic Relations

The flow and pressure in the oil film between journal and pad are solely determined by the position and motion of the journal relative to the pad. Therefore, dynamic analysis can first be made in a relative coordinate system rigidly attached to the pad. The result can then be transformed into tilting pad properties[8].

If small perturbations are given to the journal about its steady state equilibrium position (Fig. 1), the pressure and clearance distributions in the oil film can be looked upon as the sum of their static and perturbated components, with the latter much smaller in order of magnitude than the former[9]. For linear problems of small unbalance response and stability threshold of rotor-bearing systems, the perturbated components can be taken as simple harmonic. Expressed in complex forms, the P and H distributions are respectively:

$$P(\phi, Z, T) = P_o(\phi, Z) + Q(\phi, Z)e^{i\Omega T} \tag{1}$$

$$H(\phi, T) = H_o(\phi) + M(\phi)e^{i\Omega T} \tag{2}$$

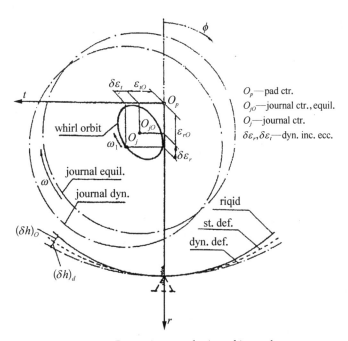

Fig. 1 Dynamic perturbation of journal

The last term contains the increment of H directly due to journal perturbation and that due to incremental deformation caused by incremental P, viz.

$$M(\phi) = (\Delta\varepsilon_t \sin\phi + \Delta\varepsilon_r \cos\phi + H_d) \tag{3}$$

The P and H distributions are mutually influential, and the relationship between them can be expressed by the non-stationary Reynolds, the deformation, and the geometric equations:

$$\frac{\partial}{\partial\phi}\left(H^3 \frac{\partial P}{\partial\phi}\right) + \left(\frac{d}{l}\right)^2 \frac{\partial}{\partial Z}\left(H^3 \frac{\partial P}{\partial Z}\right) = 6\frac{\partial H}{\partial\phi} + 12\frac{\partial H}{\partial T} \tag{4}$$

$$\delta H = (\delta H)_O + H_d e^{i\Omega T} = \iint_A P(\phi_1, Z, T) f(\phi, \phi_1) d\phi_1 dZ \tag{5}$$

$$H = 1 + (\varepsilon_{tO} + \Delta\varepsilon_t e^{i\Omega T})\sin\phi + (\varepsilon_{rO} + \Delta\varepsilon_r e^{i\Omega T})\cos\phi \tag{6}$$

where A denotes the oil film domain.

Equations (1)–(3) are substituted into equations (4)–(6), and small terms of higher orders are neglected, to get a system of steady state equations:

$$\frac{\partial}{\partial\phi}\left(H_O^3 \frac{\partial P}{\partial\phi}\right) + \left(\frac{d}{l}\right)^2 \frac{\partial}{\partial Z}\left(H_O^3 \frac{\partial P_O}{\partial Z}\right) = 6\frac{dH}{d\phi} \tag{7}$$

$$(\delta H) = \iint_A P_O f \cdot d\phi_1 dZ \tag{8}$$

$$H_O = 1 + \varepsilon_{tO}\sin\phi + \varepsilon_{rO}\cos\phi + (\delta H)_O \tag{9}$$

and a system of complex amplitude equations:

$$\frac{\partial}{\partial\phi}\left(H^3 \frac{\partial Q}{\partial\phi}\right) + \left(\frac{d}{l}\right)^2 \frac{\partial}{\partial Z}\left(H_O^3 \frac{\partial Q}{\partial Z}\right)$$
$$= -18\frac{M}{H_O}\frac{dH_O}{d\phi} + 6\frac{dM}{d\phi} + 12i\Omega M - 3H_O\left(H_O \frac{dM}{d\phi} - M \frac{dH_O}{d\phi}\right)\frac{\partial P_O}{\partial\phi} \tag{10}$$

$$H_d = \iint_A Q f \cdot d\phi_1 dZ \tag{11}$$

$$M = \Delta\varepsilon_t \sin\phi + \Delta\varepsilon_r \cos\phi + H_d \tag{12}$$

Equations (7)–(9) together with the usual boundary condition of pressure constitute the steady state EHL problem[7]. Dynamic equations (10)–(12) are partially differentiated to the two independent amplitudes of journal perturbation and, to get a system of equations of the derivatives of complex amplitudes of pressure and film thickness:

$$\frac{\partial}{\partial\phi}\left(H_O^3 \frac{\partial Q_s}{\partial\phi}\right) + \left(\frac{d}{l}\right)^2 \frac{\partial}{\partial Z}\left(H_O^3 \frac{\partial Q_s}{\partial Z}\right) = -18\frac{M_s}{H_O}\frac{dH_O}{d\phi} + 6\frac{dM_s}{d\phi} + 12i\Omega M_s$$
$$-3H_O\left(H_O \frac{dM_s}{d\phi} - M_s \frac{dH_O}{d\phi}\right)\frac{\partial P_O}{\partial\phi} \tag{13}$$

$$(H_d)_s = \iint_A Q_s f \cdot d\phi_1 dZ \tag{14}$$

$$M_s = \partial(\Delta\varepsilon_t \sin\phi + \Delta\varepsilon_r \cos\phi)/\partial S + (H_d)_s \tag{15}$$

Equations (13) – (15) with homogeneous boundary condition of Q_s constitute the dynamic EHL problem of tilting pad under small harmonic oscillations. Distributions Q_s, M_s and $(H_d)_s$ can be solved from them, and the eight dynamic coefficients of "fixed pad" can be calculated with the following formulae:

$$K_{ts}^* + i\Omega B_{ts}^* = -\iint_A Q_s \sin\phi \, d\phi \, dZ \text{ and } K_{rs}^* + i\Omega B_{rs}^* = -\iint_A Q_s \cos\phi \, d\phi \, dZ \quad (s = \Delta\varepsilon_t, \Delta\varepsilon_r) \tag{16}$$

Pad inertia is neglected, and the radial stiffness and damping of tilting pad are calculated from the dynamic properties of the "fixed pad"[8]:

$$K_{rr} = K_{rr}^* - \frac{K_{tt}^*(K_{rt}^* K_{tr}^* - \Omega^2 B_{rt}^* B_{tr}^*) + \Omega^2 B_{tt}^*(K_{rt}^* B_{tr}^* + K_{tr}^* B_{rt}^*)}{K_{tt}^{*2} + \Omega^2 B_{tt}^{*2}} \tag{17}$$

$$B_{rr} = B_{rr}^* - \frac{B_{tt}^*(\Omega^2 B_{rt}^* B_{tr}^* - K_{rt}^* K_{tr}^*) + K_{tt}^*(K_{rt}^* B_{tr}^* + K_{tr}^* B_{rt}^*)}{K_{tt}^{*2} + \Omega^2 B_{tt}^{*2}} \tag{18}$$

Relying on the above relations, nondimensional values of pad properties ($\bar{F}_r$, $H_{\min}$, $\bar{F}_t$, $\bar{Q}_s$, K_{rr}, B_{rr}) in dependence on pad radial eccentricity ratio ε_{rO} and pad deformation factor K_δ can be obtained and form a database for individual pad with given geometric parameters l/d, β, and γ.

The static deformation of pivot can be taken into consideration in the following way. It can be seen from Fig. 2 that, when the journal centre has an absolute displacement e'_r in the radial direction of a certain pad, the pivot deforms and makes the pad displace δ_p in the radial direction, hence the actual radial eccentricity of the journal relative to the displaced pad centre O' is:

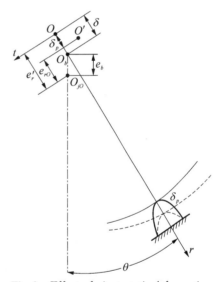

Fig. 2 Effect of pivot static deformation

$$e_{rO} = e'_r - \delta_p \text{ where } \delta_p = F_r/k_p \tag{19}$$

We can also write the above relationship in nondimensional form:

$$\varepsilon'_r = \varepsilon_{rO} + \alpha_p \bar{F}_r \tag{20}$$

Since $\bar{F}_r$ is determined by ε_{rO}, the value ε'_r is also determined by ε_{rO}, therefore, the pad nondimensional properties can also be directly linked to the absolute eccentricity ratio ε'_r.

Besides the effect of pivot static deformation in modifying the eccentricity, the effect of pivot dynamic deformation has yet to be considered, which can be done by replacing the pad complex stiffness by that of pad and pivot in series. The nondimensional values of the serial stiffness and damping are:

$$K_s = \frac{1}{\alpha_p} \cdot \frac{K_{rr}(K_{rr}+1/\alpha_p)+(\alpha B_{rr})^2}{(K_{rr}+1/\alpha_p)^2+(\Omega B_{rr})^2}; \quad B_s = \frac{B_{rr}/\alpha_p^2}{(K_{rr}+1/\alpha_p)^2+(\Omega B_{rr})^2} \quad (21)$$

Thus, all the nondimensional properties of pad with consideration of static and dynamic deformations of pivot can be directly related to ε_r'. The assembling of bearing properties is thus made very convenient, since ε_r' is directly linked with ε_b.

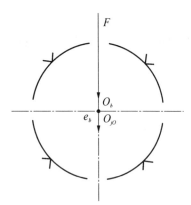

Fig. 3 Symmetric location of pivots

If we limit ourselves to the case where the pivots are located symmetrical to the load direction (Fig. 3), the steady state eccentricity of journal relative to the bearing centre will always be in the load direction. The absolute radial eccentricities related to pad k with pivot angle θ_k can be calculated as:

$$\varepsilon_{r,k'} = m + \varepsilon_b(1-m)\cos\theta_k \quad (22)$$

With this value known, all the pad properties can be interpolated from the pad database, and the bearing properties can then be assembled from them:

$$So_b = \sum_{k=1}^{n} \bar{F}_{r,k}\cos\theta_k(1-m) \quad (23)$$

$$H_{\min} = \min(H_{\min,k})/(1-m) \quad (24)$$

$$\bar{F}_{t,b} = \sum_{k=1}^{n} \bar{F}_{t,k}(1-m) \quad (25)$$

$$\bar{Q}_{s,b} = \sum_{k=1}^{n} \bar{Q}_{s,k}/(1-m) \quad (26)$$

$$K_{ij,b} = \sum_{k=1}^{n} K_{rr,k}u_i u_j (1-m)^3 \text{ and } B_{ij,b} = \sum_{k=1}^{n} B_{rr,k}u_i u_j (1-m)^3 \quad (i,j=x,y) \quad (27)$$

where $u_x = \sin\theta_k$ and $u_y = \cos\theta_k$. The assembling of dynamic properties are done separately for each whirl ratio.

The model of the rotor-bearing system considered in this paper is shown in Fig. 4. The synchronous vibration of this system can be calculated using bearing properties corresponding to unity whirl ratio[9]. The effect of pad deformation on unbalance response can be shown from calculations of this sort.

As for the effect of pad deformation on stability behavior of the system, threshold values of destabilizing factors will be used to represent the stability margin, since here only threshold analysis is possible in view of the fact that we have based our bearing analysis on assumption of small harmonic oscillations. Two forms of idealized

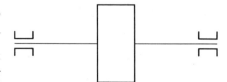

Fig. 4 Rotor-bearing system model

destabilizing factors are considered here, namely, plane isotropic cross-coupling stiffness of positive sense and negative isotropic damping, both of which are applied to the middle plane of the rotor.

When the threshold value k_{st} of cross-coupling stiffness is reached, we have the following equations of motion:

$$m_r \ddot{x}_r + k_r(x_r - x_j) - k_{st} y_r = 0 \tag{28}$$

$$m_r \ddot{y} + k_r(y_r - y_j) + k_{st} x_r = 0 \tag{29}$$

$$k_r(x_r - x_j) = 2(k_{xx,b} x_j + b_{xx,b} \dot{x}_j + k_{xy,b} y_j + b_{xy,b} \dot{y}_j) \tag{30}$$

$$k_r(y_r - y_j) = 2(k_{yx,b} x_j + b_{yx,b} \dot{x}_j + k_{yy,b} y_j + b_{yy,b} \dot{y}_j) \tag{31}$$

For symmetric locations of pivots, all the cross-coupling bearing coefficients are zero. From the condition for existence of non-trivial harmonic oscillation, the nondimensional values K_{st} and $\bar{\omega}_1$ can be deduced to be:

$$K_{st} = 2\alpha \left[\frac{\bar{\omega}_1^2 \alpha^2 B_{xx} B_{yy}}{(\beta_x^2 + \bar{\omega}_1^2 B_{xx})(\beta_y^2 + \bar{\omega}_1^2 B_{yy})} - \left(\frac{\bar{\omega} K_{xx} \beta_x + \bar{\omega}_1^2 B_{xx}^2}{\beta_x^2 + \bar{\omega}_1^2 B_{xx}^2} - \bar{\omega}_1^2 \right) \cdot \right.$$
$$\left. \left(\frac{\bar{\omega} K_{yy} \beta_y + \bar{\omega}_1^2 B_{yy}^2}{\beta_y^2 + \bar{\omega}_1^2 B_{yy}^2} - \bar{\omega}_1^2 \right) \right]^{1/2} \tag{32}$$

$$\bar{\omega}_1 = \sqrt{(-b + \sqrt{b^2 - 4ac})/(2a)} \tag{33}$$

where $\alpha = 2So_k/(f/c_{\min})$; $\beta_x = \alpha + \bar{\omega} K_{xx}$; $\beta_y = \alpha + \bar{\omega} K_{yy}$; $a = B_{xx} B_{yy}(B_{xx} + B_{yy})$; $b = B_{xx} \beta_y^2 + B_{yy} \beta_x^2 - a$; $c = -\bar{\omega}(K_{xx} B_{yy} \beta_x + K_{yy} B_{xx} \beta_y)$.

The calculation of K_{st} should be preceded by an iterative process to find the threshold whirl ratio, so that dynamic coefficients corresponding to it can be used to calculate K_{st}.

In a similar way, the threshold value of negative damping can be deduced as

$$B_{st} = \min(B_{st,x}, B_{st,y}) \text{ where } B_{st,i} = \alpha^2 B_{ii}/(\beta_i^2 + \bar{\omega}_{1,i}^2 B_{ii}^2) \quad (i = x, y) \tag{34}$$

and the nondimensional whirl ratio should also be first iterated by:

$$\bar{\omega}_{1,i} = \sqrt{(-b_i + \sqrt{b_i^2 - 4a_i c_i})/(2a_i)} \quad (i = x, y) \tag{35}$$
$$\text{where } a_i = B_{ii}; \ b_i = \beta_i^2 - a_i; \ c_i = -\bar{\omega} K_{ii} \beta_i$$

3 Results of Calculation and Discussion

The pressure and dynamic pressure increment distributions are solved by FDM and SOR. The deformation and dynamic deformation of pad are calculated using a flexibility matrix based on plane strain curved beam model. Global iterations with under-relaxation between pressure and deformation are performed to get their simultaneous solutions. Numerical integrations are used to get the properties of individual pad, and build up the pad database. All the other calculations have been stated above, and are self-explanatory.

A wide range of pad parameters have been covered in this work, but only the result concerning a four-pad bearing with the following parameters will be shown here for illustrative purpose: $l/d=0.5$, $\beta=75$ deg., $\gamma=0.6$, and $m=0.5$. The results without pad deformation are compared with [10] (Fig. 5), and the almost complete coincidence is thought to confirm the validity of the analysis.

Fig. 5 shows that, under given bearing load as represented by So_b, the minimum film thickness is significantly reduced by pad elastic deformation. The bearing stiffness is reduced significantly in light load region (low Sommerfeld number), but somewhat enhanced in high load region, while the bearing damping decreases significantly in the whole region with increase of pad deformation factor.

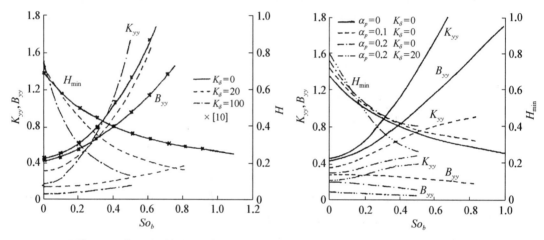

Fig. 5 Influence of pad elastic deformation on bearing properties

Fig. 6 Influence of pivot elastic deformation on bearing properties

The effect of pivot deformation on bearing properties is shown in Fig. 6. Its effect on H_{min} is beneficial although not very significant. Bearing stiffness, and especially bearing damping are significantly reduced. Simultaneous existence of pad and pivot deformations causes an combination of their effects.

This change of bearing properties will undoubtedly influence the dynamic behavior of the rotor-bearing system. It is only natural to expect the critical speed to vary somewhat due to variation of bearing stiffness, and the resonant amplitude to drastically rise due to the significant decrease of bearing damping. This has been evidenced by all of our numerous calculations. The speed-amplitude curves in Fig. 7 are shown to illustrate such effect of pad deformation. The results of resonant amplitude from a series of calculations of this sort are collected in Fig. 8, and the resulting map shows more clearly and quantitatively the adverse effect of pad elastic deformation on rotor dynamics. Practice shows that, high speed rotating machinery employing tilting pad bearings may exhibit very pronounced resonant amplitude. Pad and pivot deformation is without doubt a possible cause worthy of consideration.

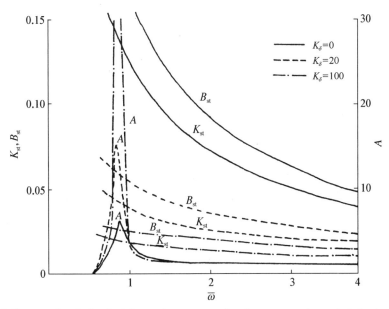

Fig. 7 Influence of pad elastic deformation on rotor-bearing system dynamic behavior

Fig. 7 also shows the threshold values K_{st} and B_{st} under different speeds. It can be seen that elastic deformation also significantly impairs the stability margin of the system, though by itself it will not cause instability.

4 Conclusions

(1) A pad database with consideration of pad elastic deformation on the basis of plane strain curved beam model can be built up by solving the steady state and dynamic EHL problems of individual pad. This database provides an improved basis for effecting theoretical calculations of deformable tilting pad bearings and the rotor systems supported on them.

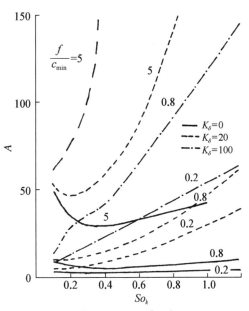

Fig. 8 Resonant amplitude map

(2) The bearing stiffness is affected, and the bearing damping drastically lowered by pad and pivot elastic deformations.

(3) The critical speed of the rotor-bearing system shifts somewhat due to pad and pivot elastic deformations.

(4) The resonant amplitude rises drastically due to pad and pivot elastic deformations.

(5) The stability margin of the rotor-bearing system as represented by threshold

values of cross-coupling stiffness and negative damping is lowered by pad and pivot elasticity.

(6) In view of the generally adverse, and under circumstance serious effect of pad and pivot elastic deformation on rotor dynamic behavior, notice should be given in design stage to provide the pads and pivots with sufficient rigidity. Alternatively, the elastic deformation effect should be accounted for when calculating bearings with large factors of deformation.

(7) The pivots have been taken as linear elastic members to facilitate the investigation of the general trend of its effect. The treatment on the static deformation of pivot should be properly modified, if the non-linearity of pivot elasticity must be considered. The dynamic character of the pivot can still be taken as linear, since assumption of small oscillation has been made in dynamic analysis.

References

[1] Caruso WJ et al. Application of recent rotor dynamics developments to mechanical drive turbines. Proceedings of the 11th Turbomachinery Symposium, 1983: 1-17.
[2] Lund JW, Pederson LB. The influence of the pad flexibility on the dynamic coefficients of a tilting pad journal bearing. Trans. ASME, J. Trib., 109(Jan), 1987: 65-70.
[3] Li X, Zhu J. Effect of thermo-elastic deformation on properties of large tilting pad bearings. Proceedings of the 4th National Tribology Conference of China, 1985.
[4] Chen X, Zhang Z. Dynamic properties of tilting pad journal bearings with consideration of elastic dynamic deformation. Journal of Shanghai University of Technology, No. 4, 1989: 310-316.
[5] Nilsson LRK. The influence of bearing flexibility on the dynamic performance of radial oil film bearings. Proceedings of the 5th Leeds-Lyon Symposium on Tribology, 1969: 311-319.
[6] Brugier D, Pascal MT. Influence of elastic deformations of turbo-generator tilting pad bearings on the static behavior and on the dynamic coefficients in different designs. Trans. ASME, J. Trib., 111(Apr), 1989: 364-371.
[7] Zhen Z et al. Effect of pad deformation on static performance of journal bearing tilting pad. Proceedings of the International Conference on Hydrodynamic Bearing-rotor System Dynamics, Xi'an, 1990: 60-66.
[8] Zhang Z et al. Theory of Hydrodynamic Lubrication of Sliding Bearings. Higher Education Ed., 1986.
[9] Zhang Z et al. The effect of dynamic deformation on dynamic properties and stability of cylindrical journal bearings. Proceedings of 13th Leeds-Lyon Symposium on Tribology, Sept. 1986: 363-366.
[10] Glienicke J, Leonhard M. Stabilitaetsprobleme bei Lagerung schnellaufender Wellen, Universitaet Karlsruhe, Juni 1981: 136.

System Damping Factor of an Elastic Rotor Supported on Tilting-Pad Bearings with Elastic and Damped Pivots*

Abstract: The system damping factors of a symmetric one-mass elastic rotor supported on the 4-tilting-pad bearings were calculated, with consideration of pivot elastic and damping properties in the tilting direction and radial direction. The influences of these parameters on system damping values were analyzed. It may be seen from the results that suitable values of elastic and damping properties of the pivots in the radial direction are very effective in improving the system damping value, but those in the tilting direction will in general reduce it while they may be advantageous in suppressing pad instability. A method of developing the determinant of the eigenmatrix into ϵ polynomial of high order is also presented in the appendix.

1 Introduction

A rotor supported on tilting-pad bearings can work stably under very high rotating speeds, but it usually has low stability reserve as characterized by system damping factor[1] under its working speed, which will moreover be influenced by various practical factors. The influence of pad inertia on system damping factor was discussed in [2] and [3]. The system damping factor of a rotor supported on air-lubricated tilting-pad bearings with radially elastic and damped pivots was touched upon in [4]. The present paper studies the influence of pad inertia and pivot elasticities and dampings in both radial and tilting directions, which might come from pivot geometry, contact deformation and surrounding medium, or might otherwise be specially designed in.

2 Model of Pivot

Only the case of the cylindrical whirl of a symmetrical one-mass rotor (Fig. 1) supported on two identical 4-pad bearings (Fig. 2) is studied as an example. All the pivots are of the line-contact type, and have the same elasticity and damping properties. Fig. 3 shows the model of a pivot and a pad. Cross-coupling terms in pivot dynamic properties are excluded for simplicity. Therefore, the increments of the radial reation and reactional moment in the tilting direction acting on the pad from the pivot due to the motion of the pad in the neibourhood of its static equilibrium position will be respectively

---

* Reprinted from *Proceedings of JSLE International Tribology Conference*, 1985:541-546.

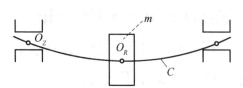
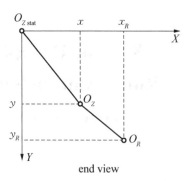

O_Z — center of journal
O_R — center of rotor
$O_{Z\,stat}$ — static equilibrium position of center of journal

end view

Fig. 1

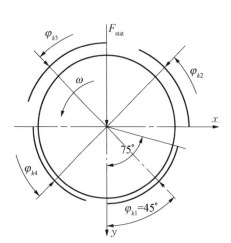
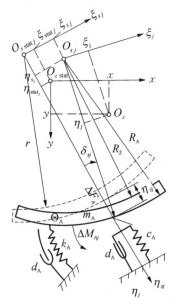

$O_{s\,statj}$ — equilibrium position of pad center
$O_{z\,stat}$ — equilibrium position of journal center
O_{sj} — instantaneous position of pad center

Fig. 2   Fig. 3

$$\Delta F_{Aj} = c_A \eta_{sj} + d_A \dot{\eta}_{sj} \tag{1}$$

$$\Delta M_{Aj} = k_A \delta_{sj} + q_A \dot{\delta}_{sj} \tag{2}$$

3 Equations of Motion and System Damping Factor

The equations of motion of the rotor in x and y directions are

$$-m\ddot{x}_R = c(x_R - x) = 2\Delta F_x$$
$$-m\ddot{y}_R = c(y_R - y) = 2\Delta F_y$$

respectively, or

$$mu\bar{x} + 2\Delta \ddot{F}_x/\omega_k^2 + 2\Delta F_x = 0$$
$$m\bar{y} + 2\Delta \ddot{F}_y/\omega_k^2 + 2\Delta F_y = 0 \tag{3}$$

where the increments of the oil film force of a bearing in x and y directions are

$$\Delta F_x = \sum_{j=1}^{4}(\Delta F_{\xi j}\cos\varphi_{kj} + \Delta F_{\eta j}\sin\varphi_{kj})$$
$$\Delta F_y = \sum_{j=1}^{4}(-\Delta F_{\xi j}\sin\varphi_{kj} + \Delta F_{\eta j}\cos\varphi_{kj}) \tag{4}$$

respectively, and the increments of the oil film force of a pad j are

$$\Delta F_{\xi j} = c_{\xi\xi j}\Delta\xi_j + c_{\xi\eta j}\Delta\eta_j + d_{\xi\xi j}\dot{\xi}_j + d_{\xi\eta j}\dot{\eta}_j$$
$$\Delta F_{\eta j} = c_{\xi\eta j}\Delta\xi_j + c_{\eta\eta j}\Delta\eta_j + d_{\xi\eta j}\dot{\xi} + d_{\eta\eta j}\dot{\eta}_j \tag{5}$$

respectively.

For a rigid rotor, the equations of motion of the rotor are

$$m\ddot{x} + 2\Delta F_x = 0, \quad m\ddot{y} + 2\Delta F_y = 0 \tag{3a}$$

The equations of motion of the pad as it tilts on its pivot is

$$\Delta F_{\xi j}R_A - \Delta M_{Aj} - m_A\ddot{\eta}_{sj}a - \Theta_s\ddot{\delta}_{sj} = 0$$

which may be written as

$$\Delta F_{\xi j} - k_A\xi_{sj} - q_A\dot{\xi}_{sj} - \frac{m_A a}{R_A}\ddot{\eta}_{sj} - \Theta_s\ddot{\xi}_{sj} = 0 \tag{6}$$

since $\delta_{sj} \approx \xi_{sj}/R_A$

The equation of motion of the pad as it moves radially is

$$\Delta F_{\eta j} - \Delta F_{Aj} - m_s(\ddot{\eta}_{sj} + a\ddot{\delta}_{sj}) = 0 \tag{7}$$

The relation between the journal displacements (x, y), the displacements $(\Delta\xi_j, \Delta\eta_j)$ of the journal center O_x relative to the pad center O_{sj} and the pad center displacements (ξ_{sj}, η_{sj}) is

$$\begin{bmatrix}\varepsilon_{sj}\\\eta_{sj}\end{bmatrix} = -\begin{bmatrix}\Delta\xi_j\\\Delta\eta_j\end{bmatrix} + \begin{bmatrix}\cos\varphi_{kj} & -\sin\varphi_{kj}\\\sin\varphi_{kj} & \cos\varphi_{kj}\end{bmatrix}\begin{bmatrix}x\\y\end{bmatrix} \tag{8}$$

since δ_{sj} is very small.

These equations are nondimentionalyzed into:

motion of rotor

$$\frac{S_0}{\mu}\left(\frac{\omega}{\omega_k}\right)^2 X'' + \left(\frac{\omega}{\omega_k}\right)^2 \Pi''_x + \Delta\Pi_x = 0$$
$$\frac{S_0}{\mu}\left(\frac{\omega}{\omega_k}\right)^2 Y'' + \left(\frac{\omega}{\omega_k}\right)^2 \Pi''_y + \Delta\Pi_y = 0 \tag{9}$$

where "'" denotes $d/d\phi$, and $\phi = \omega t$, or for a rigid rotor

$$S_0\left(\frac{\omega}{\omega_0}\right)^2 X'' + \Delta\Pi_x = 0, \quad S_0\left(\frac{\omega}{\omega_0}\right)^2 Y'' + \Delta\Pi_y = 0 \tag{9a}$$

motion of pad j

$$\begin{aligned}\Delta\Pi_{\xi j} - \alpha_\xi \xi''_{sj} - \beta_\xi \xi'_{sj} - \gamma_\xi \xi_{sj} - \alpha_\xi \eta''_{sj} = 0 \\ \Delta\Pi_{\eta j} - \alpha_\eta \xi''_{sj} - \alpha_\eta \eta'_{sj} - \beta_\eta \eta'_{sj} - \gamma_\eta \eta_{sj} = 0\end{aligned} \tag{10}$$

increments of film force of pad j

$$\begin{aligned}\Delta\Pi_{\xi j} = (\gamma^*_{\xi\xi j}\xi_j + \gamma^*_{\xi\eta j}\eta_j + \beta^*_{\xi\xi j}\xi'_j + \beta^*_{\xi\eta j}\eta'_j)/\psi_r^2 \\ \Delta\Pi_{\eta j} = (\gamma^*_{\eta\xi j}\xi_j + \gamma^*_{\xi\eta j}\eta_j + \beta^*_{\eta\xi j}\xi'_j + \beta^*_{\eta\eta j}\eta'_j)/\psi_r^2\end{aligned} \tag{11}$$

increments of film force of a bearing

$$\begin{aligned}\Delta\Pi_x = \sum_{j=1}^4 (\Delta\Pi_{\xi j}\cos\varphi_{kj} + \Delta\Pi_{\eta j}\sin\varphi_{kj}) \\ \Delta\Pi_y = \sum_{j=1}^4 (-\Delta\Pi_{\xi j}\sin\varphi_{kj} + \Delta\Pi_{\eta j}\cos\varphi_{kj})\end{aligned} \tag{12}$$

geometric relation

$$\begin{bmatrix}\bar{\xi}_{sj} \\ \bar{\eta}_{sj}\end{bmatrix} = -\begin{bmatrix}\bar{\xi}_j \\ \bar{\eta}_j\end{bmatrix} + \begin{bmatrix}\cos\varphi_{kj} & -\sin\varphi_{kj} \\ \sin\varphi_{kj} & \cos\varphi_{kj}\end{bmatrix}\begin{bmatrix}X \\ Y\end{bmatrix} \tag{13}$$

After elimination of $\bar{\xi}_{sj}$, $\bar{\eta}_{sj}$ and all the $\Delta\Pi_s$, the above system of equations may be expressed in X, Y, $\bar{\xi}_j$ and $\bar{\eta}_j (j=1, 2, \cdots, 4)$ Substituting with

$$W = W_0 e^{\lambda\phi} (W = X, Y, \bar{\xi}_j, \bar{\eta}_j) \ (j=1, 2, \cdots, 4)$$

we get a system of algebraic equations which may be expressed in matrix form as

$$\underline{K} \cdot \underline{X}^0 = \underline{O} \tag{14}$$

where $\underline{X}^0 = (X_0, Y_0, \bar{\xi}_{10}, \bar{\eta}_{10}, \cdots, \bar{\xi}_{40}, \bar{\eta}_{40})^T$; $\underline{K}$—matrix of eoefficients.

Existence of nontrivial solution calls for zero value of the determant of the matrix $\underline{K}$

$$|\underline{K}| = 0 \tag{15}$$

which may be developed into the characteristic polynomial equation

$$|\underline{K}| = A_n\lambda^n + A_{n-1}\lambda^{n-1} + \cdots + A_1\lambda + A_0 = 0 \tag{16}$$

Since n may be as high as 22, a numerical method (see Appendix) is used to determine $A_0, A_1, \cdots, A_n$.

All the complex roots $\lambda_m (m=1, 2, \cdots, n)$ are found by the Newton-Bairstow's method. The nondimentional system damping factor with reference to ω_k (or ω_0) are found from the negative real parts of λ_m

$$u/\omega_k = -\text{Re}(\lambda_m) \cdot \omega/\omega_k \text{ or } u/\omega_0 = -\text{Re}(\lambda_m) \cdot \omega/\omega_0 \tag{17}$$

The stability reserve of the whole system may be judged from the lowest value of u/ω_k (or u/ω_0).

In the calculations, the bearing eccentricity ε under the respective working condition is first determined from the relation:

$$\sum_{j=1}^{4} So_{kj} \cos\varphi_{kj}/\psi_r^2 = So = So_k \omega_k/\omega (\text{或 } So_0 \omega_0/\omega) \quad (18)$$

Next, the effective eccentricity $\varepsilon_{\text{eff}j}$[2] of each pad j is determined as

$$\varepsilon_{\text{eff}j} = 1 - (1 - \varepsilon\cos\varphi_{kj})/\psi_r \quad (19)$$

which is used to interpolate the dynamic characteristics γ_{ikj}^* and β_{ikj}^* ($i, k = \xi, \eta$) ($j = 1$, 2, ⋯, 4) from a table of the dynamic properties of "fixed" pads.

All the calculations were done for So_k (or So_0)=0.2, $\mu=1$ (or 0) and $\psi_r=2$.

4 Results and Discussion

The influence of coefficient of tilting stiffness $\gamma_{\xi k}$ is shown in Fig. 4, where it is taken that $\alpha_\eta = \alpha_\xi = \alpha_\kappa = \beta_\eta = \beta_\xi = 0$ and $\gamma_\eta = \infty$ (pivot rigid in the radial direction). It may be seen from the figure that the greater the value of $\gamma_{\xi k}$, the lower the system damping factor in the whole range of speed is. This is quite easy to imagine, since a tilting-pad bearing with greater $\gamma_{\xi k}$ loses more of its inherent stability. This tendency also shows itself in the presence of small values of coefficient of tilting damping $\beta_{\xi k}$, such as when $\beta_{\xi k} = 0.1$.

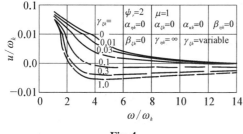
Fig. 4

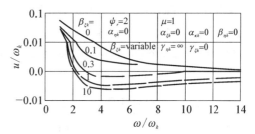
Fig. 5

The influence of coefficient of tilting damping $\beta_{\xi k}$ is shown in Fig. 5, where it is taken that $\alpha_\eta = \alpha_\xi = \alpha_\kappa = \beta_\eta = \gamma_\xi = 0$ and $\gamma_{\eta k} = \infty$. It is seen that with an increase of $\beta_{\xi k}$ from 0 to 0.1, the system damping factor decreases in the range of $\omega/\omega_k = 1$ to 9.5 but increases in the range of $\omega/\omega_k > 9.5$. But it decreases in the whole range with further increase of $\beta_{\xi k}$. It is interesting to note that with $\beta_{\xi k} = 0.3$, the system becomes unstable at $\omega/\omega_k \approx 3$, but regains stability at $\omega/\omega_k \approx 12$. The author is at a loss to say if such a regaining of stability is of practical meaning. In spite of the general tendency of the negative influence of $\beta_{\xi k}$, a small value of it might be advantageous in the presence of pad inertia. Furthermore, there exists an optimum value of $\beta_{\xi k}$ in the presence of tilting stiffness $\gamma_{\xi k}$. For example, with $\gamma_{\xi k} = 0.1$, the optimum value of $\beta_{\xi k}$ is about 0.1, as shown in Fig. 6.

The influence of coefficient of pad inertia $\alpha_{\xi k}$ is shown in Fig. 7, where it is teken that $\alpha_\eta = \alpha_\kappa = \beta_\eta = \beta_\xi = \gamma_\xi = 0$ and $\gamma_{\eta k} = \infty$. It may be seen that the system damping factor decreases with increasing α_ξ. With $\alpha_{\xi k}=0.005$, although the system damping factor of the system in the 1st mode (so called "rotor determined" mode) is positive in the whole calculated speed range, that in the 2nd mode (so called "pad determined" mode) passes into a negative value at $\omega/\omega_k \approx 12.5$. The 2nd mode loses its stability earlier with greater $\alpha_{\xi k}$, while the 1st mode has also lower system damping factor. With $\alpha_{\xi k}=0.04$, the 1st mode loses stability at a speed much lower than that of the 2nd mode. But, with $\alpha_{\xi k}<0.001$, the influence of pad inertia is negligible.

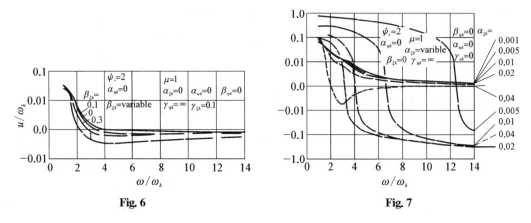

Fig. 6　　　　　　　　　　　Fig. 7

This tendency of the influence of $\alpha_{\xi k}$ shows itself also in the presence of small values of tilting stiffness $\gamma_{\xi k}$ and tilting damping $\beta_{\xi k}$.

When pad inertia is taken into consideration, small values of tilting stiffness or damping can increase the stability of the 2nd mode, and might also be advantageous to the 1st mode, as may be seen from Figs. 8 and 9 which show the results of calculation for a rigid rotor.

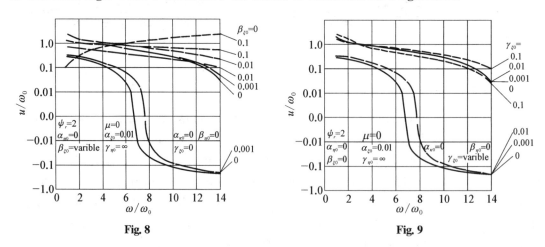

Fig. 8　　　　　　　　　　　Fig. 9

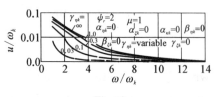

Fig. 10

The influence of coefficient of pad radial stiffness $\gamma_{\eta k}$ is shown in Fig. 10, where it is taken that $\alpha_\eta=\alpha_\xi=\alpha_\kappa=\beta_\eta=\beta_\xi=\gamma_\xi=0$. It may be seen that the system damping factor is the highest with radially rigid pivot ($\gamma_{\eta k}=\infty$), and decreases with decreasing $\gamma_{\eta k}$ but never becomes negative.

The influence of coefficient of pivot radial damping $\beta_{\eta k}$ for an radially elstic pivot is shown in Figs. 11 and 12, where it is taken that $\alpha_\eta = \alpha_\xi = \alpha_\kappa = \beta_\xi = \gamma_\xi = 0$ and $\gamma_{\xi k} = 3$ and 0.3 respectively. It may be seen that there exists an optimum value of $\beta_{\eta k}$, and it is interesting to note that the optimum ratio of $\beta_{\eta k}/\gamma_{\eta k}$ for the two calculated cases are both approximately 1.

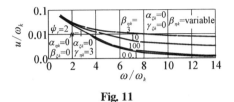
Fig. 11

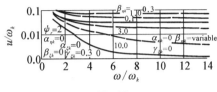
Fig. 12

A very attractive effect of $\beta_{\eta k} = \gamma_{\eta k} = 0.3$ is shown in Fig. 13 in comparison with a "common" tilting-pad bearing with rigid pivots. The system damping factor is significantly increased in the whole calculated range of speed.

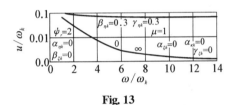
Fig. 13　　　　Fig. 14

The influence of coefficient of pad mass $\alpha_{\eta k}$ for elastic pivots is shown in Fig. 14, where it is taken that $\alpha_\xi = \alpha_\zeta = \beta_\eta = \beta_\xi = \gamma_\xi = 0$ and $\gamma_{\xi k} = 0.3$. It may be seen that the system damping factor decreases with increasing $\alpha_{\eta k}$, although it never becomes negative. But such effect is negligible for $\alpha_{\eta k} < 0.01$.

A few calculations have also been made to show the influence of coefficient of offset of pad mass center α_κ for elastic pivots and rigid rotor (Fig. 15), taking $\beta_\eta = \beta_\xi = \gamma_\xi = 0$ and $\gamma_{\eta 0} = 3$. α_ξ/α_η is taken as 0.1514, which corresponds approximately to a pad thickness about 1/10 of pad outer radius R_A. It may be seen that for $\alpha_{\eta 0} < 0.01$ the influence of $\alpha_\kappa/\alpha_\eta \approx 0.125$ is negligible. But with large pad mass ($\alpha_{\eta 0} = 0.05$), a positive value of $\alpha_{\eta 0}(\alpha_\kappa/\alpha_\eta = 0.125)$ is advantageous.

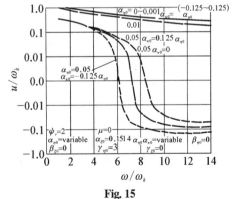
Fig. 15

5　Conclusions

Pad inertia and mass have always negative influence on the system damping factor. In general, the tilting stiffness and damping of pivot heve also negative influence, but small values of them can increase the system damping factor of the 2nd mode, compensating the negative influence of pad inertia. Radial elasticity of pivot without accompanying damping is disadvantageous, but a good match of the two can greatly increase the system damping factor.

Nomenclature

a = Offset of center of mass of pad
B = bearing width
c = bending stiffness of shaft
c_A = radial stiffness of pivot
c_{ikj} ($i, k = \xi, \eta$) = stiffness coefficients of oil film of "fixed" pad j
D = bearing diameter
d_A = radial damping of pivot
d_{ikj} = damping coefficients of oil film of fixed pad j ($i, k = \xi, \eta$)
e = distance between journal center and bearing center
F_{stat} = static bearing load
F_x, F_y = oil film forces of a bearing in x and y directions
$F_{\xi j}$, $F_{\eta j}$ = oil film forces of pad j in ξ and η direotions
f = static deflection of shaft
k_A = angular stiffness coefficient of pivot
m = mass of rotor
m_A = mass of pad
q_A = angular damping coefficient of pivot
$\Delta R = D/2 - r$, smallest bearing clearance
R_A = outer radius of pad
R_S = inner radius of pad
$\Delta R_A = R - r$, pad clearance
r = radius of journal
$So = F_{stat} \psi^2 / (BD\eta_0 \omega)$, Sommerfeld number of bearing
So_k, So_0 = Sommerfeld numbers with reference to ω_k and ω_0 respectively
So_{sj} = Sommerfeld number of pad
X, Y = dimensionless values of x and y: $X = x/\Delta R$; $Y = y/\Delta R$
x, y = horizontal and vertical distances of journal center to its static equilibrium position

x_R, y_R = the same, of middle point of rotor
$\alpha_\kappa = \dfrac{\psi^3 \omega}{2B\eta_0} m_A \dfrac{a}{R_A}$, dimensionless offset of center of mass of pad
$\alpha_\xi = \dfrac{\psi^3 \omega}{2B\eta_0} \cdot \dfrac{\Theta_s}{R_A^2}$, dimensionless moment of inertia of pad
$\alpha_\eta = \dfrac{\psi^3 \omega}{2B\eta_0} m_s$, dimensionless mass of pad
$\beta_{ik}^* = \dfrac{\psi^3}{2B\eta_0} d_{ik}$ ($i, k = \xi, \eta$), dimensionless damping coefficients of "fixed" pad j
$\beta_\xi = \dfrac{\psi^3}{2B\eta_0} \cdot \dfrac{q_A}{R_A^2}$, dimensionless angular damping coefficient of pivot
$\beta_\eta = \dfrac{\psi^3}{2B\eta_0} d_A$, dimensionless radial damping coefficient of pivot
$\gamma_{ik}^* = \dfrac{\psi^3}{2B\eta_0 \omega} c_{ik}$ ($i, k = \xi, \eta$), dimensionless stiffness coefficients of "fixed" pad
$\gamma_\xi = \dfrac{\psi^3}{2B\eta_0 \omega} \cdot \dfrac{k_A}{R_A^2}$, dimensionless angular stiffness coefficient of pivot
$\gamma_\eta = \dfrac{\psi^2}{2B\eta_0 \omega} c_A$, dimensionless radial stiffness coefficient of pivot
= angle of tilting of pad j relative to its static equilibrium position
$\varepsilon = e/\Delta R$, relative eccentricity
Θ_A = moment of inertia of pad around its pivot
$\mu = f/\Delta R$, relative elasticity of shaft
$\Pi_i = F_i \psi^2 / (BD\eta_\xi \omega)$ ($i = x, y, \xi, \eta$), dimensionles oil film forces
φ_{kj} = angle between y axis and pivot j
$\psi = \Delta R/r$, clearance ratio of bearing
$\psi_r = \Delta R_s / \Delta R$, relative clearance of pad
η_0 = dynamic viscosity of lubricant under reference

temperature

ξ_j, η_j = coordinates fixed on pad j, η_j is directed from pad center to pivot, and ξ_j is perpendicular to η_j

$\Delta\xi_j$, $\Delta\eta_j$ = displacements of journal center from its static position of equilibrium in (ξ_j, η_j) coordinates

$$\Delta\xi_j = \xi_j - \xi_{\text{statj}}$$
$$\Delta\eta_j = \eta_j - \eta_{\text{statj}}$$

$\bar{\xi}_j$, $\bar{\eta}_j$ = dimensionless values of $\Delta\xi_j$ and $\Delta\eta_j$

$$\bar{\xi}_j = \Delta\xi_j/\Delta R$$
$$\bar{\eta}_j = \Delta\eta_j/\Delta R$$

ξ_{sj}, η_{sj} = space-fixed coordinates, coinciding with ξ_j and η_j in their static equilibrium positions; also, displacements of pad center from its static equilibrium position in these directions

ω = angular speed of shaft

$\omega_k = \sqrt{c/m}$, natural angular frequency of rigidly supported rotor

$\omega_0 = \sqrt{g/\Delta R}$, reference angular frequency for rigid rotor

References

[1] Glienicke, J.: Fortsch.-Ber. VDI-Z Reihe 11 Nr. 13, Duesseldorf 1972.

[2] Han, D. C.: Dissertation, Universitaet Karlsruhe (TH), 1979.

[3] Streetz, W.: Konstruktion 31 (1979) H. 8, S. 321–324.

[4] Glienicke, J., Ehinger, M. und Hunger, H.: MTZ Motortechnische Zeitschrift 42 (1981) 12, S. 531–536.

Appendix

A Numerical Method of Finding the Coefficients of the Characteristic Polynomial

Taking $\lambda=0$ and calculating the value of the determinant $\underline{K}$, we get A_0:

$$A_0 = |\underline{K}(0)|$$

Take successively

$$\lambda_\xi = S e^{q(2\pi/\omega)} \quad (q=0, 1, 2, \cdots, n/2)$$

where S—a real value, n—always an even number in our problem. We get correspondingly complex values of $|\underline{K}|$, which we denote by D_q:

$$|\underline{K}(\lambda_q)| = D_q$$

Let $R_q = D_q - A_0$ and

$$E_q = \text{Re}(R_q), \quad F_q = \text{Im}(R_q)$$

We can get all the coefficients $A_1 \sim A_n$ from the following equations:

$$A_j = \frac{1}{nS^j} \Big\{ E_0 + (-1)^j E_{n/2} + 2 \sum_{q=1}^{\frac{n}{2}-1} [E_q \cos(2\pi jq/n) + F_q \sin(2\pi jq/n)] \Big\} \quad (j=1, 2, \cdots, n/2-1)$$

$$A_{n-2} = \frac{1}{nS^{n/2}} \Big[E_0 + (-1)^{n/2} E_{n/2} + 2 \sum_{q=1}^{\frac{n}{2}-1} (-1)^q E_q \Big]$$

$$A_{n-j} = \frac{1}{nS^{n-j}} \left\{ E_0 + (-1)^j E_{n/2} + 2 \sum_{q=1}^{\frac{n}{2}-1} [E_q \cos(2\pi jq/n) - F_q \sin(2\pi jq/n)] \right\} \quad (j = 1, 2, \cdots, n/2 - 1)$$

$$A_n = \frac{1}{nS^n} \left(E_0 + E_{n/2} + 2 \sum_{q=1}^{\frac{n}{2}-1} E_q \right)$$

The recommended value of S is the geometric mean of all the roots of the characteristic equation, which is

$$M = (|A_0/A_n|)^{1/n}$$

It is usually very easy to calculate the value of A_n. Otherwise, it is also easy to perform 1 or 2 steps of iteration to get a value of S sufficiently near M.

The coefficients calculated with this S are generally accurate enough.

Acknowledgements

The author wish to thank Professor Dr.-Ing. Jaochim Glienicke for his stimulation, support and valuable advices to this work, and Dr.-Ing. D. C. Han for his supplying the table of dynamic properties of "fixed" tilting pad and advices.

The Effect of Dynamic Deformation on Dynamic Properties and Stability of Cylindrical Journal Bearings*

Abstract: Reynolds equation and deformation equation are derived in complex form for the dynamic increment of oil film pressure and bearing surface dynamic deformation. The complex amplitudes of dynamic pressure and deformation differentiated with respect to journal motions are solved by numerical methods with iterations between them, and the whirl ratio at stability threshold is found by further iterations, wherefrom the dynamic properties relevant to stability threshold are calculated and from them the stability characteristics. It is concluded that the effect of dynamic deformation on bearing dynamic properties and stability is unnegligible for softer bearing materials. It is also shown that the two cross damping coefficients differ from each other, instead of being equal as predicted by the theory for stationary contours.

1 Introduction

Static properties of journal bearings under EHL conditions have been investigated by various researchers in the last twenty years[1,2]. But dynamic properties have mostly been calculated taking into consideration only of static deformation of the bearing[2-4]. Linear deformation model was used in [5] to account for the effect of dynamic deformation on dynamic properties, but its effect on stability threshold is still to be investigated, and more elaborate deformation model might also be applied. These are attempted in this paper.

1.1 Notation

B	Nondimensional damping coefficient, $B = b\psi^3/(\mu l)$
b	Damping coefficient
d	Bearing diameter
c	Radial clearance
$\bar{E}$	Nondimensional elasticity modulus, $\bar{E} = E\psi/(2\mu\omega_0)$
E	Young's modulus
H	Nondimensional film thickness, $H = h/c$
h	Film thickness
K	Nondimensional stiffness coefficient, $K = k\psi^3/(\mu\omega_0 l)$

---

* In collaboration with Q. Mao and H. Hu. Reprinted from *Proceedings of the 13th Leeds-Lyon Symposium on Tribology*, 1986, Paper XI(iv): 363–366.

k	Stiffness coefficient
l	Bearing width
M_{cr}	Nondimensional critical mass of rigid rotor, $M_{cr} = m_{cr}\omega_0\psi^3/(\mu l)$
m_{cr}	Critical mass of rigid rotor
P	Nondimensional pressure, $P = p\psi^2/(2\mu\omega_0)$
p	Oil film pressure
T	Nondimensional time or angle of rotation, $T = t\omega_0$
t	Time
z	Axial coordinate measured from middle plane of bearing width
ε_0	Eccentricity ratio at static equilibrium
λ	Nondimensional axial coordinate, $\lambda = z/(l/2)$
θ_0	Attitude angle at static equilibrium
μ	Dynamic viscosity of lubricant
φ	Angular coordinate, measured from maximum clearance
ψ	Clearance ratio, $\psi = c/(d/2)$
Ω	Ratio of excitation frequency to running frequency, $\Omega = \omega/\omega_0$
ω	Excitation frequency
ω_0	Running frequency

2 Theory

The characteristic motion of the journal center on stability threshold is a stationary whirl on a closed elliptical orbit around the static equilibrium position of the journal center, with a definite whirl ratio expressing the ratio of whirl frequency to running frequency. In order to be able to predict the threshold speed, the eight dynamic coefficients of the oil film corresponding to such a characteristic motion should first be calculated, wherefore all the basic relationships concerning the dynamic film pressure and the dynamic deformation of the bearing should be derived relevant to such a state of motion.

The nondimensional form of Reynolds equation may be written as:

$$\frac{\partial}{\partial \varphi}\left(H^3 \frac{\partial P}{\partial \varphi}\right) + \left(\frac{d}{l}\right)^2 \frac{\partial}{\partial \lambda}\left(H^3 \frac{\partial P}{\partial \lambda}\right) = 3\frac{\partial H}{\partial \varphi} + 6\frac{\partial H}{\partial T} \quad (1)$$

Fig. 1 shows the journal at its static equilibrium position relative to the bearing. If the journal is excited into a simple harmonic motion of small amplitude, the instantaneous eccentricity ratio and attitude angle may be expressed respectively as:

$$\varepsilon = \varepsilon_0 + \Delta\varepsilon\, e^{i\Omega T} \text{ and } \theta = \theta_0 + \Delta\theta\, e^{i\Omega T} \quad (2)$$

where $\Delta\varepsilon$ and $\Delta\theta$ are complex amplitudes of eccentricity

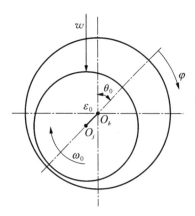

Fig. 1 End view of bearing and journal

ratio and attitude angle respectively. The corresponding dynamic increment of film thickness consists of two components, one of which is that caused by the variation of the geometric position of the journal center:

$$H_d = (\Delta\varepsilon\cos\varphi + \varepsilon_0\Delta\theta\sin\varphi)e^{i\Omega T} \tag{3}$$

the other is the dynamic deformation caused by the dynamic increment of film pressure:

$$H_\delta = \iint_A Q_\delta(\varphi_1, \lambda_1) f(\varphi, \lambda, \varphi_1, \lambda_1) d\varphi_1 d\lambda_1 \tag{4}$$

where $f(\varphi, \lambda, \varphi_1, \lambda_1)$ is the flexibility coefficient expressing the radial deformation at point (φ, λ) caused by unit pressure at point (φ_1, λ_1); $Q_\delta(\varphi_1, \lambda_1)$ is the dynamic increment of film pressure at point (φ_1, λ_1); A denotes the pressure domain. Under small simple harmonic vibration, Q_δ can also be expressed in simple harmonic form:

$$Q_\delta = Q e^{i\Omega T} \tag{5}$$

The instantaneous value of film thickness may therefore be expressed as:

$$H = H_0 + H_d + H_\delta = H_0 + (\Delta\varepsilon\cos\varphi + \varepsilon_0\Delta\theta\sin\varphi + H_\Delta)e^{i\Omega T} \tag{6}$$

where H_0 is the static film thickness including the static deformation of the bearing:

$$H_0 = 1 + \varepsilon_0\cos\varphi + \iint_A P_0(\varphi_1, \lambda_1) f(\varphi, \lambda, \varphi_1, \lambda_1) d\varphi_1 d\lambda_1 \tag{7}$$

and H_Δ is the complex amplitude of the dynamic deformation:

$$H_\Delta = \iint_A Q(\varphi_1, \lambda_1) f(\varphi, \lambda, \varphi_1, \lambda_1) d\varphi_1 d\lambda_1 \tag{8}$$

The inclusion of the dynamic deformation H_δ in equation (6) makes the expression more complete than if only static deformation is taken into consideration.

The instantaneous value of the resultant film pressure is then expressed as:

$$P = P_0 + Q_\delta = P_0 + Q e^{i\Omega T} \tag{9}$$

where P_0 denotes the static component of the film pressure.

Substituting equations (6) and (9) into equation (1), separating the time dependent part from the stationary one, and neglecting small terms of higher order, we obtain

$$\text{Reyn}(P_0) = 3\frac{\partial H_0}{\partial \varphi} \tag{10}$$

and

$$\text{Reyn}(Q) = -9\frac{M}{H_0}\frac{\partial H_0}{\partial \varphi} + 3\frac{\partial M}{\partial \varphi} + 6i\Omega M - 3H_0\left[\left(H_0\frac{\partial M}{\partial \varphi} - M\frac{\partial H_0}{\partial \varphi}\right)\frac{\partial P_0}{\partial \varphi}\right.$$
$$\left. + \left(\frac{d}{l}\right)^2\left(H_0\frac{\partial M}{\partial \lambda} - M\frac{\partial H_0}{\partial \lambda}\right)\frac{\partial P_0}{\partial \lambda}\right] \tag{11}$$

where the operator Reyn( ) denotes

$$\frac{\partial}{\partial \varphi}\left[H_0^3 \frac{\partial(\ )}{\partial \varphi}\right] + \left(\frac{d}{l}\right)^2 \frac{\partial}{\partial \lambda}\left[H_0^3 \frac{\partial(\ )}{\partial \lambda}\right]$$

and $M = \Delta\varepsilon \cos\varphi + \varepsilon_0 \Delta\theta \sin\varphi + H_\Delta$.

For each eccentricity ratio, simultaneous equations (7) and (10) are solved to obtain the static pressure P_0, static film thickness H_0 and attitude angle θ_0. In order to calculate the dynamic coefficients of the oil film, equations (11) and (8) are first partially differentiated with respect to perturbation $\Delta\varepsilon$, to obtain

$$\text{Reyn}(Q_\varepsilon) = -9\frac{\cos\varphi + H_{\Delta\varepsilon}}{H_0}\frac{\partial H_0}{\partial \varphi} + 3\left(-\sin\varphi + \frac{\partial H_{\Delta\varepsilon}}{\partial \varphi}\right)$$
$$+ 6i\Omega(\cos\varphi + H_{\Delta\varepsilon}) - 3H_0\left\{\left[H_0\left(-\sin\varphi + \frac{\partial H_{\Delta\varepsilon}}{\partial \varphi}\right)\right.\right.$$
$$\left.- (\cos\varphi + H_{\Delta\varepsilon})\frac{\partial H_0}{\partial \varphi}\right]\frac{\partial P_0}{\partial \varphi} + \left(\frac{d}{l}\right)^2\left[H_0\frac{\partial H_{\Delta\varepsilon}}{\partial \lambda}\right.$$
$$\left.\left.- (\cos\varphi + H_{\Delta\varepsilon})\frac{\partial H_0}{\partial \lambda}\right]\frac{\partial P_0}{\partial \lambda}\right\} \quad (12)$$

and

$$H_{\Delta\varepsilon} = \iint_A Q_\varepsilon(\varphi_1, \lambda_1) f(\varphi, \lambda, \varphi_1, \lambda_1) d\varphi_1 d\lambda_1 \quad (13)$$

where $Q_\varepsilon = \partial Q/\partial \varepsilon$; $H_{\Delta\varepsilon} = \partial H_\Delta/\partial \varepsilon$.

The complex distribution Q_ε is solved from the simultaneous equations (12) and (13) by iterations. Stiffness coefficients K_{ee}, $K_{\theta e}$ and damping coefficients B_{ee}, $B_{\theta e}$ are then calculated by integrations:

$$-\iint_A Q_\varepsilon \cos\varphi d\varphi d\lambda = K_{ee} + i\Omega B_{ee}$$
$$-\iint_A Q_\varepsilon \sin\varphi d\varphi d\lambda = K_{\theta e} + i\Omega B_{\theta e} \quad (14)$$

Similarly, partial differentiations with respect to $\Delta\theta$ are made to get

$$\text{Reyn}(Q_\theta) = -9\frac{\sin\varphi + H_{\Delta\theta}}{H_0} + 3\left(\cos\varphi + \frac{\partial H_{\Delta\theta}}{\partial \varphi}\right)$$
$$+ 6i\Omega(\sin\varphi + H_{\Delta\theta}) - 3H_0\left\{\left[H_0\left(\cos\varphi + \frac{\partial H_{\Delta\theta}}{\partial \varphi}\right)\right.\right.$$
$$\left.- \frac{\partial H_0}{\partial \varphi}(\sin\varphi + H_{\Delta\theta})\right]\frac{\partial P_0}{\partial \varphi} + \left(\frac{d}{l}\right)^2\left[H_0\frac{\partial H_{\Delta\theta}}{\partial \lambda}\right.$$
$$\left.\left.- \frac{\partial H_0}{\partial \lambda}(\sin\varphi + H_{\Delta\theta})\right]\frac{\partial P_0}{\partial \lambda}\right\} \quad (15)$$

$$H_{\Delta\theta} = \iint_A Q_\theta(\varphi_1, \lambda_1) f(\varphi, \lambda, \varphi_1, \lambda_1) d\varphi_1 d\lambda_1 \quad (16)$$

where $Q_\theta = \dfrac{1}{\varepsilon_0}\dfrac{\partial Q}{\partial \theta}$.

The complex distribution Q_θ is solved from equations (15) and (16). Stiffnesses $K_{e\theta}$, $K_{\theta\theta}$ and dampings $B_{e\theta}$, $B_{\theta\theta}$ are calculated by integrations:

$$-\iint_A Q_\theta \cos\varphi \, d\varphi \, d\lambda = K_{e\theta} + i\Omega B_{e\theta}$$
$$-\iint_A Q_\theta \sin\varphi \, d\varphi \, d\lambda = K_{\theta\theta} + i\Omega B_{\theta\theta} \tag{17}$$

The dynamic coefficients are then transformed into Cartesian coordinates to obtain

$$\begin{bmatrix} K_{yy} & K_{yx} \\ K_{xy} & K_{xx} \end{bmatrix} = \begin{bmatrix} \cos\theta_0 & -\sin\theta_0 \\ \sin\theta_0 & \cos\theta_0 \end{bmatrix} \begin{bmatrix} K_{ee} & K_{e\theta} \\ K_{\theta e} & K_{\theta\theta} \end{bmatrix} \begin{bmatrix} \cos\theta_0 & \sin\theta_0 \\ -\sin\theta_0 & \cos\theta_0 \end{bmatrix} \tag{18}$$

Principally, the dynamic coefficients will depend on the excitation frequency ratio Ω. The procedure of calculation of the stability threshold speed should therefore be a little different from that for nondeformable bearings. A value of Ω should first be estimated, the eight dynamic coefficients are then calculated, from which the equivalent stiffness coefficient K_{eq} and whirl ratio γ_{st} at stability threshold are obtained[6]:

$$K_{eq} = \frac{K_{xx}B_{yy} + K_{yy}B_{xx} - K_{xy}B_{yx} - K_{yx}B_{xy}}{B_{xx} + B_{yy}} \tag{19}$$

$$\gamma_{st}^2 = \frac{(K_{eq} - K_{xx})(K_{eq} - K_{yy}) - K_{xy}K_{yx}}{B_{xx}B_{yy} - B_{xy}B_{yx}} \tag{20}$$

From them, the nondimensional critical mass M_{cr} representing also the nondimensional threshold speed of a rigid rotor is obtained:

$$M_{cr} = K_{eq}/\gamma_{st}^2 \tag{21}$$

If the calculated value of whirl ratio does not coincide with the estimated excitation ratio, a new estimation should be made and the calculation repeated, until coincidence is obtained to a sufficient accuracy. It has been seen in the calculated cases that the values of the dynamic coefficients vary but little with excitation ratio between 0.5 and 1, so that the above said procedure converges rapidly. The values of the eight dynamic coefficients, the equivalent stiffness and whirl ratio at stability threshold are thus obtained, and they can be used to predict the stability threshold speed.

In the calculations, the oil film domain is discretized, the Reynolds equations are substituted by systems of finite difference equations and solved by SOR method. The flexibility coefficient $f(\varphi, \lambda, \varphi_1, \lambda_1)$ is obtained in discrete form by applying SAP5 with subsequent reduction. Iterations between the Reynolds equations and the deformation calculations are performed to obtain the simultaneous solutions for oil film pressure (or its derivatives) and bearing deformation (or its derivatives).

3 Results of Calculations

Calculations have been made for a cylindrical bearing with $l/d=0.6$, $\varepsilon_0=0.1-0.95$, $\bar{E}=200$ and 20 (corresponding approximately to bronze and nylon respectively), Poisson's ratio $\nu=0.3$, and ratio of outer to inner diameter of the bush $D/d=1.6$. The bearing bush is supposed to be uniformly and rigidly supported on its outer surface.

Results have been obtained both when only static deformation is considered and when dynamic deformation is also included. The results in the first case correspond well with [4]. The results of the two cases differ significantly from each other for the softer material ($\bar{E}=20$), they are therefore shown here.

Fig. 2 shows the comparison of the four stiffness coefficients. Fig. 3 shows that of the four damping coefficients, where it may be seen that the deviations are considerable. It is noteworthy that the two cross dampings differ from each other significantly when dynamic deformation is taken into consideration, in contrast to the theoretical predictions for stationary bearing contours[7], which might well be one of the chief causes of the unequal values of the two cross dampings determined experimentally under certain circumstances.

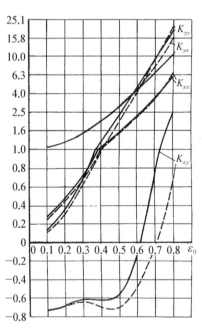

Fig. 2 Nondimensional stiffness coefficients
—— static and dynamic deformation
- - - - static deformation only

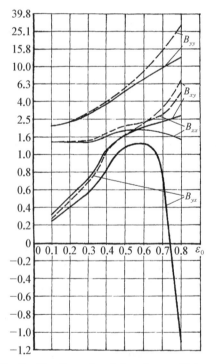

Fig. 3 Nondimensional damping coefficients
—— static and dynamic deformation
- - - - static deformation only

Fig. 4 shows that of the equivalent stiffness coefficient K_{eq}, the whirl ratio γ_{st} and the nondimensional critical mass M_{cr} of a rigid rotor. It may be seen that significant underestimation of stability threshold speed may result if the effect of dynamic deformation is neglected, especially if the material is soft and the eccentricity ratio is large.

4 Conclusions

(1) The effect of dynamic deformation of bearing on dynamic properties and stability threshold speed is unnegligible for softer bearing materials.

(2) The deviations of the damping coefficients from neglecting dynamic deformation is more pronounced than that of the stiffness coefficients.

(3) Cross damping coefficients differ from each other when dynamic deformation is taken into consideration.

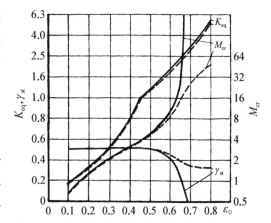

Fig. 4 Nondimensional equivalent stiffness, threshold whirl ratio and critical mass of rigid rotor

—— static and dynamic deformation
– – – static deformation only

Acknowledgement

The authors gratefully acknowledge the assistance of Mr. Yu Li in formulation of the computer programs and his valuable advices.

References

[1] Higginson, G. R. The theoretical effects of elastic deformation of the bearing liner on journal bearing performance. Proc. Symposium on Elastohydrodynamic Lubrication, Inst. Mech. Engrs., 1965, 180 (Part 3B): 31–38.

[2] Oh, K. P. and Hueber, K. H. Solution of the elastohydrodynamic finite journal bearing problem. Trans. ASME, J. Lub. Techn., 1973, 95: 342–352.

[3] Jain, S. C., Sinhasen, R. and Singh, D. V. A study of elastohydrodynamic lubrication of a centrally loaded 120° are partial bearing in different flow regimes. Proc. Inst. Mech. Engrs., 1983, 197(Part C): 97–108.

[4] Mao, Q., Han, D. C. and Glienicke, J. Stabilitaetseigenschaften von Gleitlagern bei Beruecksichtigung der Lagerelastizitaet. Konstruktion 35 (1983) H. 2, S. 45–52.

[5] Nilsson, L. R. K. The influence of bearing flexibility on the dynamic performance of radial oil film bearings. Proc. Fifth Leeds-Lyon Symposium on Tribology, 1978, Paper IX(i): 311–319.

[6] Zhang, Z. Theory of lubrication of hydrodynamic bearings. Feb. 1979, Jiaotong University, Shanghai (in Chinese).

[7] Shang, L. A matrix method for computing the dynamic characteristic coefficients of hydrodynamic journal bearings. Journal of Zhejiang University, 1984, 18(3): 126–135 (in Chinese).

复套式转子-滑动轴承系统动力学计算*

摘　要：本文介绍一种计算复套式转子-滑动轴承系统固有复频率的方法.用正向和反向隐式传递矩阵法建立特征行列式,用有限复平面轨迹显示及三角元折半法搜索特征根,形成了相应的有效方法和保证计算精度的程序.用锤击模型试验验证了计算,获得满意的符合.

一、前言

现代某些超大功率齿轮箱中,为了增大输入、输出轴的扭转柔度以减小冲击和振动的传递,同时又不加长轴以节省传动装置的轴向地位,采用了复套式转子(图1).其中细而长的挠性传动轴穿过空心的齿轮轴,内外轴在右端相连,内轴的左端则与外界相连传递扭矩.外轴有两个滑动轴承,内轴左端有一个滑动轴承.这样就使转子有较大的扭转柔度,套轴又有足够的弯曲刚度来支承齿轮以确保良好啮合,整体结构紧凑.

现有的传递矩阵法转子动力学程序[1],多数不能直接用来计算这种转子-轴承系统的固有复频率.为此,作者拟订了相应的计算方法和程序,并在简单的模型上作了实验验证.

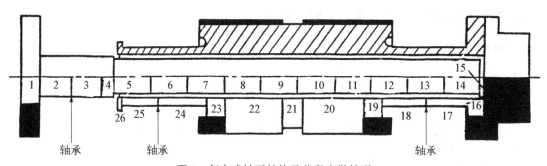

图1　复套式转子结构及分段离散情况

二、计算方法

1. 特征行列式的建立

将转子离散化为若干段等直径的分段,并将其质量集中置于该分段两端节点上.每个节点上,有八个分量构成的状态矢量:

$$R_j = (x, \phi, M, S, y, \psi, N, Q)_j^T \tag{1}$$

---

* 本文合作者:王云根.原发表于《第二届全国转子动力学学术会议论文集》,1989:89-95.

式中(x,ϕ,S,M)和(y,ψ,N,Q)分别为水平和垂直平面内的(位移,转角,弯矩,切力).

在内轴左端面,弯矩和切力均为零,可记为:

$$R_0 = B_0 \cdot (x_0,\phi_0,y_0,\psi_0)^T = B_0 \cdot R_0^* \tag{2}$$

式中

$$B_0 = \begin{bmatrix} 1 & 0 & 0 & 0 \\ 0 & 1 & 0 & 0 \\ 0 & 0 & 0 & 0 \\ 0 & 0 & 0 & 0 \\ 0 & 0 & 1 & 0 \\ 0 & 0 & 0 & 1 \\ 0 & 0 & 0 & 0 \\ 0 & 0 & 0 & 0 \end{bmatrix} \tag{3}$$

$$R_0^* = (x_0,\phi_0,y_0,\psi_0)^T$$

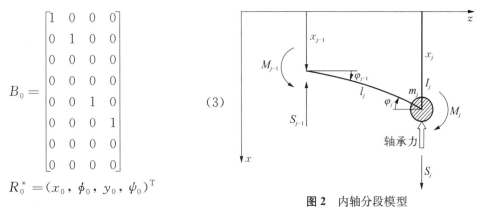

图 2 内轴分段模型

传递过程从内轴左端开始,逐个分段,用该段左端状态量 R_{j-1} 表达右端状态量 R_j. 到内轴右端后,进入外轴反向传递,直到其左端为止.对于每一分段,有:

$$R_j = A_j \cdot R_{j-1} \quad (j = 1 \sim n) \tag{4}$$

式中 n——转子分段数.内轴分段(图2)的传递矩阵为:

$$A_j = \begin{bmatrix} 1 & l_j & b_j & -c_j & 0 & 0 & 0 & 0 \\ 0 & 1 & a_j & -b_j & 0 & 0 & 0 & 0 \\ 0 & I_j\nu^2 & 1+a_jI_j\nu^2 & -l_j-b_jI_j\nu^2 & 0 & 0 & 0 & 0 \\ F_j & l_jF_j & b_jF_j & 1-c_jF_j & G_j & l_jG_j & b_jG_j & -c_jG_j \\ 0 & 0 & 0 & 0 & 1 & l_j & b_j & -c_j \\ 0 & 0 & 0 & 0 & 0 & 1 & a_j & -b_j \\ 0 & 0 & 0 & 0 & 0 & I_j\nu^2 & 1+a_jI_j\nu^2 & -l_j-b_jI_j\nu^2 \\ H_j & l_jH_j & b_jH_j & -c_jH_j & K_j & l_jK_j & b_jK_j & 1-c_jK_j \end{bmatrix}$$

(5)

其中 ν——复特征值,即复频率;l_j——分段长;I_j——轴质惯矩;$a_j = l_j/(EJ_j)$;$b_j = l_j^2/(2EJ_j)$;$c_j = l_j^3/(6EJ_j)$;E——弹性模量;J_j——截面轴惯矩;m_j——集中质量(包括轴段的及附加的);$G_j = b_{xy}\nu + k_{xy}$;$F_j = m_j\nu^2 + b_{xx}\nu + k_{xx}$;$H_j = b_{yx}\nu + k_{yx}$;$K_j = m_j\nu^2 + b_{yy}\nu + k_{yy}$;$b_{ij}(i,j = x,y)$——轴承阻尼系数;$k_{ij}(i,j = x,y)$——轴承刚度系数.

在内轴的末一分段右端(图3),计入内、外轴的联接部分结构后,有外悬的集中质量(外悬距 e),且发生传递方向的改变.其传递矩阵为:

$$A_j = \begin{bmatrix} 1 & l_j & b_j & -c_j \\ 0 & 1 & a_j & -b_j \\ m_j e\nu^2 & (I_j + m_j e^2 + m_j l_j e)\nu^2 & 1 + (a_j I_j + b_j m_j e)\nu^2 & -l_j - (b_j I_j + c_j m_j e)\nu^2 \\ F_j & l_j F_j + m_j e\nu^2 & b_j F_j + a_j m_j e\nu^2 & 1 - c_j F_j - b_j m_j e\nu^2 \\ 0 & 0 & 0 & 0 \\ 0 & 0 & 0 & 0 \\ 0 & 0 & 0 & 0 \\ H_j & l_j H_j & b_j H_j & -c_j H_j \\ 0 & 0 & 0 & 0 \\ 0 & 0 & 0 & 0 \\ 0 & 0 & 0 & 0 \\ G_j & l_j G_j & b_j G_j & -c_j G_j \\ 1 & l_j & b_j & -c_j \\ 0 & 1 & a_j & -b_j \\ m_j e\nu^2 & (I_j + m_j e^2 + m_j l_j e)\nu^2 & 1 + (a_j I_j + b_j m_j e)\nu^2 & -l_j - (b_j I_j + c_j m_j e)\nu^2 \\ K_j & l_j K_j + m_j e\nu^2 & b_j K_j + a_j m_j e\nu^2 & 1 - c_j K_j - b_j m_j e\nu^2 \end{bmatrix}$$

(6)

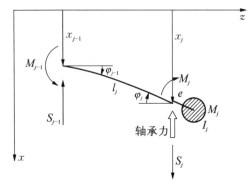 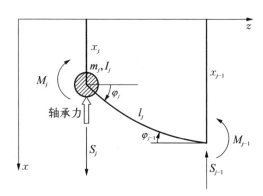

图 3 内轴末分段模型　　　　**图 4** 外轴分段模型

外轴轴段(图 4)则为由右向左传递,其传递矩阵为:

$$A_j = \begin{bmatrix} 1 & -l_j & -b_j & -c_j & 0 & 0 & 0 & 0 \\ 0 & 1 & a_j & b_j & 0 & 0 & 0 & 0 \\ 0 & I_j \nu^2 & 1 + a_j I_j \nu^2 & l_j + b_j I_j \nu^2 & 0 & 0 & 0 & 0 \\ F_j & -l_j F_j & -b_j F_j & 1 - c_j F_j & G_j & -l_j G_j & -b_j G_j & -c_j G_j \\ 0 & 0 & 0 & 0 & 1 & -l_j & -b_j & -c_j \\ 0 & 0 & 0 & 0 & 0 & 1 & a_j & b_j \\ 0 & 0 & 0 & 0 & 0 & I_j \nu^2 & 1 + a_j I_j \nu^2 & l_j + b_j I_j \nu^2 \\ H_j & -l_j H_j & -b_j H_j & -c_j H_j & K_j & -l_j K_j & -b_j K_j & 1 - c_j K_j \end{bmatrix}$$

(7)

这样,任一节点的状态量均可由 R_0^* 来表达:

$$R_1 = A_1 \cdot B_0 \cdot R_0^* = B_1 \cdot R_0^* ; R_2 = A_2 \cdot B_1 \cdot R_0^* = B_2 \cdot R_0^* ; \cdots ;$$
$$R_n = A_n \cdot B_{n-1} \cdot R_0^* = B_n \cdot R_0^*.$$

每次运算均为 8 行 8 列的矩阵 A_j 与 8 行 4 列的矩阵 B_{n-1} 相乘. 矩阵的每个元素都是 ν 的多项式. 随着传递过程的进行,B_j 矩阵元素的多项式阶数便越来越高.

在外轴左端,弯矩和切力均为零. 由此,可选出 B_n 矩阵中与 M_n, S_n, N_n, Q_n 相应的第三、四、七、八行,构成 4 行 4 列的特征行列式:

$$D = \begin{bmatrix} b_{31} & b_{32} & b_{33} & b_{34} \\ b_{41} & b_{42} & b_{43} & b_{44} \\ b_{71} & b_{72} & b_{73} & b_{74} \\ b_{81} & b_{82} & b_{83} & b_{84} \end{bmatrix} \tag{8}$$

式中 b_{ij} 表示 B_n 矩阵中第 i 行第 j 列的元素.

由 $D=0$ 的条件,确定 ν 的各个复根,就得到了系统的各个复固有频率.

2. 复特征值的寻求

以 u 和 v 分别表示 ν 的实部和虚部:$\nu = u + iv$. 以转动角速度 ω 为参考量,则其相对值为 $U = u/\omega$, $V = v/\omega$. 我们对 V 值的感兴趣范围,通常在 0~3 之间,因为共振和倍频共振相应于 $V=1$ 和 2,而油膜稳定性问题则一般总在 $V<1$ 的区域内. U 的感兴趣范围取在 -2 ~ 2 之间似亦足够了,因 $U<-2$ 时系统阻尼和对数衰减率极大,这种复模态不致引起麻烦,而 $U>2$ 则意味着系统早已丧失稳定性,如果我们从必定是稳定的低角速 ω 开始计算,逐步提高 ω 算到开始失去稳定为止,一般可避免发生 $U>2$ 的极少见情况.

在 $U + iV$ 复平面的这个限定区域内,以 $\Delta U (=0.1$ 或 $0.2)$ 和 ΔV(同 ΔU)等分为网格,计算每一网点上的 D 复值. 将 Re(D) 和 Im(D) 的符号分布情况打印出来,就得到了 Re(D)=0 和 Im(D)=0 的粗略轨迹(图 5). 这两种轨迹的交点上,Re(D)=Im(D)=0,相应的 $U + iV$ 即为一复根.

为了以足够精度算出 U 和 V 的根值(我们取为 0.0001),将含有两种轨迹交点的小区域选出,然后用三角元折半法逼近到规定精度. 根据每一小单元四角网点上 Re(D) 符号和 Im(D) 符号是否均存在"+"号及"—"号,即可认为该小单元是否应属"可疑",对此即应进一步分割搜索. 这种判断法偶尔亦会误判,即漏过了含交点的小单元而抓出了其邻近单元,这是由于网点分布不很密之故. 进一步地搜索,在错抓出的小单元中将得不到根. 好在已打印了 $U + iV$ 的复平面中 Re(D) 和 Im(D) 的轨迹图,不难判断出漏检,并在旁邻区域内将交点搜出.

三角元折半搜索,如下进行(图 6). 先将抓出的小矩形单元分成两个三角形. 根据每个三角形的三个角点上 Re(D) 和 Im(D) 是否均存在"+"号及"—"号,判定其是否可疑. 将可疑三角形的斜边中点与直角顶点相连,即划分成两个三角形. 补充计算斜边点上的 D 复值,又可按上述法分别判断这两个分三角形是否可疑. 以此类推,不断进行到分三角形的边长小于规定精度,取可疑三角形中点作为复根值.

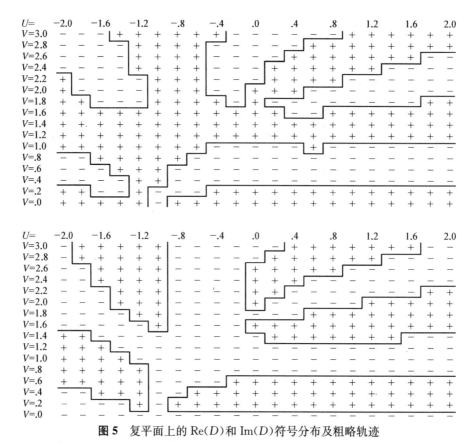

图 5 复平面上的 Re(D) 和 Im(D) 符号分布及粗略轨迹

三角元折半搜索法的优点,是每次折半细分只需补充计算一个 D 复值.

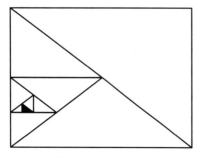

图 6 三角元折半法

3. 对称转子-轴承系统的节约处理

本文的计算法和程序,亦适用于普通非复套式的转子系统. 对于对称的普通转子系统,可以在中央截开,只取右半部计算. 这样可减少分段,大大降低 D 所含的 V 阶数,减少计算时间而提高精度.

此时应分别按"柱型"模态(左、右对称)和"锥型"模态(左、右反对称)作不同计算.

对于柱型计算,中央截分节点上 ϕ, S, ψ, Q 均为零,此时将 R_0^* 改为 $(x, M, y, N)^T$,相应的 B_0 按式(9). 对于锥型计算,中央截分节点上 x, M, y, N 均为零,此时将 R_0^* 改为 $(\phi, S, \psi, Q)^T$,相应的 B_0 按式(10).

$$B_0 = \begin{bmatrix} 1 & 0 & 0 & 0 \\ 0 & 0 & 0 & 0 \\ 0 & 1 & 0 & 0 \\ 0 & 0 & 0 & 0 \\ 0 & 0 & 1 & 0 \\ 0 & 0 & 0 & 0 \\ 0 & 0 & 0 & 1 \\ 0 & 0 & 0 & 0 \end{bmatrix} \tag{9}$$

$$B_0 = \begin{bmatrix} 0 & 0 & 0 & 0 \\ 1 & 0 & 0 & 0 \\ 0 & 0 & 0 & 0 \\ 0 & 1 & 0 & 0 \\ 0 & 0 & 0 & 0 \\ 0 & 0 & 1 & 0 \\ 0 & 0 & 0 & 0 \\ 0 & 0 & 0 & 1 \end{bmatrix} \tag{10}$$

其余计算均照旧.

4. 计算程序

为了保证计算精度,基本上采用双精度运算.过程中,将建立的特征行列式 D 记存文件中,备补充计算时调用.

经多次运算(分段数不大于30),本程序功能稳定可靠,切合实用.

三、模型试验验证

试验模型如图 7 所示.支承在三个同样刚度的弹性支座上.支座只有垂直方向的弹性,且可调节其弹性大小.支座的刚度均用静态标定.为简单起见,系统中未设阻尼.

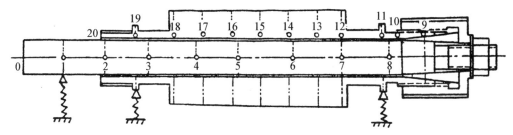

图 7 复套式模型及其离散化分段情况

用锤击法激起该系统的各阶固有振动,测试系统如图 8 所示.

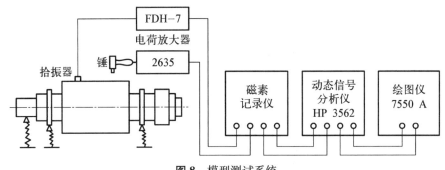

图 8 模型测试系统

图 9 示一组典型的实验结果,图(a)和(b)是敲击点与拾振点互换以作对比.这组结果表明:在 0～1 000 Hz 范围内有四个较明显的固有频率:20 Hz,42.5～45 Hz,172.5～175 Hz,

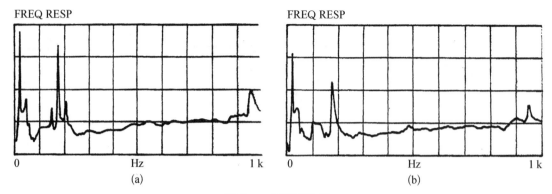

图 9 一组测试结果

950～951.2 Hz. 各组实验结果如表 1 所示.

为了排除支座可能带来的一些不确定因素,还将转子模型用绳子悬吊起来进行锤击测试,以获得它在自由—自由情况下的固有频率. 所得数据可更单纯地验证正、反传递矩阵法的可用性. 其结果亦列于表 1.

表 1 模型系统的固有频率测试结果和计算结果

支承刚度 $k/(\mathrm{N} \cdot \mathrm{m}^{-1})$	测得的固有频率/Hz	算出的固有频率/Hz	
		根据 k	设 k 与 5×10^5 N/m 串连
0 (即自由—自由)	197.5 993.5	199.4 1 082	
46 900	15.0	14.54	13.90
		29.65	28.36
	172.6	178.2	177.9
	935.0	996.3	996.3
86 900	17.5	19.78	18.26
	35.1	40.20	37.15
	176.3	180.9	180.0
	945.0	996.6	996.5
117 000	20.0	22.99	20.69
	44.0	46.58	42.01
	174.0	182.9	181.4
	951.0	996.8	996.7
168 000	23.7	27.47	23.77
	43.7	55.41	48.13
	183.7	186.3	183.5
	909.0	997.1	996.9
496 000	35.0	47.23	33.47
	71.2	92.65	67.83
	182.0	207.8	191.7
	987.0	999.3	997.7
2.15×10^6	54.95	98.15	42.74
	88.10	175.0	84.45
	180.0	299.2	202.0
	954.3	1 011	998.7

从测试结果可知,各组数据中的末两个固有频率总是大致等于自由—自由情况下测得的频率值(由于模型支座与转子模型直接连接,支座的一部分质量随该处转子振动,使在支

座上测得的这两个固有频率反而略低于自由—自由情况下的频率.至于支承在轴承油膜上的转子,则将有不同情况).各组数据中头上两个频率则随支座刚度强烈变化.前者可称为"转子主导"型,后者可称为"轴承主导"型.前者数值由于受支座刚度影响小,更能用来验证正、反传递法的计算结果.后者数值则很大程度上依赖于支承动刚度的准确性.

用本文所述方法,对模型系统作了计算.共分20段,如图7所示.在相邻段直径阶差很大的地方,利用45°影响锥的概念,适当计入了粗段端部弯曲刚度的减弱效应.支座与转子直接连接的托架质量作为转子该处的附加质量纳入计算.轴段参数见表2.计算结果如表1所列.

表2 复套式模型的分段参数

段 号	外径/mm	内径/mm	长度/mm	附加质量/kg	轴质惯矩/(kg·m^2)
1	30	0	42.5	0.221	0
2	30	0	40	0	0
3	30	0	40	0	0
4	30	0	40	0	0
5	30	0	40	0	0
6	30	0	40	0	0
7	30	0	40	0	0
8	30	0	40	0	0
9	30	0	50	1.37*	0
10	49	35	26	0	0
11	47.25	35	13	0.357	0
12	50	35	41.5	0.332 9	0
13	85	35	23.667	0	0
14	85	35	23.667	0	0
15	85	35	23.667	0	0
16	85	35	23.667	0	0
17	85	35	23.667	0	0
18	85	35	23.667	0.332 9	0
19	50	35	41.5	0.357	0
20	50	35	32.5	0	0

*附加质量离传递方向反逆节点的外悬距 $e=-4.6$ mm

对比测试结果与计算结果,可以看到:在自由—自由情况下,二者满意地符合.支承在弹性支座上时,"转子主导"型的测试和计算结果亦能满意地符合.这证实了正、反传递矩阵法的可用性.

至于"轴承主导"型,则在支座刚度小时,亦能很好符合.但支座刚度大时,却有较大偏差.仔细观察可知:测得的第一阶固有频率并不大致随支座刚度的平方根而增减,而是作缓慢得多的变化.这说明:模型支座在小振动时实际动刚度并不完全符合它在静态大变形情况下标定的值,而是不同程度地更小.考虑到支座上V形槽与轴颈接触处可能不太清洁以及不太密贴或有夹尘,接触处可能有微小变形,各接合面间可能有微小相对运动等因素(这些附加影响因素在实际轴承油膜中并不存在或完全不同),对于试验模型转子与支承静刚度间设想为串联有另一附加弹性环节.如果取该附加串联刚度为 5×10^5 N/m,则计算结果的"轴承主导"型频率亦与实测值满意地符合,如表1所示.

四、结论

用正、反传递矩阵法建立复套式转子—滑动轴承系统的弯曲固有振动特征行列式,用复平面上显示轨迹及三角元折半搜索法计算复频率;对于对称系统作节约处理. 由此建立了一套适用于复套式、普通式及对称式转子—滑动轴承系统的固有复频率计算方法和程序. 实例运算(分段数不大于 30)表明它稳定可靠.

用简单复套式模型系统的锤击试验验证了这套计算法是切实可用的.

需指出的是:在内、外轴接合处,本文是作为完善的整体来处理的. 对于不是非常紧固的内、外连接情况这样处理显然会带来误差,需要今后加以更完善的处理.

致谢

本工作由南京高速齿轮箱厂资助完成,该厂并提供了模型系统以供测试验证. 该厂李钊刚工程师对本工作给予了热情支持和关心,测试工作并得到了上海工业大学沈沛涛和奚风丰的帮助,特此一并致谢.

参 考 文 献

[1] J. Glienicke. Dynamik gleitgelagerter Wellen-Programme zur Berechnung der selbst-und unwuchterregton Schwlngungen allgemeiner gleitgelagerter (und wälzgelagerter) Rotoren mit Zusatzeinflüssen, Forschungsheft, Forschungsvereinigung Antriebstechnik E. V. , Heft 61, 1978.

Effect of Bush Viscoelasticity on Journal Bearing Dynamic Properties and Stability*

Abstract: This paper investigates the possibility of improving journal bearing dynamic properties and stability by using bearing bush possessing viscoelasticity. Reynolds equation is solved simultaneously with static deformation equation, wherefrom static performances of the bearing are calculated. Simple harmonic oscillations of journal within bearing clearance is assumed for deriving perturbated Reynolds and deformation equations, the simultaneous solutions of which are used for obtaining bearing dynamic coefficients under whirling frequencies either synchronous to running speed or at stability threshold. Numerical results show that if a rotor is supported on bearings with appropriate viscoelasticity properties, its stability and vibration suppressing ability can be significantly improved. Thus the theory confirms the noteworthy positive effect of a suitable viscoelastic bush on journal bearing dynamic behavior.

Nomenclature

B_{ij} = Nondimensional damping coefficients, $B_{ij} = b_{ij}\psi^3/(\mu l)(i, j = x, y)$

b_{ij} = Damping coefficients

c = Radial clearance

d = Bearing diameter

D = Outer diameter of bearing bush

$\bar{E}$ = Nondimensional elasticity modulus, $\bar{E} = E\psi^3/(2\mu\Omega)$

E = Young's modulus

E^*, E' and E'' = Nondimensional viscoelastic complex modulus, nondimensional storage modulus, and nondimensional loss modulus respectively, with the same relative unit as that of $\bar{E}$

f = Static flexibility of rotor, $f = mg/k$

f^* = Relative complex flexibility

g = Gravitational acceleration

H = Nondimensional film thickness, $H = h/c$

h = Film thickness

K_{ij} = Nondimensional stiffness coefficients, $K_{ij} = k_{ij}\psi^3/(\mu\Omega l)(i, j = x, y)$

k_{ij} = Stiffness coefficients

k = Static stiffness of rotor

l = Bearing width

m = Mass of rotor

P = Nondimensional pressure, $P = p\psi^2/(2\mu\Omega)$

p = Oil film pressure

r = Bearing radius

So = Sommerfeld number, $So = \psi^2 W/(2\mu\Omega rl)$

T = Nondimensional time, $T = t\Omega$

t = Time

x = Circumferential coordinate, $x = r\varphi$

W = Load

Z = Nondimensional axial coordinate, $Z = z/(l/2)$

z = Axial coordinate measured from middle plane of bearing width

β = Amplifying factor at resonance

μ = Dynamic viscosity of lubricant

ϕ = Angular coordinate measured from steady state film force direction

φ = Angular coordinate measured from maximum clearance, $\varphi = \phi - \theta$

---

* In collaboration with X. Jiang. Reprinted from *Proceedings of CSME Mechanical Engineering Forum* 1990, Vol. 1: 177-182.

ψ=Clearance ratio, $\psi = c/(d/2)$
Ω=Angular velocity of rotor
Ω_{st}=Stability threshold speed
ω=Frequency of whirling or oscillation
$\bar{\omega}$=Nondimensional whirling frequency, $\bar{\omega} = \omega/\Omega$
ω_k=Natural frequency of rotor with rigid suports
ε=Steady state eccentricity ratio
θ=Steady state attitude angle

1 Introduction

A well known method of improving stability and vibration suppressing ability of journal bearings comprises introduction of elastic-damping supports outside of bearing or shell[1-5]. The general principle underlying this method tells us that the dynamic behavior of the bearing system can be effectively improved by combining elastic and damping properties with the supporting structure in series with the lubricating film, either a little distant from the latter as in external bearing dampers, or more adjacent to it as in corrugated foil bearings. With this view in mind, it is but logical to suppose that if the bearing bush itself is given appropriate elastic and damping properties, resulting in the most intimate serial conjunction of the lubricating film with the additional elastic-damping mechanism, a significant improvement of dynamic behavior could be expected for a very simple structure. The aim of this paper is therefore to make a theoretical investigation on the possibility brought forth by such an idea so as to see if it is justified to make further experimental verification and research on bush materials in this direction. Linear theory of simple harmonic vibration will be based upon for making dynamic calculations, with interest focused on onset of instability and unbalance-induced synchronous whirl of small amplitude.

2 Analysis

For an isoviscous incompressible lubricant, the pressure field under isothermal laminar flow condition in the clearance space of a journal bearing is governed by the nonstationary Reynolds equation:

$$\partial(h^3 \partial p/\partial x)/\partial x + \partial(h^3 \partial p/\partial z)/\partial z = 6\mu\Omega r \partial h/\partial x + 12\mu \partial h/\partial t \quad (1)$$

or its nondimensional form:

$$\partial(H^3 \partial P/\partial \varphi)/\partial \varphi + (d/l)^2 \partial(H^3 \partial P/\partial Z)/\partial Z = 3\partial H/\partial \varphi + 6\partial H/\partial T \quad (1a)$$

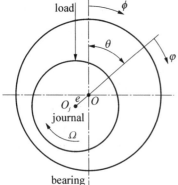

Fig. 1 Steady state equilibrium position of journal within bearing

Fig. 1 shows the equilibrium position of a journal within a cylindrical bearing under steady state. Assuming small arbitrary harmonic whirling of the journal within the bearing with frequency ω, we can express the instantaneous position of the journal centre as

$$\varepsilon_i = \varepsilon + \varepsilon_a e^{i\omega t}$$
$$\theta_i = \theta + \theta_a e^{i\omega t} \qquad (2)$$

where ε and θ express the eccentricity ratio and attitude angle of the journal centre at its steady state equilibrium position, and the mutually independent parameters ε_a and θ_a express the complex amplitudes of eccentricity ratio and attitude angle respectively. The corresponding film pressure and film thickness are composed of their steady state parts and dynamic increments. Neglecting errors of higher order, they can be expressed in nondimensional form as

$$P = P_0 + Q e^{i\bar{\omega}T} \qquad (3)$$

$$H = H_0 + H_j e^{i\bar{\omega}T} + \delta_v e^{i\bar{\omega}T} \qquad (4)$$

where H_0 is the distribution of nondimensional steady state film thickness including nondimensional steady state deformation δ_a of the bush, $H_0 = 1 + \varepsilon\cos\varphi + \delta_a$; P_0 is the distribution of steady state pressure; Q is the nondimensional complex amplitude of dynamic pressure increment; $\bar{\omega}$ is the relative whirl frequency of the journal, $\bar{\omega} = \omega/\Omega$, while Ω is the angular rotational speed of shaft; δ_v is the nondimensional complex amplitude of bush viscoelastic deformation; H_j is the distribution of nondimensional amplitude of dynamic increment of film thickness due to small dynamic displacement of journal, $H_j = \varepsilon_a \cos\varphi + \varepsilon \cdot \theta_a \sin\varphi$.

Substituting Eqs. (3)–(4) into (1a), we get the governing equation of pressure perturbation[6], taking into account bush viscoelastic deformation under small harmonic oscillation of the journal in nondimensional form, as

$$\begin{aligned}
&\partial(H_0^3 \partial Q/\partial\varphi)/\partial\varphi + (d/l)^2 \partial(H_0^3 \partial Q/\partial Z)/\partial Z \\
&= -9(M/H_0)\partial H_0/\partial\varphi - 3H_0[(H_0 \partial M/\partial\varphi \\
&\quad - M\partial H_0/\partial\varphi)\partial P_0/\partial\varphi + (d/l)^2 \partial(H_0 \partial M/\partial Z \\
&\quad - M\partial H_0/\partial Z)\partial P_0/\partial Z] + 3\partial M/\partial\varphi + 6i\bar{\omega}M
\end{aligned} \qquad (5)$$

where the variable M stands for $\varepsilon_a\cos\varphi + \varepsilon \cdot \theta_a\sin\varphi + \delta_v$.

Steady state harmonic viscoelasticity problems can be solved in the same manner as that for the corresponding elasticity problems. According to the elastic-viscoelastic correspondence principle of isothermal, linear viscoelasticity theory[7], steady state harmonic elastic solutions can be converted to the corresponding viscoelastic solutions by replacing the elastic moduli with the complex viscoelastic moduli. To be consistent with elastic problems, the relations of complex moduli are given by the following nondimensional equations:

$$\lambda^*/G^* = 2\nu^*/(1-2\nu^*) \quad E^* = 2(1+\nu^*)G^* \qquad (6)$$

where λ^* and G^* are complex Lama constants of the viscoelastic body, E^* is the complex modulus, and ν^* is complex Poisson's ratio. All the variables have been expressed in their nondimensional forms. If Poisson's ratio is assumed constant, that is $\nu^* = \nu = $ const., it is apparent from (6) that the viscoelastic solutions can be achieved from the corresponding elastic solutions simply through replacing the elastic modulus $\bar{E}$ by the complex modulus E^*.

The complex modulus E^* consists of a real part E' and an imaginary part E'' which are usually called the storage and the loss moduli, respectively.

$$E^* = E' + iE'' \tag{7}$$

Assume strain to be a harmonic function of time

$$\xi(t) = \xi_0 e^{i\omega t} \tag{8}$$

where ξ_0 is the amplitude and ω the frequency of vibration. The corresponding stress is also a harmonic function with the same frequency:

$$\sigma(t) = |E^*| \xi_0 e^{i(\omega t + \alpha)} \tag{9}$$

where $\alpha = \tan^{-1}(\eta)$, with $\eta = E''/E'$, and $|E^*|$ is the magnitude of E^*. α is usually called the loss angle of the viscoelastic material, and η is called the loss factor. Their magnitudes dominate the damping effect of the material. For a given material, E^*, E', E'', α and η are usually dependent of the exciting frequency.

In calculation of the three-dimensional elastic deformation by the finite element method, the elastic modulus can be seperated from the stiffness matrix $[K]$ so that the elements of the latter are only relative to shape function and Poisson's ratio:

$$\bar{E}[K]\{\delta_0\} = \{F\} \tag{11}$$

where $[K]$, $\{\delta_0\}$, and $\{F\}$ are nondimensional global stiffness matrix, displacement vector and load vector, respectively. The steady state deformation δ_0 is calculated by this equation with nodal loads calculated from the steady state pressure distribution. When calculating the viscoelastic displacement vector caused by the dynamic increment of film pressure, we write

$$\{\delta_v\} = [K]^{-1}\{F\}/E^* \tag{12}$$

Substituting (7) into (12), we get

$$\{\delta_v\} = (f' - if'')\{\delta_0\} \tag{13}$$

where f' and f'' are the real and imaginary parts of the relative complex flexibility respectively. $(f' - if'')$ may be termed the complex relative flexibility, and the following relationships hold

$$f^* = f' - if'' = 1/(E^*/\bar{E}) \tag{14}$$

$$f' = (E'/\bar{E})/[(E'/\bar{E})^2 + (E''/\bar{E})^2] \tag{15}$$

$$f'' = (E''/\bar{E})/[(E'/\bar{E})^2 + (E''/\bar{E})^2] \tag{16}$$

The oil film dynamic coefficients for a given value of frequency ratio $\bar{\omega}$ are calculated by the partial differential method. Derivatives of complex amplitudes of pressure and deformation distributions with respect to whirl amplitude ε_a or θ_a are first solved simultaneously, then the film stiffness coefficients and damping coefficients are obtained by integrations of the derived distributions in the same way as when film resultant forces are calculated. It is evident that the dynamic coefficients thus obtained are dependent of the assumed whirl ratio $\bar{\omega}$. A value of unity is given to $\bar{\omega}$ when the dynamic coefficients to be used in calculating synchronous vibrations are aimed at.

For stability calculations, the equivalent stiffness coefficient K_{eq} and the whirl ratio γ_{st} at stability threshold are calculated as [8]:

$$K_{eq} = (K_{xx}B_{yy} + K_{yy}B_{xx} - K_{xy}B_{yx} - K_{yx}B_{xy})/(B_{xx} + B_{yy}) \quad (17)$$

$$\gamma_{st}^2 = [(K_{eq} - K_{xx})(K_{eq} - K_{yy}) - K_{xy}K_{yx}]/(B_{xx}B_{yy} - B_{xy}B_{yx}) \quad (18)$$

The values of K_{eq} and γ_{st} thus obtained are also dependent of the assumed value of relative frequency $\bar{\omega}$. At stability threshold, the whirl ratio γ_{st} should keep in coincidence with the exciting frequency under which the dynamic coefficients are caulculated. This is arrived at by an iterative procedure.

For cylindrical whirling of a symmetrical single mass rotor supported on a pair of identical journal bearings, the threshold speed can be calculated by the following nondimensional equation[8]:

$$\Omega_{st}/\omega_k = 1/[\gamma_{st}\sqrt{1 + So_{st}/(K_{eq} \cdot f/c)}] \quad (19)$$

where Ω_{st}/ω_k is the relative threshold speed, $\omega_k = \sqrt{k/m}$ is the natural frequency of the rotor when rigidly supported, f/c represents the relative flexibility of the rotor, $f = mg/k$ denotes the static deflection of the rotor caused by its own weight, c is the radial clearance of the bearing, $So_{st} = \psi^2 W/(2\mu\Omega_{st}rl)$ is the Sommerfeld number of the bearing under threshold speed.

From a multitude of calculated results of this sort, a "stability map" can be deduced, showing the relationship between Ω_{st}/ω_k and So_k, with f/c as a parameter. A higher curve means a better stability. In the above, $So_k = \psi^2 W/(2\mu\omega_k rl)$ is the Sommerfeld number with reference to ω_k.

Response of the rotor system to residual unbalance is also influenced by the viscoelastic property of the bearing bush. The major semiaxis of the elliptical whirling orbit of the rotor centre under unit unbalance can be calculated from the corresponding nonhomogeneous equations of motion, using the dynamic coefficients of the bearing calculated under unit whirl ratio. A "resonance map" can be deduced from a multitude of such calculations, showing the resonant amplifying factor β and the relative critical speed Ω_{cr}/ω_k in relationship with So_k, with f/c as a parameter. An improvement of vibration suppressing ability is linked with a lowering of the resonant amplifying factor.

3 Methods of Solution

In the calculations, the oil film domain is discretized, and nondimensional Reynolds equation and its perturbated forms are replaced by their corresponding finite difference equations, and solved by successive over-relaxation (SOR) method. Under-relaxation iterations between Reynolds equations and deformation calculations are performed to obtain the simultaneous solutions of nondimensional oil film pressure (or the derivatives of its complex amplitudes) and bearing deformation (or the derivatives of its complex amplitudes).

The finite element method is used for calculating the elastic deformation of the bearing. For cylindrical journal bearings, where the bearing structure and boundary condition are axially symmetrical, the computational and storage amount can be significantly reduced in the following manner.

Fig. 2 shown the oil film domain discretized (16×80), the position of a node is denoted by (i, j), P_{kl} is the average pressure on the subdomain A_{kl}. Let $[\Delta]$ represent the radial displacement matrix of bearing surface composed of nodal displacements δ_{ij}, and $[F]_{kl}$ the flexibility coefficient matrix expressing radial deformations at all nodes caused by unit pressure at region A_{kl}.

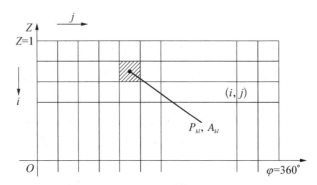

Fig. 2 Discretization of the oil film domain

$$\begin{array}{l}[\Delta]=[\delta_{ij}] \quad i=1, 2, \cdots, 17; \, k=1, 2, \cdots, 16 \\ [F]_{kl}=[f_{ij}]_{kl} \quad j=1, 2, \cdots, 81; \, l=1, 2, \cdots, 80\end{array} \quad (20)$$

From principle of superposition,

$$[\Delta]=\sum_{k=1}^{16}\sum_{l=1}^{80}\bar{P}_{kl} \cdot [F]_{kl} \cdot \Delta A \quad (21)$$

or

$$\delta_{ij}=\sum_{k=1}^{16}\sum_{l=1}^{80}\bar{P}_{kl} \cdot [f_{ij}]_{kl} \cdot \Delta A \quad (22)$$

where $(f_{ij})_{kl}$ is the element (i, j) of the flexibility matrix (k, l), $\Delta A = \Delta \psi \cdot \Delta Z$, $\Delta \varphi$ and

ΔZ are step widths in circumferential and axial diretions respectively.

Because of the axial symmetry of cylindrical bearings, it is unnecessary to obtain 16×80 flexibility matrixes for all the small areas A_{kl}. Only 16 matrices for a single column in the axial direction are needed. The matrices of all other columns can be formed by corresponding transpositions of these 16 matrices.

4 Results and Discussion

Calculations have been done for a cylindrical bearing with $l/d = 0.6$, $D/d = 1.6$, nondimensional complex modulus $E' = 20$ or 2, loss factor $\eta = 0.5$ or 1.0 and $E = 200$, $\eta = 0$ (corresponding approximately to bronze) for comparison, and Poisson's ratio $\nu = 0.3$. The bearing bush is assumed to be uniformly and rigidly supported on its outer surface.

Figs. 3 and 4 represent the comparisons between various stiffness coefficients. Figs. 5 and 6 show those between damping coefficients. It can be seen that significant increases in damping coefficients are caused by increases in η and So within the calculated range. The two cross damping coefficients differ from each other when dynamic elastic or viscoelastic deformation is taken into account.

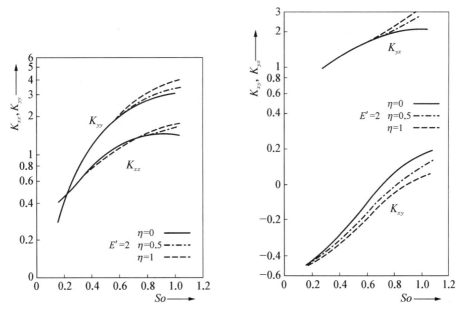

Fig. 3 Stiffness coefficients K_{xx}, K_{yy} **Fig. 4** Cross coupling stiffness coefficients K_{xy}, K_{yx}

It has been noticed from our numerous results of calculations with η value ranging from 0 up to 2 that, for a definite nondimensional storage modulus E', the optimal value of loss factor is approximately 1, since an η value of unity is always accompanied by the highest stability threshold speed and the lowest amplifying factor at resonance. Accordingly, the following discussions are limited to the case with unity η value.

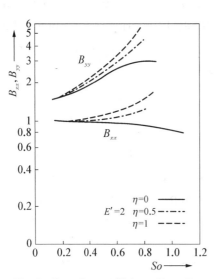
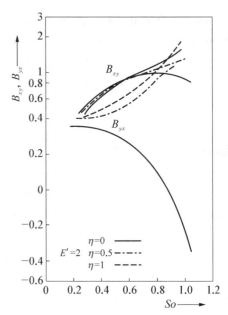

Fig. 5 Damping coefficients B_{xx}, B_{yy} **Fig. 6** Cross coupling damping coefficients B_{xy}, B_{yx}

Fig. 7 shows stability threshold speeds of a flexible rotor supported on journal bearings made of materials of different E' modulus, with unity loss factor. It can be seen that the threshold speed increases significantly for viscoelastic bush as compared to rigid bush. For example, for $f/c = 2.0$, $So_k = 1.6$, when compared to the rigid bush (no. 1), the threshold speed for the second material (no. 2) increases by 4.7%, while for the third one increases by 12.4%.

A comparison of the results of β and Ω_{cr}/ω_k is given in Fig. 8. With viscoelastic bush, β decreases to some extent, which implies that it can also give some effect in suppressing the resonant amplitude of rotors. Again for example, with $f/c = 2.0$ and $So_k = 0.8$, compared to the rigid bush, the resonant amplifying factor of the second material reduces by 2.9%, while that of the third one by 4.6%.

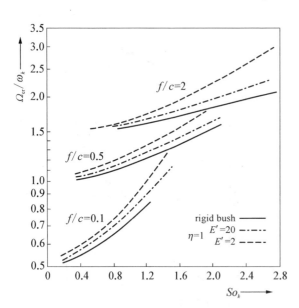

Fig. 7 Stability threshold speeds

5 Conclumions

The present study exhibits that if a rotor is supported on bearings with bush made from suitable viscoelastic materials, its stability and vibration suppressing ability can be

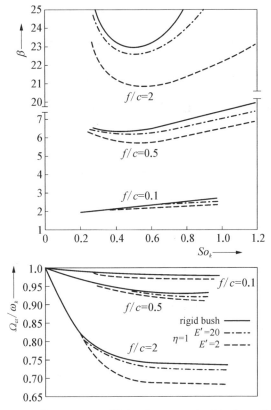

Fig. 8 Resonant amplifying factors and critical speeds

significantly improved, thus the theory confirms the suggested concept of improving journal bearing stability and vibration suppressing quality by using bearing bushes made of viscoelastic materials.

References

[1] Glienicke, J., and Stanski, U., External bearing damping — means of preventing dangerous shaft vibrations on gas turbines and exhaust turbo-chargers. Proc. 11th CIMAC, 1975: 287–311.

[2] Heshmat, H., Walowit, J. A., and Pinkus, O. Analysis of gas lubricated foil journal bearings. Trans. ASME, Journal of Lubrication Technology, 1983, 105: 638–646.

[3] Holmes, R., and Dogan, R. Investigation of a rotor bearing assembly incorporating a squeeze-film damper bearing. Journal of Mechanical Engineering Science, 1982, 24: 129–137.

[4] Nikolajsent, J. L., and Holmes, R. Investigation of sqeeze-film isolators for the vibration control of a flexible rotor. Journal of Mechanical Engineering Science, 1979, 21: 247–252.

[5] Zhang, Z. System damping factor of an elastic rotor supported on tiltingpad bearings with elastic and damped pivots. Proceedings of 1985 Tokyo JSME International Tribology Conference, 1985: 541–546.

[6] Zhang, Z., Mao, Q. and Xu. H. The effect of dynamic deformation on dynamic properties and stability of cylindrical journal bearings. Proceedings of 13th Leeds-Lyon Symposium on Tribology, 1986, Paper XI(iv): 363–366.

[7] Christensen, R. M. Theory of viscoelasticity, an introduction. 2nd ed. Academic press Inc., New York, 1982.

[8] Zhang, Z., Zhang, Y., Chen, Z., Xie, Y., Qiu D., and Zhu, J. Theory of hydrodynamic lubrication of sliding bearings (in Chinese), Zhang, Z. ed., Chinese High Education Publication Corp., Beijing, 1986: 94–95.

Calculation of Hydrodynamic Lobe Type Journal Bearings Aided by Database of Properties of Single Bush Segment*

Abstract: Design of rotors supported by hydrodynamic journal bearings usually necessitates numerous, time-consuming calculations of lobe type bearings. It is proposed to replace these calculations by searching corresponding databases of properties of single bush segment and summing them up to get the bearing properties. Systematic investigation has been done to set up three kinds of compact databases of single bush segment: static and dynamic coefficients, transient oil film forces, and properties of deformable tilting pads. Sample calculations show that, the time needed for calculating bearing property values is shortened to 1/100, and that for calculating a journal dynamic locus to 1/400 approximately. Systematic investigation of the effect of pad and pivot deformations on tilting pad bearing properties and rotor system dynamic behavior is also easily and comprehensively done.

1 Introduction

In designing rotating machines, the rotor-bearing system is demanded to meet the following criteria: the bearings should have adequate load carrying capacity, the bearing temperature should not be higher than permissible, and the rotor-bearing system should have satifactory dynamic behavior—the critical speeds sufficiently far away from the working speed, small synchronous amplitude due to residual unbalance, low resonant peak when crossing the critical speeds, and good stability of the oil film against self-excited whirl and large external excitations. The designers usually have to make a lot of bearing calculations to assess various bearing designs, especially if design optimization in a wide range is needed. For example, the calculation of the static and dynamic coefficients of a multi-lobe bearing (Fig. 1 shows a 3-lobe bearing as an example) is a common practice in rotating machinery design, which usually follows the flow chart shown in Fig. 2. Although many general purpose programs of bearings are available, such highly time consuming calculations are still causing designers headach. A number of researchers are still of the opinion that more efficient methods of solution of Reynolds equation are needed. Handbooks and books providing curves and tables of bearing characteristics are being published from time to time, but they contain only limited data of the few most commonly

---

* In collaboration with Z. Zheng, X. Wu., Z. Li and W. Wang. Reprinted from *Proceedings of the 1st International Symposium on Tribology*, 1993.

used configurations of bearings, and with rather narrow ranges of bearing parameters. The wide use of computers has naturally led to the appearance of various special bearing databases, but it has been deemed "impractical" to intend to set up comprehensive bearing databases applicable in wide ranges of bearing configuration and parameters.

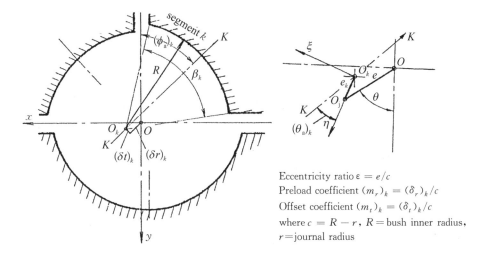

Fig. 1 A 3-lobe bearing

In order to minimize the volume of the database, bearing property data are usually given in non-dimensional form. The aim of the database is to provide non-dimensional values of various bearing properties in dependence on the determinant parameters, which determine the geometric and dynamic similarity of the bearing. Even the very simple bearings have a large number of determinant parameters. An arbitrary 3-lobe bearing, for example, has 14 geometric determinant parameters: l/d, ε, β_k, $(\Phi_a)_k$, $(m_r)_k$ and $(m_t)_k$ ($k=1, 3$). So many parameters, some of which have rather wide ranges, are hardly bearable for a compact database, to say nothing of the more complicated configurations such as bearings cosisting of both fixed bushes and tilting pads. It therefore seems to be with reason to say that a comprehensive database of journal bearing is impractical.

But if we scrutinize the flow chart in Fig. 2, it can be seen that the bearing calculation is actually based on calculations of each bush segment as the elementary cells. For the calculation of a single bush segment, there are much fewer determinant parameters, and only four are needed for geometric similarity: l/d, ε_k, β_k and $(\theta_a)_k$. It becomes entirely practical to set up a database, if it deals only with a single bush segment. Aided by such a database, the bush segment calculation can be replaced by quick searching and interpolating this database to find the segment properties, without the necessity of solving the 2-dimensional 2nd order Reynolds partial differential equations and performing various integrations. Naturally will the computing speed be drastically enhanced.

Based on this idea, three different kinds of database of bush segment have been set up. They are: 1. Static and Dynamic Coefficients; 2. Non-Stationary Oil Film Forces; 3. Properties of Deformable Tilting Pad.

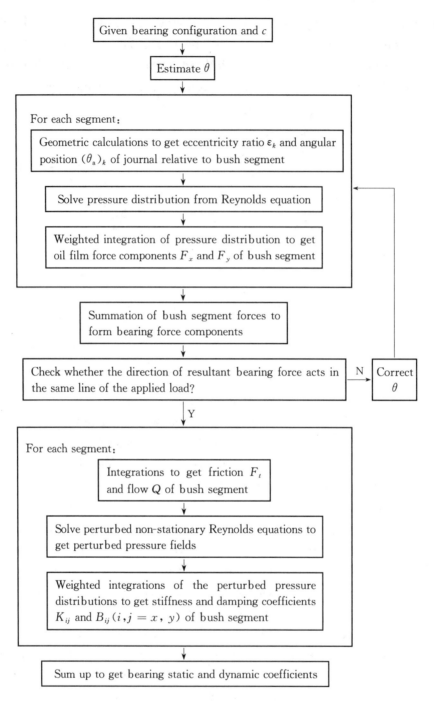

Fig. 2 Flow chart of calculation of bearing static and dynamic coefficients of a multi-lobe bearing

2 Database of Static and Dynamic Coefficients of Bush Segment

This is used for calculating bearing load carrying capacity, frictional loss, oil flow rate, stiffness and damping coefficients as described above. A 4-dimensional database relating non-dimensional properties of the bush segment to the 4 determinant parameters has been built by systematic calculations. A comparison among 8 existing efficient algorithms for solving the Reynolds equation is made. The block matrix method of Castelli[1] has been chosen as the most suitable for the particular purpose of building the database, from combined considerations regarding the efficiency and accuracy exhibited in test calculations.

When calculating a bearing, the eccentricity ratio and attitude angle of the journal relative to each bush segment are calculated from the eccentricity and attitude angle relative to the bearing center. The database is then searched and interpolated to get the property coefficients of each bush segment. These are summed up, either scalarly or vectorially, to form the bearing property coefficients.

In order to make the database applicable to high speed bearings, effects of turbulent flow regime have also been considered. The effects of depth of turbulence on various properties have been systematically investigated, using the model proposed in [2]. This leads to the idea of taking the turbulence effects into account in a condensed manner by using "corrected values of l/d" to search the database and then multiplying the outputs by "global turbulence coefficients". Both the correction on l/d and the global turbulence coefficients are determined by "effective Reynolds numbers". The correction of l/d ratio is introduced to take care of the different effects of turbulence on the pressure flow rates in the circumferential and axial directions. The global turbulence coefficients are used to transform the property values from the database into corresponding values in turbulent flow regime. In this way, the necessity of supplementing the nominal Reynolds number to the group of determinant parameters is avoided. The database keeps its compactness without being enlarged into an inconvenient 5-dimensional one. Only a small database has to be supplemented to assist in determining a "characteristic" film thickness which is based upon to determine the effective Reynolds number. These corrections result in bearing properties very near those obtained by solving the "turbulent Reynolds equation" taking full account of the non-uniformly distributed turbulence coefficients.

Effects of oil inlet pressure in an axial groove on bearing properties are taken into account in a simpler way. Since hydrodynamic bearings usually have low feed pressure, the effects of feed pressure on various properties can be adequately expressed in linear relations to the feed pressure. Systematic investigations show that the proportionality of effect can be expressed by a subordinate database relating it to l/d, ε_k, β_k and $(\theta_a)_k$. This database is very small since the value ranges of l/d and ε_k need only be very coarsely discretized.

Parallel computations have been made by performing calculation aided by the database and that using a conventional bearing program, to compare their computation speeds. It comes out that the present technique requires only about 1/100 of computational time of the conventional method, with practically the same results as shown in Fig. 3 for an example.

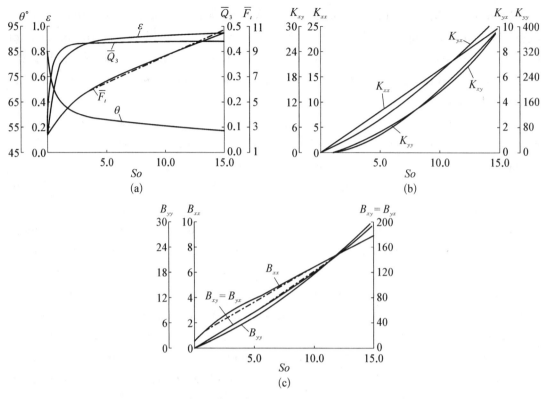

Fig. 3 Comparison of results on a 2×150 deg cylindrical bearing
 —·—· by conventional method ——— aided by database

3 Database of Non-Stationary Oil Film Forces of Bush Segment

When a rotor-bearing system is acted upon by large excitations such as seismic waves or loss of blades, or when the oil film loses its stability, the amplitude of transient motion of the journal within the bearing clearance will be comparable in order of magnitude with the radial clearance. It will then be inappropriate to express the strongly non-linear oil film force through the linearized stiffness and damping coefficients, when calculating the journal locus. It is conventional to solve the non-stationary Reynolds equation at each moment, calculate the oil film force corresponding to the current position (e and θ) and motion of the journal (ω, $\dot{\theta}$ and $\dot{e}$), and perform a step of time integration to find the journal position and motion at the next moment. Thousands of steps are needed to get a dynamic locus with adequate information. The time spent is naturally not short. Up to now, most of the analytical researchers studied only rotors supported by cylindrical bearings. [3] can be quoted as a rare example of study of a rotor supported on 2-lobe bearings, and that has been done with simplified boundary condition to reduce the computational time to a bearable degree. This situation can also be greatly improved by using database of non-stationary oil film forces of bush segment.

The non-dimensional determinant parameters of this database are l/d, ε_k, β_k, $(\theta_a)_k$

and δ which is the squeeze-rotation ratio and equal to $\dot{e}_k/e_k(\omega-2\dot{\theta}_k)$. The database has to provide values of non-dimensional oil film forces $(F_x)_k$ and $(F_y)_k$ in dependence on these 5 parameters. But the range of variation of δ is infinite, so it is impossible to use finite number of nodes to cover the entire range in the conventional way. This problem is solved by dividing the range into 3 parts: (1) $|\dot{e}_k| \leqslant |e_k(\omega-2\dot{\theta}_k)|$, where $|\delta| \leqslant 1$; (2) $\dot{e}_k = e_k(\omega-2\dot{\theta}_k) = 0$, where $(F_x)_k = (F_y)_k = 0$; and (3) $|\dot{e}_k| > |e_k(\omega-2\dot{\theta}_k)|$, where $1/\delta$ is used as the determinant parameter instead of δ, and $|1/\delta| < 1$. Infinite values of the determinant parameter is thus avoided.

Castelli's block matrix algorithm is used to make systematic calculations to set up this database. Aided by this database, the journal dynamic locus can be calculated in a time about 1/400 of that consumed by the conventional way. Fig. 4 shows a comparison of the journal dynamic locus calculated by the present technique and the practically identical result reproduced from [4].

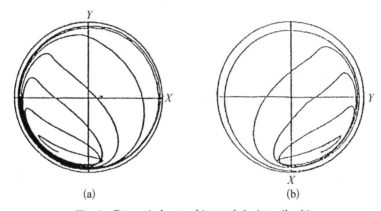

Fig. 4 Dynamic locus of journal during oil whip

Fig. 5 shows journal dynamic loci within preloaded 3-lobe and 4-lobe bearings rather easily obtained by the present technique. Such results could be obtained by the conventional method only at the expense of very long computation time.

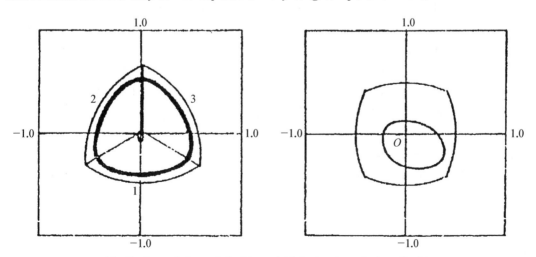

Fig. 5 Journal dynamic loci in multi-lobe bearings during oil whip

4 Database of Deformable Tilting Pad

Tilting pad bearings are widely used to support high speed rotors. The configuration of tilting pads inevitably makes them apt to be elastically deformed under load, the more so if the pads are made of material of low modulus of elasticity. The pad deformation consists of a static component due to the steady state load, and a dynamic component due to the dynamic load during vibration (Fig. 6). The contact between pad and pivot is also easily deformed, both statically and dynamically, due to the very small contact area. Since it is possible that such deformations will affect the bearing and system behavior severely[5], researchers have made investigations on their effects. Analytical work consists mainly of solving the static and dynamic EHL problems of the tilting pads to get the static and dynamic properties of the pads, coupling them with the deformation effects of the pivots and finding the bearing properties, and finally coupling the bearings with the rotor to get the system dynamic behaviors.

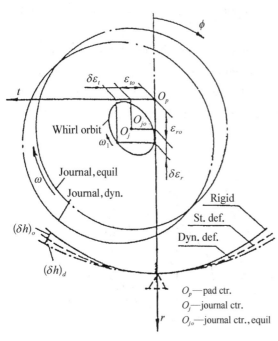

Fig. 6 Tilting pad deformations

The EHL problems of a tilting pad can only be solved at the expense of rather long computing times, even if a more or less simplified model is used for the pad deformation, such as the cantilever curved beam. Therefore, if systematic investigations are to be made over the effects of pad and pivot deformations on bearing and system dynamic behaviors, the computing time would be inconveniently long. On the other hand, a practical designer would also find it inconveniet to solve a concrete problem of this sort. It is therefore thought appropriate to single out the most time consuming part of the analytical work, viz., the solving of the EHL problems of the tilting pads, make systematic calculations with a specially compiled pad-EHL program, and store the results to form a database of properties of deformable tilting pad[6]. Using this database to supply ready values of properties of deformable tilting pads, it would be an easy thing to perform the rest parts of the analytical calculation, either to solve concrete problems, or to make systematic investigations.

For a tilting pad of uniform thickness, based on the cantilever curved beam model for its deformation, 5 parameters are needed to determine its similarity: l/d, pad span angle

β, pivot offset factor γ, radial eccentricity ratio of pad relative to journal ε_r, and pad deformation factor K_d defined as: $K_d = 0.75(1-\nu^2)\eta \cdot \omega/(T_1 \cdot E \cdot \psi^3 \cdot \alpha_J)$, where ν denotes Poisson's ratio; η the dynamic viscosity; ω the angular speed; T_1 the ratio of pad thickness to pad neutral diameter; E the modulus of elasticity; ψ the relative clearance of pad; and $\alpha_J = 1 + 3T_1^2/5 + 3T_1^4/7 + \cdots$. To avoid setting up a very large database, only two values of l/d (0.5, 0.8), three values of β (60, 75, 100 degrees), and three values of γ (0.5, 0.6, 0.65) have been covered. ε_r ranges from some negative value depending on β and γ as the lower limit to 0.95 as the upper limit. K_d ranges from 0 to 100.

The properties stored are the non-dimensional values of the film force F_r, the minimum film thickness H_{min}, the frictional force F_t, the volumetric flow Q, the dynamic properties expressed as frequency spectra of stiffness K_{rr} and damping B_{rr} in dependence on whirl ratio Ω, with Ω ranging from 0 to 1. A short program is written to search and interpolate the database to find these properties corresponding to the determinant parameters ε_r and K_d, under given l/d, β, and γ.

The effect of pivot static deformation is to reduce the effective eccentricity ε_r. This is taken into account when searching the database.

The effect of pivot dynamic deformation can be taken into consideration by replacing the pad dynamic coefficients with the serial dynamic coefficients of pad and pivot. The bearing dynamic properties will be summed up from the serial dynamic coefficients.

This method has been applied to a single mass symmetric rotor (Fig. 7) supported on a pair of identical tilting pad bearings (Fig. 8), to study the effects of pad and pivot elastic deformations. Some of the results of bearing properties and rotor behavior affected by pad and pivot deformations are shown in Figs. 9–11.

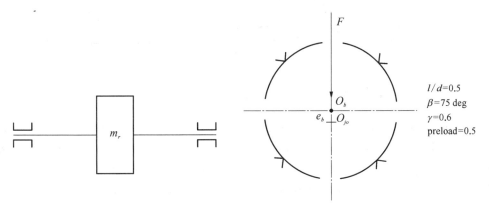

Fig. 7 Single mass rotor-bearing system　　　**Fig. 8** Symmetric bearing

5　Conclusions

Hydrodynamic journal bearing is a typical example the design or analysis of which needs large amount of complicated calculations. Databases of single bush segment are

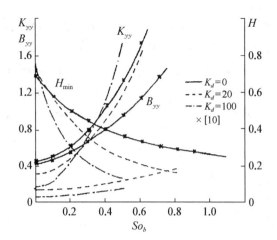
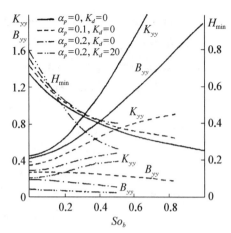

Fig. 9 Effect of pad deformation on bearing properties

Fig. 10 Effect of pivot deformation on bearing properties

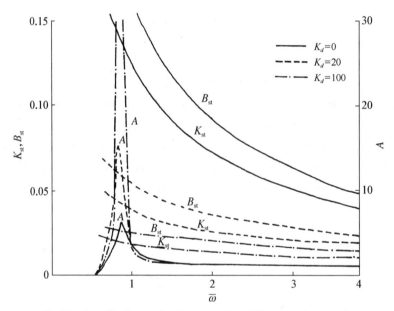

Fig. 11 Amplitude-speed relations and stability margin variation

found to be very effective in reducing the time and effort needed for obtaining bearing properties. Three kinds of bush segment databases have been set up by systematic calculations. Aided by them, the computation time for bearing properties can be shortened to 1/100, that for dynamic locus of journal to 1/400 approximately, and systematic studies over effects of pad and pivot elastic deformations on tilting pad bearing and rotor system behaviors can also be done without difficulty. Similar techniques could also be useful for improving the situation in other fields of engineering.

References

[1] Castelli, V. and Shapiro, W. ASME J. of Lub. Tech., 1967,89: 211.

[2] Ng, C. W. and Pan, C. H. T. ASME Ser. D, 1965,87: 675.

[3] Kato, T. et al. Proc. International Conference on Vibrations in Rotating Machinery, Sept., Bath, UK., 1992.
[4] Hori, Y. and Kato, T. ASME J. of Vib. and Ac., 1990,112: 160.
[5] Caruso, W. J. et al. Proceedings of the 11th Turbomachinery Symposium, 1983: 1.
[6] Zhang, Z. et al., Chinese Journal of Mechanical Engineering, 1993,6(1): 15.

Calculation of Journal Dynamic Locus Aided by Database of Non-stationary Oil Film Force of Single Bush Segment*

Abstract: A database of non-stationary oil film force of bush segment is set up and used to provide bush segment non-stationary oil film forces for assembling up the bearing non-stationary forces during calculation of journal non-linear dynamic locus within lobe type hydrodynamic bearings. A parameter representing the squeeze-rotation ratio is introduced and appropriately treated to cover the entire range of journal motion parameters. Replacing the solving of Reynolds equation and integrating in the conventional way of calculation, searching and interpolating the database to get the non-stationary oil film forces results in drastically shortening the computation time for a journal dynamic locus to 1/372. Several examples of such results are given to demonstrate the validity and noteworthy usefulness of the present technique.

Nomenclature

c: difference between bush and journal radii, $c = R - r$
FC, FS: segment forces in the ε and θ directions respectively
H: non-dimensional oil film thickness, $H = h/c$
k: shaft bending stiffness
l/d: length to diameter ratio
m: rotor mass
P: non-dimensional pressure, $P = 2p\psi^2/(\eta\omega)$
Z: non-dimensional axial coordinate, $Z = 2z/l$
ε: eccentricity ratio of journal relative to segment
ϕ: angular coordinate
η: lubricant dynamic viscosity
ψ: relative clearance, $\psi = c/r$
ω: angular speed of rotor
ω_c: 1st critical angular speed of rotor-bearing system
ω_n: natural frequency of rigidly supported rotor, $\omega_n = \sqrt{k/m}$
$'$: derivative with respect to non-dimensional time $T(=\omega t)$

1 Introduction

Non-linear dynamics of rotor-bearing systems is attracting increasing attention from researchers, engaged in researches over their responses to large transient excitation such as seismic waves[1,2] and to large unbalance, and when the linear stability threshold has been crossed, etc. The general way of such study is to discretize the time into fine steps, and perform numerical time-integration of the motion parameters of the journal center to get its

---

* In collaboration with Wen Wang. Reprinted from *Proceedings of Asia-Pacific Vibration Conference*, 1993: 365 – 369.

dynamic locus. The dynamic behavior of the system will then be assessed from the dynamic locus of the journal and/or the disks mounted on the shaft. During each step of time-integration, the knowledge of the oil film forces corresponding to the instantaneous motion parameters of the journal is needed. Solving the non-stationary Reynolds equation and integrating the instantaneous pressure distribution to get the oil film forces makes the calculation of a journal dynamic locus highly time-costing, since usually hundreds or even thousands of time steps must be performed before a locus with sufficient information can be obtained. Simplifying representations of the non-stationary oil film forces, such as short bearing forces and various field mapping[3] or approximate formulae[4-6], are often not accurate enough, especially when stability problems are concerned. They are also inconvenient when the rotor is supported on arbitrary multi-lobe bearings. Since lobe type bearings are the most widely used journal bearings, and the bearing forces are formed by vectorial summation of forces of all the bush segments, a database of non-stationary oil film forces of single lobe type bush segment will be most useful. Aided by such a database, it is only needed to search and interpolate the database to get accurate values of the instantaneous oil film forces of each bush segment, and then sum them up vectorially to get the bearing forces at each time step. In this way, not only can the computation time be drastically reduced, but also can arbitrary multilobe bearings be easily handled. Such a database will also be compact and easily realizable on micro-computers.

2 Inputs and Outputs of the Database

The database is set up in the non-dimensional form, to be as compact as possible. The non-stationary Reynolds equation is non-dimensionalized:

$$\frac{\partial}{\partial \varphi}\left(H^3 \frac{\partial P}{\partial \varphi}\right)+\left(\frac{d}{l}\right)^2 H^3 \frac{\partial^2 P}{\partial Z^2}=-3\varepsilon(1-2\theta')\sin\varphi+6\varepsilon'\cos\varphi \tag{1}$$

In order that the database is universally applicable, it should cover all possible ranges of the "rotation" parameter $\varepsilon(1-2\theta')$ and the "squeeze" parameter ε'. Since they are unlimited, the following treatments are done:

1. When both parameters $\varepsilon(1-2\theta')$ and ε' are zero, oil film forces are also zero:

$$FS=0.0, \ FC=0.0$$

2. When $\varepsilon|1-2\theta'|\geqslant 2|\varepsilon'|$, define $P_1=\dfrac{P}{\varepsilon|1-2\theta'|}$, and the equation becomes

$$\frac{\partial}{\partial \varphi}\left(H^3 \frac{\partial P_1}{\partial \varphi}\right)+\left(\frac{d}{l}\right)^2 H^3 \frac{\partial^2 P_1}{\partial Z^2}=-3\mathrm{sign}(1-2\theta')\sin\phi+3q_1\cos\phi$$

where q_1 is the squeeze-rotation ratio: $q_1=2\varepsilon'/\varepsilon|1-2\theta'|$. The range of q_1 is: $-1\leqslant q_1 \leqslant 1$.

3. When $\varepsilon|1-2\theta'|\leqslant 2|\varepsilon'|$, define $P_2=\dfrac{P}{2|\varepsilon'|}$, and the equation becomes

$$\frac{\partial}{\partial \varphi}\left(H^3 \frac{\partial P_2}{\partial \varphi}\right) + \left(\frac{d}{l}\right)^2 H^3 \frac{\partial^2 P_2}{\partial Z^2} = -3q_2 \sin\phi + 3\mathrm{sign}(\varepsilon')\cos\phi$$

where q_2 is the rotation-squeeze ratio: $q_2 = \varepsilon(1-2\theta')/2|\varepsilon'|$. The range of q_2 is also: $-1 \leqslant q_2 \leqslant 1$.

In this way, all the possible combinations of the rotation and squeeze parameters are covered by the above three cases. For case 2, the range of q_1 is discretized into a series of nodal values. For each nodal value of q_1, the distribution P_1 is solved, and integrated to give the "unit forces" FC_1 and FS_1 of the bush segment. They are stored in the database. The actual forces FC and FS will be obtained by multiplying them with $\varepsilon|1-2\theta'|$. Case 3 is treated in a similar way.

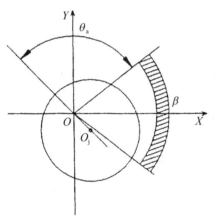

O—bush segment center O_j—journal center

Fig. 1 A bush segment

For given values of l/d and segment angular span β, the similarity of the film is determined by ε, θ_a(Fig. 1), and q_1 or q_2. After investigating into the dependence of the segment forces FC and FS on them, it is found appropriate to discretize the ranges of these three parameters into 40, 25, and 20 unevenly distributed nodal values respectively, and the structure of the database is thus set up. Castelli's block matrix method[8] is used to solve Reynolds equation, numerical integration is used to get the force values, and Reynolds boundary condition for film rupture is used. For each bush segment contained in the bearing under consideration, a database will be built in this way, and this is quickly done. These databases will not only provide segment force values for the problem currently under consideration, but also be kept for further application on segments of similar geometry. A small subprogram is written to search and interpolate the database depending on the input values ε, θ_a, q_1 or q_2, and the sense of the rotation or squeeze as the case may be. The values FC and FS are given as output values.

3 Accuracy and Speed of Calculation

The accuracy of the value of oil film force obtained from the database has been checked by comparing it to that calculated by solving the Reynolds equation and integrating. Fig. 2 shows some typical results of comparison, where $l/d=0.5$, $\beta=120$ deg., $\varepsilon=0.5$, $q_1=0.5$, $1-2\theta'>0$. It can be seen that the present technique is very satisfactory.

The accuracy of calculating the journal dynamic locus aided by the database has also been checked on an example extracted from [1]. A single mass symmetric rotor is supported on a pair of identical cylindrical bearings. $l=0.05$ m, $d=0.1$ m, $\eta=5.6 \times$

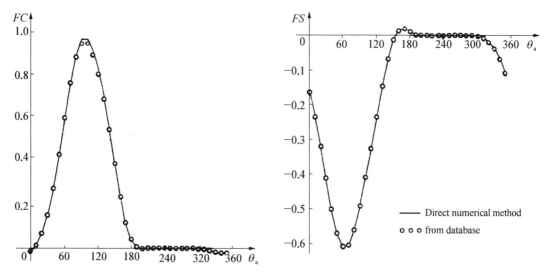

Fig. 2 Force values obtained from database compared with direct calculation

10^{-3} Pa·s, $c = 5 \times 10^{-4}$ m, $m = 62.42$ kg, $k = 1.223 \times 10^6$ N/m, $\omega_n = 140$ s^{-1}, $\omega = 448$ s^{-1}. The journal is excited from its static equilibrium position by a complete cycle of a sinusoidal acceleration of the bearing pedestal in the x-direction:

$$x_b = \begin{cases} A\sin(\omega_e t) & \text{for } 0 \leqslant t \leqslant 2\pi/\omega_e, \ \omega_e = 0.8\omega_n \\ & A = 0.3 \text{ g} \\ 0 & \text{for } t > 2\pi/\omega_e \end{cases}$$

Runge-Kutta method of the 4th order with variable step is used to calculate the journal locus. A Compaq-386/20e microcomputer is used. The values of the non-stationary oil film forces are either directly calculated (Fig. 3a) or obtained from the database (Fig. 3b). The results are practically identical. But the time needed for obtaining this locus when aided by the database is only 1/372 of the other.

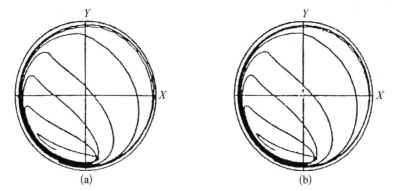

Fig. 3 Journal dynamic locus under a cycle of sinusoidal excitation, $A = 0.3$ g

Comparisons with the results in [1] are also made. The readers should be reminded of the different coordinate systems used in [1] and the present paper. Fig. 4 shows identical results when $A = 0.1$ g. Fig. 5 shows the results when $A = 0.3$ g. The insignificant difference between them is estimated to be attributable to the fact that quasi-Reynolds boundary condition has been used in [1], while full Reynolds condition has been used in the present work.

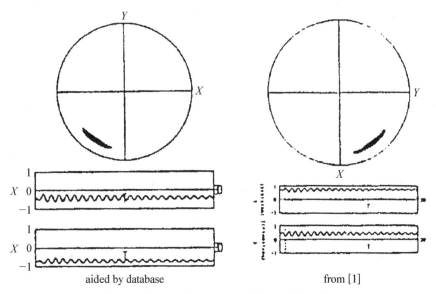

Fig. 4 Journal dynamic locus under a cycle of sinusoidal excitation, $A=0.1$ g

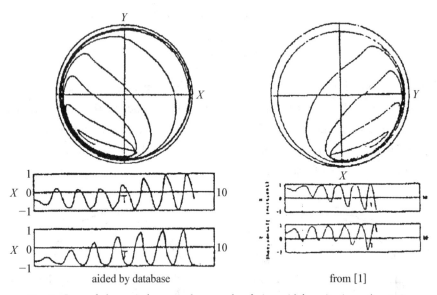

Fig. 5 Journal dynamic locus under a cycle of sinusoidal excitation, $A=0.3$ g

4 An Example of Application

The dynamic response to sudden application of unbalance of a Jeffcot rotor supported on a pair of elliptical bearings is investigated. The parameters have the following values: $l=0.05$ m, $d=0.1$ m, $\eta=5.6\times10^{-3}$ Pa·s, $k=3.0575\times10^5$ N/m, $m=15.605$ kg, $c=5.0\times10^{-4}$ N/m, $\omega_n=140$ s^{-1}, preload(ellipticity)$=0.5$. The dynamic loci of the journal are calculated under a series of speed values: $\omega = 70, 140, 210, 280, 350, 385, 395, 400 \text{ s}^{-1}$. The value of the unbalance is taken as an eccentricity of the mass center of 0.1c.

As part of the calculated results, the responses of the journal within the time interval 0.5–1.012 sec and the interval 9.4–9.912 sec are shown in Fig. 6. The frequency spectra of the responses in x-direction are also shown. The last spectrum ($\omega = 400 \text{ s}^{-1}$) has been reduced to 1/5, as the amplitude has grown very large due to oil whip.

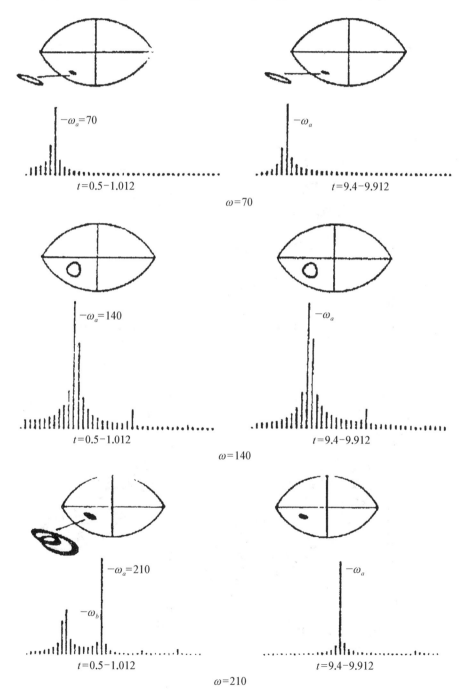

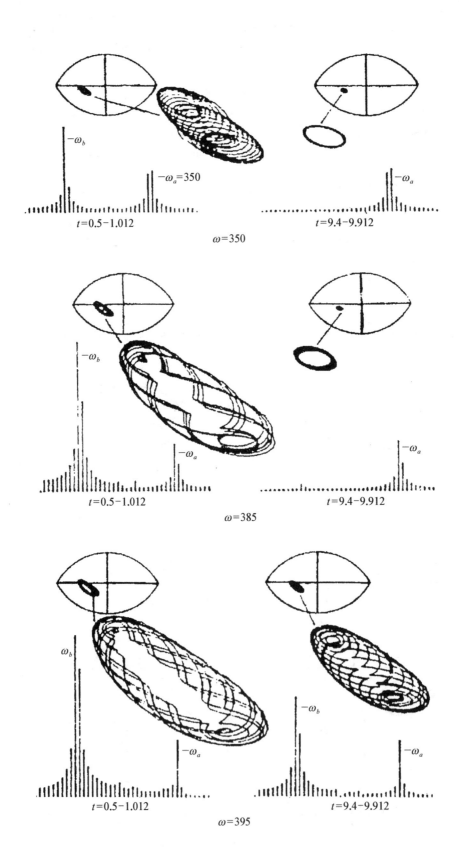

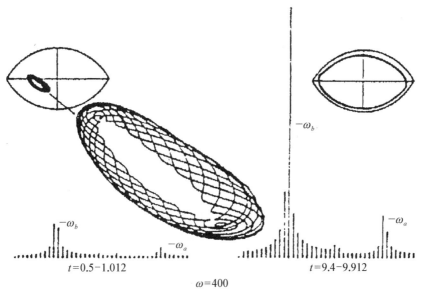

Fig. 6 Journal responses to suddenly applied unbalance

It can be seen from the journal loci that: (1) under low speeds ($\omega=70$ and $140\ s^{-1}$), the journal quickly settles to a closed orbit; (2) under speed values $\omega=210$ and $350\ s^{-1}$, the low frequency component is attenuated very slowly, and its frequency keeps a value $\omega_b=127\ s^{-1}$ which is not very far from the natural frequency of the rotor ω_n, and does not vary with the speed; (3) under $\omega=385\ s^{-1}$, the low frequency component does not completely perish even after the unbalance has been applied for nearly 10 seconds; (4) under $\omega=395\ s^{-1}$, the low frequency component persists; (5) under $\omega=400\ s^{-1}$, rather than decrease, the low frequency component gradually develops into a strong oil whip; (6) when the running speed coincides with the natural frequency $\omega=\omega_n=140\ s^{-1}$, the journal orbit deviates from a circle as would be predicted by the linear theory, and explainable by the exsistence of significant non-synchronous components, especially the second and third harmonics which can be seen in the frequency spectrum.

5 Conclusions

The technique of using a compact database of non-stationary oil film force of bush segment to aid the journal dynamic locus calculation in analyzing the dynamic behavior of rotor-bearing systems with full account of the non-linearity of the bearing forces has proven to be effective in drastically reducing the computation time, and readily applicable to all kinds of lobe type bearings.

Acknowledgments

The authors would like to thank Natural Science Foundation of China and Science Foundation of Shanghai High Education Bureau for sponsoring this research. They also wish to thank Mr. Z. Li for his valuable advices.

References

[1] Hori, Y. Proceedings of the IMechE 4th Intl. Conf. on Vibration in Rotating Machinery, Edinburgh, England, Sept. (1988), C318/88, pp. 1-8.
[2] Hori, Y., and Kato, T. J. of Vibration and Acoustics, Tran. of ASME, Vol. 112, (1990), pp. 160-165.
[3] Shapiro, W. Fluid-Film Bearing Response to Dynamic Loading, Bearing Design Problems, Technology and Future, ed. by Aderson, W. J., New York, (1980).
[4] Jakeman, R. W. Tribology International, Vol. 22, No. 1, (1989), pp. 3-10.
[5] Parszewski, Z. A., and Carter, B. M. Proc. 13th Leeds-Lyon Symp. on Trib., Sep. (1986), pp. 579-581.
[6] Hashish, E. A. Improved Mathematical Models and Dynamic Analysis of Light Rotor-Bearing Systems Unbalance and Stochastic Excitation, Ph. D Dissertation, Concordia University, (1981), Montreal, Canada.
[7] Choy, F. K., Braun, M. J., and Hu. Y. Journal of Tribology, Vol. 113, (1991), pp. 555-570.
[8] Castelli, V. and Shapiro, W. Trans. of ASME, J. of Lub. Tech., Vol. 89, (1967), pp. 211-218.

Calculation of Tilting Pad Bearings with Elastically Deformable Pads and Pivots Aided by Database of Single Deformable Tilting Pad*

Abstract: This paper investigates the effects of pad and pivot elastic deformations on tilting pad journal bearing properties and dynamic behavior of rotors supported on them. A corrected plane strain curved beam model is used for calculating the pad deformation. Small harmonic oscillations are assumed in deriving the dynamic EHL equations of single pads and for defining the pad dynamic properties. Numerical solutions of the EHL equations and the dynamically perturbed equations and numerical integrations are used to get pad properties. A database of properties of elastically deformable tilting pad is set up by such calculations. It is further transformed to account for the effects of pivot elastic deformation. When calculating the properties of a tilting pad bearing with deformable pads and pivots, values of non-dimensional properties of each pad are searched out from the database, and an "Assembling" process is used to get the properties of the bearing. Sample calculations show that, the minimum film thickness of the bearing is significantly reduced by pad deformation. The stiffness, and especially bearing damping, are also seen to be significantly affected by pad and pivot elastic deformations. In accordance with these, effects of pad and pivot elastic deformations on rotor-bearing system dynamic behavior are shown to be adverse, in the sense that resonant amplitude rises significantly, and stability margin decreases due to these deformations, although the latter alone will not cause instability. It is concluded that due consideration should be given to the pad and pivot stiffnesses during design stage, or the deformation effects should be accounted for if the deformation factors are significant.

Nomenclature

A: oil film domain

B: oil film span

b_{ij}, B_{ij}: dampings and non-dimensional dampings, $B_{ij} = b_{ij}\psi^3/(2\eta l)$ ($i, j = x, y$ or r, t)

c: difference between pad and journal radii

$c_{\min}$: minimum radial clearance of bearing

d: bearing diameter

E: modolus of elasticity

F_r, F_ϕ: non-dimensional pad force components in radial and perpendicular directions, with the relative unit $\eta\omega dl/\psi^2$

F_t: non-dimensional pad frictional force, with the relative unit $\eta\omega dl/\psi$

f: flexibility coefficient of pad

fo: static deflection of rotor due to its own weight

h, H: film thickness and its non-dimensional value, $H = h/c$ for pad; $H = h/c_{\min}$ for bearing

δH: variation of H due to deformation

K_d: pad deformation factor, $K_d = 0.75(1-\nu^2)\eta \cdot \omega/(T_1 \cdot E \cdot \psi^3 \cdot a_J)$

k_{ij}, K_{ij}: stiffnesses and non-dimensional stiffnesses,

---

* In collaboration with Z. Zheng and X. Wu. Reprinted from *Proceeding of Asia-Pasific Vibration Conference '93*, 1993, 2: 365 – 369.

$K_{ij} = k_{ij}\psi^3/(2\eta l\omega)$ ($i, j = x, y$ or r, t)

k_p: stiffness of pivot

k_r: bending stiffness of rotor

l: bearing length

m: bearing preload factor

m_r: rotor mass

p, P: pressure and non-dimensional pressure, $P = p\psi^2/(\eta\omega)$

Q_s: derivative of complex amplitude of pressure with respect to complex amplitude of parameter s ($s = \Delta\varepsilon_r, \Delta\varepsilon_t$)

q_z, Q_z: volumetric side leakage and its non-dimensional value, $Q_z = q_z/[6(l/d)c^3\omega/\psi^2]$

So, So_k: Sommerfeld numbers, $So = F\psi^2/(\eta\omega dl)$; $So_k = F\psi^2/(\eta\omega_k dl)$

T: non-dimensional shear stress, with the relative unit $\eta\omega/\psi$

T_1: ratio of pad thickness to neutral diameter

x_r, y_r: coordinates at rotor middle

x_j, y_j: coordinates at journal

z, Z: axial coordinate and its non-dimensional value, $Z = 2z/l$

α_p: pivot deformation factor, $\alpha_p = 2\eta\omega l/(\psi^3 k_p)$

α_J: factor of pad curvature, $\alpha_J = 1 + 3T_1^2/5 + 3T_1^4/7 + \cdots$

β: angular span of pad

γ: offset coefficient of pivot

$\varepsilon_r, \varepsilon_t$: radial and tangential eccentricity ratios of journal relative to pad, $\varepsilon = e/c$

η: dynamic viscosity

ϕ: angular coordinate

ψ: relative clearance, $\psi = 2c/d$

ν: Poisson's ratio

Ω: whirl ratio, $\Omega = \omega_1/\omega$

ω: angular speed

ω_1: angular frequency of whirl

ω_k: $\omega_k = \sqrt{k_r/m_r}$

1 Introduction

Tilting pad bearings are widely used to support high speed rotors, due to their high stability. The configuration of a tilting pad inevitably makes it apt to be elastically deformed under load, the more so if the pad is made of material of low modulus of elasticity, such as copper alloys. The contact between pad and pivot is also very easily deformed, due to the very small contact area. Since it is possible that such deformations will affect the bearing and rotor system behavior severely[1], researchers have made investigations on their effects. [2] used a plane strain curved beam model to calculate the pad static and dynamic deformations, but only continuity of deformation along the pad span has been observed in this model, with the demand on smoothness being violated[3]. [4] also based its analysis on curved beam, but only the effect of average curvature has been considered. [5] used FEM to calculate the pad deformation, but only the static component of deformation has been considered. Due to the complexity of the problem, if general conclusions are to be drawn from systematic analyses, it seems that one has to be satisfied with an easy-to-use but still adequately effective model, such as the plane strain curved beam. It is therefore the aim of the present work to use this model to make systematic calculations of the static and dynamic EHL problems of the tilting pad, and to arrange the results into a rather comprehensive database of properties of elastically deformable tilting pad. This database is then relied upon to make systematic investigations over the effects of pad and pivot elastic deformations on dynamic behaviors of tilting pad bearings and rotor systems supported on them.

2 Properties of a Tilting Pad

Fig. 1 shows a tilting pad of journal bearing deformed statically and dynamically. When it is acted upon by the oil film pressure, both its right and left wings will bend as cantilever curved beams. The bending deformation of the pad influences the film thickness distribution, feeding back to influence the film pressure distribution. The simultaneous solution of the pressure, pad deformation, and film thickness forms the static EHL of a tilting pad. From this, static properties of a tilting pad, such as load carrying capacity, friction, and flow rate, can be deduced.

When the journal is whirling around its static equilibrium position, simple harmonic motions with the same frequency in both horizontal and vertical directions will be assumed, if only small amplitude vibrations are considered. The journal whirl is accompanied by corresponding variation of the oil film distribution. The

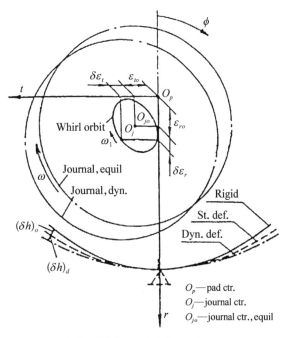

Fig. 1 Tilting pad deformations

latter causes a dynamic variation of the pad deformation. This forms a constituent of the dynamic variation of the film thickness distribution, while another constituent comes directly from the journal motion. The dynamically perturbed pressure, dynamic deformation of pad, and dynamic variation of film thickness are simultaneously solved in the dynamic EHL of a tilting pad. Whence the dynamic properties of a deformable tilting pad can be deduced, such as the stiffness and damping coefficients under a series of values of whirl ratio ranging from 0 to 1.

The governing equations of the static EHL of tilting pads are[6]

1. the lubrication equation:

$$\frac{\partial}{\partial \phi}\left(H_0^3 \frac{\partial P_0}{\partial \phi}\right)+\left(\frac{d}{l}\right)^2 \frac{\partial}{\partial Z}\left(H_0^3 \frac{\partial P_0}{\partial Z}\right)=6 \frac{\partial H_0}{\partial \phi} \qquad (1)$$

2. the deformation equation:

$$(\delta H)_0 = \iint_A P_0 f \, \mathrm{d}\phi_1 \mathrm{d}Z \qquad (2)$$

3. the geometric equation of film thickness:

$$H_0 = 1 + \varepsilon_{t_0}\sin\phi + \varepsilon_{r_0}\cos\phi + (\delta H)_0 \tag{3}$$

The flexibility coefficient f denotes the pad deflection at ϕ caused by unit force at ϕ_1. Its value is derived from the plane strain curved beam model[3].

This system of equations are solved by iterations. The static properties of the pad are calculated as:

$$F_r = -\iint_A P_0 \cos\phi\, d\phi\, dZ \tag{4}$$

$$F_\phi = -\iint_A P_0 \sin\phi\, d\phi\, dZ = 0 \tag{5}$$

$$F_t = \iint_A T\, d\phi\, dZ \tag{6}$$

$$Q_z = \int_B \left.\frac{\partial P}{\partial Z}\right|_{z=1} \cdot d\phi \tag{7}$$

The governing equations of the dynamic EHL of tilting pads are
1. the complex lubrication equation:

$$\frac{\partial}{\partial \phi}\left(H_0^3 \frac{\partial Q_s}{\partial \phi}\right) + \left(\frac{d}{l}\right)^2 \frac{\partial}{\partial Z}\left(H_0^3 \frac{\partial Q_s}{\partial Z}\right) = -18\frac{M_s}{H_0}\frac{dH_0}{d\phi} + 6\frac{dM_s}{d\phi}$$
$$+ 12i\Omega M_s - 3H_0\left(H_0\frac{\partial M_s}{\partial \phi} - M_s\frac{dH_0}{\partial \phi}\right)\frac{\partial P_0}{\partial \phi} \tag{8}$$

2. the complex deformation equation:

$$(H_d)_s = \iint_A Q_s f\, d\phi_1\, dZ \tag{9}$$

3. the complex geometric equation:

$$M_s = \frac{\partial(\Delta\varepsilon_t \sin\phi + \Delta\varepsilon_r \cos\phi)}{\partial s} + (H_d)_s \tag{10}$$

These equations are also solved by iterations, and the dynamic coefficients of pad are calculated as:

$$K_{rr} = K_{rr}^* - [K_{tt}^*(K_{rt}^*K_{tr}^* - \Omega^2 B_{rt}^* B_{tr}^*) + \Omega^2 B_{tt}^*(K_{rt}^* B_{tr}^* + K_{tr}^* B_{rt}^*)]/(K_{tt}^{*} + \Omega^2 B_{tt}^{*}) \tag{11}$$

$$B_{rr} = B_{rr}^* - [B_{tt}^*(\Omega^2 B_{rt}^* B_{tr}^* - K_{rt}^*K_{tr}^*) + K_{tt}^*(K_{rt}^* B_{tr}^* + K_{tr}^* B_{rt}^*)]/(K_{tt}^{*} + \Omega^2 B_{tt}^{*}) \tag{12}$$

where

$$K_{ts}^* + i\Omega B_{ts}^* = -\iint_A Q_s \sin\phi\, d\phi\, dZ \tag{13}$$

$$K_{rs}^* + i\Omega B_{rs}^* = -\iint_A Q_s \cos\phi \, d\phi \, dZ \tag{14}$$

As said above, when solving the dynamic EHL problems, a series of Ω values within the range 0-1 are taken, so that a spectrum of dynamic coefficients within this whirl ratio range is obtained.

3 Database of Properties of Deformable Tilting Pad

For given values of l/d, β and γ, the values of F_r, $H_{\min}$, F_t, Q_z, $K_{rr}(\Omega)$ and $B_{rr}(\Omega)$ in dependence on ε_{r0} and K_d are obtained as results of the static and dynamic EHL analyses of tilting pad. They are arranged to form a database. A short program is written to search and interpolate the database to give out deformable tilting pad properties corresponding to input values of ε_{r0} and K_d.

4 Effects of Pivot Deformation[7]

The effect of pivot static deformation under the steady state pad load F_r is to reduce the effective eccentricity of journal relative to the pad (Fig. 2):

$$\varepsilon_{r0} = \varepsilon_r' - \alpha_p F_r \tag{15}$$

In other words, a greater global eccentricity of journal relative to the pad before its pivot deforms is needed to induce a given effective eccentricity:

$$\varepsilon_r' = \varepsilon_{r0} + \alpha_p F_r \tag{16}$$

The effect of pivot dynamic deformation can be accounted for by the serial dynamic coefficients of pad and pivot:

$$K_s = \frac{1}{\alpha_p} \frac{K_{rr}(K_{rr} + 1/\alpha_p) + (\Omega B_{rr})^2}{(K_{rr} + 1/\alpha_p)^2 + (\Omega B_{rr})^2} \tag{17}$$

$$B_s = \frac{B_{rr}/\alpha_p^2}{(K_{rr} + 1/\alpha_p)^2 + (\Omega B_{rr})^2} \tag{18}$$

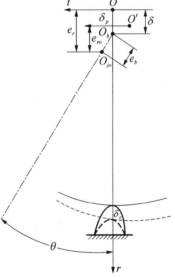

Fig. 2 Pivot deformation

Therefore, when α_p is given, the pad database can be transformed into another one relating F_r, $H_{\min}$, F_t, Q_z, $K_s(\Omega)$ and $B_s(\Omega)$ to ε_r' and K_d. The global eccentricity ratio ε_r' relative to each pad is determined from the journal eccentricity ratio relative to the bearing center by geometric relation. This value is then relied upon to search and interpolate the pad database for the serial dynamic coefficients.

5 Assembling Bearing Properties

For the case where the locations of the pivots are symmetric to the load line as shown in Fig. 3, the steady state eccentricity of the journal relative to the bearing center is always along the load line. Therefore, the eccentricity ratio of journal relative to pad k can be calculated by the relation:

$$\varepsilon'_{r,k} = m + \varepsilon_b(1-m)\cos\theta_k \tag{19}$$

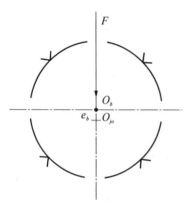

Fig. 3 Symmetric locations of pivots

The non-dimensional properties of all the pads k ($k = 1-n$) given by the database of deformable tilting pad with consideration of pivot deformation are assembled to form the non-dimensional static properties of the bearing:

$$So_b = \sum_{k=1}^{n} F_{r,k}\cos\theta_k(1-m)^2 \tag{20}$$

$$H_{\min,b} = \min(H_{\min,k})/(1-m) \tag{21}$$

$$F_{t,b} = \sum_{k=1}^{n} F_{t,k}(1-m) \tag{22}$$

$$Q_{z,b} = \sum_{k=1}^{n} Q_{z,k}/(1-m) \tag{23}$$

and the non-dimensional dynamic properties for each whirl ratio:

$$K_{ij,b} = \sum_{k=1}^{n} K_{rr,k} U_i U_j (1-m)^3 \quad (i,j=x,y) \tag{24}$$

$$B_{ij,b} = \sum_{k=1}^{n} B_{rr,k} U_i U_j (1-m)^3 \quad (i,j=x,y) \tag{25}$$

where $U_x = \sin\theta_k$; $U_y = \cos\theta_k$.

6 Dynamic Behavior of Rotor-bearing System

A single mass symmetric rotor supported on a pair of identical tilting pad bearings (Fig. 4) is taken for investigating the effects of pad and pivot elastic deformations.

When analyzing the synchronous vibration due to residual unbalance, the role of the bearings can be expressed by their dynamic coefficients of unity whirl ratio. The dynamic coefficients of other values of whirl ratio are used for investigating the behavior at stability threshold. For example, the threshold value of some destabilising factor can be calculated for the rotor under the

Fig. 4 Single mass rotor-bearing system

working speed, and used to assess the stability margin of the system. The present work takes planar isotropical cross-coupling stiffness and negative damping applied on the rotor middle as the destabilising factors[8] in such analysis.

Critical speeds, resonant amplification factor and unbalance response can be calculated using the method described in [9].

As for the calculation of the threshold values of destabilising factors, the homogeneous equations of motion will be based upon. In the case when crosscoupling stiffness is applied on the rotor and reaches its threshold value k_{st}, we have:

$$m_r \ddot{x}_r + k_r(x_r - x_j) - k_{st} y_r = 0 \tag{26}$$

$$m_r \ddot{y}_r + k_r(y_r - y_j) + k_{st} x_r = 0 \tag{27}$$

$$k_r(x_r - x_j) = 2(k_{xx,b} x_j + b_{xx,b} \dot{x}_j + k_{xy,b} y_j + b_{xy,b} \dot{y}_j) \tag{28}$$

$$k_r(y_r - y_j) = 2(k_{yx,b} x_j + b_{yx,b} \dot{x}_j + k_{yy,b} y_j + b_{yy,b} \dot{y}_j) \tag{29}$$

The following can be assumed:

$$[x_r, y_r, x_j, y_j]^T = [\hat{x}_r, \hat{y}_r, \hat{x}_j, \hat{y}_j]^T \cdot e^{i\omega_1 t} \tag{30}$$

From the above equations, the threshold value of non-dimensional cross-coupling stiffness and the corresponding whirl ratio can be derived to be[7]:

$$K_{st} = 2\alpha \left[\frac{\omega_1 \alpha^2 \beta_{xx} B_{yy}}{(\beta_x^2 + \omega_1^2 B_{xx}^2)(\beta_y^2 + \omega_1^2 B_{yy}^2)} - \left(\frac{\omega K_{xx} \beta_x + \omega_1^2 B_{xx}^2}{\beta_x^2 + \omega_1^2 B_{xx}^2} - \omega_1^2 \right) \right.$$
$$\left. \left(\frac{\omega K_{yy} \beta_y + \omega_1^2 B_{yy}^2}{\beta_y^2 + \omega_1^2 B_{yy}^2} - \omega_1^2 \right) \right]^{1/2} \tag{31}$$

$$\omega_1 = [(-b + \sqrt{b^2 - 4ac})/(2a)]^{1/2} \tag{32}$$

where $\alpha = 2So_k/(f_0/c_{min})$; $\beta_x = \alpha + \omega K_{xx}$; $\beta_y = \alpha + \omega K_{yy}$; $a = B_{xx} B_{yy}(B_{xx} + B_{yy})$; $b = B_{xx} \beta_y^2 + B_{yy} \beta_x^2 - a$; $c = -\omega(K_{xx} B_{yy} \beta_x + K_{yy} B_{xx} \beta_y)$.

In the case when negative damping is applied to the rotor and reaches its threshold value b_{st}, the equations of motion are:

$$m_r \ddot{x}_r + k_r(x_r - x_j) - b_{st} \dot{x}_r = 0 \tag{33}$$

$$m_r \ddot{y}_r + k_r(y_r - y_j) - b_{st} \dot{y}_r = 0 \tag{34}$$

$$k_r(x_r - x_j) = 2(k_{xx,b} x_j + b_{xx,b} \dot{x}_j) \tag{35}$$

$$k_r(y_r - y_j) = 2(k_{yy,b} y_j + b_{yy,b} \dot{y}_j) \tag{36}$$

The threshold value of non-dimensional negative damping can be derived to be:

$$B_{st} = \min(B_{st,x}, B_{st,y}) \tag{37}$$

where

$$B_{st,i} = \frac{\alpha^2 B_{ii}}{\beta^2 + \omega_{1,i}^2 B_{ii}^2} \quad (i = x, y) \tag{38}$$

and the corresponding whirl ratio is:

$$\omega_{1,i} = [(-b_i + \sqrt{b_i^2 - 4a_i c_i})/(2a_i)]^{1/2} \tag{39}$$

where $a_i = B_{ii}^2$; $b_i = \beta_i^2 - a_i$; $c_i = -\omega K_{ii}\beta_i$ ($i = x, y$).

7 Results of Calculation

The 4-tilting pad bearing shown in Fig. 3 has been taken as a typical example for illustrating the effects of pad and pivot elastic deformations on bearing and rotor dynamic behaviors. The parameters of the bearing are: $l/d = 0.5$; $\beta = 75$ deg.; $\gamma = 0.6$; $m = 0.5$.

A part of the results of bearing properties is shown in Fig. 5. The results without pad and pivot deformations check very well with [6], which can be taken as a proof of the validity of the mathmatical treatment.

It can be seen from Fig. 5 that, under the same load as expressed by So, the pad deformation significantly reduces H_{min}. The bearing stiffness K_{yy} is affected by pad deformation in a rather complicated manner, while the bearing damping B_{yy} decreases significantly with pad deformation.

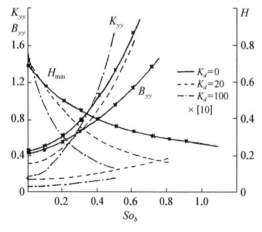

Fig. 5 Effect of pad deformation on bearing properties

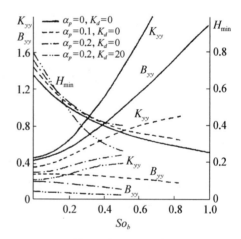

Fig. 6 Effect of pivot deformation on bearing properties

Fig. 6 shows the effects of pivot deformation on bearing properties. H_{min} is little affected, but both the bearing stiffness and damping decrease significantly. When both pad and pivot deformations are considered, their effects show in a combined manner.

From the change of bearing properties, it is easy to expect that the the critical speed will vary somewhat, and the resonant amplitude will rise seriously. Such effects have invariably been witnessed in the calculated results. The speed-amplitude curves shown in Fig. 7 display such effect of pad deformation. As known from practice, the employment of tilting pad bearings in high speed machinery sometimes causes high resonant amplitude. It ought to be worthwhile to pay attention to possible deformation effects in these cases.

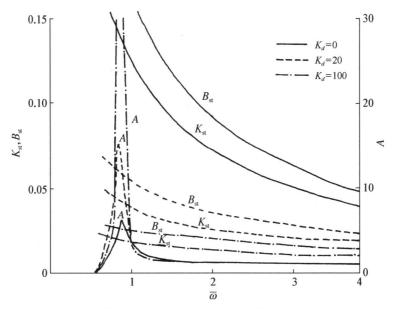

Fig. 7 Amplitude-speed relations and stability margin variation

Fig. 7 also shows the variation of the stability margin represented by K_{st} and B_{st}. It can be seen that, pad deformation is disadvantageous to the stability margin of the rotor-bearing system, although pad deformation alone will not cause a cross over to instability. The tendencies of variation of K_{st} and B_{st} are similar.

8 Conclusions

1. A database of properties of deformable tilting pad is very effective in reducing the time needed for calculation of tilting pad bearings with deformable pads and pivots.

2. Pad elastic deformation reduces the minimum film thickness of the bearing, changes the bearing stiffness, and significantly reduces the bearing damping. Pivot deformation reduces both the bearing stiffness and damping.

3. Pad and pivot elastic deformations enhances the resonance amplitude of the rotor-bearing system significantly.

4. Pad and pivot elastic deformations reduces the stability margin of the rotor-bearing system.

5. Attention should be paid in the design stage to ensure sufficient stiffnesses of pads and pivots. Alternatively, effects of elastic deformations should be accounted for if high deformation factors cannot be avoided.

Acknowledgements

The authors would like to thank The Science Foundation of Shanghai High Education

Bureau for sponsoring this work.

References

[1] Caruso, W. J. et al. , Proceedings of the 11th Turbomachinaery Symposium, 1983: 1 - 17.
[2] Nilsson, L. R. K. Proceedings of the 5th Leeds-Lyon Symposium on Tribology, 1969: 311 - 319.
[3] Zheng, Z. et al. Journal of Shanghai University of Technology(in Chinese), 1991, 12: 213 - 222.
[4] Lund, J. W. and Pederson, L. B. Trans. ASME, J. of Trib. , 1987,109: 65 - 70.
[5] Brugier, D. and Pascal, M. T. , Trans. ASME, J. of Trib. , 1989,111: 364 - 371.
[6] Zheng, Z. et al. Journal of Shanghai University of Technology (in Chinese), 1992,13: 189 - 196.
[7] Zhang, Z. et al. Journal of Shanghai University of Technology (in Chinese), 1992,13: 303 - 310.
[8] Zhang, Z. and Yu, L. , Journal of Shanghai University of Technology (in Chinese), 1985,6: 11 - 20.
[9] Glienicke, J. and Leonhard, M. Stabilitaetsprobleme bei der Lagerung schnellaufender Wellen, Bericht, Universitaet Karlsruhe, 1981.

Nonlinear Simulation of Lateral Vibration of an Experimental Rotor-Journal Bearing System[*]

Abstract: Simulation of Oscar Pinkus' well known experiment of rotor-journal bearing system dynamic behavior[1] has been done by applying a software package of nonlinear rotor dynamics aided by database of nonstationary oil film force of bearing bush. Effects of distributed rotor unbalance, rotor mass and flexibility, bearing nonlinear force and coupling mass are taken into consideration. Results compare satisfactorily with the experimental results published by Pinkus, with respect to vibration level, whirl frequency and dynamic locus of journal, within a significantly large range of running speed beginning from a low value and extending to above the 2nd critical. The effects of the magnitude and distribution of the unbalance on whirl amplitude and especially on shifting of dominant whirl frequency are clearly exhibited and noteworthy.

1 Introduction

A software package for numerical simulation of the dynamic behavior of rotors supported on journal bearings pertaining to their lateral vibration or whirl, either externally or self-excited, has been developed and applied to various cases, ranging from Jeffcot rotors to multi-span ones of large turbine-generator sets[2,3]. In particular, special behavior of rotor vibration under combined effects of unbalance and self-excitation when the speed surpasses the linear stability threshold has been exhibited, and different influences of unbalance on dynamic stability has been displayed. These phenomena have long been noticed in experiments or practice, but their mechanism has not yet been fully explained, although qualitative reasonings can usually be made more or less vaguely. A handy nonlinear simulation facility in the form of a software package makes it possible to study them quantitatively, to get at their more thorough explanation, and to find possible ways to improve the relevant design rules which are up to now based on linear modeling of oil film force. With this as aim, O. Pinkus' widely known and well noticed experiment[1] is chosen as the object of simulation, as Pinkus' paper contains detailed and rather full recordings of the above said complex behaviors. The published geometric and working parameters of the rotor-bearing system have been taken over in the calculations, with a few ones which are lacking in the paper postulated from the recorded values of critical

---

[*] In collaboration with Zhigang Li. Reprinted from *Proceedings of the 6th International Symposium on Transport Phenomena and Dynamics of Rotating Machinery*, Vol. 1, 1996: 97–105.

speeds. Values of residual unbalance and its distribution can naturally not be found from the paper, but are needed in simulation calculations. They are postulated in such a way that the calculated resonant amplitudes roughly correspond with the measured ones.

2 Methods of Calculation

Since the methods employed in the present paper have been fully published in [2,3], only a brief description will be given here.

The shaft is discretized and modeled by beam elements with lumped parameters. The linear and angular displacements in the vertical and horizontal planes at each node are used to form a state vector of that node. The system motion equations are expressed in matrix form with all these nodal state vectors as unknown quantities. The bearing forces, unbalance forces and gyroscopic forces, if any, are included in the system equation as excitation terms. The shaft can therefore be two identical, much smaller componental matrices successively for the horizontal and vertical planes. This results in significantly enhanced calculational accuracy and efficiency. The componental matrix is arranged into a tri-diagonal matrix with all its elements formed by block matrices. A new method of elimination and back sweeping is developed to solve it. During the solution process, the matrices which need be inversed are only these small matrices with 2 rows and 2 columns, and therefore direct inversion can easily be used, without the necessity of using numerical inversion as is usually done in rotor calculations. These inversed matrices, once obtained for the simulated rotor, are instored and repetitively used throughout the entire simulation calculation. The calculational accuracy and efficiency are thus further improved.

The bearing instantaneous forces are obtained by assembling the componental forces of all the included bushes. The latter are directly obtained from specially developed databases of non-stationary oil film forces of bearing bush. Each database is created beforehand for one of the combinations of length-to-diameter ratio and bush span angle present on the included bushes. This is done by systematic calculation, based on solving the non-stationary Reynolds equation within the full ranges of eccentricity ratio and attitude angle of the journal centre relative to the bush and the rotation/squeeze ratio. These parameter ranges are economically discretized while keeping adequate accuracy of interpolation, so that compact databases result. Reynolds' condition is adopted as the condition for film rupture in these calculations. The instantaneous values of the eccentricity ratio, the attitude angle, and the rotation/squeeze ratio pertaining to the bush are relied upon for interpolating the database to get the instantaneous values of non-dimensional horizontal and vertical force components of the bush. In this way, the nonlinearity and accuracy of the bearing forces are taken full care of within the scope of the present model. The resulting accuracy and efficiency of calculation are very helpful to the soundness and abundance, and therefore the usefullness, of the result.

Time marching integration of the system motion equation is performed to obtain the transient dynamic response of the rotor during a process of gradual rising of the rotor speed, beginning from a value significantly lower than the 1st critical, then surpassing the linear stability threshold, and finally ending at a value well above the 2nd critical. Houbolt's integration is adopted for this purpose, supplemented by a specially developed method of pre-estimation, to accelerate the necessary iterational correction while keeping up the stability of the time marching process. The instantaneous displacements at the main nodes of the shaft are recorded and later processed to get the required information, such as the vibration level represented by peak-to-peak excursion, whirl obit obtained by combining the vertical and horizontal displacements, and frequency spectrum obtained by applying FFT to the recorded displacement-time siries.

3 The Simulated Rotor-bearing System

Fig. 1 shows the simulated rotor schematically.

The parameters of the rotor-bearing system are taken from [1]. They are as follows.

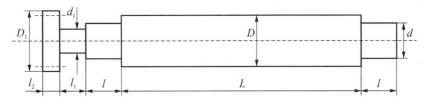

Fig. 1 The simulated rotor

(1) The bearings: halved cylindrical, double-side inlet;
  bearing length $l=50.8$ mm;
  journal diameter $d=50.8$ mm;
  radial clearance $c=0.0635$ mm;
  lubricant 150 Saybolt second/100°F;
  working temperature 120°F- 140°F.
(2) The rotor: total mass $m=29.03$ kg;
(3) The first critical speed $n_1=6\ 100$ r/min;
  the second critical speed $n_2=23\ 000$ r/min;
  the stability threshold speed $n_{st}=11\ 650$ r/min.

The other parameters needed for the simulation are postulated from the above values. They are: diameter and length of the main part of the rotor respectively $D=63.5$ mm; $L=1\ 099.2$ mm;

diameter and length of the overhang part of the rotor respectively $d_1=38.1$ mm; $l_1=50.8$ mm;

diameter and width of the coupling disk respectively $D_1=76.2$ mm; $l_2=25.4$ mm.

The density and elasticity modulus of the shaft material are taken respectively as:

$$\rho = 7\,550 \text{ kg/m}^3;$$
$$E = 2.1 \times 10^{11} \text{ Pa}.$$

The dynamic viscosity of the lubricant is taken as:

$$\mu = 0.018 \text{ Pa} \cdot \text{s}.$$

Residual unbalance is unavoidable on every rotor. Values of unbalances applied at 3 equally spaced nodes along the length of the rotor are adjusted so that the calculated 1st and 2nd resonant peaks appear to be similar to the measured ones. They are fixed in this way to be:

2.75×10^{-4} kg $\cdot$ m with phase angle $0°$ at 1/4 length from the rotor end;
4.395×10^{-4} kg $\cdot$ m with phase angle $120°$ at midplane of rotor;
2.75×10^{-4} kg $\cdot$ m with phase abgle $240°$ at 3/4 length.

A very small ellipticity $m = 0.1$ is assumed for the bearings, in view of the shape of the measured journal resonance orbit which has a rather pronounced ellipticity compared to the orbit calculated when no ellipticity of bearing is assumed. The possible existence of a small ellipticity is thus guessed at, and a value 0.1 appears to be appropriate after trials.

4 Results of Simulation and Discussion

The calculation is started from an angular speed of 500 rad/s. The time marching is done with a time step which is equal to 1/512 of the current period of revolution. The angular speed of the rotor is increased in each time step by a value 0.01 rad/s when the running speed is below 12 000 r/min, and 0.005 rad/s when the running speed is over 12 000 rad/s. The calculation is carried on until the rotor angular speed reaches 2 800 rad/s, so that the whole speed range of the simulated experiment is covered.

The journal displacements in both the vertical and horizontal directions are recorded after every 4 time steps. Fig. 2 shows a passage extracted from the recorded displacements when the rotor is crossing the 2nd critical speed. Lissajous' diagram and frequency spectrum can be obtained from the recordings, as exemplified in Figs. 3 and 4.

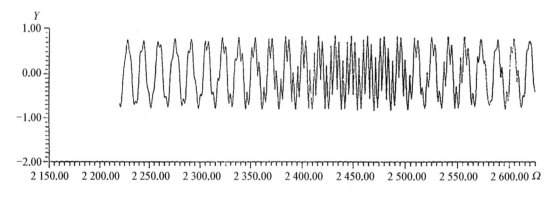

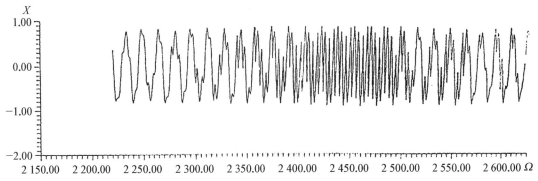

Fig. 2 A passage of the recorded journal displacement when corossing the 2nd critical speed

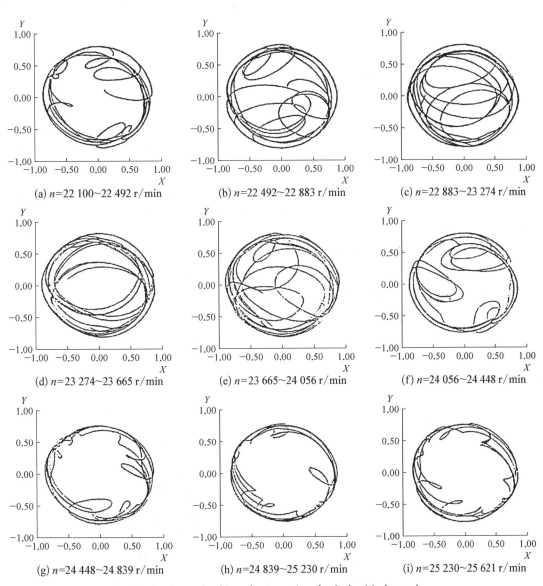

Fig. 3 Journal orbits when crossing the 2nd critical speed

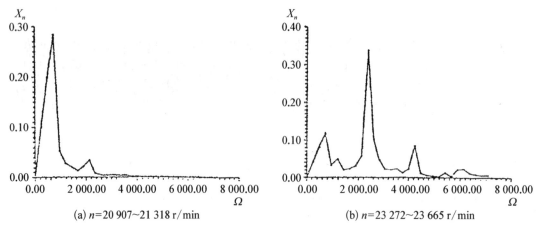

(a) $n=20\,907\sim21\,318$ r/min   (b) $n=23\,272\sim23\,665$ r/min

Fig. 4 Frequency spectra before and during crossing the 2nd critical speed

Fig. 5 shows the value of peak-to-peak excursion of the journal in vertical direction in relation to angular speed. It can be seen that the rotor system crosses its 1st critical at an angular speed about 650 rad/s. After this, the amplitude of vibration falls, but rises again at about 1 250 rad/s, where a half-frequency component appears and grows rapidly, as exhibited by the corresponding frequency spectrum. This self-excited component is also the cause of the excursion curve's quick pulsation here. With the increase of speed, the self-excited component rises drastically and quickly becomes the dominant component and makes the synchronous component relatively insignificant, so that the excursion curve becomes smoother again. Within a fairly wide range of speed, the frequency of the self-excited component keeps a value very near the 1st critical, viz., 650 rad/s, which is typical of oil whip or resonant whirl. The crossing through the 2nd critical at 2 430 rad/s shows up in the shifting of the dominant frequency from the excited to the syschronous and then back, rather than a rise and drop of the amplitude. Again, the frequency spectrum obtained by FFT, and the pulsations before and after the critical speed, are useful for exhibiting the shifts of the dominant frequency.

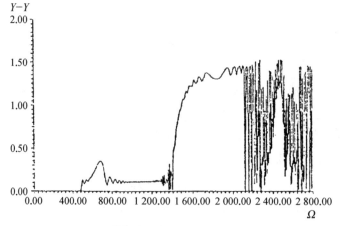

Fig. 5 Calculated excursion-time curve

Fig. 6 reproduces the measured amplitude quoted from [1]. Similarity between the measured and the calculated results, in both the vibration level and the shiftings of the dominant frequency within the whole speed range is evident. There is only one exception. A test rig resonance at 15 000 r/min is recorded in the experiment, accompanied by the corresponding shiftingsof the dominant frequency. This has not been included in the simulation, where the bearings are assumed to be mounted on rigid foundation.

Comparison between the calculated journal whirl orbits (Fig. 7) and the measured ones quoted from [1] (Fig. 8) can also be looked upon as satisfactory.

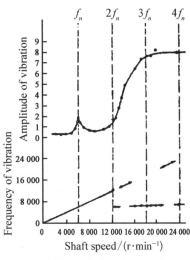

Fig. 6 Measured vibration amplitude curve, quoted from [1]

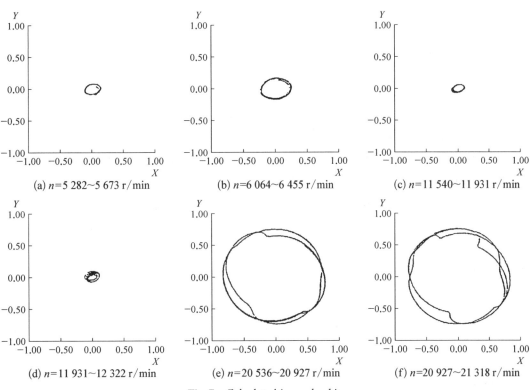

Fig. 7 Calculated journal orbits

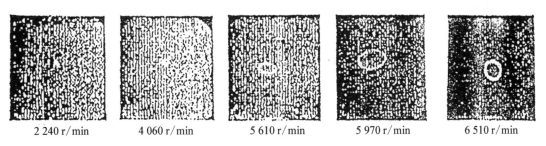

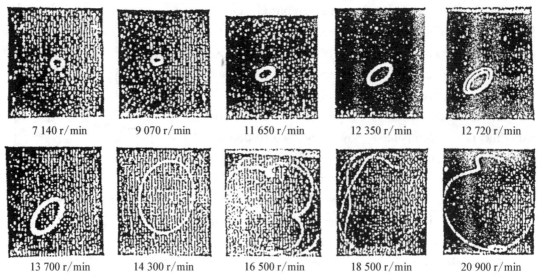

Fig. 8 Measured journal orbits, quoted from [1]

5 Conclusions

Dynamic simulation of lateral vibration of rotor-journal bearing system is capable of simultaneous consideration of nonlinear bearing force, residual unbalance and self-excited vibration. Interesting phenomenon of dominant frequency shifting, peculiar to crossing a critical speed after surpassing the stability threshold, can be realistically exhibited. This sort of simulation can therefore be very useful for explaining the seemingly complicated behavior of rotor-journal bearing systems induced by the co-existing effects nonlinearly combined.

Acknowledgement

The research project related with the present paper is sponsored by The National Natural Science Foundation of China and Foundation of Education Committee of Shanghai Municipality.

References

[1] Pinkus, O. Experimental Investigation of Resonant Whip, Trans. ASME, vol. 78, July, 1956.
[2] Wang, W. and Zhang, Z. Calculation of Journal Dynamic Locus Aided by Database of Non-Stationary Oil Film Force of Single Bush Segment. Proceedings of Asia-Pacific Vibration Conference '93, Kitakiushu, Japan, Nov. 1993, vol. 2, 365-369.
[3] Li, Z. Nonlinear Dynamic Analysis of Multi-Span Rotor-Journal Bearing Systems, PhD. Dissertation, Shanghai University, Shanghai, PR China, May 1995 (in Chinese).

Analysis of Crankshaft Bearings in Mixed Lubrication Including Mass Conserving Cavitation[*]

Abstract: The effects of two sided purely longitudinal, transverse and isotropic roughness on dynamically loaded finite journal bearings in mixed lubrication are studied. Using Christensen's stochastic model of hydrodynamic lubrication of rough surfaces and considering the running-in effect. Results show that the effect of contact loads is much smaller compared to that of the hydrodynamic film loads, and that the effects of roughness are closely tied up with the roughness texture and structure, features of nominal geometry, journal mass, and operating conditions.

Nomenclature

A = area of bearing surface, $A = BD$

A_c = total area of interference zones

B = bearing length

C = nominal radial clearance

D = bearing diameter

E' = composite elastic modulus, $\frac{1}{E'} = \frac{1}{2}\left(\frac{1-\nu_1^2}{E_1} + \frac{1-\nu_2^2}{E_2}\right)$

$\ddot{e}_x, \ddot{e}_y$ = accelerations of the journal in the x, y directions

F_{oilx}, F_{oily} = fluid film forces in the x, y directions

F_x, F_y = bearing loads in the x, y directions

h = nominal film thickness

h_T = total film thickness, $h_T = h - \delta_1 - \delta_2$

M = equivalent mass of the journal

p = film pressure

p_c = contact pressure

R = bearing radius

t = time

U_1, U_2 = tangential surface velocities of the bush and journal, $U = U_1 + U_2$

χ, z = coordinate of circumferential and axial directions

λ = length/diameter ratio, B/D

δ_1, δ_2 = asperity height measured from the nominal level of the bush and journal surfaces

ε = eccentricity ratio

η = surface density of asperity peaks

β = radius of curvature at the peak

ν_1, ν_2 = Poisson's ratio of the bush and journal

Φ^* = Gaussian probability density of the sum of peak distributions of both sufaces

σ = standard deviation of roughness height distribution

σ^* = standard deviation of Φ^*

μ = fluid viscosity

1 Introduction

The nominal minimum film thickness in dynamically loaded journal bearings, such as engine bearings, is of the same order of magnitude as the surface roughness and it often

---

[*] In collaboration with CHAO ZHANG. Reprinted from *Proceedings of Proc. IC-HBRSD'97*, 1997: 19-24.

becomes such thin that the surface asperities interfere, resulting in bearing operating in the mixed lubrication. The hydrodynamic pressure and the asperity contact pressure, in this case, will carry the applied load together.

The theories for partial hydrodynamic lubrication were established by Christensen[1], Tonder[2], Patir and Cheng[3], and for the surface contact pressure of nominally flat surfaces by Greenwood and Williamson[4], Whitehouse and Archard[5], Nayak[6], Greewood and Tripp[7], and Aramaki et al.[8] Realistic bearing surfaces may possess different texture orientations due to the machining process and running-in, while two sided longitudinal, isotropic, and transverse roughness can characterize three lypicai surface textures well. Rhow and Elrod[9], Prakash[10], Chang and Qiu[11,12] extended the Christensen's concept[13] to two sided roughness in hydrodynamic lubrication.

A bar written above a variable denotes the expected value Boedo and Booker[14] used the "Averaged Flow Model" of Patir and Cheng[4] to study surface roughness effects on partial hydrodynamic lubrication of dynamically loaded journal bearings for the Gaussian roughness cases.

Our objective is to investigate the effects of these three types of running-in roughness on partial hydrodynamic lubrication of dynamically loaded journal bearings including mass conserving cavitation. For these effects, no result has yet been available in literature.

2 Governing Equations

Using Christensen's stochastic model of hydrodynamic lubrication of rough surfaces[13], the Reynolds equation of three types of roughness texture in the hydrodynamic area for Newtonian constant viscosity and density is:

$$\frac{\partial \rho \bar{h}_T}{\partial t} + \frac{\partial}{\partial x}\left(\rho \varphi_A - \frac{\rho \varphi_B}{12\mu}\frac{\partial \bar{p}}{\partial x}\right) + \frac{\partial}{\partial z}\left(-\frac{\rho \varphi_C}{12\mu}\frac{\partial \bar{p}}{\partial z}\right) = 0 \tag{1}$$

where φ_i is orderly for the longitudinal, isotropic, and transverse textures as follows:

$$\varphi_A \text{——} \frac{U\bar{h}_T}{2},\ \frac{U\bar{h}_T}{2},\ \frac{U}{2}\frac{\overline{h_T^{-2}}}{\overline{h_T^{-3}}} - \frac{U_1\overline{\delta_1 h_T^3} + U_2\overline{\delta_2 h_T^3}}{\overline{h_T^{-3}}};$$

$$\varphi_B \text{——} \overline{h_T^3},\ \overline{h_T^3},\ 1/\overline{h_T^3};\ \varphi_C \text{——} 1/\overline{h_T^3},\ \overline{h_T^3},\ \overline{h_T^3}$$

where $h_T = h - \delta_1 - \delta_2$ and

$$\overline{(\ )} = \int_{-\infty}^{\delta_2'}\int_{-\infty}^{\delta_1'}(\ )f_1(\delta_1)f_2(\delta_2)d\delta_1 d\delta_2$$

$f_1(\delta_1)$ and $f_2(\delta_2)$ denote the probability density functions of δ_1 and δ_2. For full hydrodynamic lubrication, $\delta_1' = \delta_{1\max}/C$ and $\delta_2' = \delta_{2\max}/C$, while for partial lubrication, δ_1' and δ_2' are obtained according to the elastic modulus ratio of the bush and journal. The deduction for Eq. (1) is similar to that given by Chang and Qiu[11,12].

By modifying Elrod's algorithm[15], a void fraction, λ, and a cavitation index, g, are defined as follows:

Full film zone ($\lambda \geqslant 0$): $\lambda = \bar{p}$; $g = 1$

Cavitated zone ($\lambda \leqslant 0$): $\lambda = \theta - 1$; $g = 0$

where θ is the fractional film content. Inserting λ and g into Eq. (1), we get:

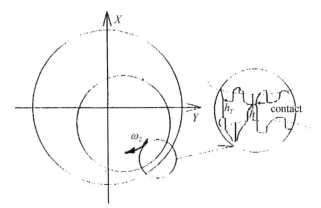

Fig. 1 Journal bearing configuration

$$\frac{\partial}{\partial t}\{[1+(1-g)\lambda]\bar{h}_T\} + \frac{\partial}{\partial x}\left\{\varphi_A[1+(1-g)\lambda] - \frac{\varphi_B g}{12\mu}\frac{\partial \lambda}{\partial x}\right\} + \frac{\partial}{12\mu\partial z}\left\{-\varphi_C g \frac{\partial \lambda}{\partial z}\right\} = 0 \tag{2}$$

The first term in Eq. (2) can be discretized as

$$\frac{\partial}{\partial t}\{[1+(1-g)\lambda]\bar{h}^T\} = \frac{[1+(1-g_{i,j})\lambda_{i,j}]\bar{h}_i^T - [1+(1-g_{i,j}^*)\lambda_{i,j}^*]h_i^{T*}}{\Delta t} \tag{3}$$

The shear and pressure induced flow terms in Eq. (2) are, respectively,

$$\frac{\partial}{\partial x}\{\varphi_A[1+(1-g)\lambda]\} = -\frac{1}{\Delta x}\Big\{\varphi_{Ai-1}(1-g_{i-1,j})[1+(1-g_{i-1,j})\lambda_{i-1,j}]$$
$$-\varphi_{Ai}(1-g_{i,j})[1+(1-g_{i,j})\lambda_{i,j}]$$
$$+\frac{g_{i-1,j}\varphi_{Ai-1}}{2}(2-g_{i,j}) + \frac{g_{i,j}\varphi_{Ai}}{2}(g_{i-1,j}-2+g_{i+1,j})$$
$$-\frac{g_{i+1,j}g_{i,j}\varphi_{Ai+1}}{2}\Big\} \tag{4}$$

$$\frac{\partial}{\partial z}\left\{-\frac{\varphi_C g}{12\mu}\frac{\partial \lambda}{\partial z}\right\} = -\frac{\varphi_C}{12\mu\Delta z^2}(g_{i,j+1}\lambda_{i,j+1} - 2g_{i,j}\lambda_{i,j} + g_{i,j-1}\lambda_{i,j-1}) \tag{5}$$

$$\frac{\partial}{\partial x}\left\{-\frac{\varphi_B g}{12\mu}\frac{\partial \lambda}{\partial x}\right\} = -\frac{1}{12\mu\Delta x^2}[\varphi_{Bi+1/2}g_{i+1,j}\lambda_{i+1,j} - (\varphi_{Bi+1/2}+\varphi_{Bi-1/2})g_{i,j}\lambda_{i,j}$$
$$+ \varphi_{Bi-1/2}g_{i-1,j}\lambda_{i-1,j}] \tag{6}$$

The nominal contact pressure given by Greewood and Tripp[8] is utilized here as follows:

$$p_c(\bar{h}/\sigma^*) = K'E'F_{5/2}(\bar{h}/\sigma^*) \tag{7}$$

the total contact load and the total area of interence zones can be calculated, respectively, from the following expressions:

$$F_c = 2BRK'E' \int_{\vartheta'}^{180} F_{5/2}(\bar{h}/\sigma^*)\cos(180-\vartheta)\mathrm{d}\vartheta \qquad (8)$$

$$A_c = 2BR\pi^2(\eta\beta\sigma^*)^2 \int_{\vartheta'}^{180} F_2(\bar{h}/\sigma^*)\mathrm{d}\vartheta \qquad (9)$$

where

$$K' = \frac{8\sqrt{2}}{15}\pi(\eta\beta\sigma^*)^2\sqrt{\frac{\sigma^*}{\beta}}$$

$$F_n(u) = \int_u^{180}(s-u)^n\Phi^*(s)\mathrm{d}s$$

$$\vartheta' = \arccos\left(\frac{3\sigma^* + \Delta_s - C}{C\varepsilon}\right) \qquad (10)$$

where Δ_s is the distance between centre lines of the roughness height and peak distributions and can be obtained from the following equation:

$$3\sigma^* + \Delta_s = \delta_{1\max} + \delta_{2\max} \qquad (11)$$

Referring to Figure 1, the equations of motion for the journal are:

$$M\ddot{e}_x = F_{oilx} + F_{ex} + F_{ex} \qquad (12)$$

$$M\ddot{e}_y = F_{oily} + F_{ey} + F_{ey} \qquad (13)$$

where F_{oilx} and F_{oily} are given by

$$F_{oilx} = -\int_A \bar{p}\cos\theta\mathrm{d}A \qquad (14)$$

$$F_{oily} = -\int_A \bar{p}\sin\theta\mathrm{d}A \qquad (15)$$

Eq. (2) is solved by the finite difference method with SOR scheme. 80 grid points are used in circumferentially and 24 axially across half the bearing width because this mesh has been found to be generally most suitable[16]. The convergence tolerance value (i. e. relative error) used for the calculation of pressure and degree of filling is 0.000 05. Eqs. (12) and (13) are integrated simultaneously using a fourth order Runge-Kutta numerical scheme with a time step equivalent to 1 degree rotation of the journal. The integrations to determine F_{oilx}, F_{oily}, F_{ex} and F_{ey} are performed using Simpson's Rule. The dynamic problem is solved with arbitrary initial conditions until periodicity is achieved.

3 Results and Discussion

The operation conditions are given in Table 1. The probability density function shown in Fig. 2 measured from the bush surface of an engine main bearing after running-in is used as that of bush and journal. K' and Δ_s are such chosen that the contact load effect can be demostrated better since K' lies in the range 0.000 03—0.003 for engineering surfaces and $\Delta_s \geqslant 0$. The loads acting on the bearing are shown in Fig. 3.

Table 1 Operating conditions

B	2.1×10^{-2} m	D	7.2×10^{-2} m
C	5.0×10^{-4} m	μ	3.0×10^{-3} Pa·s
M	400 kg	E'	46.5 GPa
K'	0.003	Δ_s	0
U_1	0	U_2	13.56 m·s^{-1}
Oil supply	Half groove(0 - 180)	p_0	3.5×10^5 N·m^{-2}

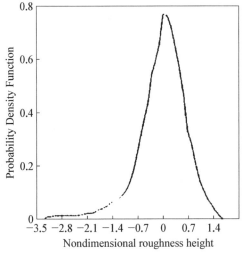

Fig. 2 Probability function

Fig. 3 Main bearing load

Table 2 Relation between σ_i and $\delta_{i\max}$ (A—running-in surface, B—Gaussian rough surface)

σ_1	σ_2	$\delta_{1\max}$		$\delta_{2\max}$		$\delta_{1\max}+\delta_{2\max}$	
		A	B	A	B	A	B
0.7	0.1	1.17	2.10	0.16	0.30	1.28	2.40
0.4	0.4	0.64	1.20	0.64	1.20	1.28	2.40

Table 3 and Figs. 4 – 9 indicate the following:

As compared to a smooth bearing, roughnesses always decrease the minimum nominal film thickness $h_{\min}$. In the three roughness textures, the longitudinal roughness always has the biggest $h_{\min}$. The isotropic roughness has bigger $h_{\min}$ than the transverse roughness when considering contact loads.

The same roughness structure ($\sigma_1=\sigma_2=0.4$ μm) has smaller $h_{\min}$ than the different roughness structure ($\sigma_1=0.7$ μm, $\sigma_2=0.1$ μm) when considering contact loads.

The contact loads have significant effects on $h_{\min}$ in the isotropic case and the

Table 3 The effects of roughness for main bearings

$h_{min}/\mu m$						$\bar{P}_{max}/MPa$					
I	I_c	L	L_c	T	T_c	I	I_c	L	L_c	T	T_c
$h_{min}(\sigma_1 = \sigma_2 = 0) = 0.67$						$\bar{P}_{max}(\sigma_1 = \sigma_2 = 0) = 189$					
$\sigma_1 = \sigma_2 = 0.4\ \mu m$											
0.37	0.65	0.88	0.86	0.60	0.61	346	231	100	167	493	740
$\sigma_1 = 0.7\ \mu m,\ \sigma_2 = 0.1\ \mu m$											
0.57	0.73	0.93	0.92	0.45	0.71	94	94	97	141	422	163

I, L, T — isotropic, longitudinal and transverse without the contact load effects.
I_c, L_c, T_c — I, L, T with the contact load effects.

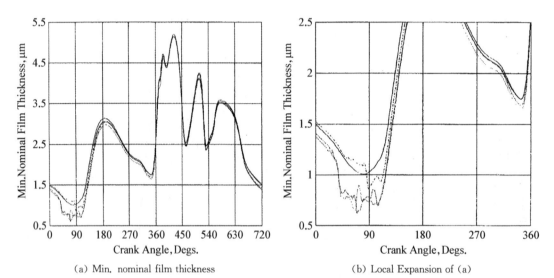

(a) Min. nominal film thickness   (b) Local Expansion of (a)

Fig. 4 Effects of roughness textures on h_{min}

——— S, ········ I, —·—· L, —··— T, $\sigma_1 = \sigma_2 = 0.4\ \mu m$

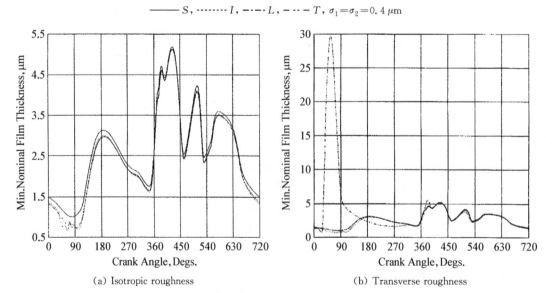

(a) Isotropic roughness   (b) Transverse roughness

Fig. 5 Effect of roughness structure on h_{min}

——— S, ········ $\sigma_1 = \sigma_2 = 0.4\ \mu m$, —··— $\sigma_1 = 0.7\ \mu m,\ \sigma_2 = 0.1\ \mu m$

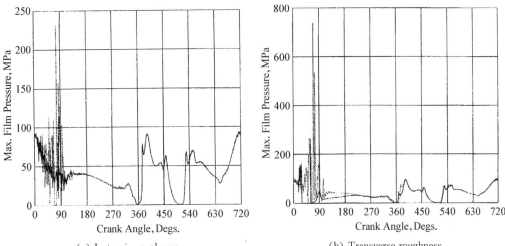

(a) Isotropic roughness (b) Transverse roughness

Fig. 6 Effects of roughness structure on $\bar{P}_{max}$

——— $\sigma_1=0.7\ \mu m, \sigma_2=0.1\ \mu m$, ········ $\sigma_1=\sigma_2=0.4\ \mu m$

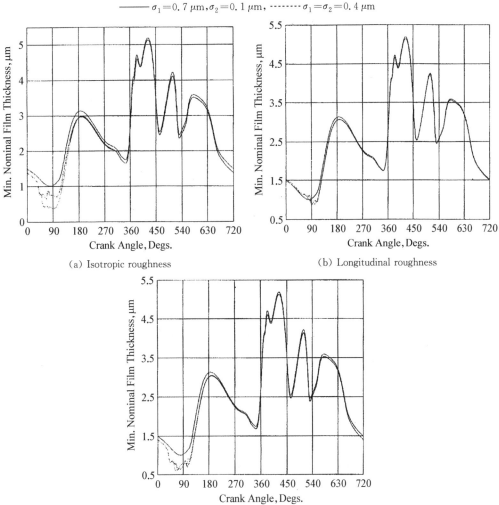

(a) Isotropic roughness (b) Longitudinal roughness

(c) Transverse roughness

Fig. 7 Effects of contact load on h_{min}

——— S, ········ without contact load, – ·· – with contact load, $\sigma_1 = \sigma_2 = 0.4\ \mu m$

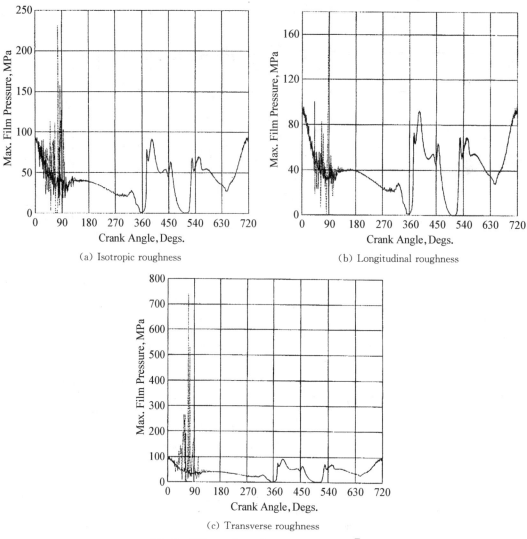

Fig. 8 Effects of roughness textures on $\bar{P}_{max}$

——— Smooth, ········ $\sigma_1=\sigma_2=0.4\ \mu m$

transverse case with the different structure, while its effects are insignificant in the longitudinal case and the transverse case with the same structure.

As compared to a smooth bearing, roughness usually induces osillations of maximum film pressure $\bar{P}_{max}$ in the contact zones. The degrees of those osillations are stronger in the case of the same roughness structure ($\sigma_1=\sigma_2=0.4\ \mu m$) than the different structure ($\sigma_1=0.7\ \mu m$, $\sigma_2=0.1\ \mu m$).

The crest values of the osillations of $\bar{P}_{max}$ in the rough cases are bigger than $\bar{P}_{max}$ in the smooth case, and these crest values in the transverse cases appear to be biggest in three roughness textures.

The roughness heighness distribution of the running-in surface deviates from the Gaussian distribution and makes its asperity peak lower than the latter as shown in Table 2 where $\delta_{i\max}$ for running-in are obtained according to Fig. 2, while $\delta_{i\max}$ for the Gaussian are calculated by the expression $\delta_{imax} = 3\sigma_i$. It means that in the mixed lubrication conditions, this deviation seriously affects the degree of contact. Therefore, the running-in effects should be considered.

4 Conclusions

Based on the results and above discussion, it can be concluded:

Roughnesses always decrease the minimum nominal film thickness and induces osillations of maximum film pressure in the contact zones. In the three roughness textures, the longitudinal roughness always has the biggest h_{min} and the transverse roughness has the biggest The crest values of the osillations of maximum film pressure. The isotropic roughness has bigger h_{min} than the transverse roughness when considering contact loads.

The effects of roughness are closely tied up with the roughness texture and tructure, features of nominal geometry, journal mass, and operation factors.

The running-in effects on roughness heighness distribution should be considered.

References

[1] Christensen, H. A Theory of Mixed Lubrication. Proc. Instn. Mech. Engrs. (London). Vol. 186, pp. 421 (1972).

[2] Tonder, K. Ssimulation of the Lubrication of Isotropically Rough Surfaces. ASLE Trans., Vol. 23, No. 3, pp. 326 - 333. (1977).

[3] Patir, N. and Cheng, H. S. An Averaged Flow Model for Determining Effects of Three-Dimensional Roughness on Partial Hydrodynamic Lubrication. Trans. ASME, F 100, pp. 12 - 17. (1978).

[4] Greewood, J. A. and Williamson, J. B. P. Contact of Nominally Flat Rough Surfaces. Proc. R. Soc. (London). Vol. A295, No. 1442, pp. 300 - 319(1966).

[5] Whitehouse, D. J. and Archard, J. F. The Pproperties of Rrandom Surfaces of Significance in Their Contact. Proc. R. Soc. (London), Vol. A316, No. 1524, pp. 97 - 121(1970).

[6] Nayak, P. R. Random Process Model of Rrough Surfaces. Journal of Lubrication Technology, Vol. 93, No. 3, pp. 398 - 407(1971).

[7] Greewood, J. A. and Tripp, J. H. The Contact of Two Nominally Flat Rough Ssurfaces. Proc. Inst. Mech. Eng. (London), Vol. 185, pp625 - 633(1970 - 71).

[8] Aramaki. H., Cheng. H. S., and Chung. Y. W. The Contact Between Rough Surfaces With Longitudinal Texture-Part I: Average Contact Pressure and Real Contact Area. ASME Journal of Tribology, Vol. 115, No. 3, pp. 419 - 424. (1993)

[9] Rhow, S. K. and Elrod, H. G. The Effects on Bearing Load-Carrying Capacity of Two-Sided Striated Roughness. ASME Journal of Lubrication Technology, Vol. 94, No. 1, pp. 554 - 560. (1974).

[10] Prakash, J. On the Lubrication of Rough Rollers. ASME Journal of Tribology, Vol. 106, No. 3, pp. 324 - 330. (1984).

[11] Chao Zhang and Zugan Qiu. Effects of Surface Roughness and Lubricant Non-Newtonian Property on the Performance of IC Engine Journal Bearings. Chinese Internal Combustion Engine Engineering, Vol. 16, No. 1, pp. 69 - 76. (1995).

[12] Chao Zhang and Zugan Qiu. Analysis of Two- Sided Roughness and Non-Newtonian Effects in Dynamically Loaded Finite Journal Bearings. Proc. of the International Tribology Conference, Yokohama, pp. 1005 - 1010 (1995).

[13] Christensen, H. Stochastic Models for Hydrodynamic Lubrication of Rough Surfaces. Proc. Instn. Mech. Engrs., Vol. 184, Part 1, No. 55, pp. 1013-1026(1969-1970).

[14] Boedo, S. and Booker, J. F. Body Force and Roughness Effects in A Mass-Conserving Model-Based Elastohydrodynamic Lubrication Model. Proc. of the International Tribology Conference, Yokohama, pp. 1061-1066 (1995).

[15] Elrod, H. G., A Cavitation Algorithm. ASME Journal of Lubrication Technology, Vol. 103, 3, pp. 350-354 (1981).

[16] Paydas, A., and Smith, E. H. A Flow-continuity approach to the analysis of hydrodynamic journal bearings. Proc Instn Mech Engrs Vol. 206, Part C, pp. 57-69(1992).

Nonlinear Dynamic Analysis of Multi-span Rotor-journal Bearing-foundation System: Part I*

Abstract: A new calculation method of nonlinear dynamics of rotor-bearing-foundation system of high degree of freedom is proposed in this paper, with the aim of studying dynamic behavior of multi-span rotor supported by journal bearings under large excitations. Elimination and sweeping algorithm of tridiagonoal block matrix is successfully applied to solve the large matrix involved; central difference pre-estimation is supplemented to Houbolt method to form a stable and efficient time marching integration method for solving the second order implicit non-linear equations; simplification of system equation by decoupling between the physical coordinates x and y further improves the calculation accuracy and speed. Also introduced in this paper are: the method of setting of suitable initial state values of the system to exclude unneeded disturbances, the method of consideration of foundation vibration, and the application of database of nonstationary oil film force. The program realized with these techniques appears to be efficient and accurate for the study of nonlinear dynamic behavior of multi-span rotor-bearings-roundation system of high degree of freedom.

Nomenclature

M_j, C_j = mass and damping matrices of the jth discrete node of rotor

X_j, F_j = displacement and force vectors of the jth discrete node of rotor

U_j, V_j, W_j = block stiffness matrices of the jth discrete node of rotor

M, C, K = mass, damping and stiffness matrices of overall rotor system

X_t, F_t = displacement and force vectors of overall rotor system at time t

$\tilde{M}$ = the equivalent mass matrix of discrete rotor system during time marching

R_t = the general force vector of rotor system at time t

Ω = rotational speed of rotor

J_{d_j} (J_{p_j}) = the diametrical and polar moments of inertia of jth discrete element

l_j, I_j = length and second moment of area of jth discrete element

$f_{x_k}(x, y, \dot{x}, \dot{y})$, $f_{y_k}(x, y, \dot{x}, \dot{y})$ = nonlinear oil film force of kth bearing in x and y directions

$\alpha_j = \dfrac{6EI_j}{l_j^2}$; $\beta_j = \dfrac{12EI_j}{l_j^3}$; $\gamma_j = \dfrac{2EJ_j}{l_j}$

1 Introduction

The rotor-bearing systems of high speed rotating machinery such as large turbo-

---

* In collaboration with Zhigang Li. Reprinted from *Proc. IC-HBRSD'97*, 1997: 277–283.

generators have high demands of dynamic properties to ensure working security. Because of the strong nonlinearity of journal bearings, linear dynamic analysis is unable to give reasonable explanation to numerous phenomena. For example, some rotors can work safely above their linear threshold speeds with harmlessly small orbits whereas oil whip may take place for some rotors working below the linear threshold speed if they are subjected to large excitations such as earthquake or blade loss. The reliability of the calculated result based on linear theory is therefore very often doubted. More and more researchers are now concentrating their interest to nonlinear rotor dynamics. A number of calculating and analysing schemes have been proposed so far. Of particular notice are the algorithm developed by M. L. Adams[1]; the finite element mothod by Joseph Padovann et al[2]; the modal synthesis method by H. D. Nelson[3]; the DT-TMM by A. Selva Kumer and T. S. Sanber[4], and Songyuan Lu[5], and the combined finite element-transient matrix method by R. Subbiah[6].

Inspite of the above mentioned significant progresses, the number of nonlinear rotordynainics research papers and the attention paid to them still do not match the importance of the problem. Most of the existing calculating algorithms still have certain inconveniences in use. For instance, analysis of a large system converted into modal coordinates needs to know a sufficiently long series of very accurate eigenvalues and eigenvectors of the overall system, which is often not easy to achieve. Besides, conversion between physical and modal coordinates at every time marching step makes the computation costly if the system is very large and experience is also needed for the appropriate truncation of the higher modes. Large computer capacity and computation time will be needed when using finite element method directly even though condensation of system matrix is made. As for DT-TMM, since the shaft displacements with the accompanying nonlinear oil film forces at every time marching step must be iterated, it is difficult to make the computation both quick and precise. Moreover, the time marching integration methods for second order implicit non-linear equations as seen in rotordynamics papers are still open to improvements, as is the method for obtaining the value of nonlinear oil film force. Because of these, the calculation scale of nonlinear dynamics of rotor-bearings systems as seen in most of the existing papers is still rather small, and comparatively few discrete nodes are often used for large rotor-bearing systems.

With the aim of studying the dynamic behavior of multi-span rotors supported on journal bearings subjected to large excitations, this paper proposes a set of new algorithms of nonlinear rotordynamics. In which, elimination and sweep algorithm of tridiagonal block matrix is applied to solve the large matrix involved; central difference pre-estimation is supplemented to Houbolt method to perform the time integration, and the system equation is simplified by decoupling between physical coordinates x and y. Also introduced in this paper are the method of evaluation of the initial state values of the system to exclude unneeded disturbances, the consideration of foundation vibration and the application of database of nonstationary oil film force to nonlinear rotordynamic analysis. The software packge realized with these techniques is capable of studying nonlinear dynamic behaviors of large rotor-bearing-foundation systems with several hundred discrete sections and more than twelve journal bearings on personal computers.

The present work will be given in two parts. The fundamental idea and main formulations will be described in the first part, and typical calculation results of nonlinear rotordynamics and comparisons with linear analyses and with experiments will be given in the second part.

2 Mechanics Model and General Dynamic Equations of the System

The shaft is discretized and modeled by beam elements and lumped parameters as shown in Fig. 1. The dynamic equation of jth discrete node of rotor and bearing at time t can be written as

$$M_j \ddot{X}_{j_t} + C_j \dot{X}_{j_t} + W_j X_{j-1_t} + U_j X_{j_t} + V_j X_{j+1_t} = F_{j_t} \tag{1}$$

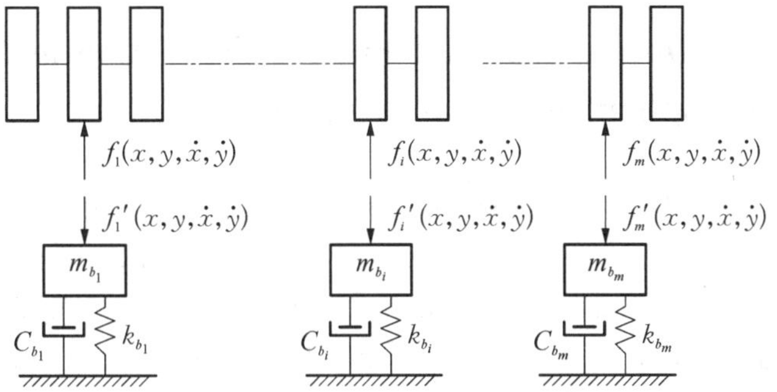

Fig. 1 Mechanics model of rotor-journal bering-foundtion system

where

$$M_j = \begin{bmatrix} m & & & \\ & m & & \\ & & J_x & \\ & & & J_y \end{bmatrix}_j ; \quad X_{j_t} = \begin{bmatrix} X \\ Y \\ \Phi \\ \Psi \end{bmatrix}_{j_t} ; \quad V_j = \begin{bmatrix} -\beta_{j+1} & 0 & \alpha_{j+1} & 0 \\ 0 & -\beta_{j+1} & 0 & \alpha_{j+1} \\ -\alpha_{j+1} & 0 & \gamma_{j+1} & 0 \\ 0 & -\alpha_{j+1} & 0 & \gamma_{j+1} \end{bmatrix} ;$$

$$U_j = \begin{bmatrix} \beta_{j+1}+\beta_j & 0 & \alpha_{j+1}+\alpha_j & 0 \\ 0 & \beta_{j+1}+\beta_j & 0 & \alpha_{j+1}+\alpha_j \\ \alpha_{j+1}-\alpha_j & 0 & 2(\gamma_{j+1}+\gamma_j) & 0 \\ 0 & \alpha_{j+1}-\alpha_j & 0 & 2(\gamma_{j+1}+\gamma_j) \end{bmatrix} ; \quad F_j = \begin{bmatrix} -f_x - P_{cx} \\ -f_y - P_{cy} \\ -M_{fy} \\ -M_{ft} \end{bmatrix}_j ;$$

$$W_j = \begin{bmatrix} -\beta_j & 0 & -\alpha_j & 0 \\ 0 & -\beta_j & 0 & -\alpha_j \\ \alpha_j & 0 & \gamma_j & 0 \\ 0 & \alpha_j & 0 & \gamma_j \end{bmatrix} ; \quad C_j = \begin{bmatrix} 0 & 0 & 0 & 0 \\ 0 & 0 & 0 & 0 \\ 0 & 0 & 0 & -J_p\Omega \\ 0 & 0 & J_p\Omega & 0 \end{bmatrix}_j$$

They are combined to give the general dynamic equation of the system as

$$M\ddot{X}_t + C\dot{X}_t + KX_t = F_t \qquad (2)$$

where, the mass and damping matrices of the overall system M and C are diagonoal block matrices formed by combinations of block mass matrices M_j and block damping matrices C_j, respectively, and the stiffness matrix of the system K is a tridiagonoal matrix formed by combining the block stiffness matrices U_j, V_j and W_j in the following way:

$$K = \begin{bmatrix} U_1 & V_1 & & & & & \\ W_2 & U_2 & V_2 & & & & \\ & \ddots & \ddots & \ddots & & & \\ & & W_j & U_j & V_j & & \\ & & & \ddots & \ddots & \ddots & \\ & & & & W_{n-1} & U_{n-1} & V_{n-1} \\ & & & & & W_n & U_n \end{bmatrix} \qquad (3)$$

3 Time Marching Integration Method with Estimation and Correction for Second Order Implicit Nonlinear Equations

Normally two kinds of method are used for solving Eq. (2) in time domain. Converting it into first order state vector and solving it by Runge-Kutta-Gill scheme is the one, and directly applying second order implicit time integration method is the other.

Eq. (2) is a large matrix with order of $4N$. If the first order state vector method is applied, inversion of state vector matrix with order of $8N$ is needed, which is quite slow when discrete number of the system N is very large, limitations from computer capacity are sometimes confronted. On the other hand, iteration of F_t and X_t is needed when using second order implicit time integration method as Eq. (2) is implicit in time domain. It appears that the methods as seen in the existing papers are still open to improvements which can be brought forth by suitable pre-estimation of X_t at each time step.

As we know, there exist two types of numerical intergration method, namely the implicit and the explicit method[7]. Their formulations are

$$\widetilde{M}X_{t+\Delta t} = R_t \qquad (4a)$$

$$\widetilde{M}X_{t+\Delta t} = R_{t+\Delta t} \qquad (4b)$$

respectively. The explicit integration method mainly includes the central difference method, while the implicit methods are several in number. Analysis about the stability and accuracy of these schemes during numerical process has been made in detail by Bathe & Wilson[7]. We can get to know from it that the central difference method is conditionally stable whereas Houbolt method is unconditionally stable, and Wilson θ method as well as Newmark β method are unconditionally stable under certain given values of θ and β.

Fig. 2(a) and Fig. 2(b) show the relative amplitude error AT and relative period error PE/T resulted from these numerical schemes after integrating one period of cosine vibration. From the point of view of numerical stability, the central difference method is by itself unsuitable for solving Eq. (2) due to the existence of a critical integration step of time Δt_{cr}, though high modes of vibration are not easily excited. However, the central difference mothod has the highest calculating precision as judged from Fig. 2. From simultaneous considerations of numerical stability and precision, it appears to be quite suitable for pre-estimating the state values of the system at each time integration step to reduce the computation time.

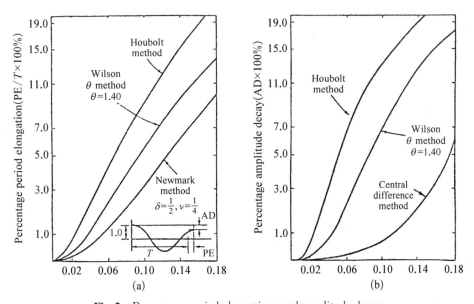

Fig. 2 Percentage period elongations and amplitude decays

Though the method of Wilson θ and Newmark β are superior to Houblot method in calculation precision as indicated in this example, the application region and convergence of them are inferior to those of Houblot method because of the limitation imposed by the assumption of linear acceleration. Therefore, it is believed that a time integration method formed by supplementing a pre-estimation by central difference to Houbolt method with correction should be superior to any existing single method for the time integration of second order implicit nonlinear equations. This conculsion has been proved by systematic calculations[8]. The combined calculating steps of estimation and correction are as follows:

　i. Solve Eq. (4a) by central difference method to obtain $X'_{t+\Delta t}$ and $\dot{X}'_{t+\Delta t}$.

　ii. Calculate $R'_{t+\Delta t}$ acconding to $X'_{t+\Delta t}$ and $\dot{X}'_{t+\Delta t}$.

　iii. Solve Eq. (4b) by Houbolt method to get $\widetilde{X}_{t+\Delta t}$ and $\dot{\widetilde{X}}_{t+\Delta t}$.

　iv. Judge whether condition of $\delta = | X'_{t+\Delta t} - \widetilde{X}_{t+\Delta t} | \leqslant \delta^* \ (\delta^* \leqslant 10^{-5})$ exsits.

　v. If $\delta > \delta^*$, then modify $X'_{t+\Delta t}$ and $\dot{X}'_{t+\Delta t}$ and repeat the steps from ii to iv. If $\delta \leqslant \delta^*$, then $X_{t+\Delta t}$ can be approximated by $\widetilde{X}_{t+\Delta t}$ and enter next time integration step.

4 Elimination and Sweeping Algorithm of Tridiagonal Block Matrix

When calculating $X'_{t+\Delta t}$ by solving Eq. (4a), it is easy to get $\tilde{M}^{-1}$ no matter how large the discrete number of rotor N is, as $\tilde{M}$ is a diagonal block matrix, inversion of which simply consists of calculating N block matrices $\tilde{M}_j^{-1}$. When calculating $\tilde{X}_{t+\Delta t}$ by solving Eq. (4b), however, it is not so easy to get $\tilde{M}^{-1}$ because $\tilde{M}$ here is a tridiagonal block matrix. If direct method is used for inversion of $\tilde{M}$, limitation by computer capacity may occur in case discrete number N is very large. Therefore, an elimination and sweeping algorithm of tridiagonal block matrix is introduced here.

The form of the matrix is given by Eq. (3). The problem to be solved can be simply expressed as

$$KX = F \qquad (5)$$

We can put it in the following form

$$K = \begin{bmatrix} O_1 & & & & & \\ P_2 & O_2 & & & & \\ & \ddots & \ddots & & & \\ & & P_j & O_j & & \\ & & & \ddots & \ddots & \\ & & & & P_{n-1} & O_{n-1} \\ & & & & & P_n & O_n \end{bmatrix} \begin{bmatrix} I_m & Q_1 & & & & \\ & I_m & Q_2 & & & \\ & & \ddots & \ddots & & \\ & & & I_m & Q_j & \\ & & & & \ddots & \ddots \\ & & & & & I_m & Q_{n-1} \\ & & & & & & I_m \end{bmatrix} \qquad (6)$$

where O_j, P_j and Q_j are block matrices of order m, and to be determined. We can get

$$\begin{cases} U_1 = O_1 \\ V_1 = O_1 Q_1 \end{cases} \qquad (7)$$

$$\begin{cases} W_i = P_i \\ U_i = P_i Q_{i-1} + O_i \\ V_i = O_i Q_i \end{cases} \qquad (8)$$

Eq. (5) is equivalent to the following:

$$\begin{cases} LY = F \\ SX = Y \end{cases} \qquad (9)$$

The steps for solving Eq. (9) are listed below.

i. Calculate vector $\{Q_i\}$ in sequence according to Eq. (8).

$$\begin{cases} Q_1 = U_1^{-1} V_1 \\ Q_i = [U_i - W_i Q_{i-1}]^{-1} V_i \end{cases} (i = 2, 3, \cdots, N) \qquad (10)$$

ii. Solve equation $LY = F$ by

$$\begin{cases} Y_1 = U^{-1} F_1 \\ Y_i = [U_i - W_i Q_{i-1}]^{-1}[F_i - W_i Y_{i-1}] \end{cases} (i = 2, 3, \cdots, N) \qquad (11)$$

iii. Solve equation $SX = Y$ by

$$\begin{cases} X_N = Y_N \\ X_i = Y_i - Q_i X_{i-1} \end{cases} (i = N-1, \cdots, 2, 1) \qquad (12)$$

The condition for applicability of this elimination and sweeping algorithm is the existence of U_i^{-1} and $[U_i - W_i Q_{i-1}]^{-1}$. However, if K^{-1} exists, the exsitence of U_i^{-1} and $[U_i - W_i Q_{i-1}]^{-1}$ is quite reasonable[9]. The computation work for Eq. (10) is only to inverse N block matrix with order of m, far less than that of inversion of matrix K with order of Nm. Moreover, the efficiency and accuracy of it will not be influenced by the discretization number N. Besides, the computation work required by Eqs. (11) and (12) is to multiply a block matrix of order m by a column vector for $3N - 2$ times only. These inversed matrices, once obtained for the simulated rotor, are instored and repetitively used throughout the entire simulation calculation. The calculational accuracy and efficiency are thus further improved.

5 Simplification of System Equation by Decoupling Between the Physical Coordinates x and y

Since Eq. (4b) is implicit in the time domain and the general force vector of system $R_{t+\Delta t}$ is dependent on the unknown vector $X_{t+\Delta t}$, they are determined by iterations with estimation and correction. It is possible to move the cross-coupling terms such as the hydroscopic ones from the left side of the equation to the right side and combine them with the general force vector $R_{t+\Delta t}$, so that the left side becomes decoupled between the physical coordinates x and y, and the system equation can be decomposed into two much simpler equations as:

$$M\ddot{X}_t + KX_t = F_{X_t} \qquad (13a)$$

$$M\ddot{Y}_t + KY_t = F_{Y_t} \qquad (13b)$$

The matrices involved will then be tridiagonal with elements in the form of block matrices of order of 2. As the inversion of these matrices can be done easily and directly, the computation amount will be reduced to only one fourth of the original, and the calculation precision will also be further improved.

6 Setting of Initial State Values of the Discrete System

When starting the simulation of a multi-span rotor-bearings system, it is very important to set suitable initial state values to the system. For a single-span rotor system with given load, the bearing loads are independent on the rotational speed. However, in a multi-span rotor-bearing system, the bearing loads are statically indeterminate and governed by the compatibility demand with consideration of the mounted bearing center heights and the journal levitation heights relative to the bearings. The latter are not only influencing but also influenced by the bearings loads, therefore further complicating the situation. It is usually necessary to use iteration in order to solve this problem. The existence of bearing lateral force components F_{x_i}, caused by posing the elastic shaft on non-uniform horizantal eccentricities ε_{x_t} in different bearings, and the cross-coupling stiffnesses of the bearings make the situation even further complicated by intimately linking the statically indeterminate problem in the horizantal plane to that in the vertical plane. The statically indeterminate problem in the horizontal plane is usually neglected in the literalure for simplification. Though this is acceptable for linear analyses, it appears somewhat too rough in nonlinear dynamic simulations, since the resulting artificially imposed loads may cause the effect like a big punch to the system at the start. The system response to such a physically aimless punch, will decay very slowly, and die out only after many periods. To get out of this entanglement, an effective calculation scheme is proposed here, in order that initial state values free of artificially imposed loads can be set at the start of simulation.

6.1 Establishing condensed system flexibility matrix A and stiffness matrix $\widetilde{K}$

Define

$$\widetilde{Y} = AP \tag{14}$$

where $A = \{a_{ji}\}_{(m-2)\cdot(m-2)}$; $P = [p_1, p_2, \cdots, p_{m-3}, p_{m-2}]^T$; $\widetilde{Y} = [\widetilde{y}_1, \widetilde{y}_2, \cdots, \widetilde{y}_{m-3}, \widetilde{y}_{m-2}]^T$. a_{ji} is the rotor deformation at location of the jth support induced by unit force acting at the ith support (see Fig. 3), and $(m-2)$ here is the number of internal supports not including the two end ones. Once the condensed flexibility matrix A is established, the condensed stiffness matrix $\widetilde{K}$ can easily be obtained as the inversion of A.

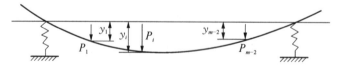

Fig. 3 Load and flexibility on rotor internal nodes

6.2 Deriving deformation equation with consideration of journal levitations

The deformation curve of the shaft centerline relative to the two end bearings is here defined as the "relative deformation curve" (see Fig. 4). Denote the relative deformations at the bearings as x_i and y_i, the arrays of which are used to form the deformation vectors

X and Y. By using Eq. (14) we can get

$$\begin{cases} X = -AF_x \\ Y = Y_0 - AF_y \end{cases} \quad (15)$$

where Y_0 is induced by the rotor weight. From Fig. 4, we can know that the relative displacements y_i' and x_i' in the ith bearing can be expressed as (take y_i' for example)

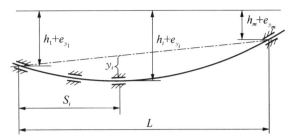

Fig. 4 Relative flexibility of rotor

$$y_i' = h_i + e_{y_i} - (h_m + e_{y_m}) - \frac{(h_1 + e_{y_1}) - (h_m + e_{y_m})}{L}(L - S_i) \quad (16)$$

The relative displacement vectors Y' and X' are then formed as

$$X' = [x_1', x_2', \cdots, x_{m-3}', x_{m-2}']^T \quad (17a)$$

$$Y' = [y_1', y_2', \cdots, y_{m-3}', y_{m-2}']^T \quad (17b)$$

Calculation routine and steps for determining the deformation vector Y and the relative displacement vector Y', as well as X and X', are as follows:

(1) Calculate the journal eccentricities e_{x_i} and e_{y_i} in each bearing by iteration, until the bearing loads f_{x_i} and f_{y_i} adequately satisfy the compatibility condition. Then calculate the linearized stiffness coefficients of each bearing $K_{l,j}$ corresponding to the calculated eccentricity values e_{x_i} and e_{y_i}, $(l, j = x, y)$.

(2) Obtain the deformation vectors X and Y by Eq. (15);

(3) Get the relative displacement vectors Y' and X' by Eq. (16);

(4) Calculate error vectors ΔX and ΔY, and combine them to form the deviation vector ΔS:

$$\begin{cases} \Delta X = X - X' \\ \Delta Y = Y - Y' \end{cases} \quad (17)$$

$$\Delta S = [\Delta x_1, \Delta y_1, \Delta x_2, \Delta y_2, \cdots, \Delta x_{m-2}, \Delta y_{m-2}]^T \quad (18)$$

(5) To make corrections in order to diminish the deviation vector ΔS, force incremental vector ΔF induced by eccentricity incremental vector $\Delta \varepsilon$ must be introduced. The relationship between ΔF and $\Delta \varepsilon$ is as follows:

$$\begin{bmatrix} k_{xx_1} & k_{xy_1} & & & & & & \\ k_{yx_1} & k_{yy_1} & & & & & & \\ & & k_{xx_2} & k_{xy_2} & & & & \\ & & k_{yx_2} & k_{yy_2} & & & & \\ & & & & \ddots & & & \\ & & & & & k_{xx_{m-2}} & k_{xy_{m-2}} \\ & & & & & k_{yx_{m-2}} & k_{yy_{m-2}} \end{bmatrix} \begin{bmatrix} \Delta e_{x_1} \\ \Delta e_{y_1} \\ \Delta e_{x_2} \\ \Delta e_{y_2} \\ \vdots \\ \Delta e_{x_{m-2}} \\ \Delta e_{y_{m-2}} \end{bmatrix} \begin{bmatrix} \Delta F_{x_1} \\ \Delta F_{y_1} \\ \Delta F_{x_2} \\ \Delta F_{y_2} \\ \vdots \\ \Delta F_{x_{m-2}} \\ \Delta F_{y_{m-2}} \end{bmatrix}$$

It can be written in short form as

$$G\Delta\varepsilon = \Delta F \tag{19}$$

(6) Due to the existence of $\Delta\varepsilon$, the actual deviation vector to be diminished is $\Delta S'$:

$$\Delta S' = \Delta S - \Delta\varepsilon \tag{20}$$

Now, expand the condensed stiffness matrix of the overall system $\widetilde{K}$ into its two-dimensional form $\bar{K}$, there exists

$$\bar{K}\Delta S' = \Delta F \tag{21}$$

(7) Making use of Eq. (19), Eq. (20) and Eq. (21) yields

$$\bar{K}[\Delta S - \Delta\varepsilon] = G\Delta\varepsilon \tag{22}$$

Thus, the eccentricity incremental vector $\Delta\varepsilon$ can be solved from

$$\Delta\varepsilon = [G + \bar{K}]^{-1}\bar{K}\Delta S \tag{23}$$

(8) Since the force vector has been increased by ΔF for modifying the deviation vector $\Delta S'$, the forces on the two reference bearings should also be increased by Δf_{x_0}, Δf_{y_0}, $\Delta f_{x_{m-1}}$ and $\Delta f_{y_{m-1}}$ to maintain the system static equilibrium, which in turn induce bearing eccentricity increments Δe_{x_0}, Δe_{y_0}, $\Delta e_{x_{m-1}}$ and $\Delta e_{y_{m-1}}$ again. These iterations should be performed through steps (2) to (7) until sufficiently high accuracy has been reached concerning the vertical assembly level, the load distribution, and the journal levitations on the bearings in both directions. As the final corrections on the two reference bearings are usually quite small as compared with ΔS, only 3 to 5 times of iterations are needed in most cases.

7 The Application of Database of Nonstationary Hydrodynamic Oil Film Force

Nonlinear dynamic simulation of rotor-bearing system is essentially a process of time marching integration. Since the governing equation of the system [Eq. (2)] is implicit in the time domain, the journal eccentricities and velocities at all the bearings and the accompanying instantaneous bearing forces have to be calculated repetitively and with iteration. Therefore, the accuracy and particularly the efficiency of the simulation calculation will largely depend on the way of obtaining the nonstationary oil film forces. The widely used ways for this purpose are: 1. solving Reynolds equation by numerical procedure and integrating, 2. using approximate values, such as the analytical solutions from the infinitely long bearing or short bearing theory, or fitted formulae based on a set of numerical calculations, ect. The first way lacks efficiency and is therefore unsuitable for cases except simple ones due to the huge amount of computation work needed. The second way may lack the accuracy necessary for realistic simulations, especially during unstable motions when the rotor speed is above its stability threshold.

To achieve both efficiency and accuracy simultaneously, the technique of aiding bearing calculations with database of properties of single bush segment[13] has been further developed and databases of nonstationary oil film force of journal bearing bushes[14] are built and applied to aid the nonlinear dynamic simulation of multi-span rotor-journal bearings-foundation system.

A tabular database is created beforhand for each given combination of length-to-diameter ratio and span angle of the involved bushes. This is done by systematic calculations, based on solving the nonstationary Reynolds equation within the full ranges of eccentricity ratio and attitude angle of the journal center relative to the bush and the rotation/squeeze ratio. These parameter ranges are economically discretized while keeping adequate accuracy of interpolation. Reynolds' oil film rupture condition is adopted. The instantaneous values of the eccentricity ratio, the altitude angle, and the rotation/squeeze ratio pertaining to the bush are relied upon for interpolating the database to get the instantaneous values of non-dimensional horizontal and vertical force components of the bush. In this way, the nonlinearity and accuracy of bearing forces are taken full care of within the scope of the present model. The application of the technique of database in nonlinear analysis of multi-span rotor-journal bearings-foundation system makes it possible to perform large scale calculation with high precision.

8 The Method of Consideration of Foundation Vibration

The mechanics model considering foundation vibration is shown by Fig. 1. The dynamic equation of foundation including the stator and the bearing can be written as

$$M_B \ddot{X}_{B_t} + C_b \dot{X}_{B_t} + K_B X_{B_t} = F'_t \tag{24}$$

where M_B, C_B and K_B are diagnoal block matrices formed by combining block matrices M_{b_j}, C_{b_j} and K_{b_j}, respectively ($j=0, 1, \cdots, m-1$). The definitions of M_{b_j}, C_{b_j} and K_{b_j} are

$$M_{b_j} = \begin{bmatrix} m_{b_j} & 0 \\ 0 & m_{b_j} \end{bmatrix}; \quad C_{b_j} = \begin{bmatrix} Cb_{xx_j} & Cb_{xy_j} \\ Cb_{yx_j} & Cb_{yy_j} \end{bmatrix}; \quad K_{b_j} = \begin{bmatrix} Kb_{xx_j} & Kb_{xy_j} \\ Kb_{yx_j} & Kb_{yy_j} \end{bmatrix};$$

$$X_B = [x_{b_0}, y_{b_0}, x_{b_1}, y_{b_1}, \cdots, x_{b_{m-1}}, y_{b_{m-1}}]^T;$$

$$F' = [f'_{x_0}, f'_{y_0}, f'_{x_1}, f'_{y_1}, \cdots, f'_{x_{m-1}}, f'_{y_{m-1}}]^T$$

where m_{b_j}, Cb_{lk_j} and Kb_{lk_j} are the equivalent coefficients of mass, damping and stiffness of the ith discrete node of foundation.

Eq. (24) is solved parallel to Eq. (2) in the time domain. The difference, compared with Eq. (2), is that the stiffness matrix K_B is a diagnoal block matrix composed of block matrices K_{b_j}, so the inversion of the equivalent mass matrix of foundation can be done directly without applying the tridiagonal matrix elimination and sweeping method.

The relative displacement of journal to bearing is shown by Eq. (25) and Fig. 5.

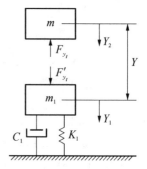

Fig. 5 Relative displacement of journal to bearing

9 Conclusions

In this paper, a new calculation procedure of non-linear dynamics of rotor-bearings-foundation system of high degree of freedom is proposed with the aim of studying the dynamic behavior of multi-span rotors supported on journal bearings. The main contents of the procedure are the expression of the space matrix in the form of a tridiagonoal block matrix and its special elimination and sweeping algorithm, the Houbolt-based time integration supplemented with central difference pre-estimation, the simplification of the system equation by decoupling between the physical coordinates x and y, the method of setting suitable initial state values of the system, the method of consideration of foundation vibration and the application of databases of nonlinear oil film force in the nonlinear dynamic calculation of rotor-bearings-foundation system. The software packdge realized with these techniques can be effectively used for efficient and accurate studies of non-linear dynamic behavior of multi-span rotor-bearings-foundation systems of high degree of freedom.

References

[1] M. L. Adams. Nonlinear Dynamics of Flexble Multi-Bearing Rotor. Journal of Sound and Vibration, Vol. 71(1), 1980, pp 129 – 144.

[2] Joseph Padovann et al. Nonlinear Transient Finite Element Analysis of Rotor-Bearing-Stator System. Computer & Structure, Vol. 18, No. 4, 1984, pp 629 – 639.

[3] H. D. Nelson. Nonlinear Analysis of Rotor-Bearing Systems Using Component Mode Systhesis. Journal of Engineering for Power, Trans. of ASME. , Vol. 105, pp 606 – 614.

[4] A. Selva Kumer, T. S. Sanber. A New Transfer Matrix Method for Response Analysis of Large Dynamic System. Computer & Structure, Vol. 23, No. 4, 1986, pp 545 – 552.

[5] Songyuan Lu. A New Transfer Matrix Method for Nonlinear Vibration Analysis of Rotor-bearing System. De-vol. 37, Vibration Analysis and Compulation, ASME, 1991.

[6] R. Rsubbian. Transient Dynamic Analysis of Rotor Using the Conbined Methodologies of Finite Elements and Transfer Matrix. Journal of Applied Mechanics, Trans. of ASME. , Vol. 55, 1988, pp 448 – 452.

[7] Bathe, Wilson. Numerical Methods in Finite Element Analysis. John Wlley Publishers.

[8] Zhigang Li. Nonlinear Dynamic Analysis of Multi-Span Rotor-Journal Bearing System. PhD. Dissertation, Shanghai University, Shanghai, PR. China. May 1995 (in Chinese).

[9] G. H. Golub, C. F. Van Loan. Matrix Computations. The Johns Hopkins University Press, 1983.

[10] J. A. George. On Block Elimination for Sparse Linear Systems. SIAM J. Num, Anal. 11, pp 585 – 603, 1974.

[11] J. M. Varah. On the Solution of Block-Tridiagonal Systems Arising from Certain Finite Difference Equations. Math. Comp. 26, pp. 859 – 868, 1972.

[12] V. Castelli, Shapiro. Improved Method for Numerical Solution of the General Incompressible Fruid Film Lubrication

Problem. J. of Lub. Tech., Trans. of ASME., Apr., 1967, pp 211-218.

[13] Zhigang Li, Zhiming Zhang. Calculation of Lobe Type Hydrodymanic Bearings Aided by Database of Properties of Single Bush Segment. Proceedings of Asia-Pacific Vibration Conference '93, Kitakyushu, Japan.

[14] Wen Wang, Zhiming Zhang. Calculation of Journal Dynamic Locus Aided by Database of Non-stationary Oil Film Force of Single Bush Segment. Proceedings of Asia-Pacific Vibration Conference '93, Kitakyushu, Japan.

Acknowledgement

The research project related with the present paper is sponsored by The National Natural Science Foundation of China and Foundation of Education Committee of Shanghai Municipality.

Nonlinear Dynamic Analysis of Multi-span Rotor-journal Bearing-foundation System: Part II[*]

Abstract: Nonlinear stability problems of multi-span rotor bearings foundation system are studied in this paper in detail, using the calculation method proposed by the authors. The problems treated consist of the effect of unbalance magnitude and distribution on stability of rotor bearing system; the investigation of the cause of non-divergence of rotor orbit for certain rotors running at a speed higher than the linear stability threshold; the analysis of the dynamic behavior of a multi-span rotor system under different kinds of large excitation. Correspondence between the simulated and the experimental results is satisfactory.

1 Stability Effect of Unbalance Distribution of Flexible Rotor-journal Bearing System

As we know, it is impossible to take the effect of unbalance-induced synchronous vibration on system stability into account by linear dynamics when calculating the stability threshold speed of a rotor-journal bearing system. Many experiments and analyses, however, show that rotor unbalance has a significant effect on system stability[1]. But so far up to now, the analyses on this are mostly limited on rigid rotors or single-mode systems, the effect of distributed unbalance on flexible rotors has not been taken into consideration, possibly due to the lack of suitable calculating tool.

Recently, effect of unbalance magnitude and distribution of rotor on system stability has been studied by the authors[2] using the calculation scheme proposed in the first part of this paper[3]. Systematically calculated results not only support the conclusion about the effect of unbalance magnitude on system stability of rigid rotor given by [1], but also show that, for flexible rotor-journal bearing systems, unbalance distribution is also an important factor, probably more important than that of unbalance magnitude. It is concluded that it would be unfavorable for system stability if the dynamic deflection shape of forced vibration caused by unbalance conforms with the first mode shape of the rotor whereas the system stability will be improved by properly increased unbalance when the dynamic deflection shape of forced vibration conforms with the second or higher vibration mode of the rotor. The explanation is that the threshold speed of a flexible rotor-journal bearing system is generally only related with the first natrual frequency of the rotor, so that self-

---

[*] In collaboration with Zhigang Li. Reprinted from *Proc. IC-HBRSD'97*, 1997: 277–291.

excited vibration will be promoted if the forced vibration shape conforms with the first natural mode of the rotor when the rotor speed is near or passing the linear threshold speed. Though there does exist some limited beneficial effect of forced synchronous vibration on stability in such ease, it appears to be not strong enough to suppress the oil whip from being initiated and promoted. On the other hand, system instability would not be easy to be promoted or would be easy to suppress if the forced vibration curve should be of higher order than the first mode of rotor. Generally speaking, rotors in practical use are balanced quite well in the first mode, while it is usual that the residual unbalance distributes mainly in the second or higher modes, and such residual unbalance distribution will benifit the stability of the rotor-journal bearing system.

Fig. 1 shows the value of peak-to-peak excursion of journal in vertical direction vs. angular speed of a single-span rotor supported on two journal bearings with different unbalance distributions on three disks mounted. The rotor configration is given by Fig. 2. The bearing type and parameters used for simulation are: $2 \times 170°$ elliptical bearing with preload factor $m = 0.1$, the minimum relative clearance $\psi_{min} = 2.7‰$, relative

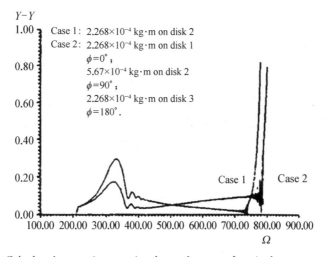

Fig. 1 Calculated excursion-rotational speed curve of a single span rotor with different unbalance distribution on three disk mounted

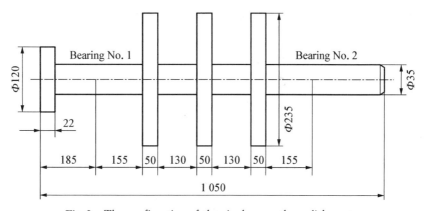

Fig. 2 The configration of the single span-three disks rotor

clearance $\psi=3‰$, lenth-to-diameter ratio $l/d=0.6$ and dynamic viscocity of lubricant $\mu=0.03$ Pa·s.

Fig. 3 shows the results of an experimental run parallel with simulating calculation, examining the effects on stability of two different kinds of unbalance distribution. The operation conditions of the test are almost identical with those of calculation. The only difference is that the bearing used in the experimental run is a cylindrical one with two axial oil grooves at both sides, the upper bush span angle is 150° and lower bush span angle 180°. The lubricant used is a 22# turbine oil with dynamic viscocity about 0.028 Pa·s at inlet oil temperature about 40 ℃. The oil pressure fed is about 0.06 MPa. Two kinds of unbalance distribution are: 1.4 g additional unbalance mass is mounted on disk 2 in case one (Fig. 3a and 3b), 1.1 g additional unbalance mass is mounted on disk 1 and 0.9g on disk 3 with 180° out of phase in case two(Fig. 3c and 3d).

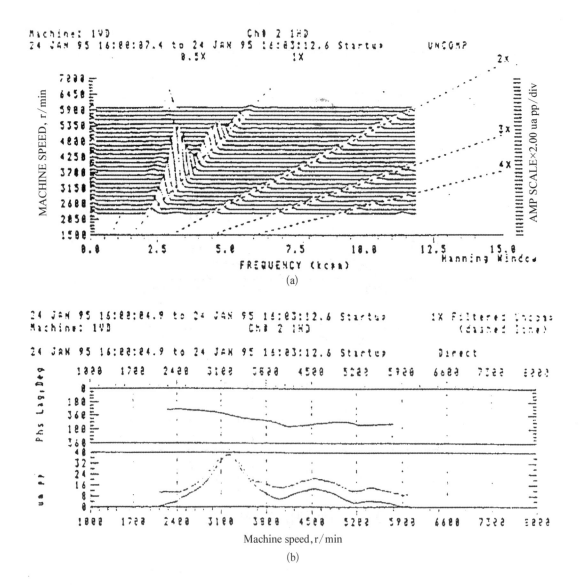

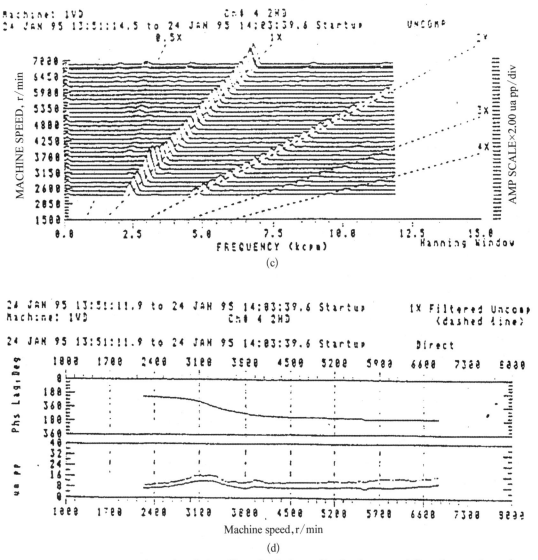

Fig. 3 The experimental results of the effcet of unbalance distribution on stability of rotor-journal bering system. (a) and (b) is the Caseade spcetrum an Bode of ease one, 1.4 g unbalance mass on disk 2. (c) and (d) is the Cascade spectrum an Bode of ease two, 1.1 g unbalance mass on disk 1 and 0.9 g on disk 3 with 180° out off phase.

Some conclusions can be obtained both from the experimental run and the simulation:

I. Unbalance distribution on the flexible rotor has marked effect on system stability. The system has lower nonlinear stability speed if the dynamic deflcetion shape of forced vibration conforms with the first vibration mode of the rotor as indicated in ease one ($n_{st} = 6\,000$ r/min). On the other hand, it will have much higher stability speed when the forced vibration shape conforms with the second vibration mode as indicated in ease two ($n_{st} \approx 6\,780$ r/min).

II. The occurrence of oil whip appears to be quite different in these two cases. Strong and sudden oil whip takes place immediate after the appearance of the subsynchronous whirl coresponding to the first vibration mode, whereas there is a noticcable intermediate process of development between the occurrence of low frequency whirl and strong oil whip in the case of the second mode of vibration, and the oil whip in case two is not so violent compared with that in case one. These phenomena have long been noticed by many researchers during their experimental or practical works, but none could be explained by linear theory. Now we can get to know that at least one mechanism will influence the process of oil whip, and that is the residual unbalance distribution of the rotor.

III. The value of the system critical speed and threshold speed as well as the shape of the curve of peak-to-peak amplitude of forced vibration vs. rotational speed obtained by numerical simulation conform with those by experiment, indicating that almost all of the important dynamic properties of rotor-journal bearing system can be deseribed not only qualitatively but also more-or-less quantitatively by careful numerical treatment.

2 The Effect of Nonlincarty of Bearing Force on System Stability

Because of the strong nonlinearity of journal bearing force, there are a number of phenomena observed in rotor-journal bearing systems which could not be explained by linear theory, the steady and non-divergence of whirl for certain rotors running at a speed higher than the linear stability threshold being the most recognized one. In this paper, a two span rotor supported by three journal bearings(Fig. 4) is studied using both linear and nonlinear methods to compare the respective stability threshold speeds predicted by them. The response to either residual unbalance during speeding-up or suddenly applied unbalance, due to blade loss for an example, is investigated using nonlinear numerical simulation method.

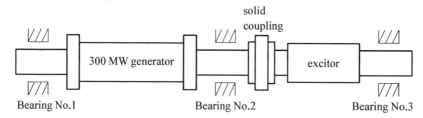

Fig. 4 A two span rotor supported by three journal bearings

2.1 Linear analysis solution

The linear analyses are performed under different calculation conditions for comparison. The bearing type and parameters used in the first case are: $2 \times 155°$ elliptical bearing for bearing No. 1 and No. 2 with preload factor $m = 0.2875$, minimum relative clearance $\psi_{min} = 1.32‰$, relative clearance $\psi = 1.85‰$, lenth-to-diameter ratio $l/d = 0.8$; $360°$ cylindral bearing for bearing No. 3 with $\psi = 1.31‰$, lenth-to-diameter ratio $l/d = 0.446$. Dynamic viscocity of lubricant for all the bearings $\mu = 0.03$ Pa·s. The calculated results are: $n_1 = 860$ r/min, and $n_{st} = 1\ 900$ r/min.

It is seen that the linear threshold speed, which is only about two times of the first critical speed, of the system appears to be rather low under these conditions. To examine the cause of such a low threshold speed, further second and third cases of studies are made. The calculation conditions are just as in the first ease except for the substitution of bearing No. 3 by a $4 \times 80°$ tilting pad bearing in the second case, or changing the preload factor to $m = 0.575$ for bearings No. 1 and No. 2 in the third case. The calculated results are: $n_1 = 860$ r/min and $n_{st} = 2\ 000$ r/min for the second case, and $n_1 = 908$ r/min, $n_1' = 1\ 000$ r/min and $n_{st} = 3\ 000$ r/min for the third case, respectively. From these results, we can see that the dominant factor affecting the system instability for this two span rotor system is the dynamic characteristics of bearings No. 1 and No. 2. It is worthwhile to note that, merely by changing the preload factor of bearings No. 1 and No. 2 from 0.287 5 to 0.575, the linear threshold speed of the system is raised about 1 000 r/min, approximately 58% from the original one.

2.2 Nonlinear analysis solution

The bearing types and parameters for the nonlinear analysis are the same as those in the first case of the linear analysis. The contents of the nonlinear analysis are simulations of the dynamic behaviors during the process of speeding-up or suddenly applied unbalance. In order to excite both the first and the second modes of vibration, the following distributed unbalances are postulated, either in the ease of residual unbalance or of suddenly applied unbalance: 0.14 kg·m at the center of the first span with zero phase angle and 0.05 kg·m at the middle of the second span with 180° out of phase, respectively. The other parameters needed for the simulation of rotor speeding-up are: the starting rotational speed $n_s = 700$ r/min; the time marching step $\Delta t = T_n/256$; the increment of rotational speed corespond in each time step $\Delta \Omega = 0.01$ rad/s, where $T_n = 2\pi/\omega_s$, $\omega_s = \pi \eta_s/30$.

The values of peak-to-peak excursions of the journal in horizantal direction vs. the angular speed during the process of speeding-up is shown in Fig. 5. The main dynamic properties obtained are: the first and second critical speeds are $\omega_1 = 102$ rad/s and $\omega_2 = 290$ rad/s; the stability threshold speed of the system is $\Omega'_{st} = 350$ rad/s. The responce of the system to the suddenly applied unbalance is shown in Fig. 6, where the coordinate r indicates the number of revolutions. It can be seen from Fig. 6 that instability occours at about 2 500~2 600 r/min of rotational speed under the excitation of suddenly applied unbalance. It seems that, at the rotational speed of 2 600 r/min, the intensity of subsynchronous whirl and the ambilitudes of vibration on bearings No. 1 and No. 2 increase steadfastly until instability occurs, while at the speed of 2 500 r/min the subsynchronous whirl decreases gradually indicating the convergence of the system vibration. From all these calculated data and figures, it can be found that:

(1) The system instability does not occur simultaneously at all the journal bearings or

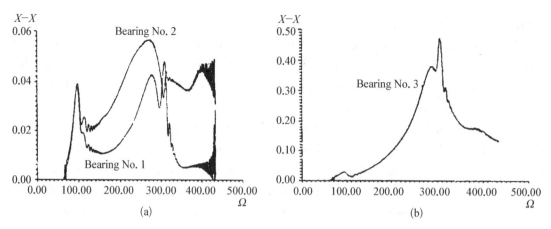

Fig. 5 Calculated excursion-rotational speed curve of the two span rotor supported by three journal bearings

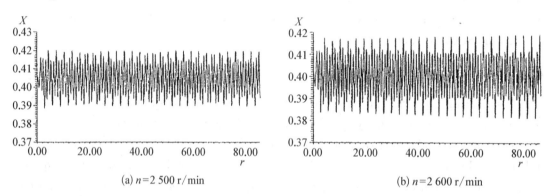

Fig. 6 The sudden unbalance responce of the two span rotor at rotational speed $N=2\,500$ r/min and $N=2\,600$ r/min, respcetively

rotor spans. On the contrary, the onset of self-excited subsynchronous whirl occurs first at a certain bearing or rotor span, and then propagates in succession to the nearby bearings or rotor spans until it contaminates the whole system. As seen from Fig. 5, for instance, instability occurs at bearing No. 1 at the angular speed of 350 rad/s, at bearing No. 2 at 370 rad/s, and at bearing No. 3 at 410 rad/s, respectively.

(2) The stability threshold speed appears to be quite different as calculated by different methods for the same rotor-bearing system. The one obtained by the present nonlinear simulation technique is much higher than by the linear analysis either in the case of speeding-up or in that of suddenly applied unbalance, while the latter case has a stability threshold speed lower than that in the former case, attributable to the time span needed for the self-excited whirl to develop up to a noticeable level. It is also exhibited that the rotor can whirl self-excitedly with a small and steady orbit without further divergence above the linear threshold speed for some rotor-bearing systems.

For a lightly loaded elliptical bearing with a small preload factor, the equilibrium position of the journal within the bearing is characteristic of small eccentricity ratio ε and almost 90 degrees of attitude angle θ when the rotational speed reaches the linear stability threshold speed. The relative eccentricity ratios ε_i and attitude angles θ_i of the journal to various bush segments are vary with time when the journal whirls under the unbalance excitation so that the oil film spans in various bush segments are different from their steady state values and vary with time, as predicted by Reynolds' film rupture condition adopted for the calculation of the non-stationary oil film forces. In particular, the upper bush has a more significant hydrodynamic and squeeze overall effect in horizontal direction than that predicted by the linear theory. The net effect of all these differences results as increasing bearing preload ratio and thus in favor of the system stability.

3 The Dynamic Behavior of Mutti-span Rotor-bearings System under Large Excitations

The phenomenon, which could not be explained by the linear dynamics, has been noticed by many researchers that oil whip may take place for some rotors working below the linear threshold speed under large excitation such as earthquake or blade loss[4,5]. Taking a test model for a 600 MW steam generater as object of simulation, the authors examined the dynamic behavior of a multi-span rotor system under different kinds of large excitations. The rotor is discretized into 276 subdivisions and supported by 11 elliptical bearings with parameter values $m=0.5$, $\psi_{min}=0.002$ and $l/d=0.6$, the dynamic viscocity of the lubricant for all the bearings is taken as $\mu=0.018$ Pa·s.

Fig. 7 shows the value of peak-to-peak excursion of journal in horizontal direction vs. angular speed during the process of speeding-up (one bearing is taken from each span for demonstration). The main dynamic behaviors of the system obtained are the first critical speed $\omega_1 \approx 285$ rad/s and the threshold speed $\Omega'_{st}=780$ rad/s. It is clearly exhibited that oil whip starts at the generater-span and propagates to the other spans one by one successively and finally brings the whole system into the state of instability.

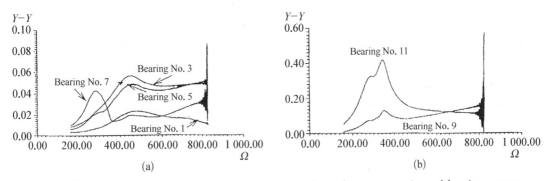

Fig. 7 The calculated excursion-rotational speed curve of a multi-span rotor-journal bearings system

Fig. 8 shows the rotor response to an excitation which was proposed in [4] as capable of representing the effect of earthquack. The rotational speed is taken as 6 000 r/min. The first critical speed of the system is taken as the excited vibration frequency and $A=0.2$ taken as the strenth of the excitation[4]. In this case, the value of Ω/ω_1 is about 2.204 6, far less than that of Ω_{st}/ω_1. Fig. 9 shows the dynamic loci of the journals in various bearings when unbalance is suddenly applied at the middle of each span at the speed of 6 000 r/min. The system exhibit different responses to these two kinds of excitation.

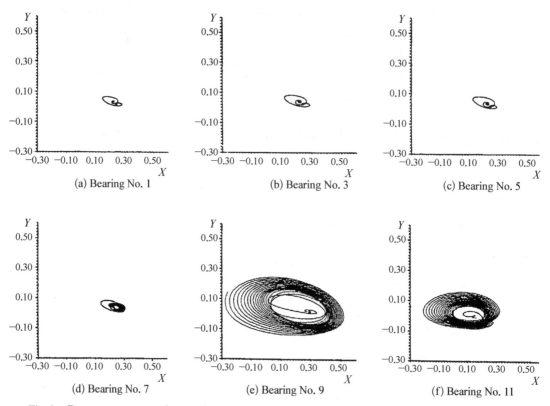

Fig. 8 Rotor responce to the simulated earthquack excitation at rotational speed of 6 000 r/min, excited frequency $\omega_{ex}=\omega_1$, and strenth of the excitation $A=0.2$[4]

The first mode of natural vibration is easily stimulated if the excitation frequency is equal or near to the first critical speed, and system instability will be initiated if the intensity of excitation is strong enough, even though the value of Ω/ω_1 is far less than that of Ω_{st}/ω_1. On the other hand, the system will keep stable if the excitation is synchronous, such as the unbalance effect after its sudden application on the system, if the value of Ω/ω_1 is far less than that of Ω_{st}/ω_1. All the modes of natural vibrations are stimulated at the start of the application of excitation, but the non-synchronous vibrations will die out gradually in the time to follow (Fig. 10). However, sufficiently large suddenly applied unbalace may cause the occurence of oil whip if the rotational speed is near the stability threshold speed.

Another difference of system response to these two kinds of excitation is that the excitation of "earthquack wave" is short-durationed and stochastic, so that, after a time span following the application of excitation, only those spans whose natural frequencies are equal or near to the excitation frequency will still vibrate with their respective natural frequencies, while the other spans with far different natural frequencies will only whirl with much lower frequency as affected by the transient vibrations of the stimulated spans,

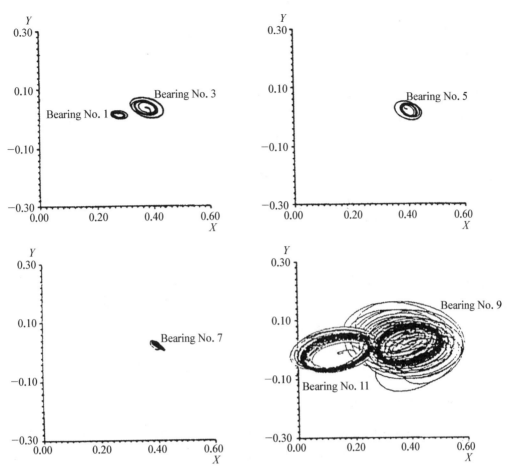

Fig. 9 Dynamic locus of shaft in each bearing when sudden change of unbalance excited on the middle of each span at the speed of 6 000 r/min

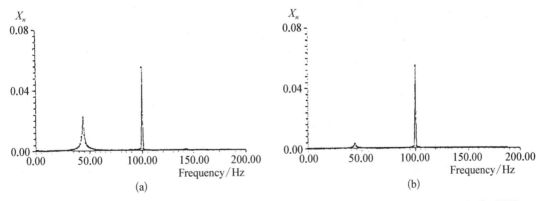

Fig. 10 FFT analysis of spectra at bearing No. 9 when sudden unbalance excited on, (a) the FFT of former 60 revolutions, (b) the FFT of latter 60 revolutions

whereas all spans of the system vibrate with the same frequency in the case of synchronous excitation.

4 Simulation of Oscar Pinkus' well Known Experiment of Rotor-journal Bearing System

The literature related to rolor-journal bearing phenomena is very rich. These phenomena have long been noticed in experiments or practice, but their mechnisms have not yet been fully explained, although qualitative reasonings can usually be made more or less vaguely by linear dynamics. With the aim of analying them quntitatively, O. Pinkus' widely known and well noticed experiment[6] is chosen as an object of nonlinear simulation here, as Pinkus' paper contains detailed and rather full recordings of the phenomena within a fairly great speed range, including speed values far surpassing the linear threshold speed. All the published geometric and working parameters of the rotor-bearing system have been taken over into the simulation in order to make meaningful comparison between the simulated and the experimental results, with a few ones, which are lacking in Pinkus' paper, postulated from the recorded values of critical speeds. Values of residual unbalance and its distribution can naturally not be found from his paper, but are needed in simulation calculations. They are postulated in such a way that the calculated resonant amplitudes roughly correspond with the measured ones.

4.1 The simulated rotor-bearing system

Fig. 11 shows the simulated rotor schematically, the parameters of the rotor-bearing system are taken from [6]. The are as follows:

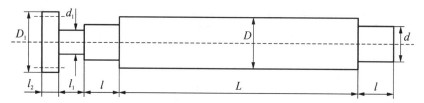

Fig. 11 The simulated rotor of O. Pinkus' experiment

(1) The bearings: halved cylindrical, with length $l=50.8$ mm, journal diameter $d=50.8$ mm and radial clearence $C=0.0635$ mm. Double-side inlet of lubricant with viscosity of 150 sybolt second/100°F and working temperature 120°F~140°F;

(2) The rotor mass: total mass $m=29.03$ kg;

(3) The first and second critical speeds $n_1=6,100$ r/min and $n_2=23,000$ r/min, the stability threshold speed $n_{st}=11,650$ r/min.

The other parameters needed for the simulation are postulated from the above values. They are the diameters and lengths respectively of the main span of the rotor $D=63.5$ mm and $L=1\,099.2$ mm; the diameter and length of the overhang part of the rotor $d_1=38.1$ mm and $l_1=50.8$ mm; and the diameter and width of the coupling disk $D_1=76.2$ mm and $l_2=25.4$ mm. The density and elasticity modulus of the shaft material are taken as $\rho=7\,550$ kg/m^3 and $E=2.1\times10^{11}$ Pa, respectively. The dynamic viscosity of the lubricant is taken as $\mu=0.018$ Pa·s.

Residual unbalance is unavoidable on every rotor. Values of unbalances applied at 3 equally spaced nodes along the rotor are adjusted until the calculated 1st and 2nd resonant peaks appear to be similar to the measured ones. They are finally fixed in this way as 2.75×10^{-4} kg·m with phase angle 0° at 1/4 length from the rotor end, 4.395×10^{-4} kg·m with phase angle 120° at midspan of the rotor and 2.75×10^{-4} kg·m with phase angle 240° at 3/4 length, respectively.

A very small ellipticity ratio $m=0.1$ is assumed for the bearings, in view of the shape of the measured journal resonance orbit which has a rather pronounced ellipticity compared to the orbit calculated when no ellipticity of bearing is assumed. The possible existence of a small ellipticity is thus guessed at, and a value 0.1 appears to be appropriate after trials.

4.2 Results of simulation and discussion

The calculation is started from an angular speed of 500 rad/s. Time marching step is 1/512 of the current period of revolution. The angular speed of the rolor is increased in each time step by a value 0.01 rad/s when the running speed is below 1 200 rad/s, and 0.005 rad/s when the speed is over 1 200 rad/s. The calculation is carried on until the angular speed reaches 2 800 rad/s, so that the whole speed range of the simulated experiment is covered.

Fig. 12 shows a passage extracted from the recorded displacements when the rotor is crossing the 2nd critical speed. The corresponding Lissajous' diagrams and frequency

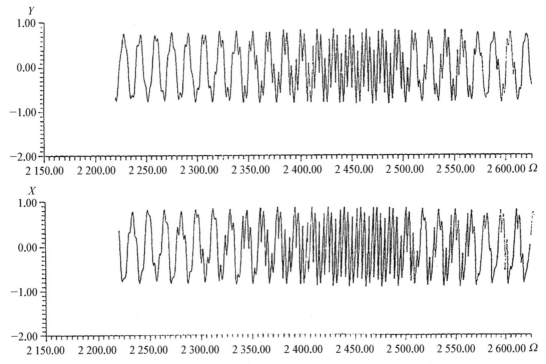

Fig. 12 A passadge of the recorded journal displacement when crossing 2nd critical speed

spectra are shown in Figs. 13 and 14.

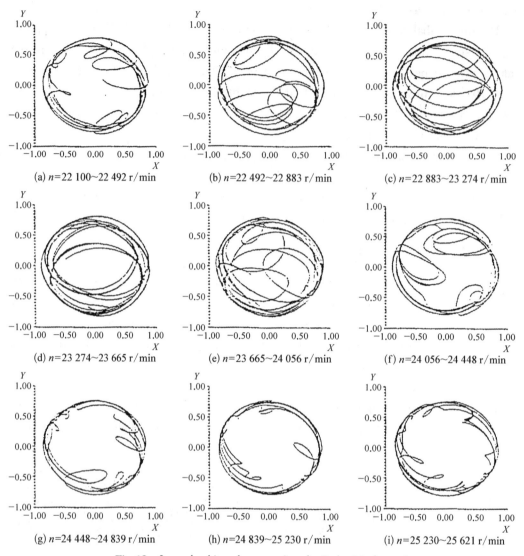

Fig. 13 Journal orbits when crossing the 2nd critical speed

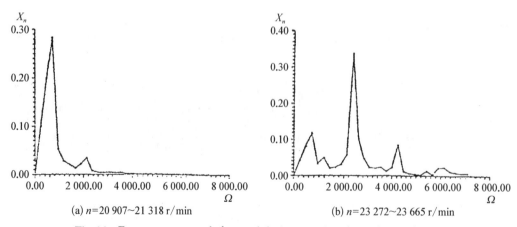

Fig. 14 Frequency spectra before and during crossing the 2nd critical speed

Fig. 15 shows the value of peak-to-peak excursion of the journal in vertical direction vs. the angular speed. It can be seen that the rotor system crosses its 1st critical at an angular speed about 650 rad/s. After this, the amplitude of vibration falls, but rises again at about 1 250 rad/s, where a half-frequency component appears and grows rapidly, as exhitbited by the corresponding frequency spectra. This self-excited component is also the cause of the excursion curve's quick pulsation here. With the increase of speed, the self-excited componet rises drastically and quickly becomes the dominant component and makes the synchronous component relatively insignificant, so that the excursion curve becomes smoother again. Within a fairly wide range of speed, the frequency of the self-excited component keeps a value very near the 1st critical, viz., 650 rad/s, which is typical of oil whip or resonant whirl. The crossing through the 2nd critical at 2 430 rad/s shows up in the shifting of the frequency from the self-excited to the synchronous and then back, rather than a rise and drop of the amplitude. Again, the frequency spectra obtained by FFT, and the pulsations before and just at the critical speed, are useful for exhibiting the shifts of the dominant frequency.

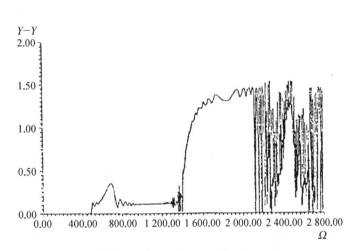

Fig. 15 Calculated excusion-rotational speed curve

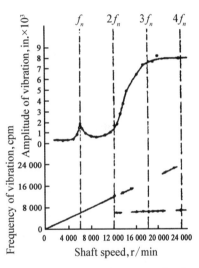

Fig. 16 Measured vibration amplitude curve, quoted from [6]

Fig. 16 reproduces the measured apmlitude quoted from [6]. Simularity between the measured and calculated results, in both the vibration level and the shiftings of the dominant frequency within the whole speed range is evident. There is only one exception. A test rig resonance at 15 000 r/min is recorded in the experiment, accompanied by the corresponding shiftings of the dominat frequency. This has not been inclued in the simulation, where the bearing are assumed to be mounted on rigid foundations.

Comparison between the calculated journal whirl orbits (Fig. 17) and the measured ones[6] can also be looked upon as satisfactory.

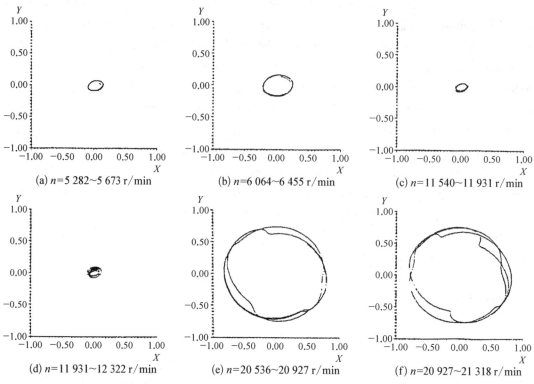

Fig. 17 Calculated journal orbits of simulated rotor of O. Pinkus' experiment

5 Conclusions

Dynamic simulation of lateral vibration of multi-span rotor-journal bearing system is capable of simutanenous consideration of nonlinear bearing force, magnitude and distribution of the residual unbalance, externally applied excitation and self-induced excitation. Interesting phenomena of the effect of unbalance magnitude and distribution on system stability, small orbit of the rotor without further divergence for certain rotors running at a speed higher than the linear stability threshold, dynamic behavior of multi-span rotor system under different kinds of large excitations and of dominant frequency shifting peculiar to crossing a critical speed after surpassing the stability threshold, can be realistically exhibited. This sort of simulation can therefore be very useful for explaining the seemingly complicated behavior of rotor-journal bearing system induced by the numerous co-existing effects nonlinearly combined.

References

[1] X. Yang. The Investigation of the Stability Marging, the Non-linear Vibration and Optimization of Dynamic Properties of Rotor. PhD. Dissatation, Xian Jiaotong University, Xian, PR. China, May 1994 (in Chinese).

[2] Z. Li. Nonlinear Dynamic Analysis of Multi-Span Rotor-Journal Bearing System. PhD. Dissatation, Shanghai University, Shanghai, PR. China, May 1995 (in Chinese).

[3] Z. Li, Z. Zhang. Nonlinear Dynamic Analysis of Multi-Span Rotor-Journal Bearing-Foundation System, Part 1, to

be published.

[4] Y. Hori, T. Kato. Earthquack-Induced Instability of a Rotor Supported by Oil Film Bearing, Journal of Vibration and Acoustics, Transaction of the ASME, Vol. 112, pp 160-165,1990.

[5] W. Zhang. The Investigation of the stability and Analying Method of Rotor-Journal Bearing System. PhD. Dissatation, Xian Jiaotong University, Xian, PR. China, Sep. 1994 (in Chinese)

[6] O. Pinkus. Experimental Investigtion of Resonant Wipe. Transaction of the ASME, Vol. 78, July, 1965.

Acknowledgement

The research project related with the present paper is sponsored by The National Seience Foundation of China and Foundation of Education Committee of Shanghai Municipality.

A General Matrix Solution for the Elastic Quarter Space*

Elastic quarter space situation is quite common in practical mechanical systems such as the contacts of rail-wheels, cam-followers, gears and roller bearings. The common characteristic of these contacts is that there are free end surfaces near the contact or loading region. Existing theoretical methods for the solution of the elastic quarter space can be roughly classified into: (1) direct numerical iteration of two overlapped, mutually orthogonal half spaces(Hetenyi[2]); (2) numerical solution of Fourier transformation of the two simultaneous integral equations of half spaces and subsequent back-transformation (Sneddon[3] and Keer et al.[4]); (3) direct numerical solution of the above-mentioned integral equations(Hanson et al.[6]); (4) Ritz-based numerical solution(Guenfoud et al.[8]). The method introduced in the present paper also utilizes the concept of overlapped half spaces, but differs from the others in employing simultaneous matrix equations at the very start instead of integral ones, and benefits from the flexibility of matrix operations to gain an explicit general solution of the elastic quarter space in the matrix form.

As a first stage of applying the present method to attack this problem, the singularity caused by edge loading is not considered. This does not affect its applicability to problems without edge loads, such as elasto-hydrodynamic lubrication where the boundary condition excludes any non-zero boundary pressure.

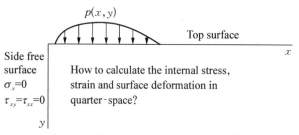

Fig. 1 Basic elastic quarter space problem

Fig. 1 shows the basic form of elastic quarter space problem considered in the present paper. The top surface is loaded by distributed normal pressure, and the side surface is free of any load.

Hetenyi[2] was the first to approach the elastic quarter space by overlapping two mutually orthogonal, symmetrically loaded half spaces. Figs. 2 and 3 show such a horizontal and a vertical half space respectively.

In both of these half spaces, the symmetrical loads cause only normal stresses in the respective mid-sections. The magnitudes of the mid-section normal stresses are related to

---

* In collaboration with W. Wang. Reprinted from *The Third International Symposium on Computational Mechanics* (ISCM III) & *Second Symposium on Computational Structural Engineering* (CSE II), 2011: 294-295.

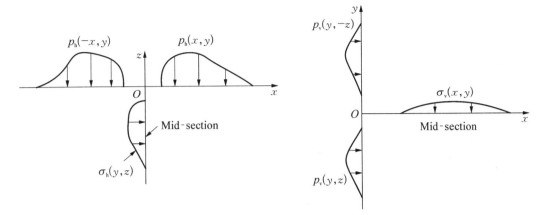

Fig. 2 Horizontal half space (H) symmetrically loaded

Fig. 3 Vertical half space (V) symmetrically loaded

the loads by matrix equations, where the elements of the matrices, or the "reflecting matrices", are defined by Love's solutions when the surfaces are discretized into rectangles and the distributed loads approached by piece-wise distributions. In general, the matrix equations can be written as:

$$\boldsymbol{S}_h = \boldsymbol{M} \cdot \boldsymbol{P}_h \quad \text{and} \quad \boldsymbol{S}_v = \boldsymbol{N} \cdot \boldsymbol{P}_v$$

where $\boldsymbol{S}_h$ and $\boldsymbol{S}_v$ are vectors formed by arrays of mid-sectional stresses; $\boldsymbol{P}_h$ and $\boldsymbol{P}_v$ are vectors formed by arrays of surface loads; and $\boldsymbol{M}$ and $\boldsymbol{N}$ are the reflecting matrices.

The left part of H-space is overlapped on the lower part of V-space. The stresses on the boundaries of the resulting quarter space must fulfill the given conditions, and two simultaneous equations are therefore obtained

$$-\boldsymbol{P}_h + \boldsymbol{S}_v = -\boldsymbol{P}_h + \boldsymbol{N} \cdot \boldsymbol{P}_v = -\boldsymbol{P} \quad \text{and} \quad -\boldsymbol{P}_v + \boldsymbol{S}_h = -\boldsymbol{P}_v + \boldsymbol{M} \cdot \boldsymbol{P}_h = 0$$

where $\boldsymbol{P}$ is the vector formed by the array of the given load on the top surface meshes of the quarter space.

The above simultaneous equations are easily decoupled, and explicit general solutions for the loads on the two overlapping half spaces are readily obtained:

$$\boldsymbol{P}_h = (\boldsymbol{I} - \boldsymbol{N} \cdot \boldsymbol{M})^{-1} \cdot \boldsymbol{P} \quad \text{and} \quad \boldsymbol{P}_v = \boldsymbol{M} \cdot (\boldsymbol{I} - \boldsymbol{N} \cdot \boldsymbol{M})^{-1} \cdot \boldsymbol{P} \tag{1}$$

With $\boldsymbol{P}_h$ and $\boldsymbol{P}_v$ known, the stress and deformation distributions within the quarter space can be calculated by applying Love's solutions again.

It is also interesting to note that the above solution is actually the limit of Hetenyi's iterations carried out indefinitely. To show this, the matrix form is applied to Hetenyi's process. The nth iterative results are:

$$\boldsymbol{P}_h^{(n)} = \left[\boldsymbol{I} + \sum_{i=1}^{n} (\boldsymbol{N} \cdot \boldsymbol{M})^i\right] \cdot \boldsymbol{P} \quad \text{and} \quad \boldsymbol{P}_v^{(n)} = \boldsymbol{M} \cdot \left[\boldsymbol{I} + \sum_{i=1}^{n} (\boldsymbol{N} \cdot \boldsymbol{M})^i\right] \cdot \boldsymbol{P}$$

The bounded character and convergence of such iterations were proven by Hetenyi[2]. It is evident that the limits of infinite iterations are identical to Eqs. (1).

The reflecting matrices involved in the present paper are only related to the mesh structure and Poisson's ratio, and un-related to the loading. Once generated, these matrices can be stored and employed in different loading cases of the same Poisson's ratio and geometrically similarly meshed structures of the surfaces. A special case has been selected from [6], and results obtained by the present solution correlate satisfactorily with those of [6]. Thus, the present solution should be conveniently applicable in practical elastic quarter space problems where edge loading is non-existent or of insignificant effect.

References

[1] Zhu D, Ren N, Wang QJ, "Pitting life prediction based on a 3D line contact mixed EHL analysis and subsurface von Mises stress calculation", J. of Trib., ASME, Vol. 131, No. 4 (2009), 1-8.

[2] Hetenyi M., "A general solution for the elastic quarter space", J. of Appl. Mechanics, ASME, 37, Ser E(1) (1970), 70-76.

[3] Sneddon I. N., "Fourier transformation solution of the quarter plane problem in elasticity", File PSR-99/6 (1971) Appl. Math. Res. Group, North Carolina State University.

[4] Keer L. M., Lee J. C. and Mura T., "Hetenyi's elastic quarter space problem revisited", Inlt. J. Solids and Structures, Vol. 19, No. 6, (1983), 497-508.

[5] Keer L. M., Lee J. C. and Mura T., "A contact problem for the elastic quarter space", Intl. J. Solids and Structures, Vol. 20, No. 5, (1984), 513-524.

[6] Hanson M. T. and Keer L. M, "A simplified analysis for an elastic quarter-space", Mech. Appl. Mats, Vol. 43, Pt. 4, (1990), 561-587.

[7] Yu C. C., Keer L. M., and Moran B., "Elastic-plastic rolling-sliding contact on a quarter space", Wear, 191 (1996), 219-225.

[8] Guenfoud S., Bosakov S. V., and Laefer D. F., "A Ritz's method based solution for the contact problem of a deformable rectangular plate on an elastic quarter-space", Intl. J. of Solids and Structures, Vol. 47, No. 14-15 (2010), 1822-1829.